INTRODUCTION TO ELECTRIC CIRCUITS AND MACHINES

COLIN D. SIMPSON
George Brown College

PRENTICE-HALL CANADA INC.
SCARBOROUGH, ONTARIO

Canadian Cataloguing in Publication Data
Simpson, Colin D. (Colin David), date
Introduction to electric circuits and machines

Includes index.
ISBN 0-13-473257-X

1. Electric circuits. 2. Electric machines.
I. Title.

TK3001.S55 1992 621.319 C91-094315-X

Prentice-Hall, Inc., Englewood Cliffs, New Jersey
Prentice-Hall International, Inc., London
Prentice-Hall of Australia, Pty., Ltd., Sydney
Prentice-Hall of India Pvt., Ltd., New Delhi
Prentice-Hall of Japan, Inc., Tokyo
Prentice-Hall of Southeast Asia (Pte.) Ltd., Singapore
Editora Prentice-Hall do Brasil Ltda., Rio de Janeiro
Prentice-Hall Hispanoamericana, S.A., Mexico

ISBN 0-13-473257-X

Acquisitions Editor: Jacqueline Wood
Developmental Editor: Linda Gorman
Copy Editor: Barbara Zeiders & Associates
Production Editor: Elizabeth Long
Production Coordinator: Florence Rousseau
Cover Design and Page Layout: Derek Chung Tiam Fook
Technical Art: Greg Dorosh
Typesetting: Compeer Typographic Services Ltd.

1 2 3 4 5 AG 96 95 94 93 92

Printed and bound in the U.S.A. by Arcata Graphics

CONTENTS

PREFACE

The primary objective of this book is to provide a comprehensive treatment of traditional topics in dc and ac circuit analysis, suitable for two- and four-year programs in electrical and electronic engineering technology. The level of content is suitable for community colleges, technical institutes, and as a first year text in four-year university programs. Although some of the subject matter, particularly ac and dc machines, is intended for students in electrical programs, there are numerous examples of electronic circuits throughout all chapters. Topics such as semiconductor devices, electronic control of servomotors, variable frequency drives, audio communication circuits and filters are presented with an emphasis on practical application.

In the field of electronics, it is particularly important that the student have an understanding of electric-circuit analysis, and every effort has been made to meet this requirement. The hundreds of worked examples and problems in this book have been selected and written in such a way as to illustrate fundamental concepts essential to troubleshooting and design of electrical and electronic circuits.

Students using this book should have an understanding of algebra up to quadratic equations, and of trigonometry to the simple properties of triangles. Some knowledge of logarithms will also help the student in understanding the information on filters. The availability of a scientific calculator is assumed. The calculator must handle trigonometric and exponential functions as a minimum. A calculator capable of rectangular to polar conversions will prove most useful. Basic principles, theorems, circuit behavior, and problem-solving procedures are presented so that the average student can get a clear understanding of essential concepts. The large number of example problems and exercises should also make this book very useful for self-study, review, and in support of more advanced courses.

The introductory chapters cover the basics of electricity, including atomic structure, electric charges, Ohm's law, Kirchhoff's laws, energy, and power. Emphasis has been placed on introductory circuit analysis techniques early in the book to assist the student in developing an approach to solving problems related to electric circuits. In chapters 6 and 10, a thorough treatment of transient circuit analysis is presented. Thévenin's theorem, Norton's theorem, superposition theorem, maximum power transfer theorem and analysis of dependent sources are covered in Chapter 5. These concepts are reinforced in Chapter 15, when ac circuit analysis is discussed.

In addition to electric circuit analysis, the book also covers the principles of magnetism, measuring instruments, generators, transformers, motors, and three-phase power systems. Conventional current flow is used throughout the text and, in accordance with the Institute of Electrical and Electronic Engineers (IEEE), the SI system of measurement is used wherever practical.

ACKNOWLEDGEMENTS

This book could never have emerged in its present form without the help of many people. I would like to thank my colleagues past and present at George Brown College, particularly Ray Davis, Leon Harris, George Danac, and Bill Houghton. I would also like to thank the editorial staff at Prentice-Hall, especially Yolanda de Rooy, Jackie Wood, Elizabeth Long, Linda Gorman, and Marta Tomins. A special thanks goes to Barbara Zeiders whose copyediting skills contributed greatly to the accuracy of this text. I also wish to thank the following individuals who offered many helpful suggestions in the early stages of the manuscript's development:

Fraser Cooper, Conestoga College
Joseph Thomas, University of Wisconsin
J.N. Tompkin, British Columbia Institute of Technology
Richard Parker, Seneca College
Robert I. Eversoll, Western Kentucky University
Doug Fuller, Humber College
Bud Skinner, Applied Physics Specialties
Leonard Sokoloff, DeVry Institute of Technology
Stanley W. Lawrence, Utah Technical College
Mike Goulding, Fanshawe College
H. Hayre, University of Houston
Bernard H. Moisey, Northern Alberta Institute of Technology
O.S. Zemlak, Algonquin College

Colin D. Simpson
1992

CHAPTER

Introduction

LEARNING OBJECTIVES

Upon completion of this chapter you will be able to:

- Define the basic units of measurement.
- Describe the SI system of measurement.
- Understand the usefulness and applications of conversion factors.
- Be able to express numbers in scientific notation.
- Convert from one power of 10 to another.
- Define engineering notation.
- Describe basic atomic structure.
- Define electric charge.
- Express Coulomb's law.

1-1 UNITS

There are at least three quantities that can be proven to exist without the use of any measuring instruments. These three quantities are **length**, **mass**, and **time**. These are known as **fundamental units**. Quantities that are obtained from the fundamental units are referred to as **derived units**.

The basic standard of length is the **meter**. The meter was originally defined in the eighteenth century as exactly one ten-millionth of the distance from the earth's pole to its equator. The distance was marked off by two lines on a platinum–iridium bar that had been cooled to 0°Celsius. Until 1960, the meter was measured against this standard platinum–iridium bar, which was kept at the International Bureau of Weights and Measures in France. The meter's length is now determined by atomic radiation, resulting in a value virtually identical to the platinum–iridium bar. The meter is defined as exactly 1,650,763.73 wavelengths of one of the spectral lines of the krypton isotope Kr^{86}.

The basic unit of mass is the **kilogram**. This unit was defined in 1901 as the mass of a platinum block kept with the standard meter bar at the International Bureau of Weights and Measures in France. In 1960, the basic unit of mass was redefined as being approximately equal to 1000 times the mass of 1 cm^3 of pure water at 4°C.

The basic unit of time is the **second**, which was defined prior to 1956 as 1/86,400 of an average solar day. At that time, the second was determined as 1/31,556,925.9747 of the tropical year 1900. In 1964, the second was defined more carefully as 9,192,631,770 periods of the transition frequency between the hyperfine levels $F = 4$, $m_F = 0$ and $F = 3$, $m_F = 0$ of the ground state $^2S_{1/2}$ of the atom of cesium 133, unperturbed by external fields.

Over the years measurement of these three quantities has led to at least three systems of measurement: the MKS system, the CGS system, and the English system of units. The SI system represents a rationalized selection of the MKS system. In recent history, the systems of units most commonly used were English and metric, as shown in Table 1-1. The MKS system, representing meters–kilograms–seconds, and the CGS system, with centimeters–grams–seconds, contain unrelated standards and are therefore grouped under the same heading. The English system of units developed in a rather natural way based on physical objects used to convey the dimensional information. English units of length consist of the foot, rod, and chain, while English units of mass consist of the stone and slug.

TABLE 1-1 *English and Metric System of Units*

Quantity	English	Metric	
		MKS	CGS
Length	yard	meter	centimeter
Mass	slug	kilogram	gram
Time	second	second	second
Force	pound	newton	dyne
Energy	foot-pound	newton-meter	dyne-centimeter

1-2 THE SI SYSTEM OF MEASUREMENT

In October 1965, the Institute of Electrical and Electronics Engineers (IEEE) adopted the **International System of Units (SI)** as a standard for all engineering and scientific literature. The SI system uses unit prefixes to form multiples and submultiples of a unit. The International System of Units has seven base units, listed as follows:

1. *Length/meter:* defined in terms of wavelengths of a particular radiation from the krypton-86 atom. This orange-red line has a wavelength of 6057.802×10^{-10} m. 1 meter (m) = 39.370 079 inches.
2. *Mass/kilogram:* defined as equal to the mass of the international prototype kept at Sèvres, France. 1 kilogram = 2.204 622 pounds.

3. *Time/second:* defined in terms of the duration of a specific number of periods of a particular radiation from the cesium-133 atom.

4. *Electric current/ampere:* defined as the constant electric current in two infinite parallel conductors of negligible cross section exactly 1 meter apart that produces a force of 2×10^{-7} newton (N) per meter length of wire. The symbol for current is I, and the abbreviation for amperes is A.

5. *Temperature/kelvin:* defined as the fraction 1/273.16 of the thermodynamic temperature of water.

6. *Luminous intensity/candela:* defined as the light intensity of the freezing point of platinum under specified conditions. 1 candela (cd) will produce a luminous flux of 1 lumen (lm) within a solid angle of 1 steradian (sr).

7. *Molecular substance/mole:* defined as the amount of substance that contains as many elementary entities as there are atoms in 0.012 kg of carbon 12.

There are many situations when base or supplementary units of measure are not suitable and derived units are required. The SI derived units are formed from the previously defined SI base units. For example, the base unit of time has the following derived units: frequency, velocity, and acceleration. Units derived from the base unit of mass deal with work, force, power, pressure, and other physical science measurements. There are 16 derived units, of which 12 are considered essential for the study of electric circuits and machines. Table 1-2 lists these 12 derived units with special names.

TABLE 1-2 ***Partial Listing of SI Derived Units***

Quantity	Quantity symbol	Unit	Unit symbol
Capacitance	C	farad	F
Conductance	G	siemens	S
Electric charge	Q	coulomb	C
Electromotive force	E	volt	V
Energy, work	W	joule	J
Force	F	newton	N
Frequency	f	hertz	Hz
Inductance	L	henry	H
Magnetic flux	Φ	weber	Wb
Magnetic flux density	B	tesla	T
Power	P	watt	W
Resistance	R	ohm	Ω

1-3 CONVERSION OF UNITS

In some situations it is advantageous to convert from one set of units to another by use of a **conversion factor**, which is a constant that relates to two different units. Table 1-3 is a partial listing of some common conversion factors. It should be noted that conversion factors can be used only to change the units of a given physical quantity. They cannot be used to change the physical quantity itself. Also, only one conversion factor is needed for any given physical quantity for conversion between systems of units. For example, if we know that 1 in. = 2.54 cm, then 1 cm must equal the reciprocal, or 1/2.54 in.

TABLE 1-3 *Common Conversion Factors*

Quantity	Conversion
Length	1 inch (in.) = 2.54 centimeters (cm)
	1 foot (ft) = 0.3048 meter (m)
	1 meter = 39.37 in.
	1 kilometer (km) = 0.6213 mi
Area	1 square inch ($in.^2$) = 6.452 cm^2
	1 square foot (ft^2) = 0.093 m^2
Volume	1 quart = 0.946 liter (L)
	1 liter = 1000 cm^3
	1 cubic inch ($in.^3$) = 16.39 cm^3
Mass	1 pound (lb) = 453.6 grams (g)
	1 kilogram (kg) = 2.2046 lb
Force	1 newton (N) = 0.2248 pound force (lbf)
Energy	1 ft·lb = 1.3549 joules (J)
	1 Btu = 1055 J
	1 watthour (Wh) = 3600 J
Power	1 horsepower (hp) = 746 watts (W)
	1 hp = 550 ft·lb/s

EXAMPLE 1-1 Convert 4.37 feet to meters.

Solution

$$1 \text{ ft} = 0.3048 \text{ m}$$

$$4.37 \text{ ft} \times \frac{0.3048 \text{ m}}{\text{ft}} = 1.332 \text{ m}$$

EXAMPLE 1-2 Convert 0.648 hp to its equivalent in watts.

Solution

$$0.648 \text{ hp} \times \frac{746 \text{ W}}{\text{hp}} = 483.41 \text{ W}$$

1-4 SCIENTIFIC NOTATION

Numbers that are very large or very small can be more conveniently written as a number multiplied by 10 and raised to a power. This method of writing numbers is known as **scientific notation**. A number written in scientific notation is expressed as the product of a number greater than or equal to 1 and less than 10, and is a power of 10. To express a number in scientific notation the decimal point is moved until there is one significant digit to the left of the decimal point. The result is then multiplied by the appropriate power of 10 to return the quantity to its original value.

EXAMPLE 1-3 Express the following numbers in scientific notation: (a) 72,300; (b) 0.0057.

Solution (a) $72{,}300 = 7.23 \times 10^4$; (b) $0.0057 = 5.7 \times 10^{-3}$.

Since some of these powers of 10 appear frequently, various multiples and submultiples are assigned prefixes and symbols. Those that are most frequently used in electrical calculations are listed in Table 1-4.

TABLE 1-4 *Prefixes for Use with SI Units*

Prefix	Symbol	Scientific notation	Value
tera	T	10^{12}	1 000 000 000 000
giga	G	10^{9}	1 000 000 000
mega	M	10^{6}	1 000 000
kilo	k	10^{3}	1 000
milli	m	10^{-3}	0.001
micro	μ	10^{-6}	0.000 001
nano	n	10^{-9}	0.000 000 001
pico	p	10^{-12}	0.000 000 000 001
femto	f	10^{-15}	0.000 000 000 000 001

EXAMPLE 1-4 Express 5600 grams in a more appropriate form using SI unit prefixes.

Solution Moving the decimal point three places to the left results in a figure of 5.6, which must now be multiplied by 1000 or 10^3 to return the quantity to its original value.

$$5600 \text{ g} = 5.6 \times 10^3 \text{ g}$$

From Table 1-4, the value of 10^3 can be replaced by using the prefix *kilo* on the root unit:

$$5600 \text{ g} = 5.6 \times 10^3 \text{ g} = 5.6 \text{ kg}$$

To convert a fraction to scientific notation, it is first necessary to divide the fraction out into decimal form. The decimal point is now moved to the right until there is one significant digit to the left of the decimal.

EXAMPLE 1-5 Express 1/500 meter in scientific notation.

Solution Divide the fraction out into decimal form.

$$\frac{1}{500} \text{ m} = 0.002 \text{ m}$$

The decimal point is now moved three places to the right, and the result multiplied by 10^{-3}.

$$0.002 \text{ m} = 2.0 \times 10^{-3} \text{ m}$$

or, using SI unit prefixes,

$$2.0 \times 10^{-3} \text{ m} = 2.0 \text{ mm}$$

To change a number from scientific notation to ordinary notation, the procedure is reversed. The following examples illustrate this method.

EXAMPLE 1-6 Convert 7.84×10^5 to ordinary notation.

Solution The decimal point must be moved five places to the right. Therefore, additional zeros must be included for proper location of the decimal point.

$$7.84 \times 10^5 = 784{,}000$$

To add or subtract numbers expressed in scientific notation, it is necessary to convert numbers to a common power of 10.

EXAMPLE 1-7 Calculate the sum $(3.4 \times 10^5) + (5.9 \times 10^6)$.

Solution $(0.34 \times 10^6) + (5.9 \times 10^6) = 6.24 \times 10^6$

EXAMPLE 1-8 Find the difference $(8.4 \times 10^3) - (4.7 \times 10^2)$.

Solution $(8.4 \times 10^3) - (0.47 \times 10^3) = 7.93 \times 10^3$

To multiply numbers expressed in scientific notation, the exponents are added; to divide, the exponents are subtracted.

EXAMPLE 1-9 Calculate the product $(4 \times 10^5) \times (2.2 \times 10^2)$.

Solution

$$\begin{aligned}(4 \times 10^5) \times (2.2 \times 10^2) &= 4 \times 2.2^{5+2} \\ &= 8.8 \times 10^7\end{aligned}$$

EXAMPLE 1-10 Find the result of $\dfrac{6 \times 10^6}{3 \times 10^2}$.

Solution

$$\begin{aligned}\frac{6 \times 10^6}{3 \times 10^2} &= \frac{6}{3} \times 10^{6-2} \\ &= 2 \times 10^4\end{aligned}$$

Numbers that are written in powers-of-10 notation having exponents that are multiples of 3 are written in **engineering notation**. This technique calls for the use of numbers between 1.0 and 999 times the appropriate third power of 10. The values shown in Table 1-3 would be considered to be in engineering notation since these prefixes all represent powers-of-10 notation that have exponents with multiples of 3. The following rules are followed to express a decimal number in engineering notation:

1. Express the number in scientific notation.
2. If required, make the exponent a multiple of 3 by subtracting 1 or 2 from the exponent.
3. The decimal point is moved one or two places to the right to correspond to the change in exponent.

With some calculators, it is possible to fix a readout to either SCI or ENG notation, as well as selecting the number of decimal places required.

EXAMPLE 1-11 Write 645 000 in engineering notation.

Solution First, express the number in scientific notation.

$645\,000 = 6.45 \times 10^5$

Subtract 2 from the exponent.

$6.45 \times 10^{5-2}$

Move the decimal point two places to the right.

645×10^{3}

The number 645 000 could also be expressed in engineering notation by converting the number to scientific notation and moving the decimal point one place to the left.

$$\begin{aligned} 645\,000 &= 6.45 \times 10^{5} \\ &= 0.645 \times 10^{6} \end{aligned}$$

1-5 MOLECULES AND ATOMIC STRUCTURE

Mass is defined as the quantity of inertia possessed by a substance that occupies space. Anything that occupies space and has mass is called **matter**. All matter is composed of small particles called *atoms*. There are presently over 100 known types of atoms, referred to as *elements*. An element is a substance that cannot be chemically decomposed. Atoms can combine to form molecules. For example, one molecule of water contains two atoms of hydrogen and one atom of oxygen.

Prior to the twentieth century, there was considerable speculation concerning the nature of the atom. A discovery in 1911 by an English scientist named Rutherford established an atomic structure. This model was based on the principle of the atom being primarily an open space with all the mass concentrated in a central core, called the **nucleus**.

Rutherford's model of the atom was further extended by Niels Bohr. In 1913 Bohr proposed the theory that the simplest atom, hydrogen, consisted of a nucleus with a positively charged particle, or **proton**, and a planetary negative electron revolving in a circular orbit. The charge of a proton is equal in magnitude to that of an electron. Since the number of protons in the nucleus is equal to the electrons orbiting the nucleus, the atom is considered to be electrically neutral in charge.

Bohr's model of a hydrogen atom with one revolving electron is shown in Figure 1-1. This planetary atom theory may seem rather naive compared with more recent quantum-mechanical theories, but in fact, considering what it was designed to do, Bohr's theory is an excellent example of a successful physical model.

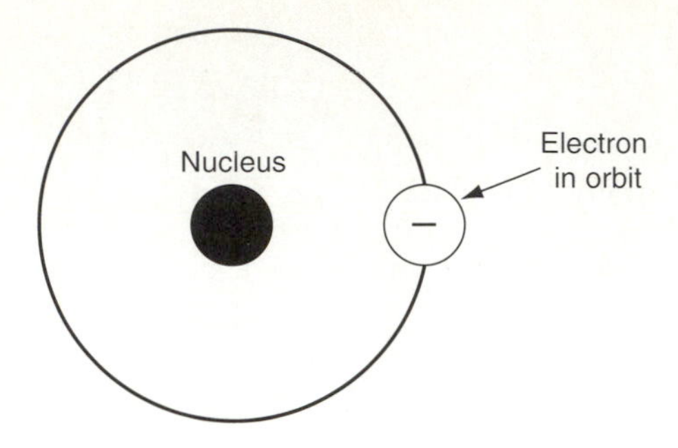

FIGURE 1-1 Bohr model of hydrogen atom.

As each electron is constantly in motion about the nucleus, it has energy of motion, or **kinetic energy**. Electrons travel at very high velocity and have charges that are equal, but opposite, to the charge of a proton. It is this attraction between the proton and electron that causes equilibrium with the centrifugal force acting on it.

Neutrons are particles of mass that do not have an electrical charge. The name is derived from the word *neutral*, meaning in this case neither positive nor negative. Neutrons are about the same size as protons and have approximately the same mass. Most of the mass of an atom is located in the nucleus since each proton has 1837 times as much mass as an electron. Because neutrons are neither positively nor negatively charged, the net charge of the nucleus is positive, due to the presence of protons. The neutron does not contribute to the flow of electricity, but it does contribute significantly to the mass of an atom.

In addition to having kinetic energy, the electron also has **potential energy**. This energy is the result of an attraction between the nucleus and the electron, which tends to "draw" the electron toward the nucleus. The potential energy increases as the distance from the nucleus increases. The closer an electron is to the nucleus, the greater its kinetic energy. However, the farther an electron is from the nucleus, the greater its potential energy. The potential energy of an electron is opposed to its kinetic energy. The net energy of an electron is the difference in magnitudes between the two energies.

The electrons very close to the nucleus are tightly held in their orbit and are called **bound electrons**. The outermost orbit, or shell, is known as the **valence orbit**. Since electrons in the valence orbit are farther from the nucleus, the attracting forces are not as great as the inner orbit. Figure 1-2 is a diagram of a copper atom with a single valence electron in the outer shell. By adding energy, such as heat or light, it is possible for a valence orbit electron to absorb this energy and move into the valence orbit of a neighboring atom. An electron freed from its valence state is called a **free electron**. Since a single cubic inch of copper contains approximately 1.4×10^{24} free electrons, a veritable cloud of free electrons moves about in a piece of copper. An atom that has lost one or more valence electrons has a net positive charge and is called a **positive ion**. This concept of free electrons is extremely important, as it is through them that an electrical current is possible.

As an electron moves from the valence orbit of one atom into the valence orbit of another atom, it leaves behind a space that it used to occupy. This space is often referred to as a hole. By definition, a **hole** is the absence of a negative electron and cannot be electrically neutral. Consequently, holes are considered to be positively charged and equal in magnitude to the negative charge of an electron.

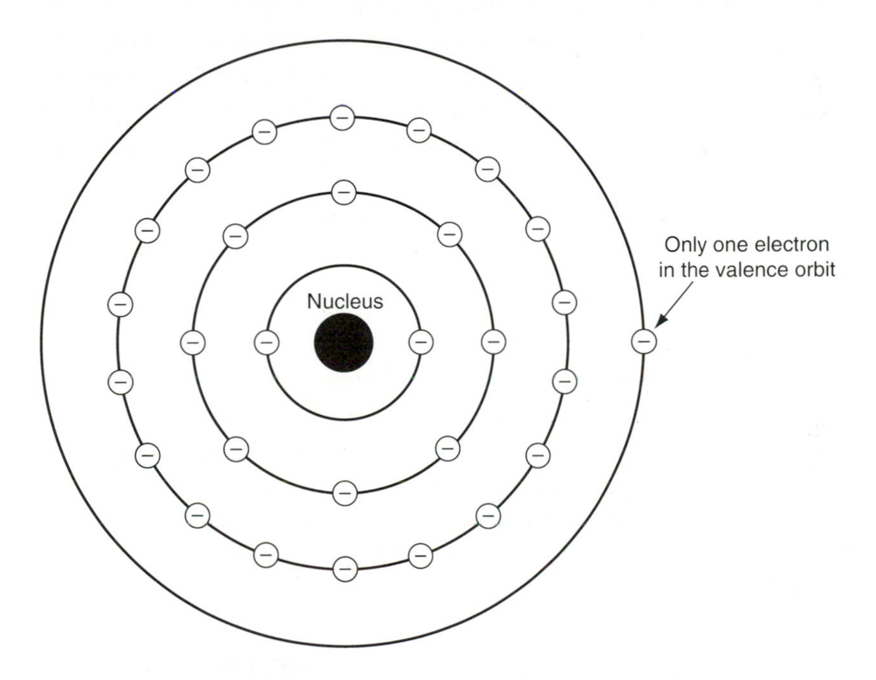

FIGURE 1-2 Copper atom with 29 electrons in nucleus and 29 orbital electrons.

1-6 ELECTRIC CHARGES

If a neutral atom were to acquire an extra electron, it would become negatively charged. Conversely, if it were to lose an electron, it would become positively charged. Charge is one of the fundamental properties of matter. Whenever one atom exchanges charge with another, the charge gained by one is exactly equal to the charge lost by the other. Because of this, the total net charge of any isolated system never changes. This fact is expressed in terms of the **law of**

conservation of electric charge, which is stated as follows:

> The algebraic sum of all electric charges in any isolated system is a constant.

It is possible to transfer an electric charge by causing friction between such materials as glass and silk. In these types of materials, charges do not move as freely as they would in metal. For this reason they are referred to as **static charges**, and their movement is called **static electricity**. A piece of glass rubbed with silk obtains a positive charge that does not easily escape, due to the glass being a good insulator. If a copper rod is also rubbed with silk, it will become charged, but the charge will rapidly be conducted away, due to the poor insulation qualities of copper.

A field of force, or electrostatic field, surrounds any charged body. The strength of the electric charge will determine the extent of this field. If an electron were placed within the electrostatic field of a positively charged body, the electron would be drawn toward the positive charge. If an electron were placed within a negative electrostatic field, the lines of force surrounding the charged body would repel the electron. In summary, *like charges repel each other, and unlike charges attract each other*.

1-7 COULOMB'S LAW

The French physicist Charles Augustine de Coulomb used a torsion balance, as shown in Figure 1-3, to provide direct experimental evidence for the inverse-square law, which was suggested by Joseph Priestly. The torsion balance consists of a horizontal balanced insulating rod which is suspended by a thin silver wire. The wire twists when a force is exerted on the end of the rod. This twisting effect can be used as a measure of the force. Coulomb determined that *the force of attraction or repulsion between two charged bodies is directly proportional to the square of the distance between them*.

Coulomb verified this law by attaching a charged body at point A of Figure 1-3, and placed another charged body at point B. The electrical force exerted on A by B caused the wire to twist. By measuring the twisting effect for different separations between the centers of spheres A and B, Coulomb was able to measure the force between the spheres. Mathematically, **Coulomb's law** can be expressed as

$$F = \frac{kQ_1Q_2}{d^2} \tag{1-1}$$

where F = force between the bodies, in newtons

k = a constant whose value depends on what material fills the volume of space in which the bodies are located (i.e., air, gas, water, etc.)

Q = charge of each body, in coulombs

d = distance between the two charges, in meters

The proportionality constant, k, called the **Coulomb constant**, is often expressed as $k = 1/(4\pi\epsilon_0)$, where ϵ_0 is referred to as the **permittivity of free space**. In SI units, the coulomb (C) is defined as the rate of flow of charge through a conductor. A total of 6.24×10^{18} electrons is required to amount to a charge of 1 C. Since the SI units of force, distance, and charge do not rely on Coulomb's law, the Coulomb constant, k, is determined as $k = 9.0 \times 10^9 \text{N} \cdot \text{m}^2/\text{C}^2$.

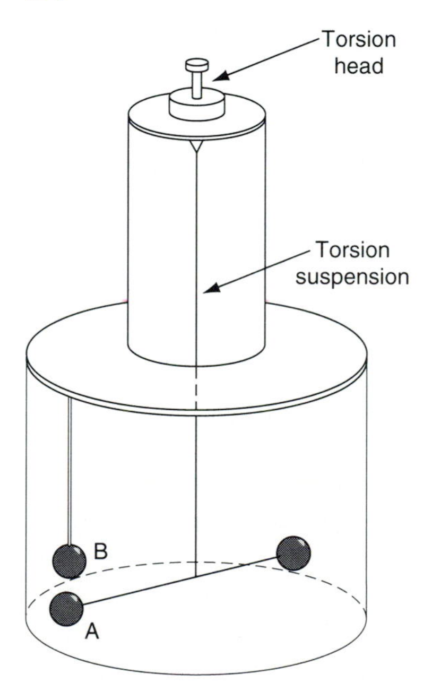

FIGURE 1-3 Coulomb's torsion balance.

Electric force is a vector quantity, with equation 1-1 giving only its magnitude. The direction of the vector **F** is always along the line joining Q_1 and Q_2. Therefore, when a charge is acted on by more than one force, the resultant force is the vector sum of the separate forces.

EXAMPLE 1-12 Two charges, each having a magnitude of 4 μC, are 6 cm apart. Determine the force of repulsion between the two charges.

Solution $F = \frac{kQ_1Q_2}{d^2} = \frac{(9.0 \times 10^9)(4 \times 10^{-6})^2}{(0.06\,\text{m})^2} = 40\text{ N}$

EXAMPLE 1-13 7.5×10^{12} electrons are transferred from body *A* to body *B*. Calculate the charge on each body.

Solution $1\text{ C} = 6.24 \times 10^{18}$ electrons

$$Q = \frac{7.5 \times 10^{12}}{6.24 \times 10^{18}} = 1.2\,\mu\text{C}$$

KEY TERMS

Length
Mass
Time
Fundamental units
Derived units
Meter
Kilogram
Second
International System of Units (SI)
Conversion factor
Scientific notation
Engineering notation
Mass
Matter
Nucleus
Proton
Kinetic energy
Neutron
Potential energy
Bound electron
Valence orbit
Free electron
Positive ion
Hole
Law of conservation of electric charge
Static charges
Static electricity
Coulomb's law
Coulomb constant
Permittivity of free space

PROBLEMS

1-1 Convert 7.2 in. to an equivalent length in centimeters.

1-2 Convert 8.5 hp to watts.

1-3 Convert 325 Btu to watthours.

1-4 Given that 0.3048 m = 1 ft. Calculate the following conversion factors in ratio form: (a) ft/m; (b) ft/cm; (c) m/ft.

1-5 Express the following numbers as powers of 10: (a) 0.0001; (b) 100,000; (c) 0.000 000 01; (d) 100,000,000.

1-6 Put the following numbers into scientific notation form: (a) 220,000; (b) 0.00318; (c) 4,570,000; (d) 0.0002755.

1-7 Express 1/2750 m in scientific notation.

1-8 Express the following voltages in engineering notation: (a) 5000 V; (b) 60,000 V; (c) 0.0000025 V; (d) 0.072 V.

1-9 Convert the following numbers to ordinary notation: (a) 6.77×10^6; (b) 3.25×10^{-3}; (c) 8.46×10^8; (d) 1.73×10^{-4}.

1-10 Find the sum of the following numbers.
(a) $(8.25 \times 10^3) + (3.57 \times 10^4)$
(b) $(9.75 \times 10^6) + (1.33 \times 10^5)$
(c) $(6.25 \times 10^3) - (0.38 \times 10^3)$
(d) $(4.38 \times 10^6) - (2.65 \times 10^5)$

1-11 Determine the products as powers of 10.
(a) $0.001 \times 1,000,000$
(b) 1000×100
(c) $0.01 \times 0.0001 \times 10^{-3}$
(d) $100 \times 10,000 \times 10^6$
(e) $10^6 \times 10^5 \times 10^{-8}$

1-12 Find the quotients as powers of 10.
(a) $\frac{1000}{0.01}$
(b) $\frac{0.0001^2}{10^2}$
(c) $\frac{1000^2}{0.01}$
(d) $\frac{0.00001^3}{0.0001^2}$

1-13 Calculate the product of the following numbers.

(a) $(4.38 \times 10^6) \times (7.75 \times 10^3)$
(b) $(1.77 \times 10^5) \times (6.11 \times 10^{-8})$
(c) $(9.51 \times 10^{-3}) \times (2.28 \times 10^{-6})$
(d) $(8.75 \times 10^9) \times (3.15 \times 10^6)$

1-14 Find the quotient of the following numbers.

(a) $\dfrac{3.25 \times 10^6}{1.61 \times 10^4}$

(b) $\dfrac{8.35 \times 10^2}{2.44 \times 10^5}$

(c) $\dfrac{3.71 \times 10^{-5}}{7.52 \times 10^3}$

(d) $\dfrac{6.25 \times 10^{-3}}{9.75 \times 10^{-6}}$

1-15 Determine the electric force between two very small particles in free space if they have charges of 3.5×10^{-7} C and -1.8×10^{-6} C and they are 52 cm apart.

1-16 How many coulombs are represented by each of the following quantities of electrons?
(a) 1.248×10^{20}; (b) 1.56×10^{18}; (c) 1.872×10^{19}; (d) 3.12×10^{19}.

1-17 Two charges, $Q_l = -12\ \mu C$ and $Q_2 = 16\ \mu C$, are 10 cm apart in air. Determine the resultant force on a third charge if it is placed in the center of Q_1 and Q_2 and has a value of $-6\ \mu C$.

CHAPTER

Current, Voltage, and Resistance

LEARNING OBJECTIVES

Upon completion of this chapter you will be able to:

- Define electric current.
- Differentiate between conductors and insulators.
- Define conventional current.
- Develop an understanding of potential energy.
- Describe five different voltage sources.
- Understand the difference between a voltage source and current source.
- Differentiate between a dependent source and an independent source.
- Define resistance and conductance.
- Express the relation between temperature and resistance.
- Calculate the resistance of various lengths of wire.
- Describe the difference between a rheostat and a potentiometer.
- Utilize the resistor color code.

2-1 ELECTRIC CURRENT

Solid materials are composed of large numbers of atoms. In certain materials, called **conductors**, it is very easy to remove some of the electrons from individual atoms. A conductor is a material that has the ability to transfer charge from one object to another. Materials that are poor conductors have electrons that are tightly bound to individual atoms. These materials are called **insulators**. An insulator is a material that resists the flow of charge. A class of materials called **semiconductors** is intermediate between conductors and insulators in its ability to transfer charge.

Electric current is defined as the rate of flow of charged particles through a conductor in a specified direction. The charged particle may be an electron, a positive ion, or a negative ion. In a solid such as copper wire, the charged particle is the electron. In a copper wire, the ions are rigidly held in place by the atomic structure of the material. Therefore, ions cannot be current carriers in solid material. In a semiconductor, the current is a movement of electrons in one direction and a movement of positively charged holes in the opposite direction.

Electric current, or rate of flow of charge, is measured in the units of coulombs per second. The SI unit of electric current is the **ampere**. It can then be stated that 1 ampere is equal to 1 coulomb of electric charge passing a certain point in an electric circuit in 1 second.

Amperes and coulombs per second express the same unit. The symbol for electric current is I. In equation form, current can be represented by the statement

$$I = \frac{Q}{t} \tag{2-1}$$

where I = current, in amperes

Q = charge transferred, in coulombs

t = time, in seconds

Conventional current is defined as the direction in which positive charge carriers flow through a circuit. If electrons flow one way in a material, the conventional electric current is in the opposite direction. This conventional flow is taken to imply that the electric current flows from the positive terminal of the power supply, then travels through the external circuit, and returns to the negative terminal of the power supply. The conventional direction of current flow is used in this book.

A current is considered to be unidirectional if it always maintains the same direction of flow, and bidirectional if it changes direction. When a unidirectional current is unchanging or changes negligibly, it is referred to as **direct current**. This type of current is a constant value that does not change with time or cross the zero axis. If the magnitude and direction of the current vary with time, the current is referred to as **alternating current**. These two types of current are shown in Figure 2-1.

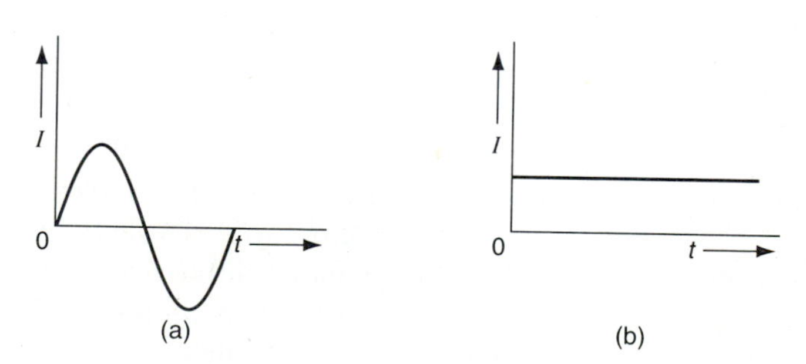

FIGURE 2-1 (a) Alternating current; (b) direct current.

2-2 ELECTRIC POTENTIAL AND VOLTAGE

Potential energy is defined as energy possessed by a system by virtue of position. For example, a lifted object or a compressed spring both have the potential for doing work because of position or condition. It is energy that can be stored for long periods of time in its present form. Potential energy is capable of doing work when it is converted from its stored form into another form, such as kinetic energy. The potential energy of a charge at a given point is directly proportional to the charge itself. Whenever a positive charge is moved against an electric field, the potential energy increases; whenever a negative charge moves against an electric field, the potential energy decreases.

Work is the amount of energy converted from one form to another as a result of motion or conversion of energy from potential to kinetic. For work to be done, three things must occur:

1. A force must be applied.
2. The force must act through a certain distance.
3. The force must have a component along the displacement.

In the SI system, the unit of work is the **joule**, abbreviated J. One joule is equal to the work done by a force of 1 N acting through a distance of 1 m. In equation form,

$$W = Fd \tag{2-2}$$

where W = work, or energy, in joules

F = force, in newtons

d = distance through which the force acts, in meters

The electric potential, or potential difference, between two points in an electric circuit is the amount of work required to move a unit charge from one point to another. In other words, electric potential is a measure of work per unit charge. The units of electric potential are expressed in joules per coulomb. Because this quantity is so frequently employed, its units have been given the name **volts** (V). One volt is the potential difference between two points in an electric circuit, when 1 joule of energy is required to move 1 coulomb of electric charge from one point to the other. The relationship between voltage, energy, and charge can be expressed as

$$E = \frac{W}{Q} \tag{2-3}$$

where E = potential difference, in volts

W = energy, in joules

Q = quantity of electric charge, in coulombs

EXAMPLE 2-1 What value of charge is moved between two points if 0.44 J of energy is used and a potential of 2.6 V is developed?

Solution

$$E = \frac{W}{Q}$$

$$Q = \frac{W}{E} = \frac{0.44}{2.6} = 0.169\,\text{C}$$

To maintain a constant current in a conductor, it is necessary to maintain a steady potential difference across a conductor. This potential difference can be supplied only if some device converts another form of energy into electric energy. Such a device is called a source of **electromotive force** (emf). As charges pass through a source of electrical energy, work is done on them. The emf of the source is the work done per coulomb on the charges. A typical car battery has an emf of 12 V, which means that 12 J of work is done on each coulomb of charge that passes through the battery.

To distinguish between a source of voltage in a circuit and a voltage loss across a dissipative component, the following notation will be used in this book: E = voltage source (volts) and V = voltage drop (volts).

2-3 VOLTAGE AND CURRENT SOURCES

A voltage source is a device capable of converting one form of energy into electrical potential energy. A voltage **cell** is considered to be a voltage source since it is a source of stored chemical energy that may be converted to electrical energy as required. A **battery** consists of one or more cells connected in series. A **primary cell** provides electrical energy to the limit of its chemical energy. Primary cells are nonrenewable voltage sources. A **secondary cell** is capable of being recharged. The electrical energy returned to the secondary cell during its charging process restores the chemicals in the battery to their original state.

There are five principal types of voltage sources:

1. *Chemical sources*. These sources convert chemical energy into electrical energy. Examples of such sources are primary cells, secondary cells, and fuel cells. Fuels cells operate on the principle that when hydrogen and

oxygen are combined, water is produced and electrical energy is released. Hydrogen is available from hydrocarbon sources such as ammonia, petroleum, propane, and natural gas.

Primary cells

Source: Courtesy of Duracell Canada Inc. Duracell and Dynacharge are registered trademarks of Duracell Canada Inc., 1991

2. *Solar and photovoltaic cells.* The photovoltaic effect, which is the basis of all solar cells, was discovered in 1893 by Edmond Becquerel. Using a pair of electrodes immersed in electrolyte, Becquerel observed a flow of current when the cell was illuminated with sunlight. A solar cell is a semiconductor device consisting of a thin layer of heavily doped p-type silicon on a heavily doped n-type silicon wafer. Incident light passes through the thin layer of p-type silicon to the junction region, where it creates a large number of free electron-hole pairs. A strong electric field causes the free electrons to flow to the negative terminal, while free holes move to the positive output terminal. The net result is a potential difference created in the device when it is irradiated with light.

Secondary cells

Source: Courtesy of Duracell Canada Inc. Duracell and Dynacharge are registered trademarks of Duracell Canada Inc., 1991

3. *Thermoelectric generation*. Thermoelectric generation is based on the principle that if a metal rod is heated at one end, negatively charged electrons flow from the hot end to the cooler end to reduce their energy. Examples of such sources are thermocouples and semiconductor thermoelectric engines.

4. *Electromagnetic generation*. This type of device is capable of converting mechanical energy to electrical energy. A generator, or dynamo, is a device in which mechanical energy is used to rotate conductors in a magnetic field to produce an emf. Examples of such sources are dc generators.

Solar cell

Source: Anthony Sara

5. *Electrical conversion*. A **power supply** is a device that converts one type of electric potential or current to another. A dc power supply converts an alternating signal into one of a fixed magnitude. In addition to dc power supplies, other examples of electrical conversion sources are rotary converters and motor-generator sets.

An ideal voltage source provides a constant voltage across its terminals, regardless of the size of load connected to it. This type of source is also referred to as an **ideal independent voltage source** or **constant voltage source**.

A **current source** is similar to a voltage source. An ideal or constant current source provides a specified value of current through its terminals regardless of the voltage across the terminals. This type of source is also referred to as an ideal independent current source. Ideal voltage and current sources are generally represented as a circle with an arrow symbol for a current source, and a circle with a polarity marking for a voltage source.

A **dependent source**, also referred to as a **controlled source**, produces a voltage or current as a function of a voltage or current at some other point in a circuit. Dependent sources are not usually purchased and installed in a circuit. Instead, they often appear in circuit diagrams as parts of models that operate similarly to transistors and other electronic components. There are four types of dependent sources:

1. *Voltage-dependent voltage sources (VDVS)*. Produces a voltage as a function of a voltage elsewhere in the circuit.
2. *Current-dependent voltage sources (CDVS)*. Produces a voltage as a function of a current elsewhere in the circuit.
3. *Voltage-dependent current sources (VDCS)*. Produces a current as a function of a voltage elsewhere in the circuit.
4. *Current-dependent current sources (CDCS)*. Produces a current as a function of a current elsewhere in the circuit.

Dependent current sources are usually represented by a diamond-shaped symbol with an arrow inside; dependent voltage sources are indicated by a diamond with a polarity symbol. Dependent sources are often associated with electronic components, such as transistors. For example, a bipolar junction transistor (BJT) has three terminals (emitter, base, and collector), and each terminal has a value of current flowing. The current flowing between the emitter and base can be expressed as a specific fraction of the current flowing between the emitter and collector. This dependence of the emitter-collector current on the emitter-base current results in the model of a bipolar junction transistor.

2-4 RESISTANCE

The ease with which electric current flows through a material depends on whether there are relatively large numbers of free electrons. A material with few electrons per unit volume would have a substantial opposition to current flow. If an electric potential is applied to a conductor, the electrons are given increased kinetic energy and collide more frequently with atoms. This friction caused by atoms colliding increases the temperature of the conductor. Therefore, when electric current flows in a conductor, some of the electrical potential energy is converted to heat energy. Because of this heat dissipation, **resistance** is not only an opposition to current flow but is also a producer of heat energy in a conductor.

The unit of measurement of resistance is the ohm, for which the symbol is Ω, the capital Greek letter omega. The standard international ohm is defined as the resistance at zero degrees Celsius of a column of mercury of uniform cross section having a mass of 14.4521 g and a length of 106.3 cm.

The **resistance** of any material with a uniform cross-sectional area is determined by the following four factors:

1. Type of material
2. Length
3. Cross-sectional area
4. Temperature

As the cross-sectional area of the conductor increases, its resistance decreases. As the length of the current path through the conductor increases, its resistance increases. This statement can be expressed in the form of the following basic rule: *The resistance of a conductor is directly proportional to its length and inversely proportional to its cross-sectional area.*

By using a constant of proportionality, ρ (the lowercase Greek letter rho), this rule can be expressed by the equation

$$R = \frac{\rho L}{A} \tag{2-4}$$

where R = resistance of material, in ohms

ρ = constant of proportionality between R and L/A

L = length of the current path through the conductor, in meters

A = cross-sectional area of the conductor, in square meters

The resistance of a conductor having unit length and unit cross-sectional area is defined as the specific resistance, or **resistivity**. The constant of proportionality in equation 2-4 would represent the resistivity of the conductor. Resistivity is used to compare the inherent resistance characteristics of different materials. A material with the lowest resistivity will be the best conductor and the poorest insulator. Conductors are defined as materials having resistivities from 10^{-6} to 10^{-8} $\Omega \cdot m$. Semiconductors are defined as materials with resistivities between 10^{7} and 10^{-6} $\Omega \cdot m$.

Since the SI base unit of length and cross section is the meter, the resistivity of a material is equal to the resistance of a piece of the material 1 m in length and 1 m^2 in cross-sectional area. A table of resistivities for common conducting materials is given in Table 2-1.

TABLE 2-1 *Resistivities of Common Conducting Materials*

Material	Resistivity ($\Omega \cdot$m at 20°C)	Resistivity ($\Omega \cdot$circular mils per foot at 20°C)
Aluminum	2.83×10^{-8}	17.0
Brass	7×10^{-8}	42.0
Copper, annealed	1.72×10^{-8}	10.37
Copper, hard-drawn	1.78×10^{-8}	10.7
Gold	2.45×10^{-8}	14.7
Lead	22.1×10^{-8}	132.0
Nichrome	100×10^{-8}	600.0
Nickel	7.8×10^{-8}	47.0
Platinum	10×10^{-8}	60.2
Silver	1.64×10^{-8}	9.9
Tungsten	5.52×10^{-8}	33.2

EXAMPLE 2-2 Find the resistance of a uniform copper conductor 1000 m in length and 0.0005 m^2 in cross-sectional area.

Solution

$$R = \frac{\rho L}{A} = \frac{1.72 \times 10^{-8}\,\Omega\cdot\text{m} \times 1000\,\text{m}}{0.0005\,\text{m}^2} = 0.034\,\Omega$$

EXAMPLE 2-3 A round copper wire 0.8 cm in diameter has a length of 1200 m. Calculate its resistance at ambient temperature.

Solution

$$A = \frac{\pi}{4} \times (0.008)^2 = 0.785 \times 64 \times 10^{-6} = 5.03 \times 10^{-5}\,\text{m}^2$$

Therefore,

$$R = \frac{\rho L}{A} = \frac{1.72 \times 10^{-8}\,\Omega\cdot\text{m} \times 1200\,\text{m}}{5.03 \times 10^{-5}\,\text{m}^2} = 0.411\,\Omega$$

For a circular wire, it is convenient to use circular units to denote the area of a material. The unit chosen to represent the relatively small diameter of a wire is the mil (0.001 in.). The **circular mil** (CM) is the area of a circular cross section having a diameter of 1 mil. The standard of reference for resistivity is therefore defined as the resistance of 1 circular mil-foot of conductor. The equivalent area of 1 circular mil in square inches is

$$1 \text{ CM} = 7.854 \times 10^{-7} \text{ in.}^2$$

To express the cross-sectional area of a circular conductor in circular mils, we simply multiply the inch value of the diameter by 1000 and square the result. In equation form, the area represented by a circular mil is given as

$$\text{area (CM)} = d^2 \tag{2-5}$$

EXAMPLE 2-4

How many circular mils in the circular cross section of a bar 2 in. in diameter?

Solution

$$\begin{aligned} 2 \text{ in.} &= 2 \text{ in.} \times 1000 \text{ mils/in.} \\ &= 2 \times 1000 \text{ mils} \\ &= 2 \times 10^3 \text{ mils} \end{aligned}$$

$$\begin{aligned} A = d^2 &= (2 \times 10^3)^2 \\ &= 4 \times 10^6 \text{ CM} \end{aligned}$$

2-5 RELATION BETWEEN TEMPERATURE AND RESISTANCE

By increasing the temperature of a conductor, an increase in atomic motion results in the conductor. As free electrons flow through a metallic conductor, their motion is impeded when they interact with the atomic ion cores. When a conductor is heated, its atoms increase their energy and vibrate with larger amplitudes. The higher the temperature, the faster the electrons move about between atoms, and the more the atoms vibrate. Consequently, the higher the temperature of a conducting material, the more likely it is that collisions will occur between the electrons drifting along the conductor and the atoms of the material. If the number of collisions increase, the flow of current will decrease.

Generally, the resistance of a conductor will increase with temperature. When pure metals are used within normal temperature ranges, the resistance increases in a fairly linear progression. For copper, the resistance at 100°C is nearly 43% greater than at 0°C, and the resistance increases proportionately between these temperatures.

Due to the linear relationship between the resistance and temperature of a conductor, the slope $\Delta R/\Delta T$ is constant, as shown in Figure 2-2. The increase

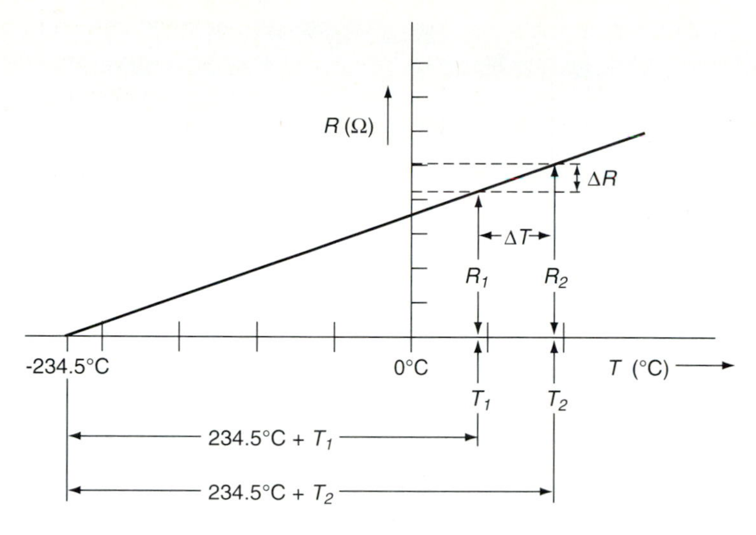

FIGURE 2-2 Variation of resistance with temperature for copper.

in resistance for most metals is approximately linear when compared with temperature changes. The proportional change in resistance to the change in temperature can be expressed in terms of the original resistance R as follows:

$$R = \frac{\Delta R}{\alpha \Delta T}$$

where ΔR represents the change in resistance and ΔT the change in temperature. The proportionality constant α, called the **temperature coefficient of resistance**, is defined as the change in resistance per unit resistance per degree change in temperature. Since the resistance of all pure metals increases as the temperature increases, metals are said to have a *positive temperature coefficient*.

The temperature intercepts and coefficients for common conducting materials are shown in Table 2-2. The temperature of $-234.5°C$ is called the **inferred zero-resistance temperature** of copper. Nichrome wire has a very small temperature coefficient, which means that its straight-line graph of resistance versus temperature has a very small slope.

When temperature is measured in degrees above inferred absolute temperature, the resistance is directly proportional to temperature. This can be expressed in equation form as

$$\frac{234.5 + t_1}{R_1} = \frac{234.5 + t_2}{R_2}$$

TABLE 2-2 *Temperature Coefficients of Resistance and Conductor Materials*

Material	Temperature coefficient (Ω·°C/Ω at 0°C)	Inferred zero-resistance temperature (°C)
Aluminum	0.0039	−236
Brass	0.002	−480
Copper, annealed	0.00393	−234.5
Copper, hard-drawn	0.00382	−242
German silver	0.0004	−2480
Iron	0.005	−180
Lead	0.0041	−224
Mercury	0.00089	−1100
Nichrome	0.0004	−2480
Nickel	0.006	−147
Platinum	0.003	−310
Silver (99.98% pure)	0.0038	−243
Steel, soft	0.0042	−218
Tin	0.0042	−218
Tungsten	0.0045	−200

Within normal ranges of temperature, the resistance (R_2) of a conductor at any temperature (t) is given in terms of the resistance (R_1) at a given temperature. This can be expressed by the equation

$$R_2 = R_1(1 + \alpha \Delta T)$$
$$\Delta T = t_2 - t_1 \qquad (2\text{-}6)$$

EXAMPLE 2-5 What is the resistance of a copper wire at 30°C if the resistance at 20°C is 4.31 Ω and if $\alpha(20°C) = 0.00393$?

Solution

$$\begin{aligned} R_2 &= R_1(1 + \alpha \Delta T) \\ &= 4.31\ \Omega\ [1 + 0.00393(30°C - 20°C)] \\ &= 4.31\ \Omega\ (1 + 0.0393) \\ &= 4.48\ \Omega \end{aligned}$$

2-6 CONDUCTANCE

Conductance can be defined as the reciprocal of resistance: that is, the ease with which current flows through a circuit. The SI unit of measurement is siemens (S) and is represented by the symbol G:

$$G = \frac{1}{R} \tag{2-7}$$

EXAMPLE 2-6 A given conductor has a resistance of 200 Ω. What is its conductance?

Solution

$$G = \frac{1}{200}$$
$$= 0.005\ \text{S} \quad \text{or} \quad 5\ \text{mS}$$

The reciprocal of resistivity is conductivity, or specific conductance, which is assigned the symbol σ. The relationship between resistivity and conductivity is expressed by

$$\sigma = \frac{1}{\rho} \tag{2-8}$$

Sometimes the conductivities of different materials are compared. In making a comparison, the conductivity of standard annealed copper is taken as 100%. The percent conductivity of a given material is expressed as a ratio of that material, with the annealed copper as a standard.

$$\%\ \text{Conductivity} = \frac{\text{conductivity of material}}{\text{conductivity of annealed copper}} \times 100$$
$$= \frac{\sigma}{\sigma_{\text{std}}} \times 100 \tag{2-9}$$

EXAMPLE 2-7 Determine the percent conductivity of silver.

Solution

$$\text{Silver (Table 2-1)} = 9.9\ \text{CM}\cdot\Omega/\text{ft}$$
$$\text{Copper (annealed)} = 10.37\ \text{CM}\cdot\Omega/\text{ft}$$
$$\%\ \text{Conductivity} = \frac{1/\rho(\text{silver})}{1/\rho(\text{copper})} \times 100$$
$$= \frac{0.101}{0.096} \times 100$$
$$= 105.2\%$$

2-7 WIRE SIZES

Wires are manufactured in sizes numbered according to the **American Wire Gage** (AWG). By analyzing Table 2-3, the following statements can be made. A lower AWG number indicates a greater cross-sectional area in circular mils. A decrease of one gage number represents an increase in the cross-sectional area by approximately 25%. An increase in three gage numbers multiplies the resistance by a factor of 2 and decreases the area by a factor of 2. A decrease in three gage numbers doubles the area and halves the resistance. A change of 10 wire gage numbers represents a 10:1 change in the cross-sectional area.

For such considerations as cost, weight, and bulk, it is desirable to use wire of the smallest diameter consistent with the minimum resistance that can be tolerated. Table 2-3 lists the information for copper wire only. Wire is also available in many other metals, such as aluminum, silver, platinum, and iron.

As Table 2-3 illustrates, the wire diameters become smaller as the gage numbers become larger. The largest wire size shown in the table is 0000 (pronounced "4 naught"), and the smallest number is 40 AWG. Larger and smaller sizes of conductors other than what is shown by the table are available. Very large conductors are measured in thousand circular mils (kCM). The Greek symbol M is also used to express 1000 CM. For example, 250,000 circular mils could be expressed as either 250 MCM or 250 kCM.

Table 2-3 can be utilized when making calculations of a conductor's resistance, as shown below.

EXAMPLE 2-8 Calculate the resistance of 600 ft of No. 18 AWG wire at 20°C.

Solution

$$R = \text{length} \times \frac{\Omega}{1000\text{ ft}}$$
$$= 600\text{ ft} \times \frac{6.385\ \Omega}{1000\text{ ft}}$$
$$= 3.83\ \Omega$$

TABLE 2-3 *American Wire Gage (AWG) Wire Sizes of Commercial Solid Copper Conductors at 20°C*

	Diameter		Area		Resistance	
AWG	mil	mm	CM	mm²	Ω/kft	Ω/km
0000	460.0	11.68	211,600	107.2	0.04901	0.160
000	409.6	10.40	167,810	85.01	0.0618	0.203
00	364.8	9.266	133,080	67.43	0.0780	0.255
0	324.9	8.252	105,530	53.49	0.0983	0.316
1	289.3	7.348	83,694	42.41	0.1240	0.406

Table 2-3 continued

	Diameter		Area		Resistance	
AWG	mil	mm	CM	mm²	Ω/kft	Ω/km
2	257.6	6.543	66,373	33.62	0.1563	0.511
3	229.4	5.827	52,634	26.67	0.1970	0.645
4	204.3	5.189	41,742	21.15	0.2485	0.813
5	181.9	4.620	33,102	16.77	0.313	1.028
6	162.0	4.115	26,250	13.30	0.395	1.29
7	144.3	3.665	20,816	10.55	0.498	1.63
8	128.5	3.264	16,509	8.367	0.628	2.06
9	114.4	2.9065	13,094	6.631	0.792	2.59
10	101.9	2.588	10,381	5.261	0.998	3.27
11	90.74	2.305	8,234	4.170	1.260	4.10
12	80.81	2.0525	6,529.9	3.310	1.588	5.20
13	71.96	1.828	5,178.4	2.630	2.003	6.55
14	64.08	1.628	4,106.8	2.08	2.525	8.26
15	57.07	1.450	3,256.7	1.650	3.184	10.4
16	50.82	1.291	2,582.9	1.310	4.016	13.1
17	45.26	1.150	2,048.2	1.040	5.064	16.6
18	40.30	1.024	1,624.3	0.823	6.385	21.0
19	35.89	0.912	1,288.1	0.653	8.051	26.3
20	31.96	0.812	1,021.5	0.519	10.15	33.2
21	28.46	0.723	810.1	0.412	12.80	41.9
22	25.35	0.644	642.4	0.324	16.14	52.8
23	22.57	0.573	509.45	0.259	20.36	66.7
24	20.10	0.5105	404.01	0.205	25.67	83.9
25	17.90	0.4547	320.40	0.162	32.37	106
26	15.94	0.405	254.1	0.128	40.81	134
27	14.20	0.3607	201.50	0.102	51.47	168
28	12.64	0.321	159.79	0.0804	64.90	213
29	11.26	0.286	126.72	0.0647	81.83	267
30	10.03	0.255	100.50	0.0507	103.2	337
31	8.92	0.227	79.70	0.0401	130.1	425
32	7.95	0.202	63.21	0.0324	164.1	537
33	7.08	0.180	50.13	0.0255	206.9	676
34	6.30	0.160	39.75	0.0201	260.9	855
35	5.61	0.1426	31.52	0.0159	329.0	1071
36	5.00	0.127	25.00	0.0127	414.8	1360
37	4.45	0.113	19.83	0.0103	523.1	1715
38	3.96	0.101	15.72	0.00811	659.6	2147
39	3.53	0.0897	12.47	0.00621	831.8	2704
40	3.14	0.0799	9.89	0.00487	1049.0	3422

2-8 RESISTORS

Resistors are devices that conduct electricity but also dissipate electric energy as heat. By adding resistance, supply voltage may be reduced, or current limited. There are two general types of resistors: composition resistors and wire-wound resistors.

Carbon composition resistors are the most common of the composition resistors. They are relatively inexpensive and are available in power ratings of 1/10 to 5 W. Figure 2-3 shows the basic construction of a fixed molded carbon composition resistor. The resistive element contains a mixture of graphite powder and silica as well as a binding compound, which is molded under a combination of heat and pressure. Since graphite is a semiconductor and silica is an insulator, higher resistance values are obtained by increasing the amount of silica in proportion to the graphite.

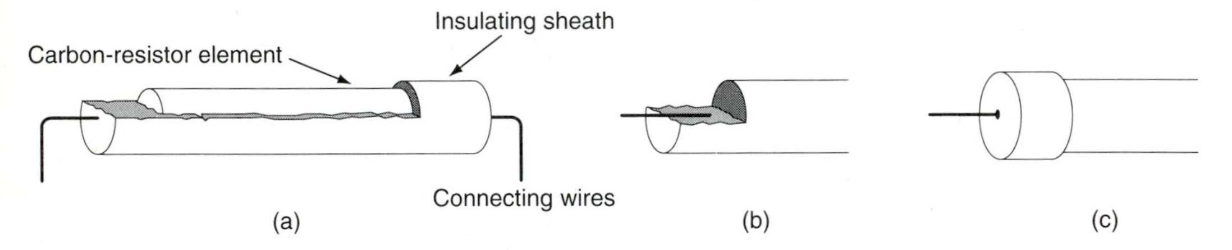

FIGURE 2-3 (a) Cutaway view of carbon composition resistor; (b) connecting wires embedded in the carbon composition; (c) connecting wires attached to end caps.

Wire-wound resistors are considerably different from composition resistors. They are made from alloys of relatively high resistivity drawn into wire with precisely controlled characteristics. This wire is then wrapped around a ceramic-core form. Different wire alloys are used to provide various resistor ranges. Wire-wound resistors are characterized by their high stability and power ratings up to 250 W. Figure 2-4 shows the schematic symbol for either a wire-wound or a composition resistor of a fixed value.

FIGURE 2-4 Symbol for a fixed resistor.

There are two basic types of variable resistors: the **rheostat** and the **potentiometer**. The rheostat is a two-terminal variable resistor that consists of a resistance element, wound around a circular insulated form. Inside this form is a contact that is rotated to vary the resistance. Rheostats are used to control

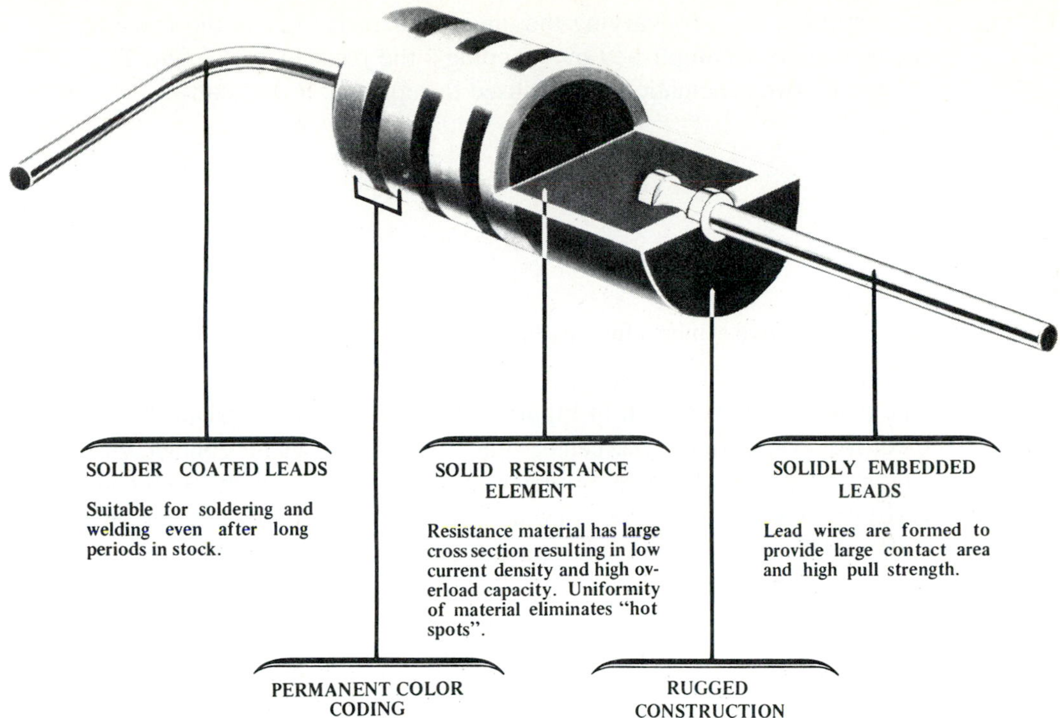

Carbon- composition resistor (hot-molded)

Source: Courtesy of Allen-Bradley Co.

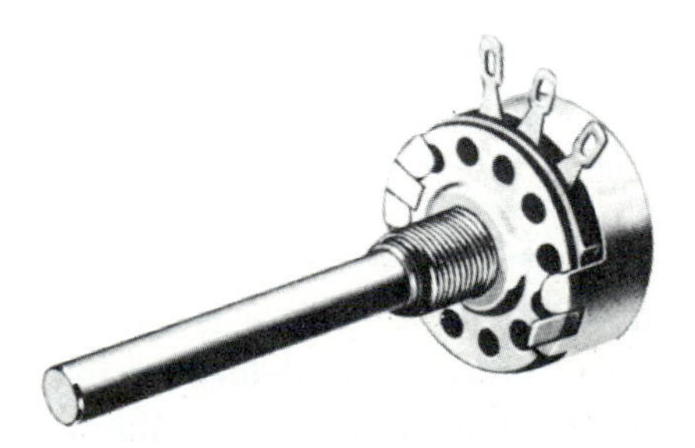

Potentiometer

Source: Prentice Hall Inc.

the circuit current by varying the amount of resistance in the resistance element. This is accomplished by "tapping" the resistance element. Figure 2-5 shows thc two schematic symbols used to represent a rheostat.

FIGURE 2-5 Alternative symbols for variable resistor or rheostat.

The potentiometer, shown in Figure 2-6(a), is a variable resistor having three electrical connections, the center one being the movable contact, or **wiper**. Figure 2-6(b) shows a potentiometer connected as a rheostat. Both ends of the resistance element of the potentiometer are connected in the circuit. In Figure 2-6(c), the wiper, or movable contact, is used to control the voltage applied across the circuit load.

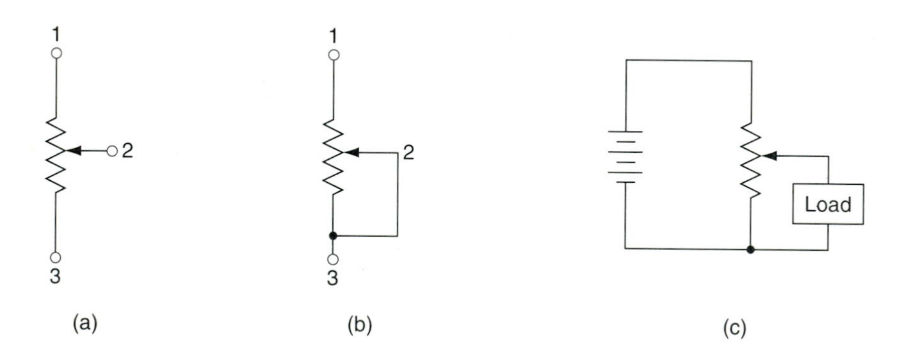

FIGURE 2-6 Symbols for (a) potentiometer, (b) potentiometer connected as a rheostat, and (c) potentiometer connected to vary circuit resistance and load voltage.

2-9 RESISTOR COLOR CODE

The resistance value of a carbon composition resistor is specified by a set of four **color-code bands** on the resistor housing. To determine the ratings of the resistor, it is held so that the bands are on the left and the reliability designation is on the right. The first and second bands produce a two-digit number, the third band indicates the multiplier, and the fourth band designates the percent tolerance, or accuracy of the resistor's rating. The four resistor color bands are shown in Figure 2-7.

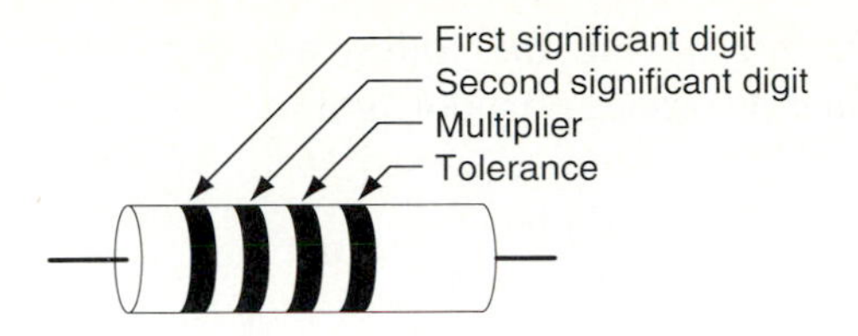

FIGURE 2-7 Resistor color bands.

Table 2-4 contains the color code for carbon composition resistors. Three color bands near one end of the resistor indicate the resistance value, and an added gold or silver band means that the accuracy is ±5% or ±10% of the amount designated by the three bands. When the resistor contains only three bands, the tolerance is taken as ±20%.

Some resistors also have a fifth color band, which indicates the **reliability factor.** The fifth band gives the percentage of failure per 1000 hours of use. For example, a 1% failure would indicate that 1 out of every 100 resistors will not lie within the tolerance range after 1000 hours of operation at its rated power.

TABLE 2-4 *Color Code for Carbon Composition Resistors*

	First band:	Second band:	Third band:	Fourth band:	Fifth band:
Color	first significant digit	second significant digit	third significant digit	tolerance (%)	reliability (%)
Black	—	0	10^0	—	—
Brown	1	1	10^1	—	1
Red	2	2	10^2	—	0.1
Orange	3	3	10^3	—	0.01
Yellow	4	4	10^4	—	0.001
Green	5	5	10^5	—	—
Blue	6	6	10^6	—	—
Violet	7	7	10^7	—	—
Gray	8	8	10^8	—	—
White	9	9	10^9	—	—
Gold	—	—	10^{-1}	±5	—
Silver	—	—	10^{-2}	±10	—
None	—	—	—	±20	—

EXAMPLE 2-9 Determine the rating of a resistor having the following four color bands: band 1, red; band 2, violet; band 3, orange; band 4, gold.

Solution

Red $= 2$ (first significant digit)
Violet $= 7$ (second significant digit)
Orange $= 10^3$ (multiplier in third band)
Gold $= \pm 5\%$ (tolerance)

$$\text{Resistor} = 27 \times 10^3\ \Omega \pm 5\%$$
$$= 27{,}000\ \Omega \pm 5\%$$

EXAMPLE 2-10 Find the rating of a resistor with the following color bands: band 1, orange; band 2, white; band 3 gold; band 4, no color.

Solution

Orange $= 3$ (first significant digit)
White $= 9$ (second significant digit)
Gold $= 0.1$ (multiplier in third band)
None $= \pm 20\%$ (tolerance)

$$\text{Resistor} = 39 \times 0.1\ \Omega \pm 20\%$$
$$= 3.9\ \Omega \pm 20\%$$

KEY TERMS

Conductor
Insulator
Semiconductor
Electric current
Ampere
Conventional current
Direct current
Alternating current
Potential energy
Work
Joule
Volt
Electromotive force
Cell
Battery
Primary cell
Secondary cell
Power supply
Ideal independent voltage source (constant voltage source)
Current source
Dependent (controlled) source
Resistance
Resistivity
Circular mil
Temperature coefficient of resistance
Inferred zero-resistance temperature
Conductance
American Wire Gage
Resistor
Rheostat
Potentiometer
Wiper
Color-code band
Reliability factor

PROBLEMS

2-1 Determine the resistance of an annealed copper bus bar that is 0.75 m long and 1.5 × 3.5 cm wide.

2-2 Find the resistance of a round annealed copper conductor 0.22 cm in diameter and 750 m long.

2-3 Calculate the resistance of 250 m of aluminum conductor which has a cross-sectional area of 3.5 mm^2.

2-4 A 1-kΩ carbon resistor has a diameter of 1.75 mm and a length of 1.55 cm. Determine the resistivity of its material.

2-5 Determine the resistance of a copper conductor at 20°C which is 50 ft long and 30 mils in diameter.

2-6 Calculate the area in circular mils for the following diameters: (a) 0.0375 in.; (b) 0.062 in.; (c) 0.166 in.; (d) 0.003 in.

2-7 Find the diameter in inches for the following areas: (a) 270 CM; (b) 5500 CM; (c) 1200 CM; (d) 820 CM.

2-8 A cable contains four conductors 0.025 in. in diameter, six conductors 0.0035 in. in diameter, and five conductors 0.055 in. in diameter. What is the number of circular mils of conductor in this cable?

2-9 Determine the resistance of a copper conductor 150 ft long that has a square cross section 75 mils on each side.

2-10 Calculate the length of a round annealed copper conductor 8 mm in diameter that has a resistance of 0.0025 Ω.

2-11 A conductor made of annealed copper has a resistance of 4.45 Ω at 20°C. What will the resistance of the conductor be at 60°C?

2-12 A length of pure silver wire has a resistance of 2.4 Ω at 20°C. Calculate its resistance at 40°C.

2-13 Determine the resistance of an aluminum conductor at 50°C if the resistance is 1.86 Ω at 20°C.

2-14 What is the resistance of a copper conductor 250 m in length with a cross-sectional area of 1.87 mm^2 at 50°C?

2-15 A copper conductor has a resistance of 220 μΩ at 70°C. If the temperature is reduced to −20°C, what is the resistance of the conductor?

2-16 What is the resistance of an aluminum conductor at −25°C if its resistance at +40°C is 2.33 Ω?

2-17 A copper conductor has a resistance of 24.5 Ω at 20°C. Find its resistance at the following temperatures: (a) 60°C; (b) −25°C; (c) 40°C.

2-18 A coil of copper wire has a resistance of 120 Ω. Determine its conductance.

2-19 Find the conductance of 150 ft of 0.06-in.-diameter copper wire.

2-20 Determine the conductance of 75 m of aluminum wire with a diameter of 1.20 mm.

2-21 Using the data in Table 2-1, determine the percent conductivity of aluminum.

2-22 Calculate the resistivity of a length of annealed copper wire in ohm-circular mils per foot if its conductivity is 92.3%.

2-23 Calculate the resistance of 400 ft of No. 16 AWG wire at 20°C.

2-24 What is the resistance of 750 ft of No. 8 wire at 20°C?

2-25 The distance between an electrical panel and a motor is 600 ft. The resistance of the circuit to energize the motor must not exceed 0.68 Ω. Calculate the correct wire gage to be used.

2-26 The distance between load and source is 100 ft. The total resistance of the conductor cannot exceed 0.037 Ω. What wire gage should be selected?

2-27 A No. 24 gage conductor has a resistance of 17.7 Ω. What is the length of the conductor?

2-28 What is the length of a No. 16 gage conductor if it has a resistance of 0.75 Ω?

2-29 Find the length of a No. 20 gage conductor with a resistance of 0.53 Ω.

2-30 Determine the range of resistance for a carbon composition resistor having color bands of red, black, orange, and gold.

2-31 What is the resistance rating of a carbon composition resistor having color bands of yellow, violet, orange, and silver?

2-32 A resistor has color bands in the order of blue, yellow, and gold. What are the ohmic value and tolerance of this resistor?

2-33 Determine the color bands for a 1.5-Ω $\pm$ 10% resistor.

2-34 Find the color bands for a 6.8-MΩ $\pm$ 5% resistor.

CHAPTER

Ohm's Law, Power, and Energy

LEARNING OBJECTIVES

Upon completion of this chapter you will be able to:

- Define Ohm's law.
- Utilize Ohm's law to determine current, voltage, or resistance.
- Describe the linear relationship between current and voltage.
- Differentiate between work and energy.
- Define power.
- Determine the efficiency of an electrical device.
- Calculate power consumption in terms of kilowatthours.

3-1 OHM'S LAW

In 1827 a German scientist named Georg Simon Ohm discovered a relationship between the voltage that existed across a simple electric circuit and the current through that circuit. He determined that the magnitude of a current is, in general, proportional to the magnitude of the emf that produces it. When he doubled the potential difference he found that the current was doubled, when he tripled the potential difference the current was tripled, and so on. Ohm discovered that as the voltage increased, the current increased in direct proportion to the applied voltage, maintaining a constant ratio of voltage to current. **Ohm's law** is stated as follows:

> The current produced in a given conductor is directly proportional to the difference of potential between its endpoints.

The linear relationship between voltage and current is illustrated in Figure 3-1. The straight line that results means that the resistance of the conductor is the same regardless of the magnitude of applied voltage. Assuming that the temperature of a conductor is constant, the slope of the curve, k, is the value of the voltage divided by the current at any given instant. This relationship can be expressed mathematically as

$$k = \frac{E}{I} \tag{3-1}$$

The slope k is resistance; therefore, Ohm's law is expressed in equation form as

$$R = \frac{E}{I} \tag{3-2}$$

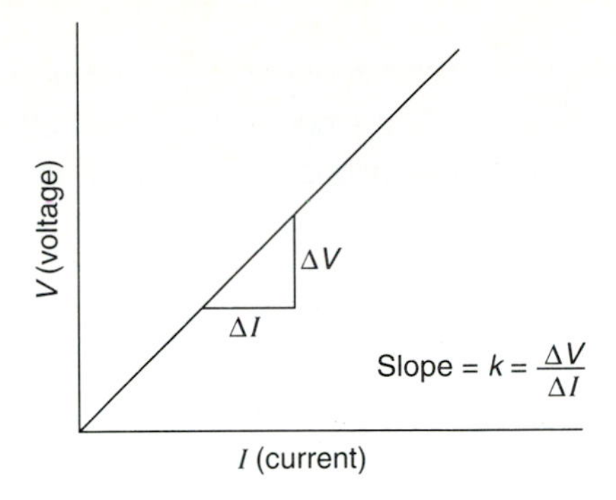

FIGURE 3-1 Current versus voltage curve.

Recall from Chapter 2 that the symbol for a voltage source is considered to be E and the symbol for a voltage drop is represented by V. The curve shown in Figure 3-1 would be applicable to either a voltage source or a voltage drop. Consequently, E and V are interchangeable in equations 3-1 and 3-2.

A semiconductor has a negative temperature coefficient of resistivity α that varies greatly with temperature. When a resistance is not constant, a graph of the current versus the voltage will result in a *curved* line, due to the fact that the resistance is *nonlinear*. By manipulating Ohm's law, any one of the three variables can be determined, if two are known.

EXAMPLE 3-1 An incandescent lamp is connected to a 120-V supply; current flow is 0.833 A. What is the resistance of the lamp?

Solution

$$R = \frac{E}{I} = \frac{120\text{ V}}{0.833\text{ A}}$$
$$= 144.06\ \Omega$$

EXAMPLE 3-2 A current of 160 mA is measured through a conductor whose resistance is 0.25 Ω. Determine the voltage drop across the conductor.

Solution

$$V = I \times R = 160\text{ mA} \times 0.25\ \Omega$$
$$= 40\text{ mV}$$

EXAMPLE 3-3 A 48-V source is connected to a circuit with a resistance of 15 Ω. Determine the current drawn from the source.

Solution

$$I = \frac{E}{R} = \frac{48\text{ V}}{15\ \Omega}$$
$$= 3.2\text{ A}$$

3-2 WORK AND ENERGY

Whenever a force acts so as to produce motion in a body, the force acts during a displacement of the body. Unless a force acts through a distance, no **work** is done, regardless of how great the force. When the force and its displacement are perpendicular to each other, no work is done, even though the force has moved. An example of this is motion parallel to the earth's surface, where the force of gravity, which pulls downward, is perpendicular to all horizontal displacements. Motion must take place in a mechanical consideration of work, such as a rotating electric motor.

Work is independent of time. The SI unit for work is the joule. One joule is the work accomplished when an object is moved a distance of 1 meter against an opposing force of 1 newton. In Chapter 2, work was expressed using the following equation:

$$W = Fd$$

where W = work, in joules

F = force, in newtons

d = distance, in meters

EXAMPLE 3-4 If it takes a force of 8 N to move an object that is pushed by this force for a distance of 1.5 m, what amount of work is done?

Solution

$$W = Fd = 8\text{ N} \times 1.5\text{ m}$$
$$= 12\text{ J} \quad \text{or} \quad \text{N}\cdot\text{m}$$

In the case of electrical machinery, work input or output is delivered through a shaft. A turning shaft that is doing work can be represented as being rotated by a force (F) operating at a radius (r). The distance through which the force operates depends on how many times the shaft rotates. For one revolution, force operates through the circumference of the circle, whose radius is measured in meters. This circumference can be expressed as $2\pi r$ meters. Therefore, the work done in rotating the shaft one revolution is

$$W = 2\pi rF \qquad (3\text{-}3)$$

where W = work, in joules

r = radius, in meters

F = force, in newtons

The portion rF in equation 3-3, called **torque**, is expressed in SI units as newton-meters. In English units, torque is expressed in foot-pounds.

Since work is performed when electrical energy is converted to mechanical energy, and when mechanical energy is converted to its electrical counterpart, the joule is also used as a unit measurement of electric energy. Consequently, the joule is used as the SI unit for both electric energy and work. The definition of the joule can be stated in electrical terms as:

> One joule of electric energy is required to raise 1 coulomb of electric charge through a potential difference of 1 volt.

When energy is converted from mechanical energy to electrical energy, work is said to be done. This implies that energy can be defined as the *ability to do work*. There are two kinds of energy: kinetic and potential. **Kinetic energy** is possessed by a body by virtue of its motion. **Potential energy** is possessed by a system by virtue of its position, or condition. Examples of kinetic energy are a moving car, or bullet. Examples of potential energy are a compressed spring or a lifted object.

3-3 POWER

Power is an indication of how much work can be accomplished in a specified amount of time. From this statement, power can be defined as the *rate at which work is done, or energy expended*. This definition is true for both mechanical power and electrical power. Since work is measured in joules (J) and time in seconds (s), power is measured in *joules per second*. The electrical unit of measurement for power is the **watt** (W). One watt is the rate of doing work when 1 joule of work is done in 1 second. Expressing this in equation form, we have

$$P = \frac{W}{t} \tag{3-4}$$

where P = power, in watts

W = work, in joules

t = time, in seconds

As stated, the unit of measurement of power is the watt, which is derived from the surname of James Watt, the developer of the steam engine. In the British Engineering System, power may be expressed in foot-pounds per second, but more often it is given in **horsepower**. The horsepower, watt, and foot-pound are related in the following manner:

$$1 \text{ horsepower} = 550 \text{ ft-lb/second} = 746 \text{ watts} \quad (3\text{-}5)$$

Electric power may also be defined as:

> One watt is the power expended when 1 ampere of direct current flows through a resistance of 1 ohm.

One watt is also the power dissipated when there is 1 ampere as the result of the application of 1 volt.

In Chapter 2 the relationship between current and voltage was discussed. Voltage was defined as being a ratio between energy and charge:

$$\text{electric potential } (E) = \frac{\text{energy (joules)}}{\text{charge (coulombs)}}$$

and electric current is given as a ratio between charge and time:

$$\text{electric current } (I) = \frac{\text{charge (coulombs)}}{\text{time (seconds)}}$$

If the current and voltage were multiplied together, the product of $E \times I$ would result in a ratio of energy/time, or joules per second, which according to equation 3-4, is the formula for power. Therefore, in a direct-current circuit the amount of power dissipated can be determined by the equation

$$P = EI \quad (3\text{-}6)$$

where P = power dissipated, in watts

E = applied voltage, in volts

I = current flow, in amperes

By direct substitution of Ohm's law, the equation for power can be obtained in two other forms:

$$\begin{aligned} P &= EI \\ &= E\frac{E}{R} \\ &= \frac{E^2}{R} \end{aligned} \quad (3\text{-}7)$$

and

$$\begin{aligned} P &= EI \\ &= (IR)I \\ &= I^2R \end{aligned} \quad (3\text{-}8)$$

EXAMPLE 3-5 What is the power delivered to a dc motor when the input current is 8 A and the supply voltage is 120 V?

Solution

$$P = EI = 120\text{ V} \times 8\text{ A} = 960\text{ W}$$

EXAMPLE 3-6 What is the power dissipated by an incandescent light bulb with a rated voltage of 120 V and a resistance of 240 Ω?

Solution

$$P = \frac{E^2}{R} = \frac{120^2\text{ V}}{240\,\Omega} = 60\text{ W}$$

EXAMPLE 3-7 An electronic device operates at 12 V and is supplied with 200 mW of power. Determine the current drawn by the device.

Solution

$$P = EI$$

Therefore,

$$I = \frac{P}{E} = \frac{200\text{ mW}}{12\text{ V}} = 16.67\text{ mA}$$

EXAMPLE 3-8 Determine the voltage across a 2-kΩ resistor receiving a power of 50 mW.

Solution

$$V = \sqrt{PR} = \sqrt{(50 \times 10^{-3}\text{ W})(2 \times 10^{3}\,\Omega)} = 10\text{ V}$$

3-4 EFFICIENCY

Efficiency is a measure of how completely the power put into a device is used as output. Some of the energy put into a machine or system is lost either in overcoming friction or in some other way. When any system or device converts one form of energy to another, some energy is lost or wasted, usually in the form of heat. The resistance of electrical conductors causes the production of heat when current flows. This energy is lost when it occurs in a power supply, such as a generator or battery. To overcome these losses of power in an electric circuit, the input power must be greater than the output power. Efficiency is defined as the ratio of useful output energy to total input energy. The letter symbol for efficiency is the lowercase Greek letter η (eta) and is expressed by the equation

$$\eta = \frac{\text{output power}}{\text{input power}} \times 100 \tag{3-9}$$

EXAMPLE 3-9 A 4-hp motor draws 4.8 kW from a power source. What is the efficiency of the motor?

Solution

$$\text{Output power } (P_O) = 4 \text{ hp}$$
$$= 4 \times 746 \text{ (W/hp)}$$
$$= 2984 \text{ W}$$
$$= 2.984 \text{ kW}$$

$$\text{Input power } (P_I) = 4.8 \text{ kW}$$
$$= \frac{P_O}{P_I} \times 100$$
$$= \frac{2.984 \text{ kW}}{4.8 \text{ kW}} \times 100$$
$$= 62.2\%$$

EXAMPLE 3-10 A motor has an efficiency 85%. What current does it draw from a 120-V source when its output is 7 hp?

Solution

$$P_O = 7 \times 746 \text{ (W/hp)}$$
$$= 5222 \text{ W}$$

$$\eta = \frac{P_O}{P_I}$$

$$\therefore \quad P_I = \frac{P_O}{\eta}$$

$$I = \frac{5222 \text{ W}}{0.85}$$
$$= 6143.53 \text{ W} \quad \text{(input power)}$$

$$P = EI$$

$$I = \frac{P}{E} = \frac{6143.53 \text{ W}}{120 \text{ V}}$$
$$= 51.2 \text{ A}$$

3-5 THE KILOWATTHOUR

The basic unit of electric energy is the joule. Because this unit is too small for typical problems, a more practical unit of electrical energy is required. Since $P = W/t$, then $W = Pt$. If P is power in watts and t is time in seconds, W must be work in watt-seconds or joules. Therefore, if P is power in kilowatts and t is time in hours, W must be work in **kilowatthours**. The watthour meter is an instrument for measuring the energy supplied to the residential or commercial user of electricity.

EXAMPLE 3-11 A generator delivers 4 kW to a customer for 60 hours. How much energy does the customer receive in this time?

Solution

$$\begin{aligned} W = Pt &= (4\text{ kW})(60\text{ h}) \\ &= 240\text{ kW}\cdot\text{h} \end{aligned}$$

EXAMPLE 3-12 At 8 cents ($0.08) per kW·h, how much will it cost to use a 60-W lamp for 20 days?

Solution

$$\begin{aligned} \text{Cost} &= \frac{8\text{ cents}}{\text{kW}\cdot\cancel{\text{h}}} \times 60\text{ W} \times 20\ \cancel{\text{days}} \times \frac{24\ \cancel{\text{h}}}{\cancel{\text{day}}} \\ &= \frac{8 \times 60 \times 20 \times 24}{1000}\text{ cents} \\ &= 230\text{ cents} \\ &= \$2.30 \end{aligned}$$

EXAMPLE 3-13 A light-emitting diode (LED) is used as an indicating light in a fire alarm panel. The LED is continuously on, drawing 30 mA from a 6-V supply. Determine the yearly cost to maintain its operation at $0.12 per kW·h.

Solution

$$P = EI = (6)(0.03) = 0.18\text{ W}$$

$$\begin{aligned} \text{Cost} &= \frac{12\text{ cents}}{\text{kW}\cdot\cancel{\text{h}}} \times 0.18\text{ W} \times 365\ \cancel{\text{days}} \times \frac{24\ \cancel{\text{h}}}{\cancel{\text{day}}} \\ &= \frac{12 \times 0.18 \times 365 \times 24}{1000}\text{ cents} \\ &= 18.92\text{ cents} \\ &= \$0.19 \end{aligned}$$

KEY TERMS

Ohm's law	Power
Work	Watt
Torque	Horsepower
Kinetic energy	Efficiency
Potential energy	Kilowatthour

PROBLEMS

3-1 What is the resistance of a lamp that draws 2.15 A from a 120-V supply?

3-2 A resistor in a television receiver must have a 2.5-V drop across it when its current flow is 4 mA. What value of resistance is required?

3-3 If the current in a circuit is 0.5 A and the resistance is 12 Ω, what is the voltage?

3-4 How much voltage must be applied across a 30-Ω load to cause a current of 4 A to flow?

3-5 An electric heater draws 12 A and its terminal voltage is 120 V. If the total resistance of the supply conductors is 0.75 Ω, what is the voltage drop in the supply conductors?

3-6 A radio receiver has an ''internal'' resistance of 225 Ω. How much current will the receiver draw when 120 V is applied?

3-7 If the voltage across a resistance doubles, what is the effect on the current flowing through the resistance?

3-8 A 1-MΩ resistor in a television receiver has a voltage drop of 5 V. When the resistor heats up, its resistance changes to 1.12 MΩ. What value of current flows when the resistor is heated?

3-9 Determine the amount of work done in lifting a 5-kg mass up to a height of 25 m.

3-10 How much work is done in lifting an 80-lb weight from the floor to a platform 5 ft high?

3-11 A force of 8.37 N is applied to a mass. If the mass is displaced 2.8 ft, how much work is done in (a) ft-lb; (b) joules?

3-12 What is the power supplied to a dc motor when the input current is 6 A and the supply voltage is 240 V?

3-13 A soldering iron requires 0.33 A at 120 V. How much power is dissipated?

3-14 An electric range will draw 80 A at 220 V when on the maximum heat setting. What is the power requirement of the range?

3-15 Determine the maximum voltage and current ratings for a 2-kΩ 5-W resistor.

3-16 Find the current and resistance of a 120-V 100-W lamp.

3-17 At what rate must a 60-Ω soldering iron dissipate heat if the current through it is 0.64 A?

3-18 A 240-V 20-hp motor has a full-load current rating of 65 A. Calculate the efficiency of the motor.

3-19 A motor requires 6.4 A when operating from a 120-V source. If the efficiency of the motor is 81%, what is the horsepower output?

3-20 If the efficiency of a motor is 93%, how much current does it draw from a 220-V source if it has a rated output of 10 hp?

3-21 What is the input power in watts to a 2-hp motor operating at 96% efficiency?

3-22 One kilowatthour is equivalent to how many joules?

3-23 A microwave oven rated at 700 W operates for 20 minutes. How much electric energy is used in (a) kW·h; (b) joules?

3-24 A generator delivers 100 kW to a building for 30 days. How many kW·h were supplied to the building?

3-25 If the cost per kW·h is 6 cents, how much would it cost to supply 80 kW over a period of 15 days?

3-26 Assuming a rate of $0.08 per kW·h, determine the cost of operating the following electrical equipment for 30 days:
(a) 600-W iron for 2 h/day
(b) 260-W TV receiver for 5 h/day
(c) four 60-W, two 100-W, and three 40-W lamps for 6 h/day
(d) 750-W microwave oven for 30 min/day

3-27 A 30-hp pump motor is used for an average of 8 h/day. If the motor has an efficiency of 88%, what would be the cost of operating the motor over a 20-day period if the industrial rate is 6 cents/kW·h?

CHAPTER

Series and Parallel Circuits

LEARNING OBJECTIVES

Upon completion of this chapter you will be able to:

- Explain how voltages are distributed around a series circuit.
- Recognize the difference between series and parallel circuits.
- Understand the purpose of double-subscript notation.
- Define Kirchhoff's laws.
- Express the voltage divider rule and understand where it can be applied.
- Calculate power in a series and a parallel circuit.
- Explain how current divides in a parallel circuit.
- Determine the total resistance of a parallel circuit.

4-1 SERIES CIRCUITS

The coverage of this chapter is confined to direct-current networks: that is, to those in which there is only direct current without variation, and in which only the effects of resistance are considered. Later, in the study of alternating-current circuits, many of the principles contained in this chapter will be found to be applicable to ac circuits.

A **series circuit** can be defined as a circuit in which there is only one current path and all components are connected end to end along this path. The same amount of current must flow in every part of a series circuit. This is because the same number of electrons must pass through the source per second as pass through any other part of the series circuit in the same time interval. Mathematically, this is expressed as

$$I_T = I_1 = I_2 = I_3 \tag{4-1}$$

Figure 4-1 depicts a series circuit with three resistors connected. Conventional current is shown by the arrows.

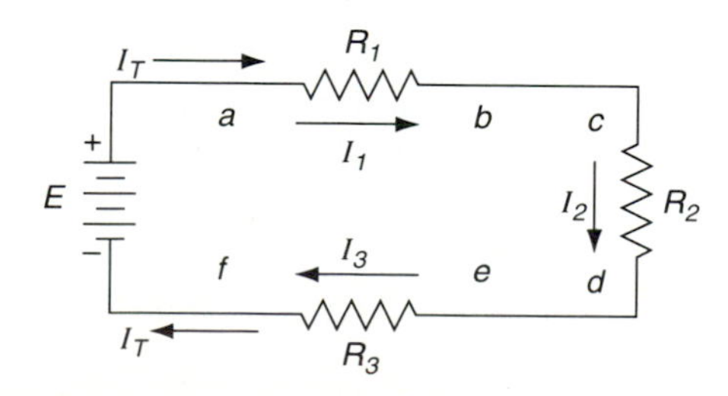

FIGURE 4-1 Series circuit.

Another characteristic of a series circuit is that the total resistance is the sum of the individual resistances. For the circuit shown in Figure 4-1, the total resistance is expressed as

$$R_T = R_1 + R_2 + R_3 \qquad (4\text{-}2)$$

Once we determine the total resistance of a series circuit, we can then solve for the common current by using Ohm's law.

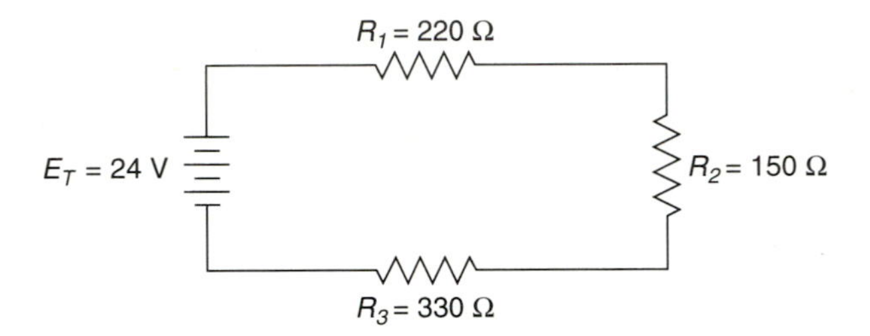

FIGURE 4-2

EXAMPLE 4-1 For the circuit shown in Figure 4-2, determine the amount of current that will flow.

Solution

$$R_T = R_1 + R_2 + R_3$$
$$= 220\ \Omega + 150\ \Omega + 330\ \Omega$$
$$= 700\ \Omega$$

$$I = \frac{E_T}{R_T}$$
$$= \frac{24\ \text{V}}{700\ \Omega}$$
$$= 34.29\ \text{mA}$$

4-2 DOUBLE-SUBSCRIPT NOTATION

In a dc circuit, the direction of current indicates the polarity of a device and whether the device is absorbing power or delivering it. In Figure 4-1, current flow through resistor R_1 is shown as left to right. This means that the left side of R_1 is more positive than the right side, or point a is more positive than point b. This must be true in order for current to flow in the conventional direction.

When current flows through a **passive element**, such as resistor R_1 in Figure 4-1, a drop of potential occurs in the direction of current. Since the current is flowing from point a to point b, the voltage drop of R_1 is referred to as V_{ab}. The notation V_{ab} has two subscripts and is called **double-subscript notation**.

A rise of potential occurs in an **active element**, such as a voltage source, when current flows from the (+) to the (−) terminals. If a single subscript such as V_a is indicated, the voltage at point a is considered to be referenced to ground. An electrical ground utilizes the earth or an equipment chassis as a reservoir of charge and it is represented schematically by the symbols shown in Figure 4-3.

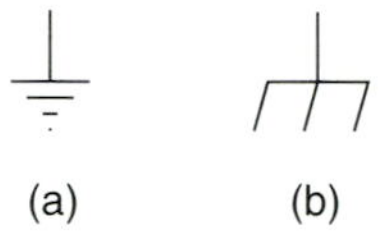

(a) (b)

FIGURE 4-3 Symbols for (a) earth ground and (b) chassis ground.

4-3 KIRCHHOFF'S VOLTAGE LAW

In 1845, Gustav Robert Kirchhoff developed a law dealing with the distribution of voltages around a circuit. **Kirchhoff's voltage law** states:

> In any closed loop, the algebraic sum of the voltage drops and rises equals zero.

As current is forced through an electric circuit, it encounters resistance. Some of the voltage of the supply source will appear across each resistance, forcing current through it. The amount of voltage required to force current through a given resistor is called the **voltage drop** of the resistor. Kirchhoff's voltage law may also be stated as:

> The algebraic sum of all the voltage drops in a circuit must equal the applied voltage.

Mathematically, Kirchhoff's voltage law is expressed as

$$E_{\text{supply}} = V_{R1} + V_{R2} + V_{R3} + \cdots + V_{RN} \tag{4-3}$$

To measure currents and voltages correctly, it is necessary to determine the polarity of the emfs and voltage drops in a circuit. As mentioned earlier, a source of emf is called an active circuit element, since it generates electric energy. Inside the voltage source, electrons flow from the (+) to the (−) terminal. In passive circuit elements (which consume electric energy), electrons flow from the (−) to the (+) terminal. Therefore, the polarity of a voltage drop in a resistance is always such that if it were an emf it would oppose the current producing it.

Figure 4-4 shows a series circuit with a voltmeter and ammeter connected. Since the ammeter's purpose is to show the magnitude of the common current, it must be connected in series with the circuit. This will allow the current being measured to flow through the device. The internal resistance of a typical ammeter is low. Ideally, the resistance of the ammeter should be zero ohms in order to have no effect on the circuit. If the internal resistance is high, it can have an effect on the current level it is measuring. A dc ammeter must always have its positive terminal connected on the positive side of the power supply. If the connections should be reversed accidentally, the meter will try to indicate current flow in the reverse direction and may become damaged.

The voltmeter shown in Figure 4-4 is connected in parallel around resistor R_2. The voltmeter is always connected across the load to be measured, never in series. It must also be connected with the positive terminal connected on the positive side of the power supply. Due to the fact that voltmeters have a very high internal resistance, they do not draw any significant amount of current from the power supply.

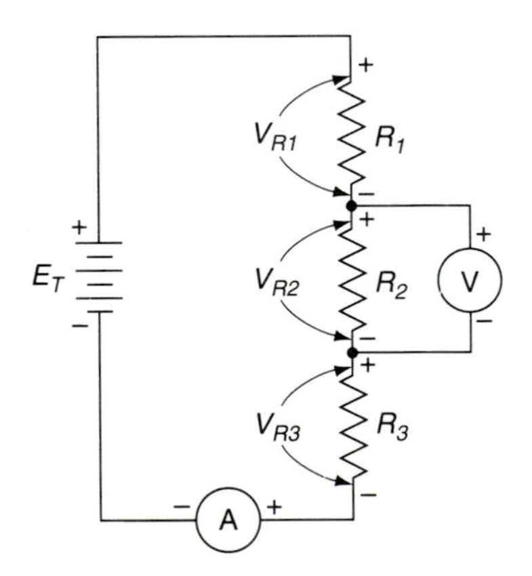

FIGURE 4-4 Ammeter and voltmeter connections.

EXAMPLE 4-2 Figure 4-4 has three series-connected resistors. Assume that $R_1 = 330\ \Omega$, $R_2 = 560\ \Omega$, $R_3 = 470\ \Omega$, and the power supply (E_T) = 100 V. What is the voltage drop across resistor R_2?

Solution

$$R_T = R_1 + R_2 + R_3 = 330\,\Omega + 560\,\Omega + 470\,\Omega$$
$$= 1360\,\Omega$$

$$I_T = \frac{E_T}{R_T} = \frac{100\text{ V}}{1360\,\Omega}$$
$$= 73.53\text{ mA}$$

$$V_{R2} = IR_2 = 73.53\text{ mA} \times 560\,\Omega$$
$$= 41.18\text{ V}$$

EXAMPLE 4-3 Two lamps with ratings of 100 W at 80 V and 60 W at 100 V are connected in series across a 120-V supply. Calculate the voltage drop of each lamp.

Solution

$$P = \frac{E^2}{R}$$

$$R = \frac{E^2}{P}$$

$$R_1 = \frac{80^2\ \text{V}}{100\ \text{W}} = 64\ \Omega$$

$$R_2 = \frac{100^2\ \text{V}}{60\ \text{W}} = 166.7\ \Omega$$

$$R_T = R_1 + R_2 = 64\ \Omega + 166.7\ \Omega = 230.7\ \Omega$$

$$I = \frac{E_T}{R_T} = \frac{120\ \text{V}}{230.7\ \Omega} = 0.52\ \text{A}$$

$$V_{R1} = IR_1 = 0.52\ \text{A} \times 64\ \Omega = 33.3\ \text{V}$$

$$V_{R2} = IR_2 = 0.52\ \text{A} \times 166.7\ \Omega = 86.7\ \text{V}$$

4-4 VOLTAGE DIVIDERS

In a series circuit, the voltage across a given resistance has the same relation to the total voltage as the resistance has to the total resistance. This proportional method of determining voltage drop is known as the **voltage divider rule**. The voltage divider rule may also be stated as:

> The ratio between any two voltage drops in a series circuit is the same as the ratio of the two resistances across which these voltage drops occur.

A useful application of the voltage divider rule is for **voltage divider circuits**. A voltage divider circuit allows a load to operate at a voltage that is different from the supply voltage. Mathematically, the voltage divider rule is stated in equation form as

$$V_{Rx} = E\frac{R_x}{R_T} \tag{4-4}$$

where V_{Rx} = voltage drop across a given resistance

E = applied voltage

R_x = given resistance

R_T = total resistance

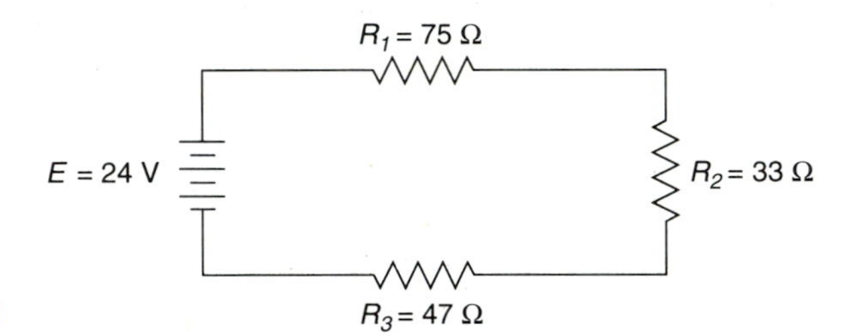

FIGURE 4-5

EXAMPLE 4-4 Use the voltage divider rule to find the voltage drops across the resistors in the circuit shown in Figure 4-5.

Solution

$$R_T = R_1 + R_2 + R_3 = 75\ \Omega + 33\ \Omega + 47\ \Omega$$
$$= 155\ \Omega$$

$$V_{R1} = E\frac{R_1}{R_T} = 24\left(\frac{75}{155}\right)$$
$$= 11.61\text{ V}$$

$$V_{R2} = E\frac{R_2}{R_T} = 24\left(\frac{33}{155}\right)$$
$$= 5.11\text{ V}$$

$$V_{R3} = E\frac{R_3}{R_T} = 24\left(\frac{47}{155}\right)$$
$$= 7.28\text{ V}$$

To check: $E = V_{R1} + V_{R2} + V_{R3} = 11.61\text{ V} + 5.11\text{ V} + 7.28\text{ V}$
$= 24\text{ V}$

4-5 POWER IN A SERIES CIRCUIT

The total power dissipated in a circuit is the sum of the individual amounts of power expended by the individual resistances. The total power in the circuit shown in Figure 4-6 is the sum of the power at each point in the network. In equation form this is expressed as

$$P_T = P_1 + P_2 + P_3 \tag{4-5}$$

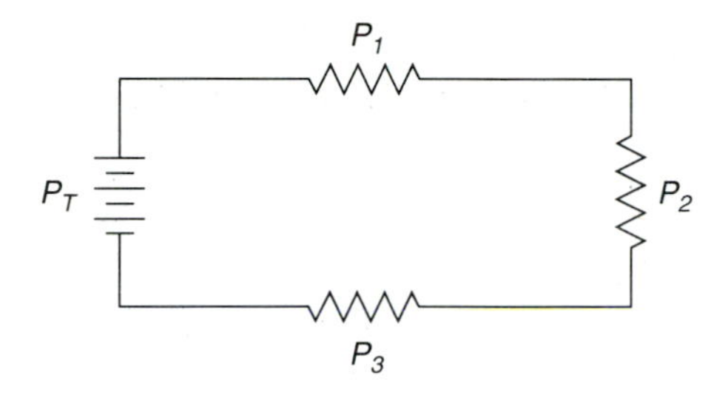

FIGURE 4-6 Series circuit.

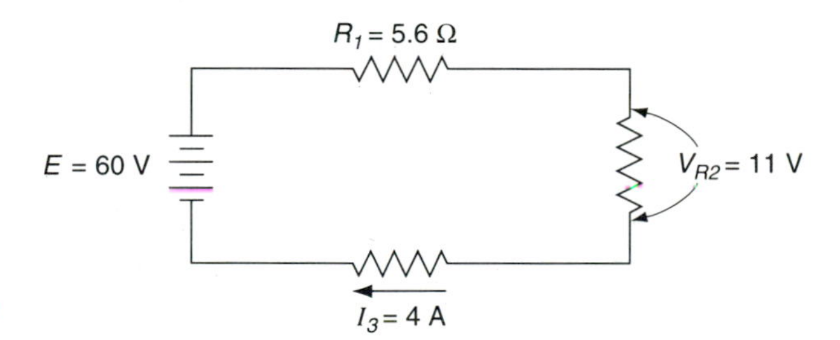

FIGURE 4-7

EXAMPLE 4-5 Calculate the total power as well as the power used in each resistor shown in Figure 4-7.

Solution

$$I_T = I_1 = I_2 = I_3$$
$$= 4\text{ A}$$

$$R_2 = \frac{V_{R2}}{I_2} = \frac{11\text{ V}}{4\text{ A}}$$
$$= 2.75\ \Omega$$

$$V_{R1} = I_{R1} \times R_1 = 4\text{ A} \times 5.6\ \Omega$$
$$= 22.4\text{ V}$$

$$V_{R3} = E_T - (V_{R1} + V_{R2}) = 60\text{ V} - (22.4\text{ V} + 11\text{ V})$$
$$= 26.6\text{ V}$$

$$R_3 = \frac{V_{R3}}{I_3} = \frac{26.6\text{ V}}{4\text{ A}}$$
$$= 6.65\ \Omega$$

$$R_T = R_1 + R_2 + R_3 = 5.6\ \Omega + 2.75\ \Omega + 6.65\ \Omega$$
$$= 15\ \Omega$$

$$P_1 = I^2R_1 = 4^2\text{ A} \times 5.6\ \Omega$$
$$= 89.6\text{ W}$$

$$P_2 = \frac{V_{R2}^2}{R_2} = \frac{11^2\text{ V}}{2.75\ \Omega}$$
$$= 44\text{ W}$$

$$P_3 = I^2R_3 = 4^2\text{ A} \times 6.65\ \Omega$$
$$= 106.4\text{ W}$$

$$P_T = P_1 + P_2 + P_3 = 89.6\text{ W} + 44\text{ W} + 106.4\text{ W}$$
$$= 240\text{ W}$$

4-6 PARALLEL CIRCUITS

A **parallel circuit** can be defined as a circuit that provides more than one current path. When resistances, or other circuit components, are connected so that they have the same pair of terminal points, or nodes, the resistances are said to be in parallel. A node is any point in a circuit where two or more circuit paths intersect. An example of a parallel circuit is shown in Figure 4-8. Since each of the resistors is connected directly across the terminals of the power supply, the parallel circuit may be described as a circuit having a common voltage across its components. The voltage across any branch of a parallel commination is equal to the voltage across each of the other branches in parallel. This fact may be expressed mathematically in the following manner:

$$E = V_{R1} = V_{R2} = V_{R3} \tag{4-6}$$

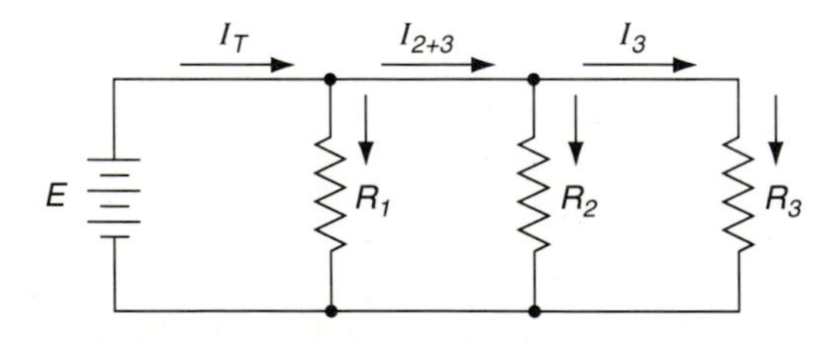

FIGURE 4-8 Circuit diagram for three parallel-connected resistors.

In a parallel circuit, the voltage measured across the load will not change if the load opens. If R_2 in Figure 4-8 were open, a voltmeter connected across R_2 would still measure the source voltage.

The current leaving the source enters a parallel circuit and divides, with a portion of the current traveling through each resistor.

> The total current flowing from the source is the sum of the individual branch currents.

Therefore, in equation form,

$$I_T = I_1 + I_2 + I_3 + \cdots + I_N \tag{4-7}$$

4-7 KIRCHHOFF'S CURRENT LAW

Kirchhoff's current law states that:

> The algebraic sum of the currents entering and leaving a node is zero.

EXAMPLE 4-6 Determine the amount of current flowing in branch R_3 of the circuit shown in Figure 4-9.

Solution

$$I_1 = \frac{E}{R_1} = \frac{12\text{ V}}{30\,\Omega} = 0.4\text{ A}$$

$$I_2 = \frac{E}{R_2} = \frac{12\text{ V}}{24\,\Omega} = 0.5\text{ A}$$

$$I_T - (I_1 + I_2 + I_3) = 0$$

$$I_3 = I_T - (I_1 + I_2) = 1.6\text{ A} - (0.4\text{ A} + 0.5\text{ A}) = 0.7\text{ A}$$

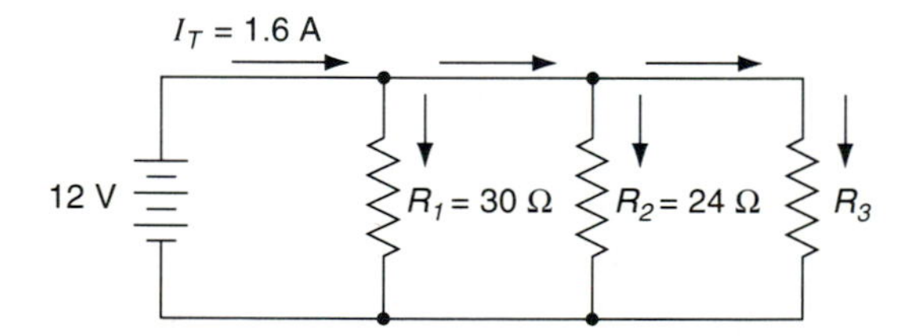

FIGURE 4-9

4-8 CURRENT DIVIDERS

It is often necessary to find the individual branch currents in a circuit from the resistances and total current, but without knowing what the applied voltage of the circuit is. This type of problem can be solved by using the fact that currents divide inversely as the resistances of its path. Therefore, the branch with the smallest resistance will have the greatest current, and the branch with the largest resistance will have the least current.

The **current divider rule** is stated as follows:

> The amount of current in one of two parallel resistances is calculated by multiplying their total current by the other resistance and dividing by their sum.

The formulas for calculating currents through two branch circuits are

$$I_1 = \frac{R_2}{R_1 + R_2} \times I_T \tag{4-8}$$

$$I_2 = \frac{R_1}{R_1 + R_2} \times I_T \tag{4-9}$$

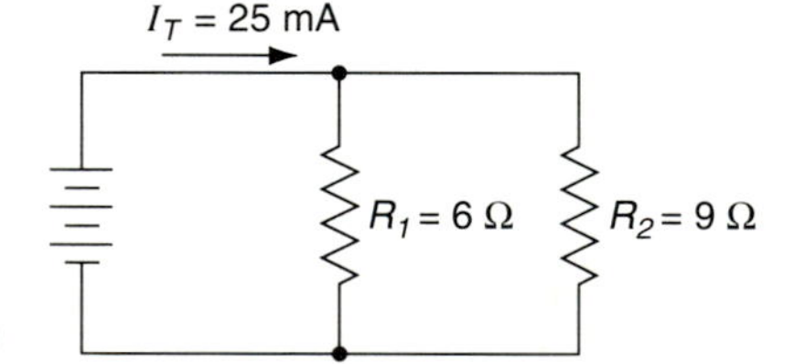

FIGURE 4-10

EXAMPLE 4-7 Using the current divider rule, calculate the currents flowing through branches R_1 and R_2 of Figure 4-10.

Solution

$$I_1 = \frac{R_2}{R_1 + R_2} I_T = \frac{9\,\Omega}{6\,\Omega + 9\,\Omega} \times 25\text{ mA}$$
$$= 15\text{ mA}$$

$$I_2 = \frac{R_1}{R_1 + R_2} I_T = \frac{6\,\Omega}{6\,\Omega + 9\,\Omega} \times 25\text{ mA}$$
$$= 10\text{ mA}$$

To check: $I_T = I_1 + I_2 = 15\text{ mA} + 10\text{ mA}$
$= 25\text{ mA}$

It should be noted that equations 4-8 and 4-9 are applicable for only *two* resistances. When a parallel circuit contains more than two resistances, the following equation may be used to find current in an individual branch:

$$I_X = \frac{R_T}{R_X} I_T$$

4-9 RESISTANCES IN PARALLEL

The total resistance of a parallel circuit can be found by Ohm's law:

> Divide the common voltage across the parallel resistances by the total current of all the branches.

$$R_T = \frac{E}{I_T}$$

where R_T = equivalent resistance of all the parallel branches

E = supply voltage

I_T = sum of all the branch currents

Conductance is a term used to describe a circuit's, or component's, ability to pass current. The SI unit of conductance is the **siemens** (S). The ease with which current flows through a circuit is said to be the circuit's conductance. As the number of resistors connected in parallel increases, the ease with which current flows also increases. Therefore, when resistances are connected in parallel, the total conductance increases, or

$$G_T = G_1 + G_2 + G_3 + \cdots + G_N \tag{4-10}$$

Resistance is the exact opposite, or inverse, of conductance:

$$G_T = \frac{1}{R_T} = \frac{1}{R_1} + \frac{1}{R_2} + \frac{1}{R_3} + \cdots + \frac{1}{R_N} \tag{4-11}$$

The conductance of a parallel circuit and the resistance of a series circuit are often called the **duals** of each other since it is possible to develop some of the relationships for the parallel circuit directly from those of a series circuit simply by interchanging R and G.

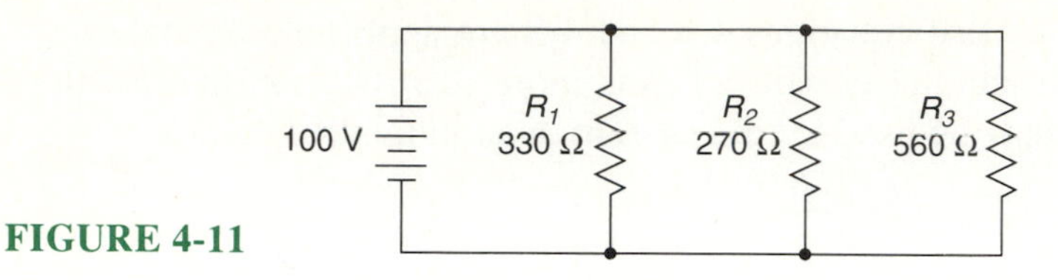

FIGURE 4-11

EXAMPLE 4-8 Calculate the total resistance of the circuit shown in Figure 4-11.

Solution

$$G_T = \frac{1}{R_T} = \frac{1}{R_1} + \frac{1}{R_2} + \frac{1}{R_3} = \frac{1}{330\,\Omega} + \frac{1}{270\,\Omega} + \frac{1}{560\,\Omega}$$
$$= 8.52 \text{ mS}$$

$$R_T = \frac{1}{0.00852}$$
$$= 117.37\ \Omega$$

Two common cases of parallel circuits are more easily solved for by using the following two rules:

1. When all the parallel resistances are of equal value, the combined resistance equals the value of one branch resistance divided by the number of resistances.

$$R_T = \frac{R_X}{N} \tag{4-12}$$

where R_T = equivalent resistance of parallel circuit

R_X = value of one resistance

N = number of resistors in circuit

2. When only two resistors are connected in parallel, the resistance of the parallel combination of two resistances is equal to the product of the individual resistances divided by their sum. This rule is also known as the **product-over-sum rule**. In equation form,

$$R_T = \frac{R_1 \times R_2}{R_1 + R_2} \tag{4-13}$$

EXAMPLE 4-9 Using equation 4-12, determine the total resistance of the circuit shown in Figure 4-12.

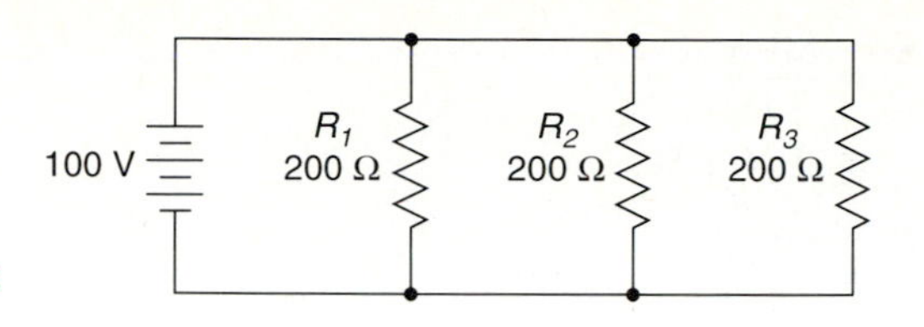

FIGURE 4-12

Solution
$$R_T = \frac{R_X}{N} = \frac{200\ \Omega}{3}$$
$$= 66.7\ \Omega$$

EXAMPLE 4-10 Using the product-over-sum rule, calculate the total resistance of the circuit of Figure 4-13.

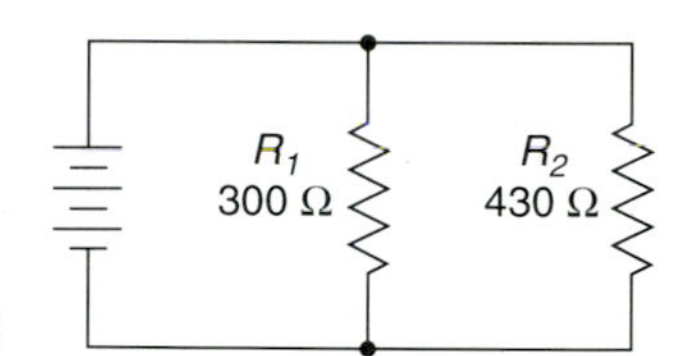

FIGURE 4-13

Solution
$$R_T = \frac{R_1 \times R_2}{R_1 + R_2} = \frac{300\ \Omega \times 430\ \Omega}{300\ \Omega + 430\ \Omega}$$
$$= 176.71\ \Omega$$

4-10 POWER IN PARALLEL CIRCUITS

As in the case of any other circuit, power dissipation in a parallel circuit is the sum of the power dissipated in the individual resistances. The power expended by each component is calculated using the previous equations:

$$P = EI = I^2R = \frac{E^2}{R}$$

EXAMPLE 4-11 Determine the power dissipated in each resistance and the total power expended in the circuit of Figure 4-14.

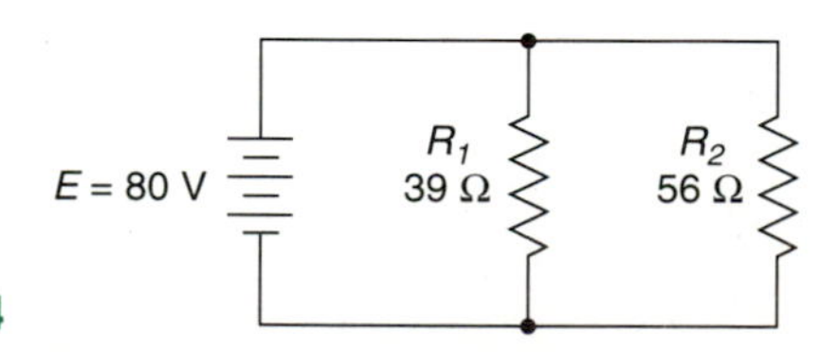

FIGURE 4-14

Solution

$$P_1 = \frac{E^2}{R} = \frac{80^2\ \text{V}}{39\ \Omega}$$
$$= 164.1\ \text{W}$$

$$P_2 = \frac{E^2}{R} = \frac{80^2\ \text{V}}{56\ \Omega}$$
$$= 114.29\ \text{W}$$

$$R_T = \frac{R_1 \times R_2}{R_1 + R_2} = \frac{39\ \Omega \times 56\ \Omega}{39\ \Omega + 56\ \Omega}$$
$$= 22.9895\ \Omega$$

$$P_T = \frac{E^2}{R_T} = \frac{80^2\ \text{V}}{22.9895\ \Omega}$$
$$= 278.39\ \text{W}$$

To check: $P_T = P_1 + P_2 = 164.1\ \text{W} + 114.29\ \text{W}$
$= 278.39\ \text{W}$

KEY TERMS

Series circuit
Passive element
Double-subscript notation
Active element
Kirchhoff's voltage law
Voltage drop
Voltage divider rule
Voltage divider circuit
Parallel circuit
Kirchhoff's current law
Current divider rule
Conductance
Siemens
Dual
Product-over-sum rule

PROBLEMS

4-1 The following resistors are connected in series across a 24-V supply: $R_1 = 30\ \Omega$, $R_1 = 100\ \Omega$, and $R_3 = 70\ \Omega$. Determine the total current in this circuit.

4-2 A 50-Ω resistor is connected in series with another resistor whose value is not known. If the current is 0.75 A when 120 V is applied, determine the unknown resistance.

4-3 A series circuit containing three resistors has a total current of 2 A when 120 V is applied. If $R_1 = 20\ \Omega$ and $R_2 = 15\ \Omega$, calculate the ohmic value of R_3.

4-4 What is the current through each resistor in a series circuit if the total resistance is 1.3 MΩ and the total voltage is 24 V?

4-5 The current in a 40-V circuit is 0.44 A. One of three series resistors has a value of 40 Ω, and another resistor has a value of 22 Ω. How much power is dissipated by the unknown resistor?

4-6 A series circuit containing four resistors has the following values: $R_1 = 2\ k\Omega$, $R_2 = 1\ k\Omega$, $R_3 = 5\ k\Omega$, and $R_4 = 1.5\ k\Omega$. The voltage across the 2-kΩ resistor is 5.5 V. Determine the total applied voltage of the circuit.

4-7 Three resistors with values of 1 kΩ, 5 kΩ, and 10 kΩ are connected in series across a 120-V supply. What power is dissipated by each resistor, and what is the total power dissipated by the circuit?

4-8 Using the voltage divider rule, determine the voltage drop across the 51-Ω resistor for the circuit shown in Figure 4-15.

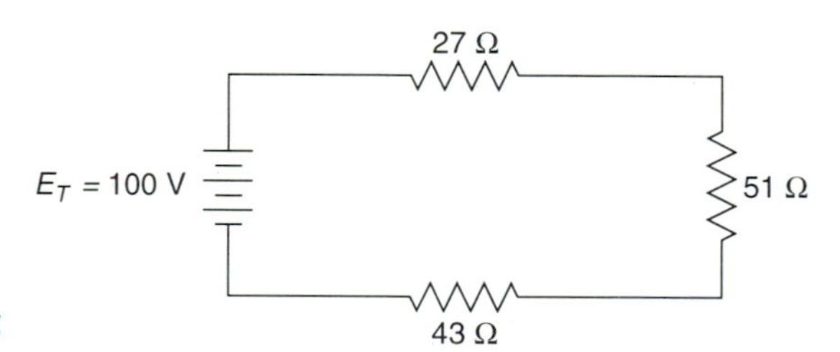

FIGURE 4-15

4-9 The following values have been measured in a series circuit containing three resistors: $R_1 = 11.5\ \Omega$, $I_{R2} = 3.2$ A, and $V_{R3} = 18$ V. The applied voltage of the circuit is 80 V. Calculate the total power as well as the power used in each resistor.

4-10 In a series circuit, R_1 is to have one-third the voltage drop of R_2. If R_2 has a resistance of 1000 Ω, what value is R_1?

4-11 A certain series electronic circuit requires a current-limiting resistor to lower the supply voltage to an acceptable level. The load voltage is 8 V and load current is 20 mA. The supply voltage is 18 V. Calculate the ohmic value and power rating of the resistor required for this circuit.

4-12 A bank of 10 series-connected light-emitting diodes are to be supplied by a 24-V source. Each LED has a voltage drop of 1.6 V and an internal resistance of 600 Ω. Calculate the resistance and power rating of the necessary voltage dropping resistor.

4-13 Two resistances in series have ohmic values that result in the power dissipated by R_1 being four times greater than the power dissipated by R_2. What is the voltage ratio of R_1 and R_2?

4-14 What is the total resistance of three 470-Ω resistors connected in parallel?

4-15 Find the equivalent resistance of a 20-Ω resistor in parallel with a 68-Ω resistor.

4-16 How much resistance must be connected in parallel with a 10-kΩ resistor to have a total resistance of 5 kΩ?

4-17 What value of resistance must be connected in parallel with a 470-Ω resistor to have an equivalent resistance of 220 Ω?

4-18 Determine the amount of current flowing in branch R_1 of the circuit shown in Figure 4-16.

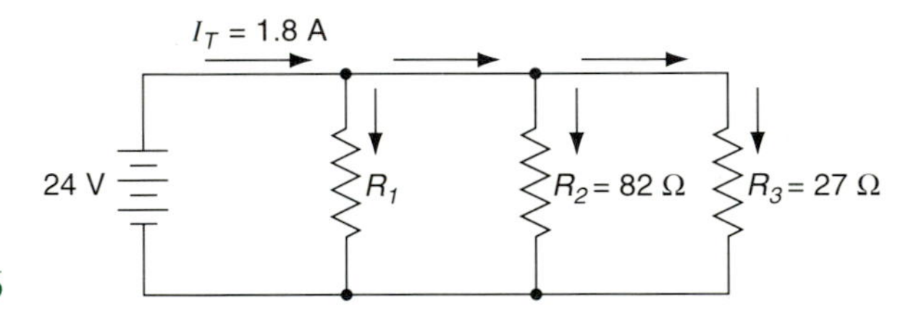

FIGURE 4-16

4-19 Two parallel-connected resistors have values of R_1 = 220 Ω and R_2 = 470 Ω. The supply voltage is 12 V. Calculate the current flowing through each resistor as well as the total current supplied by the source.

4-20 Two resistors connected in parallel have values of R_1 = 30 Ω and R_2 = 60 Ω. The resistors are supplied from a current source rated at 20 A. Use the current divider rule to find the value of current in each branch.

4-21 Use the current divider rule to determine the branch currents in a circuit containing the following three parallel-connected resistors: R_1 = 1 kΩ, R_2 = 2.2 kΩ, and R_3 = 3 kΩ. The supply current for the circuit is 40 mA.

4-22 Two circuit devices are connected in parallel. One component has a conductance of 480 μS and the other has a conductance of 800 μS. Find the total resistance of the circuit.

4-23 Three components with values of 1500 μS, 200 Ω, and 2500 μS are connected in parallel. If the total current is 81 mA, determine the voltage drop across the parallel components as well as the current through each branch.

4-24 Use the product-over-sum rule to find the equivalent resistance of the parallel combination of 40 Ω and 90 Ω.

4-25 Four 10-kΩ resistors are connected in parallel. What is the total equivalent (a) resistance; (b) conductance?

4-26 Two resistors, R_1 = 20 Ω and R_2 = 8 Ω, are connected in parallel. If the circuit is supplied by a 24-V source, what is the amount of power dissipated by each resistance?

4-27 Three 60-W lamps rated at 120 V are connected in parallel. What is the total resistance of these lamps?

4-28 A 12-Ω resistor and a 30-Ω resistor are connected in parallel. If the applied voltage is 12 V, determine the power dissipated by each resistor and the total power dissipated.

4-29 Three resistors, R_1 = 1 kΩ, R_2 = 2 kΩ, and R_3 = 4 kΩ, are connected in parallel to a 24-V supply. Find (a) total resistance; (b) total current; (c) current through each resistor; (d) total power; (e) power dissipated by each resistor.

4-30 Four 120-V lamps are connected in parallel to a 120-V source. The lamps are rated as follows: L_1 = 25 W, L_2 = 60 W, L_3 = 40 W, and L_4 = 100 W. Find (a) the amount of current flowing through each lamp; (b) the total resistance of the circuit; (c) the total power dissipated.

4-31 Three resistors, R_1 = 2.2 kΩ, R_2 = 4.7 kΩ, and R_3 = 10 kΩ, are connected in parallel. If each resistor is rated at ¼ W, find the maximum total current that would not cause any resistor to overheat.

4-32 Two resistors, R_1 = 1 kΩ and R_2 = 2.2 kΩ, are connected in parallel. The current in the first resistor is 5 mA. Find (a) the power dissipated by each resistor; (b) the total power drawn from the supply.

CHAPTER

DC Circuit Analysis

LEARNING OBJECTIVES

Upon completion of this chapter you will be able to:

- Identify series and parallel configurations in a series–parallel circuit.
- Define practical voltage and current sources.
- Convert voltage sources to current sources, and vice versa.
- Apply loop analysis to dc circuits.
- Understand Norton's theorem and its application to circuit analysis.
- Describe three-wire distribution circuits.
- Understand the application of determinants in circuit analysis.
- Define Thévenin's theorem and understand how to use it to reduce a dc circuit to a simple equivalent.
- Apply Norton's theorem to dc circuits.
- Use Millman's theorem to reduce multiple voltage sources in parallel to a single equivalent voltage source.
- Apply superposition to a circuit with more than one voltage or current source.
- Define the maximum power transfer theorem.
- Analyze dependent voltage and current sources.

5-1 INTRODUCTION

The primary objective of circuit analysis is to determine the response of a circuit to a given voltage or current applied to the circuit. In most cases, circuit analysis techniques involve determining the performance of an isolated portion of a complex circuit. In these situations it is necessary to replace the remainder of the circuit with a simplified equivalent network. Often, the equivalent circuit is reduced to a source and a resistor in series or parallel with the source. Thévenin's theorem and Norton's theorem enable us to do this.

A complex circuit is defined as any network that combines series and parallel elements in various interconnections. Solving for circuit parameters in complex networks is simplified by establishing a systematic approach for setting up and solving several circuit equations. Two of the most common methods used are loop analysis and nodal analysis.

In this chapter we examine linear networks, that is, circuits made up of resistors and driven by sources of constant voltage and current. An understanding of algebra is the only mathematical tool required for analyzing these types of circuits. The advantage of using algebraic methods of circuit analysis is that the same basic approach may be used on any circuit, regardless of the circuit complexity.

In the first section of this chapter we deal with solving series–parallel networks. In many cases a complete series–parallel combination can be replaced by a single resistor that is electrically equivalent to the entire network.

5-2 SERIES–PARALLEL RESISTORS

Circuits that contain both series and parallel combinations are called **series–parallel circuits**. A series–parallel circuit may be defined as a network in which some portions of the circuit have the characteristics of a simple series circuit and other portions have the characteristics of simple parallel circuits.

Series circuit laws and relations are applied to the components of a circuit that are connected in series, and parallel circuit laws and relations are applied to the components of a circuit that are connected in parallel. For example, the equivalent resistance of a series circuit is the sum of the resistance values of the components. In a parallel circuit, the reciprocal of the value of the equivalent resistance equals the sum of the reciprocals of the resistance values of the branches.

A series–parallel circuit can be analyzed using the following steps:

1. Reduce all parallel circuits to series equivalent resistances.
2. Combine all the branches containing more than one resistance in series into a single resistance.
3. Redraw the resulting equivalent circuit.
4. When the series–parallel circuit has been simplified, solve for total resistance, current, or voltage, as required.
5. To obtain a complete solution for a series–parallel circuit, use values obtained in the equivalent circuit and apply them to the original circuit. Calculate voltage, current, and power, as required.

There are, essentially, two basic methods in which series and parallel networks can be combined. The circuit shown in Figure 5-1 consists of two series networks connected in parallel. Each branch current in the circuit is equal to the voltage across the branch divided by the total series resistance in the branch. The total current, I_T, is the sum of the branch currents I_A and I_B.

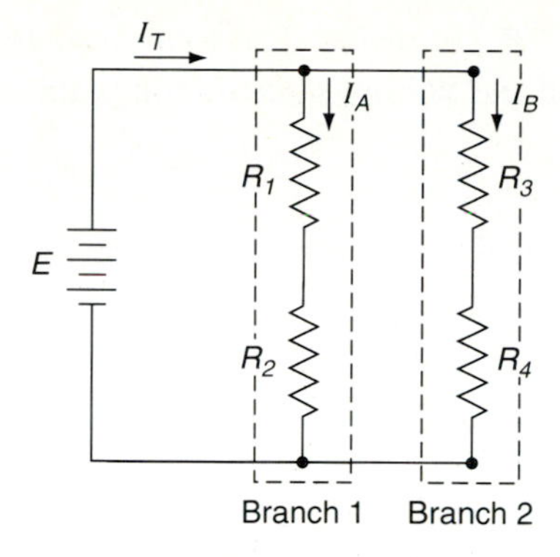

FIGURE 5-1 Series–parallel circuit.

For any individual resistance in a branch, the current in the branch multiplied by the resistance equals the *IR* voltage drop across that particular resistance. The sum of the series *IR* drops in the branch equals the voltage drop across the entire branch. For the circuit of Figure 5-1,

$$E_T = V_{R1} + V_{R2} \qquad \text{also,} \quad E_T = V_{R3} + V_{R4}$$

Figure 5-2 shows a parallel network connected in series. The two groups of parallel resistances (R_1, R_2, R_3 and R_4, R_5, R_6) are called *banks of resistance*. Bank *A* is in series with bank *B*, since the total current of bank *A* must go through bank *B*. Therefore, the total current, I_T, in each of the parallel banks is the same ($I_T = I_1 + I_2 + I_3$ and $I_T = I_4 + I_5 + I_6$). The voltage drop across either bank of resistors in Figure 5-2 can be found by multiplying the total current, I_T, by the equivalent resistance of the bank. The voltage across each resistor in either of the parallel banks is the same (V_A is the voltage across R_1,

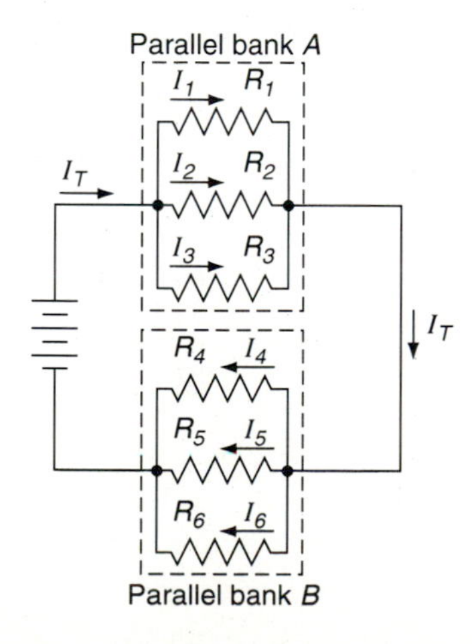

FIGURE 5-2 Series–parallel circuit.

R_2, and R_3; V_B is the voltage across R_4, R_5, and R_6). The total voltage across the series–parallel network is the sum of the voltages across the parallel banks ($E_T = V_A + V_B$).

EXAMPLE 5-1 Determine the values of currents I_A and I_B for the circuit shown in Figure 5-3.

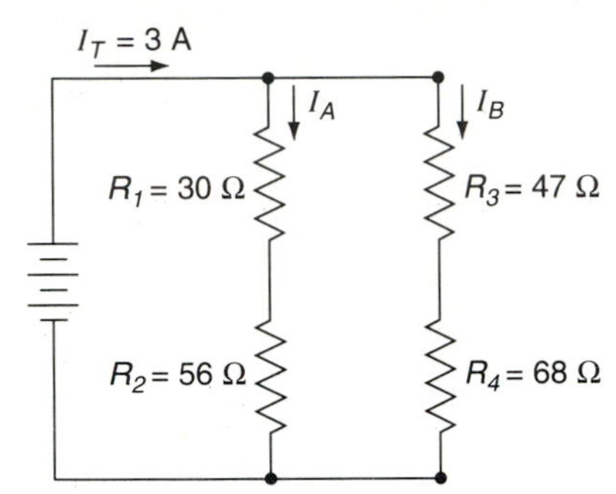

FIGURE 5-3

Solution

$$R_A = R_1 + R_2 = 30\ \Omega + 56\ \Omega = 86\ \Omega$$

$$R_B = R_3 + R_4 = 47\ \Omega + 68\ \Omega = 115\ \Omega$$

See Figure 5-4 and use the current divider rule:

$$I_A = I_T \frac{R_B}{R_A + R_B} = 3\ \text{A} \times \frac{115\ \Omega}{86\ \Omega + 115\ \Omega} = 1.72\ \text{A}$$

$$I_B = I_T \frac{R_A}{R_A + R_B} = 3\ \text{A} \times \frac{86\ \Omega}{86\ \Omega + 115\ \Omega} = 1.28\ \text{A}$$

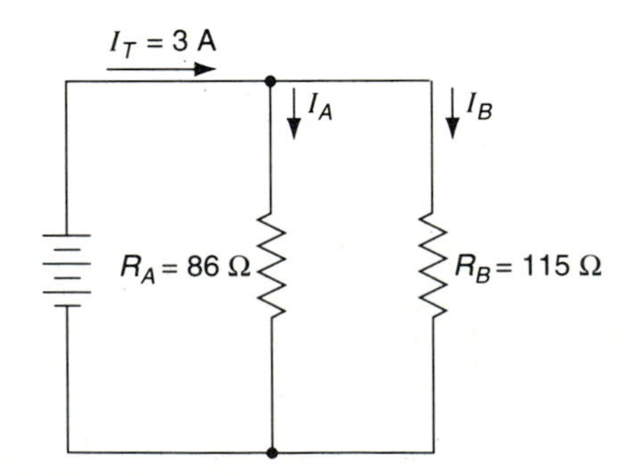

FIGURE 5-4

EXAMPLE 5-2 For the circuit shown in Figure 5-5, resistors R_1, R_2, and R_3 form one group and R_4, R_5, and R_6 form another group. Find the current through resistor R_2 and the voltage drop across each group.

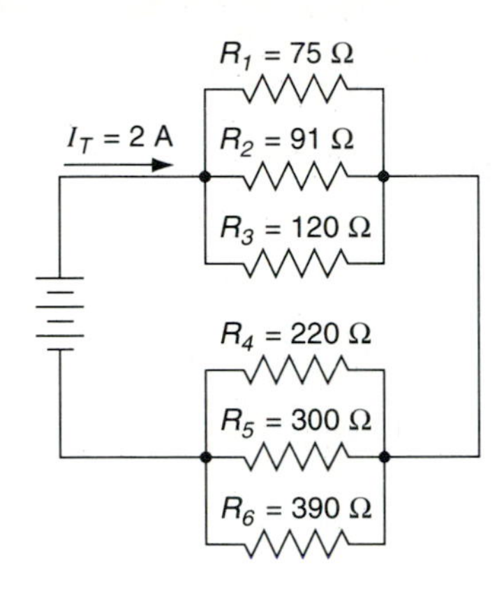

FIGURE 5-5

Solution

$$\frac{1}{R_A} = \frac{1}{R_1} + \frac{1}{R_2} + \frac{1}{R_3} = \frac{1}{75\,\Omega} + \frac{1}{91\,\Omega} + \frac{1}{120\,\Omega}$$

$$R_A = 30.62\,\Omega$$

$$\frac{1}{R_B} = \frac{1}{R_4} + \frac{1}{R_5} + \frac{1}{R_6} = \frac{1}{220\,\Omega} + \frac{1}{300\,\Omega} + \frac{1}{390\,\Omega}$$

$$R_B = 95.76\,\Omega$$

The circuit is now redrawn as shown in Figure 5-6.

$$E_{RA} = I_T R_A = 2\text{ A} \times 30.62\,\Omega = 61.24\text{ V}$$

$$E_{RB} = I_T R_B = 2\text{ A} \times 95.76\,\Omega = 191.52\text{ V}$$

$$I_2 = \frac{E_{RA}}{R_2} = \frac{61.24\text{ V}}{91\,\Omega} = 0.67\text{ A}$$

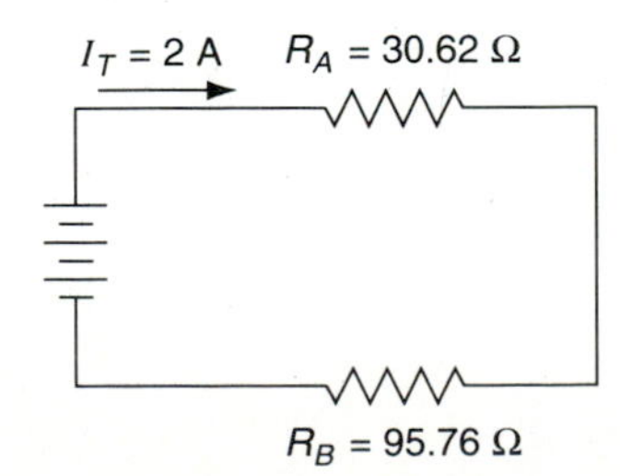

FIGURE 5-6

EXAMPLE 5-3 What is the voltage at point A in the circuit of Figure 5-7?

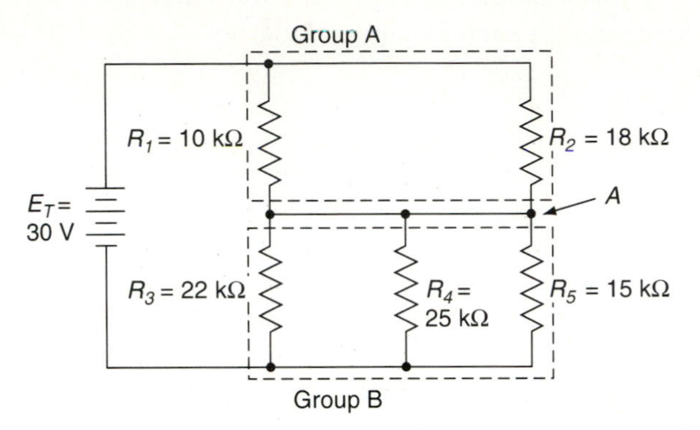

FIGURE 5-7

Solution R_A represents group A, and R_B represents group B:

$$R_A = \frac{R_1 R_2}{R_1 + R_2} = \frac{10\text{ k}\Omega \times 18\text{ k}\Omega}{10\text{ k}\Omega + 18\text{ k}\Omega}$$
$$= 6.43\text{ k}\Omega$$

$$\frac{1}{R_B} = \frac{1}{R_3} + \frac{1}{R_4} + \frac{1}{R_5} = \frac{1}{22\text{ k}\Omega} + \frac{1}{25\text{ k}\Omega} + \frac{1}{15\text{ k}\Omega}$$

$$R_B = 6.57\text{ k}\Omega$$

The circuit is now redrawn as shown in Figure 5-8. Using the voltage divider rule gives

$$V_{RB} = E_T \frac{R_B}{R_A + R_B} = 30\text{ V} \times \frac{6.57\text{ k}\Omega}{6.43\text{ k}\Omega + 6.57\text{ k}\Omega}$$
$$= 15.16\text{ V}$$

The voltage at point A with respect to ground is 15.16 V.

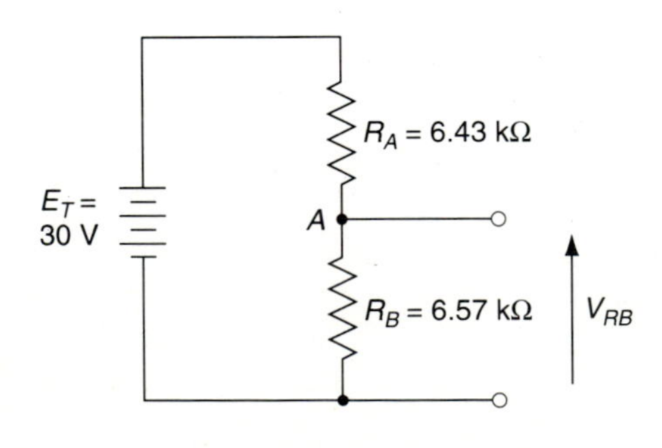

FIGURE 5-8

EXAMPLE 5-4 Calculate the total resistance and total current for the circuit shown in Figure 5-9.

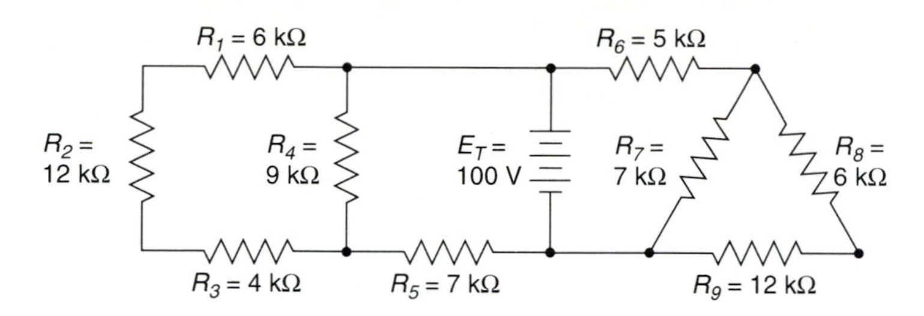

FIGURE 5-9

Solution The circuit can first be redrawn in the simplified diagram of Figure 5-10. There are now, basically, two groups of resistors in the circuit. The equivalent resistors can now be calculated:

$$R_A = \frac{22\,\text{k}\Omega \times 9\,\text{k}\Omega}{22\,\text{k}\Omega + 9\,\text{k}\Omega} + 7\,\text{k}\Omega$$
$$= 13.39\,\text{k}\Omega$$

$$R_B = \frac{18\,\text{k}\Omega \times 7\,\text{k}\Omega}{18\,\text{k}\Omega + 7\,\text{k}\Omega} + 5\,\text{k}\Omega$$
$$= 10.04\,\text{k}\Omega$$

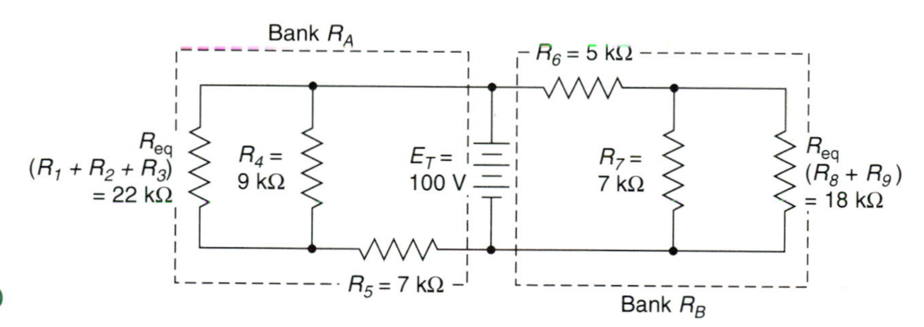

FIGURE 5-10

The circuit is now simplified further, as shown in Figure 5-11.

$$R_T = \frac{R_A \times R_B}{R_A + R_B} = \frac{13.39\,\text{k}\Omega \times 10.04\,\text{k}\Omega}{13.39\,\text{k}\Omega + 10.04\,\text{k}\Omega}$$
$$= 5.74\,\text{k}\Omega$$

$$I_T = \frac{E_T}{R_T} = \frac{100\,\text{V}}{5.74\,\text{k}\Omega}$$
$$= 17.42\,\text{mA}$$

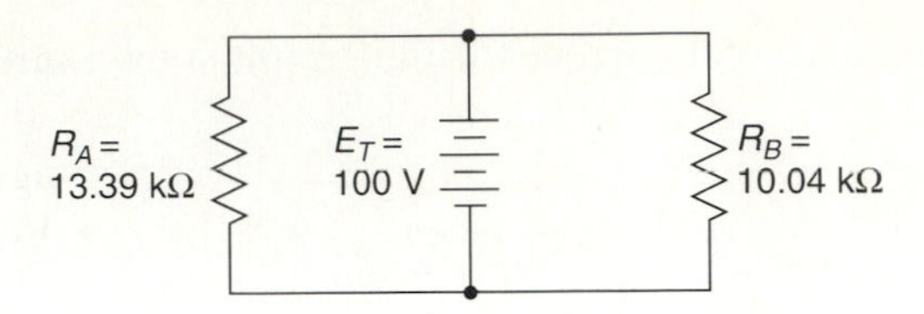

FIGURE 5-11 Circuit containing equivalent resistors R_A and R_B.

The transistor circuit shown in Figure 5-12 is essentially a series–parallel circuit. The bipolar junction transistor (BJT) is a three-terminal device with an emitter (E), collector (C), and base (B). The BJT can be thought of as two resistors, R_{CB} and R_{BE}. If the bipolar junction transistor is made of silicon, the voltage drop between the base and emitter, V_{BE}, is typically considered to be 0.7 V. In the following example, the emitter current (I_E) and collector current (I_C) can be taken as approximately equal to each other, although I_E is actually the sum of $I_C + I_B$.

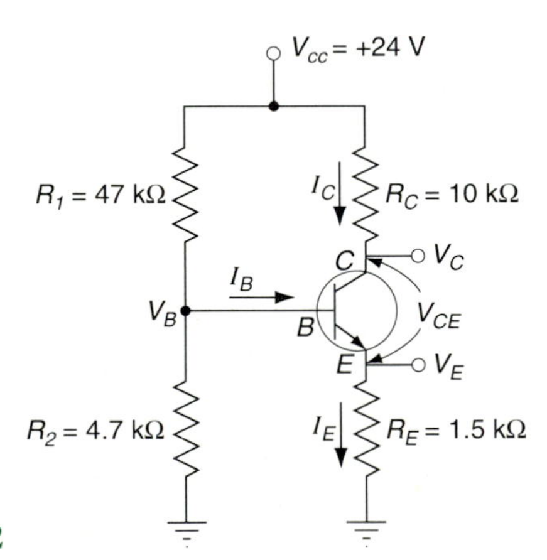

FIGURE 5-12

EXAMPLE 5-5 For the circuit shown in Figure 5-12, assume that I_B is negligible and $V_{BE} = 0.7$ V. Determine (a) V_B; (b) V_E; (c) I_E; (d) V_C; (e) V_{CE}.

Solution (a) The voltage at the base of the transistor (with respect to ground) is equal to the voltage drop across R_2. The voltage divider rule is used to solve for V_B [Figure 5-13(a)].

$$V_B = V_{R2} = \frac{R_2}{R_1 + R_2} V_{cc} = \frac{4.7\,\text{k}\Omega}{47\,\text{k}\Omega + 4.7\,\text{k}\Omega} \times 24\text{ V}$$
$$= 2.18\text{ V}$$

(b) If the circuit of Figure 5-12 is redrawn as the equivalent resistor network of Figure 5-13(b), it can be seen that the voltage drop V_E can be solved by Kirchhoff's voltage law. Since V_B is equal to $V_E + V_{BE}$, V_E must equal $V_B - V_{BE}$.

$$V_B = V_E + V_{BE}$$
$$V_E = V_B - V_{BE} = 2.18 - 0.7$$
$$= 1.48 \text{ V}$$

(c) The current through the emitter resistor can now be solved by Ohm's law.

$$I_E = \frac{V_E}{R_E} = \frac{1.48}{1500}$$
$$= 0.99 \text{ mA}$$

(d) Assuming that I_C and I_E are approximately equal, the voltage V_C (with respect to ground) can be found by Kirchhoff's voltage law [Figure 5-13(c)].

$$I_C \approx I_E$$

$$V_C = V_{cc} - I_C R_C = 24 - (0.99 \times 10^{-3})(10 \times 10^3)$$
$$= 14.1 \text{ V}$$

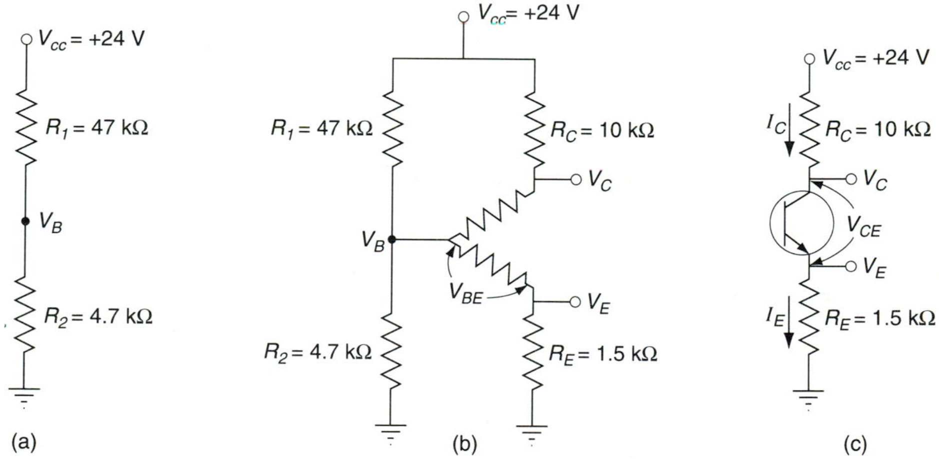

FIGURE 5-13

(e) The voltage between the collector and emitter of the transistor is the difference between V_C and V_E.

$$V_{CE} = V_C - V_E = 14.1 \text{ V} - 1.48 \text{ V}$$
$$= 12.62 \text{ V}$$

V_{CE} can also be solved for by using another KVL equation for the circuit:

$$V_{cc} = I_E R_E + V_{CE} + I_C R_C$$

Therefore,

$$V_{CE} = V_{cc} - (I_E R_E + I_C R_C) = 24 \text{ V} - (1.48 + 9.9)\text{V}$$
$$= 12.62 \text{ V}$$

5-3 NONIDEAL VOLTAGE AND CURRENT SOURCES

In Chapter 2 an ideal independent voltage source was defined as a source of potential energy that maintained a specified voltage regardless of the current drawn from it. In practice, however, as the current drawn from the source increases, the terminal voltage of the source decreases. This is due to an internal resistance of the source, called an **internal voltage drop**. Ideal voltage sources have no internal resistance. Sources that contain internal resistance as referred to as **nonideal**, or **practical sources**. Figure 5-14(a) shows the schematic symbol used to represent a practical dc source of voltage such as a battery or power supply. The long line on the source indicates the **anode** or positive terminal, and the short line indicates the **cathode** or negative terminal. The internal resistance of the source is shown by the resistor R_i, which is in series with the source. A source of electromagnetic generation is shown in Figure 5-14(b). The positive and negative terminals of the generator are shown on the device. The internal resistance is identified on this diagram, although it is often not indicated in circuit problems dealing with dc generators.

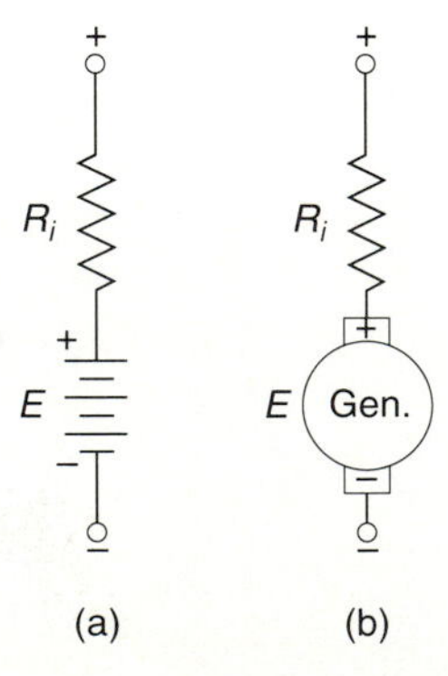

FIGURE 5-14 Practical voltage sources.

If a nonideal voltage source is connected to a load resistor, as shown in Figure 5-15, the current I is solved for by Ohm's law.

$$I = \frac{E}{R_i + R_L}$$

Using the voltage divider rule, the output voltage developed across R_L can be found:

$$V_{RL} = E\frac{R_L}{R_i + R_L}$$

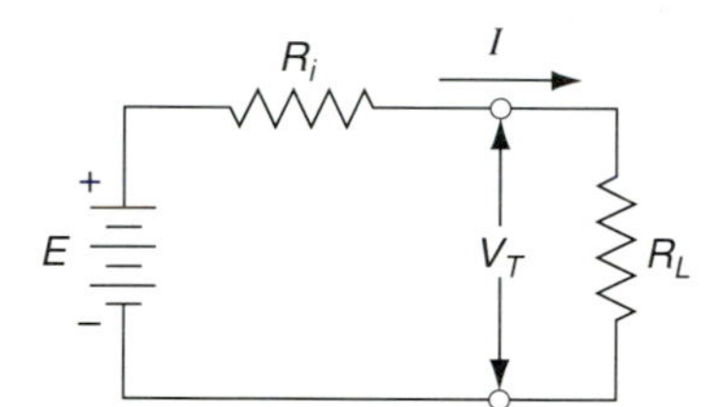

FIGURE 5-15

Since V_{RL} is the potential difference at the terminals of the voltage source, the load voltage is considered equal to the terminal voltage, V_T. The terminal voltage of a practical voltage source under load must always be less than the open-circuit voltage due to the voltage drop across the internal resistance.

$$V_T = E - IR_i$$

From the equations above it is apparent that the terminal voltage of a practical source is dependent on the internal resistance, open-circuit emf, and load resistance of the circuit. If the load resistance decreases, the current drawn from the source increases. As the current increases, the internal voltage drop increases. An increase in the internal voltage drop causes a decrease in the terminal voltage.

A **practical current source** is the dual of a practical voltage source. The symbol for a nonideal independent current source with internal resistance is shown in Figure 5-16. This type of device has its internal resistance shown in parallel with the ideal source. In Chapter 2 an ideal independent current source was defined as a device capable of providing a constant current regardless of the size of load connected across it. The internal resistance of an ideal source is considered to be infinite.

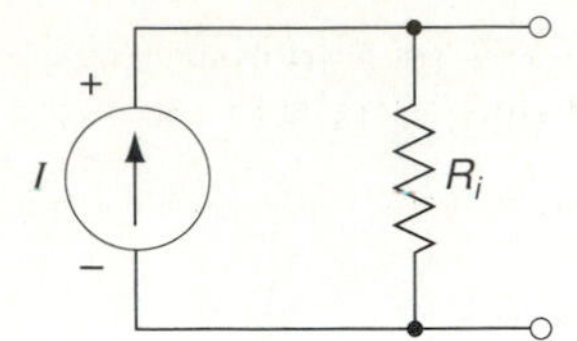

FIGURE 5-16

The current, I_L, that will flow through the load in the circuit of Figure 5-17 is determined by

$$I_L = I - \frac{V_T}{R_i}$$

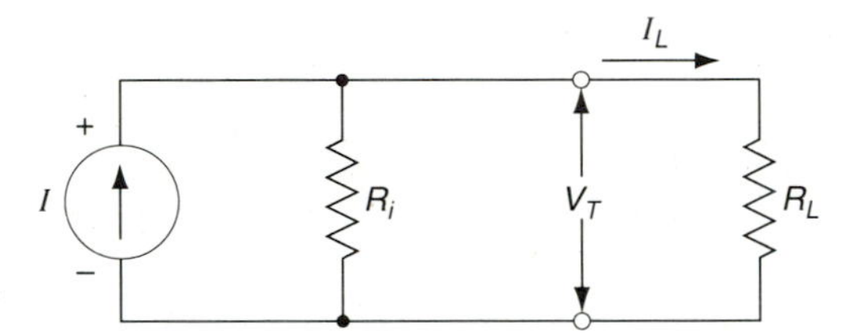

FIGURE 5-17

If V_T is the terminal voltage of the current source, the value of V_T when under load is determined as

$$V_T = IR_i - I_L R_i$$

When the load resistance is considerably smaller than the source resistance, a practical current source is assumed to have an infinite value of source resistance, and all the current supplied by the source is considered to flow through the load. If the practical current source is open-circuited, the open-circuit voltage V_{oc} is then

$$V_{oc} = IR_i$$

5-4 SOURCE CONVERSIONS

When solving circuit problems it is often advantageous to replace a current source with a voltage source or a voltage source with a current source. This replacing of sources is called **source conversion**. In Section 5-3 the terminal voltage of a practical voltage source was determined by the equation

$$V_T = E - IR_i$$

and the terminal voltage of a practical current source was found by

$$V_T = IR_i - I_L R_i$$

A voltage source and a current source are **equivalent** if they produce identical currents in the same resistive load. Therefore, if the internal resistances of both sources are equal, the load current for a current source can be found by the current divider rule,

$$I_L = I\frac{R_i}{R_L + R_i}$$

The load current for a voltage source is found by Ohm's law:

$$I_L = \frac{E}{R_i + R_L}$$

In many practical applications, the internal resistance of a source is ignored. For this reason, some examples and problems discussed in this book may not have internal resistances shown with sources. However, it must be noted that to convert one type of source to another, the internal resistances must be identified.

EXAMPLE 5-6 Transform the voltage source of Figure 5-18 to a current source, and determine the load current for each source.

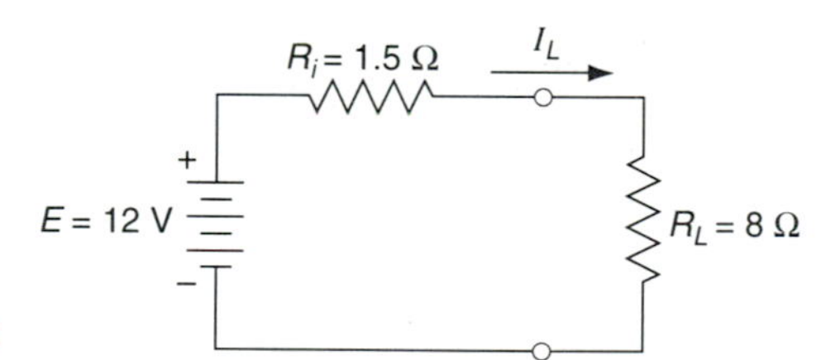

FIGURE 5-18

Solution As a voltage source, the load current is found by Ohm's law:

$$I_L = \frac{E}{R_i + R_L} = \frac{12\text{ V}}{1.5\,\Omega + 8\,\Omega} = 1.26\text{ A}$$

As a current source, the supply current is found by Ohm's law and the load current is found by the current divider rule:

$$I = \frac{E}{R_i} = \frac{12\text{ V}}{1.5\,\Omega} = 8\text{ A}$$

$$I_L = I\frac{R_i}{R_L + R_i} = 8\text{ A} \times \frac{1.5\,\Omega}{8\,\Omega + 1.5\,\Omega}$$
$$= 1.26\text{ A}$$

The equivalent current source is shown in Figure 5-19.

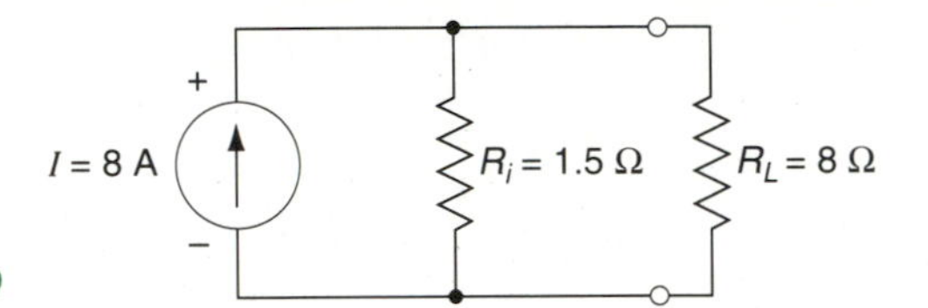

FIGURE 5-19

5-5 LOOP ANALYSIS

Kirchhoff's voltage and current laws can be applied as an alternative method of solving series–parallel networks. These laws provide a means of solving a problem without reducing the circuit to a simple series or parallel network. **Loop analysis** is a method of solving circuit problems by use of Kirchhoff's voltage law. Mathematically, this method of analysis results in a number of simultaneous equations in which the unknown quantities are the currents in various parts of the circuit. The three methods of solving simultaneous equations are substitution, determinants, and matrices.

Any closed path is called a **loop**. There is usually, but not always, a current flowing in a loop. This current is referred to as a **loop current**. An electric circuit may be analyzed by counting the unknown branch currents and unknown node voltages in the circuit. It is then necessary to write simultaneous equations to obtain a solution for the circuit. For a given network, the number of simultaneous equations to be solved may be determined by counting the minimum number of loops required.

The steps taken in solving a circuit problem by loop analysis are as follows:

1. Assign an arbitrary direction of current flow in each branch of the network, as shown in Figure 5-20.

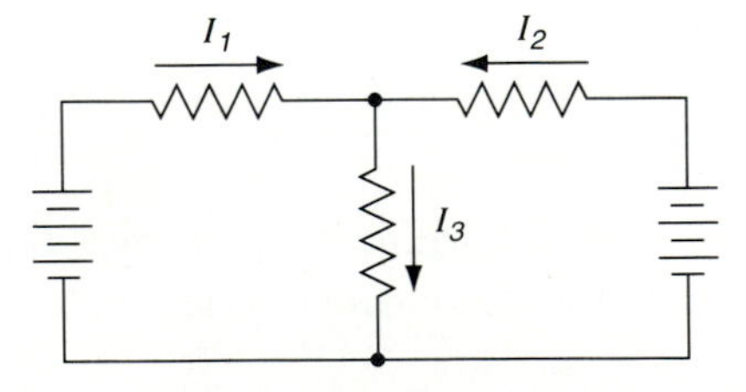

FIGURE 5-20 Loop analysis of circuit.

2. Using the arrows as a guide, polarize each resistor for each current. Draw loops (Figure 5-21).

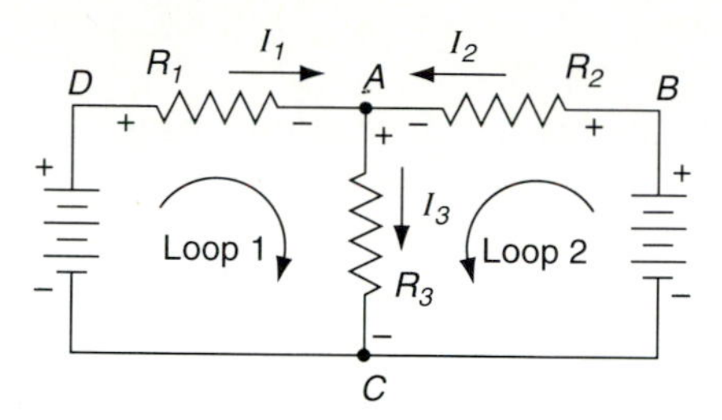

FIGURE 5-21 Properly labeled circuit for loop equations.

3. Write the Kirchhoff voltage law (KVL) equation for each loop. If the direction of the loop is from plus to minus across a component, *add* the voltage in the KVL equation. If the direction of the loop is from minus to plus, *subtract* the voltage in the KVL equation.

4. Solve the simultaneous equations for the unknown loop currents.

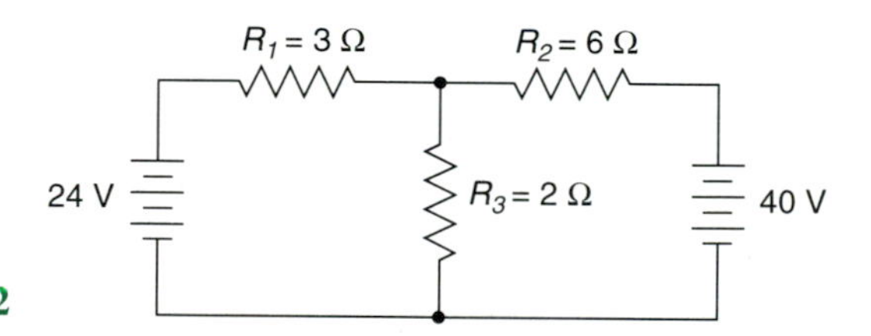

FIGURE 5-22

EXAMPLE 5-7 For the circuit shown in Figure 5-22, solve for the currents in each of the three resistors.

Solution Assign direction of current flow, indicate polarity of resistors, and draw loops, as shown in Figure 5-23. Write loop equations. Using point *a* as a reference, loops 1 and 2 are written as follows:

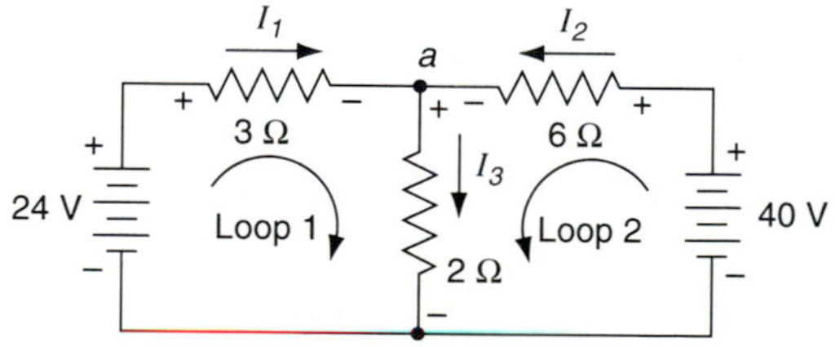

FIGURE 5-23

Loop 1: $2I_3 - 24 + 3I_1 = 0$

$3I_1 + 2I_3 = 24$

Loop 2: $2I_3 - 40 + 6I_2 = 0$

$6I_2 + 2I_3 = 40$

At this point, there are two equations in the three resistor currents. A third equation is derived by applying Kirchhoff's current law at node *a*.

$$I_3 = I_1 + I_2$$

Substituting the equation above into the two loop equations results in

$$3I_1 + 2(I_1 + I_2) = 24$$

$$6I_2 + 2(I_1 + I_2) = 40$$

which simplifies to

$$5I_1 + 2I_2 = 24$$

$$2I_1 + 8I_2 = 40$$

The loop equations are now solved by eliminating an unknown. This is done by multiplying one or more equations by an appropriate constant and adding or subtracting the resulting equations. In this example the number 4 is chosen as the constant for the first equation.

$$\begin{array}{r} 20I_1 + 8I_2 = 96 \\ \underline{2I_1 + 8I_2 = 40} \\ 18I_1 \qquad\quad = 56 \end{array}$$

Therefore,

$$I_1 = \frac{56}{18} = 3.11 \text{ A}$$

With I_1 known, it is substituted into either equation to obtain I_2.

$$5(3.11) + 2I_2 = 24$$

$$15.55 + 2I_2 = 24$$

$$I_2 = \frac{8.45}{2} = 4.23 \text{ A}$$

I_3 is now found by Kirchhoff's current law.

$$I_3 = I_1 + I_2 = 3.11 \text{ A} + 4.23 \text{ A} = 7.34 \text{ A}$$

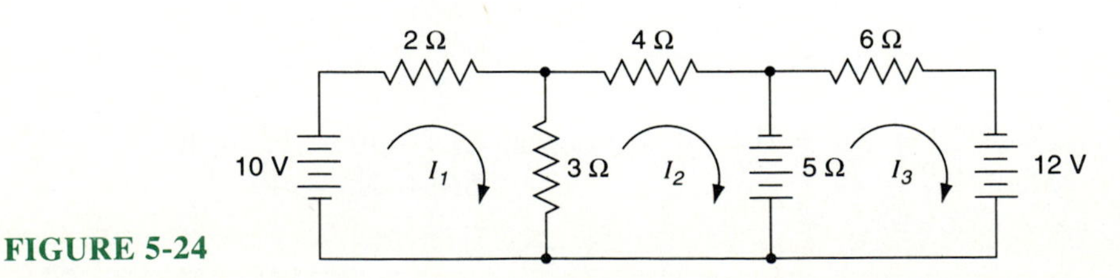

FIGURE 5-24

EXAMPLE 5-8 For the circuit of Figure 5-24, solve for the loop currents I_1, I_2, and I_3.

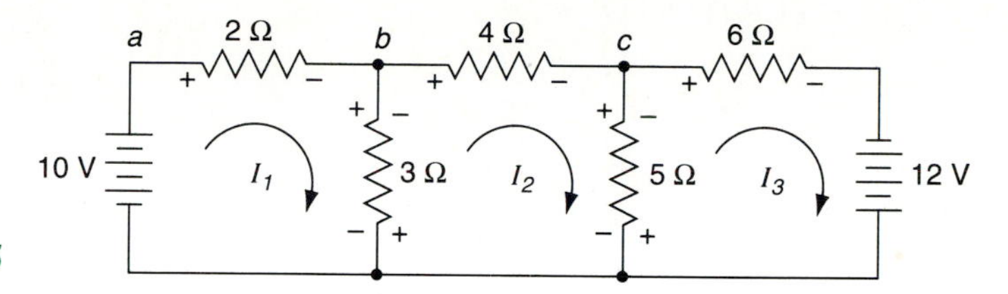

FIGURE 5-25

Solution Identify polarity of resistors, and establish reference points, as shown in Figure 5-25. Using points *a*, *b*, and *c* as references for each loop, the following equations are derived:

Loop 1: $2I_1 + 3I_1 - 3I_2 - 10 = 0$

$$5I_1 - 3I_2 = 10$$

Loop 2: $4I_2 + 5I_2 - 5I_3 + 3I_2 - 3I_1 = 0$

$$-3I_1 + 12I_2 - 5I_3 = 0$$

Loop 3: $6I_3 - 12 + 5I_3 - 5I_2 = 0$

$$-5I_2 + 11I_3 = 12$$

Combine loop 1 and loop 2. Multiply loop 1 by 3, and loop 2 by 5, and add the equations.

$$\begin{array}{rrrl} 15I_1 & - 9I_2 & & = 30 \\ -15I_1 & + 60I_2 & - 25I_3 & = 0 \\ \hline & 51I_2 & - 25I_3 & = 30 \end{array}$$

Combine the result of loops 1 and 2 with loop 3. Multiply the top equation by 11 and loop 3 by 25. Add the equations.

$$\begin{array}{rrl} 561I_2 & - 275I_3 & = 330 \\ -125I_2 & + 275I_3 & = 300 \\ \hline 436I_2 & & = 630 \end{array}$$

$$I_2 = \frac{630}{436}$$

$$= 1.44 \text{ A}$$

I_2 is now substituted into loop 1,

$$5I_1 - 3(1.44) = 10$$

$$5I_1 = 14.32$$

$$I_1 = \frac{14.32}{5}$$

$$= 2.86 \text{ A}$$

and into loop 3,

$$
\begin{aligned}
-5(1.44) + 11I_3 &= 12 \\
11I_3 &= 19.2 \\
I_3 &= \frac{19.2}{11} \\
&= 1.75\text{ A}
\end{aligned}
$$

5-6 NODAL ANALYSIS

Nodal analysis is a technique where equations based on Kirchhoff's current law (KCL) are written, and unknown voltages solved for. A **node** is a junction between two or more components. A **node voltage** is a voltage at a node with respect to a common reference point. The reference point is generally chosen as a node in the given circuit, although it can be referenced to a point outside the circuit. The number of node voltages in a circuit is always one less than the total number of nodes.

When using nodal analysis to solve simultaneous equations, the reference, or ground, node is often chosen as the node with the largest number of components connected. Many electronic circuits are built on a metallic chassis, which is the practical choice for the ground node. In other circuits, such as electric power systems, the ground node is the earth. In either case, chassis ground or earth ground is assumed to be at zero volts, and node voltages will be at potentials above zero.

The steps taken in solving circuit problems by nodal analysis are as follows:

1. Identify the number of nodes. The circuit of Figure 5-26(a) contains four nodes. They are identified as points a, b, c, and d.

2. Use one node as a reference point. In this case, node b is chosen as the reference. The voltages from c to b and d to b are called **dependent nodes**. Node a is referred to as an **independent node**. The voltage between the independent node and the reference node is the unknown voltage, which is labeled V_x in Figure 5-26(b).

3. Identify and assign current direction at each independent node. The currents flowing into and out of node a are shown in Figure 5-26(c).

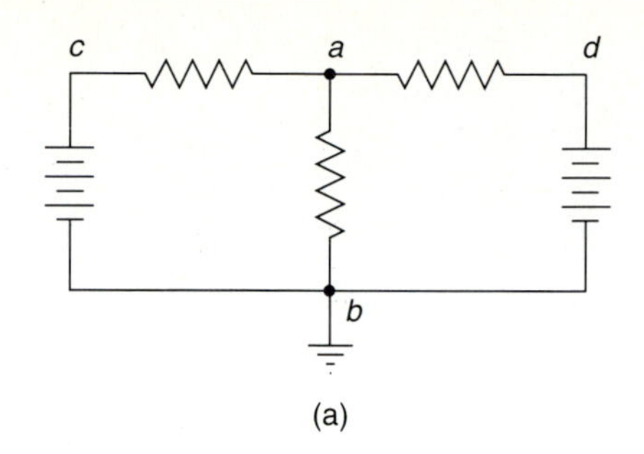

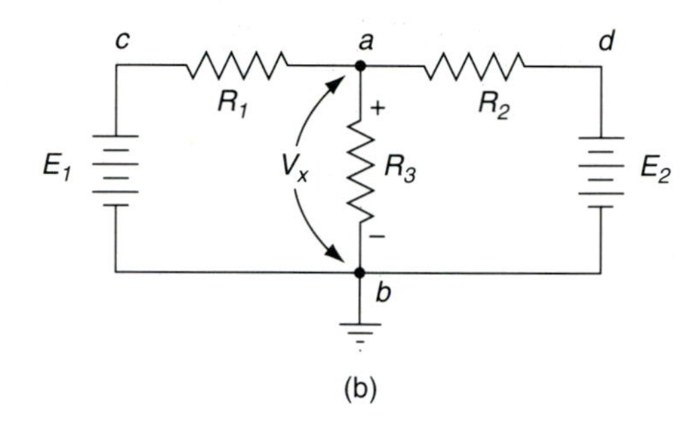

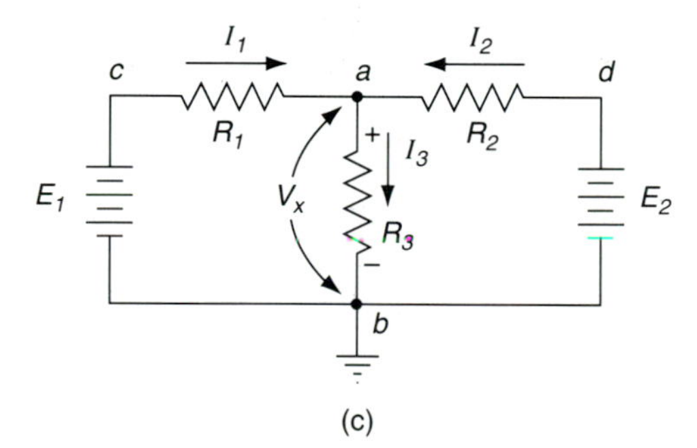

FIGURE 5-26

4. Use Kirchhoff's current law to write an equation for each independent node.

$$I_1 + I_2 + I_3 = 0$$

5. Use Ohm's law to express each current equation in terms of voltages.

$$I_1 = \frac{V_{R1}}{R_1} = \frac{V_x - V_c}{R_1}$$

$$I_2 = \frac{V_{R2}}{R_2} = \frac{V_x - V_d}{R_2}$$

$$I_3 = \frac{V_x}{R_3}$$

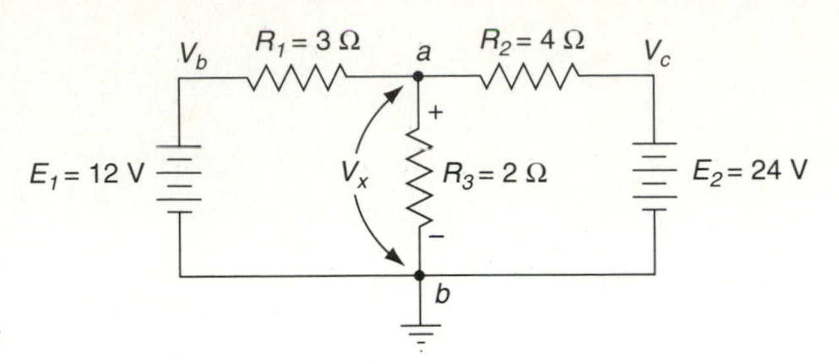

FIGURE 5-27

EXAMPLE 5-9 Use nodal analysis to find the voltage V_x for the circuit shown in Figure 5-27.

Solution Apply KCL at node a.

$$I_1 + I_2 + I_3 = 0$$

Apply Ohm's law.

$$I_1 = \frac{V_{R1}}{R_1} = \frac{V_x - V_b}{R_1} = \frac{V_x - 12}{3}$$

$$I_2 = \frac{V_{R2}}{R_2} = \frac{V_x - V_c}{R_2} = \frac{V_x - 24}{4}$$

$$I_3 = \frac{V_{R3}}{R_3} = \frac{V_x}{2}$$

The KCL equation is now substituted with values obtained in Ohm's law calculations.

$$I_1 + I_2 + I_3 = 0$$

$$\frac{V_x - 12}{3} + \frac{V_x - 24}{4} + \frac{V_x}{2} = 0$$

The equation above is now solved in terms of V_x.

$$\left(\frac{1}{3} + \frac{1}{4} + \frac{1}{2}\right)V_x = \frac{12}{3} + \frac{24}{4}$$

$$\frac{13}{12V_x} = \frac{120}{12}$$

$$(1.083)V_x = 10$$

$$V_x = \frac{10}{1.083}$$

$$= 9.23 \text{ V}$$

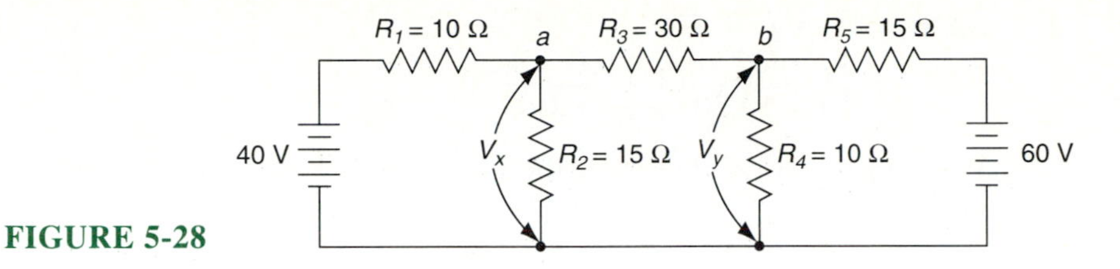

FIGURE 5-28

EXAMPLE 5-10 Use nodal analysis to solve for the unknown voltages V_x and V_y in the circuit of Figure 5-28.

Solution Redraw circuit, assign current direction, and identify independent nodes, as shown in Figure 5-29. Write KCL equations at independent nodes.

Node a: $I_1 = I_2 + I_3$

Node b: $I_4 = I_3 + I_5$

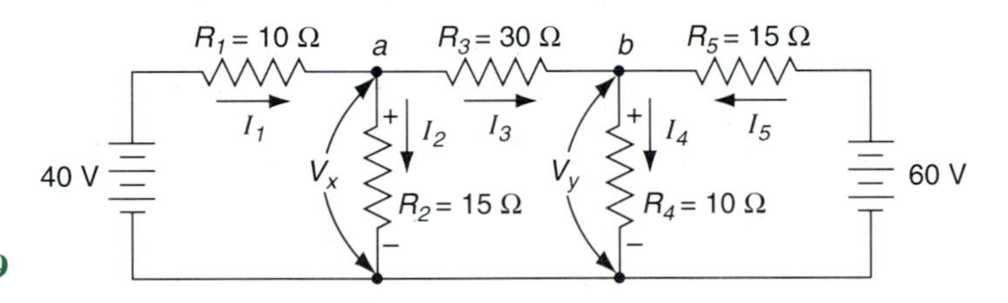

FIGURE 5-29

Use Ohm's law to express each current in terms of voltage.

$$I_1 = \frac{V_{R1}}{R_1} = \frac{40 - V_x}{10} \qquad I_4 = \frac{V_y}{R_4} = \frac{V_y}{10}$$

$$I_2 = \frac{V_x}{R_2} = \frac{V_x}{15} \qquad I_5 = \frac{V_{R5}}{R_5} = \frac{60 - V_x}{15}$$

$$I_3 = \frac{V_{R3}}{R_3} = \frac{V_x - V_y}{20}$$

Substitute KCL equations with Ohm's law values:

Node a: $I_1 = I_2 + I_3$

$$\frac{40 - V_x}{10} = \frac{V_x}{15} + \frac{V_x - V_y}{30}$$

Node b: $I_4 = I_3 + I_5$

$$\frac{V_y}{10} = \frac{V_x - V_y}{20} + \frac{60 - V_x}{30}$$

The number 30 is chosen as the common denominator for both equations.

Node a: $3(40 - V_x) = 2V_x + V_x - V_y$

Node b: $3V_y = V_x - V_y + 2(60 - V_x)$

which is simplified further as follows:

Node a: $120 - 3V_x = 2V_x + V_x - V_y$

$$-6V_x + V_y = 120$$

Node b: $3V_y = V_x - V_y + 120 - 2V_x$

$$V_x + 4V_y = 120$$

Multiply the equation for node a by 4 and subtract node b from node a.

$$\begin{array}{r} -24V_x + 4V_y = 480 \\ V_x + 4V_y = 120 \\ \hline -25V_x \qquad\quad = 360 \end{array}$$

$V_x = -14.4$ V

Substitute V_x into either node equation.

$$-6(-14.4) + V_y = 120$$

$$86.4 + V_y = 120$$

$$V_y = 33.6\text{ V}$$

5-7 THREE-WIRE DISTRIBUTION CIRCUITS

Three-wire distribution circuits transmit power at 240 V and utilize it at 120 V. This system consists of a positive line, a negative line, and a neutral wire, as shown in Figure 5-30. The loads are connected between the positive line and the neutral and between the negative line and the neutral. When the loads are unbalanced, the neutral wire carries a current equal to the difference in the currents in the negative and positive line.

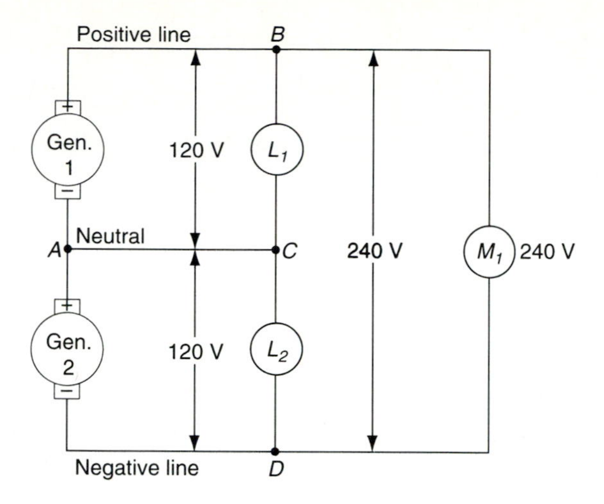

FIGURE 5-30 Three-wire distribution circuit.

When L_1 equals L_2, and the potential difference across L_1 equals the voltage across L_2, the current in L_1 will be the same as in L_2, and no current will flow in the neutral wire. When the current through L_1 is greater than the current through L_2, current will flow through the neutral from point C to A. When the current through L_1 is less than the current through L_2, the current in the neutral will flow from A to C. This statement can easily be verified by applying Kirchhoff's voltage law at junction C. The motor M_1 is connected between the positive line and the negative line, and will affect only the currents flowing in these two lines. The currents and voltages in three-wire distribution systems may be determined by means of Kirchhoff's laws, provided that the applied voltages and resistances of the system are known.

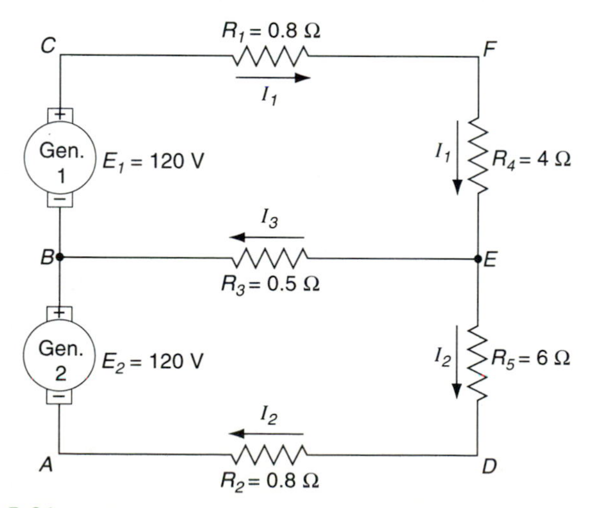

FIGURE 5-31

EXAMPLE 5-11 For the circuit of Figure 5-31, determine the current in each line and load of the system.

Solution Draw a loop around the closed path *B–C–F–E–B*, and apply Kirchhoff's voltage law:

$$E_1 - I_1R_1 - I_1R_4 - I_3R_3 = 0$$

$$120 - 0.8I_1 - 4I_1 - 0.5I_3 = 0$$

$$120 - 4.8I_1 - 0.5I_3 = 0$$

Draw a second loop around the closed path *A–B–E–D–A*, and apply Kirchhoff's voltage law.

$$E_2 + I_3R_3 - I_2R_5 - I_2R_2 = 0$$

$$120 + 0.5I_3 - 6I_2 - 0.8I_2 = 0$$

$$120 + 0.5I_3 - 6.8I_2 = 0$$

Applying Kirchhoff's current law to point B, the equation is

$$I_3 + I_2 - I_1 = 0$$

Therefore,

$$I_2 = I_1 - I_3$$

The value of I_2 can now be substituted into the second closed loop:

$$120 + 0.5I_3 - 6.8I_1 + 6.8I_3 = 0$$

$$120 + 7.3I_3 - 6.8I_1 = 0$$

$$I_2 = \frac{120 + 7.3I_3}{6.8}$$

Solving for the first closed loop, the equation for I_1 becomes

$$I_1 = \frac{120 - 0.5I_3}{4.8}$$

The next step requires eliminating I_1 and solving for I_3 by comparing the equations for I_1 in both loops:

$$\frac{120 + 7.3I_3}{6.8} = \frac{120 - 0.5I_3}{4.8}$$

$$(120 + 7.3I_3)(4.8) = (120 - 0.5I_3)(6.8)$$

$$576 + 35I_3 = 816 - 3.4I_3$$

$$35I_3 + 3.4I_3 = 816 - 576$$
$$38.4I_3 = 240$$
$$I_3 = \frac{240}{38.4}$$
$$= 6.25\text{ A}$$

Substituting the value of I_3 into either loop equation for I_1 allows the value of I_1 to be determined:

$$I_1 = \frac{120 - 0.5(6.25)}{4.8} = 24.35\text{ A}$$

Substituting the values of I_1 and I_3 into the original Kirchhoff current loop for point B gives

$$I_2 = I_1 - I_3 = 24.35\text{ A} - 6.25\text{ A}$$
$$= 18.1\text{ A}$$

The values of I_1, I_2, and I_3 can be checked for accuracy by substituting the values into the original Kirchhoff voltage loops:

$$E_1 - I_1R_1 - I_1R_4 - I_3R_3 = 0$$
$$120\text{ V} - (24.35\text{ A})(0.8\,\Omega) - (24.35\text{ A})(4\,\Omega) - (6.25\text{ A})(0.5\,\Omega) = 0$$
$$120\text{ V} - 19.48\text{ V} - 97.4\text{ V} - 3.12\text{ V} = 0$$

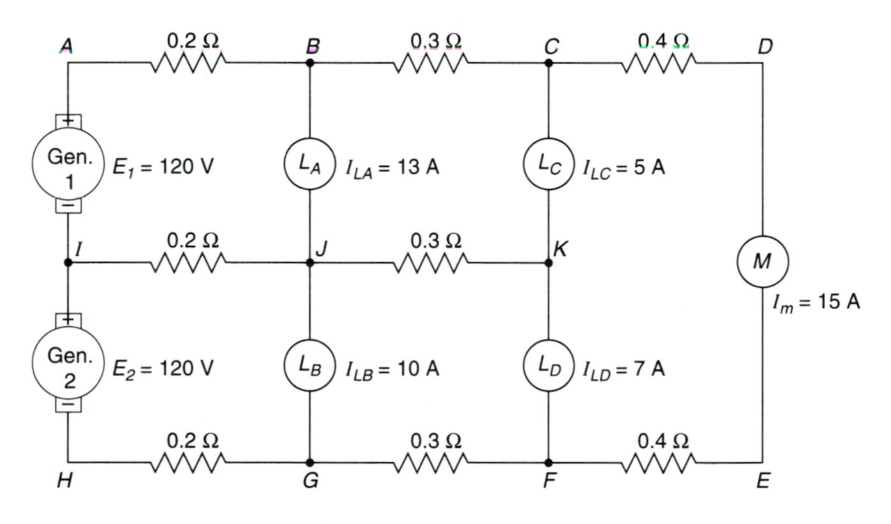

FIGURE 5-32

EXAMPLE 5-12 For the circuit shown in Figure 5-32, determine the following voltages: (a) V_m; (b) V_{LA}; (c) V_{LB}; (d) V_{LC}; (e) V_{LD}.

Solution First determine the various line currents:

$$
\begin{aligned}
I_{AB} &= I_{LA} + I_{LC} + I_m \\
&= 13\text{ A} + 5\text{ A} + 15\text{ A} \\
&= 33\text{ A}
\end{aligned}
$$

$$
\begin{aligned}
I_{FG} &= I_{LD} + I_m \\
&= 7\text{ A} + 15\text{ A} \\
&= 22\text{ A}
\end{aligned}
$$

$$
\begin{aligned}
I_{BC} &= I_{LC} + I_m \\
&= 5\text{ A} + 15\text{ A} \\
&= 20\text{ A}
\end{aligned}
$$

$$I_{EF} = I_m = 15\text{ A}$$

$$I_{CD} = I_m = 15\text{ A}$$

$$
\begin{aligned}
I_{GH} &= I_{LB} + I_{LD} + I_m \\
&= 10\text{ A} + 7\text{ A} + 15\text{ A} \\
&= 32\text{ A}
\end{aligned}
$$

$$
\begin{aligned}
I_{JK} &= I_{FG} - I_{BC} \\
&= 22\text{ A} - 20\text{ A} \\
&= 2\text{ A}
\end{aligned}
$$

$$
\begin{aligned}
I_{IJ} &= I_{AB} - I_{GH} \\
&= 33\text{ A} - 32\text{ A} \\
&= 1\text{ A}
\end{aligned}
$$

The voltage drops of the line can now be calculated:

$$
\begin{aligned}
V_{AB} &= I_{AB}\,R_{AB} \\
&= (33)(0.2) \\
&= 6.6\text{ V}
\end{aligned}
$$

$$
\begin{aligned}
V_{FG} &= I_{FG}\,R_{FG} \\
&= (22)(0.3) \\
&= 6.6\text{ V}
\end{aligned}
$$

$$
\begin{aligned}
V_{BC} &= I_{BC}\,R_{BC} \\
&= (20)(0.3) \\
&= 6\text{ V}
\end{aligned}
$$

$$
\begin{aligned}
V_{EF} &= I_{EF}\,R_{EF} \\
&= (15)(0.4) \\
&= 6\text{ V}
\end{aligned}
$$

$$
\begin{aligned}
V_{CD} &= I_{CD}\,R_{CD} \\
&= (15)(0.4) \\
&= 6\text{ V}
\end{aligned}
$$

$$
\begin{aligned}
V_{IJ} &= I_{IJ}\,R_{IJ} \\
&= (1)(0.2) \\
&= 0.2\text{ V}
\end{aligned}
$$

$$
\begin{aligned}
V_{GH} &= I_{GH}\,R_{GH} \\
&= (32)(0.2) \\
&= 6.4\text{ V}
\end{aligned}
$$

$$
\begin{aligned}
V_{JK} &= I_{JK}\,R_{JK} \\
&= (2)(0.3) \\
&= 0.6\text{ V}
\end{aligned}
$$

The load voltages can now be solved using Kirchhoff's voltage loops:

(a) $$
\begin{aligned}
V_m &= (E_1 + E_2) - (V_{AB} + V_{BC} + V_{CD} + V_{EF} + V_{FG} + V_{GH}) \\
&= 240\text{ V} - (6.6\text{ V} + 6\text{ V} + 6\text{ V} + 6\text{ V} + 6.6\text{ V} + 6.4\text{ V}) \\
&= 202.4\text{ V}
\end{aligned}
$$

(b) $V_{LA} = E_1 - V_{AB} - V_{IJ} = 120\text{ V} - 6.6 - 0.2 = 113.2\text{ V}$

(c) $V_{LB} = E_2 - V_{GH} + V_{JK} = 120\text{ V} - 6.4 + 0.2 = 113.8\text{ V}$

(d) $$
\begin{aligned}
V_{LC} &= V_{LA} - V_{BC} + V_{JK} = 113.2\text{ V} - 6\text{ V} + 0.6\text{ V} \\
&= 107.8\text{ V}
\end{aligned}
$$

(e) $V_{LD} = V_{LB} - (V_{JK} + V_{FG}) = 113.8 \text{ V} - (0.6 \text{ V} + 6.6 \text{ V})$
$= 106.6 \text{ V}$

5-8 DETERMINANTS

It is often necessary to solve for more than two simultaneous equations in circuit analysis. When three or more simultaneous equations are required to be solved, it is often convenient to use **determinants**. A determinant is simply a matrix, or square array of elements arranged between two vertical lines.

The determinant method of solving a system of three simultaneous equations in three unknowns consists of arranging the equations in the following manner:

$$\begin{aligned} A_1x + B_1y + C_1z &= D_1 \\ A_2x + B_2y + C_2z &= D_2 \\ A_3x + B_3y + C_3z &= D_3 \end{aligned} \qquad (5\text{-}1)$$

The equations can now be set up in determinant form, where x, y, and z will represent the solution for each of the three simultaneous equations.

$$x = \frac{\begin{array}{|ccc|cc} D_1 & B_1 & C_1 & D_1 & B_1 \\ D_2 & B_2 & C_2 & D_2 & B_2 \\ D_3 & B_3 & C_3 & D_3 & B_3 \end{array}}{\begin{array}{|ccc|cc} A_1 & B_1 & C_1 & A_1 & B_1 \\ A_2 & B_2 & C_2 & A_2 & B_2 \\ A_3 & B_3 & C_3 & A_3 & B_3 \end{array}}$$

$$y = \frac{\begin{array}{|ccc|cc} A_1 & D_1 & C_1 & A_1 & D_1 \\ A_2 & D_2 & C_2 & A_2 & D_2 \\ A_3 & D_3 & C_3 & A_3 & D_3 \end{array}}{\begin{array}{|ccc|cc} A_1 & B_1 & C_1 & A_1 & B_1 \\ A_2 & B_2 & C_2 & A_2 & B_2 \\ A_3 & B_3 & C_3 & A_3 & B_3 \end{array}} \qquad (5\text{-}2)$$

$$z = \frac{\begin{vmatrix} A_1 & B_1 & D_1 & A_1 & B_1 \\ A_2 & B_2 & D_2 & A_2 & B_2 \\ A_3 & B_3 & D_3 & A_3 & B_3 \end{vmatrix}}{\begin{vmatrix} A_1 & B_1 & C_1 & A_1 & B_1 \\ A_2 & B_2 & C_2 & A_2 & B_2 \\ A_3 & B_3 & C_3 & A_3 & B_3 \end{vmatrix}}$$

Once the simultaneous equations have been arranged in the form of equation 5-2, the solution for the denominator is found by using the pattern shown in Figure 5-33. The product of the diagonal to the extreme left ($A_1 B_2 C_3$) is added to the products of the next two diagonals:

$$A_1 B_2 C_3 + B_1 C_2 A_3 + C_1 A_2 B_3$$

The products of the diagonals with arrows pointing upward are assigned negative values and subtracted from the first three diagonals. The total solution for the denominator is the sum of the first three diagonals minus the last three diagonals:

$$A_1 B_2 C_3 + B_1 C_2 A_3 + C_1 A_2 B_3 - A_3 B_2 C_1 - B_3 C_2 A_1 - C_3 A_2 B_1$$

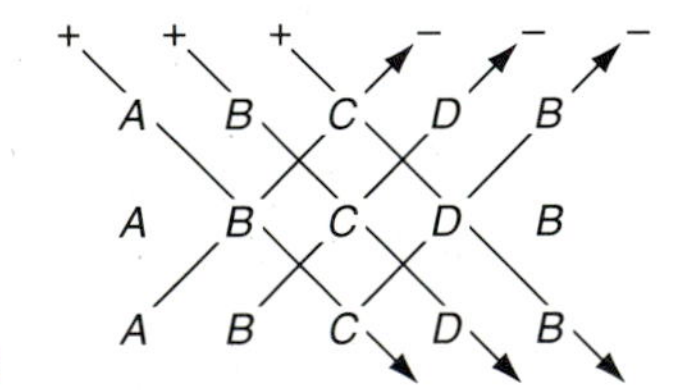

FIGURE 5-33

The numerator for the determinant is solved in the same manner. For determinant x, the solution for the numerator is

$$D_1 B_2 C_3 + B_1 C_2 D_3 + C_1 D_2 B_3 - D_3 B_2 C_1 - B_3 C_2 D_1 - C_3 D_2 B_1$$

Therefore, to solve for determinant x, the equation would be

$$x = \frac{D_1 B_2 C_3 + B_1 C_2 D_3 + C_1 D_2 B_3 - D_3 B_2 C_1 - B_3 C_2 D_1 - C_3 D_2 B_1}{A_1 B_2 C_3 + B_1 C_2 A_3 + C_1 A_2 B_3 - A_3 B_2 C_1 - B_3 C_2 A_1 - C_3 A_2 B_1}$$

EXAMPLE 5-13

Solve for x, y, and z.

$$3x + 2y - 5z = -1$$

$$2x - 3y - z = 11$$

$$5x - 2y + 7z = 9$$

Solution

$$x = \frac{\left|\begin{array}{rrr|rr} -1 & 2 & -5 & -1 & 2 \\ 11 & -3 & -1 & 11 & -3 \\ 9 & -2 & 7 & 9 & -2 \end{array}\right.}{\left|\begin{array}{rrr|rr} 3 & 2 & -5 & 3 & 2 \\ 2 & -3 & -1 & 2 & -3 \\ 5 & -2 & 7 & 5 & -2 \end{array}\right.}$$

$$= \frac{21 - 18 + 110 - 135 + 2 - 154}{-63 - 10 + 20 - 75 - 6 - 28}$$

$$= \frac{-174}{-162} = 1.07$$

$$y = \frac{\left|\begin{array}{rrr|rr} 3 & -1 & -5 & 3 & -1 \\ 2 & 11 & -1 & 2 & 11 \\ 5 & 9 & 7 & 5 & 9 \end{array}\right.}{-162}$$

$$= \frac{231 + 5 - 90 + 275 + 27 + 14}{-162}$$

$$= \frac{462}{-162} = -2.85$$

$$z = \frac{\left|\begin{array}{rrr|rr} 3 & 2 & -1 & 3 & 2 \\ 2 & -3 & 11 & 2 & -3 \\ 5 & -2 & 9 & 5 & -2 \end{array}\right.}{-162}$$

$$= \frac{-81 + 110 + 4 - 15 + 66 - 36}{-162}$$

$$= \frac{48}{-162} = -0.30$$

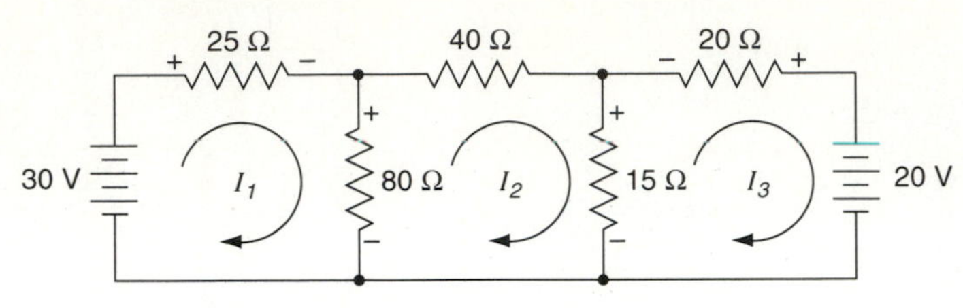

FIGURE 5-34

EXAMPLE 5-14 Solve for the currents I_1, I_2, and I_3 in the circuit of Figure 5-34.

Solution Develop three simultaneous equations:

$$\text{Loop } I_1\text{: } 25I_1 + 80I_1 - 80I_2 + 0I_3 = 30$$
$$105I_1 - 80I_2 + 0I_3 = 30$$

$$\text{Loop } I_2\text{: } 40I_2 + 15I_2 - 15I_3 + 80I_2 - 80I_1 = 0$$
$$-80I_1 + 135I_2 - 15I_3 = 0$$

$$\text{Loop } I_3\text{: } 15I_3 - 15I_2 + 20I_3 = -20$$
$$0I_1 - 15I_2 + 35I_3 = -20$$

The simultaneous equations can now be placed in the determinant grid:

$$x = \frac{\begin{vmatrix} 30 & -80 & 0 \\ 0 & 135 & -15 \\ -20 & -15 & 35 \end{vmatrix} \begin{matrix} 30 & -80 \\ 0 & 135 \\ -20 & -15 \end{matrix}}{\begin{vmatrix} 105 & -80 & 0 \\ -80 & 135 & -15 \\ 0 & -15 & 35 \end{vmatrix} \begin{matrix} 105 & -80 \\ -80 & 135 \\ 0 & -15 \end{matrix}}$$

$$= \frac{111{,}000}{248{,}500}$$
$$= 0.447$$

$$y = \frac{\begin{vmatrix} 105 & 30 & 0 \\ -80 & 0 & -15 \\ 0 & -20 & 35 \end{vmatrix} \begin{matrix} 105 & 30 \\ -80 & 0 \\ 0 & -20 \end{matrix}}{248{,}500}$$

$$= \frac{52{,}500}{248{,}500}$$
$$= 0.211$$

$$z = \frac{\begin{vmatrix} 105 & -80 & 30 \\ -80 & 135 & 0 \\ 0 & -15 & -20 \end{vmatrix} \begin{matrix} 105 & -80 \\ -80 & 135 \\ 0 & -15 \end{matrix}}{248{,}500}$$

$$= \frac{-119{,}500}{248{,}500}$$

$$= -0.481$$

$$I_1 = 0.447 \text{ A} \qquad I_2 = 0.211 \text{ A} \qquad I_3 = -0.481 \text{ A}$$

The values of I_1, I_2, and I_3 can now be checked for accuracy by substituting these numbers into the original simultaneous equations:

$$105(0.447) - 80(0.211) = 30$$

$$-80(0.447) + 135(0.211) - 15(-0.481) = 0$$

$$-15(0.211) + 35(-0.481) = -20$$

5-9 THÉVENIN'S THEOREM

In the study of electric circuits, it is often desirable to determine the effect that a change in a single resistance will have on the currents and voltages of a circuit, while all other resistances remain unchanged. This type of problem may be solved by loop equations and determinants, but the solution by using Thévenin's theorem is an easier method. It is particularly useful when studying the relationship between a load of some kind and the system of sources that supplies the load, as the theorem makes it possible to replace the entire supply system by one equivalent source in series with one resistance. This can be done regardless of the complexity of the supplying system.

Thévenin's theorem is stated as follows:

> Any two-terminal circuit made up of fixed-value resistances and of voltage and current sources can be replaced by a single voltage source in series with a single resistance which will produce the same effects at the terminals.

Figure 5-35(a) shows a two-terminal linear network which may consist of fixed resistance and any combination of constant current and voltage sources. A linear network is made up of components that have a directly proportional relationship between voltage and current. The Thévenin equivalent circuit for Figure 5-35(a) is shown in Figure 5-35(b). The Thévenin equivalent voltage, V_{TH}, is in series with the Thévenin equivalent resistance, R_{TH}.

The Thévenin voltage of Figure 5-35(a) could be measured by placing a high-resistance voltmeter between terminals A and B. It is possible to measure the equivalent resistance of Figure 5-35(a) by connecting an ohmmeter between points A and B. For the ohmmeter reading to be accurate, any voltage sources in the circuit would have to be short-circuited and any current sources open-circuited.

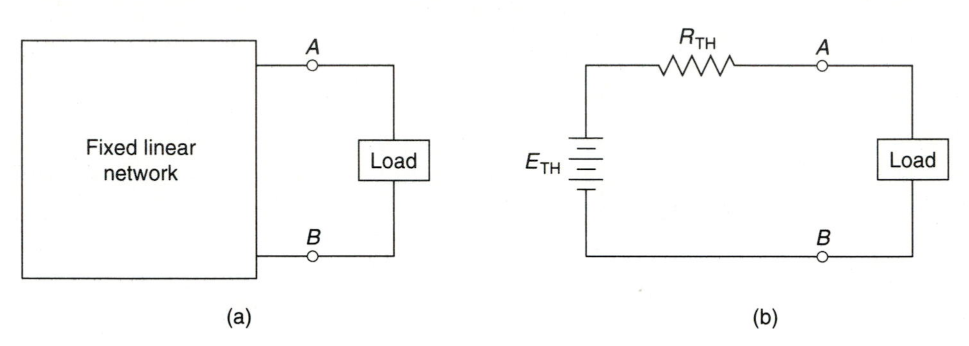

FIGURE 5-35 (a) Fixed linear network; (b) Thévenin equivalent circuit.

The steps taken in applying Thévenin's theorem are as follows:

1. Remove the portion of the circuit where the Thévenin equivalent circuit is to be determined.
2. Determine the Thévenin equivalent resistance, R_{TH}, by short-circuiting voltage sources and open-circuiting current sources.
3. Determine the Thévenin voltage, E_{TH}, across the portion of the circuit removed in step 1. Return all sources to their original position.
4. Draw the Thévenin equivalent circuit. Place the resistance removed in step 1 across the terminals of the Thévenin equivalent circuit. The current through this resistor can now be found by Ohm's law.

FIGURE 5-36

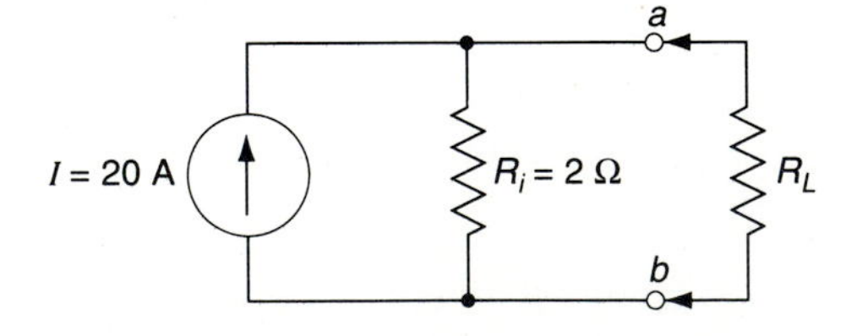

EXAMPLE 5-15 Determine the Thévenin equivalent circuit for the network to the left of terminals a–b in Figure 5-36.

Solution Disconnect the load and replace the current source with an open circuit, as shown in Figure 5-37(a). The resistance between points a and b is 2 Ω. This is the equivalent resistance of the circuit.

$$R_{TH} = R_i = 2\ \Omega$$

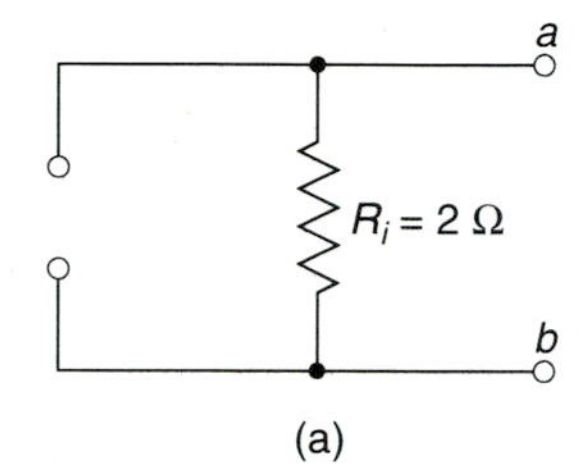

FIGURE 5-37 (a)

Calculate E_{TH} by returning the current source to its original position, as shown in Figure 5-37(b). With a–b open, the entire current of the circuit will flow through the internal resistance R_i. Therefore, the voltage at terminals a–b will be the product of the current through R_i and the ohmic value of R_i.

$$E_{TH} = I \times R_i = 20\ \text{A} \times 2\ \Omega = 40\ \text{V}$$

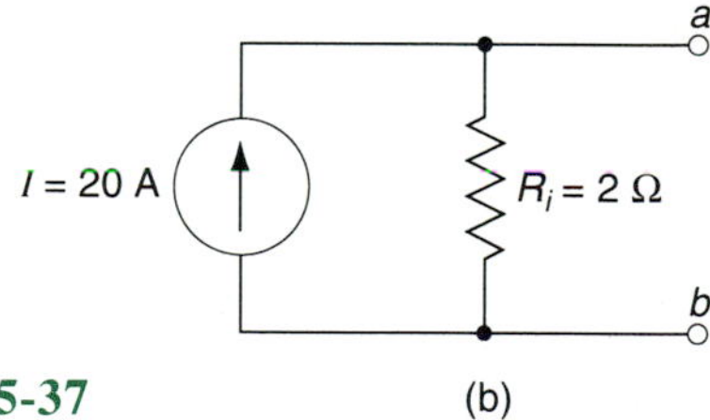

FIGURE 5-37 (b)

The Thévenin equivalent circuit is shown in Figure 5-37(c).

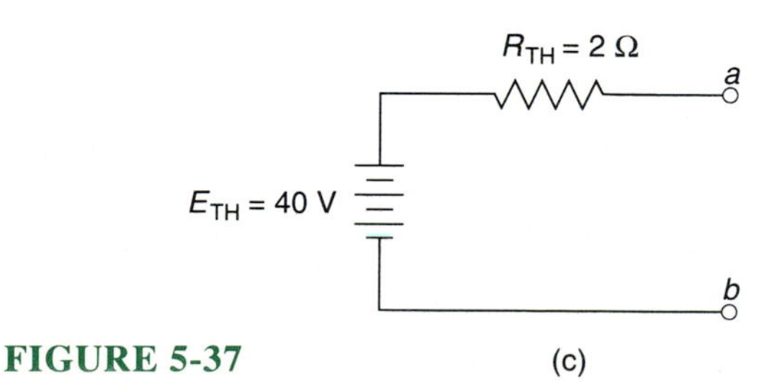

FIGURE 5-37 (c)

A very common problem in electronic circuits is to determine the value of current flowing in a circuit such as the one shown in Figure 5-38. This type of circuit will often have a transistor connected where resistor R_3 is shown, and the value of current flowing into the transistor would be required. The following example illustrates how to solve this type of problem using Thévenin's theorem.

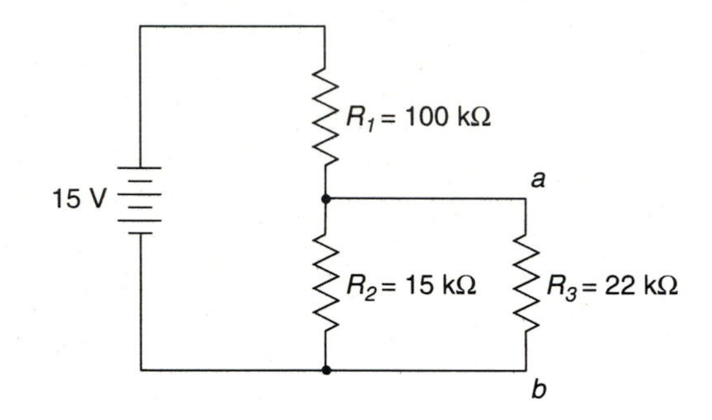

FIGURE 5-38

EXAMPLE 5-16 For the circuit of Figure 5-38, determine the current through resistor R_3 using Thévenin's theorem.

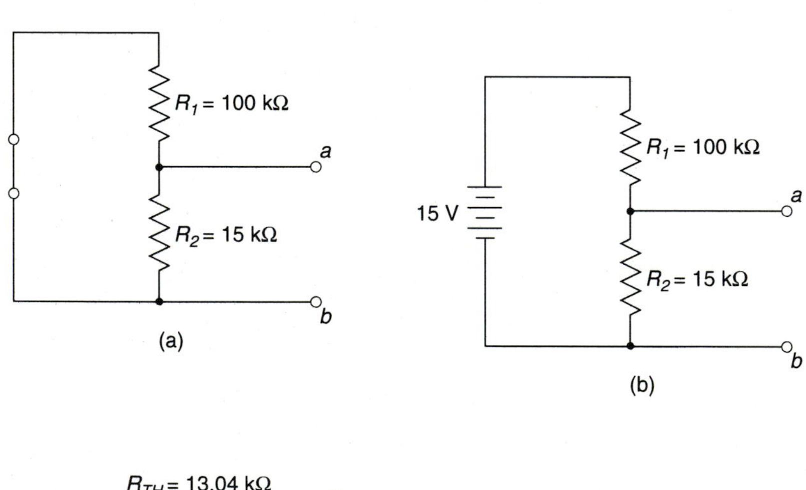

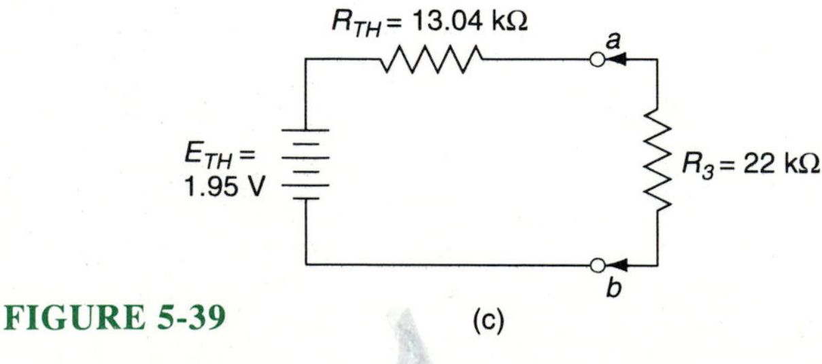

FIGURE 5-39

Solution Disconnect resistor R_3, and short circuit the voltage source, as shown in Figure 5-39(a). The Thévenin equivalent resistance is now determined.

$$R_{TH} = \frac{R_1 \times R_2}{R_1 + R_2} = \frac{(100 \times 10^3) \times (15 \times 10^3)}{(100 \times 10^3) + (15 \times 10^3)}$$
$$= 13.04\ \text{k}\Omega$$

Insert the voltage source back in the circuit, as shown in Figure 5-39(b). Determine the current flow in the circuit with R_3 still disconnected.

$$I = \frac{E}{R_1 + R_2} = \frac{15\ \text{V}}{100\ \text{k}\Omega + 15\ \text{k}\Omega}$$
$$= 0.13\ \text{mA}$$

The Thévenin equivalent voltage across terminals a–b is the voltage dropped across resistor R_3.

$$E_{TH} = V_{ab} = (15\ \text{k}\Omega)(0.13\ \text{mA})$$
$$= 1.95\ \text{V}$$

The Thévenin equivalent circuit can now be drawn [Figure 5-39(c)]. The current through the 22-kΩ resistor is found by Ohm's law.

$$I_3 = \frac{E_{TH}}{R_{TH} + R_3} = \frac{1.95\ \text{V}}{13.04\ \text{k}\Omega + 22\ \text{k}\Omega}$$
$$= 55.65\ \mu\text{A}$$

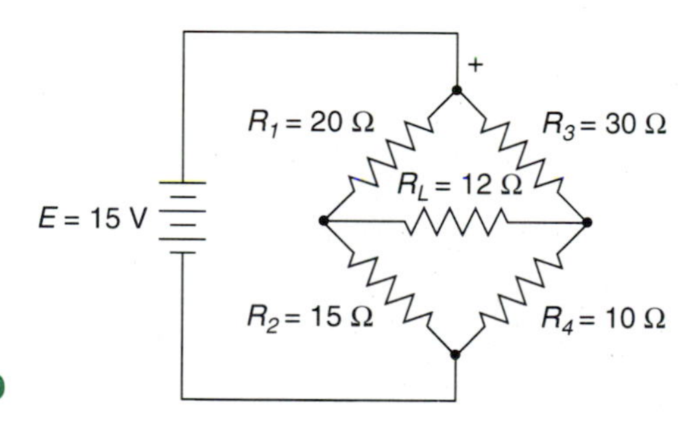

FIGURE 5-40

EXAMPLE 5-17 For the circuit shown in Figure 5-40, calculate the current through resistor R_L using Thévenin's theorem.

Solution Disconnect R_L and short circuit the voltage source, as shown in Figure 5-41(a). It is apparent that resistors R_1 and R_2 are joined at point a, while their opposite ends are connected across the shorted voltage source. The same is true at point b for resistors R_3 and R_4. If the circuit is redrawn, as in Figure 5-41(b), the equivalent resistance of the circuit is found. For the circuit of Figure 5-41(b), the resistance between points a and b consists of the parallel-connected resistors R_1 and R_2, which are in series with the parallel resistors R_3 and R_4. Therefore, the equivalent resistance is determined as follows:

$$R_{TH} = \frac{R_1 \times R_2}{R_1 + R_2} + \frac{R_3 \times R_4}{R_3 + R_4} = \frac{20\,\Omega \times 15\,\Omega}{20\,\Omega + 15\,\Omega} + \frac{30\,\Omega \times 10\,\Omega}{30\,\Omega + 10\,\Omega}$$
$$= 16.07\,\Omega$$

Insert the voltage source back in the circuit of Figure 5-41(a), as shown in Figure 5-41(c). Use the voltage divider rule to find the voltages across resistors R_2 and R_4.

$$V_{R2} = \frac{R_2}{R_1 + R_2}E = \frac{15\,\Omega}{20\,\Omega + 15\,\Omega} \times 15\text{ V}$$
$$= 6.43\text{ V}$$
$$V_{R4} = \frac{R_4}{R_3 + R_4}E = \frac{10\,\Omega}{30\,\Omega + 10\,\Omega} \times 15\text{ V}$$
$$= 3.75\text{ V}$$

The potential difference between points a and b is the difference in voltage between V_{R2} and V_{R4}.

$$E_{TH} = V_{ab} = V_{R2} - V_{R4} = 6.43\text{ V} - 3.75\text{ V}$$
$$= 2.68\text{ V}$$

The Thévenin equivalent circuit can now be drawn [Figure 5-41(d)]. The current through R_L is now solved by Ohm's law.

$$I_{RL} = \frac{E_{TH}}{R_{TH} + R_L} = \frac{2.68\text{ V}}{16.07\,\Omega + 12\,\Omega}$$
$$= 95.48\text{ mA}$$

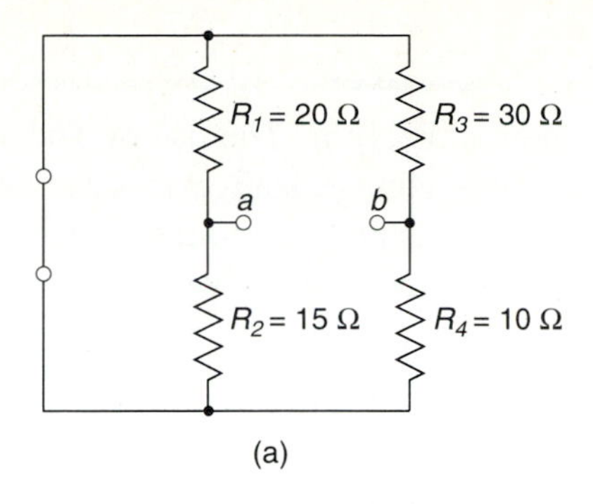

(a)

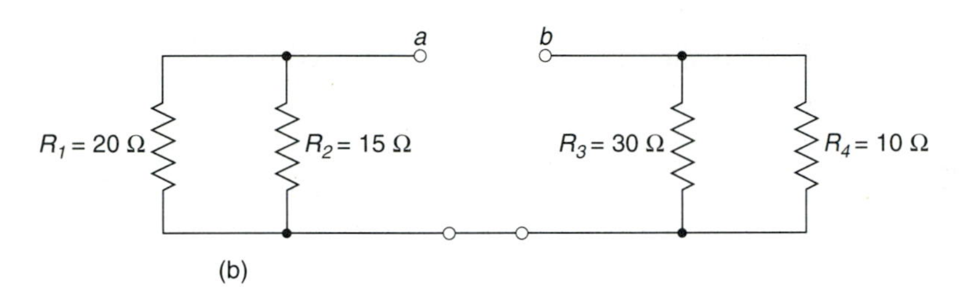

(b)

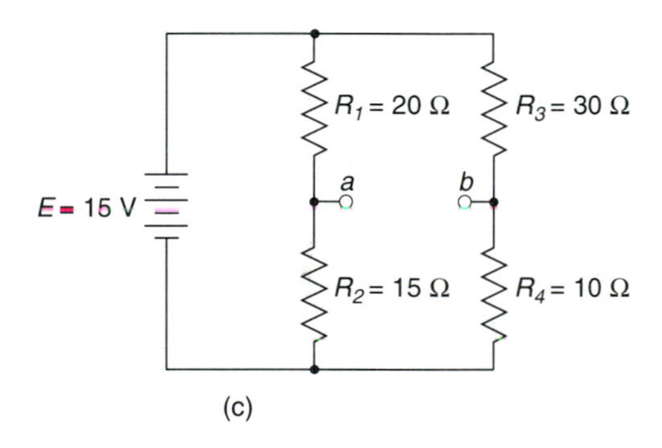

(c)

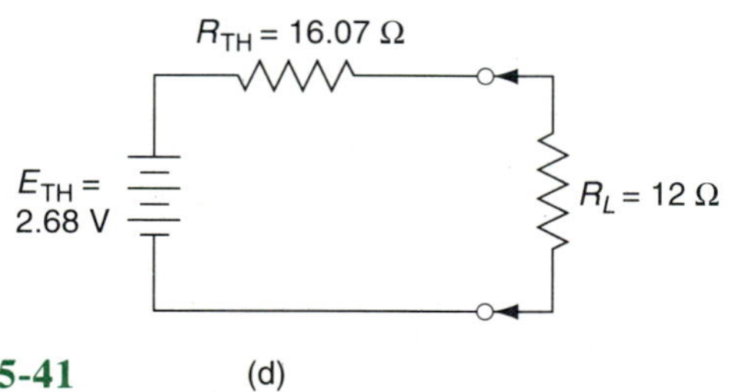

(d)

FIGURE 5-41

5-10 NORTON'S THEOREM

Nortons's theorem is the dual of Thévenin's theorem. The use of Thévenin's theorem is limited to the identification of a voltage source in series with a resistance value. Norton's theorem is implemented to reduce a circuit to a current source and a parallel resistance value. **Norton's theorem** is stated as follows:

> Any two-terminal circuit can be replaced by an equivalent current source in parallel with an equivalent resistance. The current source determines the maximum possible current that would flow if the terminals of the original network are short-circuited. The equivalent parallel resistance is the resistance value measured looking back into the original circuit.

The term *looking back* refers to the resistance of a circuit if voltage sources are removed and replaced by their internal resistance values, and if all current sources are open-circuited.

When applying Norton's theorem, the current and resistance must be determined to form a **Norton equivalent circuit**, as shown in Figure 5-42. The **Norton equivalent current**, I_N, is the short-circuit current between two points in a circuit. When the load terminals are short-circuited, the **Norton equivalent resistance**, R_N, is calculated. Figure 5-42(a) consists of a fixed linear network. In Figure 5-42(b) the Norton equivalent current is found by shorting terminals *a* and *b*. The equivalent resistance is the resistance seen from terminals *a–b* when the independent sources of the fixed network are disconnected.

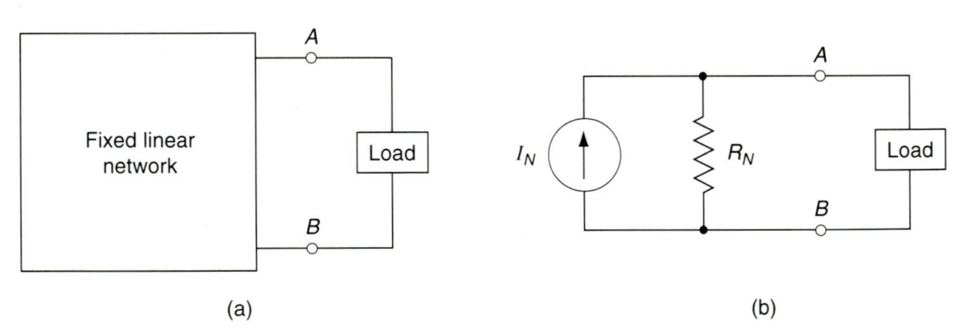

FIGURE 5-42 (a) Fixed linear network; (b) Norton equivalent circuit.

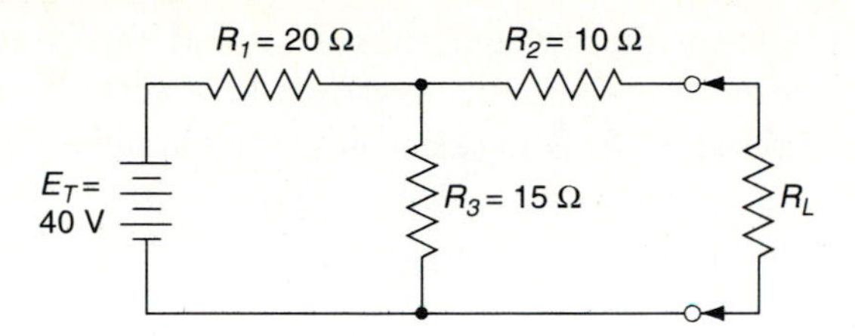

FIGURE 5-43

EXAMPLE 5-18 For the circuit shown in Figure 5-43, find the Norton equivalent circuit.

Solution Short-circuit the load as shown in Figure 5-44(a). Resistors R_2 and R_3 are now in parallel. The total resistance seen by the 40-V source is given by

$$R_T = \frac{R_2 \times R_3}{R_2 + R_3} + R_1 = \frac{10\,\Omega \times 15\,\Omega}{10\,\Omega + 15\,\Omega} + 20\,\Omega$$

$$= 26\,\Omega$$

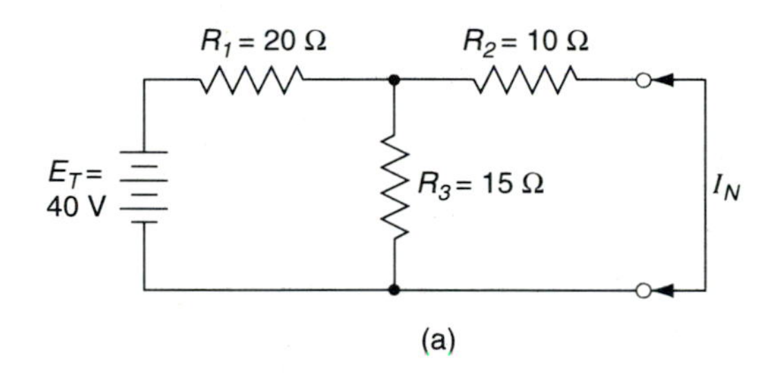

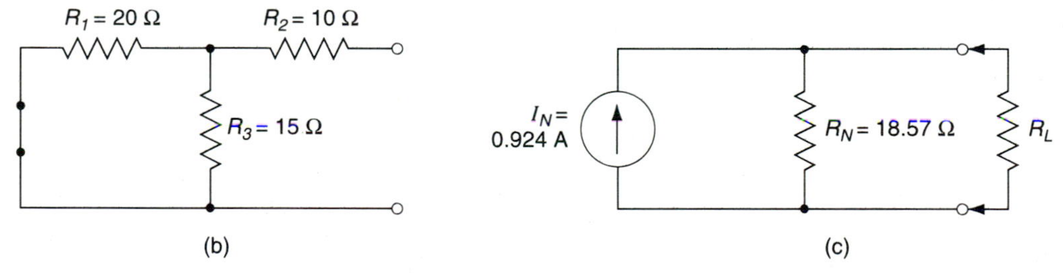

FIGURE 5-44

The Norton equivalent circuit is shown in Figure 5-44(c).

The total current is found by Ohm's law.

$$I_T = \frac{E_T}{R_T} = \frac{40\text{ V}}{26\,\Omega}$$

$$= 1.54\text{ A}$$

Short-circuit the voltage source to find the Norton equivalent resistance of the circuit, as shown in Figure 5-44(b). R_1 and R_3 are in parallel; R_2 is in series with the parallel connection.

$$R_N = \frac{R_1 \times R_3}{R_1 + R_3} + R_2 = \frac{20\,\Omega \times 15\,\Omega}{20\,\Omega + 15\,\Omega} + 10\,\Omega$$
$$= 18.57\,\Omega$$

The Norton equivalent current is now found by using the current divider rule.

$$I_N = \frac{R_3}{R_2 + R_3} I_T = \frac{15\,\Omega}{10\,\Omega + 15\,\Omega} \times 1.54\text{ A}$$
$$= 0.924\text{ A}$$

When current sources are in parallel, the total current is the sum of the individual current sources. The following example illustrates the use of Norton's theorem for a two-source network.

EXAMPLE 5-19 Using Norton's theorem, solve for the current in resistor R_3 for the circuit of Figure 5-45.

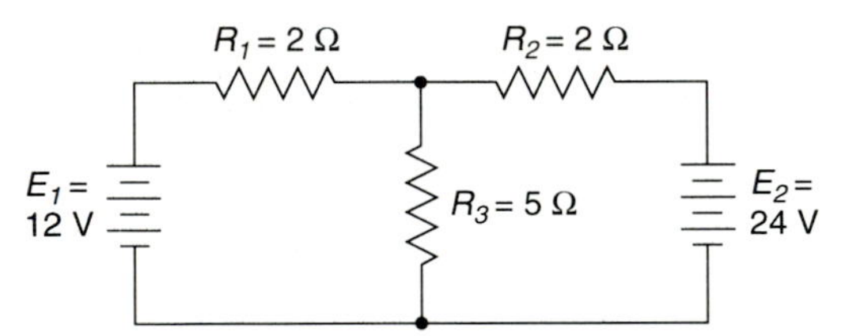

FIGURE 5-45

Solution Short-circuit resistor R_3, and draw arrows indicating current flow, as shown in Figure 5-46(a). Convert voltage sources to current sources [Figure 5-46(b)].

$$I_{N1} = \frac{E_1}{R_1} = \frac{12\text{ V}}{2\,\Omega}$$
$$= 6\text{ A}$$

$$I_{N2} = \frac{E_2}{R_2} = \frac{24\text{ V}}{2\,\Omega}$$
$$= 12\text{ A}$$

The two current sources of Figure 5-46(b) can be combined as a single current source by adding the currents of the sources together. The equivalent resistance can be found by the product-over-sum rule. The Norton equivalent circuit is shown in Figure 5-46(c).

$$I_N = I_{N1} + I_{N2} = 6\text{ A} + 12\text{ A}$$
$$= 18\text{ A}$$

$$R_N = \frac{R_1 \times R_2}{R_1 + R_2} = \frac{2\,\Omega \times 2\,\Omega}{2\,\Omega + 2\,\Omega}$$
$$= 1\,\Omega$$

The current through R_3 is now determined by the current divider rule.

$$I_3 = \frac{1\,\Omega}{1\,\Omega + 5\,\Omega} \times 18\text{ A}$$
$$= \frac{R_N}{R_N + R_3} \times I_N$$
$$= 3\text{ A}$$

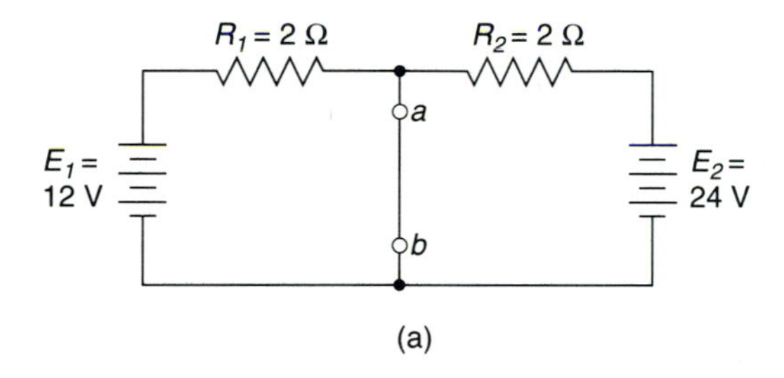

(a)

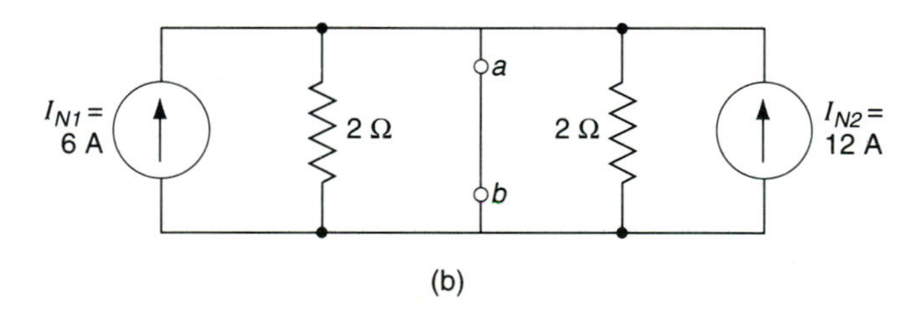

(b)

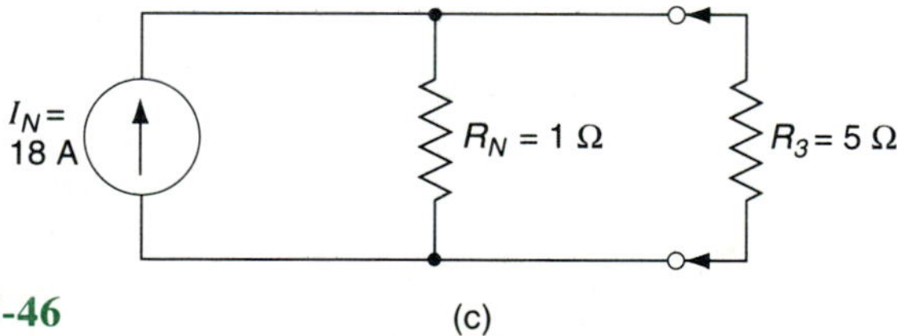

FIGURE 5-46

(c)

5-11 MILLMAN'S THEOREM

Millman's theorem is used for circuits having more than one voltage or current source in parallel. Essentially, Millman's theorem combines the source transformation theorem with both the Thévenin and Norton theorems. Millman's theorem applies only to *sources connected directly in parallel*. If there are resistors between the sources, this theorem cannot be applied. To utilize Millman's theorem, all voltage sources must first be converted to current sources. Once the equivalent current source is determined, it can be transformed to a voltage source if desired. **Millman's theorem** is stated as follows:

> Any number of constant current sources in parallel may be combined into a single current source in which the equivalent current source is the algebraic sum of the individual source currents, and the resistance of the equivalent source is the equivalent resistance of the original parallel-connected resistors.

If all the sources in a circuit were constant voltage sources, the following equation would determine the Millman equivalent voltage source, E_M:

$$E_M = \frac{\pm G_1E_1 \pm G_2E_2 \pm G_3E_3 \pm \cdots \pm G_NE_N}{G_1 + G_2 + G_3 + \cdots + G_N}$$

The plus and minus signs in the equation above are used to indicate whether the sources are supplying energy in the same direction. If the sources have the same polarity, they are added. Conversely, if the sources have opposite polarity, they are subtracted.

The Millman equivalent resistance R_M for parallel-connected voltage sources is found as

$$R_M = \frac{1}{G_1 + G_2 + G_3 + \cdots + G_N}$$

where E_1, E_2, E_3, . . . are the voltages of individual voltage sources and G_1, G_2, G_3, . . . are the conductances of individual voltage sources.

EXAMPLE 5-20 Use Millman's theorem to find the voltage across R_L and the current through R_L in the circuit of Figure 5-47.

Solution First, calculate the Millman equivalent voltage. Since the 15-V source is connected with its polarity opposite the 10- and 12-V sources, the 15-V source is subtracted.

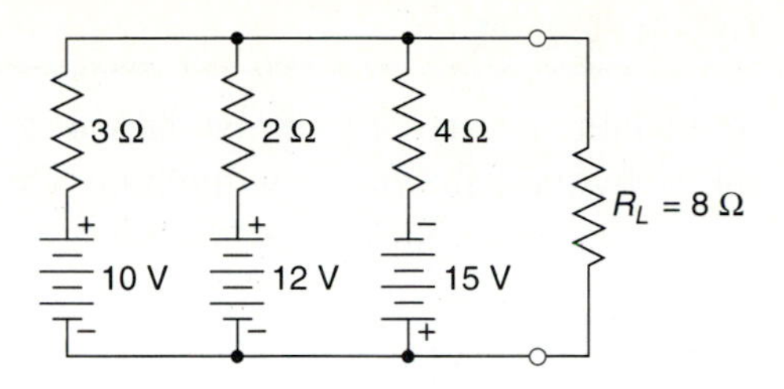

FIGURE 5-47

$$E_M = \frac{G_1E_1 + G_2E_2 - G_3E_3}{G_1 + G_2 + G_3} = \frac{\left(\frac{1}{3}\right)(10) + \left(\frac{1}{2}\right)(12) - \left(\frac{1}{4}\right)(15)}{\frac{1}{3} + \frac{1}{2} + \frac{1}{4}}$$

$$= 5.16 \text{ V}$$

Next, determine the Millman equivalent resistance.

$$R_M = \frac{1}{G_1 + G_2 + G_3} = \frac{1}{\frac{1}{3} + \frac{1}{2} + \frac{1}{4}}$$

$$= 0.923 \ \Omega$$

The Millman equivalent circuit can now be drawn (Figure 5-48). The voltage across R_L is found by the voltage divider rule, and the current through R_L is found by Ohm's law.

$$V_{RL} = \frac{R_L}{R_M + R_L} E_M = \frac{8 \ \Omega}{0.92 \ \Omega + 8 \ \Omega} \times 5.16 \text{ V}$$

$$= 4.63 \text{ V}$$

$$I_{RL} = \frac{V_{RL}}{R_L} = \frac{4.63 \text{ V}}{8 \ \Omega}$$

$$= 0.58 \text{ A}$$

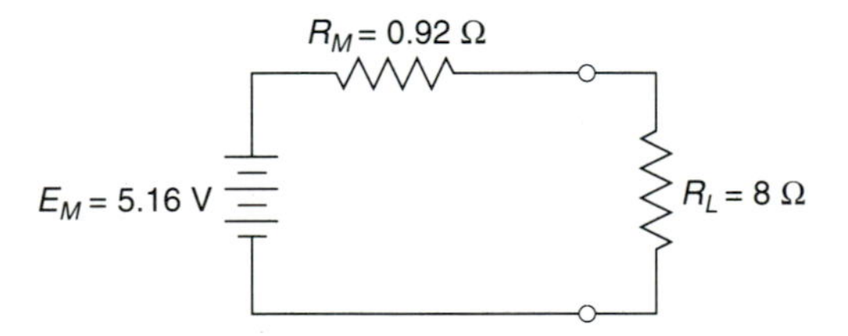

FIGURE 5-48

5-12 SUPERPOSITION THEOREM

Circuits containing more than one source of emf may be solved by the **superposition theorem**, as well as by the Kirchhoff law method. The superposition theorem, as applied to dc circuits, may be stated as follows:

> The current that flows at any point in a circuit, or the potential difference between any two points in a circuit, resulting from more than one source of emf connected in the circuit is the algebraic sum of the separate currents or voltages at these points. These values are the voltages and currents that would exist if each source of emf were considered separately and if each of the other sources were replaced by a unit of equivalent internal resistance.

To consider the effects of each source, the voltage sources must be short-circuited (zero resistance) and the current sources open-circuited (infinite resistance). When a voltage or current source contains an internal resistance, it must be taken into account when the source is removed.

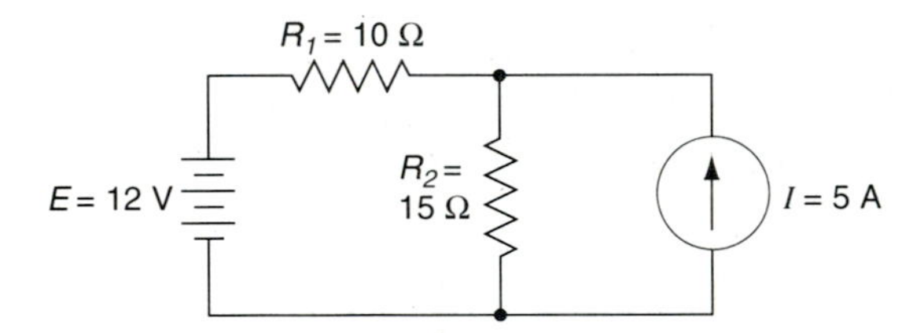

FIGURE 5-49

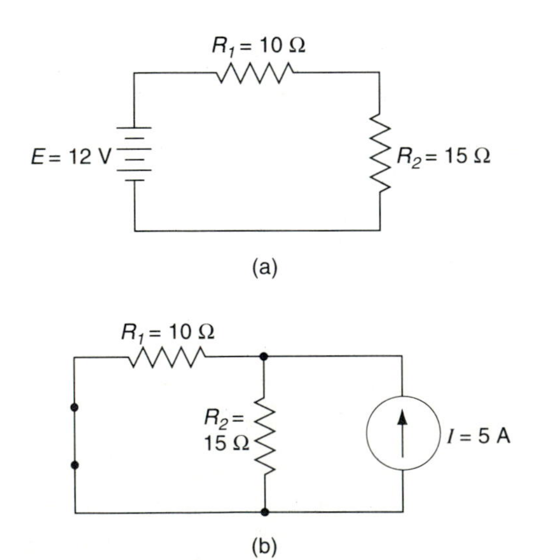

FIGURE 5-50

EXAMPLE 5-21 Use the superposition theorem to find the current through resistor R_2 in the circuit shown in Figure 5-49.

Solution The first step is to remove the current source, as shown in Figure 5-50(a). The component current through resistor R_2 with the current source removed, I', can now be found by using Ohm's law:

$$I' = \frac{E}{R_1 + R_2} = \frac{12\text{ V}}{10\,\Omega + 15\,\Omega} = 0.48\text{ A}$$

The next step involves short-circuiting the voltage source, as shown in Figure 5-50(b). The component current through R_2 with the voltage source removed, I'', can now be determined by using the current divider rule:

$$I'' = \frac{R_1}{R_1 + R_2} \times I = \frac{10\,\Omega}{10\,\Omega + 15\,\Omega} \times 5\text{ A} = 2\text{ A}$$

Since the current I' is the same direction as the current I'', the current through resistor R_2 is the sum of these two currents.

$$I_{R2} = I' + I'' = 0.48 + 2 = 2.48\text{ A}$$

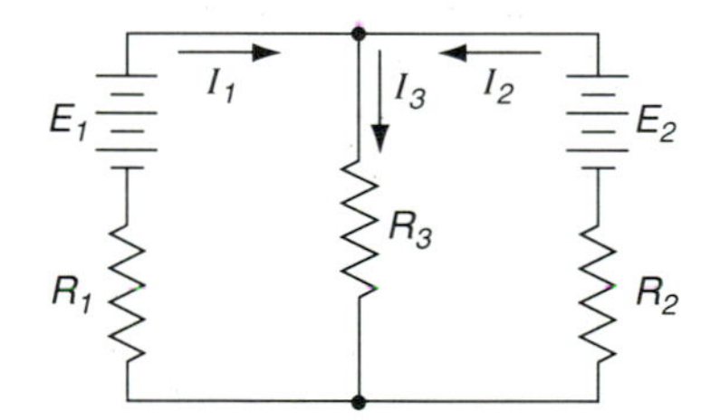

FIGURE 5-51

EXAMPLE 5-22 For the circuit shown in Figure 5-51, use superposition to determine currents I_1, I_2, and I_3. The values of the network are as follows: $R_1 = 5\,\Omega$, $R_2 = 15\,\Omega$, $R_3 = 20\,\Omega$, $E_1 = 60$ V, and $E_2 = 100$ V.

Solution The circuit is redrawn, as shown in Figure 5-52(a), with voltage source E_1 short-circuited.

$$R_{1\|3} = \frac{R_1 \times R_3}{R_1 + R_3} = \frac{5\,\Omega \times 20\,\Omega}{5\,\Omega + 20\,\Omega} = 4\,\Omega$$

$$R_T = R_{1\|3} + R_2 = 4\,\Omega + 15\,\Omega = 19\,\Omega$$

$$I_2' = \frac{E_2}{R_T} = \frac{100\text{ V}}{19\,\Omega}$$
$$= 5.263\text{ A}$$

$$E_{1\|3} = R_{1\|3} \times I_2' = 4\,\Omega \times 5.263\text{ A}$$
$$= 21.052\text{ V}$$

$$I_1' = -\frac{E_{1\|3}}{R_1} = \frac{21.052\text{ V}}{5\,\Omega}$$
$$= -4.21\text{ A}$$

$$I_3' = \frac{E_{1\|3}}{R_3} = \frac{21.052\text{ V}}{20\,\Omega}$$
$$= 1.053\text{ A}$$

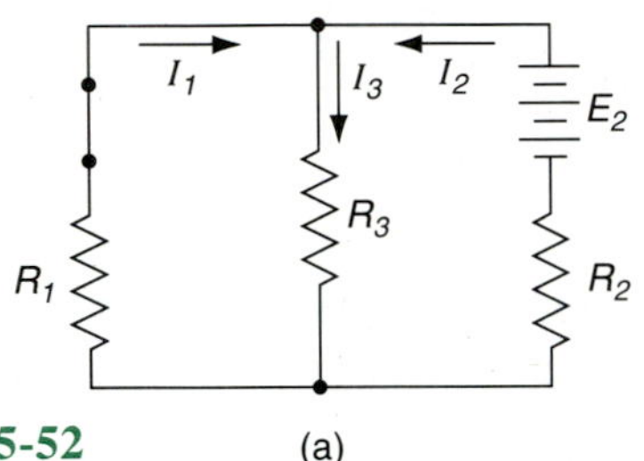

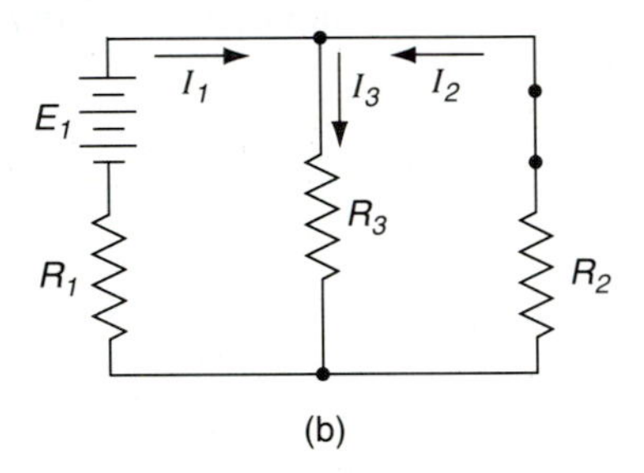

FIGURE 5-52 (a) (b)

The circuit is now redrawn, as shown in Figure 5-52(b), with source E_2 short-circuited.

$$R_{2\|3} = \frac{R_2 \times R_3}{R_2 + R_3} = \frac{15\,\Omega \times 20\,\Omega}{15\,\Omega + 20\,\Omega}$$
$$= 8.571\,\Omega$$

$$R_T = R_{2\|3} + R_1 = 8.571\,\Omega + 5\,\Omega$$
$$= 13.571\,\Omega$$

$$I_1'' = \frac{E_1}{R_T} = \frac{60\text{ V}}{13.571\,\Omega}$$
$$= 4.421\text{ A}$$

$$E_{2\|3} = R_{2\|3} \times I_1'' = 8.571\,\Omega \times 4.421\text{ A}$$
$$= 37.892\text{ V}$$

$$I_3'' = \frac{E_{2\|3}}{R_3} = \frac{37.892\text{ V}}{20\,\Omega}$$
$$= 1.895\text{ A}$$

$$I_2'' = -\frac{E_{2\|3}}{R_2} = -\frac{37.892\text{ V}}{15\,\Omega}$$
$$= -2.526\text{ A}$$

The total current is now found by superimposing the results:

$$I_1 = I_1' + I_1'' = (-4.21\text{ A}) + 4.421\text{ A}$$
$$= 0.211\text{ A}$$

$$I_2 = I_2' + I_2'' = 5.263\text{ A} + (-2.526\text{ A})$$
$$= 2.737\text{ A}$$

$$I_3 = I_3' + I_3'' = 1.053\text{ A} + 1.895\text{ A}$$
$$= 2.948\text{ A}$$

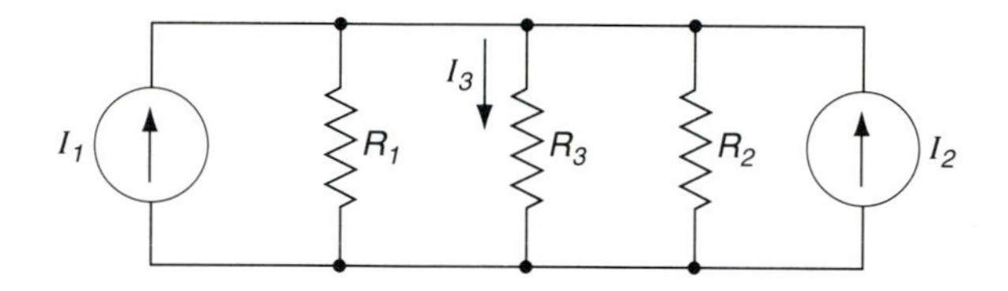

FIGURE 5-53

EXAMPLE 5-23 Use the superposition theorem to find the current through resistor R_3 in the circuit shown in Figure 5-53. The values of the network are $R_1 = 20\ \Omega$, $R_2 = 30\ \Omega$, $R_3 = 15\ \Omega$, $I_1 = 5$ A, and $I_2 = 10$ A.

Solution Redraw the circuit, as shown in Figure 5-54(a), with the current source I_2 open-circuited.

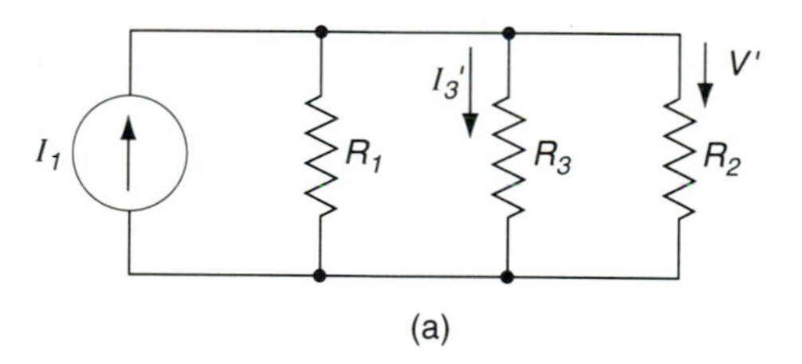

FIGURE 5-54

$$\frac{1}{R_T} = \frac{1}{R_1} + \frac{1}{R_3} + \frac{1}{R_2}$$
$$= 0.15\text{ S}$$

$$R_T = 6.67\ \Omega$$

$$V' = I_1 \times R_T = 5\text{ A} \times 6.67\ \Omega$$
$$= 33.35\text{ V}$$

$$I_3' = \frac{V'}{R_3} = \frac{33.35\text{ V}}{15\ \Omega}$$
$$= 2.22\text{ A}$$

The circuit is now redrawn, as in Figure 5-54(b), with current source I_1 disconnected.

$$V'' = I_2 \times R_T = 10\text{ A} \times 6.67\ \Omega$$
$$= 66.7\text{ V}$$

$$I_3'' = \frac{V''}{R_3} = \frac{6.67\text{ V}}{15\ \Omega}$$
$$= 4.45\text{ A}$$

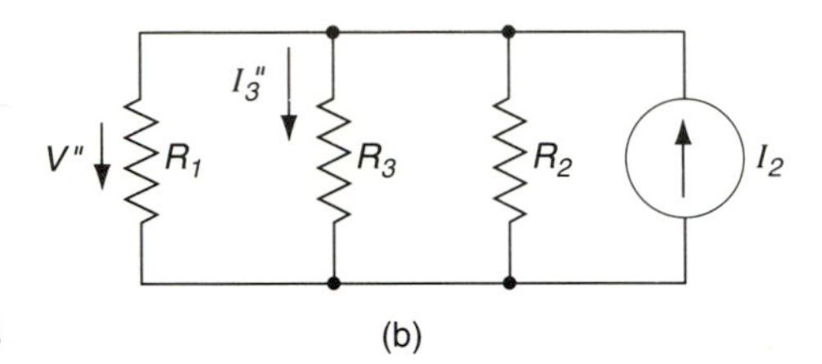

FIGURE 5-54 (b)

The result is then found by superimposing the two currents for I_3.

$$I_3 = I_3' + I_3'' = 2.22\text{ A} + 4.45\text{ A}$$
$$= 6.67\text{ A}$$

5-13 MAXIMUM POWER TRANSFER THEOREM

The **maximum power transfer theorem** can be stated in the following manner:

> Maximum power is drawn from a source when the load resistance equals the internal resistance of the source.

The Thévenin resistance of a circuit is comparable to the internal resistance of the source, since it absorbs some of the available power from the source. The effect of the source resistance, R_i, on the power output of a dc source may be shown by an analysis of the circuit in Figure 5-55. The current for any value of load resistance, R_L, is found by Ohm's law:

$$I = \frac{E}{R_i + R_L}$$

When the variable load resistor is set at the zero-ohms position (equivalent to a short circuit), the current is limited only by the internal resistance of the source. The maximum current that may be drawn from the source would be found by

$$I = \frac{E}{R_i}$$

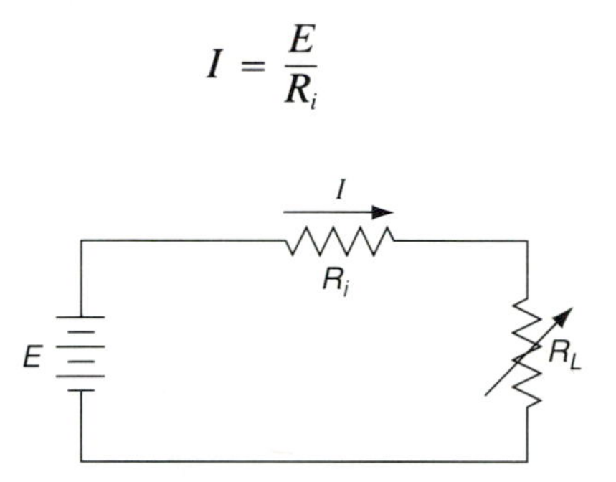

FIGURE 5-55 Maximum power transfer. Thévenin equivalent of a circuit with a variable load resistance.

If the load resistance is increased, current drawn from the source will decrease. At the same time, the terminal voltage applied across the load will increase and will approach a maximum value as the current approaches zero. To calculate the power delivered to the load, the following formula may be used:

$$P_L = I^2R_L = \frac{E^2R_L}{(R_i + R_L)^2} \tag{5-3}$$

The load power is dependent on both R_i and R_L; however, R_i is considered to be constant for any circuit. The ratio of output power to input power, or the power transfer, from the source to the load increases as the load resistance is increased. The efficiency of power transfer approaches 100% as the load resistance approaches a relatively large value compared with the source resistance. The efficiency of power transfer is expressed by the equation

$$\eta = \frac{P_L}{P_T} = \frac{I^2R_L}{I^2R_L + I^2R_i} \tag{5-4}$$

where P_L is the load power and P_T is the power developed by the source.

In electronic communication systems, maximum power transfer is extremely important when analyzing signal sources such as antennas. When the load resistance and the source resistance are equal, the circuit is said to be **matched**, and maximum power transfer occurs between the antenna and load.

The graph of Figure 5-56 shows the relationship between load power and efficiency with a load resistance that varies between 0 and 200 Ω. When the load resistance is 100 Ω, the maximum load power of 100 W is obtained. This occurs when the resistance of the source is equal to the load resistance. However, the efficiency of power transfer is only 50% at the maximum power transfer resistance of 100 Ω and approaches zero efficiency at relatively low values of load resistance compared with that of the source.

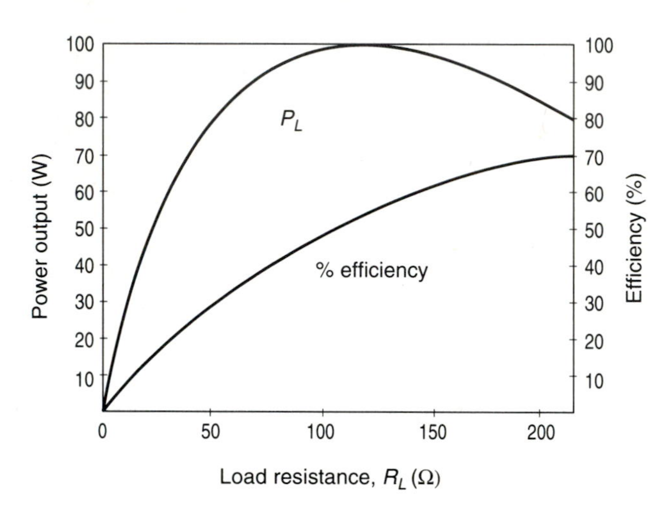

FIGURE 5-56 Effect of load resistance on power output.

The problem of obtaining high efficiency and maximum power transfer is resolved as a compromise between the low efficiency of maximum power transfer and the high efficiency of the high-resistance load. Where circuits deal with large values of power and efficiency is critical, the load resistance is made large in proportion to the source resistance. Unfortunately, this is not practical in situations such as power transmission by power companies, although it is feasible to a certain extent in electronic circuits, where maximum power is required from a small power source. The source and load resistance are matched for maximum power, and the efficiency of the circuit is ignored.

EXAMPLE 5-24 A dc power supply has a no-load voltage of 40 V and delivers 80 W at a full-load current of 3 A. Find (a) the internal resistance of the supply; (b) the value of load resistance when full-load current flows; (c) the load resistance for maximum power transfer; (d) the efficiency of the source at full load.

Solution (a) At full load the load voltage is found by dividing the rated power by the rated current. The difference between the no-load voltage and the full-load voltage is the internal voltage drop. The internal resistance is then found by Ohm's law.

$$V_L = \frac{P_L}{I_L} = \frac{80\text{ W}}{3\text{ A}} = 26.67\text{ V}$$

$$V_i = V_{NL} - V_{FL} = 40\text{ V} - 26.67\text{ V} = 13.33\text{ V}$$

$$R_i = \frac{V_i}{I} = \frac{13.33\text{ V}}{3\text{ A}} = 4.44\ \Omega$$

(b) $$R_L = \frac{V_L}{I_L} = \frac{26.67\text{ V}}{3\text{ A}} = 8.89\ \Omega$$

(c) For maximum power transfer, $R_L = R_i = 4.44\ \Omega$.

(d) $$\eta = \frac{P_L}{P_T} \times 100 = \frac{P_L}{V_{NL} \times I_L} \times 100 = \frac{80\text{ W}}{40\text{ V} \times 3\text{ A}} \times 100 = 66.67\%$$

5-14 ANALYSIS OF DEPENDENT SOURCES

Independent sources are current or voltage sources that operate independently of the network to which they are connected. A **dependent**, or **controlled, source** has the property that its voltage or current is dependent on the voltage or current in another branch of the circuit.

By definition, a simple circuit component is the mathematical model of a two-terminal electrical or electronic device. Mathematical models are used by engineers, technologists, and technicians to analyze the behavior of a component in a circuit. Dependent sources appear in the models of electronic devices such as transistors and operational amplifiers (op-amps).

A dependent source is represented by a diamond-shaped symbol. Figure 5-57(a) shows a dependent voltage source. A dependent current source is shown in Figure 5-57(b). The polarity of both sources is specified within the diamond shape.

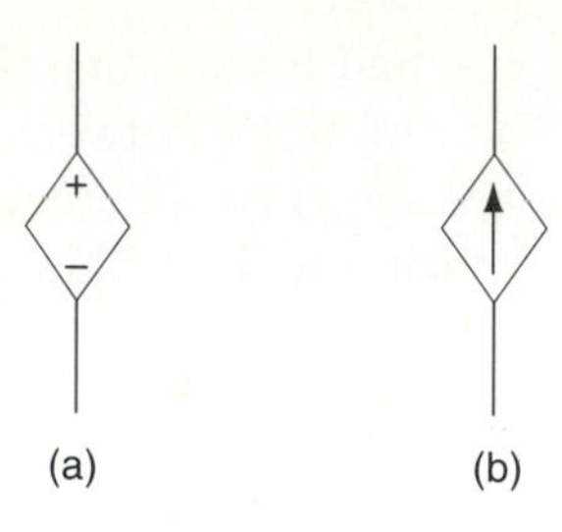

FIGURE 5-57 Schematic symbols for (a) dependent voltage source and (b) dependent current source.

Dependent and independent voltage and current sources are **active elements**, which means that they are capable of delivering power to an external device. A network that contains at least one active element is called an **active network**. A circuit that does not contain any active elements is called a **passive network**.

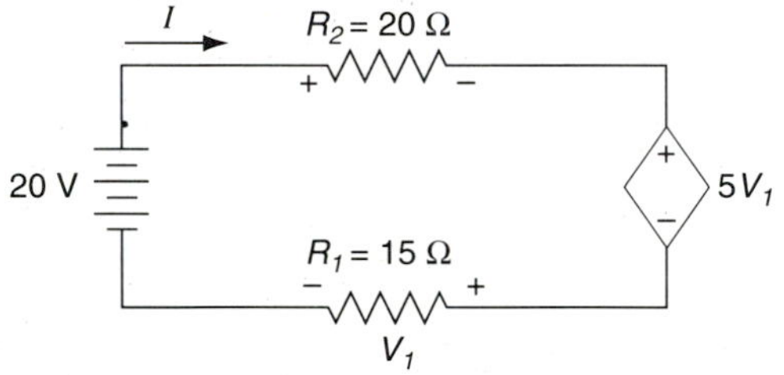

I
$R_2 = 20\ \Omega$
20 V
$5V_1$
$R_1 = 15\ \Omega$
V_1

FIGURE 5-58

In Figure 5-58, two voltage sources are shown. The independent source supplies 20 V, and the dependent source is given as $5V_1$. This means that the dependent source is supplying five times the voltage that is being dropped across resistor R_1. The following steps are taken to find the current, I, in the circuit of Figure 5-58:

1. Apply Kirchhoff's voltage law around the circuit.

$$20I + 5V_1 + 15I = 20$$

or

$$35I + 5V_1 = 20$$

$$V_1 = 15I$$

2. This value of V_1 can now be substituted into the KVL equation and solved by Ohm's law.

$$35I + 5V_1 = 20$$

$$35I + 5(15I) = 20$$

$$110I = 20$$

$$I = \frac{20}{110}$$

$$= 181.8\,\text{mA}$$

Proof:

$$V_1 = 0.1818\text{ A} \times 15\ \Omega = 2.727\text{ V}$$
$$V_2 = 0.1818\text{ A} \times 20\ \Omega = 3.636\text{ V}$$
$$5V_1 = 5 \times 2.727\text{ V} = 13.635\text{ V}$$
$$\text{Total} = 2.727\text{ V} + 3.636\text{ V} + 13.635\text{ V} \approx 20\text{ V}$$

EXAMPLE 5-25 Determine the voltage, V, for the circuit shown in Figure 5-59.

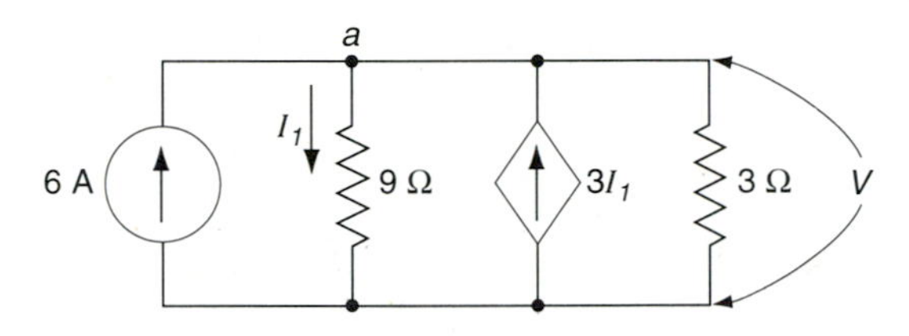

FIGURE 5-59

Solution Apply Kirchhoff's current law at node a.

$$-6 + I_1 - 3I_1 + \frac{V}{3} = 0$$

Also, by Ohm's law, the equation for I_1 is

$$I_1 = \frac{V}{9}$$

Combining both equations results in

$$-6 + \frac{V}{9} - \frac{3V}{9} + \frac{V}{3} = 0$$
$$-6 - \frac{2V}{9} + \frac{V}{3} = 0$$
$$\frac{V}{3} - \frac{2V}{9} = 6$$
$$V = 54\text{ V}$$

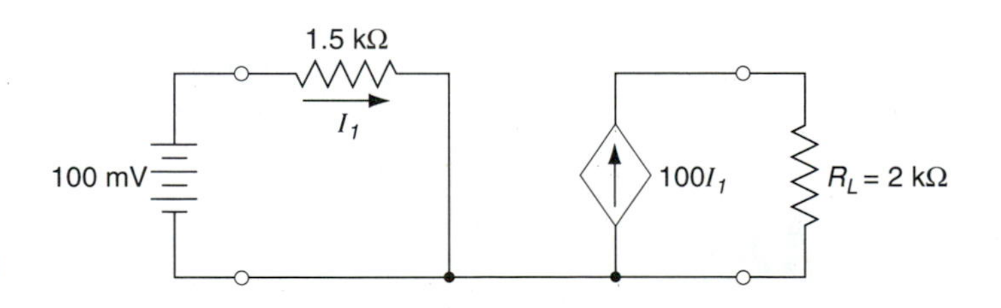

FIGURE 5-60

EXAMPLE 5-26 The circuit shown in Figure 5-60 is a simplified model of a transistor. Determine the voltage across the load resistor R_L.

Solution The current from the dependent source is 100 times the value of the current flowing through the 1.5-kΩ resistor. The first step in solving this problem is to write the KVL equation for the loop with the voltage source.

$$(1.5 \times 10^3\,\Omega)I_1 = 100\,\text{mV}$$

$$I_1 = \frac{100 \times 10^{-3}\,\text{V}}{1.5 \times 10^3\,\Omega} = 66.67\,\mu\text{A}$$

Next, write the KVL equation for the loop with the current source.

$$V_{RL} = 100I_1R_L = 100(66.67 \times 10^{-6}\,\text{A})(2 \times 10^3\,\Omega)$$
$$= 13.33\text{ V}$$

KEY TERMS

Series–parallel circuit
Internal voltage drop
Nonideal (practical) source
Anode
Cathode
Practical current source
Source conversion
Equivalent
Loop analysis
Loop
Loop current
Node
Node voltage
Dependent node
Independent node
Three-wire distribution circuit
Determinant
Thévenin's theorem
Norton's theorem
Norton equivalent circuit
Norton equivalent current
Norton equivalent resistance
Millman's theorem
Superposition theorem
Maximum power transfer theorem
Matched circuit
Dependent (controlled) source
Active element
Active network
Passive network

PROBLEMS

5-1 Calculate the equivalent resistance of the circuit shown in Figure 5-61.

5-2 If the applied voltage in Figure 5-61 is 24 V, what would be the current through each branch?

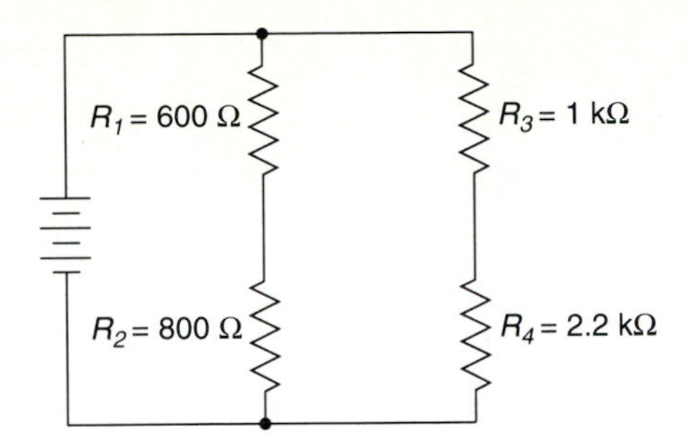

FIGURE 5-61

5-3 For the circuit shown in Figure 5-62, calculate the total resistance.

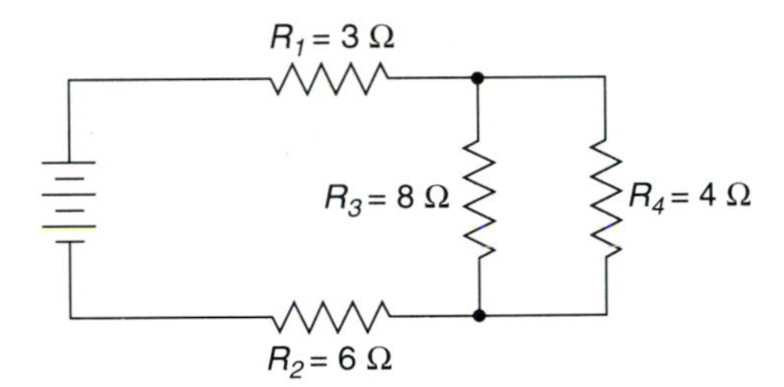

FIGURE 5-62

5-4 If the total current in the circuit shown in Figure 5-62 is 3.43 A, what value of current flows in resistor R_4?

5-5 In the circuit of Figure 5-62, find the current through R_3 if the applied voltage is 24 V.

5-6 Calculate the total resistance of the circuit shown in Figure 5-63.

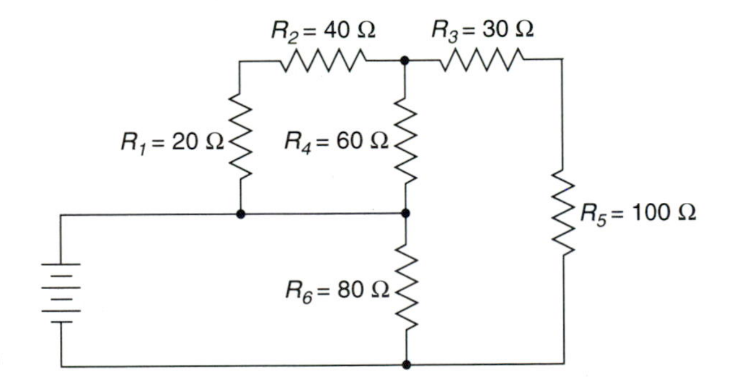

FIGURE 5-63

5-7 In the circuit of Figure 5-63, find the current through R_5 if the applied voltage is 120 V.

5-8 If the total current flowing in Figure 5-63 is 400 mA, what is the voltage drop across resistor R_4?

5-9 What is the total power dissipated by the circuit of Figure 5-63 if the applied voltage is 24 V?

5-10 If the total current flowing in the circuit of Figure 5-63 is 1.8 A, how much power is dissipated by resistor R_6?

5-11 Calculate the total resistance of the circuit shown in Figure 5-64.

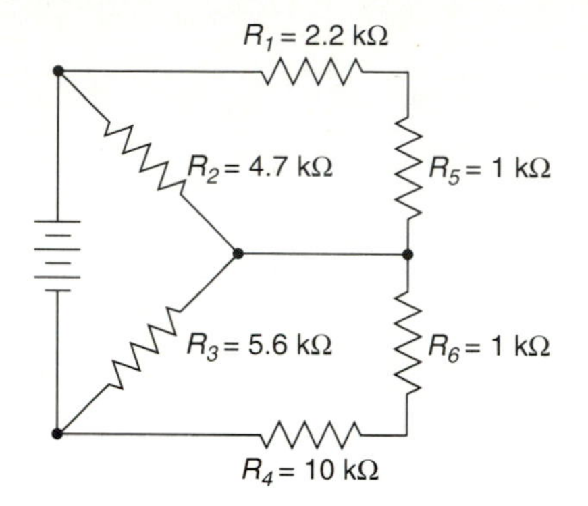

FIGURE 5-64

5-12 In the circuit of Figure 5-64, find the current through R_3 if the applied voltage is 6 V.

5-13 If the total current flowing in Figure 5-64 is 2 mA, find the voltage drop across R_3.

5-14 Find the total power dissipated by the circuit shown in Figure 5-64 if the applied voltage is 12 V.

5-15 Solve for currents I_1 and I_2 in the circuit shown in Figure 5-65.

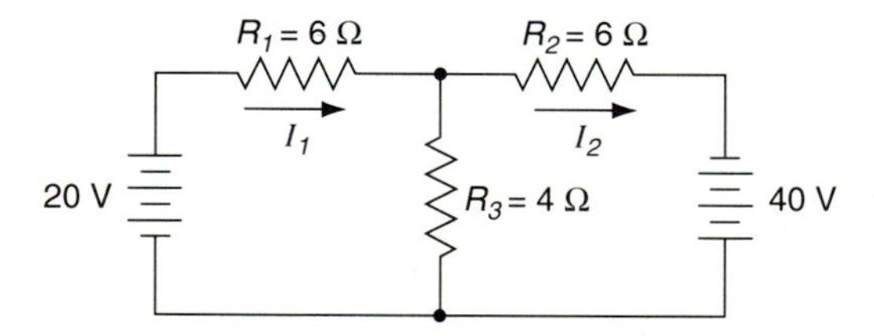

FIGURE 5-65

5-16 Find the voltage drops across resistors R_1, R_2, and R_3 in Figure 5-65.

5-17 Use loop analysis to solve for the three currents flowing in the circuit shown in Figure 5-66.

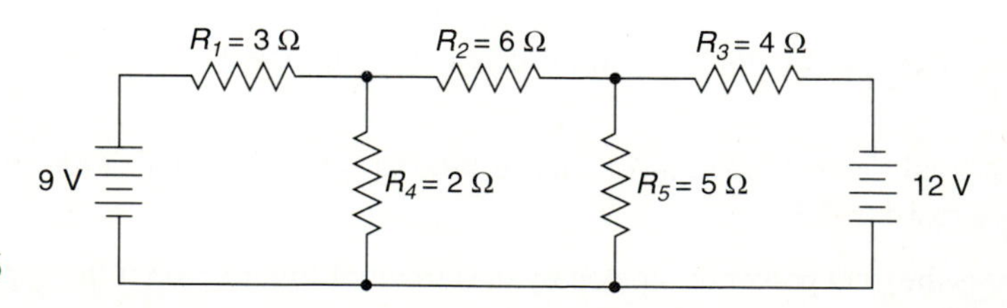

FIGURE 5-66

5-18 Find the total power dissipated by the circuit of Figure 5-66.

5-19 Find the value of current flowing through resistor R_5 in Figure 5-67.

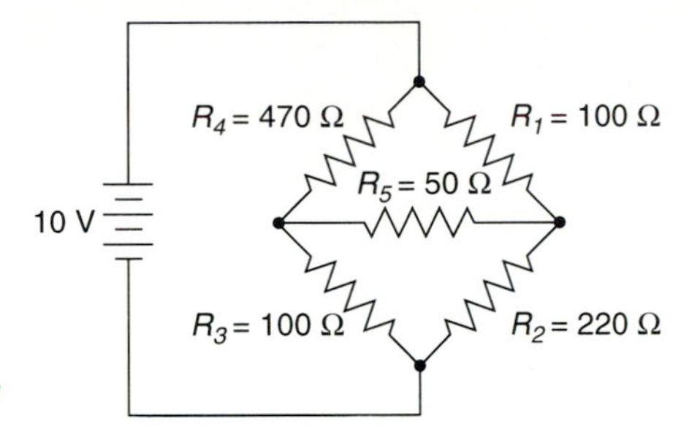

FIGURE 5-67

5-20 Solve for all currents in the circuit shown in Figure 5-68.

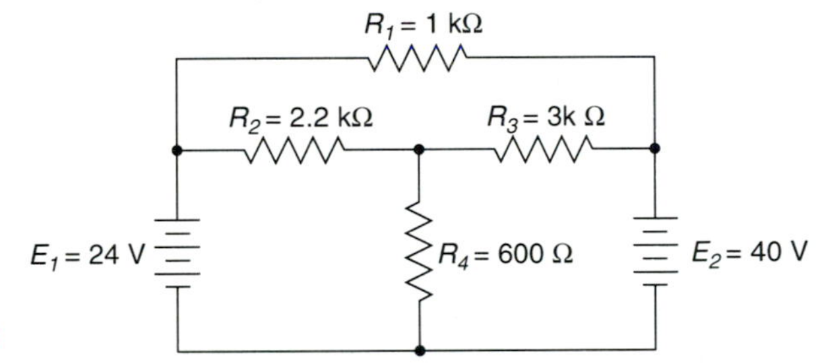

FIGURE 5-68

5-21 Use nodal analysis to find the voltage V_x for the circuit shown in Figure 5-69.

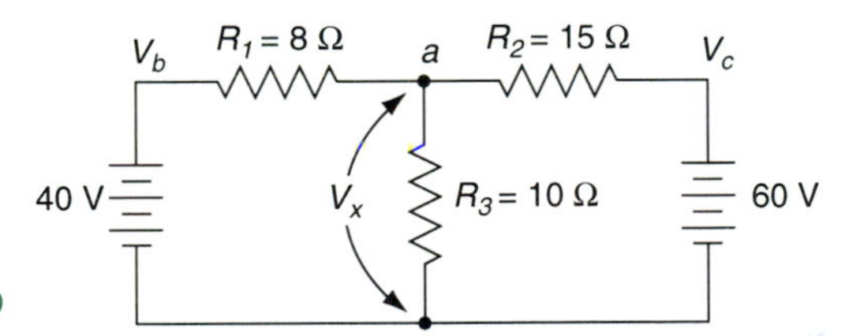

FIGURE 5-69

5-22 Solve for the five branch currents shown in Figure 5-70 using nodal analysis.

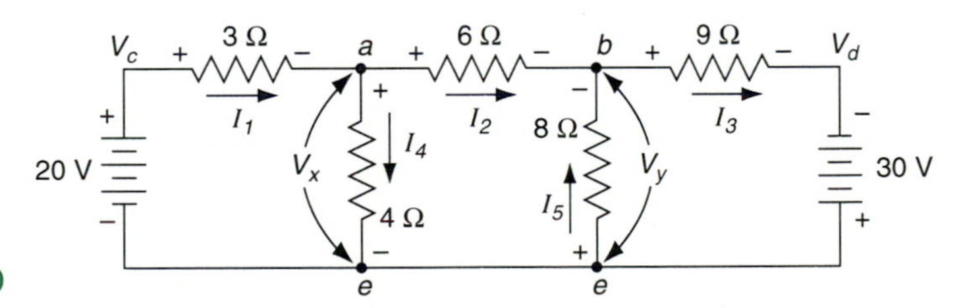

FIGURE 5-70

5-23 The circuit shown in Figure 5-71 has the following loads: $R_A = 2$ kW, $R_B = 1$ hp, and $R_C = 1$ kW. Determine the line and neutral currents.

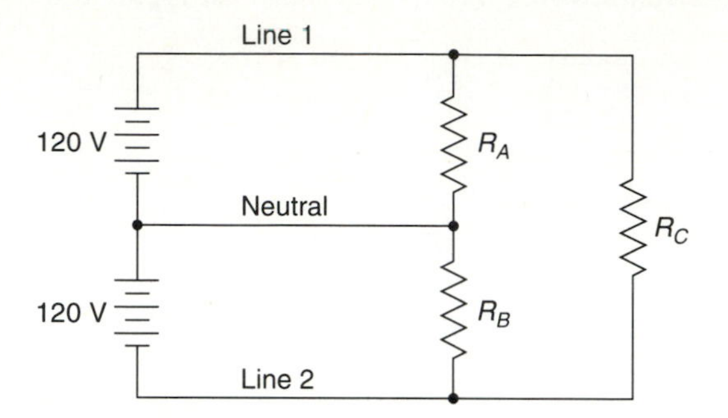

FIGURE 5-71

5-24 If the neutral became disconnected in Figure 5-71, what value of current would flow through load R_B?

5-25 The circuit shown in Figure 5-71 has line resistances $R_1 = 0.5\ \Omega$, $R_2 = 0.5\ \Omega$, and $R_N = 0.5\ \Omega$. The load resistances are as follows: $R_A = 3\ \Omega$, $R_B = 2\ \Omega$, and $R_C = 4\ \Omega$. Determine the value of current flowing in each load of the system.

5-26 The load resistors in Figure 5-71 are $R_A = 1$ kΩ, $R_B = 2$ kΩ, and $R_C = 3$ kΩ. If the line resistances are $R_1 = 100\ \Omega$, $R_2 = 100\ \Omega$, and $R_N = 80\ \Omega$, find the currents flowing in each load.

5-27 Calculate the total power loss in the three line conductors shown in the circuit of Figure 5-72.

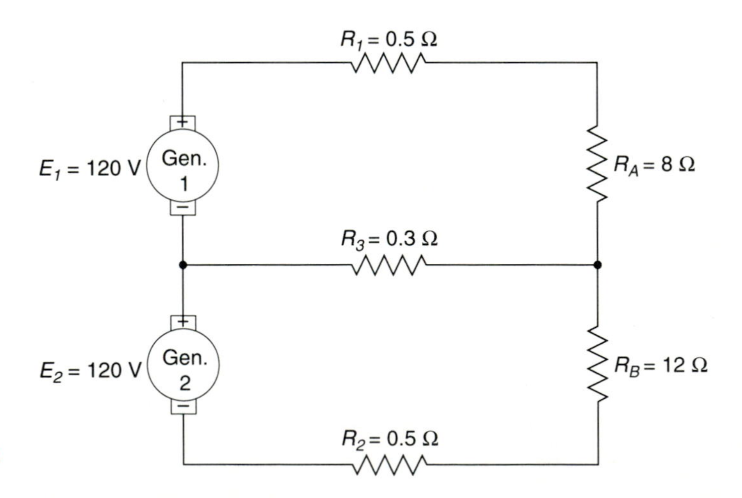

FIGURE 5-72

5-28 Solve for the currents I_1, I_2, and I_3 in the circuit of Figure 5-73.

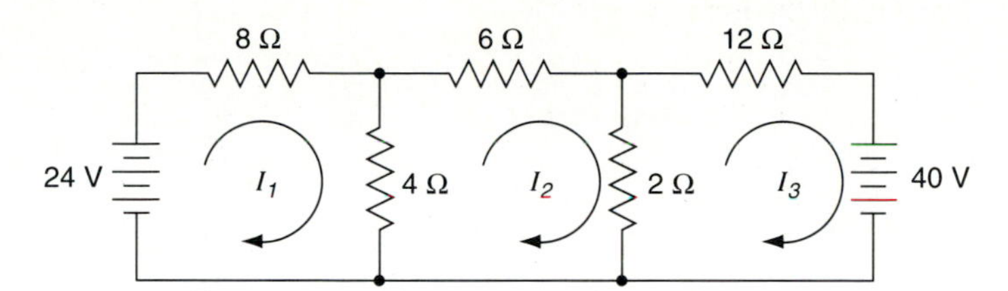

FIGURE 5-73

5-29 Find the current through resistor R_3 in Figure 5-74 using Thévenin's theorem.

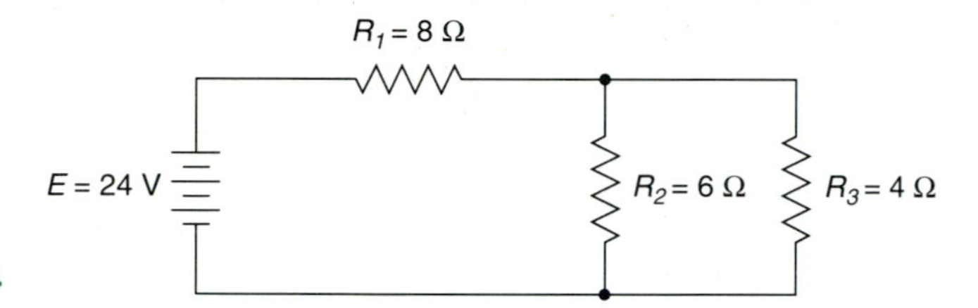

FIGURE 5-74

5-30 Find the voltage and current for load resistor R_L in the circuit shown in Figure 5-75.

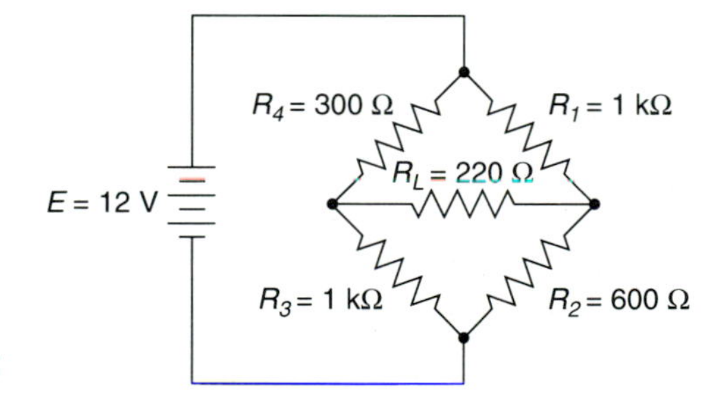

FIGURE 5-75

5-31 Using Thévenin's theorem, find the current flowing through resistor R_3 in the circuit shown in Figure 5-76.

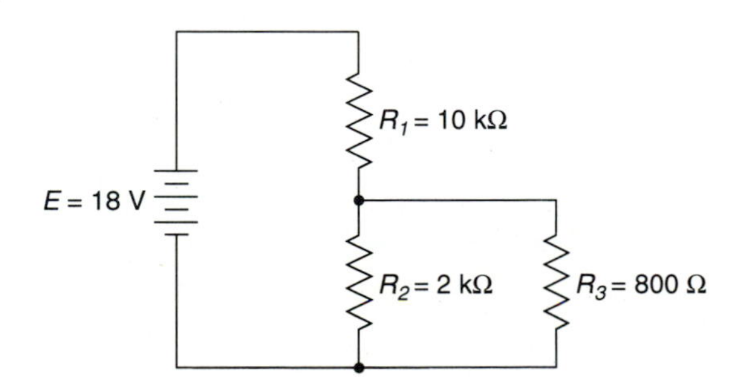

FIGURE 5-76

5-32 For the circuit of Figure 5-77, find the current flowing through resistor R_3 by use of Thévenin's theorem.

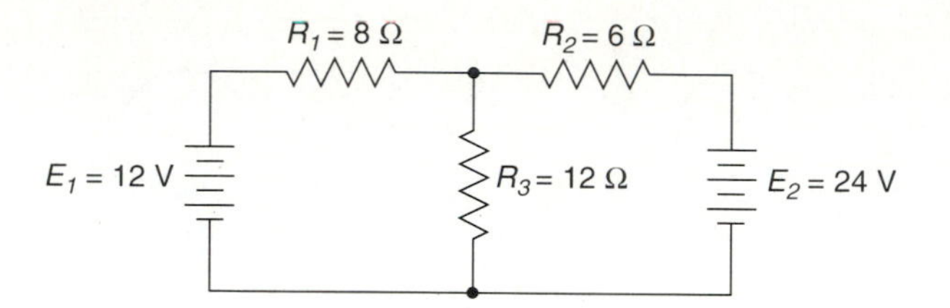

FIGURE 5-77

5-33 Use Thévenin's theorem to find the equivalent voltage source for the circuit shown in Figure 5-78.

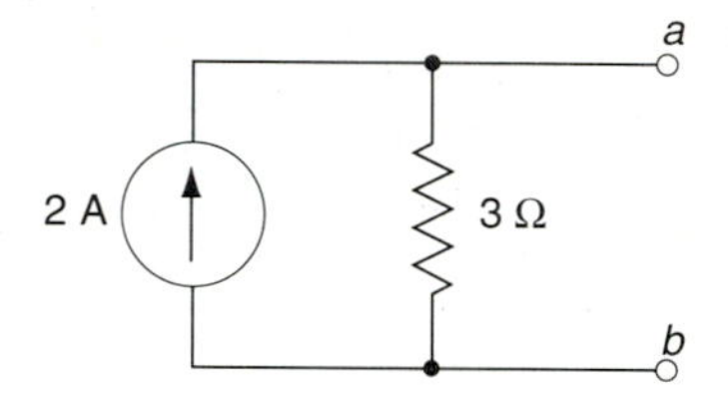

FIGURE 5-78

5-34 Use Norton's theorem to find the equivalent current source for the circuit shown in Figure 5-79.

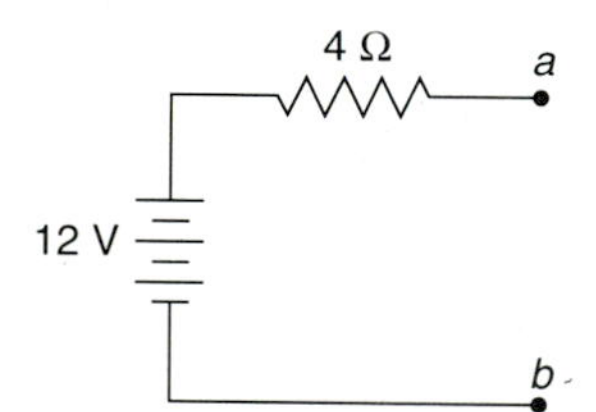

FIGURE 5-79

5-35 For the circuit of Figure 5-80, calculate the load current and load voltage using Norton's theorem if the load resistance is (a) 47 kΩ; (b) 68 kΩ.

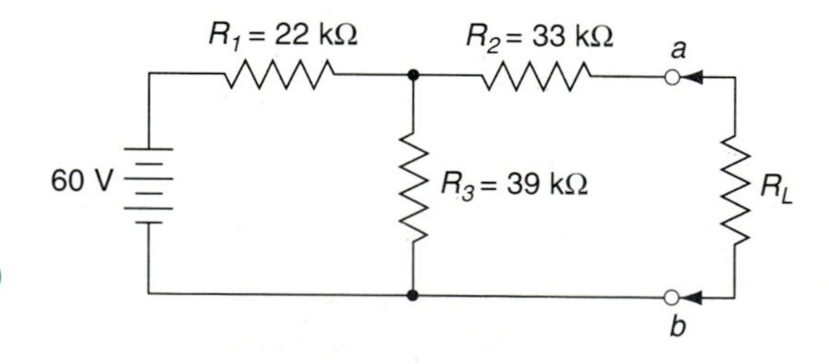

FIGURE 5-80

5-36 Figure 5-81 shows a battery-generator charging circuit. Determine the load current and load voltage by using Norton's theorem.

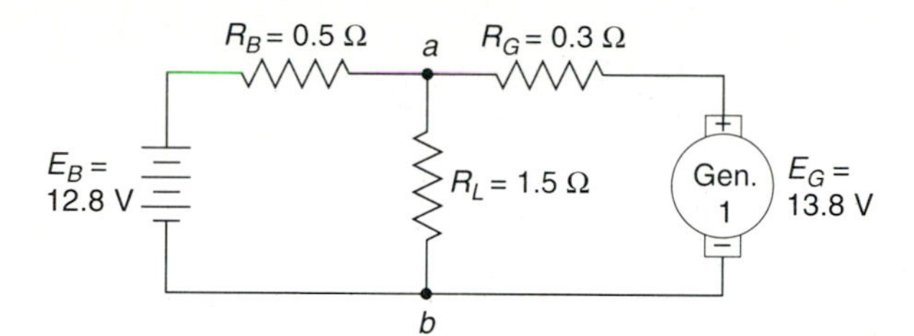

FIGURE 5-81

5-37 Use the superposition theorem to determine the value of current flowing in resistor R_3 in the circuit shown in Figure 5-82.

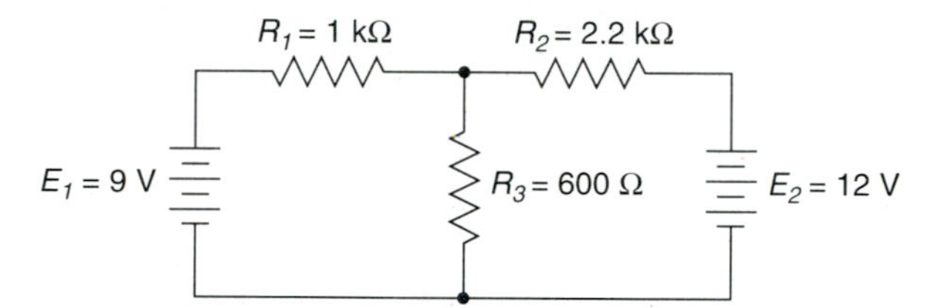

FIGURE 5-82

5-38 Find the values of the three currents flowing through the resistors shown in Figure 5-83 using the superposition theorem.

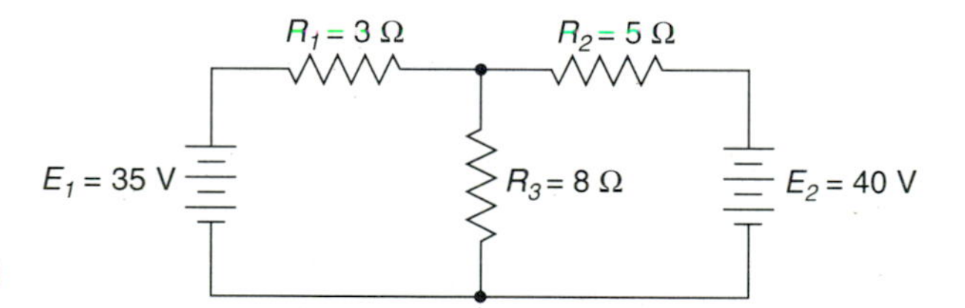

FIGURE 5-83

5-39 Using superposition, determine the current through resistor R_2 in the circuit shown in Figure 5-84.

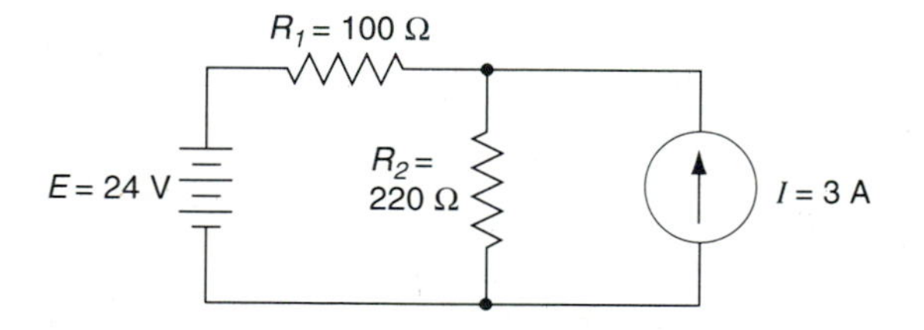

FIGURE 5-84

5-40 Find the values of the three currents in the circuit of Figure 5-85 using superposition. The values of the circuit components are as follows: $R_1 = 6\ \Omega$, $R_2 = 10\ \Omega$, $R_3 = 12\ \Omega$, $E_1 = 9$ V, and $E_2 = 18$ V.

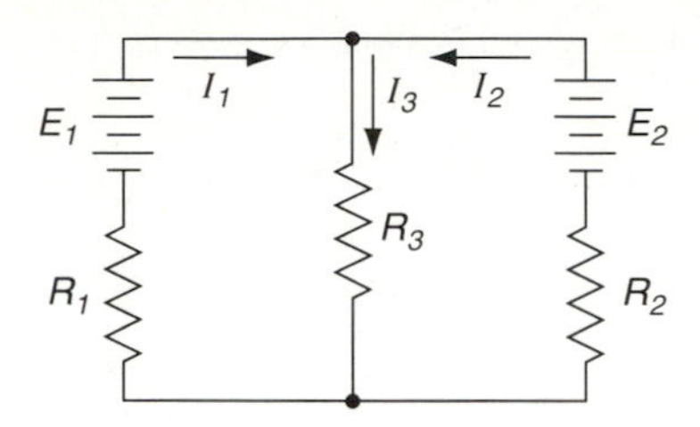

FIGURE 5-85

5-41 For the circuit shown in Figure 5-86, use the maximum power transfer theorem to find (a) the ohmic value of R_L for maximum power; (b) the maximum power that can be delivered to R_L.

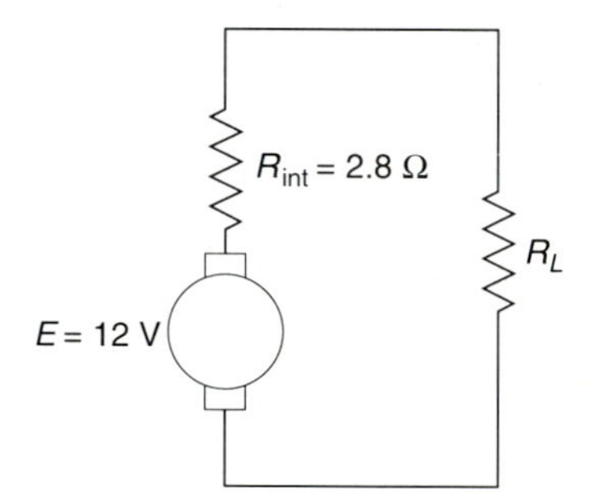

FIGURE 5-86

5-42 For the circuit of Figure 5-86, determine the power for the following values of load resistance: (a) 20 Ω; (b) 100 Ω; (c) 4 Ω.

5-43 A dc power supply delivers 24 V to its output terminals when no-load is connected. When the rated full-load current of 5 A is drawn from the supply, it provides 100 W to the load. Calculate (a) the power supply's internal resistance; (b) the value of load resistance for full-load current; (c) the value of load resistance for maximum power transfer; (d) the full-load efficiency of the power supply.

CHAPTER

Capacitance and Capacitors

LEARNING OBJECTIVES

Upon complction of this chapter you will be able to:

Describe the electrostatic field between two charged surfaces.

Determine the flux density of a capacitor.

Define relative permittivity and dielectric strength.

Understand the relationship between dielectric constants and common insulating materials.

Express the capacitance of a device in terms of charge and potential difference.

List the three factors that determine the capacitance of a capacitor.

Understand the terms *leakage current* and *leakage resistance*.

Describe various types of capacitors used in electrical and electronic circuits.

Utilize the capacitor color code.

Understand transients in *RC* circuits.

Describe the universal time constant curve.

Define a relaxation oscillator.

Determine the energy stored by a capacitor.

Understand the relationship between capacitors connected in series and in parallel.

6-1 INTRODUCTION

When two conducting surfaces are separated by an insulator such as a dielectric, and a difference of potential is applied between the conductors, a state of stress will be established in the dielectric. If the potential difference is sufficient, the stress will cause the insulation to break down, and a spark will pass between the conducting surfaces, puncturing the insulation.

In the dielectric, a field of electric force, or an electrostatic field, is established. Contained in this electrostatic field is stored energy. Figure 6-1 shows two parallel plates of a conducting material, separated by a dielectric, and connected through a switch to a battery. If the parallel plates are initially uncharged and the switch is left open, no net positive or negative charge will exist on either plate. When the switch is closed, charges from the source will distribute themselves on the plates; that is, a current will flow. There will be a surge of current at first, limited in magnitude by the resistance present. As more charge is accumulated, and more voltage developed across the plates, the accumulated charge tends to oppose the further flow of charge. This action creates a net positive charge on the top plate. Electrons are being repelled by the negative terminal through the lower conductor to the bottom plate at the same rate that they are being drawn to the positive terminal. Finally, when enough charge has been transferred from one plate to the other, a voltage equal to the applied emf will have been developed across the plates. The final result is a net positive charge on the top plate and a negative charge on the bottom plate. Figure 6-2 shows the two plates as charged bodies and an electric field set up in the dielectric.

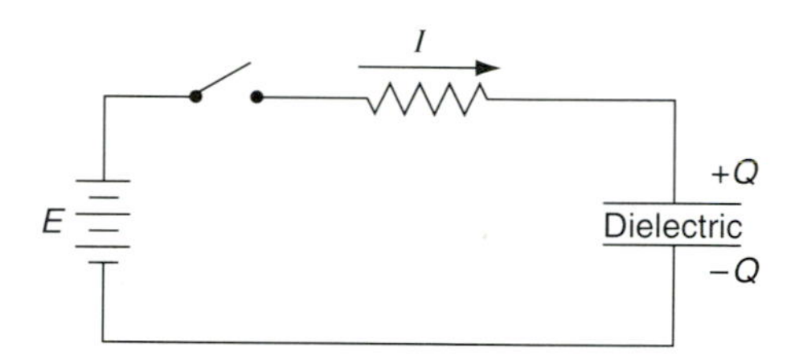

FIGURE 6-1 Dielectric circuit.

The element shown in Figure 6-2, constructed of two plates separated by an insulating material, is called a **capacitor**. Capacitance is a measure of a capacitor's ability to store charge on its plates. The total strength of the electric field is represented by the total number of lines of force, or **dielectric flux**. The lowercase Greek letter ψ (psi) is used to represent this flux.

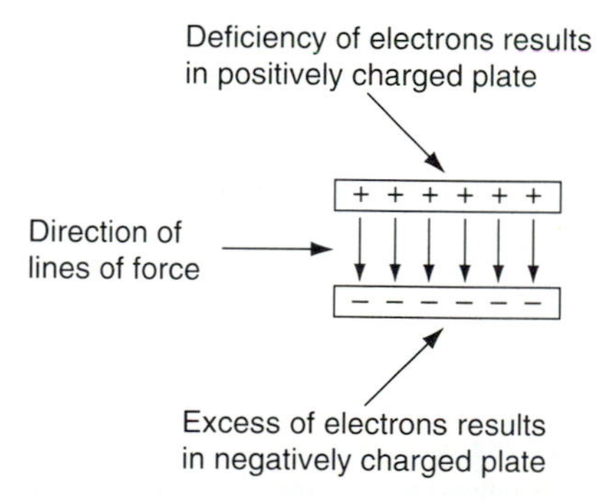

FIGURE 6-2 Electrostatic field between two charged surfaces.

Electric charge is measured in coulombs, and the coulomb is also the unit of electric flux. Therefore, a capacitor having a certain number of coulombs of charge will have a similar amount of electric flux between its plates:

$$\text{electric flux, } \psi = Q \tag{6-1}$$

The larger the charge, Q, in coulombs, the greater the number of lines of flux extending or terminating per unit area. Twice the charge will produce twice the flux per unit area.

The charge-inducing capability of an electric field is called its **electric flux density**, D. Equation 6-1 implies that the electric flux density will be constant whenever the flux is evenly distributed, as it would be across the plates of a capacitor. The flux density that results from charge distribution across the surface of a capacitor is

$$D = \frac{\psi}{A} = \frac{Q}{A} \tag{6-2}$$

where D = flux density, in coulombs per square meter

A = area of plate, in square meters

Q = charge, in coulombs

If the distance between the plates shown in Figure 6-2 were decreased, the reaction of the energy field of each plate on the other is increased. The positively charged plate will attract electrons to the negatively charged plate with greater strength. This means that the flux increases without the applied voltage increasing. Therefore, the flux can be said to be inversely proportional to the distance between plates. In equation form,

$$\psi = \frac{1}{d} \tag{6-3}$$

Elastance can be defined as the opposition to the setting up of electric lines of force in an electric insulator or dielectric. The letter symbol for elastance is S. Elastance increases when the distance between plates increases. If the area of the plates were increased, the elastance would decrease.

6-2 RELATIVE PERMITTIVITY (DIELECTRIC CONSTANT)

The elastance of any dielectric circuit depends on the material of the path. The **relative permittivity**, or **dielectric constant**, is a measure of how good a material is for the production of dielectric flux. The symbol for relative permittivity is ϵ_r.

The dielectric constant for a vacuum is taken as unity. For any other material, the constant will be more than 1, depending on how many more lines of force would be produced if the material were substituted for a vacuum as the path between the plates. The relative permittivity is dimensionless, since it is a ratio of the **absolute permittivity**, ϵ, of a material to the **absolute permittivity**, ϵ_0, of a vacuum. This ratio is expressed by the equation

$$\epsilon_r = \frac{\text{flux produced with material as dielectric}}{\text{flux produced with a vacuum as dielectric}} = \frac{\epsilon}{\epsilon_0} \tag{6-4}$$

The choice of ϵ_0 as the proportionality constant is the result of Gauss's law, which applies to any closed hypothetical surface (called a *Gaussian surface*). **Gauss's law** is stated as follows:

> The net number of electric lines of force crossing any closed surface in an outward direction is numerically equal to the net total charge within that surface.

The value of ϵ_0, also referred to as the *permittivity of free space*, is $8.85 \times 10^{-12}\ \text{C}^2/\text{N}\cdot\text{m}^2$.

The dielectric constant of a material will vary depending on the processing method in its manufacture. Some common values are given in Table 6-1.

TABLE 6-1 *Dielectric Constants*

Material	Typical dielectric constant
Vacuum	1.0
Air	1.0006
Ceramic (low loss)	5.0–570
Ceramic (high loss)	600–10,000
Glass	4.4–10.0
Mica (typical)	5.5
Mylar	3.0
Paper	4.0–6.0
Paraffin	2.1–2.5
Plastics	2.1–4.5
Porcelain	5.7
Rubber	2.5
Water	81.0
Wood	2.5–7.7

6-3 DIELECTRIC STRENGTH

Dielectric strength can be defined as the voltage per unit thickness at which breakdown occurs. The dielectric strength therefore corresponds to the field intensity required for breakdown. As the field intensity is increased, the polarization of the dielectric atoms becomes more pronounced. Finally, a value of field intensity may be reached at which so much force is exerted on the orbital electrons that they are torn free from their orbits. When breakdown occurs, the capacitor has characteristics very similar to those of a conductor. A typical example of breakdown is lightning, which occurs when the potential between the clouds and the earth is so high that charge can pass from one to the other through the atmosphere, which acts as a dielectric. The dielectric strengths of selected materials are listed in Table 6-2.

TABLE 6-2 *Dielectric Strengths*

Material	Strength (volts per millimeter)
Air	76
Bakelite	150–500
Ebonite	30–100
Fiber	50
Glass (commercial)	760–3800
Mica	760–5600
Oil	100–500
Paper (kraft, dry)	250–600
Porcelain	100–250
Rubber	400–1270
Vinyl (plastic)	15,800
Water (distilled)	380
Wood	25–75

Capacitance exists between any two conductors separated by a dielectric. The conductors do not have to be plates. They may be wires, grids, or conductors of any shape. The dielectric may be air or any other material that is an insulator. In any two-wire cable, there will be capacitance between the two wires. Capacitance will also exist between circuit wiring and metal chassis, between adjacent or opposite conductors on a printed circuit board, or between the collector, base, and emitter of a transistor. Capacitance resulting from these and other unwanted sources is referred to as **stray capacitance**. Stray capacitance is most easily minimized by keeping conductors as far apart as possible.

A capacitor has a capacitance of 1 farad if 1 coulomb of charge is deposited on the plates by a potential difference of 1 volt across the plates. The farad is named after Michael Faraday, a nineteenth-century English chemist and physicist. Since the farad is generally too large a measure of capacitance for most practical applications, the microfarad is more commonly used. Expressed as an equation, capacitance is determined by

$$C = \frac{Q}{V} \tag{6-5}$$

where C = capacitance, in farads

Q = charge, in coulombs

V = potential difference, in volts

EXAMPLE 6-1 What is the capacitance in which 200 V stores 3 C?

Solution

$$C = \frac{Q}{V} = \frac{3\text{ C}}{200\text{ V}} = 0.015\text{ F}$$

EXAMPLE 6-2 A capacitance of 40 μF is connected for considerable length of time across a 600-V source. How much charge is stored?

Solution

$$Q = CV = (40 \times 10^{-6}\text{ F})(600\text{ V}) = 0.024\text{ C}$$

EXAMPLE 6-3 A 1200-μF capacitor holds a charge of 0.016 C. What is the voltage across it?

Solution

$$V = \frac{Q}{C} = \frac{0.016\text{ C}}{1200 \times 10^{-6}\text{ F}} = 13.33\text{ V}$$

There are essentially three factors that determine the capacitance of a capacitor:

1. Effective area of the plates. The larger the area, the greater the capacitance.
2. Distance between plates.
3. Nature of the dielectric.

A relationship can be derived from these three factors and expressed in the equation

$$C = \frac{\epsilon_0 \epsilon_r A}{d} \qquad (6\text{-}6)$$

where C = capacitance, in farads

ϵ_0 = permittivity of a vacuum

ϵ_r = dielectric constant (relative permittivity)

A = area of plates, in square meters

d = distance between plates, in meters

ϵ_0, the permittivity of a vacuum, is 8.85×10^{-12} F/m. For air, the value is only negligibly greater. The capacitance will be greater if the area of the plates is increased, or the distance between plates is decreased, or the dielectric is changed so that ϵ_r is increased.

EXAMPLE 6-4 What is the capacitance of the capacitor shown in Figure 6-3?

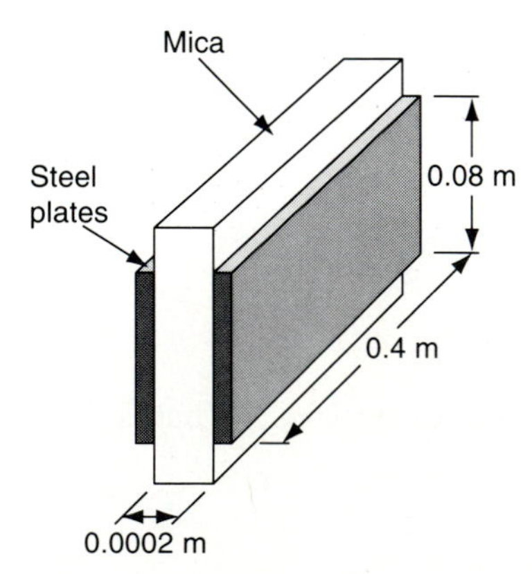

FIGURE 6-3

Solution $A = l \times w = 0.08 \times 0.4 = 0.032 \text{ m}^2$

ϵ_r (from Table 6-1) $= 5.5$

$$C = 8.85 \times 10^{-12} \times \epsilon_r \times \frac{A}{d}$$
$$= (8.85 \times 10^{-12})(5.5)\left(\frac{0.032}{0.0002}\right)$$
$$= 7.788 \times 10^{-9} \text{ F}$$
$$= 7.788 \text{ nF}$$

EXAMPLE 6-5 A 500-pF capacitor is constructed using a porcelain dielectric between two plates, each having an area of 3×10^{-4} m. What is the thickness of the dielectric?

Solution From Table 6-1, $\epsilon_r = 5.7$.

$$C = \epsilon_0 \epsilon_r \frac{A}{d}$$

$$d = \frac{\epsilon_0 \epsilon_r A}{C}$$
$$= \frac{(8.85 \times 10^{-12})(5.7)(3 \times 10^{-4})}{500 \times 10^{-12}}$$
$$= 30.267 \times 10^{-6} \text{ m}$$
$$= 30.267\ \mu\text{m}$$

6-4 LEAKAGE CURRENT

The **leakage current** in a capacitor is defined as the dc current that flows through the capacitor due to imperfections in the dielectric or to surface paths from one plate to another. This very small value of current varies in inverse proportion to the insulation resistance of the dielectric.

Ideally, the dielectric insulating material has an infinite resistance. In reality, all dielectric materials have a finite resistance, called the **leakage resistance**, R_p. The leakage resistance will have a very large ohmic value and is shown in Figure 6-4 as being in parallel with the capacitor. The leakage resistor is the reason it is impossible for a capacitor to maintain a charge indefinitely. If a charge is placed on a capacitor and not **refreshed**, it will eventually **leak off** through this leakage resistance. In most electric and electronic circuits, R_p is very large in comparison to other resistors in the circuit. Consequently, the effect of R_p on the circuit is considered to be negligible in most situations.

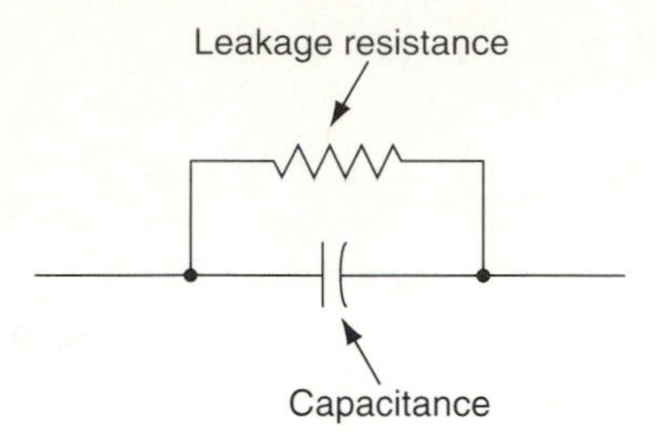

FIGURE 6-4 Leakage resistance of a capacitor.

The leakage current of a capacitor multiplied by the applied voltage represents a power loss. A high value of leakage current will cause not only a rapid loss of charge, but will also result in the capacitor overheating. The **dissipation factor** of a capacitor is determined by the capacitor losses. These losses include the power loss caused by leakage current as well as by **dielectric hysteresis**. Dielectric hysteresis is defined as an effect in a dielectric material caused by changes in orientation of electron orbits in the dielectric, such as the rapid reversals of the polarity of the line voltage.

If the losses are negligible and the capacitor returns the total charge to the circuit, it is considered to be a perfect capacitor. Therefore, the dissipation factor of a capacitor is a measure of its efficiency. The dissipation factor of commercial capacitors varies between 0.0001 and 0.025.

6-5 TYPES OF CAPACITORS

Capacitors are available in many shapes and sizes. Generally, they are designated by their dielectric material. All capacitors can also be included under two headings: fixed and variable. The schematic symbols for fixed and variable capacitors are shown in Figure 6-5. The curved line of the capacitor symbol often represents the plate of the capacitor that is connected to the point of lower potential. If its capacitance cannot be deliberately controlled, a capacitor is called *fixed*. Many types of fixed capacitors are available today. Some of the most common are paper, plastic, mica, ceramic, glass, and electrolytic.

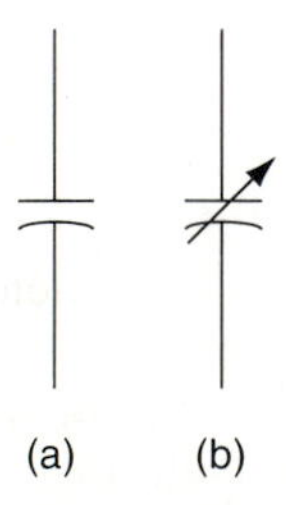

FIGURE 6-5 Capacitor symbols: (a) fixed; (b) variable.

The paper capacitor consists of aluminum foil and kraft paper dielectric rolled together and impregnated with wax or resin to exclude moisture. Paper capacitors commonly range from about 0.0005 to 2 μF. A variation of the paper capacitor is the oil-filled capacitor. These capacitors are generally of a higher capacitance value (1.0 μF and up), with voltage ratings from 400 to 5000 V. They are usually mounted in metal cases and, in the case of high voltage ratings, the terminals are brought out through stand-off ceramic insulators.

The plastic film capacitor is very similar in construction to the paper capacitor. It is, however, much denser than paper, and contaminating foreign particles are virtually nonexistent. Plastics can withstand higher temperatures and are considerably more stable than paper. The disadvantage of plastic film capacitors is that for a given capacitance and voltage rating, plastic film capacitors are either bulkier or more costly than paper capacitors.

Mica capacitors are used mainly in the RF circuits of receivers and transmitters. Mica is one of the best natural insulators known. Radio transmitters use these capacitors since the voltage rating and current may go as high as 30 kV and 100 A. Due to its high cost, mica capacitors are seldom found with capacitance values greater than 0.05 μF. A variation of this kind of capacitor is the silver–mica capacitor. A thin layer of silver is deposited on the surface of the mica, and the resulting capacitor has excellent stability and tolerance. Silver–mica capacitors are used in frequency-selective (tuned) circuits, and particularly for temperature (drift) compensation.

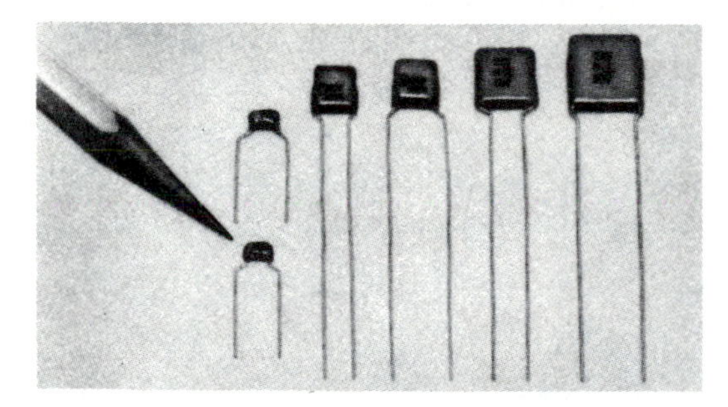

Mica capacitors

Source: Prentice Hall Inc. / Sprague Electric Co.

Ceramic capacitors consist of a ceramic disk with silver plates attached to each flat surface. The leads are then attached through electrodes to the plates. An insulating coating of ceramic is then applied over the plates and dielectric. These types of capacitors have values ranging from a few picofarads to about 2 μF. Another type of ceramic capacitor is the multilayer type, which consists of metal plates stacked alternately with ceramic dielectrics then molded into a single block. This construction is called *monolithic*.

Glass was first used as a capacitor dielectric in the early 1950s. Capacitors of this type are characterized by extremely low losses, excellent stability, and reliability. Glass capacitors are available in capacitance values up to 0.01 μF and with voltage ratings up to 6000 V.

The electrolytic capacitor is used most commonly in situations where capacitances between 1 and several thousand microfarads are required. They are designed primarily for use in circuits where only dc voltages will be applied across the capacitor. An electrolytic capacitor consists of two plates separated by an electrolyte and a dielectric. One of the plates is oxidized and it is this oxide that forms the dielectric. These capacitors depend on chemical action within them to produce the dielectric. Current must flow through the

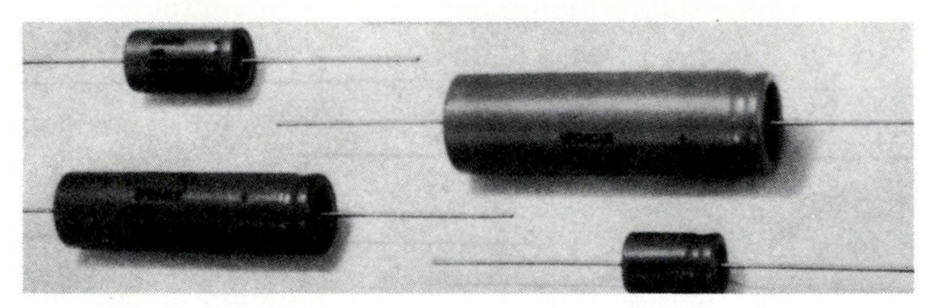

Electrolytic capacitors

Source: Prentice Hall Inc. / Sprague Electric Co.

capacitor to maintain this dielectric. To maintain the dielectric film, electrolytic capacitors are connected into a circuit with proper observation of the polarity. For this reason, the terminals of this type of capacitor are marked positive and negative. The applied voltage must be connected positive to positive and negative to negative. If the capacitor is connected incorrectly, the current flowing through the device will be opposite in direction to the current that formed the dielectric, causing the dielectric oxide to be destroyed and short-circuiting the capacitor. This can often cause the electrolytic capacitor to explode.

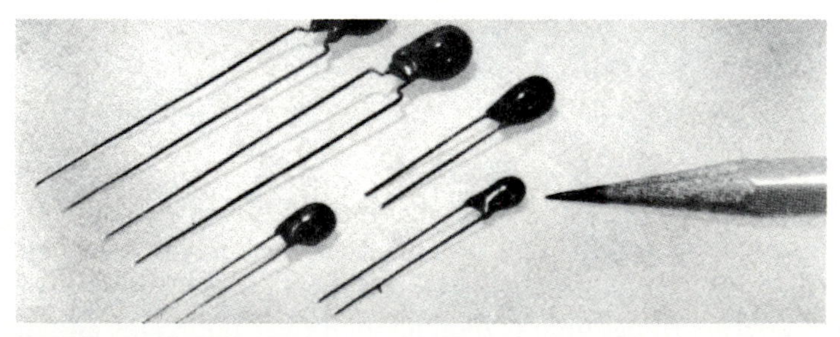

Tantalum capacitors

Source: Prentice Hall Inc. / Sprague Electric Co.

The tantalum capacitor is essentially another type of electrolytic capacitor. Tantalum powder of high purity is pressed into cylindrical shape. The capacitor is then sintered (baked) at very high temperatures, resulting in a porous

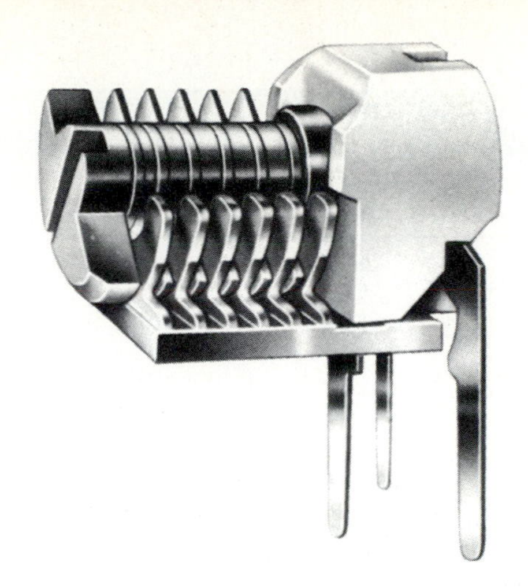

Variable capacitor

Source: Prentice Hall Inc. / E.F. Johnson Company, Components Division

material. When this material is immersed in a container of electrolyte, the electrolyte is absorbed in the tantalum. A layer of tantalum oxide forms on the inner surface of the positive plate, and this oxide layer acts as the dielectric.

Variable capacitors generally use air as the dielectric and have one set of plates meshing between a second set of plates. The capacitance is altered by turning the shaft at one end to vary the common area of the movable and fixed plates. The greater the common area, the larger the capacitance. The fixed plate assembly is called the *stator*, and the movable portion is called the *rotor*. Due to a fringing effect, the minimum value of capacitance occurring when the plates are enmeshed is not zero, but is a finite value. In high-voltage applications, the spacing between the plates of a variable capacitor is increased.

Another type of variable capacitor is the trimmer, or padding capacitor. This type of capacitor consists of two or more plates separated by a dielectric of mica. A screw is mounted so that tightening the screw compresses the plates more tightly against the dielectric. By applying compression to the dielectric, the thickness of the dielectric is reduced and the capacitance increases.

6-6 CAPACITOR COLOR CODE

Some capacitors, such as Mylar and molded mica, have color codes to indicate the voltage, tolerance, and value in picofarads of the capacitor. The method of determining the value of a capacitor by use of color code is very similar to that of the resistor color code. The first two digits indicate the first and second significant digits, the third digit is the multiplier, the fourth digit is the tolerance, and the fifth digit, if any, is the rated voltage of the capacitor. Table 6-3 shows the chart for determining capacitance values based on color code.

TABLE 6-3 *Capacitor Color Code*

Color	Significant digit	Decimal multiplier	Tolerance (%)	Voltage rating
Black	0	10^0	±20	
Brown	1	10^1	±1	100
Red	2	10^2	±2	200
Orange	3	10^3	±3	300
Yellow	4	10^4	±4	400
Green	5	10^5	±5	500
Blue	6	10^6	±6	600
Violet	7	10^7	±12.5	700
Gray	8		±30	800
White	9		±10	900
Gold		10^{-1}	±5	1000
Silver		10^{-2}	±10	2000
No color			±20	

EXAMPLE 6-6 What is the value of a capacitor with the following color bands: brown, black, orange, red, and gold?

Solution The significant numbers are 10, the multiplier is 1000, the tolerance is ±2%, and the voltage rating is 1000 V. Therefore, the capacitor has a rating of 10,000 pF ±2% at 1000 V.

6-7 TRANSIENTS IN *RC* CIRCUITS

A **transient** is the part of the change in a variable that disappears when going from one steady-state condition to another. At **steady state**, the capacitor's charge and voltage across it are constant and do not change with time. Since the potential of a capacitor cannot change instantaneously, a time period is required for the transition of the capacitor from the uncharged state to the charged state. The **transient state** of a capacitor would be the state of the capacitor between being fully charged and fully discharged. Although a device such as a switch would have a transient state, the term is usually applied to voltages and currents which increase or decrease in an exponential manner. Two typical exponential curves are shown in Figure 6-6, where x represents the horizontal component, such as time or frequency, and y represents the vertical component, such as current or voltage. In Figure 6-6(a), the exponential decay of the vertical component is given by the equation

$$y = e^{-x} \tag{6-7}$$

where y = vertical component, such as voltage or current

e = 2.71828, the base of natural logarithms

x = a function of time, in seconds

The exponential growth of the curve shown in Figure 6-6(b) is expressed by the equation

$$y = 1 - e^{-x}$$

Although both curves of Figure 6-6 begin at a definite point, they never reach a final value. However, in most cases the curve can be considered to rise to 100% and to fall to 0%.

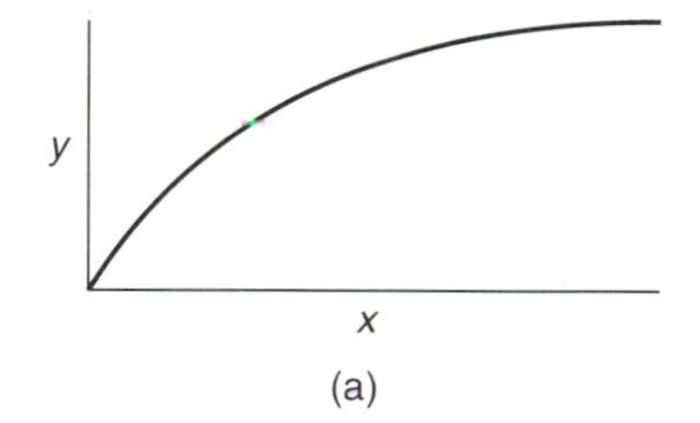

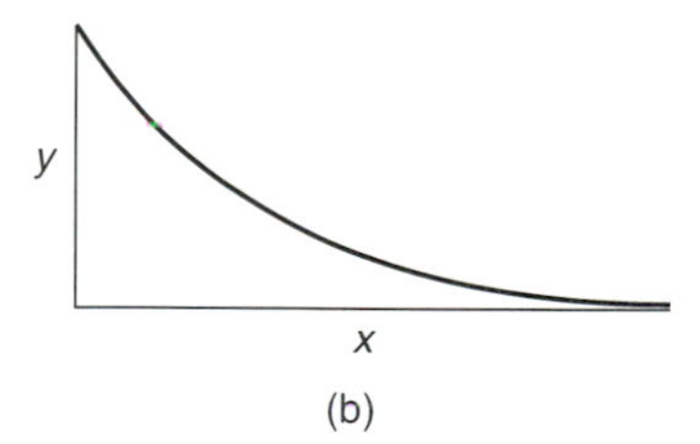

FIGURE 6-6

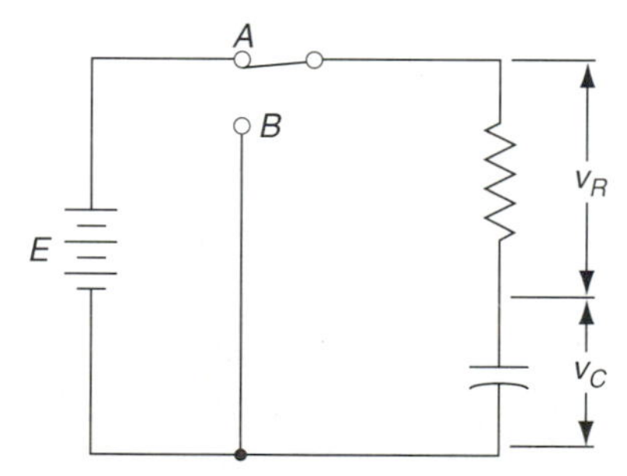

FIGURE 6-7

Figure 6-7 shows a capacitor and a resistor connected through a switch to a dc power source. When the switch is in position A the capacitor will charge, and when it is in position B the capacitor will discharge. If it is in the open position, there is no potential difference between the plates and no charge on the capacitor. When the switch is at position *A*, the capacitor begins to charge. Initially (time = 0), the only limitation to the flow of current is the resistance (*R*), so the current will immediately rise to its maximum value. The instant the switch closes, electrons from the negative terminal of the power supply accumulate on the lower plate of the capacitor. The lower plate becomes negative, while the upper plate becomes positively charged from the positive terminal of the supply. A difference of potential now exists between the plates, which will oppose the applied voltage of the circuit.

Although the net voltage of the circuit is reduced, a slightly lower current continues to flow, which increases the counter emf and charge of the capacitor. This is a cumulative process; by increasing the charge on a capacitor the counter emf increases, while the charging current decreases. When the counter emf is equal to the electromotive force of the dc supply, the current ceases to flow.

The exponential curves of Figure 6-8 illustrate the values of the counter electromotive force as the capacitor is charged (curve A) and as the capacitor is discharged (curve B). On curve A, the time required for the counter emf to reach approximately 63% of its maximum value is equal to 1 time constant. On curve B, the time required for the counter emf to fall to approximately 37% of its maximum value is also equal to 1 time constant. The value of 1 time constant is represented by the lowercase Greek letter τ (tau).

From the chart of Figure 6-8 it can be seen that after 5 time constants a capacitor has been charged to within 1% of its maximum or discharged to within 1% of its minimum value. Since the graph represents an exponential

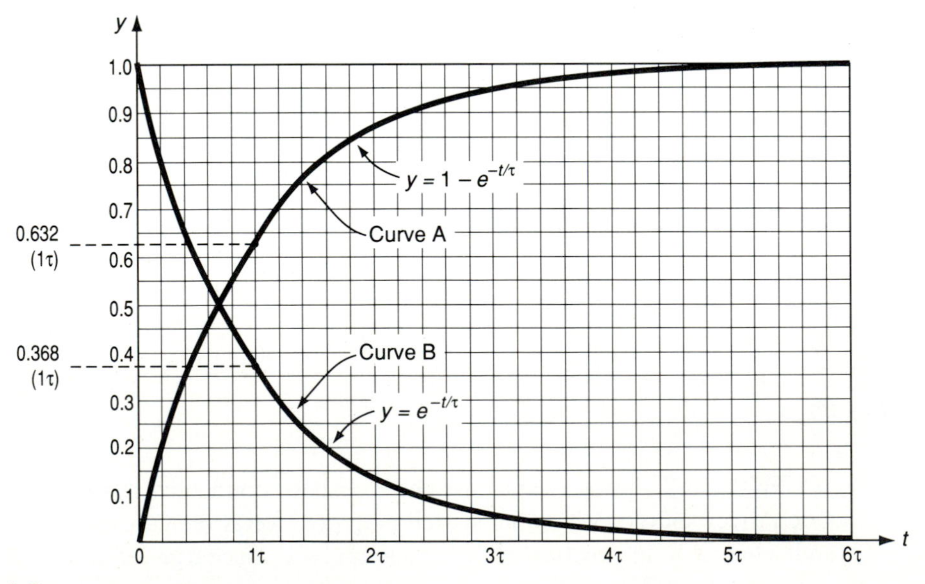

FIGURE 6-8 Universal time constant curves.

curve, the curve never reaches 100% or 0%. Even after 6 time constants the capacitor is at 99.75% on the charging curve, 0.25% on the discharge curve. The value of the time constant in seconds is equal to the product of the circuit resistance in ohms and its capacitance in farads. In equation form,

$$\tau = RC \tag{6-8}$$

where τ = time, in seconds

R = resistance, in ohms

C = capacitance, in farads

EXAMPLE 6-7 What would be the value of 1 time constant in a series circuit containing a 0.001-μF capacitor and a 47-kΩ resistor?

Solution

$$\tau = RC = (47 \times 10^3)(0.001 \times 10^{-6})$$
$$= 47\ \mu s$$

Instantaneous voltage is the value of voltage at any specific instant in time. The IEEE recommends the use of lowercase letter symbols to symbolize quantities that change with respect to time. The equation for the instantaneous voltage on a capacitor when charging in a circuit containing resistance and capacitance can be derived by differential calculus:

$$v_C = E(1 - e^{-t/RC}) \tag{6-9}$$

where v_C = capacitor voltage at time t
E = supply voltage
e = exponential constant = 2.71828
t = time in seconds, from charge commencing
R = resistance of circuit, in ohms
C = capacitance of circuit, in farads

Since τ represents RC, it can be substituted into the foregoing equation.

EXAMPLE 6-8 For the circuit shown in Figure 6-9, assume that the capacitor is initially discharged. Determine the voltage across the capacitor after the switch has been closed for 81.6 ms.

Solution

$$\tau = RC = (0.68 \times 10^{-6})(1 \times 10^{6}) = 0.68\ \text{s}$$
$$v_C = E(1 - e^{-t/RC}) = 110(1 - e^{-0.0816/0.68})$$
$$= 12.44\ \text{V}$$

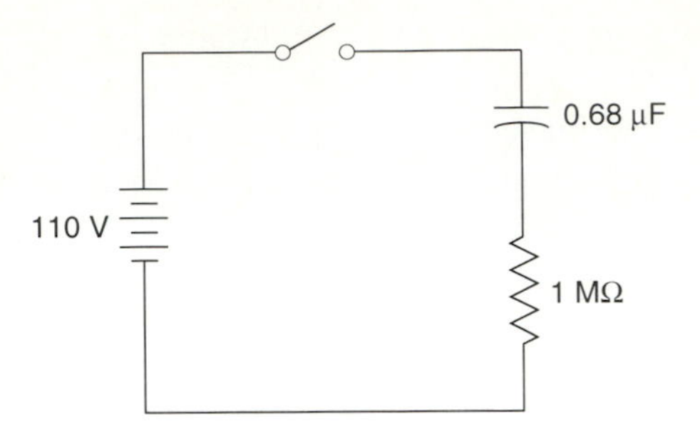

FIGURE 6-9

The following equation is used to calculate the length of time required for the voltage across a capacitor to rise to a certain value:

$$t = RC \times \ln\left(\frac{E}{E - v_C}\right) = \tau \ln\left(\frac{E}{E - v_C}\right) \tag{6-10}$$

EXAMPLE 6-9 Given the circuit of Figure 6-9, where $R = 1\ \text{M}\Omega$, $C = 0.68\ \mu\text{F}$, and $E = 110$ V, determine the length of time required after the switch is closed for voltage across the capacitor to rise to 83.5 V.

Solution

$$\tau = RC = (0.68 \times 10^{-6})(1 \times 10^{6}) = 0.68\ \text{s}$$

$$t = \tau \ln\left(\frac{E}{E - v_C}\right) = 0.68 \ln\left(\frac{110}{110 - 83.5}\right)$$

$$= 0.97\ \text{s}$$

Figure 6-10(a) shows a **UJT relaxation oscillator** circuit that is useful in many electronic control applications. An oscillator is a circuit that is used to create an electronic signal voltage. A UJT relaxation oscillator utilizes the characteristic of the charge and discharge of a capacitor in an *RC* circuit to control the frequency of oscillations. In Figure 6-10(a), a unijunction transistor (UJT) is used to generate voltage pulses across resistor R_3. When the voltage at the emitter, *E*, of the UJT reaches a certain value the UJT switches on and current flows from the emitter through R_3 to ground. Current continues to flow through the UJT until the voltage at the emitter falls below a value required to maintain conduction, V_v. At this point the UJT switches off.

Figure 6-10(b) shows the voltage waveform at the emitter of the UJT. When the supply voltage, V_{BB}, is applied, the capacitor begins charging through R_E. Eventually, the capacitor reaches the trigger voltage, V_p, for the UJT. When V_p is reached, the UJT switches on, and the capacitor is rapidly discharged through R_3. When the current flowing through the device falls below the minimum required to maintain conduction, the UJT switches off

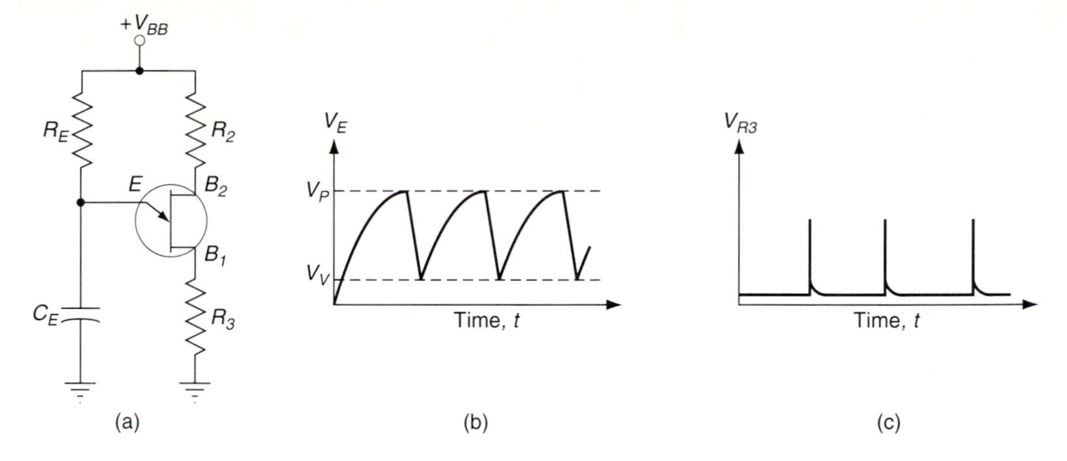

FIGURE 6-10 (a) Relaxation oscillator circuit; (b) sawtooth waveform at the emitter, E; (c) pulse waveform at base B_1.

and the cycle repeats. Figure 6-10(c) shows the voltage drop that will appear across resistor R_3 during the time that the UJT is conducting. The waveforms of Figure 6-10(b) and (c) represent an exponential sawtooth at the emitter terminal and a pulse waveform at the B_1 terminal.

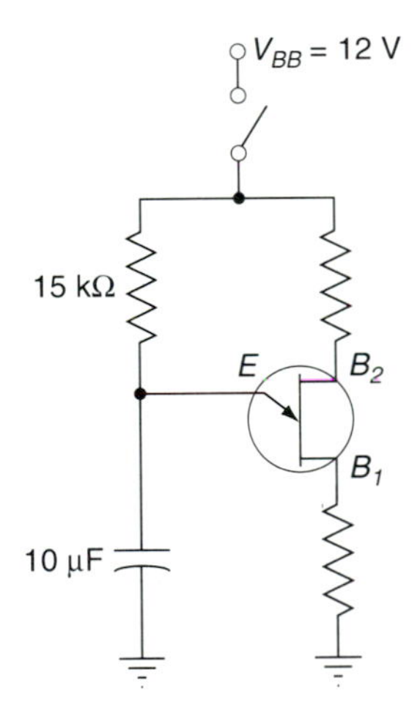

FIGURE 6-11

EXAMPLE 6-10 Figure 6-11 shows a relaxation oscillator. When the voltage at the emitter (E) of the UJT reaches 3.2 V, the UJT will switch on and current will flow through the device. Determine the length of time required after the switch is closed for the voltage at the emitter to reach 3.2 V.

Solution

$$\tau = RC = (15 \times 10^3)(10 \times 10^{-6}) = 0.15 \text{ s}$$

$$t = \tau \ln\left(\frac{E}{E - v_C}\right) = 0.15 \ln\left(\frac{12}{12 - 3.2}\right)$$
$$= 46.52 \text{ ms}$$

In most practical applications, a capacitor is considered to be fully charged after 5 time constants. This is because

$$v_C = E(1 - e^{-5}) = E \times 0.993$$

which is considered to be close enough to 100% for the majority of electrical and electronic circuits. The charging current in a resistive capacitive circuit is determined by

$$i_C = \frac{E}{R} e^{-t/RC} \qquad (6\text{-}11)$$

EXAMPLE 6-11 A 56-kΩ resistor is connected in series with a 20-μF capacitor and a switch. If the circuit is connected across a 40-V supply and the capacitor is completely discharged initially, calculate the following when the switch is closed: (a) circuit time constant, τ; (b) capacitor voltage at time $t = 2.5$ s; (c) charge on the capacitor at time $t = 2.5$ s; (d) time required for voltage to rise to 27.3 V; (e) capacitor circuit current at $t = 2.5$ s; (f) voltage drop across 56-kΩ resistor at $t = 2.5$ s.

Solution

(a) $\tau = RC = (56 \times 10^3 \ \Omega)(20 \times 10^{-6} \text{ F}) = 1.12 \text{ s}$

(b) $v_C = E(1 - e^{-t/RC}) = 40(1 - e^{-2.5/1.12})$
$= 35.71 \text{ V}$

(c) $$C = \frac{Q_C}{v_C}$$

$$\therefore \ Q_C = Cv_C = (20 \times 10^{-6} \text{ F})(35.71 \text{ V})$$
$$= 714.2 \ \mu\text{C}$$

(d) $$t = \tau \ln\left(\frac{E}{E - v_C}\right) = 1.12 \ln\left(\frac{40 \text{ V}}{40 \text{ V} - 27.3 \text{ V}}\right)$$
$$= 1.285 \text{ s}$$

(e) $$i_C = \frac{E}{R} e^{-t/RC} = \left(\frac{40 \text{ V}}{56 \times 10^3 \ \Omega}\right) e^{-2.5/1.12}$$
$$= 76.64 \ \mu\text{A}$$

(f) $$v_R = i_C R = (76.64 \times 10^{-6} \text{ A})(56 \times 10^3 \ \Omega)$$
$$= 4.29 \text{ V}$$

In Figure 6-12, when the switch is at position A, the capacitor is part of the load being supplied with energy from the source. When the switch is at position B, the capacitor is the source, and the resistor its load. Essentially, the capacitor is functioning like a battery. When the switch is in position A, the battery is charging, and in position B the battery is discharging.

Although the polarity of the capacitor remains the same during both charging and discharging, the polarity of the resistor will reverse when the switch is changed from position A to position B. After 5 time constants, the capacitor is considered to be completely discharged. If the capacitor was fully charged and the switch changed to position B, the equation for the decaying voltage across the capacitor would be

$$v_C = Ee^{-t/RC}$$

The capacitor circuit current will also decrease with time. The value of i_C for a discharging capacitor is determined by

$$i_C = \frac{E}{R}e^{-t/RC}$$

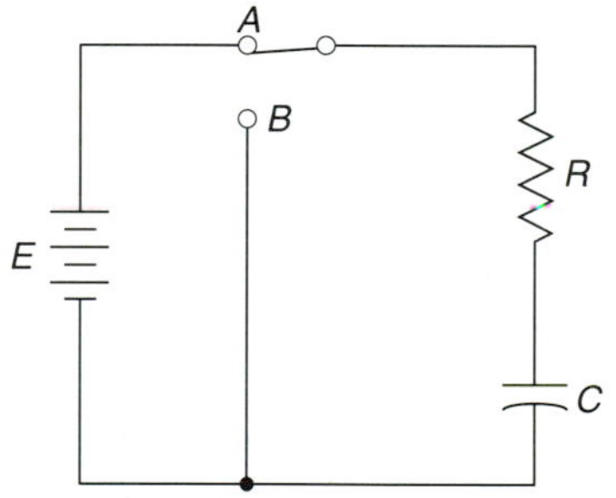

FIGURE 6-12

EXAMPLE 6-12 The circuit shown in Figure 6-12 has a capacitance of 100 μF, a resistance of 5.6 kΩ, and a supply voltage of 24 V. Assume that the capacitor is fully charged and the switch is changed to position B. Determine (a) the voltage across the capacitor; (b) the capacitor circuit current after the switch has been at position B for 1.05 s.

Solution

$$\tau = RC = (5.6 \times 10^3\ \Omega)(100 \times 10^{-6}\ \text{F}) = 0.56\ \text{s}$$

(a)
$$v_C = Ee^{-t/RC} = 24e^{-1.05/0.56} = 3.68\ \text{V}$$

(b)
$$i_C = \frac{E}{R}e^{-t/RC} = \left(\frac{24\ \text{V}}{5.6 \times 10^3\ \Omega}\right)e^{-1.05/0.56} = 0.66\ \text{mA}$$

The length of time required for a voltage to decay to a certain value in an *RC* circuit is determined by the equation

$$t = \tau \times \ln\left(\frac{V_0}{V_1}\right) \qquad (6\text{-}12)$$

where V_0 is the initial value of voltage and V_1 is the voltage at any point on the decay curve for which time is desired.

EXAMPLE 6-13 A 10-μF capacitor is charged to a value of 40 V and is then discharged into a 1-kΩ resistor. Determine the length of time required for the voltage to decrease from 21 V to 13 V.

Solution

$$\tau = RC = (1 \times 10^3\ \Omega)(10 \times 10^{-6}\ \text{F})$$
$$= 0.01\ \text{s}$$

$$t = \tau \times \ln\left(\frac{V_0}{V_1}\right) = 0.01 \times \ln\left(\frac{21\ \text{V}}{13\ \text{V}}\right)$$
$$= 4.8\ \text{ms}$$

6-8 ENERGY STORED BY A CAPACITOR

When an ideal capacitor is connected to a voltage source, the source must expend energy to charge the capacitor. If the charging source is disconnected from the capacitor, energy will remain stored between its plates. The energy being stored in the capacitor is in the form of an electrostatic field.

If a capacitor is charged with a constant current, the voltage between its plates at any given time will increase at a uniform rate. The total energy transferred to a capacitor during a given charging time will be determined by the average voltage. Since the voltage increases at a uniform rate with the charge current, the average voltage is one-half the final voltage. The energy stored by a capacitor can be found using the equation

$$W = \frac{CV_C^2}{2} \qquad (6\text{-}13)$$

where W = energy stored, in joules

C = capacitance, in farads

V_C = potential difference across capacitor, in volts

EXAMPLE 6-14 Calculate the energy stored by a 100-μF capacitor when 24 V is applied across the capacitor's plates.

Solution $$W = \frac{CV_C^2}{2} = \frac{(100 \times 10^{-6})(24)^2}{2} = 0.029 \text{ J}$$

6-9 CAPACITORS IN SERIES AND PARALLEL

Capacitors may be connected in series or parallel to give resultant values, which may be either the sum of the individual values (in parallel) or a value less than that of the smallest capacitance (in series). A circuit consisting of a number of capacitors in series is similar in some respects to a circuit containing several series-connected resistors. In a series capacitive circuit the same current flows through each part of the circuit and the applied voltage will divide across the individual capacitors.

When capacitors are connected in a series configuration, the magnitude of the charge on each plate must be the same. In Figure 6-13, the electrons that produce the negative charge on C_1 must come from the plate of C_2 and leave it with a positive charge. The battery maintains a potential difference between the positive plate of C_1 and the negative plate of C_2, transferring electrons from one to the other. The charge cannot pass between the plates of a capacitor. Therefore, the charge (Q) is the same in all parts of the circuit. In equation form,

$$Q_T = Q_1 = Q_2 \cdots Q_N$$

Also, $$C = \frac{Q}{V} \quad \text{and} \quad V = \frac{Q}{C}$$

Then $$V_T = \frac{Q_T}{C_T} = \frac{Q_T}{C_1} = \frac{Q_T}{C_2}$$

Dividing this last equation by Q results in

$$C = \frac{1}{1/C_1 + 1/C_2} \tag{6-14}$$

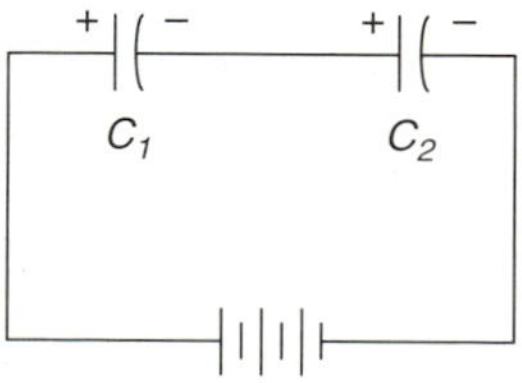

FIGURE 6-13 Capacitors in series.

Equation 6-8 for series capacitors is similar to that for the total resistance of parallel resistors. For two capacitors in series, the total capacitance is found by the product-over-sum relationship:

$$C_T = \frac{C_1 \times C_2}{C_1 + C_2} \tag{6-15}$$

When capacitors are connected in parallel, one plate of each capacitor is connected directly to one terminal of the supply, while the other plate of each capacitor is connected to the other terminal of the supply. In Figure 6-14, since all the negative plates of the capacitors are connected together and all the positive plates are connected together, *C* appears as a capacitor with a plate area equal to the sum of all the individual plate areas. Connecting capacitors in parallel effectively increases the plate area; consequently, the capacitance increases.

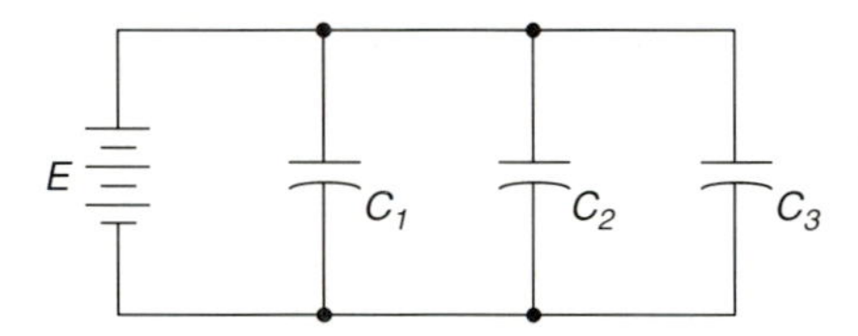

FIGURE 6-14 Capacitors in parallel.

For capacitors connected in parallel, the total charge is the sum of all the individual charges.

$$Q_T = Q_1 + Q_2 + Q_3 + \cdots + Q_N$$

since $Q_1 = C_1V_1$, $Q_2 = C_2V_2$, and $Q_3 = C_3V_3$, where $V_T = V_1 = V_2 = V_3$. The total charge is then

$$Q_T = C_TV_T = Q_1 + Q_2 + Q_3 = C_1V_T + C_2V_T + C_3V_T$$

Dividing the last equation by V_T, we obtain an expression for the total parallel capacitance. This expression is given by the equation

$$C_T = C_1 + C_2 + C_3 + \cdots + C_N \tag{6-16}$$

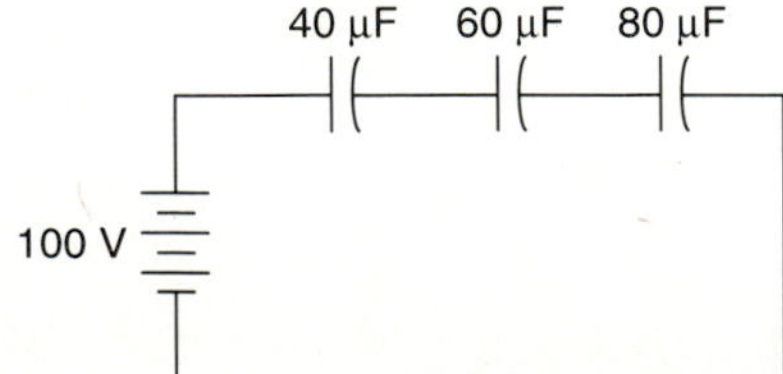

FIGURE 6-15

EXAMPLE 6-15 For the circuit of Figure 6-15, find (a) the total capacitance; (b) the voltage across each capacitor.

Solution (a) The capacitance is calculated from equation 6-8.

$$\frac{1}{C_T} = \frac{1}{C_1} + \frac{1}{C_2} + \frac{1}{C_3}$$

$$= \frac{1}{40\,\mu\text{F}} + \frac{1}{60\,\mu\text{F}} + \frac{1}{80\,\mu\text{F}}$$

$$= 0.025 \times 10^6 + 0.0167 \times 10^6 + 0.0125 \times 10^6$$

$$= 54.17 \times 10^3$$

$$C = \frac{1}{54.17 \times 10^3}$$

$$= 18.46\,\mu\text{F}$$

(b) $$Q_1 = Q_2 = Q_3 = Q_T = C_T V_T = 100(18.46 \times 10^{-6})$$

$$= 18.46 \times 10^{-4}\,\text{C}$$

$$V_1 = \frac{Q_1}{C_1} = \frac{18.46 \times 10^{-4}}{40 \times 10^{-6}}$$

$$= 46.15\,\text{V}$$

$$V_2 = \frac{Q_2}{C_2} = \frac{18.46 \times 10^{-4}}{60 \times 10^{-6}}$$

$$= 30.77\,\text{V}$$

$$V_3 = \frac{Q_3}{C_3} = \frac{18.46 \times 10^{-4}}{80 \times 10^{-6}}$$

$$= 23.08\,\text{V}$$

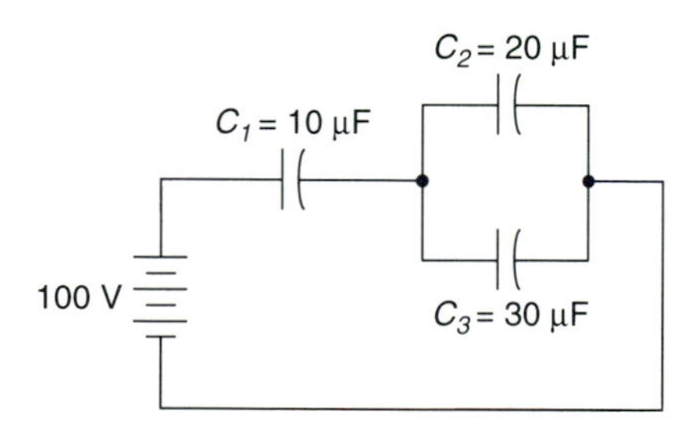

FIGURE 6-16

EXAMPLE 6-16 For the circuit of Figure 6-16, solve for (a) the total capacitance; (b) the charge on each capacitor; (c) the voltage across each capacitor.

Solution $20\,\mu\text{F} + 30\,\mu\text{F} = 50\,\mu\text{F}$

(a) $C_T = \dfrac{10\,\mu\text{F} \times 50\,\mu\text{F}}{10\,\mu\text{F} + 50\,\mu\text{F}} = \dfrac{500\,\mu\text{F}}{60\,\mu\text{F}}$
$= 8.33\,\mu\text{F}$

(b) $Q_T = C_T V_T = (8.33 \times 10^{-6})(100)$
$= 8.33 \times 10^{-4}\text{ C}$

This is the value of charge on C_1. To solve for C_2 and C_3, it is necessary first to find the capacitor voltages.

(c) $V_1 = \dfrac{Q_1}{C_1} = \dfrac{8.33 \times 10^{-4}\text{ C}}{10 \times 10^{-6}\text{ F}}$
$= 83.3\text{ V}$

$V_2 = V_3 = V_T - V_1 = 100\text{ V} - 83.3\text{ V}$
$= 16.7\text{ V}$

The charge on capacitors C_2 and C_3 can now be found:

$Q_2 = C_2 V_2 = (20 \times 10^{-6}\text{ F})(16.7\text{ V})$
$= 3.3 \times 10^{-4}\text{ C}$

$Q_3 = C_3 V_3 = (30 \times 10^{-6}\text{ F})(16.7\text{ V})$
$= 5.0 \times 10^{-4}\text{ C}$

KEY TERMS

Capacitor
Dielectric flux
Electric flux density
Elastance
Relative permittivity
Dielectric constant
Absolute permittivity
Gauss's law
Dielectric strength
Capacitance
Stray capacitance
Leakage current
Leakage resistance
Refreshed
Leak-off
Dissipation factor
Dielectric hysteresis
Transient
Steady state
Transient state
UJT relaxation oscillator

PROBLEMS

6-1 A 6.8-μF capacitor is connected to a 120-V supply. How much charge will it store?

6-2 What size capacitor is required to store a charge of 2.17 C from a 250-V supply?

6-3 A 200-μF capacitor holds a charge of 0.00325 C. What voltage is applied across the capacitor?

6-4 A capacitor is marked 0.0022 μF. What is its value in picofarads?

6-5 A computer-grade 0.39-nF capacitor has 12 V across it terminals. Determine the charge in coulombs.

6-6 Find the capacitance of a mica capacitor with a dielectric constant of 5.5, a plate area of 2.7 cm^2, and a parallel-plate separation of 0.25 cm.

6-7 A capacitor has plates of 1.5 cm $\times$ 2.5 cm, separated by a ceramic dielectric ($K = 35$) 0.33 mm in thickness. Find the capacitance.

6-8 A variable capacitor with air dielectric ($K = 1$) has eight stationary and seven movable plates. When the plates are completely meshed and maximum capacitance is achieved, the area of each plate facing the dielectric is 0.00315 m^2 and the plate separation is 0.0015 m. Determine the maximum capacitance.

6-9 Determine the electric field strength between the plates of a capacitor if the parallel plates are 1.5 mm apart and 600 mV is applied across it.

6-10 How much charge is stored in a capacitor which has parallel plates with a plate area of 0.002 m^2, a plate separation of 0.01 m^2, a paper dielectric ($K = 4$), and an applied voltage across the plates of 12 V?

6-11 Find the capacitance of a capacitor made of two parallel plates with diameters of 3.09 cm^2 separated by an air gap of 0.225 cm.

6-12 A 0.0068-μF capacitor is constructed using a Mylar dielectric ($K = 3$). If the distance between plates is 1 mm, what is the area of the plates?

6-13 Find the thickness of mica dielectric ($K = 5.5$) when used in a 0.0022-μF capacitor with parallel plates each having an area of 2.5×10^{-2} m^2.

6-14 Referring to Table 6-3, what is the value of a ceramic capacitor marked red, red, orange, brown, and red?

6-15 What is the value in microfarads of a capacitor having the following color bands: blue, gray, yellow, white, and brown?

6-16 What are the color bands of a 100-μF capacitor with a tolerance of 1% and a voltage rating of 300 V?

6-17 Find the value of 1 time constant in a series circuit containing a 0.022-μF capacitor and a 1-kΩ resistor.

6-18 What size resistor would have to be connected in series with a 470-μF capacitor to produce a time constant of 0.0031 s?

6-19 A 100-Ω resistor is connected in series with a capacitor and the resulting time constant is 0.022 s. Find the size of the capacitor.

6-20 The circuit shown in Figure 6-17 has components with the following values: C = 100 μF, R = 33 kΩ, and E = 24 V. If the capacitor is initially discharged, determine the length of time required for v_C to rise to 11.5 V.

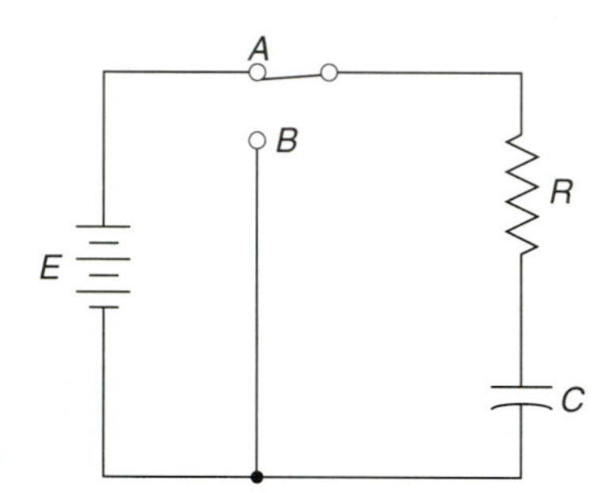

FIGURE 6-17

6-21 The capacitor shown in Figure 6-17 is fully charged. The component values are E = 15 V, C = 22 μF, and R = 100 kΩ. How long will it take for the voltage to fall to 10 V if the switch is set at point B?

6-22 The windshield wiper motor of an automobile is controlled by a UJT relaxation oscillator. The capacitor for C_E is 40 μF. The resistor for R_E is the series combination of a 51-kΩ resistor and a 500-kΩ potentiometer. What is the minimum-to-maximum range of blade strokes per minute?

6-23 Three 40-μF capacitors are connected in series. Find the total capacitance.

6-24 Find the total capacitance of three 10-μF capacitors connected in parallel.

6-25 A 20-μF capacitor is connected in series with a 10-kΩ resistor and is separated from a 24-V source by a switch. Determine (a) the current flowing at the instant the switch is closed; (b) the value of current flowing after 1 time constant.

6-26 A 50-μF capacitor is in series with a 30-kΩ resistor and is connected to a 12-V source. Determine the time required for the voltage to rise from 3 V to 9 V.

6-27 A 100-V source is applied across a 10-μF capacitor in series with a 200-kΩ resistor. What is the voltage across the capacitor after 2 s?

6-28 A 100-kΩ resistor is connected in series with a 0.2-μF capacitor. How much time does it take for the voltage across the capacitor to rise to 43 V after 120 V is applied to the circuit?

6-29 A 50-kΩ resistor and a 20-μF capacitor are connected in series to a switch and a 40-V source. If the capacitor is initially discharged, what is the voltage across the capacitor after the switch is closed for 2.75 s?

6-30 A fully charged 0.68-μF 100-V capacitor is in series with a 220-kΩ resistor and an open switch. How long will it take the capacitor to discharge to 20 V after the switch is closed?

6-31 Determine how long it will take for the voltage across a 10-μF capacitor to reach 15 V if it is connected in series with a 6-kΩ resistor and the applied voltage is 40 V. Assume that the capacitor is initially discharged.

6-32 For the circuit shown in Figure 6-18, find (a) the total capacitance; (b) the voltage across each capacitor in the circuit.

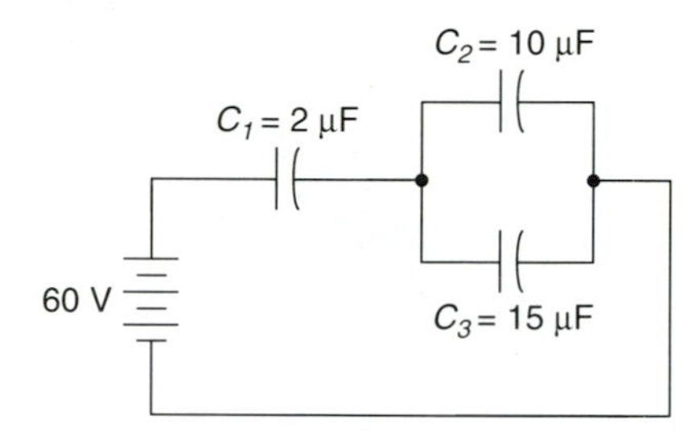

FIGURE 6-18

6-33 Determine the charge on each capacitor for the circuit shown in Figure 6-18.

CHAPTER

Magnetism

LEARNING OBJECTIVES

Upon completion of this chapter you will be able to:

- Express Weber's theory.
- Define the term *domain*.
- Understand the principle of the magnetic field.
- List four characteristics of magnetic lines of force.
- Describe the three laws of magnetic attraction and repulsion.
- List the three classifications of magnetic materials.
- Describe the field around a current-carrying conductor.
- Define the right-hand rule.
- List the three factors affecting the strength of an electromagnetic field.

7-1 INTRODUCTION

The phenomenon called **magnetism** has been known and investigated for a very long time. At some point in the sixth century B.C. it was discovered that a lodestone would point itself in a particular direction if it were freely suspended. If a lodestone is placed on a pile of iron filings, the filings will tend to move so that they are concentrated around two regions on the stone. These regions are called the **poles** of the stone. When the lodestone is suspended, a line drawn through the poles is found to point roughly north and south. The magnetic pole that points northward is called the **north pole**, and the other is referred to as the **south pole**.

Eventually, this magnetic effect was found to be caused by the presence of iron in the lodestone. More recent discoveries have shown that this property is shared by a group of materials known as **ferromagnetic materials**. These include iron, cobalt, and nickel, as well as certain alloys. Magnets made of these materials can exhibit the effect to a much greater degree than can the original lodestone.

William Gilbert undertook the first serious study of magnets around A.D. 1600. From his studies, Gilbert theorized that the earth is a huge magnet with its magnetic north pole near the geographical north pole, and the magnetic south pole located near the geographical south pole.

7-2 NATURE OF MAGNETISM

A popular theory of magnetism is known as **Weber's theory**. According to this theory, the molecules of magnetic material, such as iron, are tiny magnets, each with a north and south magnetic pole and with a surrounding magnetic field. All unmagnetized materials have the magnetic forces of its molecular magnets neutralized by adjacent molecular magnets, thereby eliminating any magnetic effect. In magnetic materials the molecular magnets align themselves so that their magnetic properties add to one another. The magnetized material will be lined up so that the north pole of each molecule points in one direction and the south pole of each molecule points in the opposite direction. When all the molecules of a material are aligned in such a manner, it is said to have one effective north pole and one effective south pole. In nonmagnetic materials, the molecular magnets are in a random configuration and do not become aligned in the presence of a magnetic material.

If a magnet is split in half, Weber's theory appears to be proven correct, since each half will possess both a north and a south pole. If the magnet is again divided, there will also be one effective north and south pole. The polarities of each subsequent division of the magnet will be in the same direction as the original magnet. Also, the magnetizing effect of a magnet can be nullified by any means that will disarrange the orderly array of the molecules, such as jarring or heating the magnet.

A more modern theory of magnetism is the modified version of Weber's molecular theory. Scientists now believe that a magnetic field is produced by a moving electric field. From the study of atomic structure it is known that the electrons of an atom rotate in concentric shells around the nucleus. Electrons are believed to spin on their axis in the same way that the earth turns on its axis, as they orbit around the nucleus. The phenomenon of magnetism seems to be associated with both the spinning and orbiting activities of electrons.

A moving electron carries a negative electrical charge. The spinning effect of the electron creates a magnetic field. The polarity of the magnetic field is determined by the direction the electron is spinning. The effectiveness of the magnetic field of an atom depends on the number of electrons spinning in each direction. If an atom has equal numbers of electrons spinning in opposite directions, the magnetic fields surrounding the electrons cancel each other out, and the atom is unmagnetized. However, if more electrons spin in one direction than the other, the magnetic fields do not cancel out completely, and the atom is said to be magnetized.

An atom such as iron has an atomic number of 26, meaning that there are 26 protons in the nucleus and 26 revolving electrons arranged in four shells. If 13 electrons are spinning in a clockwise direction and 13 electrons are spinning in a counterclockwise direction, the opposing magnetic fields will be neutralized. However, iron has 14 electrons in its third shell. Of these 14 electrons, nine spin in one direction and five in the other. The net result is an external magnetic field caused by the four electrons that do not have canceled magnetic fields.

The atoms in a given magnetic material do not act independently but are bound together in groups. When a number of such atoms are grouped together, there is an interaction between the magnetic forces of various atoms. This interaction causes a number of neighboring atoms to line up parallel to each other in such a way that their magnetic fields aid each other. An electrostatic force, referred to as **exchange interaction**, maintains these neighboring groups in parallel even when thermally agitated. Exchange interaction is effective against heat only to a certain point. When heated above the Curie point, the atomic alignment breaks down and the material becomes nonmagnetic. For example, the Curie temperature of iron is 770°C, and when iron is heated above this temperature, it is no longer a magnetic material.

In solid materials, molecules and atoms form groups of regular geometric shape called **crystals**. The crystalline structures have magnetized regions which are known as **domains**. A domain is a microscopic needle-shaped crystal which contains a very large number of spinning electrons. The domain has a net magnetic effect oriented in a given direction and is similar to the molecular magnet postulated by Weber.

The domains in any material are always magnetized to saturation but are randomly oriented throughout the material. That is, the domains are pointing magnetically in all different directions. This random configuration results in the magnetic field of each domain being neutralized by opposing magnetic forces of other domains. When an external field is applied to a magnetic material, the domains will line up with the external field. Since the domains themselves are always magnetized to saturation, the magnetic strength of a magnetized material is determined by the number of domains aligned by the magnetizing force. When the magnetizing force is removed, the amount of magnetism remaining is considered to be a measure of the material's retentivity.

7-3 MAGNETIC FIELD

If a layer of iron filings is sprinkled over a piece of cardboard, and a magnetized strip, or bar, is laid upon it, the filings will arrange themselves in a definite pattern when they are moved by a slight tap on the cardboard. This pattern may be thought of as a map of the conditions surrounding the magnet, which are said to be due to the field of the magnet. Therefore, a magnetic field has the property of exerting a force on iron filings which tends to set them into a particular pattern.

The magnetic field surrounding a bar magnet may be investigated more accurately than with iron filings. If a small magnetic compass is placed near one end of a bar magnet, the force between the poles of the magnet and the poles of the compass-needle magnet will move the compass needle away from its usual north-south direction. As long as the compass needle is small in comparison with the bar magnet, the compass will point in the direction of force exerted by the bar magnet on the poles of the compass. More exactly, the compass will point in the direction of the magnetic field surrounding the bar magnet. Since the north pole of the compass will be repelled by the north pole of the bar magnet, the compass will point away from the north pole of the bar magnet. This implies that the direction of a magnetic field at any point is the direction of the force exerted on the north pole of a magnet placed at that point.

If the compass is moved steadily in the direction in which it points at any instant, it will follow a continuous path from the north pole to the south pole of the magnet. Any number of these paths can be traced out, depending on the starting point chosen for the plotting compass. The resulting pattern, which is a replica of the pattern formed by the iron filings, represents a map of the magnetic field of a magnet. Since the lines of this map represent the direction of the force exerted on the north pole of the plotting compass, they are called **lines of force**. The pattern formed by the magnetic lines of force is shown in Figure 7-1. Some important properties of the lines of force can be deduced from the pattern of Figure 7-1. The lines are seen to emerge from the north pole of the magnet and to reenter it at the south pole. They are continuous in the sense that every line leaving the north pole will eventually arrive at the south pole. These lines of force do not begin at the north pole of a magnet and end at the south pole, but continue inside the magnet to form complete closed loops.

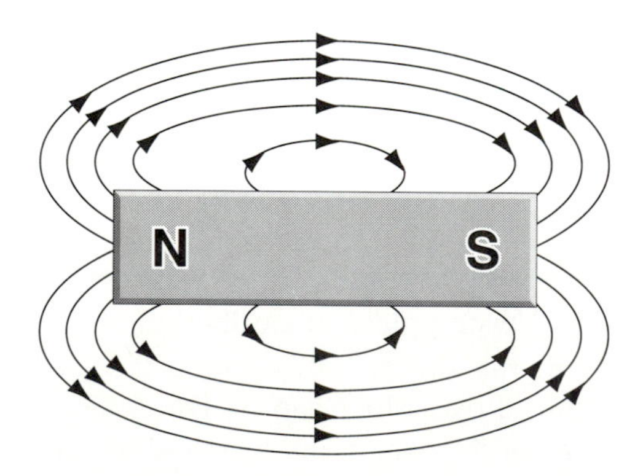

FIGURE 7-1 Magnetic field around a bar magnet.

The characteristics of magnetic lines of force can be stated as follows:

1. Magnetic lines of force are continuous and will always form closed loops.
2. Magnetic lines of force cannot intersect each other.
3. Magnetic lines of force in a field around a magnet can be plotted by the use of iron filings.
4. Magnetic lines of force that are traveling in the same direction tend to repel each other. Parallel magnetic lines of force traveling in opposite directions tend to attract each other.

The total number of lines of force in a given region is called **magnetic flux**. Flux in a magnetic circuit corresponds to current in an electric circuit. The number of lines of force per unit area is called **flux density**. The unit of flux density is the **tesla** (T). The unit of magnetic flux is the **weber**. One weber is equal to 100 million (10^8) lines of force. One tesla is equal to 1 weber per square meter (Wb/m^2). The relationship between flux density and lines of force is expressed in equation form as follows:

$$B = \frac{\Phi}{A} \tag{7-1}$$

where B = flux density, in teslas

Φ = total field flux, in webers

A = cross-sectional area, in square meters

EXAMPLE 7-1 An iron core with a cross-sectional are of 0.25 m^2 has a total flux of 750 μWb. Calculate the flux density in the core material.

Solution

$$B = \frac{\Phi}{A} = \frac{750 \times 10^{-6}}{0.25} = 0.003 \text{ T}$$

7-4 MAGNETIC ATTRACTION AND REPULSION

When the north pole of one magnet is brought near the south pole of another magnet, a force of attraction is exerted between them. However, if the north pole of the magnet is brought near the north pole of another magnet, there is a force of repulsion between them. From this, the first two **laws of attraction and repulsion** are stated as:

1. Like magnetic poles repel each other.
2. Unlike magnetic poles attract each other.

The third law of magnetic attraction and repulsion concerns the strength of the attraction and repulsion of the magnets based on their distance from each other. This third law is stated as:

3. The attraction or repulsion between magnets varies directly with the product of their strengths and inversely with the square of the distance between them.

7-5 MAGNETIC MATERIALS

When a material is easy to magnetize, it is said to have a high **permeability**. Soft iron, being relatively easy to magnetize, has a high permeability, whereas steel is much harder to magnetize and has a lower permeability than iron. Magnetic materials can be classified into one of three groups: paramagnetic, diamagnetic, and ferromagnetic.

Paramagnetic materials are those that become only slightly magnetized even though they are under the influence of a strong magnetic field. This slight magnetization is in the same direction as the magnetizing field. Materials of this type are aluminum, chromium, platinum, and air.

Diamagnetic materials exhibit a very slight opposition to magnetic lines of force. These materials are magnetized in a direction opposite to the external field that is being applied. In all known cases of materials which are diamagnetic, the diamagnetic effect is so small that very sensitive instruments are required to detect it. Examples of diamagnetic materials are copper, mercury, silicon, gold, and silver.

Materials possessing pronounced magnetic properties are said to be **ferromagnetic**. These are the materials used as permanent magnets and electromagnets. Ferromagnetic materials are easy to magnetize and are considered to have high permeability, such as iron, nickel, and cobalt.

7-6 FIELD AROUND A CURRENT-CARRYING CONDUCTOR

When an electric current is passed through a long straight conductor, a magnetic field is established in and around the conductor. In 1820, the Danish physicist Hans Christian Oersted discovered that when a compass is placed near a current-carrying conductor, the compass needle sets itself at right angles to the conductor. If the direction of current is reversed, the direction of the compass needle will also be reversed. If the current is switched off, the compass needle will return to its original north-south direction. If the compass needle is moved steadily in the direction in which it points at any instant, it will trace out a circular path around the current, with the current axis at the center of the circle. Any number of such circles can be traced out, depending on the starting radius of the plotting compass. This experiment demonstrates that the magnetic field exists in concentric circles around the conductor.

The direction of the magnetic field surrounding a conductor can be determined by what is known as the **right-hand rule**. This rule is illustrated in Figure 7-2 and is further specified by the following statement:

> If a current-carrying conductor is grasped in the right hand with the thumb pointing in the direction of conventional current, the fingers will point in the direction of the magnetic lines of force.

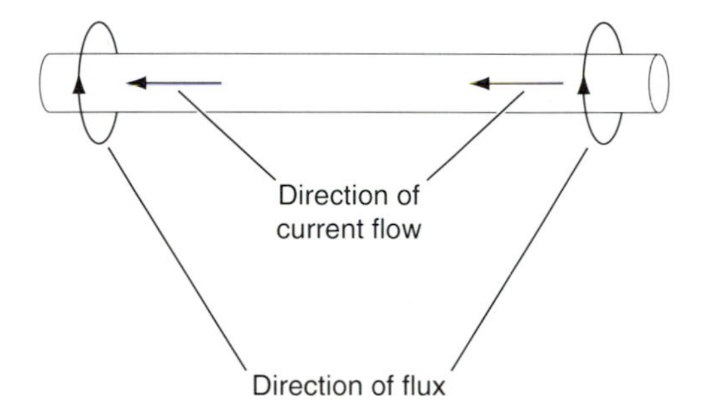

FIGURE 7-2 Right-hand rule for determining direction of flux around a conductor.

The symbol ⊙ is used in diagrams to denote a cross-sectional view of a conductor carrying current toward the reader, while the symbol ⊕ is used to indicate current flowing away from the reader. Figure 7-3 shows the use of these symbols.

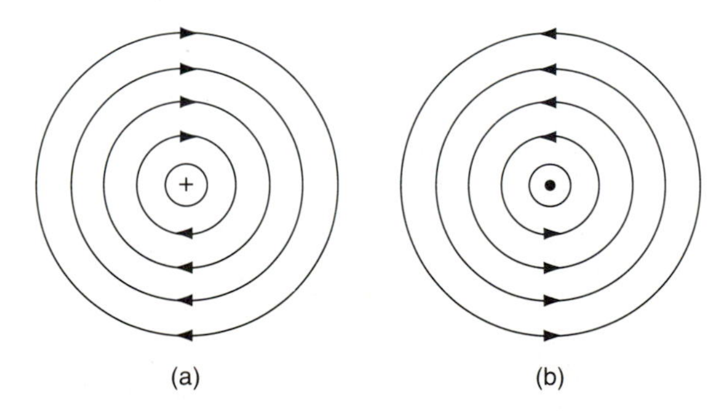

FIGURE 7-3 Direction of magnetic field surrounding current in a conductor: (a) for current direction away from reader; (b) for current direction toward reader.

When two parallel conductors carry current in the same direction, the magnetic fields tend to encircle both conductors, drawing them together with a force of attraction, as shown in Figure 7-4(a). Two parallel conductors carrying current in opposite directions are shown in Figure 7-4(b). The field around one conductor is opposite in direction to the field around the other conductor. The resulting lines of force are crowded together in the space between the conductors and tend to push the conductors apart. Therefore, two parallel adjacent conductors carrying currents in the same direction attract each other and two parallel conductors carrying currents in opposite directions repel each other.

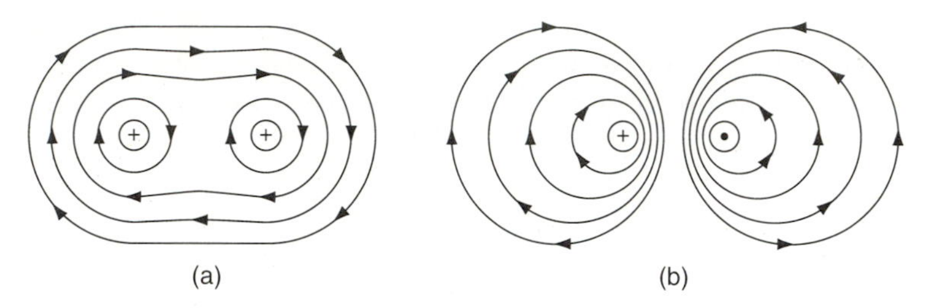

FIGURE 7-4 Magnetic field around two parallel conductors.

If the current-carrying conductor is formed into a loop, the magnetic lines of force will all pass through the center of the loop in the same direction. This is indicated approximately in Figure 7-5, where a few lines of force are sketched in as typical of those it would be possible to trace. For any one segment of the loop, the surrounding magnetic field is much the same as it is for straight wire. This implies that the number of lines of force has not changed, even though the conductor is now concentrated in a smaller physical area.

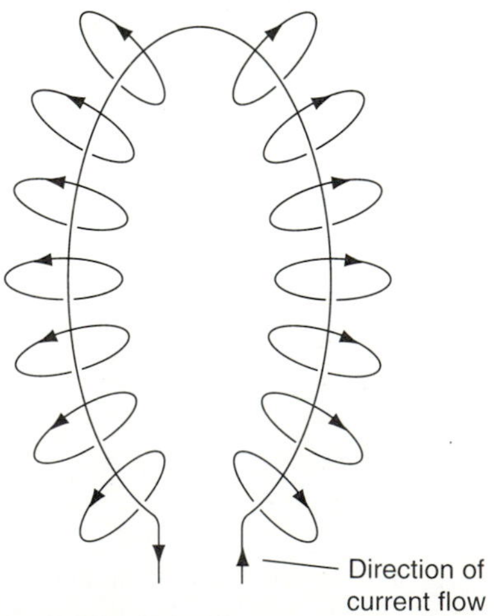

FIGURE 7-5 Concentrating the magnetic field by forming the conductor into a single-turn loop.

Figure 7-5 shows that the lines of force representing the magnetic field of the current in the loop leave the loop on one side and enter it at the other. If the current direction is assumed to be counterclockwise, the lines of force would appear to be coming toward the reader when viewed from one side of the loop. If viewed from the other side of the loop, the current direction would be clockwise, and the lines of force would appear to be going away from the reader. In other words, the current loop is acting as a magnet, which has one side as a north pole and the other side as a south pole. A simple method of remembering which current direction around the loop makes the face of the loop a north pole and which makes it a south pole is illustrated in Figure 7-6.

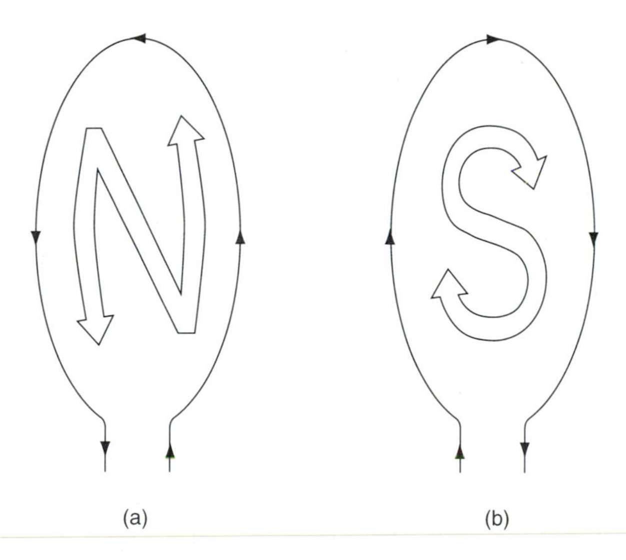

FIGURE 7-6 Magnetic field of current flowing around a loop wire: (a) direction of flux is toward reader; (b) direction of flux is away from reader.

7-7 MAGNETIC FIELD AROUND A COIL

The magnetic field around a current-carrying conductor exists in all points around its length. The magnetism associated with this conductor can be intensified by forming the conductor into a coil, or solenoid. If a current is passed through the coil, the field around each turn reinforces the field of adjacent turns. These lines of force produce a greatly strengthened magnetic field. The combined influence of all the turns produces a two-pole field similar to that of a simple bar magnet. One end of the coil will be a north pole and the other end will be a south pole.

The polarity of any coil may be found by means of the right-hand rule for a coil, which may be stated in the following manner:

> Grasp the solenoid with the right hand so that the fingers follow the conventional current direction around the circumference of the solenoid; the thumb then points in the direction of the magnetic lines of force through the solenoid.

7-8 ELECTROMAGNETS

An electromagnet is composed of a coil of wire wrapped around an iron core. When an electric current is passed through the coil, a magnetic field is developed which is strengthened by the presence of the iron core. The magnetic field will have the same polarity regardless of whether or not the iron core is present. If the current direction is reversed through the coil, the polarity of both the coil and iron core is reversed. The addition of the iron core accomplishes two things. First, the magnetic flux is greatly increased, due to the iron core being more permeable than air. Second, the flux is more highly concentrated than it would be if only air were used.

When a piece of soft iron is placed near the poles of an electromagnet, the iron is attracted to the magnet. This force of attraction between the magnet and iron is due to a property that tends to make flux contract and become as short as possible. Figure 7-7 shows a solenoid with an iron core placed near the coil. The lines of force are shown to extend through the soft iron and magnetize it. Since unlike poles attract, the piece of iron is drawn toward the coil. If the bar is free to move, it will be drawn into the coil to a position near the center where the field is strongest. If a spring were attached to the iron bar, it would return to its original position when the current through the coil was interrupted. The solenoid-and-plunger type of magnet is employed extensively in electromechanical systems.

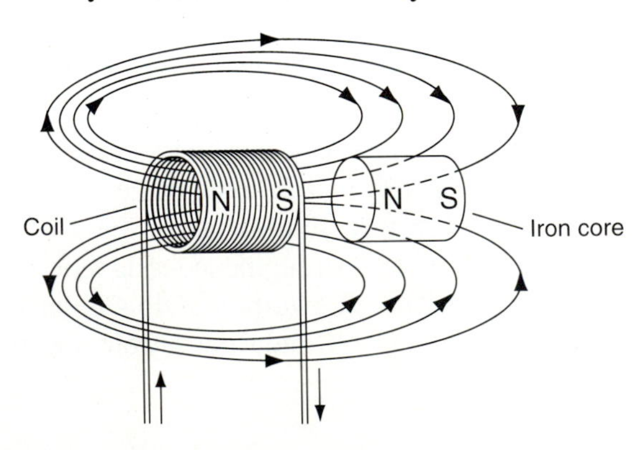

FIGURE 7-7 Solenoid with iron core.

In practice, the solenoid is used for tripping circuit breakers, operating relays, and for operating contactors in automatic motor starters. In practically all cases, a soft-iron plunger, or armature, is necessary to obtain the tractive effort required of the solenoid.

The strength of an electromagnetic field depends on the following three factors:

1. Amount of current in the coil, I
2. Number of turns of the coil, N
3. Size and material of the core

The product of the first two factors, turns $\times$ amperes (NI), is referred to as ampere-turns (At). For example, a coil of 800 turns carrying 5 A has 800 $\times$ 5, or 4000 At.

KEY TERMS

Magnetism
Pole
North pole
South pole
Ferromagnetic materials
Weber's theory
Exchange interaction
Crystal
Domain
Lines of force
Magnetic flux
Flux density
Tesla
Weber
Laws of attraction and repulsion
Permeability
Paramagnetic material
Diamagnetic material
Ferromagnetic material
Right-hand rule

PROBLEMS

7-1 An iron core with a cross-sectional area of 0.01 m^2 has a total flux of 300 μWb. Determine the flux density in the core material.

7-2 With a flux of 200×10^{-4} Wb through an area of 75 mm $\times$ 30 mm, what is the flux density?

7-3 Determine the amount of flux required to provide a flux density of 220×10^{-6} Wb/m^2 in a cross-sectional area measuring 0.0025 m $\times$ 0.00317 m.

7-4 If the flux density in an iron core is 3.35 T and the material has a cross-sectional area of 0.25 in.^{2}, find the flux through the core.

7-5 Calculate the flux density of a certain magnetic material having a cross-sectional area of 0.155 in.^{2} and a total flux of 275 μWb.

7-6 A core has a flux density of 7.2×10^3 Wb/m^2. If the flux is 26×10^2 Wb, find the cross-sectional area.

7-7 How much magnetomotive force does a current of 5 A in a 60-turn coil provide?

7-8 For a 100-turn coil to produce a magnetomotive force of 60 At, what value of current must flow?

CHAPTER

The Magnetic Circuit

LEARNING OBJECTIVES

Upon completion of this chapter you will be able to:

- Convert a magnetic quantity from SI to English units, and vice versa.
- Define magnetomotive force.
- Express magnetic reluctance in terms of magnetomotive force and magnetic flux.
- Define field intensity.
- Understand the permeability curves of common magnetic materials.
- Describe the magnetic properties of common materials.
- Define magnetic hysteresis and residual magnetism.
- Express Ampère's circuit law.
- Describe the effect of air gaps in a magnetic circuit.
- Understand the principles of series and parallel magnetic circuits.
- Determine flux in a magnetic circuit.
- Define tractive force.

8-1 INTRODUCTION

The magnetic circuit and the electric circuit have a great number of similarities. The electric circuit makes it possible to calculate voltage, resistance, and current, provided that sufficient data are given. Similarly, with the magnetic circuit it is possible to calculate magnetic quantities such as magnetomotive force, flux, and reluctance, provided that sufficient data are given. For example, magnetic flux is equal to the magnetomotive force divided by the reluctance.

In the magnetic circuit, after the flux is established, no heat is produced in the circuit itself, and therefore no energy is used in retaining this magnetism, provided that the electric current in the coil is steady. All the power used in supplying current to the coil can be calculated as I^2R losses in the coil itself. Since no energy is used in the magnetic circuit, it indicates that there is no movement of the flux or flux lines.

Table 8-1 shows a comparison of the magnetic units. Although the SI system of units is taken as the international standard, there are still many situations where both the English system and the CGS system are still in use.

TABLE 8-1 *Comparison of Magnetic Units*

Quantity	SI unit	English unit	CGS unit
Flux	weber	line	maxwell
Flux density	tesla (Wb/m^2)	$lines/in.^2$	gauss ($maxwell/cm^2$)
MMF	ampere/tesla	ampere-turn	gilbert
Field intensity	ampere/tesla/meter	ampere-turn/in.	oersted

8-2 MAGNETOMOTIVE FORCE

The amount of flux developed in a solenoid is dependent on the current, I, and the number of turns, N. The product of I and N in a magnetic circuit, called **magnetomotive force** (mmf), is a measure of the ability of a coil to produce flux. Mmf determines the degree to which electromagnetic effects are produced. Flux is the "current" of a magnetic circuit, and the magnetomotive force is the force that establishes it. Therefore, mmf is analogous to the emf of the electric circuit. Since the mmf developed is proportional to the current and to the number of turns in the coil, the unit of measurement is the ampere-turn and is found by

$$\mathcal{F} = I \times N \tag{8-1}$$

where $\mathcal{F}$ = magnetomotive force, in ampere-turns

I = current through a coil

N = number of turns of a coil

8-3 RELUCTANCE

The opposition that a magnetic path offers to magnetic flux when a magnetomotive force is applied is called **reluctance**. The symbol for reluctance is the script letter $\mathcal{R}$, and the SI unit of measurement is the ampere-turn/weber. The relationship among magnetomotive force, magnetic flux, and reluctance may be expressed by the equation

$$\mathcal{R} = \frac{\mathcal{F}}{\Phi} \tag{8-2}$$

where $\mathcal{R}$ = reluctance, in ampere-turns/weber

$\mathcal{F}$ = magnetomotive force, in ampere-turns

Φ = magnetic flux, in webers

8-4 FIELD INTENSITY

Magnetomotive force is an important consideration in a magnetic circuit, as it gives a measure of the magnetic stress imposed on a given core material for its total length. In the SI system, magnetizing force is in ampere-turns per meter. **Magnetic field intensity** is a measurement of the magnetomotive force needed to establish a certain flux density in a unit length of the magnetic circuit. Therefore, the magnetic force per unit length is defined as field intensity (H) and is expressed in mathematical form as

$$H = \frac{\mathcal{F}}{l} \qquad (8\text{-}3)$$

where H = magnetic field strength, in ampere-turns/meter

$\mathcal{F}$ = applied mmf, in ampere-turns

l = average length of magnetic path, in meters

8-5 PERMEANCE

The **permeance** of a circuit is the reciprocal of the reluctance and may be defined as that property of the circuit which permits the passage of magnetic flux, or lines of induction. Consequently, a magnetic circuit in which the path of the lines of force is almost all through iron, the permeance of the circuit is relatively high compared with the permeance of air or other nonmagnetic material. Permeance corresponds to conductance in the electric circuit. The SI unit for permeance is the henry (H), named after the American physicist Joseph Henry (1797–1878).

8-6 PERMEABILITY

The ability of a material to concentrate magnetic flux is called **permeability** (μ). The permeability of any material is a measure of the ease with which its atoms can be aligned, or the ease with which it can establish lines of force. Permeability is measured in henrys per meter (H/m). Numerical values of μ

for different materials are assigned by comparing their permeability with the permeability of air, or a vacuum. The μ value of nonmagnetic materials such as air, copper, wood, and plastic is for all practical purposes equal to unity (1). Magnetic materials such as iron, nickel, steel, and their alloys have a μ value much greater than unity. For example, soft iron has a permeability several hundred times that of air. Figure 8-1 shows a graph of four common magnetic materials and their permeability curves.

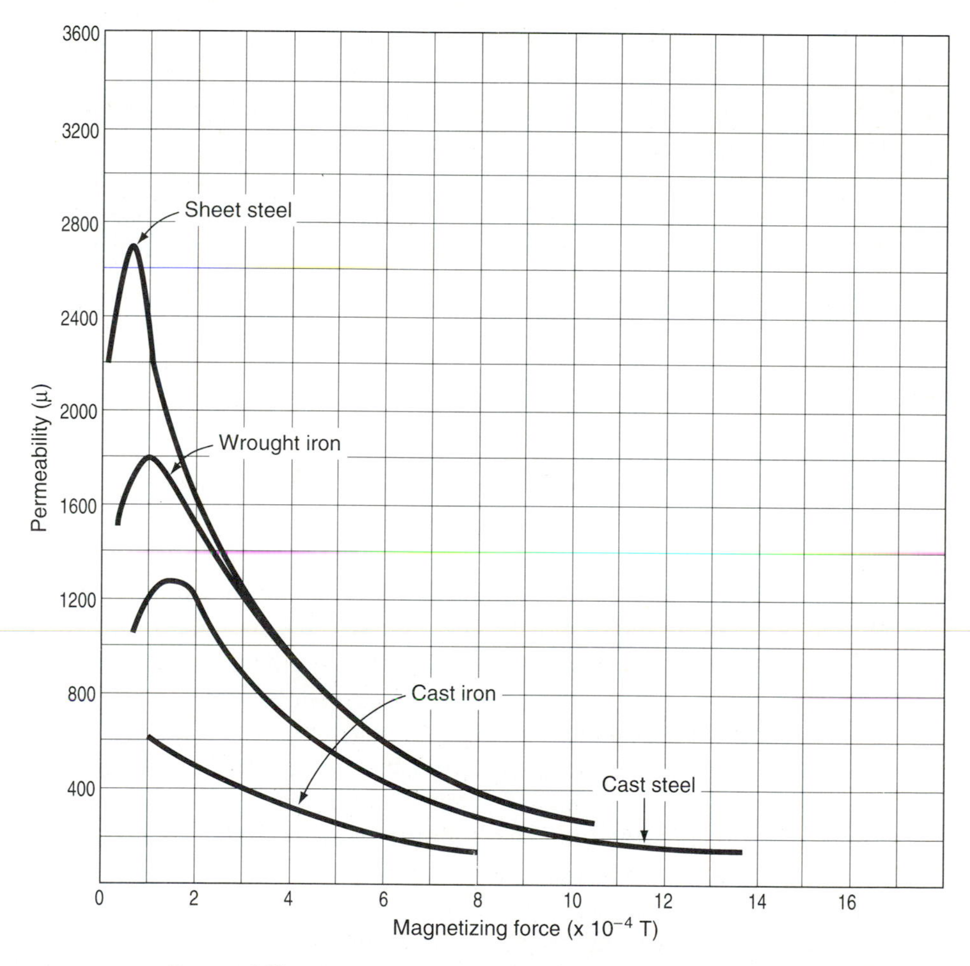

FIGURE 8-1 Permeability curves of common ferromagnetic materials.

Space permeability (μ_0) is the permeability of air, or vacuum, and is taken as the standard reference value. It is also referred to as the absolute, or free-space, permeability. The permeability of free space is $4\pi \times 10^{-7}$ H/m. The permeability of any material can be stated as the ratio of the magnetic flux density to the magnetic field intensity of a material. Expressed as an equation,

$$\mu = \frac{B}{H} \tag{8-4}$$

where μ = permeability of a material, in henrys/meter

B = flux density, in teslas

H = magnetic field strength, in ampere-turns/meter

The relative permeability (μ_r) of a material is the ratio of the flux in that material to the flux that would exist if that material were replaced with air, with the mmf on the material remaining unchanged. Relative permeability is defined by the equation

$$\mu_r = \frac{\text{number of lines produced with the material as a core}}{\text{number of lines produced with air as a core } (4\pi \times 10^{-7})} = \frac{\mu}{\mu_0} \tag{8-5}$$

8-7 MAGNETIC PROPERTIES OF MATERIALS

It was stated earlier in the chapter that for nonmagnetic materials the permeability is unity, and that for magnetic materials the permeability is dependent on the flux density. Figure 8-2 shows the typical magnetization curves for sheet steel, cast steel, and cast iron. These curves, which must be determined experimentally, show the manner in which the flux density varies with the magnetizing force. In this *B–H* graph, the slope of the curve is the change in *B* divided by the corresponding change in *H*, which is equal to the permeability of the material. Eventually, the curve reaches a point where increasing the field intensity has virtually no effect on the flux density. At this point the material is said to be saturated. The term **magnetic saturation** means simply that to produce an increase in the flux density, an extremely large increase in magnetizing force is necessary.

Since permeability is defined as $\mu = B/H$, the permeability can be constant only if the magnetization curve is in a straight line. In other words, the steeper the slope of the magnetization curve, the greater the permeability. The curves of Figure 8-2 show that the magnetic properties of cast iron are very inferior to those of cast steel. The use of cast iron instead of cast steel for the frames of direct-current machines would require a sectional area and weight twice as great as that needed if cast steel were used. For magnetic materials, the permeability varies with the flux density, as indicated in Figure 8-3. Starting from a value of $B = 0$, the permeability falls off fairly rapidly, as shown.

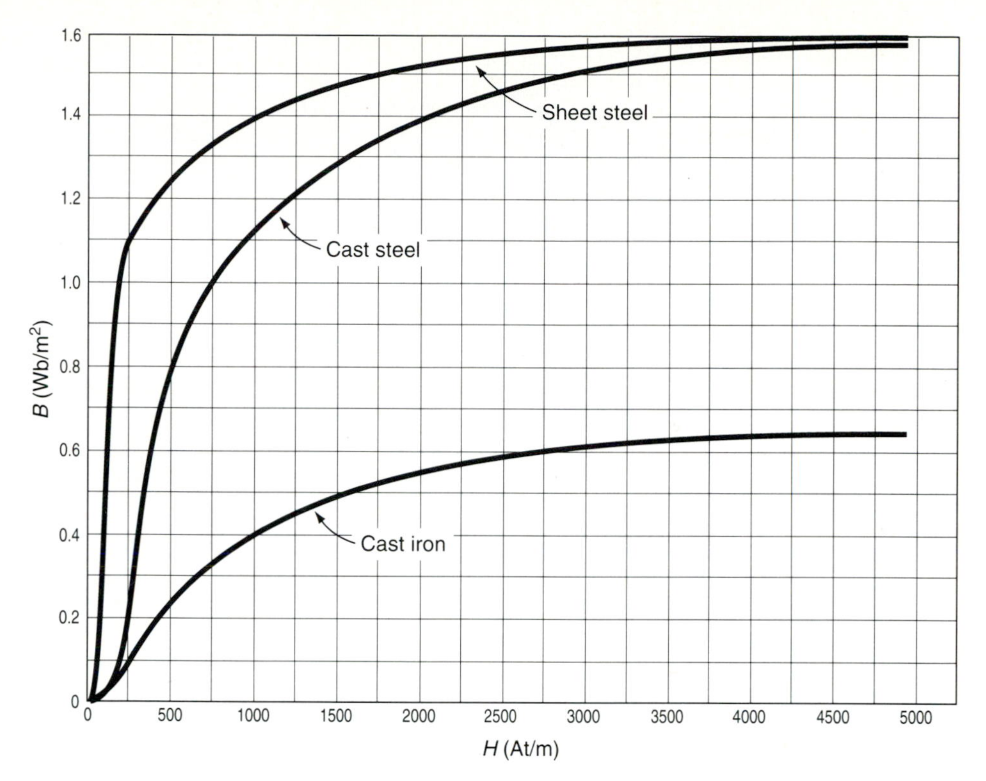

FIGURE 8-2 Magnetization and relative permeability curves.

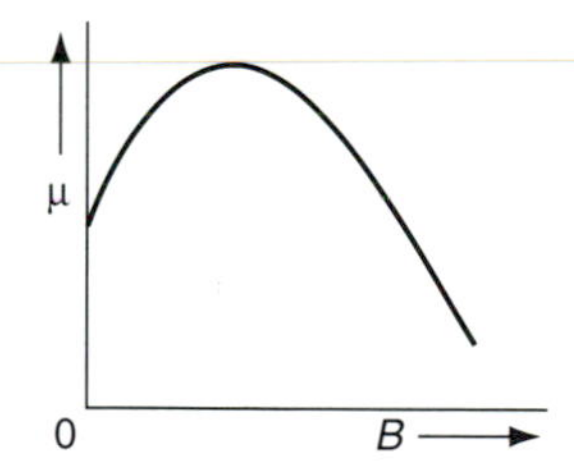

FIGURE 8-3 Manner in which the permeability of a magnetic material varies with flux density.

8-8 MAGNETIC HYSTERESIS AND RESIDUAL MAGNETISM

In Figure 8-3 it was demonstrated that the flux density increases when the magnetizing force increases, even when the curve was carried to a point beyond saturation. When the magnetizing force is decreased, the flux density will not decrease along the same line as it had increased. The reason for this will be explained using the curves shown in Figure 8-4.

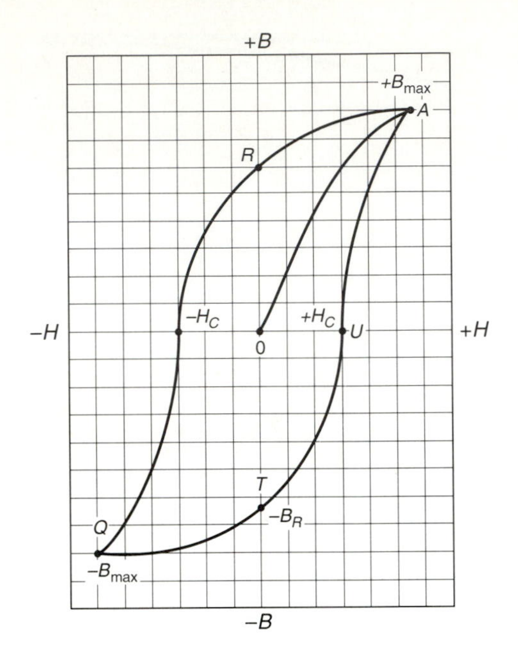

FIGURE 8-4 Hysteresis loop.

Experiment proves that iron, once magnetized, tends to retain its magnetism. Permanent magnets are all dependent on this property. If a ferromagnetic material was magnetized to a value represented in Figure 8-4 by *OA*, the core is said to become saturated along the normal magnetization curve. By the time the magnetizing force has fallen to zero, the flux density has only been reduced to *AR*, which is called the **residual flux density**, or **residual magnetism**. The residual flux density is the amount of flux density remaining in the material after the magnetizing force has been removed.

To reduce the flux to zero, the **coercive force**, $-H$, must be applied, tending to establish magnetic flux in the opposite direction. Coercive force is defined as the demagnetizing force necessary to remove the residual flux from the material.

If starting from the condition represented by the point *Q*, the magnetizing force is again gradually reversed, the magnetizing curve will be as shown by *QUA*. The curve for increasing magnetization always lies below that for decreasing magnetization. The flux density *B*, therefore, is said to lag behind the force *H*. This lagging is called **hysteresis**. Not only does magnetic hysteresis introduce a degree of uncertainty into the calculation of magnetic circuits, but in addition, due to this hysteresis, energy is wasted when the iron is subjected to alternating magnetism.

8-9 AMPÈRE'S CIRCUIT LAW

Ampère's circuit law states that the algebraic sum of the rises and drops of the mmf around a closed loop of a magnetic circuit is equal to zero, which can also be expressed as: the sum of the mmf rises are equal to the sum of the mmf drops.

When Ampère's circuit law is applied to magnetic circuits, sources of mmf are expressed by the following equations:

$$\mathcal{F} = NI$$

where $\mathcal{F}$ = mmf, in ampere-turns

N = the number of turns

I = current, in amperes

$$\mathcal{F} = \Phi\mathcal{R} \tag{8-6}$$

where $\mathcal{F}$ = mmf, in ampere-turns

Φ = the flux passing through a section of the magnetic circuit, in webers

$\mathcal{R}$ = reluctance of that section, in amperes/tesla/weber

$$\mathcal{F} = Hl \tag{8-7}$$

where $\mathcal{F}$ = mmf, in ampere-turns

H = magnetizing force on a section of a magnetic circuit, in amperes/tesla/meter

l = length of the section, in meters

8-10 AIR GAPS

In a magnetic circuit, it is quite common for the magnetic core not to be continuous. This may be due to an **air gap** in the circuit. The total flux in the ferromagnetic material must also exist in the air gap. Air gaps are often deliberately placed in magnetic circuits to increase reluctance. By increasing the total reluctance, saturation of the iron core may be prevented, allowing a larger current flow through the coil wound on the core. In these types of magnetic circuits the "air" gap may consist of a sheet of nonmagnetic material, such as fiberboard, allowing the length of the air gap to be accurately set.

When an air gap exists, as in Figure 8-5, the lines of flux passing through the gap are not limited to the projected area of the iron core. The spreading of the flux lines outside the area of the core for the air gap is known as **fringing**. The flux area increases due to fringing, resulting in a decrease in flux density. Therefore, the flux density in the air gap is slightly less than that in the iron sections of the magnetic circuit.

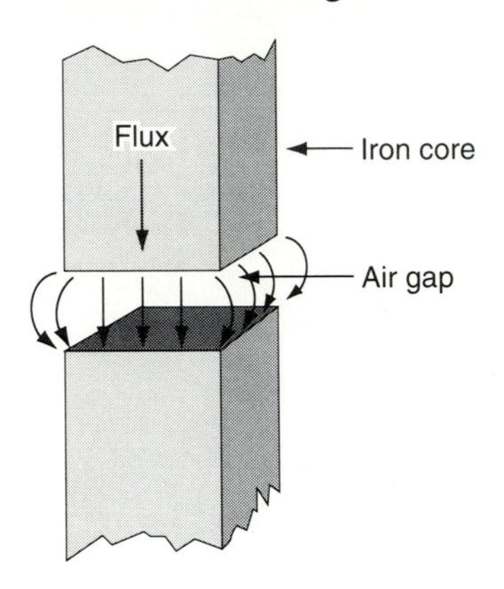

FIGURE 8-5 Fringing flux at an air gap.

If the length of the air gap is small, the resulting fringing is also small. The correcting factor which takes fringing into account is to add the length of the gap to each cross-section dimension of the adjoining material. For example, if a core 7 × 3 cm has a 0.17-cm air gap, the air gap's effective area is $A_g = (7.17) \times (3.17) = 22.73\ \text{cm}^2$. For our purposes, we shall neglect fringing unless otherwise noted.

The flux density of the air gap in Figure 8-5 is defined by the equation

$$B_g = \frac{\Phi_g}{A_g} \tag{8-8}$$

where $\Phi_g = \Phi_{\text{core}}$ and $A_g = A_{\text{core}}$. The magnetizing force of the air gap is determined in the following manner:

$$H_g = \frac{B_g}{\mu_0} \tag{8-9}$$

where μ_0 is the permeability free space and has a value of $4\pi \times 10^{-7}$ H/m. The magnetizing force of the air gap is also equal to

$$H_g = 7.97 \times 10^5\, B_g$$

A certain number of ampere-turns are required to send flux through a ferromagnetic material, and additional ampere-turns are required to send flux through the air gap of a magnetic circuit. Since the paths are in series, the total ampere-turns needed is the sum of the two.

Due to the fact that flux density is the product of flux divided by area, flux density may also be expressed by

$$B = \frac{NI}{l}\mu \tag{8-10}$$

where B = flux density, in teslas

NI = ampere-turns

l = length of the material, in meters

μ = permeability of the material

Solving for magnetomotive force, the equation may be rewritten as

$$NI = \frac{Bl}{\mu} \tag{8-11}$$

Since the permeability for air is constant at $4\pi \times 10^{-7}$, the equation for ampere-turns in an air gap would be

$$NI = \frac{Bl}{4\pi \times 10^{-7}}$$

8-11 SERIES MAGNETIC CIRCUIT

The method of calculating the magnetomotive force required to establish a given flux through a magnetic circuit consisting of a number of parts in series is very similar to the corresponding problem for an electric circuit. The total reluctance of the path is simply the sum of the values for all the individual parts. The most convenient method, however, of dealing with the problem is to calculate separately the actual magnetomotive force absorbed in establishing the flux in each part of the circuit, and then add these values.

EXAMPLE 8-1 For the series magnetic circuit of Figure 8-6, find (a) the value of current required to develop a magnetic flux of $\Phi = 7 \times 10^{-4}$ Wb; (b) the permeability of the material under these conditions.

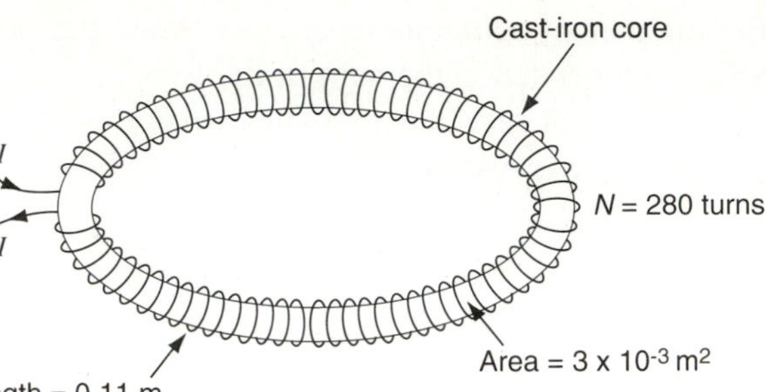

FIGURE 8-6

Solution (a) The flux density B is

$$B = \frac{\Phi}{A} = \frac{7 \times 10^{-4}}{3 \times 10^{-3}} = 0.233 \text{ T}$$

Using the B–H curves of Figure 8-2, we can determine the magnetizing force H:

$$H(\text{cast iron}) = 600 \text{ At/m}$$

From Ampère's circuit laws, the following equation is derived:

$$NI = Hl \tag{8-12}$$

Therefore,

$$\begin{aligned} I &= \frac{Hl}{N} \\ &= \frac{(600)(0.11)}{280} \\ &= 235.71 \text{ mA} \end{aligned}$$

(b) The permeability of the material can be found in the following manner:

$$\begin{aligned} \mu &= \frac{B}{H} = \frac{0.233}{600} \\ &= 3.88 \times 10^{-4} \text{ H/m} \end{aligned}$$

A coil wound on a circular core, as in Example 8-1, is called a **toroid.** A toroidally wound coil has efficient magnetic characteristics. Since its core is endless and entirely within the coil winding, it concentrates the magnetic field to the maximum degree.

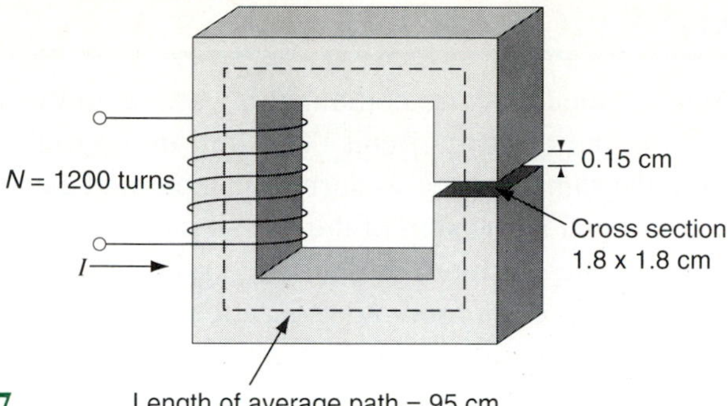

FIGURE 8-7

EXAMPLE 8-2 The core shown in Figure 8-7 is made of cast steel. The magnetizing coil has 1200 turns. How much current should be sent through this coil to produce a flux of 3×10^{-4} Wb?

Solution Area of air gap (taking fringing into account):

$$A_g = (1.95 \times 10^{-2})^2 = 38.025 \times 10^{-5}\,\text{m}^2$$

$$B = \frac{\Phi}{A} = \frac{3 \times 10^{-4}}{38.025 \times 10^{-5}} = 0.789\,\text{T}$$

$$H = \frac{B}{\mu} = \frac{0.789}{4\pi \times 10^{-7}} = 627.859 \times 10^3\,\text{At/m}$$

$$(Hl)_g = (627.859 \times 10^3)(1.5 \times 10^{-3}) = 941.789\,\text{At}$$

Area of steel:

$$A = (1.8 \times 10^{-2})^2 = 3.24 \times 10^{-4}\,\text{m}^2$$

$$B = \frac{\Phi}{A} = \frac{3 \times 10^{-4}}{3.24 \times 10^{-4}} = 0.926\,\text{T}$$

The graph of the B–H curves (Figure 8-2) reveals the following information:

$$H = 625\,\text{At/m}$$

$$Hl = (625)(9.5 \times 10^{-1}) = 593.75\,\text{At}$$

$$\mathcal{F} = (Hl)_g + (Hl) = 941.789\,\text{At} + 593.75\,\text{At} = 1535.539\,\text{At}$$

Therefore,

$$I = \frac{\mathcal{F}}{N} = \frac{1535.539}{1200} = 1.28\,\text{A}$$

8-12 PARALLEL MAGNETIC CIRCUIT

In the magnetic circuit, reluctances are combined in parallel in the same manner as resistances are in the electric circuit. The flux in a parallel magnetic circuit also behaves in the same manner as current in a parallel electric circuit. That is, the total flux is equal to the sum of the individual flux flowing in each part of the circuit. Magnetic circuits are deliberately designed with parallel magnetic paths, since this has been found to reduce leakage flux substantially. In practice, there are no truly parallel magnetic circuits. The branch on which a coil is wound is always in series with the parallel paths of the circuit, forming a series–parallel configuration.

EXAMPLE 8-3 In the magnetic circuit of Figure 8-8, the cross-sectional area of the central cast steel core is 2.2×10^{-4} m^2. The rest of the core has a cross-sectional area of 1.3×10^{-4} m^2. The path lengths of the core are $l_1 = 7$ cm and $l_2 = 12$ cm. The coil consists of 700 turns. Determine the current required to establish a flux of 1.8×10^{-4} Wb in the center branch.

Solution As the circuit is symmetrical about its centerline, $\frac{1}{2}\Phi_1 = \Phi_2 = \Phi_3$. Since the two loops are identical, only one loop equation for both outer branches is necessary. For the center branch:

$$B = \frac{\Phi}{A} = \frac{1.8 \times 10^{-4}}{2.2 \times 10^{-4}} = 0.818 \text{ T}$$

From the B–H curve for cast steel (Figure 8-2),

$$H = 520 \text{ At/m}$$

$$(Hl)l_1 = (520)(7 \times 10^{-2}) = 36.4 \text{ At}$$

For either outside loop,

$$B = \frac{\Phi}{A} = \frac{\frac{1}{2}(1.8 \times 10^{-4})}{1.3 \times 10^{-4}} = 0.692 \text{ T}$$

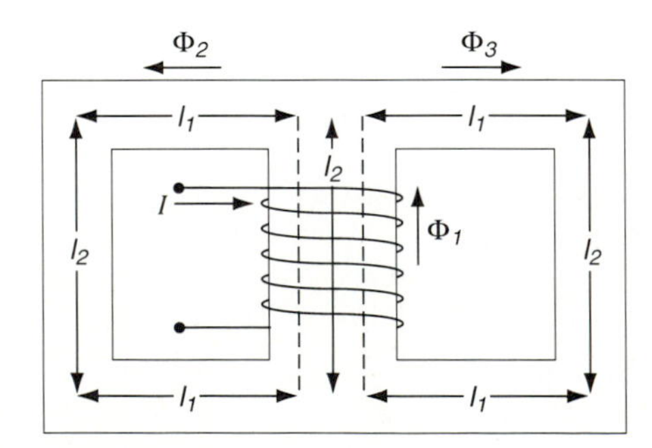

FIGURE 8-8

From the B–H curve for cast steel (Figure 8-2),

$$H = 450 \text{ At/m}$$

$$(Hl)(l_1 + l_2 + l_1) = (450)(26 \times 10^{-2}) = 117 \text{ At}$$

$$\begin{aligned}\mathcal{F} &= (Hl)l_1 + (Hl)(l_1 + l_2 + l_1)\\ &= (520)(7\text{ cm}) + (450)(7\text{ cm} + 12\text{ cm} + 7\text{ cm})\\ &= 153.4\text{ At}\end{aligned}$$

Therefore,

$$\begin{aligned}I &= \frac{\mathcal{F}}{N} = \frac{153.4}{700}\\ &= 219.14\text{ mA}\end{aligned}$$

8-13 DETERMINING FLUX IN A MAGNETIC CIRCUIT

Calculating flux is relatively straightforward if only one magnetic section, such as a toroid, is involved. It can be stated that

$$H = \frac{NI}{l}$$

and $$\Phi = BA$$

However, if the magnetic circuit is made up of different sections in series, it is not quite as easy to solve for the total flux. For magnetic circuits with more than one section, there is no set order of steps that will lead to an exact solution.

It is usual, therefore, in this kind of problem to make an estimate of the approximate flux density and note from the B–H curves how much magnetomotive force is necessary. From the information obtained by this calculated guess, we can make closer estimates, and obtain more definite information, until we bring our guess closer to the actual value of Φ. For most applications, a value within $\pm 5\%$ of the actual is acceptable. This cut-and-try method is illustrated in the following example.

EXAMPLE 8-4 The magnetic circuit of Figure 8-9 consists of a cast steel core whose cross-sectional dimensions are 6 cm × 3 cm and has an average path length of 22 cm. A coil of 600 turns carries a current of 0.7 A. The continuity of this core is broken by a single air gap that is 0.33 mm in length. Calculate the total flux in the core.

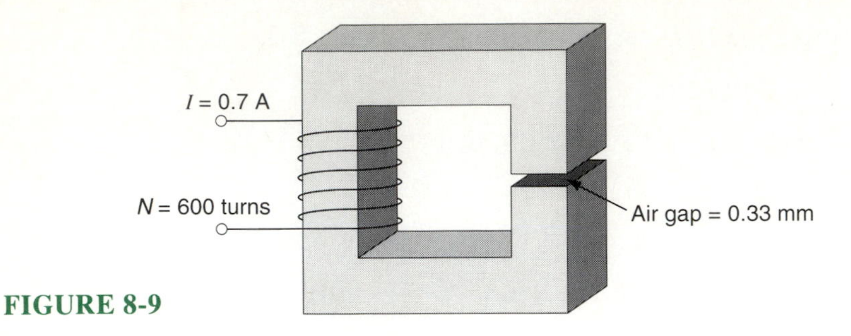

FIGURE 8-9

Solution The area of air gap (taking fringing into account) is

$$A_g = 6.033 \times 3.033 = 18.298\ \text{cm}^2$$

$$\mathcal{F} = NI = 600 \times 0.7 = 420\ \text{At}$$

Since air has a much greater reluctance than steel, we may assume that the greater part of the magnetomotive force will be used in sending the magnetic flux across the air gap. Assume 60% of the 420 ampere-turns for the air gap.

$$420 \times 0.6 = 252\ \text{At}$$

Then

$$H_g = \frac{\mathcal{F}}{l_g} = \frac{252}{0.33 \times 10^{-3}} = 763.636 \times 10^3\ \text{At/m}$$

$$B_g = \mu_0 H_g = 4\pi \times 10^{-7}\ \text{H/m} \times 763.636 \times 10^3\ \text{At/m} = 0.96\ \text{T}$$

Therefore,

$$\Phi = B_g A_g = 0.96\ \text{T} \times 18.298\ \text{cm}^2 = 1.757 \times 10^{-3}\ \text{Wb}$$

For this to be correct, the flux density in the core must be

$$B = \frac{\Phi}{A} = \frac{1.757 \times 10^{-3}}{18.298 \times 10^{-4}} = 0.96$$

From the B–H curve for cast steel (Figure 8-2),

$$B = 0.96\ \text{T}$$

$$H = 750\ \text{At/m}$$

Applying Ampère's circuital law gives

$$\begin{aligned} NI &= Hl + H_g l_g \\ &= (750 \times 0.22) + (763.636 \times 10^3)(0.33 \times 10^{-3}) \\ &= 165 + 252 \\ &= 417\ \text{At} \end{aligned}$$

Since this value of 417 At is within the $\pm 5\%$ tolerance, it can be stated that the total flux in the core is approximately 1.757×10^{-3} Wb.

8-14 TRACTIVE FORCE

The air gap in a magnetic circuit is similar to magnetic poles of opposite polarity separated by air. Due to opposite poles attracting each other, a force of attraction attempts to close this gap. The magnitude of this force can be calculated in terms of the flux density and the cross-sectional area of the air gap. By increasing the length of the air gap, the total reluctance of the magnetic circuit increases. To maintain the same amount of flux in the circuit, more current must be added. By increasing the current, extra energy is added to the magnetic circuit.

When current flows through a coil wound on the core, energy must be supplied to the coil in order to set up the flux in the air gap. Electrical energy is used to maintain this flux, and magnetic energy is stored in the air gap. The amount of stored energy is dependent on the magnetic field strength and the flux density. In equation form,

$$W = \frac{1}{2} BH \tag{8-13}$$

where W = energy stored in the air gap, in joules

B = flux density, in teslas

H = magnetic field strength, in ampere-turns/meter

Therefore, stored energy may also be expressed as

$$W = \frac{\Phi \times H \times l}{2} \tag{8-14}$$

However, $\Phi = B \times A$ and $H = B/\mu_0$. Therefore,

$$W = l\frac{B^2 A}{2\mu_0} \tag{8-15}$$

In the event of two magnetic surfaces being separated by a short distance, the mechanical energy involved in pulling them apart is determined by

$$W = F \times l \tag{8-16}$$

where W = work done, in joules

F = amount of mechanical force, in newtons

l = distance moved, in meters

Therefore, $F \times l = l \dfrac{B^2A}{2\mu_0}$

In this case, the two lengths both represent the thickness of the air gap, and they cancel out, resulting in the force between the surface being found by the equation

$$F = \frac{B^2A}{2\mu_0} \tag{8-17}$$

where F = force acting close to the air gap, in newtons

B = flux density, in teslas

A = cross-sectional area of the air gap, in square meters

μ_0 = space permeability

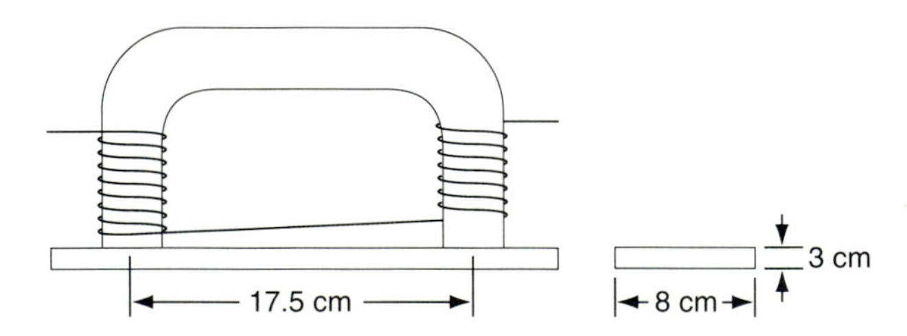

FIGURE 8-10

EXAMPLE 8-5 The electromagnet shown in Figure 8-10 has two pole pieces, each having a cross-sectional area of 24 cm². The total flux crossing each pole is 300 μWb. Calculate the maximum weight of a wrought-iron bar that can be lifted by this magnet.

Solution

$$B = \frac{\Phi}{A} = \frac{300\,\mu\text{Wb}}{24 \times 10^{-4}\,\text{m}^2} = 0.125\ \text{T}$$

$$A = \text{total cross-sectional area} = 2 \times 24 \times 10^{-4}\ \text{m}^2 = 48 \times 10^{-4}\ \text{m}^2$$

$$F = \frac{B^2A}{2\mu_0} = \frac{(0.125\ \text{T})^2(48 \times 10^{-4}\ \text{m}^2)}{2 \times (4\pi \times 10^{-7})} \quad \text{newtons}$$
$$= 29.842\ \text{N}$$

Since force is the product of mass times acceleration due to gravity,

$$F = m \times a$$

then

$$m = \frac{F}{a} = \frac{29.842\text{ N}}{9.81\text{ m/s}^2} = 3.04\text{ kg}$$

KEY TERMS

Magnetomotive force (mmf)	Residual flux density (residual magnetism)
Reluctance	Coercive force
Magnetic field intensity	Hysteresis
Permeance	Ampère's circuit law
Permeability	Air gap
Space permeability	Fringing
Magnetic saturation	Toroid

PROBLEMS

8-1 A toroidal coil consists of 300 turns and has a current of 200 mA flowing. If the length of the magnetic circuit is 8 cm, calculate the mmf and field intensity of the coil.

8-2 A solenoid has 50 turns and carries a current of 0.75 A. If the cross-sectional area of the solenoid's core is 0.035 m^2 and the flux density is 5.75 T, find the reluctance of the solenoid.

8-3 Determine the reluctance of a coil with an mmf of 72 At and a total magnetic flux of 20×10^{-6} Wb.

8-4 An iron core has a flux density of 2.15 T and a cross-sectional area of 0.13 in.^{2}. The coil around the core has 200 turns and carries a current of 0.35 A. Determine the reluctance.

8-5 If the reluctance of a magnetic path is 225×10^5 At/Wb, what value of mmf would be required for a flux of 120 μWb?

8-6 How much flux is established in the magnetic path of a toroidal coil with 20 turns carrying 2 A if the reluctance of the material is 0.38×10^4 At/Wb?

8-7 To develop a flux of 0.033 Wb, how much current must flow through a coil with 300 turns in a magnetic circuit of 2.7×10^4 At/Wb?

8-8 A solenoid has a magnetomotive force of 150 At, its magnetic path is 0.21 m, and it has a permeability of 4.35×10^{-3} H/m. Find the flux density.

8-9 What value of flux density will be produced in a coil having a permeability of 246×10^{-5} H/m and a field intensity of 760 At/m?

8-10 A toroidal coil consists of 2500 turns and carries a current of 150 mA. The coil has a cross-sectional area of 6.25 cm^2 and the length of the magnetic circuit is 22 cm. The permeability of the material is 275×10^{-6} H/m. Calculate (a) the magnetic field strength; (b) the flux density; (c) the total field flux.

8-11 Calculate the mmf required to produce a total flux of 250 μWb in an air gap 0.0035 m in length and having a cross-sectional area of 0.0017 m^2.

8-12 A magnetic circuit has an air gap 2.5 mm long and 1.5 cm^2 in cross section. If the total flux is 450 μWb, what is the mmf of the air gap?

8-13 A toroidal coil has a core cross-sectional area of 3.5 cm^2 and an average core length of 27.5 cm. If the coil has 300 turns and carries a current of 250 mA, determine (a) the mmf; (b) the magnetic field intensity.

8-14 A solenoid with an air core has an inside diameter of 2.25 cm and a length of 20 cm. The coil consists of 400 turns and carries a current of 5 A. Calculate the total flux within the solenoid.

8-15 A square, symmetrical cast-iron core has outside dimensions of $0.08 \times 0.08 \times 0.0025$ m. The inside dimensions of the core are 0.06×0.06 m. One leg of the core is wrapped with a 200-turn coil which carries 2 A. Determine (a) the field intensity; (b) the flux density; (c) the total flux.

8-16 A magnetic relay has an iron core 30 cm in length and has an air gap 0.25 cm in length. The core has an average cross-sectional area of 1.75 cm^2. The relative permeability of the core is 200 and the coil consists of 400 turns. Determine the amount of current required to produce a flux density of 0.28 T in the air gap.

8-17 A toroidal coil consists of 100 turns and carries a current of 250 mA. The length of the magnetic circuit is 5 cm. Find (a) the mmf; (b) the field intensity; (c) the flux density. Assume a permeability of 0.03.

8-18 A 200-turn toroidal coil with a cast-iron core is intended to have a flux density of 0.4 T. The core has an average diameter of 0.035 m. Determine the current required to produce this value of flux density and also calculate the relative permeability of the material.

8-19 For the circuit shown in Figure 8-11, find the value of current required to develop a magnetic flux of 2000 μWb.

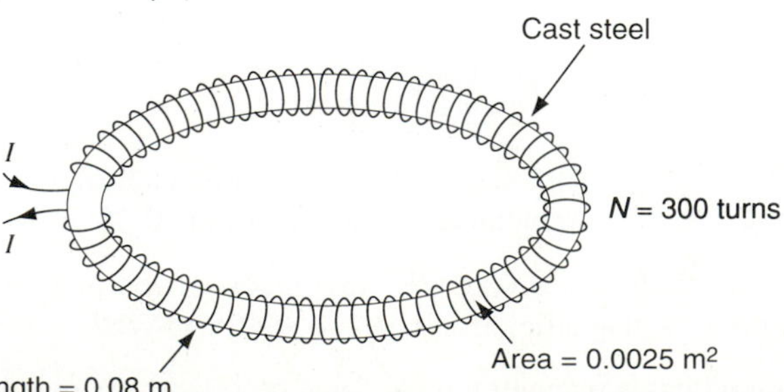

FIGURE 8-11

8-20 Find the relative permeability of the toroidal core shown in Figure 8-11.

8-21 For the circuit shown in Figure 8-12, determine the amount of current required to produce a flux of 100 μWb. Assume that the core is cast iron. Take fringing into account when doing the calculation.

8-22 The core shown in Figure 8-12 is made of cast steel and has a relative permeability of 600 and a core flux of 350 μWb. Determine the current required in the coil.

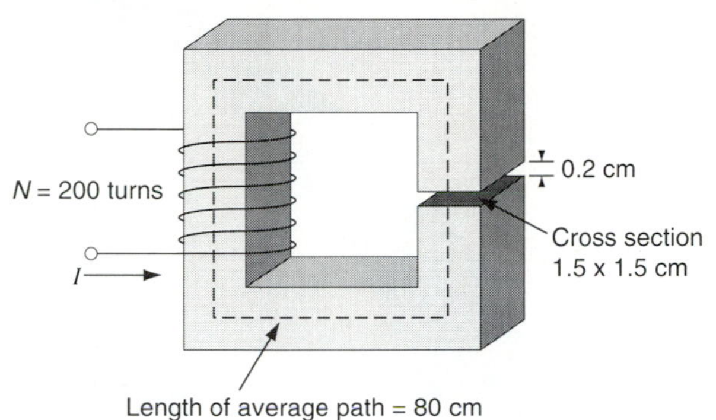

FIGURE 8-12

8-23 The magnetic circuit shown in Figure 8-13 has a cast-iron core with the central path having a cross-sectional area of 4 cm^2. The rest of the core has a cross-sectional area of 2.5 cm^2. The path lengths of the core are as follows: $l_1 = 5$ cm and $l_2 = 9$ cm. The coil consists of 400 turns. What value of current is required to produce a flux of 80 μWb in the center branch?

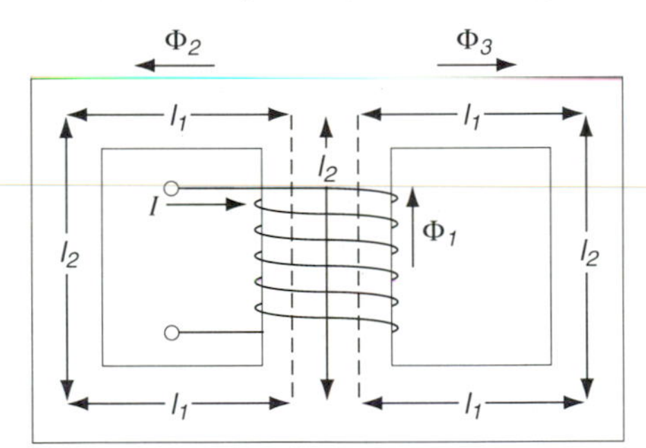

FIGURE 8-13

8-24 Assume that the circuit shown in Figure 8-13 has a 1.5-cm air gap cut into the right outside branch and a 2.5-cm air gap cut into the left outside branch. Using the information about the circuit given in Problem 8-23, calculate the flux density within each gap when a current of 2.5 A flows through the coil.

8-25 An electromagnet has two pole pieces each having a cross-sectional area of 15 cm^2. The total flux crossing each pole is 800 μWb. Determine the maximum weight that can be lifted by this magnet.

8-26 An electromagnetic relay has a 4-cm^2 air gap and a force of 20 N is required to close the gap. Determine the flux density required by this relay.

8-27 A U-shaped lifting magnet has a cross-sectional area of 0.7 m^2 for each pole and a total flux of 0.06 Wb. Determine the tractive force of this magnet.

CHAPTER

DC Measuring Instruments

LEARNING OBJECTIVES

Upon completion of this chapter you will be able to:

- Describe the D'Arsonval galvanometer.
- Explain galvanometer damping.
- Describe the difference between a Weston-type instrument and a galvanometer.
- Explain the necessity of a shunter resistor in a dc ammeter circuit.
- Describe the effects of ammeter and voltmeter loading.
- Explain the basic operation of a multirange ammeter.
- Discuss the purpose of a multiplier resistor in a dc voltmeter.
- Define voltmeter sensitivity.
- Describe the operating characteristics of the dc wattmeter.
- Understand the operation of the ohmmeter.
- Define a megger and explain the application of this type of instrument.
- Determine unknown values of resistance using the Wheatstone bridge.
- Understand the basic principles of electronic and digital multimeters.

9-1 INTRODUCTION

All moving-coil instruments operate on a principle based on the reaction between two magnetic fields. One of these fields is obtained from a moving coil, while the other is obtained from a permanent magnet. When current is fed through this moving coil, the resulting magnetic field reacts with the magnetic field of the permanent magnet and causes the coil to rotate. This is called the **D'Arsonval principle**.

Dc measuring instruments consist of a coil that translates current flowing through it into a displacement of a pointer, which will indicate the amount of displacement on a **scale**. The greater the amount of current flow through the coil, the stronger the magnetic field produced. The stronger the field produced, the greater the rotation of the coil. The effectiveness of any force to produce rotation is called **torque**. A spiral spring's angular deflection is proportional to the amount of torque applied. According to Hooke's law, the force exerted by the spring is a restoring force that always points toward the origin. If the pointer of a dc meter is restrained by a spiral spring, the movement of the coil and pointer will wind up the spring by a small amount. This action makes possible the use of a linear scale on the dc meter, since the deflection of the spring is proportional to the torque applied.

In some instruments there is no spring control; the moving component is acted on by two opposing electrical torques and it takes up a position where these torques are equal. Such instruments are often referred to as *ratio meters*, because they measure the ratio of two quantities, such as voltage to current in ohmmeters, or kVAR to kW in power-factor meters. However, in dc measuring instruments, the moving-coil meter is the most common.

9-2 D'ARSONVAL GALVANOMETER

The D'Arsonval **galvanometer** is based on the D'Arsonval principle, which was developed by the French physicist Jacques Arsene D'Arsonval in 1881. It is suitable for use in dc/ac voltmeters, dc/ac ammeters, ohmmeters, and the Wheatstone bridge. The D'Arsonval galvanometer is a very sensitive electrical instrument used in detecting and measuring extremely small currents. Figure 9-1 shows the principle of its construction. A coil of very fine copper wire is suspended between the poles of a permanent magnet by means of thin flat ribbons of phosphor bronze. These ribbons provide the conducting path for the current between the circuit under test and the movable coil.

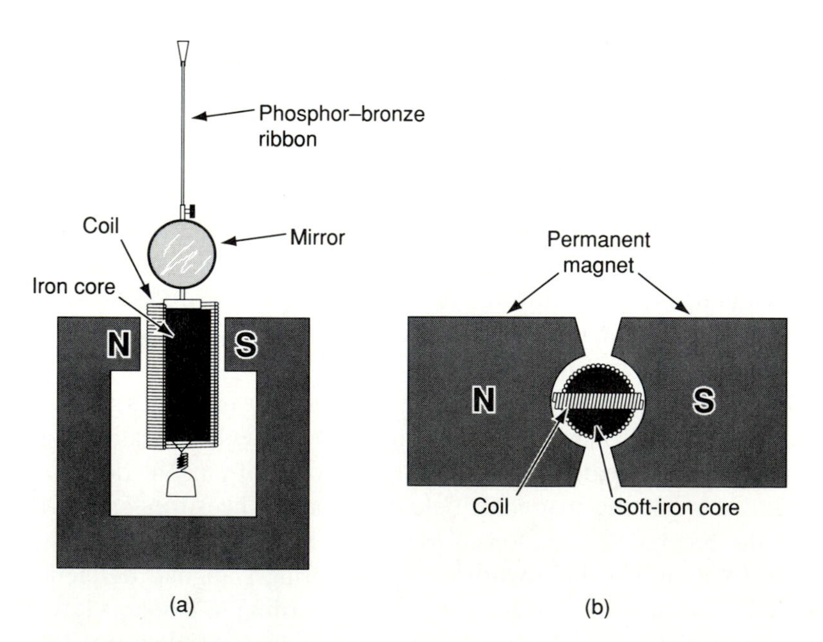

FIGURE 9-1 Essential parts of a D'Arsonval galvanometer: (a) coil suspended in magnetic field; (b) radial magnetic field resulting from cylindrical pole faces and core.

The faces of the magnetic poles are usually cylindrical in shape. A cylindrical core of soft iron is mounted between the poles and inside the moving coil, as indicated in Figure 9-1(b). By reducing the length of the air gap, the core increases the flux linking the coil. Since the core and the pole faces are concentric with each other, the flux crosses the air gap in a radial manner. The deflections of the suspended coil will, therefore, be directly proportional to the current in the coil, and the scale divisions will be uniform. When the suspended coil is deflected by a current, the turning of the coil produces torsion in the suspension ribbon. The phosphor bronze ribbons oppose the turning of the coil and attempt to restore the coil to its normal position after deflection. The torsion is referred to as the *restoring force.*

To determine the amount of current flow, a means must be provided to indicate the amount of coil rotation. Either of two methods may be used: (1) the pointer arrangement, and (2) the light and mirror arrangement. In the pointer arrangement, one end of the pointer is fastened to the rotating coil, and as the coil turns, the pointer also turns. The other end of the pointer moves across a graduated scale, indicating the amount of current flow. The light and mirror arrangement uses a small mirror mounted on the supporting ribbon [Figure 9-1(a)], which turns with the coil. An internal light source is directed to the mirror and then reflected to the scale of the meter. The moving coil turns, and so does the mirror, causing the light reflection to move over the scale of the meter.

9-3 GALVANOMETER DAMPING

If a galvanometer coil that is hung freely [as in Figure 9-(1a)] starts to swing, it will continue to swing, or oscillate, for some time. Any means employed to keep the moving parts from oscillating is referred to as **damping**. Practically all galvanometers are damped electromagnetically. A D'Arsonval type of galvanometer is damped by winding the moving coil on a very light aluminum bobbin. As the bobbin oscillates in the magnetic field, an emf is induced in it because it cuts through the lines of force. Since the induced voltage in the coil is in such a direction as to set up an mmf that tends to oppose the motion, the coil moves the pointer upscale slowly without overshooting its final deflected position. If the coil overshoots its zero position and oscillates before coming to rest, it is said to be *underdamped*. If the coil returns very slowly to its zero position, it is considered to be overdamped.

D'Arsonval galvanometers are made so sensitive that 10 nA will cause a deflection of one scale division. The deflections of the galvanometer are proportional to the current flowing in the coil. Since the resistance of the coil is constant, the voltage must vary as the current. Therefore, the deflections are also proportional to the voltage across the coil.

9-4 WESTON-TYPE INSTRUMENT

The Weston-type instrument utilizes the D'Arsonval principle and is the basis for practically all dc ammeters and voltmeters. Figure 9-2 shows the basic construction of the Weston-type instrument. The moving coil, instead of being supported by a delicate suspension wire, as in a galvanometer, is fastened to hardened steel pivots. These pivots are fitted with great precision into jeweled bearings. This method of supporting the moving coil is almost frictionless and makes the instrument portable, whereas the D'Arsonval galvanometer is not readily portable. Two flat spiral springs provide the path for current into and out of the coil. These springs also act as a controlling device for the coil and oppose any tendency of the coil to turn. The two spiral springs are wound in opposite directions and act against each other. By doing so, these springs negate any changes in temperature, since they will both coil or uncoil accordingly, without the pointer changing from its zero position.

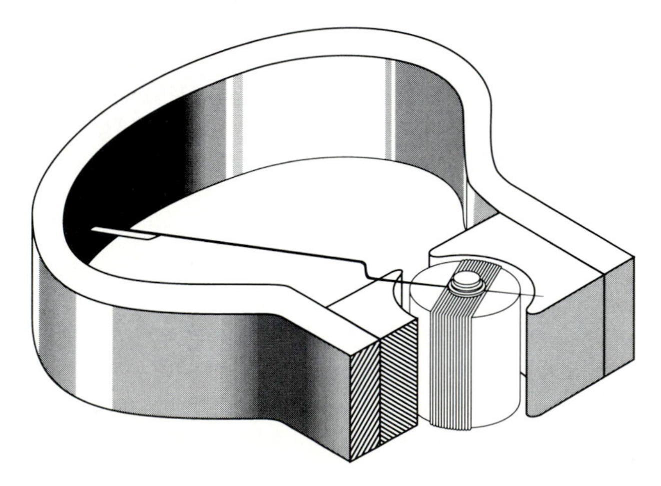

FIGURE 9-2 Magnet and movement of Weston instrument.

The pointer, which is very light and made of aluminum, is balanced by an adjustable counterweight. This helps to reduce the friction of the supportive pivots on the jeweled bearings. Due to this radial magnetic field, the deflection of the moving coil is virtually proportional to the current in the moving coil, allowing the scale divisions of the Weston-type instrument to be uniform throughout. The Weston-type instrument may be used as either an ammeter or as a voltmeter, depending on whether the resistance of the coil is large or small. When used as a voltmeter, the moving coil has a resistance connected in series with it. When used as an ammeter, the coil is provided with a parallel resistor.

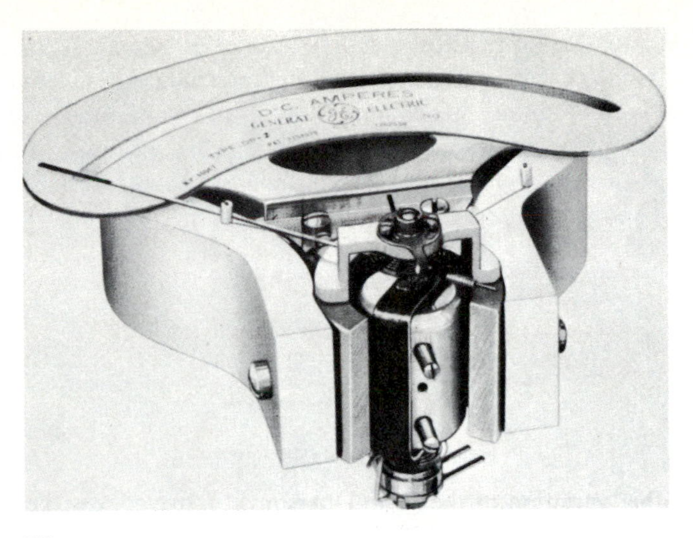

Weston movement

Source: Prentice Hall Inc. / General Electric Co.

9-5 DC AMMETER

An **ammeter** is an instrument used for measuring electric currents. The dc ammeter is almost always a Weston type of instrument. However, the moving coil of a Weston instrument cannot carry currents greater than 30 μA, so to read larger values of current, a resistor is placed in parallel with the moving coil. The current going into the instrument will divide, with part going through the coil and the rest through the resistor. This type of resistor is called a **shunt resistor**.

Direct-current ammeter

Source: Bach-Simpson

The shunt is a low-value resistor, usually made of manganin strips brazed to heavy copper blocks. This method ensures that the shunt has a low-temperature coefficient of resistance. Manganin is an alloy consisting of copper, nickel, and ferromanganese. Instead of having a resistance rating, shunts are rated in terms of current and voltage. When an instrument is made with a built-in shunt, as shown in Figure 9-3, its scale is calibrated to read the total current directly.

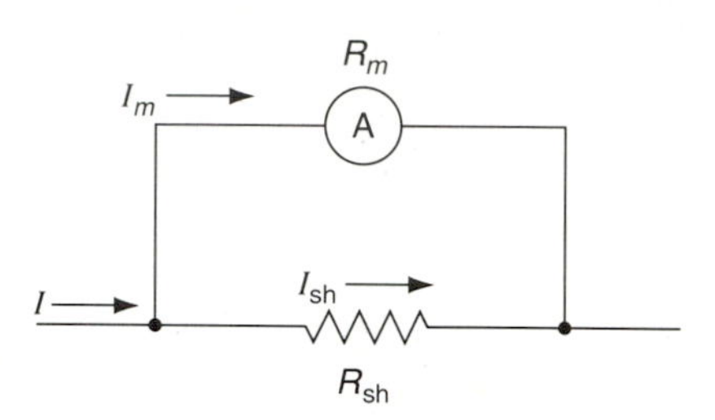

FIGURE 9-3 Current division between ammeter and shunt resistor.

The Weston-type ammeter is actually a voltmeter that measures the voltage drop across the shunt resistor. If the resistance of the shunt is constant, the voltage drop across R_{sh} is proportional to the current in the shunt. Therefore, the ammeter is just a divided circuit consisting of a millivoltmeter and a shunt, where the millivoltmeter measures the voltage drop across the shunt.

Since Figure 9-3 is a divided circuit, the currents in the shunt and in the instrument vary inversely with their resistances. Expressed in equation form,

$$\frac{I_{sh}}{I_m} = \frac{R_m}{R_{sh}} \tag{9-1}$$

where I_{sh} = shunt current

I_m = meter current

R_m = meter resistance

R_{sh} = shunt resistance

The internal resistance of an ammeter is very small. Ideally, the internal resistance would be zero. If the meter resistance were appreciable, its insertion would change the amount of current being read by the meter. This is referred to as the **loading effect** of an ammeter. Ammeter loading occurs mainly in low-voltage, low-resistance circuits.

EXAMPLE 9-1 Assume that the coil resistance is 1.25 Ω, the resistance of the shunt is 7×10^{-4} Ω, and the line current is 70 A. Determine the value of the meter current.

Solution

$$\frac{I_{\text{sh}}}{I_m} = \frac{R_m}{R_{\text{sh}}}$$

Then

$$I_m = \frac{I_{\text{sh}} \times R_{\text{sh}}}{R_m} = \frac{(70\ \text{A})(7 \times 10^{-4}\ \Omega)}{1.25\ \Omega} = 0.0392\ \text{A}$$

EXAMPLE 9-2 What shunt resistance is required to extend the range of a 0- to 10-mA movement having a meter resistance of 15 Ω, to read a total of 100 mA?

Solution

$$\frac{I_{\text{sh}}}{I_m} = \frac{R_m}{R_{\text{sh}}}$$

Then

$$R_{\text{sh}} = \frac{R_m \times I_m}{I_{\text{sh}}} = \frac{(15\ \Omega)(10 \times 10^{-3}\ \text{A})}{90 \times 10^{-3}\ \text{A}} = 1.67\ \Omega$$

A multirange ammeter can be constructed by using several values of shunt resistors, with a rotary switch to select the desired range. The ammeter shown in Figure 9-4 has a switch, Sw, which selects a shunt resistor (R_1, R_2, or R_3)

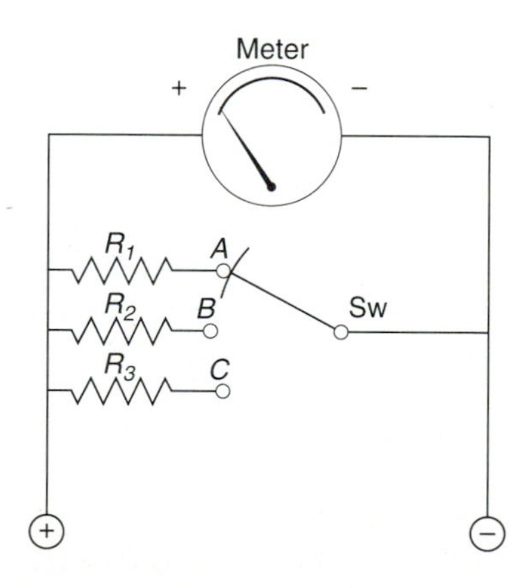

FIGURE 9-4 Multirange ammeter with individual shunts.

to be used across the coil of the meter. This curved typed of switching mechanism is known as a *make-before-break switch*. The wide-ended moving contact connects to the next terminal to which it is being moved before it loses contact with the previous terminal. This is to ensure that there is a shunt across the coil of the meter at all times. Otherwise, an open circuit may occur and the full current would flow through the coil of the meter.

The Ayrton shunt, shown in Figure 9-5, is another method used to protect the ammeter coil from excessive current. When the moving contact of the rotary switch is connected to point A, the resistances of R_2 and R_3 are added to the coil resistance, while R_1 becomes the shunt for the meter. When the moving contact is switched to point B, the shunt becomes the sum of $R_1 + R_2$, and resistor R_3 is now in series with the meter. Finally, with the moving contact at point C, the resistance of the shunt in parallel with the instrument is the sum of all three resistors, $R_1 + R_2 + R_3$. Note that there is a shunt in parallel with the instrument at all times.

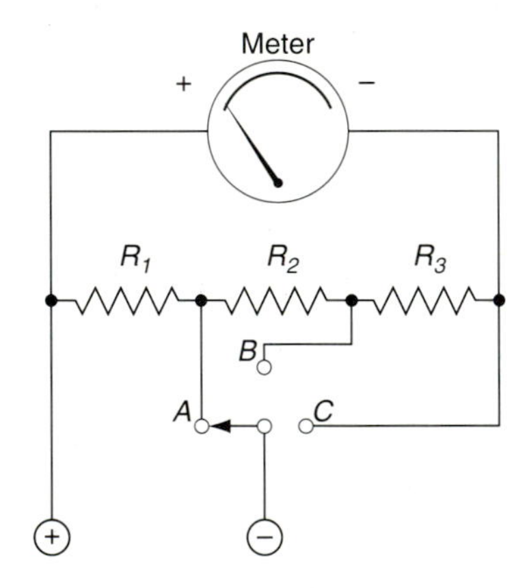

FIGURE 9-5 Ayrton shunt, or universal, ammeter.

EXAMPLE 9-3 An ammeter using an Ayrton shunt, as shown in Figure 9-5, has the following resistance values: $R_1 = 0.025\ \Omega$, $R_2 = 0.050\ \Omega$, and $R_3 = 0.075\ \Omega$. The meter resistance is 70 Ω, and the meter has a full-scale deflection when 150 μA passes through the moving coil. Calculate the full-scale range of the ammeter for the switch at point A.

Solution Switch at point A:

$$\begin{aligned} V_{\text{sh}} &= I_m \times (R_m + R_2 + R_3) \\ &= 150 \times 10^{-6}(70\ \Omega + 0.05\ \Omega + 0.075\ \Omega) \\ &= 10.519\ \text{mV} \end{aligned}$$

$$I_{sh} = \frac{V_{sh}}{R_1} = \frac{10.519 \times 10^{-3}\,\text{V}}{0.025\,\Omega}$$
$$= 420.76 \text{ mA}$$

$$I(\text{full scale}) = I_{sh} + I_m = 420.76 \text{ mA} + 150\,\mu\text{A}$$
$$= 420.91 \text{ mA}$$

EXAMPLE 9-4 For the circuit shown in Figure 9-6, find the shunt resistances for the following ranges: (a) 0 to 10 mA; (b) 0 to 100 mA; (c) 0 to 1 A. Also determine the values of R_1, R_2, and R_3.

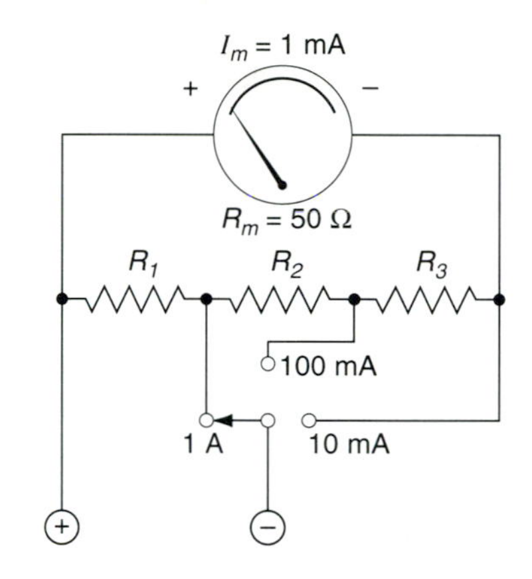

FIGURE 9-6

Solution (a) The shunt resistance for this range consists of $R_1 + R_2 + R_3$.

$$I_{sh} = I_T - I_m = 10 \text{ mA} - 1 \text{ mA} = 9 \text{ mA}$$

By Ohm's law,

$$R_{sh} = \frac{I_m R_m}{I_{sh}} = \frac{(0.001\text{ A})(50\,\Omega)}{0.009\text{ A}}$$
$$= 5.556\,\Omega$$

$$R_1 + R_2 + R_3 = 5.556\,\Omega$$

(b) 0 to 100 mA range consists of $R_1 + R_2$.

$$I_{sh} = I_T - I_m = 100 \text{ mA} - 1 \text{ mA} = 0.99 \text{ mA}$$

$$R_{sh} = \frac{I_m R_m}{I_{sh}} = \frac{(0.001\text{ A})(50\,\Omega)}{0.099\text{ A}}$$
$$= 0.505\,\Omega$$

$$R_1 + R_2 = 0.505\,\Omega$$

(c) 0 to 1 A range consists of only R_1.

$$I_{sh} = I_T - I_m = 1\text{ A} - 1\text{ mA} = 999\text{ mA}$$

$$R_{sh} = R_1 = \frac{I_m R_m}{I_{sh}} = \frac{(0.001\text{ A})(50\,\Omega)}{0.999\text{ A}}$$
$$= 0.05\,\Omega$$

To determine R_1, R_2, and R_3, substitute the results into the equations.

$$R_1 + R_2 = 0.505\,\Omega$$

$$R_1 = 0.05\,\Omega$$

$$R_2 = 0.505 - R_1 = 0.505\,\Omega - 0.05\,\Omega = 0.455\,\Omega$$

$$R_1 + R_2 + R_3 = 5.556\,\Omega$$

$$\therefore \quad R_3 = 5.56 - (R_1 + R_2) = 5.556\,\Omega - 0.505\,\Omega = 5.051\,\Omega$$

9-6 DC VOLTMETER

When the Weston-type instrument is used as a **voltmeter**, it is performing essentially the same function as when it is used as an ammeter. Instead of a shunt resistor, the voltmeter has a high value of resistance placed in series with the moving coil. The size of the series resistor used depends on the current range of the movement and the voltage range desired. The resistance connected in series with the moving coil is called a **multiplier**.

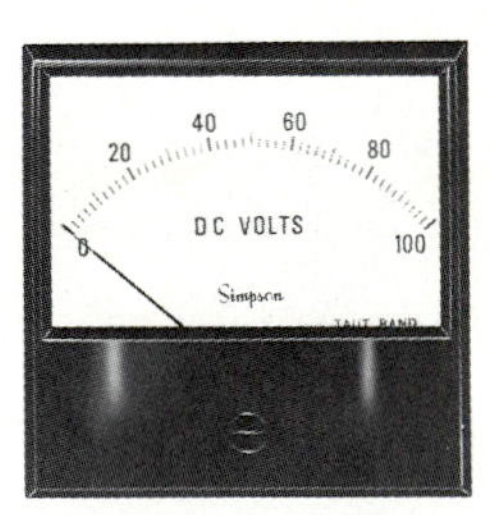

Direct-current voltmeter

Source: Bach-Simpson

Since a voltmeter is placed across, or in parallel with, a circuit, it is desirable that it take as little current as possible. Therefore, the moving coil of a voltmeter is usually wound with more turns and of finer wire than that of the ammeter. Less current is then required to develop the same torque for a full-scale deflection.

A basic voltmeter is shown in Figure 9-7. It consists of a 0- to 2-mA meter movement with a multiplier resistor in series with it. To calculate the size of multiplier for an instrument, simply apply the rules of a series electric circuit.

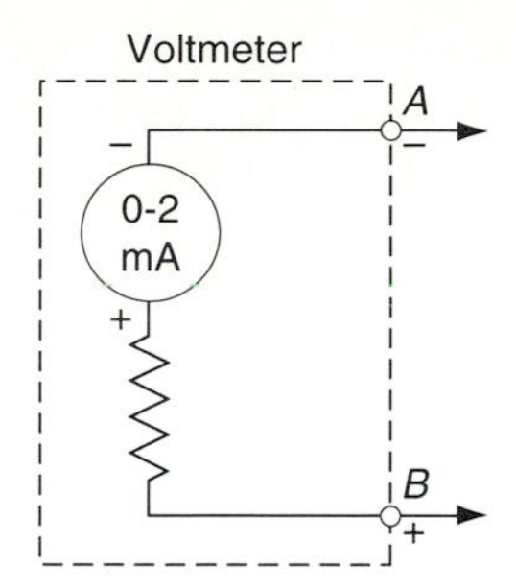

FIGURE 9-7

EXAMPLE 9-5 The 0- to 2-mA instrument in Figure 9-7 has a moving coil resistance of 50 Ω. What size of multiplier is required to convert this to a 100-V voltmeter?

Solution

$$R_T = \frac{E_T}{I_T} = \frac{100\text{ V}}{0.002\text{ A}} = 50{,}000\ \Omega$$

$$R_m(\text{multiplier}) = 50{,}000\ \Omega - 50\ \Omega \approx 50{,}000\ \Omega$$

Usually, the one-meter movement is arranged to read several different maximum voltage values by connecting resistors as shown in Figure 9-8. Unlike the ammeter, the rotary switch used with the voltmeter, as in Figure 9-8(a), should be a break-before-make type, which means that the moving contact should disconnect from one terminal before connecting to the next terminal.

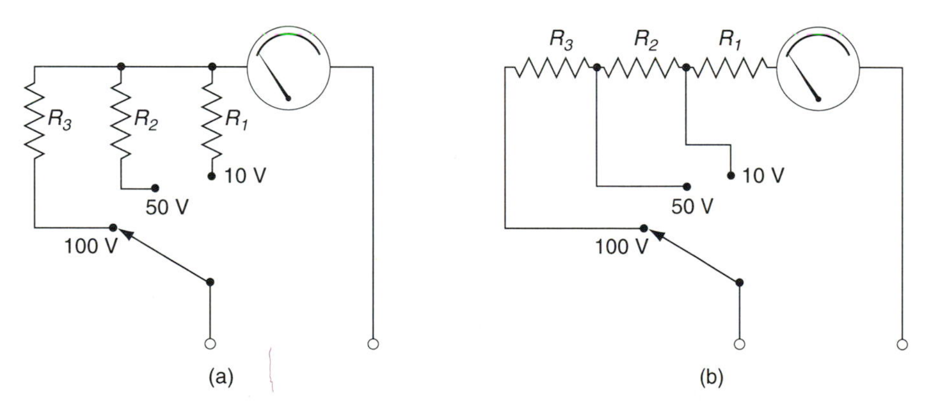

FIGURE 9-8 Switching configuration of voltmeter.

EXAMPLE 9-6 The voltmeter shown in Figure 9-8(a) has a moving-coil resistance of 25 Ω and a full-scale deflection of 2 mA. Determine the required value of multiplier resistor for each range.

Solution For the 10-V range:

$$I = \frac{E}{R_1 + R_m}$$

$$R_1 = \frac{E}{I} - R_m$$

$$= \frac{10\text{ V}}{0.002\text{ A}} - 25\ \Omega$$

$$= 4975\ \Omega$$

$$R_2 = \frac{50\text{ V}}{0.002\text{ A}} - 25\ \Omega$$

$$= 24{,}975\ \Omega$$

$$R_3 = \frac{100\text{ V}}{0.002\text{ A}} - 25\ \Omega$$

$$= 49{,}975\ \Omega$$

9-7 VOLTMETER SENSITIVITY

Since the voltmeter is connected in parallel with the circuit it measures, some of the circuit current will flow through the voltmeter. In order not to unbalance the circuit conditions, the meter current should be very small compared with the original circuit current. When voltage measurements are made in high-resistance circuits, it is necessary to use a high-resistance voltmeter to prevent this shunting action of the meter. The effect is less noticeable in low-resistance circuits because the shunting effect is less.

The sensitivity of a voltmeter is determined by dividing the resistance of the meter plus the multiplier by the full-scale deflection of the meter. Expressed in equation form, we have

$$\text{sensitivity} = \frac{R_m + R_s}{E} \tag{9-2}$$

where R_m = meter resistance

R_s = series resistance (multiplier)

E = full-scale reading, in volts

Voltmeter sensitivity is expressed in ohms/volt. This rating can also be used to determine the current sensitivity of the meter. A voltmeter is considered more sensitive if it draws less current from the circuit. Therefore, the sensitivity of a voltmeter varies inversely with the current required for full-scale deflection. Expressed mathematically,

$$\text{sensitivity} = \frac{1}{I_{FS}} \tag{9-3}$$

where I_{FS} is the current required for full-scale deflection of the meter movement. For example, the sensitivity of a 200-μA movement is the reciprocal of 0.0002 A, or 5000 Ω/V.

EXAMPLE 9-7 What is the sensitivity of a voltmeter having a total internal resistance of 250,000 Ω for a 100-V range?

Solution

$$\text{Meter sensitivity} = \frac{250{,}000}{100} = 2500\ \Omega/\text{V}$$

EXAMPLE 9-8 What is the sensitivity of a voltmeter that produces full-scale deflection when the meter current is 85 μA?

Solution

$$\text{Meter sensitivity} = \frac{1}{85 \times 10^{-6}} = 11{,}765\ \Omega/\text{V}$$

9-8 VOLTMETER LOADING

When measuring voltage, it is necessary to connect the voltmeter across the circuit under test. Since a portion of the circuit current must flow through the voltmeter, the circuit behavior is modified somewhat. By connecting a voltmeter across two points in a highly resistive circuit, the meter acts as a shunt and consequently reduces the equivalent resistance in that portion of the network. This means that the voltmeter will actually produce a lower reading of the voltage drop than what existed before the meter was connected. As we learned earlier, this is called the *loading effect* of a meter.

Often, the loading effect of a voltmeter can be ignored, especially if the meter has a high Ω/V, or sensitivity rating. However, if the meter has a low sensitivity, or the circuit under test has a high resistance, the loading effect of the meter must be taken into account. Consider the circuit shown in Figure 9-9(a). Two 10-kΩ resistors are connected in series to a 9-V source, and a voltmeter is connected across R_2. Since the resistors are of equal value, each resistor will drop one-half of the applied voltage, or 4.5 V. Therefore, we would expect the voltmeter to read 4.5 V if it was connected across either resistor. If the voltmeter has a sensitivity of 1000 Ω/V, on the 0 to 10 V range, it would have an internal resistance of 10,000 Ω. R_m is in parallel with R_2, so the equivalent resistance in that portion of the circuit is 5 kΩ, as shown in Figure 9-9(b). By connecting the voltmeter across R_2, the total resistance of the circuit has now been reduced to 15 kΩ instead of the original 20 kΩ.

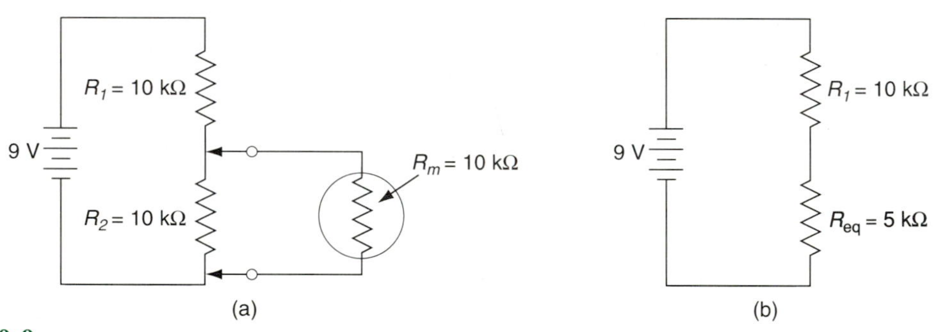

FIGURE 9-9

The voltage drop across $R_2 \| R_m$ is found by the voltage divider rule.

$$V_{R2} = \frac{5\ \text{k}\Omega}{15\ \text{k}\Omega} \times 9\ \text{V} = 3\ \text{V} \qquad \text{(with meter connected)}$$

The percent error of a voltmeter is a ratio of the true voltage minus the voltage read by the meter, divided by the true voltage:

$$\%\ \text{error} = \frac{\text{true voltage} - \text{apparent voltage}}{\text{true voltage}} \times 100$$

The meter used in the example above would have a percent error of

$$\%\ \text{error} = \frac{4.5 - 3}{4.5} \times 100 = 33.33\%$$

By increasing the sensitivity, the loading effect is minimized. For example, if a voltmeter with a sensitivity of 1 MΩ/V were to be used in the circuit of Figure 9-9(a), the internal resistance on a 0 to 10 V scale would be 10 MΩ, as shown in Figure 9-10(a). The equivalent resistance of the 10-kΩ and 10-MΩ resistance is found by the product-over-sum rule.

$$R_{eq} = \frac{R_2 \times R_m}{R_2 + R_m} = \frac{(10 \times 10^3\,\Omega)(10 \times 10^6\,\Omega)}{10 \times 10^3\,\Omega + 10 \times 10^6\,\Omega} = 9990.01\,\Omega$$

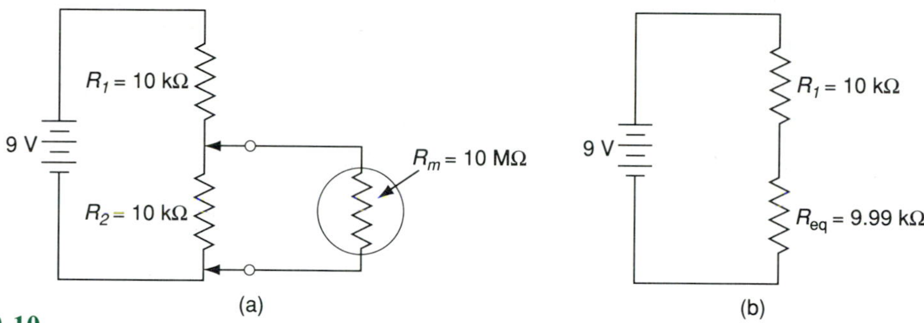

FIGURE 9-10

The equivalent resistance, R_{eq}, shown in Figure 9-10(b) represents the resistance of R_2 and R_m combined. The voltage drop across this resistance is found by the voltage divider rule.

$$V_{R2} = \frac{9990.01\,\Omega}{19{,}990.01\,\Omega} \times 9\text{ V} = 4.498\text{ V}$$

By increasing the sensitivity of the meter, the loading effect has been reduced to an inconsequential value. The percent error is now

$$\frac{4.5\text{ V} - 4.498\text{ V}}{4.5\text{ V}} \times 100 = 0.044\%$$

XAMPLE 9-9 The voltage drop across the 100-kΩ resistor in the circuit of Figure 9-11 is to be measured. The voltmeter that is to be used has a sensitivity of 20,000 Ω/V and is set in the range 0 to 10 V. Determine (a) the reading of the meter; (b) the percent error.

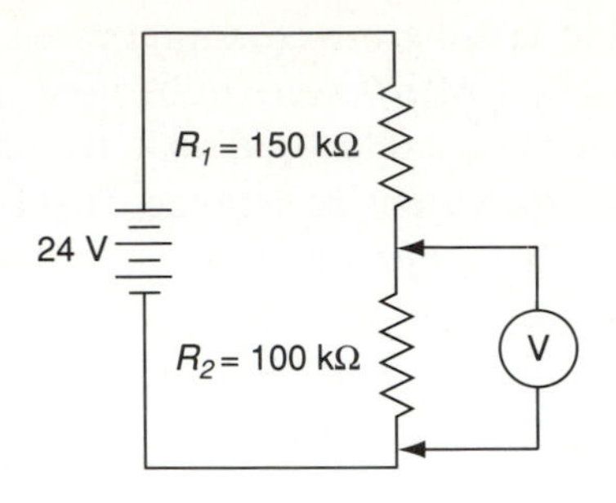

FIGURE 9-11

Solution (a) $R_m = 10 \times 20{,}000 = 200\ \text{k}\Omega$

$$R_{\text{eq}} = \frac{R_m \times R_2}{R_m + R_2} = \frac{(200 \times 10^3\,\Omega)(100 \times 10^3\,\Omega)}{200 \times 10^3\,\Omega + 100 \times 10^3\,\Omega}$$
$$= 66.667\ \text{k}\Omega$$

$$V_{R2} = \frac{R_{\text{eq}}}{\text{R}_{\text{eq}} + R_1} \times 24 = \frac{66.667\,\text{k}\Omega}{66.667\,\text{k}\Omega + 150\,\text{k}\Omega} \times 24$$
$$= 7.385\ \text{V}$$

(b) $$V_{R2}(\text{true voltage}) = \frac{R_2}{R_1 + R_2} \times 24 = \frac{100\,\text{k}\Omega}{250\,\text{k}\Omega} \times 24$$
$$= 9.6\ \text{V}$$

$$\%\ \text{Error} = \frac{\text{true voltage} - \text{measured voltage}}{\text{true voltage}} \times 100$$

$$= \frac{9.6\,\text{V} - 7.385\,\text{V}}{9.6\,\text{V}} \times 100$$
$$= 23.07\%$$

9-9 WATTMETER

Although power can be measured using a voltmeter and ammeter, it is more convenient and accurate to measure power using a **wattmeter**. The basic element of the wattmeter is the electrodynamometer mechanism. An electrodynamometer instrument consists primarily of two coils, one stationary, the other pivoted in the field set up by the stationary coil. The deflecting force is directly proportional to the field established and to the current in the fixed coil.

Figure 9-12 shows a wattmeter with an electrodynamometer mechanism. The pivoted coil, *E*, consists of a few turns of fine wire and is held in position by spiral springs, which also act as lead-in wires, as in the Weston-type voltmeters. The moving coil is connected across the line in series with a multiplier resistor in the same manner that a voltmeter is connected.

The current in the fixed coils is proportional to the current in the circuit being measured. Since the deflecting force is proportional to the product of the voltage and amperage of the circuit, the deflection of the instrument is proportional to the power being measured in the circuit.

The wattmeter is capable of measuring power in both direct-current and alternating-current circuits. However, since a high degree of precision is obtainable by the use of the voltmeter and ammeter, the wattmeter is seldom used for dc measurements. It is also affected by stray magnetic fields, and in certain cases it must be equipped with a magnetic shield. The wattmeter is used extensively for power measurement in ac circuits.

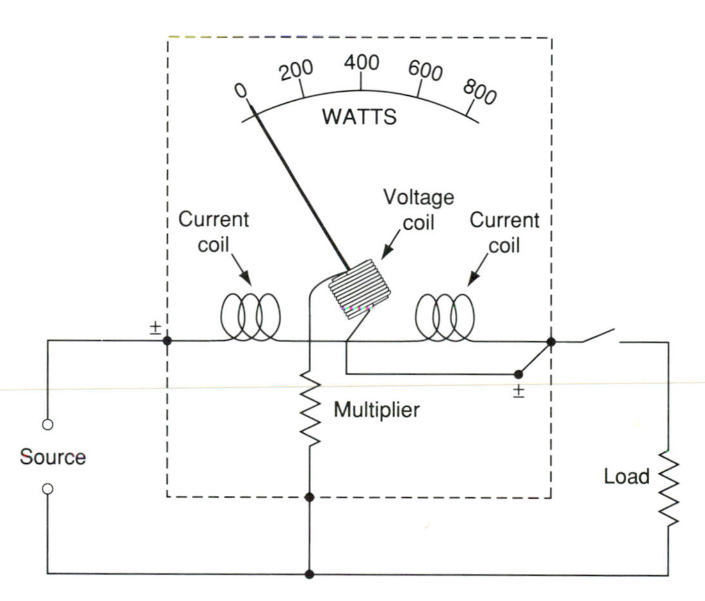

FIGURE 9-12 Electrodynamometer movement as a wattmeter.

9-10 MEASURING RESISTANCE USING AN AMMETER AND VOLTMETER

By using an ammeter to measure the current flowing through a resistor and a voltmeter to measure the voltage across it, as in Figure 9-13, the resistance can be determined by Ohm's law. However, there exists a degree of inaccuracy by using the meters in the configuration shown in Figure 9-13. The voltmeter functions properly since it is connected across R_L and measures the actual load voltage. But the ammeter will measure the load current I_L as well as the current I_V that flows through the voltmeter. As long as the resistance of the load current is small, the current I_V would be small and the percent error would be insignificant. In the case of a large load resistance, the effect of I_V could be quite substantial. Therefore, the circuit connection shown in Figure 9-13 should be used only for determining low resistance values.

One method of detecting the possibility of error in measurement is first to connect the ammeter in series with the load resistance and record the reading of the ammeter. Next, connect the voltmeter across the load resistance and check the ammeter for a change in deflection. If the ammeter is not noticeably altered with the voltmeter connected, the readings can be considered accurate. If the ammeter reading has been changed, the voltmeter should be connected on the opposite side of the ammeter.

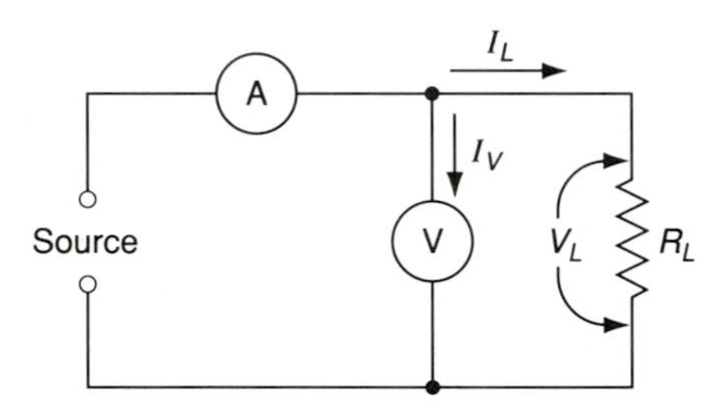

FIGURE 9-13 Resistance measurement by ammeter/voltmeter method.

9-11 OHMMETER

An **ohmmeter** is an instrument that indicates the resistance of a circuit or part of a circuit directly in ohms without any need for calculation. The principle of the ohmmeter is based on supplying a fixed value of low voltage to a circuit of high resistance. A milliammeter is used to measure the current in the circuit. Since the voltage is of a fixed value, the current varies inversely with the resistance, and the scale of the milliammeter is marked in the ohms required to give the corresponding value of current. If the external resistance connected to the meter is zero ohms, the pointer has a full-scale deflection. If the external resistance is infinity, the pointer has no deflection. The pointer will produce a midscale deflection when the external resistance being measured is equal to the internal resistance of the meter. The circuit of a simple series ohmmeter is shown in Figure 9-14. The ohmmeter's pointer deflection is controlled by the amount of battery current passing through the moving coil. When the test

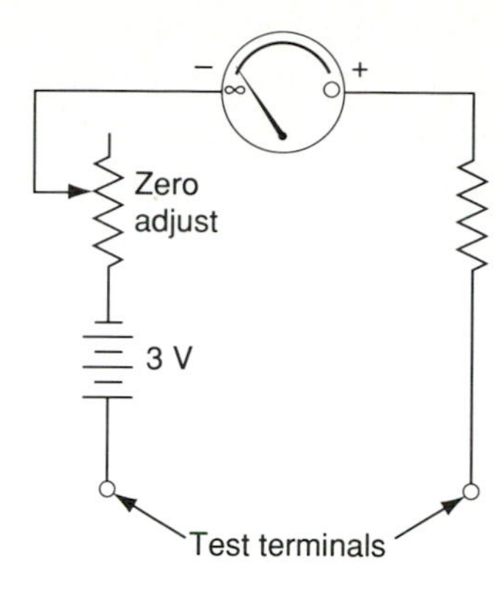

FIGURE 9-14 Simplified schematic diagram of typical ohmmeter circuit.

terminals of the ohmmeter are connected together, current will flow through the circuit. To obtain a full-scale deflection of the meter, the variable resistor is adjusted. This calibrating resistor is provided to compensate for reduction in terminal voltage in the battery due to aging.

After the ohmmeter is calibrated for zero reading, it is ready to be connected in a circuit to measure resistance. An ohmmeter of any type is *never* connected to an energized circuit. The additional voltage introduced by an energized circuit may be enough to establish a current through the moving coil much greater in magnitude than its current sensitivity. This could result in permanent damage to the ohmmeter.

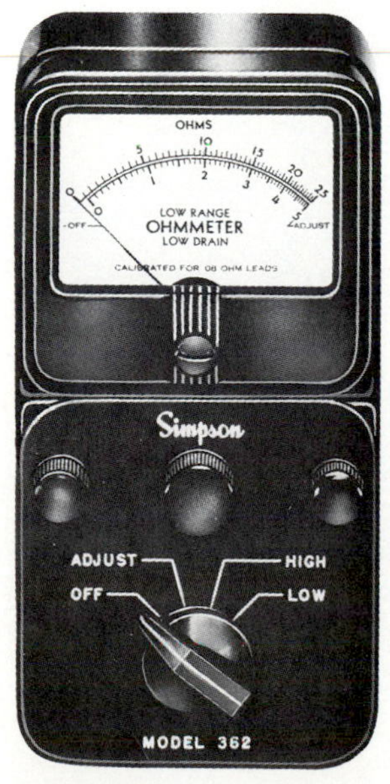

Ohmmeter

Source: Bach-Simpson

The test leads of the ohmmeter are connected across the circuit to be measured. This causes the current produced by the meter's 3-V battery to flow through the circuit being tested. The greater the resistance of the circuit, the smaller the current flow, resulting in less deflection of the pointer. If no resistance is introduced, such as the test terminals short-circuited, the reading will be full scale.

Figure 9-15 shows that the scale of an ohmmeter is not divided into uniform divisions. The scale can be said to be nonlinear due to the unequal division of units. The most accurate readings group around the center of the scale. As the accuracy of an ohmmeter is greatest around the center of the scale, ohmmeters are designed with several different ranges. These devices are referred to as *multirange ohmmeters*. The circuit for a multirange ohm-

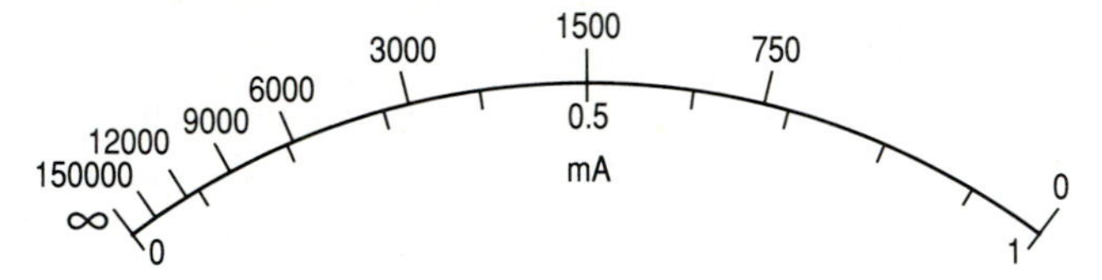

FIGURE 9-15 Ohmmeter with 1-mA movement. Ohms scale reads higher resistances from right to left.

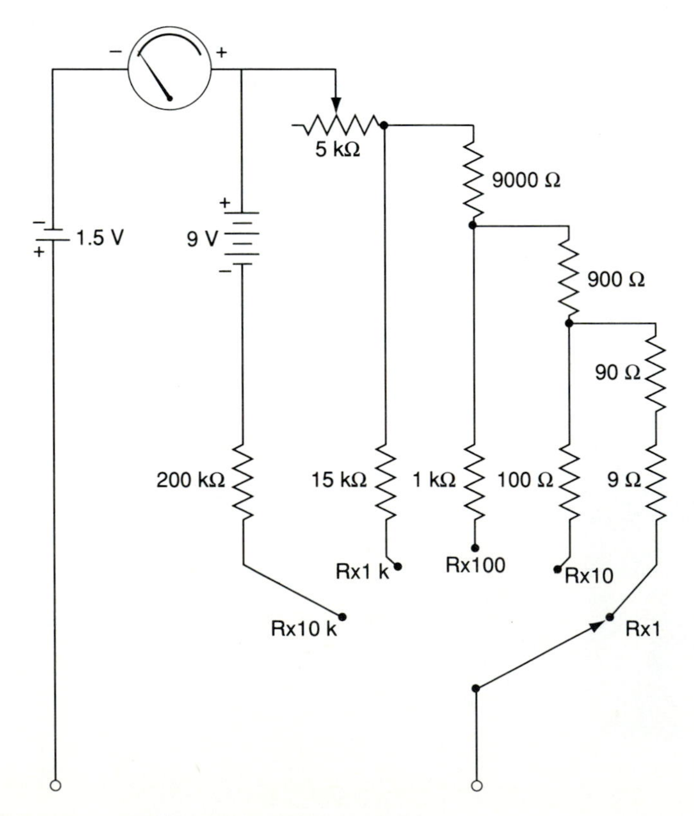

FIGURE 9-16 Multirange ohmmeter circuit.

meter is shown in Figure 9-16. The range switch of this ohmmeter has resistance ranges of $R \times 1$, $R \times 10$, $R \times 100$, $R \times 1000$, and $R \times 10{,}000$. When the range switch is on $R \times 100$, the direct reading on the meter scale is multiplied by 100 for the true reading. For high-resistance readings, such as $R \times 10\text{k}$ on this meter, an additional 9-V battery is added. Generally, the maximum resistance that can be measured by an ohmmeter increases directly with the voltage of the battery used.

9-12 MEGGER

The **megger**, or megohmmeter, is an instrument for measuring very high resistance values. An ordinary ohmmeter cannot be used for measuring the resistance of thousands of megohms, such as conductor insulation. To test adequately for insulation breakdown, it is necessary to use a much higher potential than is supplied by an ohmmeter's battery. This potential is placed between the conductor and the outside surface of the insulation.

The megger, shown in Figure 9-17, is a portable instrument consisting of a hand-driven dc generator, which supplies the required voltage for making the measurement. The generated voltage may be anything from 100 V to 2.5 kV. The instrument mechanism consists primarily of two coils, *A* and *B*, both mounted on the same moving system. The flux is provided by the pole pieces, MM, which are permanent magnets. An iron core, *C*, increases the permeance of the magnetic circuit and produces a radial field. The moving system turns in spring-supported jeweled bearings. The lead-in wires to the coils consist of very flexible conducting filaments having the least possible

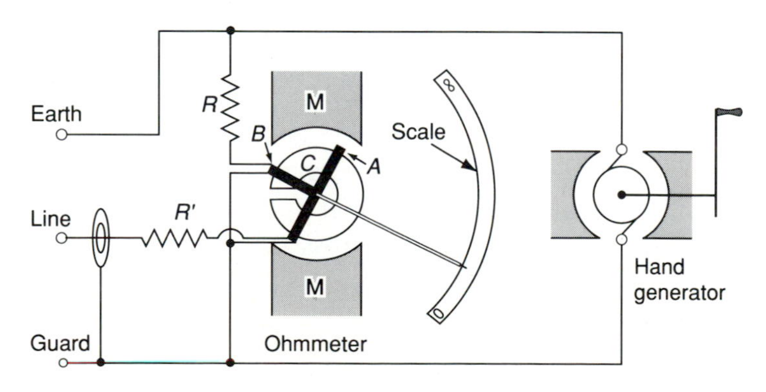

FIGURE 9-17 Megger internal circuit.

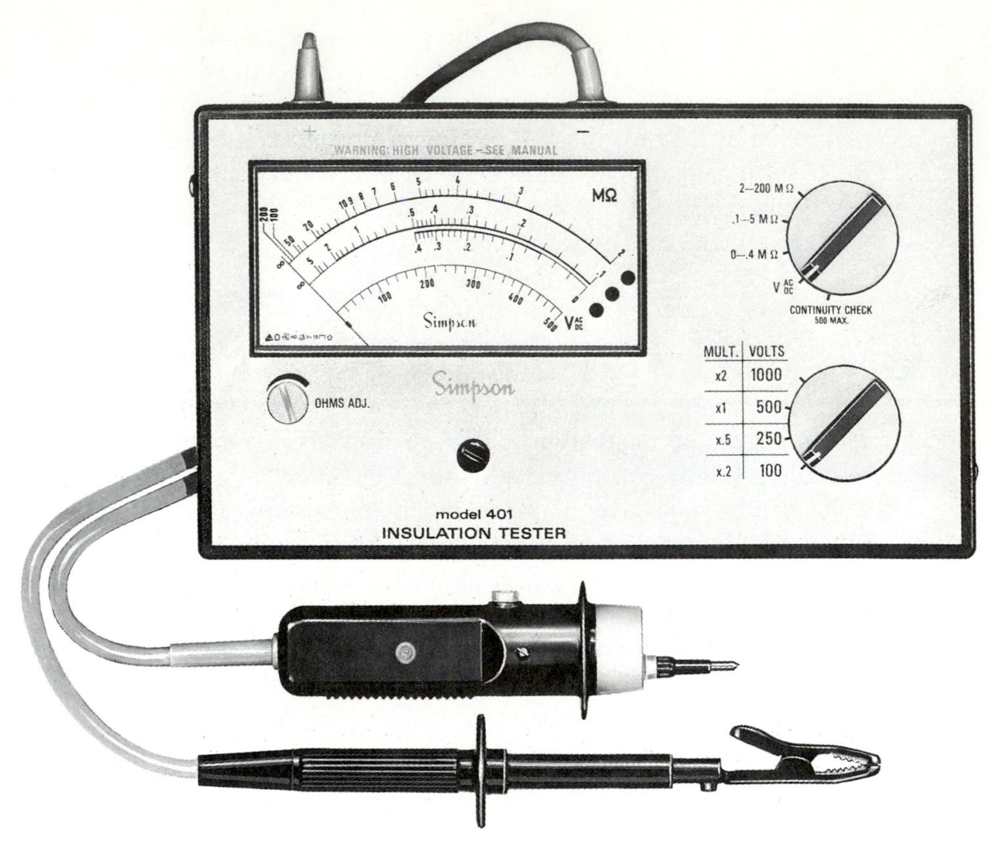

Megger

Source: Bach-Simpson

torsion. Therefore, when there is no current in the coils, the moving element is not restrained in any way, and the pointer ''floats'' over the scale.

Coil *A* of Figure 9-17 is called the *current coil*. It is connected in series with the resistance *R′*, between the negative side of the generator and the line terminal of the megger. When the resistance to be measured is connected between the earth and line terminals, the coil *A*, the resistor *R′*, and the resistance to be measured are all in series across the generator terminals. Any current in coil *A* sets up a clockwise torque, tending to move the pointer to the zero portion of the scale.

Coil B is called the *potential coil* and is connected in series with the resistance R across the generator terminals. Coil B is narrower than coil A and moves so that in some positions it encircles a portion of the C-shaped iron core. A current in coil B sets up a counterclockwise torque. The coil tends to assume a position where the least flux is threading it, which is opposite the gap in the iron core, and where the pointer will indicate infinity.

The torque set up by the current in the coils A and B is in opposite directions. If the line and earth terminals of the instrument are open circuited, no current flows in coil A and the instrument indicates infinity. If a resistance is connected between the earth and line terminals so that some current flows in the circuit of coil A, a torque is set up which tends to move the pointer away from the infinity position and into a field of gradually increasing strength. The pointer will continue to move toward zero until an equilibrium is reached between the torques of coils A and B.

9-13 WHEATSTONE BRIDGE

The **Wheatstone bridge** is an instrument that measures resistance with a much greater degree of accuracy than that obtainable from the typical ohmmeter. The accuracy of an ohmmeter is at best only 5 to 10%, while a Wheatstone bridge will measure resistance values from 0.01 to 10 MΩ with an accuracy below 1%.

The circuit of the Wheatstone bridge is shown in Figure 9-18. Three

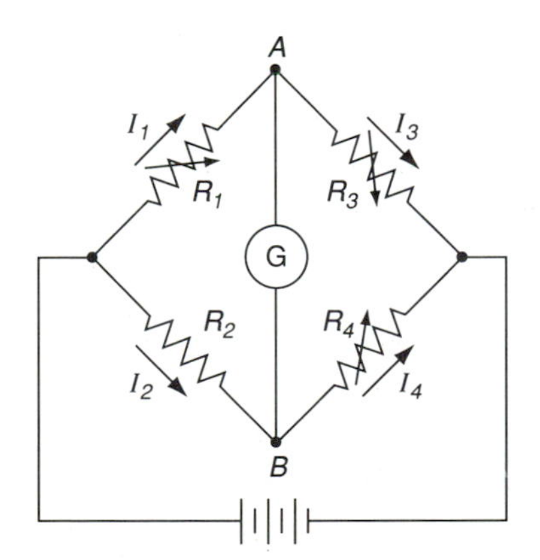

FIGURE 9-18 Wheatstone bridge.

variable resistors R_1, R_3, R_4 and the unknown resistor R_2 are connected in the form of a diamond. Connections are made from a battery to the two opposite corners of the diamond. A galvanometer is connected across the other two corners.

The resistors are adjusted until the galvanometer reads zero. This operation is known as *balancing* the bridge. When the bridge is balanced, there is no current flow through the galvanometer. Therefore, the two points, A and B, must be at the same potential, and the currents $I_1 = I_3$ and $I_2 = I_4$. Since points A and B are at the same potential, the voltage drops across R_1 and R_2 must be equal, so

$$I_1 \times R_1 = I_2 \times R_2 \tag{9-4}$$

Also,

$$I_3 \times R_3 = I_4 \times R_4 \tag{9-5}$$

The equation relating the unknown resistor, R_2, to the known resistors can now be derived by dividing the two equations and remembering current division, which results in

$$R_2 = \frac{R_1 \times R_3}{R_4} \tag{9-6}$$

EXAMPLE 9-10 In the Wheatstone bridge in Figure 9-18, $R_1 = 56\ \text{k}\Omega$ and $R_4 = 100\ \text{k}\Omega$. Determine the value of R_2 if R_3 must be adjusted to 69.66 kΩ to balance the bridge.

Solution

$$R_2 = \frac{R_1 \times R_3}{R_4} = \frac{(56 \times 10^3)(69.66 \times 10^3)}{100 \times 10^3}$$
$$= 39\ \text{k}\Omega$$

9-14 MURRAY LOOP

One type of commercial bridge is known as a **slide wire bridge**. In this mechanism, the ratio arm is a continuously variable slide wire or potentiometer, and its ratio is read on an accurately calibrated dial. The slide wire bridge offers a convenient method for locating grounds in cables and wires. The ground test using a slide wire bridge is known as the **Murray loop** and is demonstrated in Figure 9-19. There is a ground at point F in the conductor

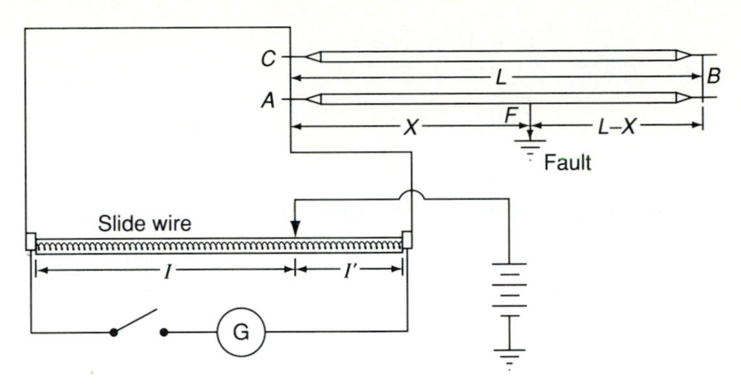

FIGURE 9-19 The Murray loop, a special Wheatstone bridge arrangement for locating a ground in a cable.

AB, caused by a defect in the conductor's insulation. *CB* is the return conductor, which is similar to *AB* except that it has no ground fault. The two conductors are "looped," or connected, at point *B*.

The slide wire is connected between points *A* and *C*, with a galvanometer connected in parallel to the slide wire. A battery is connected to the slide wire in such a manner as to prevent ground, or earth, currents from entering the galvanometer circuit and produce false balances on the meter. The distance *X* to the ground fault may be found using the following equation:

$$X = \frac{2Ll'}{l + l'} \qquad \text{(9-7)}$$

where L = length of the cable, in meters

l = movement of the slide wire to balance the bridge, in centimeters

l' = difference between the total length of the slide wire and the amount of movement required for a balanced bridge, in centimeters

Equation 9-7 assumes that the resistance per meter of both conductors, *AB* and *BC*, are identical.

EXAMPLE 9-11 A 1500-m cable consists of two conductors. At some point in the cable a conductor is grounded. A Murray loop test with a 150-cm slide wire bridge is connected to determine the location of the fault. At 105 cm the bridge is balanced. Find the location of the ground fault from where the slide wire bridge is placed.

$$X = \frac{2Ll'}{l + l'} = \frac{2 \times 1500 \times 45}{105 + 45}$$
$$= 900 \text{ m}$$

Solution

9-15 MULTIMETERS

Since ammeters, ohmmeters, and voltmeters can use the same meter movement, they are often combined into a volt-ohm-milliammeter (VOM). The number of scales used in a particular meter will vary with the ranges of voltages and currents required. The **multimeter** shown in Figure 9-20 has a separate terminal provided for each scale. This method of completing the meter circuit for test requires a large number of terminals, but when using this method there is less chance of meter burnout. By using switches in place of the terminals, much greater convenience is obtained. However, if a switch is set incorrectly at the beginning of the measurement, or if it is carelessly moved without regard for the correct meter scale, the meter may be burned out.

Multimeter

Source: Bach-Simpson

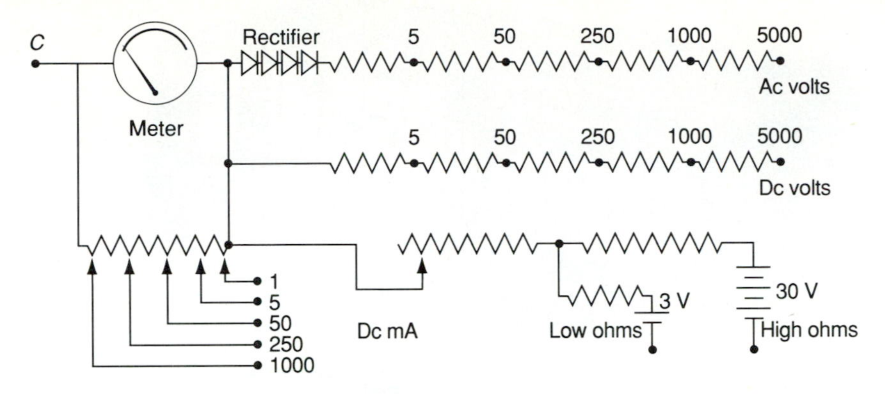

FIGURE 9-20 Multimeter with terminals for various scale readings.

9-16 ELECTRONIC METERS

The electronic meter (EVOM) has several advantages over the passive type of multimeter. By electronically processing the incoming voltage or current, much greater sensitivity with less circuit loading can be obtained. The following features of an electronic multimeter illustrate some basic differences between an EVOM and a VOM:

1. The *zero control* is used to zero the meter electronically on all ranges and functions.
2. Electronic VOMs have an *auto-polarity* function. This allows either test lead to be used on any point in a circuit and the meter will indicate the relative polarity of the voltage at the (+) input of the meter. With this type of meter there is no danger of damaging the device by connecting it backward, or reverse polarity.
3. Many electronic VOMs have a *low-*Ω setting. This setting improves the accuracy of very low ohmic value readings.

9-17 DIGITAL METERS

A **digital meter** is a solid-state electronic instrument that measures electrical quantities and displays the measured value in decimal numeric form. Numerical readout is advantageous in many applications since it reduces human reading error and eliminates **parallax error**. Parallax is caused by looking at a meter from an angle that will cause the pointer to appear left or right of the true position. Since the digital meter has no pointer, parallax is nonexistent.

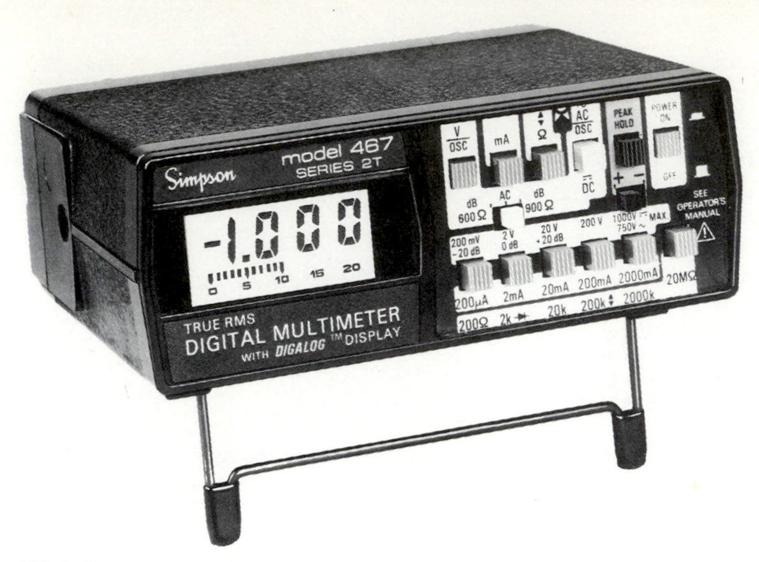

Digital meter

Source: Bach-Simpson

A digital meter is commonly referred to as a digital voltmeter (DVM), even though the meter is also capable of reading current and resistance. A typical DVM will have four digits and a decimal point. By turning a range selector, the decimal point is moved so that the full-scale voltage reading may be selected as 9.999 V, 99.99 V, or 999.9 V. When used as a voltmeter, a DVM has a very high internal resistance (typically, 10 MΩ), which is one reason the percent error of this type of meter is so low. Some DVMs are accurate to within $\pm 0.005\%$ of the true value.

The output of a DVM is indicated by **seven-segment displays**. Each segment is individually controlled so that groups of segments can be formed to represent the decimal numbers 0 through 9. The two types of seven-segment displays used by digital meters are **light-emitting diode** (LED) and **liquid-crystal display** (LCD). The LCD type requires less power but is difficult to read in poor lighting.

A digital meter converts an analog quantity into an equivalent digital readout. The basic functions required to accomplish this conversion are shown in Figure 9-21. The signal processor converts the incoming signal to a quantity that is usable by the **analog-to-digital** (A/D) **converter**. The A/D converter only responds to dc voltages of a limited range. Consequently, the processor may be required to convert alternating current (ac) to dc, or to amplify or attenuate the incoming signal.

The analog-to-digital converter essentially compares an unknown voltage with a reference voltage and indicates which of the two voltages is higher. The output state of the A/D converter is determined by the relative polarity of the

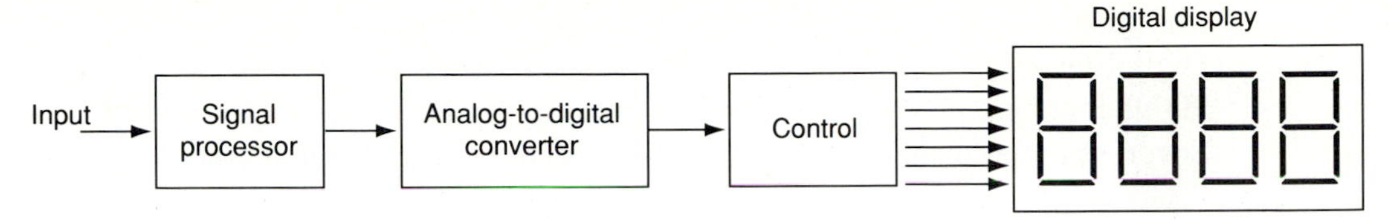

FIGURE 9-21

two input voltages. A digital signal can be in only one of two states, high or low. If input *A* is greater than input *B*, the output is high. If *B* is greater than *A*, the output is low. The A/D converter has a very high ratio of voltage amplification. As a result, an input signal will cause the output either to *saturate* or *cut off* at relatively low differential input levels.

A common A/D conversion technique in digital meters is the **dual-slope integration method**. This method consists of an integrating amplifier, a level detector (comparator circuit), and a pulse amplifier. An integrating amplifier performs the mathematical function of integration and amplifies the input signal. The rate of pulse generation is determined by the magnitude of the dc input voltage. The higher the input signal, the greater the pulse repetition rate (PRR).

An integrating converter measures the time required to charge a capacitor to a given reference voltage. By using dual-slope integration, the capacitor in the integrator is allowed to charge for a fixed period of time. At the end of this fixed charge time, the capacitor is discharged by a constant current source, the time required for discharge is counted, and the count is displayed as an electrical quantity.

The control block shown in Figure 9-21 manages the flow of information within the meter. This block contains latch, decoder, driver, and timing circuits. A latch is a digital circuit that holds data for a specified period of time. A decoder/driver converts digital information into an amplified signal that operates the seven-segment display.

KEY TERMS

D'Arsonval principle
Scale
Torque
Galvanometer
Damping
Ammeter
Shunt resistor
Loading effect
Voltmeter
Multiplier
Wattmeter
Ohmmeter
Megger
Wheatstone bridge
Slide wire bridge
Murray loop

Multimeter
Digital meter
Parallax error
Seven-segment display
Light-emitting diode
Liquid-crystal display
Analog-to-digital converter
Dual-slope integration method

PROBLEMS

9-1 A meter movement with a deflection of 10 mA has a resistance of 35 Ω. Determine the value of shunt required to change the full-scale deflection to 1 A.

9-2 A 0.2-mA meter has a resistance of 50 Ω. What size shunt is required to extend the meter's range to 100 mA?

9-3 An ammeter with an internal resistance of 0.75 Ω is connected in parallel with a 0.0015-Ω shunt. The line current flowing is 50 A. Find the value of meter current.

9-4 Determine the resistance of a shunt to be used with a 12-Ω moving coil in an ammeter that is to have a full-scale reading of 40 A. Without the shunt, the meter has a full-scale deflection of 1 A.

9-5 If the internal resistance of an ammeter is 2.16 Ω and the shunt resistance is 0.004 Ω, what value of meter current flows when the line current is 100 A?

9-6 An ammeter using an Ayerton shunt, as shown in Figure 9-22, has resistance values as follows: $R_1 = 0.017\ \Omega$, $R_2 = 0.06\ \Omega$, and $R_3 = 0.08\ \Omega$. The meter has a full-scale deflection of 100 mA and a moving-coil resistance of 22 Ω. Calculate the full-scale range of the ammeter at point *B*.

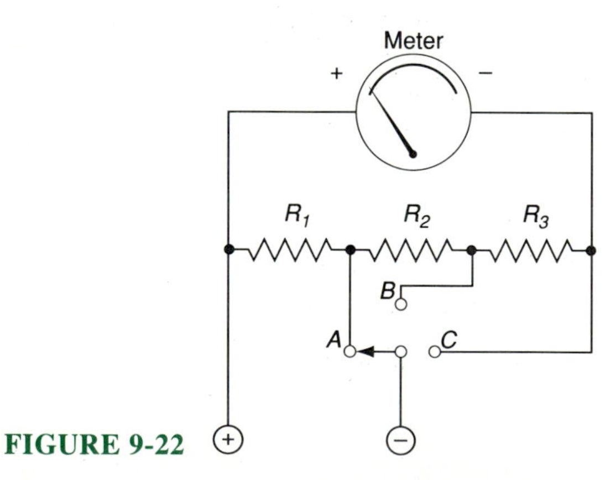

FIGURE 9-22

9-7 Using the values given in Problem 9-6, calculate the full-scale range of the meter shown in Figure 9-22 if the switch is at point *A*.

9-8 A voltmeter consists of a 45-Ω moving coil that carries 100 mA with a full-scale deflection of 300 V. Calculate the size of resistor connected in series with the coil.

9-9 A meter movement has a full-scale deflection of 10 mA and a resistance of 20 Ω. Determine the value of multiplier required to convert the movement to indicate 300 V full scale.

9-10 A meter movement with a full-scale deflection of 2500 μA and a coil resistance of 50 Ω is to be used as a voltmeter with the following ranges: 10 V, 50 V, and 100 V. Determine the value of multiplier resistor for each range.

9-11 Determine the sensitivity of a voltmeter with a total internal resistance of 4250 Ω for a 50-V range.

9-12 If a voltmeter has an internal resistance of 0.85 MΩ when the 100-V range is selected, what is the sensitivity of the meter?

9-13 Determine the sensitivity of a voltmeter that has a full-scale deflection when the meter current is 300 μA.

9-14 Calculate the voltage across a 300-Ω, 800-Ω/V meter movement if the pointer is deflected to the full-scale position.

9-15 A voltmeter is marked 10,000 Ω/V. Find the meter's resistance when used on the 600 V range.

9-16 A 600-Ω, 5-kΩ/V meter movement is deflected to midscale. Find the voltage across the movement.

9-17 Two 10-kΩ resistors are connected in series to a 100-V source. A voltmeter with a 50-V scale and a sensitivity of 1 kΩ/V is to be used to measure the voltage across the resistors. Determine the voltage indicated by the meter when connected across one resistor.

9-18 A 300-V 600-Ω/V voltmeter and an ammeter are connected to measure the ohmic value of a resistor. The voltmeter indicates 110 V and the ammeter reads 180 mA. Determine the true resistance.

9-19 What percentage of accuracy would the meters used in Problem 9-18 have?

9-20 For the circuit shown in Figure 9-23, R_1 = 100 kΩ, R_2 = 50 kΩ, and E = 40 V. The voltmeter has a sensitivity of 10,000 Ω/V and is set on the range 0 to 10 V. Determine (a) the reading of the meter; (b) the percent error.

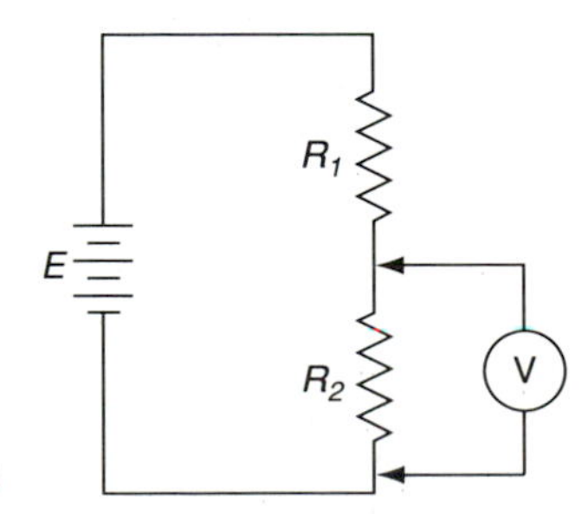

FIGURE 9-23

9-21 The voltmeter shown in Figure 9-23 has a sensitivity of 1000 Ω/V. The applied voltage is 240 V and $R_1 = 200$ kΩ. When the meter is on the 100-V setting, the voltage across R_2 is indicated as being 63.5 V. Determine the ohmic value of R_2.

9-22 Determine the midscale reading of an ohmmeter that has a 1½-V battery and a 100-μA movement.

9-23 In the Wheatstone bridge circuit shown in Figure 9-15, $R_1 = 12$ kΩ, $R_3 = 2$ kΩ, and $R_4 = 28$ kΩ. Find the value of resistor R_2.

9-24 If the Wheatstone bridge circuit of Figure 9-15 has $R_1 = 5$ kΩ, $R_4 = 10$ kΩ, and the galvanometer indicates zero when $R_3 = 1556$ Ω, what value must R_2 equal?

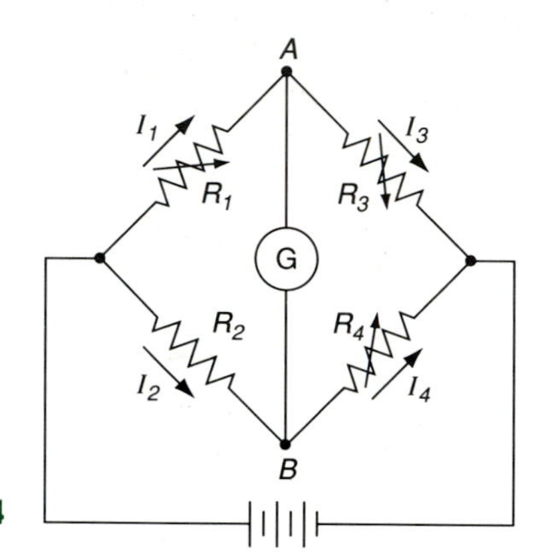

FIGURE 9-24

9-25 A 2400-m cable has two conductors. One conductor is grounded at some point in the cable. A Murray loop test with a 200-cm slide wire bridge is connected, and at 133 cm the bridge is balanced. Determine the location of the ground fault from the point where the slide wire bridge is located.

CHAPTER

Inductance

LEARNING OBJECTIVES

Upon completion of this chapter you will be able to:

- Explain electromagnetic induction and flux linkages.
- List the four basic factors that determine the magnitude of an induced emf.
- Discuss Fleming's right-hand rule and its application in determining the direction of induced emf.
- Explain Lenz's law and the principle of counter emf.
- Define self-inductance and mutual inductance.
- List various types of inductors used in electrical and electronic circuits.
- Discuss the differences between inductors connected in series and in parallel.
- Explain inductive time constants and transients in *RL* circuits.
- Discuss energy stored in a magnetic field.
- Explain the effects of arcing in inductive circuits.
- Define the term *freewheeling diode* and explain its purpose in inductive circuits.

10-1 ELECTROMAGNETIC INDUCTION

In 1831, Michael Faraday discovered that if a conductor is moved through a magnetic field so that it cuts magnetic lines of flux, a voltage will be induced across the conductor. Figure 10-1 shows a moving conductor and a stationary magnetic field. If the speed which the conductor passes through the field is increased, or if the strength of the magnetic field is increased, the amount of induced voltage will also increase.

A magnetic field always exists at right angles to the current flowing in a conductor. Conversely, a changing magnetic field always has an emf associated with it that is at right angles to the direction of the lines of flux. The emf that is associated with a changing magnetic field is referred to as **induced emf**. Induced voltages that are developed by mechanical motion between conductors and magnetic fields are called **generated voltages**.

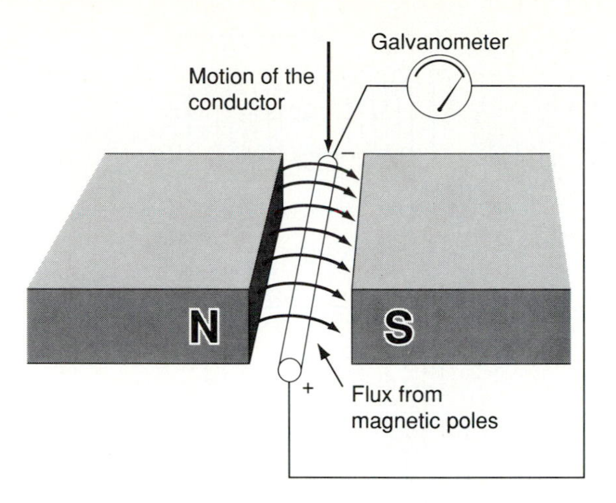

FIGURE 10-1 Conductor moving down through a magnetic field.

Electromagnetic induction is the process by which an electromotive force is induced, or generated, in an electric circuit when there is a change in the magnetic flux linking the circuit. The magnitude of the voltage induced is proportional to the time rate of change of the flux linkage.

10-2 FLUX LINKAGES

Flux linkage exists when the flux and electric circuit are connected like two links of a chain. If a current flows in a conductor, a magnetic flux is set up around the conductor. This magnetic flux completely encircles the flux. The products of the turns of conductor and the number of lines of flux linking these turns are called the linkages of the circuit.

Several examples of flux linkages are shown in Figure 10-2. In Figure 10-2(a), the cross section of a current-carrying coil illustrates magnetic linkage occurring from the flux produced by the current in the coil. In Figure 10-2(b), which shows a portion of a magnetic circuit having two coils, the flux produced by the energized coil links each coil. The linkage of the coil shown in Figure 10-2(c) has its linkage flux produced externally. In Figure 10-2(d) a fixed coil is linked by the flux produced by a U-shaped permanent magnet.

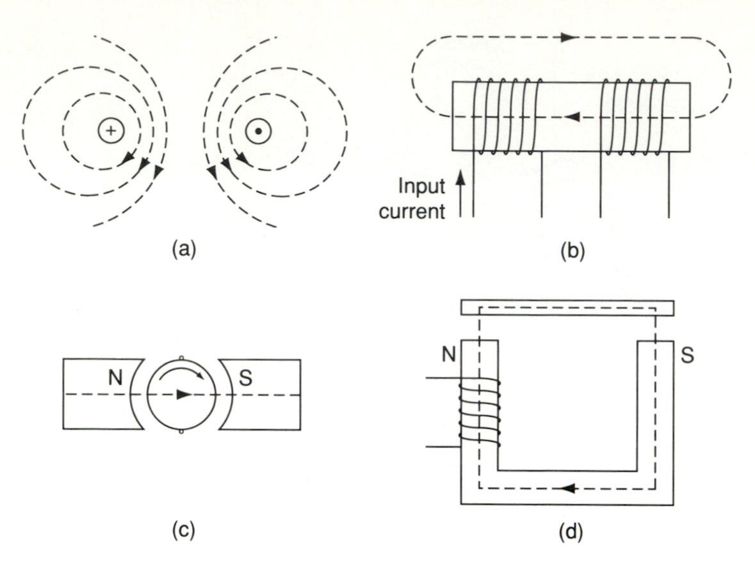

Examples of flux linkages. Dashed lines represent flux direction.

10-3 MAGNITUDE OF INDUCED VOLTAGE

Four basic factors determine the magnitude of an induced emf:

1. Number of turns in the conductor
2. Strength of the magnetic field
3. Relative speed between the coil and the magnetic field
4. Angle at which the conductor passes through the magnetic field

When the magnetic flux linking a coil changes, a voltage proportional to the rate of flux change is induced in the coil. This statement is known as **Faraday's law** and is expressed mathematically as

$$e = N\frac{d\phi}{dt} \tag{10-1}$$

where e = induced emf, in volts

N = number of turns

$d\phi/dt$ = rate of flux change, in webers per second

The induction of voltages in a conductor moving in a magnetic field is fundamental to the operation of all types of generators. Consequently, this principle is known as **generator action**.

EXAMPLE 10-1 A 205-turn coil establishes a flux of 820×10^{-3} Wb. What is the average voltage induced in the coil if the flux changes to 955×10^{-3} Wb in 1.8×10^{-4} s?

Solution The change in flux is 135×10^{-3} Wb over a time interval of 1.8×10^{-4} s.

$$e = N\frac{d\Phi}{dt} = 205 \times \frac{135 \times 10^{-3}\,\text{Wb}}{1.8 \times 10^{-4}\,\text{s}}$$
$$= 153.75\ \text{kV}$$

EXAMPLE 10-2 The flux linking a coil changes from 0 to 2.5×10^{-3} Wb in 0.02 s. If 4.5 V is developed during the flux change, how many turns are there in the coil?

Solution
$$N = e\frac{dt}{d\phi} = 4.5\ \text{V} \times \frac{0.02\ \text{s}}{2.5 \times 10^{-3}\,\text{Wb}}$$
$$= 36\ \text{T}$$

10-4 DIRECTION OF INDUCED EMF

The relation among the direction of motion of the conductor, the direction of the magnetic field, and the direction of the induced emf is expressed by **Fleming's right-hand rule**. This relation is shown in Figure 10-3 and may be stated as follows:

1. Hold the thumb, forefinger, and middle finger of the right hand at right angles to each other.
2. Point the thumb in the direction of motion of the conductor.

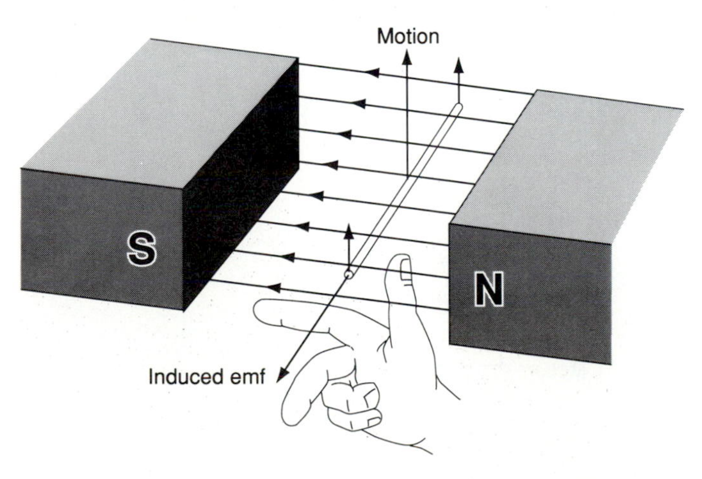

FIGURE 10-3 Fleming's right-hand rule: forefinger along lines of flux, thumb in direction of motion; middle finger shows direction of induced emf.

3. Point the forefinger in the direction of the magnetic field.

4. The middle finger will then point in the direction of the induced voltage.

10-5 LENZ'S LAW

The induced emf in a circuit always tends to oppose any change in the amount of current in that circuit. This is commonly referred to as **Lenz's law**, which may also be stated as follows:

> In all cases of electromagnetic induction, current that flows as a result of an induced emf is in such a direction that the magnetic field established by the current reacts to stop the motion that generates the emf.

This law may be demonstrated by using a bar magnet, coil, and galvanometer, as shown in Figure 10-4. When the north pole of the bar magnet is inserted in the coil [Figure 10-4(a)], an emf is induced in the windings and the galvanometer will momentarily deflect. Since the electric circuit is completed by connecting the galvanometer between the terminals of the coil, current will flow in the circuit.

The direction of this current will establish an electromagnetic field with its north pole adjacent to the north pole of the bar magnet. Therefore, the direction of the electromagnetic field will oppose the north pole of the bar magnet entering the coil. If the magnet is withdrawn from the coil, as in Figure 10-4(b), the current in the coil reverses. The galvanometer will once again deflect, but the deflection is now opposite to that of Figure 10-4(a). The

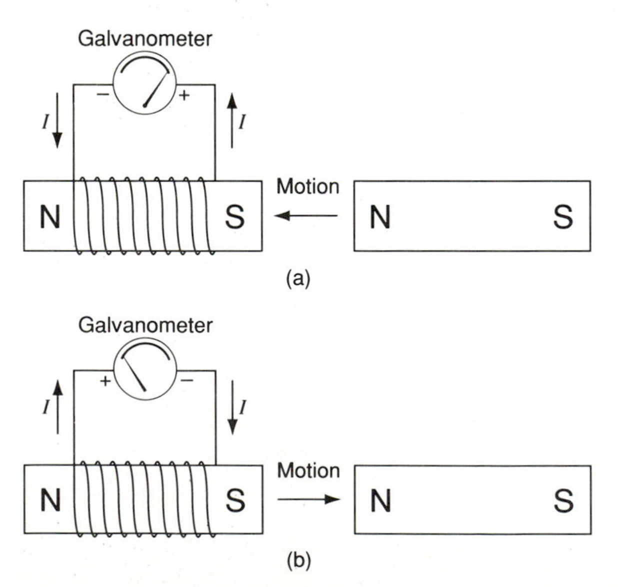

FIGURE 10-4 Induced electromotive force: (a) north pole inserted in coil; (b) north pole withdrawn.

current in the coil has now reversed and the electromagnetic field is in the same direction as the field of the bar magnet. Since a north pole tends to move in the direction of the lines of force, this mmf would oppose the withdrawal of the bar magnet from the coil. The emf in each case is momentary and will cease when the change of flux linking the coil stops.

In Figure 10-5, a coil is connected to a battery with an open switch. If the switch is closed, current will flow in the coil. As the current increases in the coil, the flux increases, which increases the flux linking the coil. This action will induce an emf whose value is determined by the rate that the flux increases and by the number of turns in the coil. The induced voltage, often referred to as **counter emf** (CEMF), opposes the increase in current and retards it, thereby preventing the current from reaching its maximum value. This emf is said to be self-induced since it is induced in the coil which the current changes. **Self-induction** may therefore be defined as

> the emf that is generated in an electric circuit by a change of current within the circuit itself.

Circuits consisting of coils in which strong magnetic fields are set up have high self-inductance. Such circuits are generally constructed by winding a great many turns of wire upon a continuous magnetic field core of high permeability.

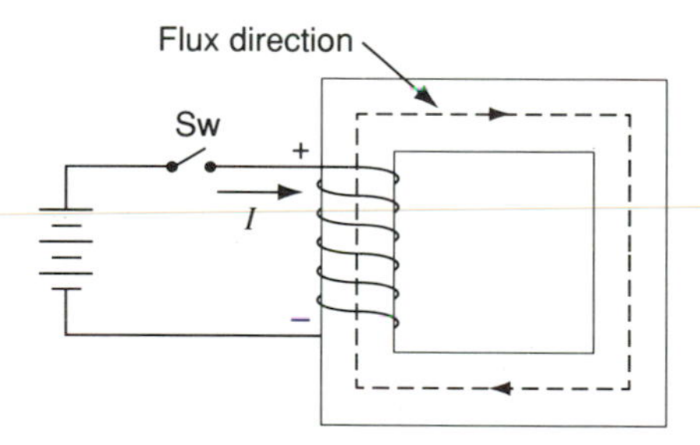

FIGURE 10-5 Principle of counter emf. Voltage induced in coil tends to oppose supply voltage.

10-6 SELF-INDUCTANCE

Inductance is defined as the property of a circuit that opposes any change in the amount of current in that circuit. The ability of an inductor to induce a voltage into itself is referred to as self-inductance. The unit of inductance is the **henry**. A circuit has a self-inductance of 1 henry when a current changing at the rate of 1 ampere per second induces an average of 1 volt. The symbol used to represent inductance is the capital letter L. Inductance may be expressed using the equation

$$L = \frac{eL}{di/dt} \tag{10-2}$$

where L = inductance of a circuit, in henrys

eL = induced voltage, in volts

di/dt = instantaneous rate of change of current, in amperes per second

EXAMPLE 10-3 The current flowing through a coil rises from 0 to 9 A in 6 s. If 11 V is produced across the device, what is its inductance?

Solution

$$L = \frac{eL}{di/dt} = \frac{11\text{ V}}{9\text{ A}/6\text{ s}}$$
$$= 7.33\text{ H}$$

When the length of a coil is much greater than its diameter and the permeability of the magnetic path is constant, the inductance of the coil can be found using the following equation:

$$L = \frac{\mu N^2 A}{l} \tag{10-3}$$

where μ = permeability of magnetic path

N = total turns of coil

A = area, in square meters

l = length of coil, in meters

EXAMPLE 10-4 1250 turns of wire are wound around a toroidal cast-steel ring. The ring has a cross-sectional area of 1.5 cm^2 and an average path length of 17.5 cm. The relative permeability at the rated current of the coil is 1000. Determine the inductance of the coil.

Solution Permeability of free space, $\mu_0 = 4\pi \times 10^{-7}$ H/m. Then

$$\mu = \mu_r \mu_0 = (1000)(4\pi \times 10^{-7})$$

$$L = \frac{\mu N^2 A}{l} = \frac{(1 \times 10^3)(4\pi \times 10^{-7})(1250)^2(0.015)^2}{0.175}$$
$$= 2.52\text{ H}$$

10-7 MUTUAL INDUCTANCE

When two windings are placed so that a change of current in one will cause its changing magnetic field to cut the turns of the other, an induced emf will be set up in the second coil. The two circuits are then said to possess **mutual inductance**. An example of mutual inductance is shown in Figure 10-6. When the switch in series with the dc supply and coil is closed, a magnetic field is established in the iron core. This magnetic field links the iron core with the secondary winding. During the time the flux is increasing from a minimum to a maximum value, an emf is induced in the secondary coil. This emf is referred to as an emf of mutual induction. When the switch is closed, current will flow in the secondary winding in a direction which will establish an mmf that opposes the increase in flux.

When the switch in the primary circuit is opened, the magnetic field will collapse. During this time interval, when the flux is decreasing from a maximum to a minimum value, an emf will also be developed in the secondary coil. This emf will cause current to flow in the direction illustrated by Figure 10-6(b).

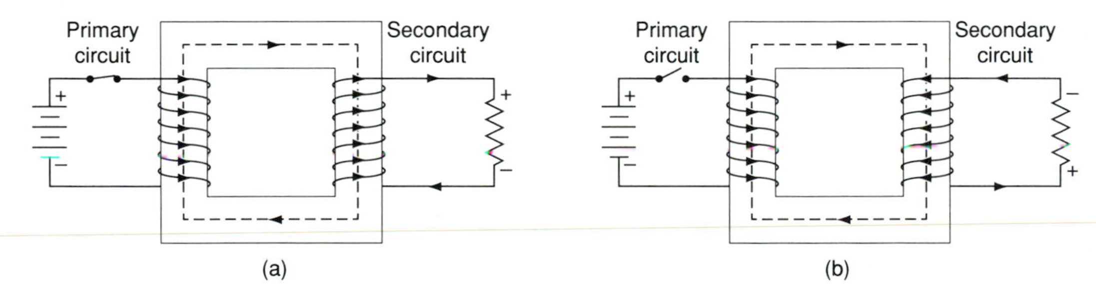

FIGURE 10-6 Direction of current in secondary winding as a result of primary current increasing and decreasing: (a) magnetic field increasing after switch is closed; (b) magnetic field decreasing after switch is opened.

The mutual inductance of two adjacent coils is dependent on such factors as the permeability of the core material, the number of turns on each coil, the physical dimensions of the coils, and the coefficient of coupling. The coefficient of coupling, k, is defined as the degree of coupling that exists between two magnetic circuits. When maximum magnetic coupling exists between two coils, virtually all the magnetic flux created by one coil passes through the other. Under this condition the coefficient of coupling is said to be 1, or unity. This degree of closeness of the coils is known as **tight coupling**.

Theoretically, the coupling is never perfect. If two coils are wound on a common closed iron core, the coefficient of coupling can be considered as unity. However, if two coils are wound on separate iron cores, the coefficient of coupling is dependent on the distance between coils and the angle between the axes of the coils. **Loose coupling** produces a small value of coupling. For example, if 2% of the lines of force from one coil were cutting the second coil, the coefficient of coupling would be 0.02. In many electronic circuits, two coils are wound on an air-core plastic form. The coefficient of coupling is then quite low (0.01 and 0.1) since the relative permeability of the coil is unity.

Figure 10-7 shows two coils on separate iron cores. Coil *B* is placed at right angle to coil *A*. The coils are centered so that equal parts of ΦA link equal numbers of turns on coil *B* but in opposite directions. The emf induced in one part of coil *B* would be equal and opposite that induced in the other part, and the net emf induced in coil *B* would be zero.

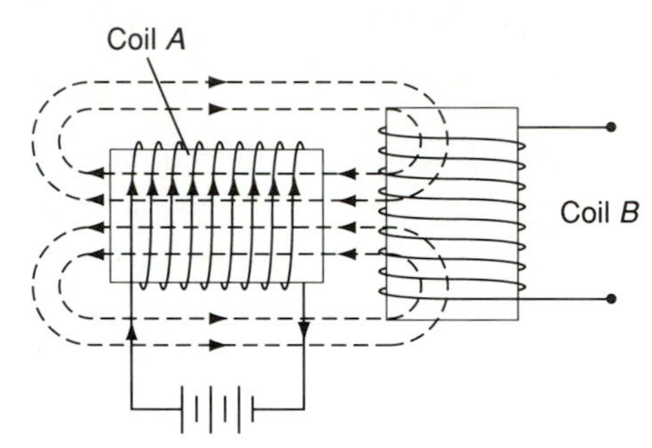

FIGURE 10-7 Effect of degree of coupling.

The mutual inductance between two coils, L_1 and L_2, may be expressed in terms of the inductance of each coil and the coefficient of coupling, k, as follows:

$$M = k\sqrt{L_1L_2} \tag{10-4}$$

where M = mutual inductance, in henrys

k = coefficient of coupling

L_1 = inductance of first coil, in henrys

L_2 = inductance of second coil, in henrys

EXAMPLE 10-5 Two coils have inductances of 7.5 H and 4.9 H, respectively. If they are coupled with a coefficient of coupling of 0.95, what is the mutual inductance?

Solution

$$M = k\sqrt{L_1L_2} = 0.95\sqrt{7.5 \times 4.9}$$
$$= 5.76 \text{ H}$$

Since mutual inductance is a measurement of the ability of a changing current in one coil to induce an emf in the other coil, mutual inductance may also be expressed as a ratio of the secondary turns multiplied by the flux that passes through it, divided by the primary current. In equation form,

$$M = \frac{N_s \Phi_s}{I_p} \tag{10-5}$$

where M = mutual inductance, in henrys

N_s = secondary turns

Φ_s = secondary flux, in webers

I_p = primary current, in amperes

EXAMPLE 10-6 A 300-turn coil and a 500-turn coil have a resistance of 80 Ω and 110 Ω, respectively. The coefficient of coupling is 0.82. The application of 150 V dc to the 300-turn coil produces a primary flux of 1.0 Wb. Calculate (a) the flux linking the secondary; (b) the mutual inductance.

Solution (a) $\Phi_s = k\Phi_p = 0.82 \times 1.0 = 0.82 \text{ Wb}$

(b) $I_p = \frac{V_p}{R_p} = \frac{150 \text{ V}}{80 \, \Omega} = 1.875 \text{ A}$

$$M = \frac{N_s \Phi_s}{I_p} = \frac{500 \times 0.82 \text{ Wb}}{1.875 \text{ A}} = 218.67 \text{ H}$$

10-8 TYPES OF INDUCTORS

A device that introduces inductance into an electric circuit is called an **inductor**. Basically, all inductors are made by winding a length of conductor around a core. The two main categories of inductors are air-core and iron-core. **Air-core inductors** contain no magnetic iron. They are usually wound on some tubular insulating material. This material may be made of treated cardboard, fiberglass, Bakelite, hard rubber, or plastic. The straight spiral coils of air-core inductors are said to be spiral wound. When a high-quality insulator is used, the losses of the coil are reduced. Large air-core inductors with high power capabilities do not require an insulated core. These coils are wound with bare copper wire of large diameter and are sufficiently rigid to be self-supporting. The schematic symbol for an air-core inductor is illustrated in Figure 10-8.

FIGURE 10-8 Air-core inductor.

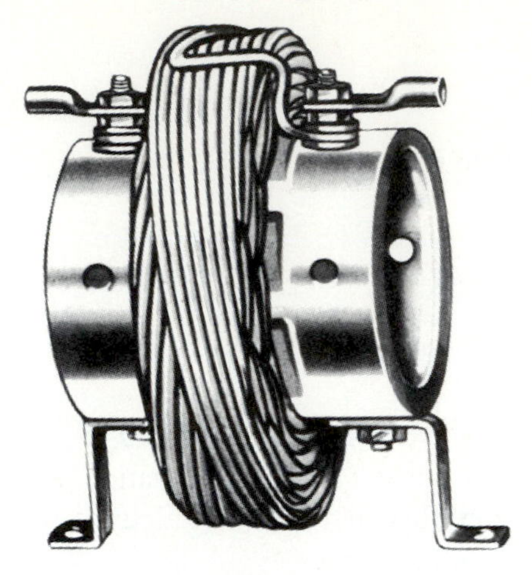

Air-core inductor

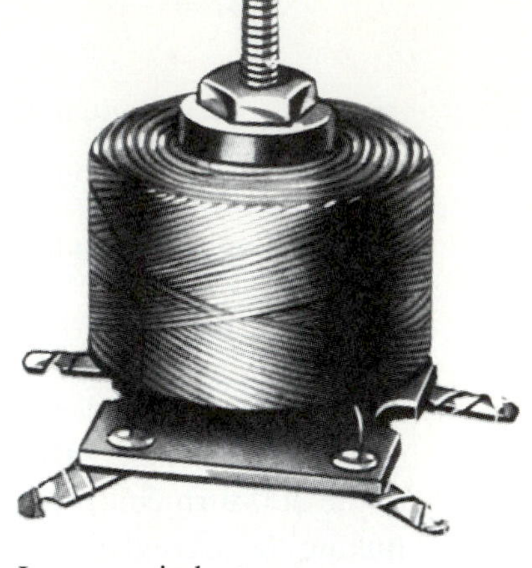

Iron-core inductor

Source: Prentice Hall Inc. / J.W. Miller Co.

Iron-core inductors have cores made of various alloys of iron and other materials to give the required characteristics for a particular application. Iron-core inductors contain all the losses associated with the iron itself.

The main cause of losses in an iron-core inductor are **eddy currents**. An eddy-current loss takes place in cores of magnetic material that are subjected to cycles of magnetization. The moving magnetic field induces an emf in the core, which in turn causes an electric current to flow. To reduce the magnitude of these currents and thereby minimize eddy-current loss, magnetic cores are constructed out of thin laminations. The surfaces of these laminations are treated so that high electrical resistance is offered to the flow of eddy currents from one lamination to another. The schematic symbol for an iron-core inductor is illustrated in Figure 10-9.

Hysteresis losses also occur in an iron-core inductor. Hysteresis losses take place in any magnetic material in which the magnetic field continually reverses direction. When the current in the inductor increases, a completely demagnetized iron core becomes magnetized in a nonlinear manner. If the inductor current is decreased to zero, the magnetization does not return to zero. Some residual magnetism is left in the core.

Both eddy-current losses and hysteresis losses are considered to be **heat losses**. Hysteresis loss is a heat loss caused by the alternating movement of magnetic domains in the iron, and eddy-current loss is a heat loss caused by an alternating current induced in the iron. Both losses are generated by the alternating magnetic field of the iron-core inductor.

FIGURE 10-9 Iron-core inductor.

Inductors can also be classified as fixed or variable. Fixed inductors are designed with set values of inductance that cannot be changed. Variable inductors have inductance that can be altered by using either a tapped coil, movable core, or slider.

Tapped coils are manufactured by placing taps at certain points in the winding. The value of inductance can be changed by using one fixed end of the winding and selecting any one of the taps.

Movable core inductors are made so that the core can be moved into and out of the winding. As the length of the air gap increases, the value of reluctance also increases. Circuits that use movable core inductors are often referred to as *permeability-tuned circuits*.

The slider inductor operates on the same principle as the slide wire bridge. The sliding contact varies the amount of inductance introduced into the circuit. The three types of variable inductors are shown in Figure 10-10.

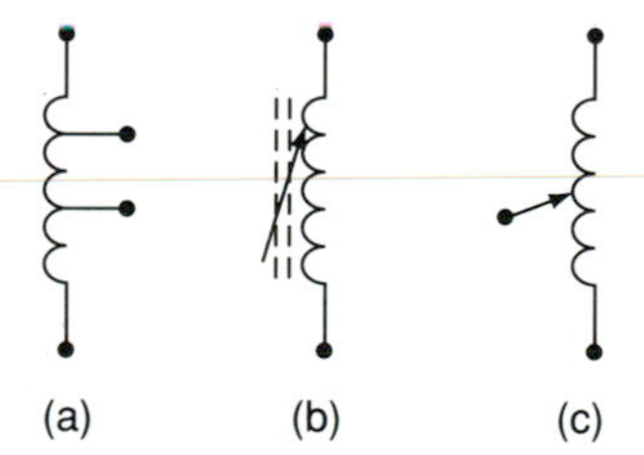

FIGURE 10-10 Inductor symbols: (a) tapped coils; (b) movable core; (c) slider.

10-9 INDUCTORS IN SERIES

When inductors are connected in series and are far enough apart, or shielded, so that there is no coupling between them, the inductors may be added in the same manner in which resistors are added. Therefore, when inductors are connected in series and have no mutual inductance, the total inductance is equal to the sum of the individual inductances. In equation form,

$$L_T = L_1 + L_2 + L_3 + \cdots + L_N \qquad (10\text{-}6)$$

When mutually coupled coils are connected in series, they can be connected with their mmfs either aiding or opposing. Figure 10-11 shows two series connected inductors with aiding mmf.

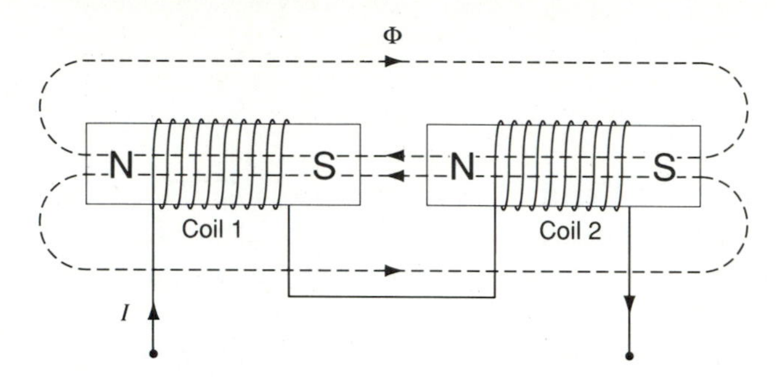

FIGURE 10-11 Two series-connected coils with aiding flux.

The inductance of two coils connected with their flux aiding each other can be calculated by adding the inductance of the first and second coils to the mutual inductance of the two coils. In equation form,

$$L_a = L_1 + L_2 + 2(k\sqrt{L_1 L_2}) \tag{10-7}$$

where L_a = total inductance of circuit

L_1 = inductance of first coil

L_2 = inductance of second coil

k = coefficient of coupling

The two inductors shown in Figure 10-12 have their flux in opposite directions, which means that they are connected series opposing.

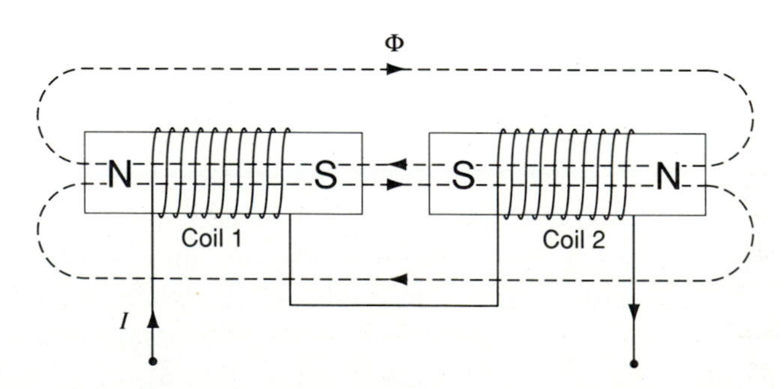

FIGURE 10-12 Two series-connected coils with opposing flux.

The inductance of two coils connected with their flux opposing each other can be calculated in the following manner:

$$L_0 = L_1 + L_2 - 2(k\sqrt{L_1L_2}) \qquad (10\text{-}8)$$

If equations 10-7 and 10-8 are combined, the result is as follows:

$$L_a - L_0 = 4M$$

Therefore, in terms of series-aiding and series-opposing connections, the mutual inductance of two series-connected coils can be determined by the following equation:

$$M = \frac{L_a - L_0}{4}$$

EXAMPLE 10-7 A 50-H inductor and a 30-H inductor are connected in series aiding. The coefficient of coupling is 0.75. Calculate the inductance of the circuit.

Solution

$$\begin{aligned} L_a &= L_1 + L_2 + 2(k\sqrt{L_1L_2}) \\ &= 50\text{ H} + 30\text{ H} + 2(0.75)\sqrt{50\text{ H} \times 30\text{ H}} \\ &= 138.09\text{ H} \end{aligned}$$

EXAMPLE 10-8 A 17-H inductor and a 15-H inductor are connected series opposing. The mutual inductance of these two coils is 12 H. What are the coefficient of coupling and the inductance of the circuit?

Solution

$$M = k\sqrt{L_1L_2}$$

$$\therefore k = \frac{M}{\sqrt{L_1L_2}} = \frac{12}{\sqrt{17 \times 15}} = 0.75$$

$$\begin{aligned} L &= L_1 + L_2 - 2(k\sqrt{L_1L_2}) \\ &= 17\text{ H} + 15\text{ H} - 2(0.75)\sqrt{17\text{ H} \times 15\text{ H}} \\ &= 8.05\text{ H} \end{aligned}$$

A convenient method for indicating the instantaneous flux direction is to use the **dot convention**. The dot system allows us to represent the direction of flux produced by a coil without having to indicate the windings and the flux path. The dot represents the tip of the flux arrow. If the current through each of the mutually coupled coils enters the dot terminals, or terminals without dots, the mutually coupled coils will be positive, or series aiding. If the arrow indicating current direction enters a dot and the other current is shown leaving a dot, the polarity of the mutually coupled coil is negative, or series opposing. Figure 10-13 illustrates the dot conventions for additively and subtractively coupled coils. Figure 10-12(a) and (b) would be represented by Figure 10-13(a) and (b), respectively.

M(+)

(a)

M(−)

(b)

FIGURE 10-13 Dot convention for coupled coils: (a) additive polarity, mutual flux aiding; (b) subtractive polarity, mutual flux opposing.

10-10 INDUCTORS IN PARALLEL

If no mutual induction exists, inductors that are connected in parallel are added in the same manner as resistors in parallel. In equation form,

$$\frac{1}{L_T} = \frac{1}{L_1} + \frac{1}{L_2} + \frac{1}{L_3} + \cdots + \frac{1}{L_N} \tag{10-9}$$

For two inductors in parallel, with no mutual inductance, the product-over-sum equation may be used.

$$L_T = \frac{L_1 \times L_2}{L_1 + L_2} \tag{10-10}$$

When mutual inductance exists between two parallel-connected inductors, the following equation may be used to find the total inductance of the circuit.

$$L_T = \frac{L_1L_2(1 - k^2)}{L_1 + L_2 - 2k\sqrt{L_1L_2}} \tag{10-11}$$

EXAMPLE 10-9 A 35-H inductor and a 25-H inductor are connected in parallel. The coefficient of coupling is 0.6. Calculate the inductance of this circuit.

Solution

$$L_T = \frac{L_1L_2(1 - k^2)}{L_1 + L_2 - 2k\sqrt{L_1L_2}} = \frac{(35)(25)[1 - (0.6)^2]}{35 + 25 - 2(0.6)\sqrt{35 \times 25}}$$
$$= 22.85 \text{ H}$$

10-11 INDUCTIVE TIME CONSTANT

When emf is first applied to an inductor, it takes a certain amount of time for current to build up and reach a constant value. A dc circuit containing inductance as well as resistance will have a gradual current change between zero and maximum. Since the buildup current in an inductive circuit starts at zero, it must always be changing as it approaches its stable value. During these changes, a relationship exists between the values of current reached and the time it takes to reach them. This relationship is expressed by a quantity called an **inductive time constant** and is defined as follows:

> The inductive time constant is the amount of time, in seconds, required for the current in an *RL* circuit to rise to 63.2% of its total value after an emf is initially applied across the series-connected inductance and resistance.

The value of the time constant is directly proportional to the inductance and inversely proportional to the resistance. As was the case with the time constant for a capacitive circuit, the time constant for an inductive circuit is represented by the symbol τ, which is the lowercase Greek letter tau. If the values of resistance and inductance are known, the time constant can be calculated using the equation

$$\tau = \frac{L}{R} \tag{10-12}$$

where τ = time for current to reach 63.2% of its final value, in seconds

L = inductance of the circuit, in henrys

R = resistance of the circuit, in ohms

EXAMPLE 10-10 A 12-Ω resistor is connected in series with a 0.36-H inductor. What time is required for the circuit to build up to 63.2% of its final value?

Solution

$$\tau = \frac{L}{R} = \frac{0.36}{12} = 0.03 \text{ s}$$

For capacitive circuits, the product of $R \times C$ was the time constant. For inductive circuits the time constant is L/R. Figure 10-14 shows the universal time constant chart, which is identical to the chart shown in Figure 6-8. It is indeed a universal chart, with the charging and discharging paths applicable to both *RC* and *RL* circuits. Curve *A* would represent the charging time of an inductive circuit and curve *B* would represent the discharge time.

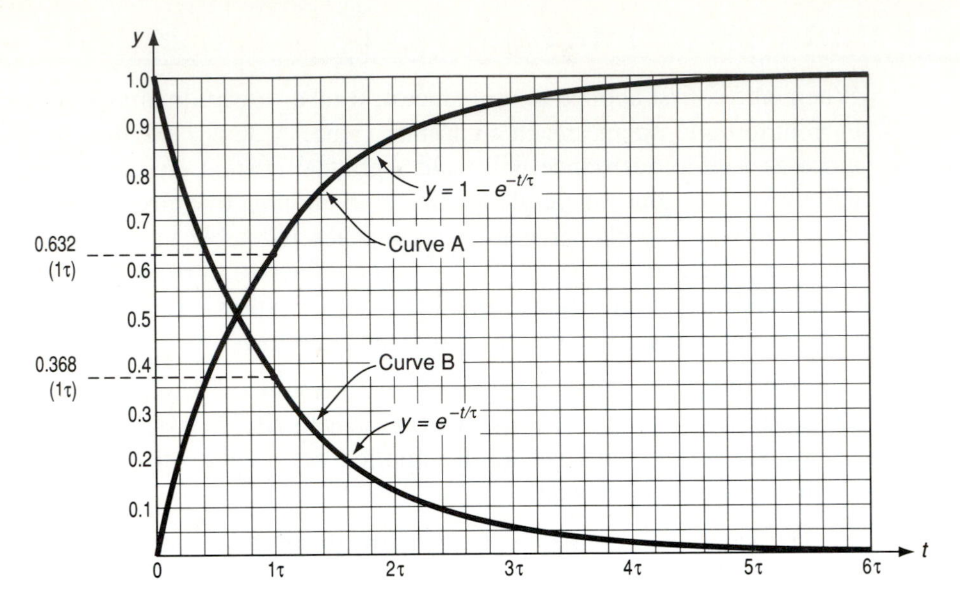

FIGURE 10-14 Universal time constant curves.

10-12 TRANSIENTS IN *RL* CIRCUITS

By using differential calculus it is possible to calculate instantaneous current in an inductive–resistive circuit. Figure 10-15 shows an *RL* circuit with a switch and a source of emf. At any time after the switch is closed, the mathematical expression for current through the coil can be found by first applying Kirchhoff's voltage law:

$$E = v_R + v_L$$

where v_R and v_L are the instantaneous voltage drops across the resistor and coil, respectively. By substituting v_R with iR, Kirchhoff's voltage law may now be expressed as

$$E = iR + v_L$$

Since the current, i, is the same for the resistor and the inductor, the equation now becomes

$$E = iL \times R + L \times \frac{diL}{dt} \tag{10-13}$$

The current iL can then be found by applying differential calculus:

$$iL = \frac{E}{R}[1 - e^{-t(R/L)}] \quad (10\text{-}14)$$

i = instantaneous current, in amperes, at time t

L = inductance of the inductor, in henrys

E = supply voltage, in volts

R = series resistance, in ohms

e = exponential constant 2.71828

t = time in seconds, from current commencement

EXAMPLE 10-11 The circuit of Figure 10-15 has an 8-H inductor and a 1-kΩ resistor connected in series to a 10-V supply. Calculate the current iL at t = 6 ms when the switch is closed.

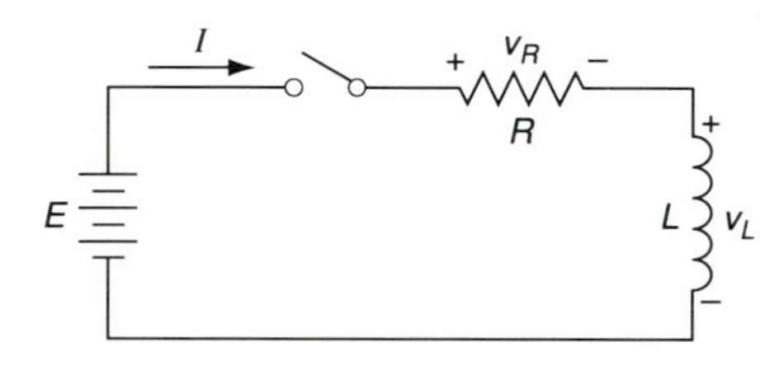

FIGURE 10-15 The *RL* circuit.

Solution

$$iL = \frac{E}{R}[1 - e^{-t(R/L)}] = \frac{10}{1 \times 10^3}[1 - e^{-0.006(1000/8)}]$$
$$= 5.28 \text{ mA}$$

In the *RC* circuit, the voltage v_C essentially reaches its final value in 5 time constants. The same is true for the current iL in an *RL* circuit. If R is held constant and L is reduced, the rise time of 5 time constants decreases. When plotting a curve of time constants for an *RL* circuit, the ratio L/R always has the same numerical value. The larger the inductance, the larger the time constant, and the longer it will take iL to reach its final value.

When a circuit that contains only resistance is opened, the current drops to zero immediately. In an inductive circuit, the current tends to drop to zero when the switch is opened; however, any change in current produces an emf that tends to keep the current flowing in the circuit. For this reason, the current through an inductor cannot change instantaneously. The equation for calculating a **discharge curve** for an *RL* circuit is

$$iL = I_m(e^{-t/\tau}) \quad (10\text{-}15)$$

where I_m is the current flowing in the coil just prior to the start of the decay transient and τ is the time constant L/R. Since the current is a decaying exponential, the time constant is the time required for the current to decay to 36.8% of its original value.

EXAMPLE 10-12 The RL circuit of Figure 10-16 consists of a 43-Ω resistor, a 0.4-H coil, and an applied voltage of 12 V. Assuming that switch S_1 has been closed for a very long time, calculate the current flowing 25 ms after switch S_1 is opened and S_2 is closed.

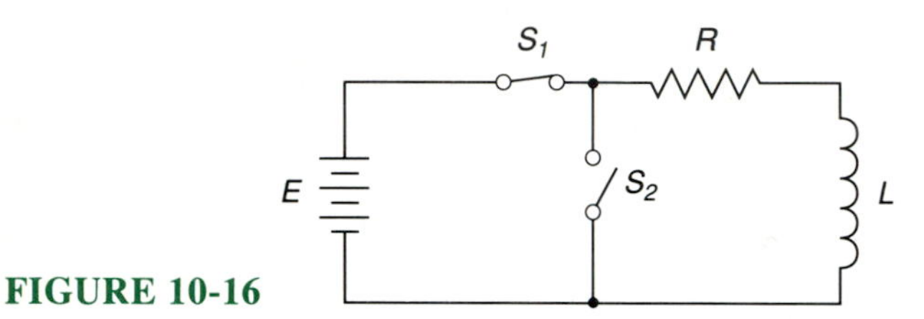

FIGURE 10-16

Solution The current I_m is found by Ohm's law.

$$I_m = \frac{E}{R} = \frac{12\text{ V}}{43\ \Omega} = 0.279\text{ A}$$

$$\tau = \frac{L}{R} = \frac{0.4\text{ H}}{43\ \Omega} = 9.302 \times 10^{-3}\text{ s}$$

$$iL = I_m(e^{-t/\tau}) = 0.0279(e^{-0.025/0.009302})$$
$$= 18.98\text{ mA}$$

The voltage across the coil in an RL circuit can change instantaneously. Figure 10-17 shows the voltages across a series-connected inductance and resistance. At the instant the switch is closed, the current is zero, due to the sudden rate of change of current, causing the counter emf to equal the applied

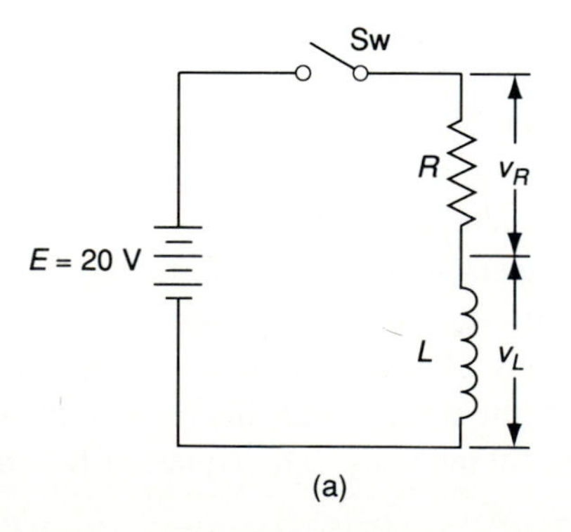

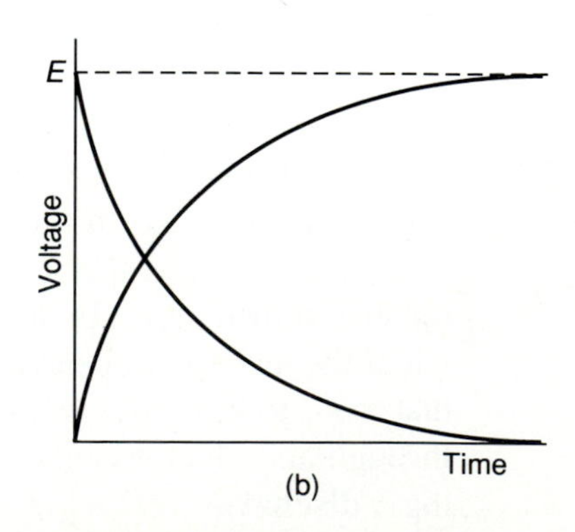

FIGURE 10-17 Voltages varying with time immediately after emf is initially applied.

emf. Since there is no current, the voltage across the resistance is zero. Therefore, the entire applied voltage appears across the inductance.

Eventually, the voltage across the inductor begins to decline, and the voltage across the resistor increases. This continues until the total applied emf appears across R and none across L. The voltage across R will be at its maximum value when the current has reached its steady state, which means there is no counter emf in L. To calculate the voltage across an inductor in an RL circuit, the following equation is used:

$$v_L = E(e^{-t/\tau}) \tag{10-16}$$

EXAMPLE 10-13 The RL circuit of Figure 10-16 consists of a 30-Ω resistor and a 2-H coil. An emf of 20 V is applied to the circuit. What is the voltage across the inductor 0.02 s after switch S_1 is closed and S_2 opened?

Solution

$$\tau = \frac{L}{R} = \frac{2\,\text{H}}{30\,\Omega} = 66.67 \times 10^{-3}\ \text{s}$$

$$v_L = E(e^{-t/\tau}) = 20(e^{-0.02/0.06667}) = 14.82\ \text{V}$$

While the voltage across the inductor is decreasing with time, the voltage across the resistor is increasing exponentially. The voltage across the resistor increases at a rate determined by the current through the circuit. The equation for calculating the changing voltage drop across a resistor in an RL circuit is as follows:

$$v_R = E(1 - e^{-t/\tau}) \tag{10-17}$$

EXAMPLE 10-14 Using the values given in Example 10-13, solve for the voltage across the resistor. Verify by using Kirchhoff's voltage law.

Solution

$$\tau = \frac{L}{R} = \frac{2\,\text{H}}{30\,\Omega} = 66.67 \times 10^{-3}\ \text{s}$$

$$v_R = E(1 - e^{-t/\tau}) = 20(1 - e^{-0.02/0.06667}) = 5.18\ \text{V}$$

$$E = v_R + v_L$$

$$v_R = E - v_L = 20\ \text{V} - 14.82\ \text{V} = 5.18\ \text{V}$$

The length of time required for a current to reach a certain value in an RL circuit is determined by equations 10-18 and 10-19. The time at any point on the curve for exponential *rise* of current is as follows:

$$t = \tau \ln \frac{I_m - I_0}{I_m - I_1} \qquad (10\text{-}18)$$

The time at any point on the curve for exponential *decay* of current is determined by

$$t = \tau \ln \frac{I_m}{I_1} = \tau \ln \frac{I_0}{I_1} \qquad (10\text{-}19)$$

where t = time required to reach a certain value, in seconds

I_m = steady-state value of current as found by E/R

I_0 = initial value of current

I_1 = current at any point on the curve where time is desired

EXAMPLE 10-15 In an RL circuit, the current rises from 0 to a final value of 45 mA. The circuit has a time constant of 3.75 ms. Determine the length of time required for the current to reach a value of 14.58 mA.

Solution

$$t = \tau \ln \frac{I_m - I_0}{I_m - I_1} = 3.75 \text{ ms} \times \ln \frac{45 - 0}{45 - 14.58} = 1.47 \text{ ms}$$

EXAMPLE 10-16 The current in an RL circuit with a time constant of 2.68 ms decays from an initial value of 6.25 mA to an instantaneous given value of 1.72 mA. Determine the time of decay.

Solution

$$t = \tau \ln \frac{I_0}{I_1} = 2.68 \text{ ms} \times \ln \frac{6.25}{1.72} = 3.46 \text{ ms}$$

The time at any point on the curve during exponential decay of voltage is the same equation for both inductors and capacitors. That is,

$$t = \tau \ln \frac{V_0}{V_1}$$

EXAMPLE 10-17 In an RL circuit, the time constant is 7.5 ms. Determine the length of time required for the voltage to decay from an initial value of 20 V to 14.57 V.

Solution

$$t = \tau \ln \frac{V_0}{V_1} = 7.5 \text{ ms} \times \ln \frac{20}{14.57} = 2.38 \text{ ms}$$

10-13 ENERGY STORED IN A MAGNETIC FIELD

During the period that a magnetic field is being established by a coil, a portion of the energy that is supplied by the source is stored in the field. The remainder of the energy supplied during the rise of the current is dissipated from the coil of the magnet as heat. After the magnetic field is established and the current has reached a steady state, all the energy input to the coil windings of the electromagnet is dissipated as heat, and no additional energy is required to maintain this magnetic field. The amount of energy stored in the magnetic field for any value of current may be expressed using the following equation:

$$W = \frac{LI^2}{2} \tag{10-20}$$

where W = energy stored, in joules

L = inductance, in henrys

I = current, in amperes

EXAMPLE 10-18 Find the energy stored in a series-connected *RL* circuit having an inductance of 5 H and a resistance of 120 Ω. The supply voltage is 100 V.

Solution

$$I = \frac{E}{R} = \frac{100\ \text{V}}{120\ \Omega} = 0.833\ \text{A}$$

$$W = \frac{LI^2}{2} = \frac{(5)(0.833)^2}{2} = 1.74\ \text{J}$$

10-14 ARCING

If a switch connected in series with a charged inductor is opened, a very high induced voltage results due to the rapidly collapsing field. Current flowing through the switch cannot cease in zero time, for this would produce infinite voltage across the inductor. The practical result is that the voltage, though not infinite, rises high enough to produce an electric **arc** through air between the switch contacts as they separate. This arc permits current to exist around the circuit, through the open switch contacts.

The arc developed by the switch opening may last for only a brief instant, but it allows a finite interval in which one current can rise and another fall without violating the laws of electricity. The tendency for a collapsing magnetic field to produce a high-voltage arc is used in automobiles to produce an arc across the gap of spark plug electrodes. It is possible to induce a voltage of greater than 20,000 V from the collapsing field of a 12-V electrical system.

In electrical systems where high-voltage inductive loads are switched, inductive arcing is an undesirable condition that causes switch contacts to become pitted and carbonized. The formation of an arc between opening contacts can be suppressed by slowing the rate of the collapsing field. Connecting a resistor, or a resistor and a capacitor, in parallel with the switch will reduce arcing substantially.

Another method that is often used to prevent contact arcing is placing a **diode** in parallel with the inductive load. A diode is made from semiconductor materials such as germanium (Ge) or silicon (Si). The schematic symbol for a diode is shown in Figure 10-18(a). It offers very high opposition to current flow in one direction, but very low opposition to current flow in the opposite direction.

The two terminals of the diode are identified as the anode and cathode. The internal resistance of the diode is very low when it is **forward-biased**. The device is considered to be forward-biased when the anode voltage is more positive than the cathode voltage, as shown in Figure 10-18(b). The internal resistance of the diode is very high when it is **reverse-biased**, that is, when the cathode voltage is more positive than the anode voltage, as shown in Figure 10-18(c). A simple electrical analogy would be that a forward-biased diode acts as a closed switch, and a reverse-biased diode acts as an open switch.

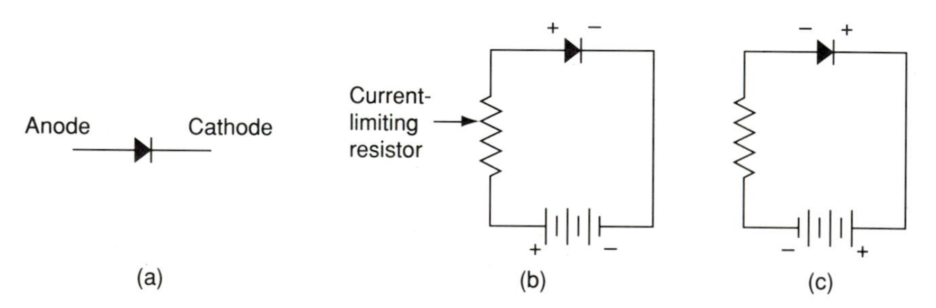

FIGURE 10-18 (a) Schematic symbol of a diode; (b) forward bias; (c) reverse bias.

In Figure 10-19, a diode is used in the circuit to provide a discharge path for the current in the coil. When the switch is opened, current will flow through the diode instead of across the open switch contacts. In this circuit, when the switch opens the polarity of the induced emf causes the diode to conduct. The forward-biased diode provides a low-resistance discharge path. The time constant of the circuit (L/R) is large and the current gradually decreases, so very little voltage is induced across the coil when the diode is forward-biased. A diode used in this manner, referred to as a **freewheeling diode**, is very common in dc motors and relay coils to suppress high induced voltage when equipment is deenergized.

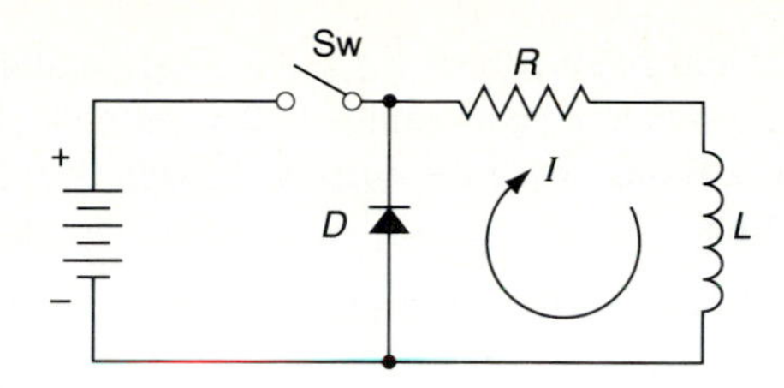

FIGURE 10-19 Freewheeling diode.

KEY TERMS

Induced emf
Generated voltage
Flux linkage
Faraday's law
Generator action
Fleming's right-hand rule
Lenz's law
Counter emf (cemf)
Self-induction
Henry
Mutual inductance
Tight coupling
Loose coupling
Inductor
Air-core inductor
Iron-core inductor
Eddy current
Hysteresis loss
Heat loss
Dot convention
Inductive time constant
Discharge curve
Arcing
Diode
Forward-biased
Reverse-biased
Freewheeling diode

PROBLEMS

10-1 A bar magnet is placed in a 400-turn coil. The magnet has a field strength of 650 μWb. Calculate the emf induced in the coil if the magnet is removed completely after 100 ms.

10-2 Find the induced voltage across a 150-turn coil located in a magnetic field that is changing at a rate of 200 mWb/s.

10-3 A 300-turn solenoid has a field strength of 2200 μWb. Find the average induced voltage if the flux changes to 2800 μWb in 8.5 ms.

10-4 The flux linking a coil changes from 0 to 15.3 mWb in 50 ms. If 6.45 V is developed during the flux change, how many turns are in the coil?

10-5 The magnetic flux of 80 mWb changes to 120 mWb in 1.77 ms. Determine the rate of flux change.

10-6 Given a 500-turn air-core solenoid 25 cm in length having a cross-sectional area of 7.5 cm^2. The solenoid has a secondary winding consisting of 10 turns. Calculate the voltage induced in the secondary winding when the current through the solenoid changes from 0 to 20 mA in 4.4 ms.

10-7 The current flowing through a coil rises from 0 to 28 mA in 5 ms. If 3.3 V is developed, what is the inductance of the coil?

10-8 A 5-H coil carries a current that increases 22.5 mA in 13.3 ms. Determine the average voltage induced in the coil.

10-9 An air-core solenoid 10 cm long has a diameter of 3 cm and consists of 200 turns. Find the inductance of the solenoid.

10-10 A 250-turn coil has an air core, a length of 0.025 m, and a radius of 0.0035 m. Determine its inductance.

10-11 Two coils have inductances of 3.3 H and 5.4 H and are coupled with a coefficient of coupling of 0.83. Determine the mutual inductance.

10-12 Find the mutual inductance of 22-mH and 35-mH coils with a coefficient of coupling of 0.91.

10-13 A 400-turn coil has a resistance of 37 Ω, and a 250-turn coil has a resistance of 26 Ω. The coils have a coefficient of coupling of 0.77. When 60 V dc is applied to the 400-turn coil, a primary flux of 220 mWb is produced. Determine (a) the flux linking the secondary; (b) the mutual inductance.

10-14 Two coils having inductances of L_1 = 350 mH and L_2 = 275 mH are tightly wound on top of each other on a ferromagnetic core. Determine the maximum mutual inductance that is possible between the coils.

10-15 The mutual inductance between two coils is 0.39 H. The coils have self-inductances of 0.35 H and 0.79 H. Calculate the coefficient of coupling.

10-16 If the primary and secondary windings of a transformer each have an inductance of 550 mH and the mutual inductance between the windings is 30 mH, what percentage of the flux from one winding reaches the other winding?

10-17 A 100-H coil and a 70-H coil are connected in series aiding. The coefficient of coupling is 0.82. Determine the inductance of the circuit.

10-18 A 200-mH coil and a 200-mH coil are connected series aiding. The mutual inductances of these two coils is 150 mH. Determine (a) the coefficient of coupling; (b) the inductance of the circuit.

10-19 A 20-H coil and a 15-H coil are connected series opposing. The coefficient of coupling is 0.77. Calculate the inductance of the circuit.

10-20 A 50-mH coil and a 35-mH coil are connected series opposing. The mutual inductance of these two coils is 28 mH. Determine (a) the coefficient of coupling; (b) the inductance of the circuit.

10-21 Two series-connected coils, each having an inductance of 6 H, have a total inductance of 14 H when connected series aiding and 11 H when connected series opposing. Determine (a) the mutual inductance; (b) the coefficient of coupling of this circuit.

10-22 Two 700-μH coils have a mutual inductance of 250 μH. Calculate the total inductance of the two coils when they are connected (a) series aiding; (b) series opposing.

10-23 A 400-mH coil and a 300-mH coil are connected in parallel. The coefficient of coupling is 0.85. Determine the inductance of the circuit.

10-24 A 300-Ω resistor is connected in series with a 20-H coil. Determine the length of time required for the circuit voltage to rise to 63.2% of its final value.

10-25 For the circuit shown in Figure 10-20, calculate the time required after the switch is closed for the circuit current to rise to 40 mA.

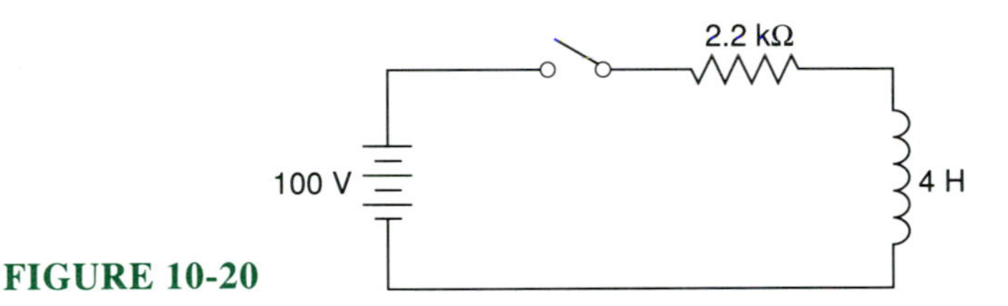

FIGURE 10-20

10-26 Determine the value of current flowing in the circuit of Figure 10-20 after the switch has been closed for 2.5 ms.

10-27 The *RL* circuit of Figure 10-21 consists of a 100-Ω resistor, a 250-mH coil, and an applied voltage of 24 V. Assuming that switch S_1 has been closed for a very long period of time, calculate the current flowing 10 ms after switch S_1 is opened and S_2 is closed.

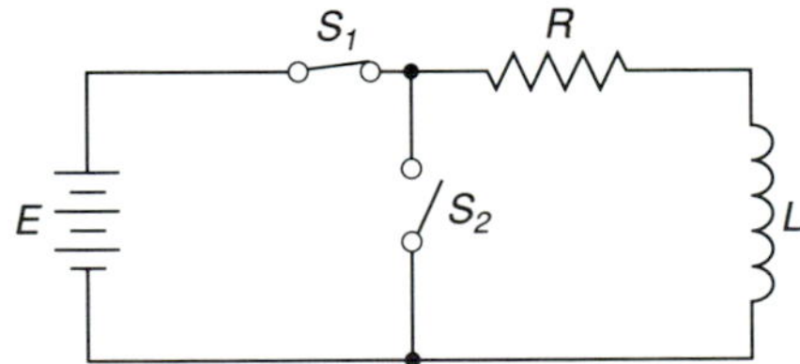

FIGURE 10-21

10-28 The *RL* circuit of Figure 10-21 has a 1-kΩ resistor and a 5-H coil. An emf of 120 V is applied to the circuit. Determine the voltage across the inductor 10 ms after S_1 is closed and S_2 opened.

10-29 In an *RL* circuit, the current rises from 0 to a final value of 16 mA. The circuit has a time constant of 4.55 ms. Determine the length of time required for the current to reach a value of 7.25 mA.

10-30 The current in an *RL* circuit with a time constant of 1.28 ms decays from an initial value of 22.5 mA to an instantaneous given value of 6.35 mA. Determine the time of decay.

10-31 In an *RL* circuit, the time constant is 12.5 ms. Determine the length of time required for the voltage to decay from an initial value of 40 V to 28.65 V.

10-32 An *RL* circuit has an initial rate of rise of 50 A per second when 60 V is applied. Determine the inductance of the circuit.

10-33 Determine the amount of energy stored in a series-connected *RL* circuit with a 200-mH coil and a 5-Ω resistor. The supply voltage is 24 V.

10-34 Find the energy store in a magnetic field when the inductance is 7.5 H and the current flowing is 1.5 A.

CHAPTER

11 Direct-Current Generators

LEARNING OBJECTIVES

Upon completion of this chapter you will be able to:

- Explain the difference between a generator and a motor.
- List the basic components of a typical generator.
- Name the two classifications of armature windings.
- Define the term *frequency*.
- Explain the principle of rectification using a commutator.
- Discuss the magnetization curve and the effect of air-gap flux on the generator voltage.
- List the two basic types of dc generators.
- Explain the voltage buildup of a shunt generator.
- Discuss armature reaction.
- Determine the percent voltage regulation of a dc generator.
- Explain the losses and efficiency of a dynamo.
- Discuss the advantages of paralleling generators.

11-1 INTRODUCTION

A **dynamo** is a dc machine that may be used either to convert mechanical energy into electrical energy or to convert electrical energy into mechanical energy. For this reason a dynamo is referred to as an **electromechanical device**. When the dynamo is mechanically driven by a prime mover such as a diesel engine, or turbine, to supply an electric current, it is called a **generator**. When the dynamo is supplied with electric energy and is used as a source of mechanical energy to drive machinery, pumps, compressors, and so on, it is called a **motor**.

11-2 GENERATOR CONSTRUCTION

As mentioned previously, a dc generator converts mechanical energy into electrical energy. This conversion is accomplished by rotating an armature, which carries conductors, in a magnetic field. The action of conductors rotating in a magnetic field induces an emf in the conductors. In most dc generators, the armature rotates and the magnetic field is stationary. A mechanical force is applied to the shaft of the rotating armature to cause the relative motion.

The basic components of a typical generator are shown in Figure 11-1, and consist of a frame, field windings, commutator, interpoles, armature, and brushes. The dc generator field frame, or **yoke**, is usually made of annealed steel. The frame provides mechanical support for the pole pieces and serves as a portion of the magnetic circuit to provide the necessary flux across the air gap. The end bells are bolted to the frame structure and support the armature shaft bearings. One end bell also supports the brush rigging and extends over the commutator.

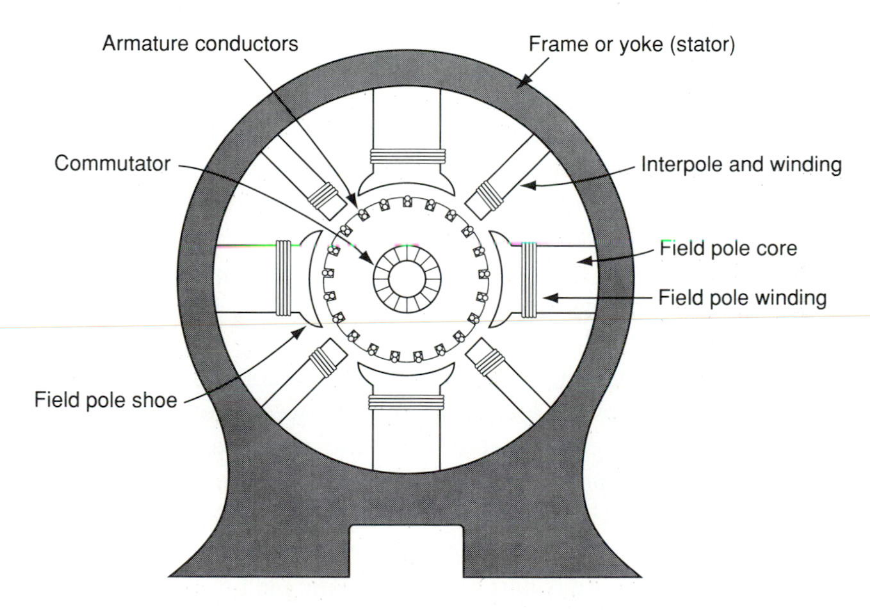

FIGURE 11-1 Sectional view of dynamo construction and electrical circuits.

The **field windings**, or coils, are connected so that they produce alternate north and south poles to obtain the correct direction of emf in the armature conductors. The field windings are connected in series to form the field circuit. Field circuits may be designed to be connected in series or in parallel with the armature circuit. When a series–parallel combination is used, it is known as a **compound field**. Parallel, or shunt, field windings have many turns of relatively small wire and are designed to produce the required ampere-turns with a small current from the line supply. Series-field windings have few turns of relatively large wire and are designed to produce the required ampere-turns with a small voltage drop.

Field poles, which support the field windings, are constructed of laminated steel and are welded, or bolted, to the yoke. The outer end of the laminated pole is called the **shoe**. The pole shoe is curved and is wider than the pole core to produce a more uniform magnetic field.

The conventional dc generator is an alternating-current machine equipped with a mechanical rectifier to convert the ac to dc. This rectifier is called a **commutator**. The commutator is composed of alternate sections of copper bars and mica separators clamped together with a mica-insulated V-ring. The number of commutator bars is determined by the number of coils in the armature and the type of winding.

The **brushes** make a rubbing contact on the commutator and conduct current to, and from, the outside circuit. Each brush is supported and held against the commutator by a brush holder, which, in turn, is clamped to a brush stud. The brushes, holders, and studs are stationary are referred to as the **brush rigging**. Brushes are made of carbon or copper graphite and are held down firmly on the commutator by a spring that exerts a pressure of about 1 to 2 pounds per square inch (psi).

The **interpole** and its windings are also mounted on the yoke of the generator. The interpoles are located midway between the main poles. The interpole winding is connected in series with the armature circuit. The purpose of the interpole is to improve commutation. In large generators there are as many interpoles as there are main poles. In small machines, to save expense, there may be only half as many of these poles as main poles.

The **armature** core is part of the magnetic path through the machine. It is usually constructed of laminated steel. The laminations are required to reduce the eddy currents due to change of flux in the core. The circumferential edge of the armature is slotted to hold the armature windings.

11-3 ARMATURE WINDINGS

The **armature windings** are the windings in which a voltage is induced. These windings consist of diamond-shaped preformed coils which are inserted into the armature slots as a unit. Each coil consists of a number of turns of wire; each turn, or loop, is insulated from the other turns and from the armature slot.

Cutaway illustration of a large dc generator

Source: Prentice Hall Inc. / Reliance Electric Company

Armature windings are classified according to the **plex** of their windings. A **simplex armature winding** is a single, complete, closed winding wound on an armature. A **duplex armature winding** consists of two complete sets of windings that are independent of each other. An armature containing a duplex winding will be connected alternately to the commutator segments. All armatures with more than one set of windings are referred to as **multiplex windings**.

The span of the armature winding should be about equal to the peripheral distance between the centers of adjacent field poles, so that the voltages generated in the two sides of the coil will be aiding, or series additive. This distance is called the **pole pitch**. If a coil covers a span of 180 electrical degrees, it is termed a **full-pitch coil**. When the span of a coil is less than the pole pitch the winding is referred to as a **fractional-pitch coil.** Fractional-pitch windings have reduced emf, as their coil-side voltages do not reach their maximum values at the same time. An armature wound with a fractional pitch is called a **chorded winding**. A coil that spans 130° would have a pitch factor of 130/180 = 0.72, or 72%. Generally, pitch factors below 80% are avoided.

There are two basic sequences of armature winding connections to commutator segments: **lap windings** and **wave windings**. A third winding, the **frog-leg winding**, is a combination of the lap and wave windings.

The simplest type of winding construction is the **simplex-lap winding**. In this type of winding, the coil ends are connected to adjacent commutator segments. When the current passes through any armature winding, it always divides into an even number of parallel paths. In a simplex-lap winding, the current divides into the number of pole paths, P. If C represents the number of coils and commutator segments present in the rotor, there will be C/P coils in each of the parallel current paths through the machine.

Figure 11-2 shows a lap armature winding coil. In a simplex-lap winding, a coil edge under one pole is directly connected to a second in a nearly identical position under the next pole. This second coil edge is then connected through the commutator segment and back to a third coil edge under the original pole, but two coil edges over from the first coil edge. With this arrangement, the coils overlap each other, in a manner similar to shingles on a roof, and the winding is made to fit readily on the armature.

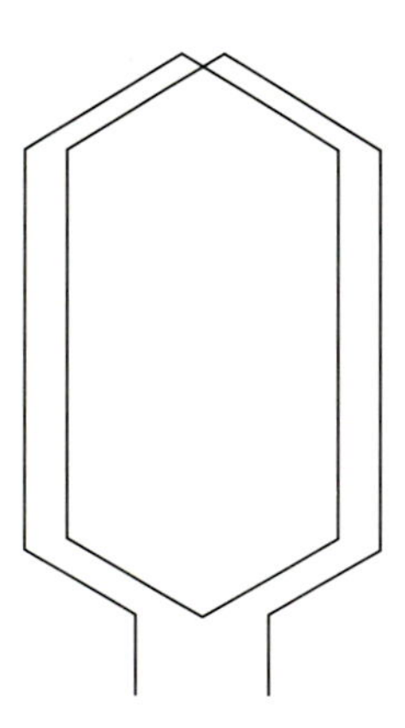

FIGURE 11-2 Lap armature winding.

Figure 11-3 gives a diagrammatic representation of a four-pole simplex-lap winding. To obtain maximum emf and current, four brush riggings must be used, the brushes being set so as to reverse connections between each coil and the brush when the conductors pass through the neutral position. The brushes are alternately positive and negative, and brushes of like polarity must be joined in parallel. In the case of Figure 11-3, there are four groups of coils which are placed in parallel by connecting the two positive brushes together

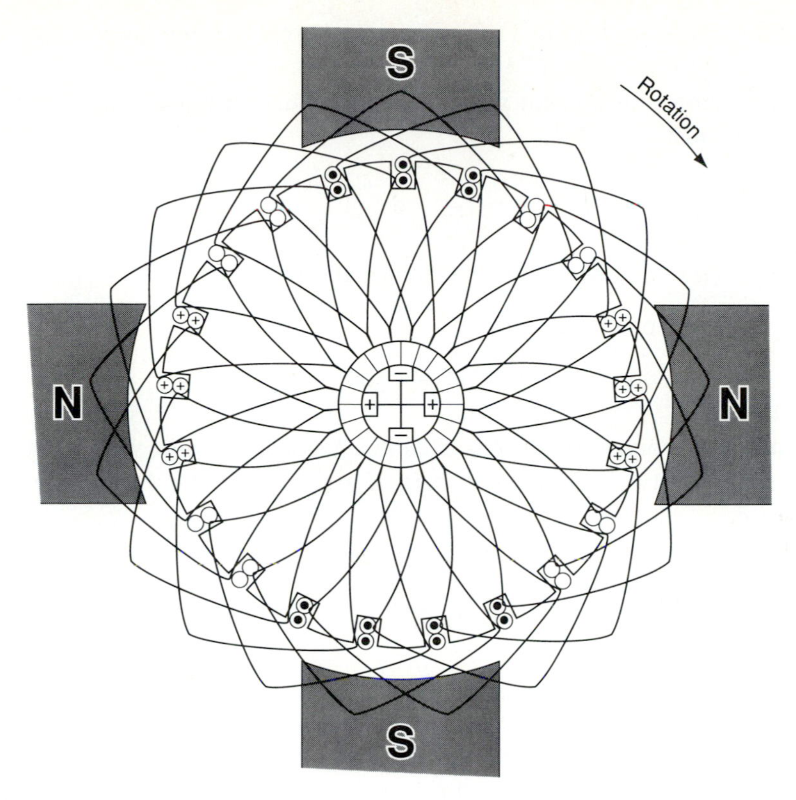

FIGURE 11-3 Simplex-lap four-pole armature winding. Shown as a two-layer winding.

and the two negative brushes together. No circulating current flows between these four parallel paths when no load is connected across the positive and negative brushes. This is due to the fact that the voltage of each group is equal and of opposite polarity to the voltages of the other groups.

For the induced voltage to be in the same direction in the two sides of a single coil, one side of a coil must be under a north magnetic pole while the other side is under a south pole. There are four groups of coils generating the same voltage between brushes of opposite polarity in Figure 11-3. These four groups of coils are placed in parallel by connecting two positive brushes together and two negative brushes together. Connecting a load between the positive and negative brushes results in current flowing through the armature in these four paths. Therefore, if the total armature output current is 200 A, each path through the armature will carry 50 A.

In all simplex-lap windings, there are as many paths through the armature as there are field poles. Also, there are as many brush positions on the commutator as there are field poles. The brushes must be equally spaced around the commutator. In Figure 11-3, since there are 21 commutator segments, the brushes must be placed at 2¼ or 5¼ segments apart.

The number of current paths in a lap-wound armature can be calculated using the equation

$$a = m \times P \tag{11-1}$$

where a = number of current paths in the armature

m = plex of the winding

P = number of poles in machines

EXAMPLE 11-1 A duplex lap-wound armature is used in a 12-pole machine with 12 brush sets, each spanning two commutator bars. Calculate the number of paths in the armature.

Solution

$$a = m \times P = 2 \times 12$$
$$= 24 \text{ paths}$$

All two-pole machines have lap windings, while most four-pole machines have **wave windings**. In dc generators with small current-handling capabilities, it is desirable to limit the number of circuits in parallel to the minimum possible value for a closed winding. In the case of multipolar machines, this involves using a winding that has similar coil sides under poles of like polarity. This type of winding is called the wave winding, and has series-connected coil sides, unlike the lap winding, which has a parallel connection. The wave winding increases the number of conductors in series between brushes without increasing the number of coils on the armature. In a simplex wave winding, there are only two current paths, and $C/2$, or ½, of the windings in each current path. The brushes in a simplex wave-wound machine will be located a full pole pitch apart from each other. Since one cycle occurs in a distance covered by a pair of poles, this distance is referred to as 360 electrical degrees. Therefore, in a simplex-wave winding, the coil ends are connected to commutator segments approximately every 360 electrical degrees.

A **multiplex wave winding** is a winding that consists of multiple independent sets of wave windings on the armature. The extra sets of windings each contain two current paths. The number of current paths is determined by

$$a = 2 \times m \tag{11-2}$$

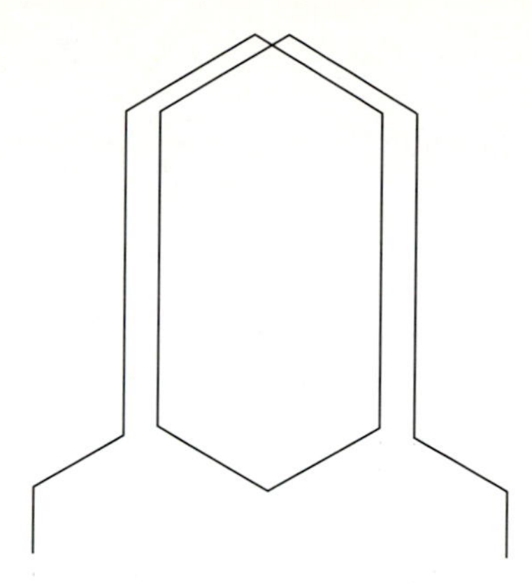

FIGURE 11-4 Wave armature winding.

EXAMPLE 11-2 A four-plex wave-wound armature is used in a 16-pole armature with four brush sets. Calculate the number of paths in the armature.

Solution $a = 2m = 2 \times 4 = 8$ paths

The frog-leg winding is a combination of the lap and wave winding. It is called a frog-leg winding because of the shape of its coils, as shown in Figure 11-5. Since the frog-leg coil is a combination "lap–wave" coil, it is wound with two conductors and two corresponding ends are brought out to be connected to adjacent commutator bars. The other corresponding ends are connected to commutator segments at some distance from each other. Frog-leg windings have as many circuits in parallel as duplex windings have. The number of current paths in a frog-leg winding may be calculated using the following equation:

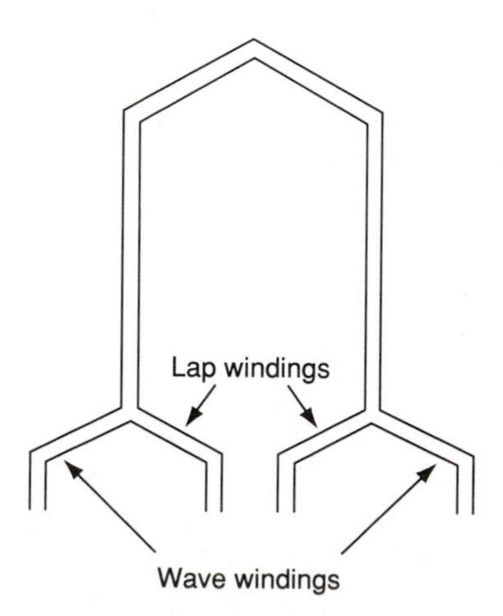

FIGURE 11-5 Self-equalizing or frog-leg winding.

$$a = 2Pm_{\text{lap}} \tag{11-3}$$

where P is the number of poles on the machine and m_{lap} is the plex of the lap winding.

11-4 GENERATOR CHARACTERISTICS

The generation of voltage in a dc machine requires the cutting of magnetic flux by moving conductors. The rate at which the flux is cut will determine the magnitude of the emf. The direction of the emf will be determined by the magnetic polarity and the direction of rotation of the armature. Since the armature has a total of z conductors divided into a parallel paths, each path will have z/a conductors in series. This relation can be expressed in the following equation:

$$E_g = \frac{z}{a} \times \Phi \times P \times \frac{\text{rpm}}{60}$$

$$= \frac{(z)(\Phi)(P)(\text{rpm})}{(60)(a)} \tag{11-4}$$

where
E_g = total generated voltage, in volts
z = total number of armature conductors
a = number of parallel paths
Φ = flux per pole, in webers
P = number of poles
rpm = speed of armature, in revolutions per minute

EXAMPLE 11-3 The flux per pole of a four-pole generator with a duplex lap winding is $50{,}000 \times 10^{-7}$ Wb. The armature has 4500 conductors and is rotating at 2400 rpm. Calculate the generated voltage.

Solution

$$a = mP = 2 \times 4 = 8$$

$$E_g = \frac{(z)(\Phi)(P)(\text{rpm})}{(60)(a)} = \frac{(4500)(50{,}000 \times 10^{-7})(4)(2400)}{(8)(60)}$$

$$= 450 \text{ V}$$

Figure 11-6 represents an elementary two-pole generator with a single-turn coil. In the two-pole single-turn coil arrangement, flux is assumed to pass from left to right, or from the north to the south pole. In Figure 11-6(a) the coil is in a vertical plane so that the conductors *ab* and *cd* are not cutting flux. At this instant no voltage is generated. When the coil is in this position, it is considered to be in the **neutral plane**.

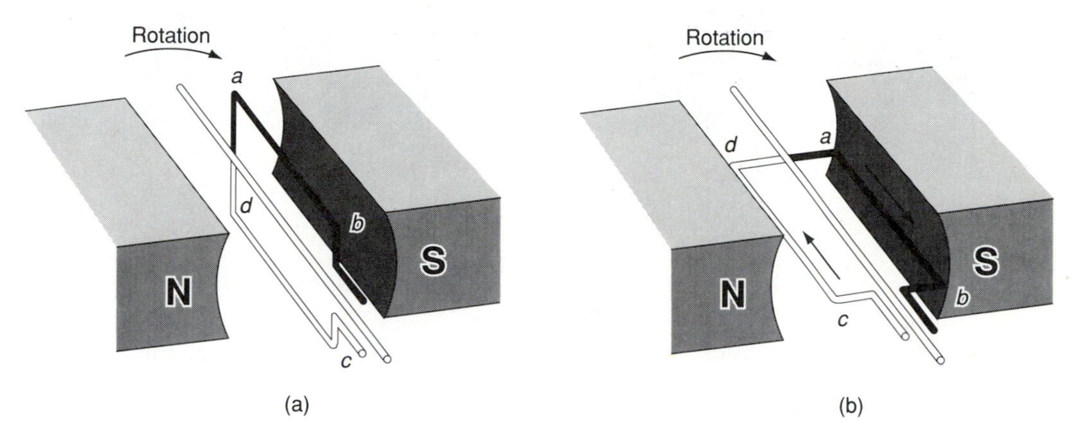

FIGURE 11-6 Voltage induced in a coil of a single turn at (a) 0° and (b) 90°.

If the coil is rotated 45° in a clockwise direction, less flux will pass through the coil. The conductor *ab* is moving down across the flux and a voltage would be produced in a direction determined by Fleming's right-hand rule. Therefore, the direction of induced emf would be from point *a* to *b* on one conductor and from *c* to *d* on the other. When the coil has rotated 90° from its neutral plane, as in Figure 11-6(b), the conductors are at right angles to the flux, and the voltage induced is at its greatest value.

As the coil continues to rotate, the voltage diminishes to zero as it approaches the vertical plane. Conductors *ab* and *cd* are once again in a position parallel to the flux, and no voltage is generated. When the coil is rotated in a clockwise direction from its position in Figure 11-7(a), the conductors again begin to cut flux. However, the sides of the coil have interchanged positions from that in Figure 11-6, and the direction of emf is now from *b* to *a* and *d* to *c*.

The emf again begins to increase in magnitude until it reaches its maximum value at 270° from the original starting point. This maximum value is at 180° to the maximum reached when *ab* passed the south pole. Therefore, these two maximum amplitudes can be said to be of opposite polarity. After the 270° mark, the emf generated decreases until it once again is zero in the neutral plane. Because of this action, as the coil revolves, a voltage is induced in one direction during one-half revolution, and in the opposite direction during the other half-revolution.

The series of values that the generated emf passes through in the coil

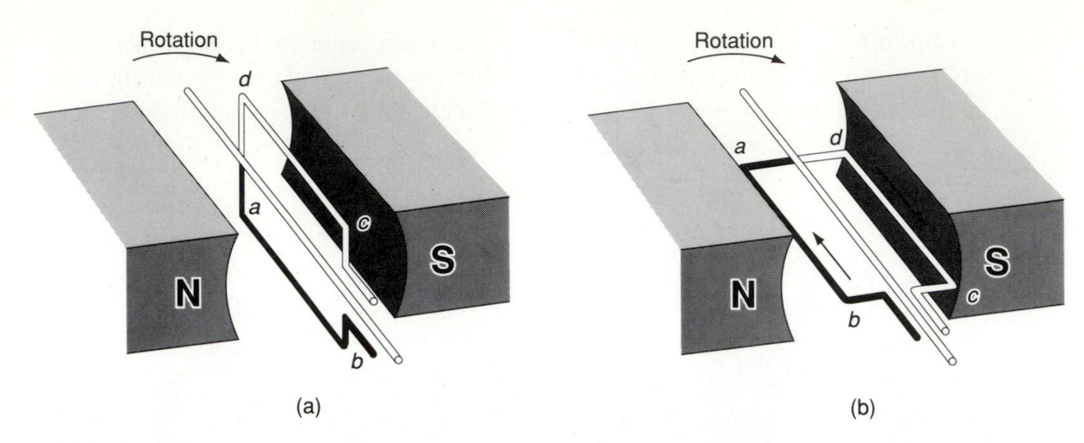

FIGURE 11-7 Voltage induced in a coil of a single turn at (a) 180° and (b) 270°.

during one revolution may be expressed in the form of a curve. Figure 11-8 illustrates the result of a voltage induced in a coil at various instants during a revolution. Since the voltage is changing every 180°, the current must also be alternating in a similar manner.

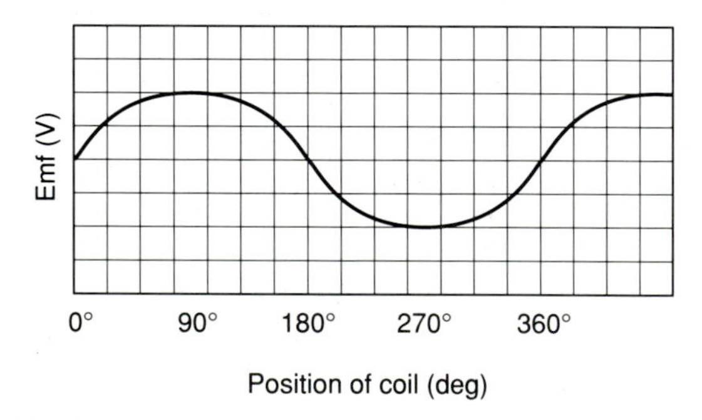

FIGURE 11-8 Voltage induced in a coil at various instants during one revolution.

From this analysis of a two-pole single-turn generator, it should be obvious that the generated voltage, as well as the current in a dc armature winding, is alternating. The number of alternations that occur every second is called **frequency** and is measured in hertz:

$$\text{frequency} = \text{cycles/second} \qquad \text{Hz} \tag{11-5}$$

A cycle is a unit consisting of 360 electrical degrees. Since the frequency of a dc generator is proportional to both the speed in revolutions per second, rpm/60, and the number of pairs of poles, $P/2$, frequency may be expressed using the following equation:

$$f = \frac{P}{2} \times \frac{\text{rpm}}{60}$$
$$= \frac{P \times \text{rpm}}{120} \qquad (11\text{-}6)$$

where f = frequency, in hertz

P = number of poles

rpm = speed of armature, in revolutions per minute

EXAMPLE 11-4 What is the frequency of the alternating emf in the armature winding of a six-pole generator that operates at a speed of 1600 rpm?

Solution

$$f = \frac{P \times \text{rpm}}{120} = \frac{6 \times 1600}{120}$$
$$= 80 \text{ Hz}$$

The dc generator develops an internal ac electromotive force. To convert this to a dc voltage, some type of rectification must be used. The rectifier in a dc generator consists of a commutator and commutator brushes. A simple split-ring commutator is illustrated in Figure 11-9. The two ends of the armature winding are each connected to a split ring. Brushes are located so that as the coil moves through the zero position, both commutator segments move out from under one brush and into contact with the other.

When conductor *ab*, in Figure 11-9, is moving downward in a clockwise direction, the split ring marked *x* will be positive. At the same instant, conductor *cd* will be moving upward, which will make the split ring *y* negative.

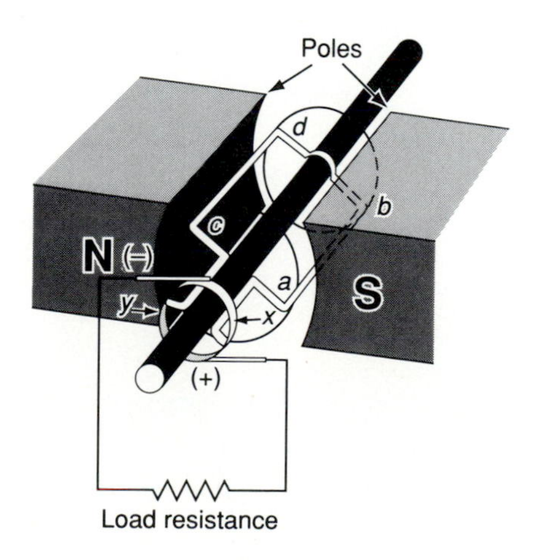

FIGURE 11-9 Basic two-pole dc generator with split-ring commutator.

When the coil rotates until conductor *ab* is under the north pole and *cd* is under the south, the generated voltage will reverse. However, the direction of current in the external circuit will remain the same.

Although the voltage across the load resistance is always in the same direction, it is not a steady emf. This is due to the fact that generated voltage varies from zero to a maximum value and back to zero twice each revolution. Figure 11-10(a) shows the generated voltage wave of a single winding armature. A voltage of this type cannot be used satisfactorily for direct-current applications, since the fluctuating voltage would produce a pulsating current, instead of a constant, steady value.

The practical dc generator consists of many coils that are joined together at a multisegment commutator. This method will develop a large voltage at the commutator brushes, and the voltage pulsations will be greatly subdued. Figure 11-10(b) shows the pulsating voltage of a two-coil armature with four commutator segments.

The more turns in each coil, the higher the value of induced voltage; the more coils and commutator segments, the steadier the voltage at the brushes. However, a dc generator can never deliver a pure direct current; there will always be a small amount of ripple voltage.

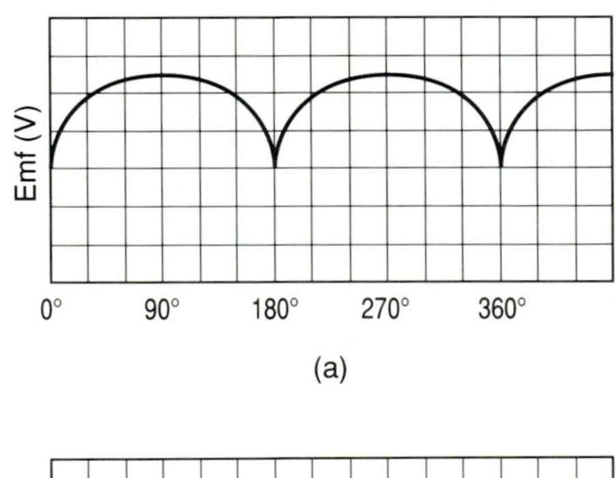

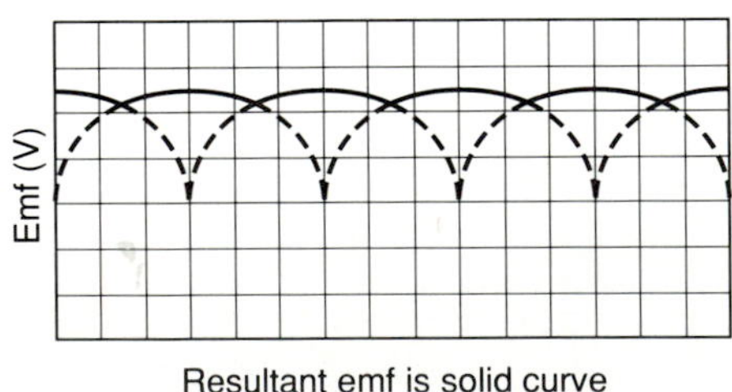

FIGURE 11-10 Output voltage for (a) a single coil and (b) a two-coil armature.

11-5 SATURATION, OR MAGNETIZATION CURVE

The generated voltage of a dc machine will be directly proportional to the air-gap flux if the machine is operated at a constant speed. Therefore, the induced emf in a generator is directly proportional to the flux and to the speed. The flux is produced by the field ampere-turns, since the turns on the field remain constant, the flux is a product of the field current. It is not directly proportional to the field current because of the varying permeability of the magnetic circuit.

Figure 11-11 shows the relation existing between the field ampere-turns and the flux per pole in a generator driven at a constant speed. The flux is not zero when the field current is zero; instead, it has some value, such as 0 A, due to the residual magnetism in the magnetic circuit. At low values of flux, most of the field ampere-turns are used in overcoming the reluctance of the air gap, and the curve is straight.

As the field poles approach saturation, the curve begins to bend at point *b*. Initially, the armature teeth become saturated, followed by other parts of the magnetic circuit, such as the armature core, the field cores, and the yoke. Because of this saturation, the rate of increase of the flux diminishes more and more as the field current increases to higher values. Beyond point *c* the magnetic circuit becomes highly saturated, resulting in a small increase in flux when there is a large increase in current.

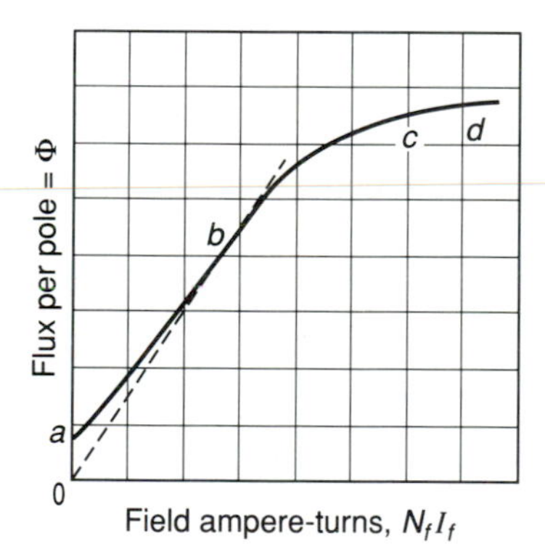

FIGURE 11-11 Saturation curve.

To obtain data to plot a **magnetization**, or **saturation curve**, a dc generator is driven at a constant speed. Then, with the field excited separately and varied over a considerable range, the emf and field current values are recorded. The data are then plotted with the generated voltage as a dependent variable, or the ordinate, versus the field current as the independent variable, or the abscissa. Saturation curves are extremely important when analyzing, predicting, and comparing the operating performance of the various types of generators.

After data have been obtained for increasing values of field current, a different set of values will be obtained as the magnetization is gradually reduced to zero. The descending curve will not match the ascending curve exactly. It will lie slightly above the ascending curve, due to the hysteresis property of the iron. Hysteresis tends to produce a lag of the magnetization with respect to the magnetizing force. Figure 11-12 shows the relative position of the two curves.

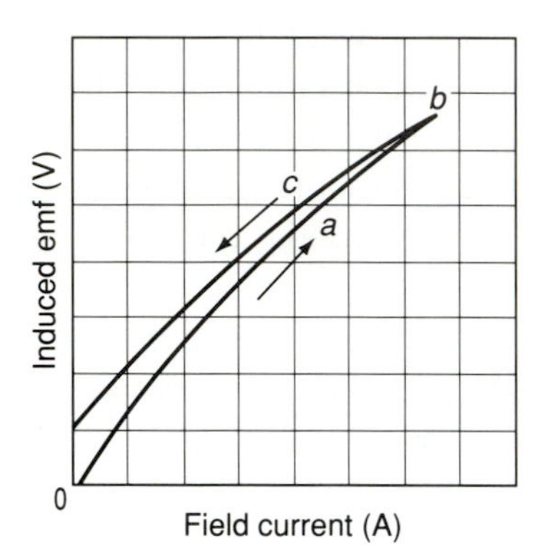

FIGURE 11-12 Magnetization curves illustrating effect of hysteresis.

To determine the saturation curve experimentally, connect the field, in series with an ammeter, across a dc source of power. This ammeter measures the field current. A voltmeter should be connected across the armature terminals to read the emf. Since the *IR* drop in the armature is negligible, the terminal volts and the induced emf are equal under these conditions. The generator is then driven at rated speed, and successive values of emf are obtained from the voltmeter readings for various values of field current. Speed must be kept constant while the readings are being taken. If the speed cannot be maintained at a constant value, corrections can be made for any variation, since the induced emf is directly proportional to the speed.

If the machine has a high-resistance field winding, it is convenient to use a field rheostat connected in series with the field winding. This rheostat enables the curve to be started from the zero value of the field current. The connections for obtaining a magnetization curve are shown in Figure 11-13. If the turns in the field winding are known, the curve can be plotted between volts and field ampere turns.

The values of field current should be increased continuously in one direction from zero to a maximum for a curve of increasing field current, and then decreased from a maximum value to zero. Otherwise, hysteresis effects will be introduced.

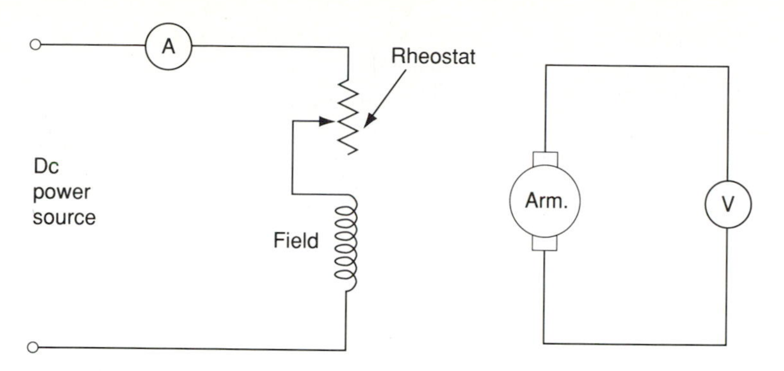

FIGURE 11-13 Circuit for obtaining saturation curve.

11-6 TYPES OF DIRECT-CURRENT GENERATORS

There are two general classifications of dc generators:

1. Separately excited generators, whose fields are energized by a source of direct current external to the machine
2. Self-excited generators, whose fields are energized by their own armatures

If the field of a self-excited generator is connected in series with the armature, the generator is called a **series generator**. When the field is connected in parallel with the armature circuit, it is referred to as a **shunt generator**. If both series and shunt fields are used, the generator is called a **compound generator**.

In the compound generator, each pole of the machine has two coils wound around it. One coil is part of the series winding and the other is part of the shunt winding. When the current in the series coil is in such a direction that it sets up a flux in the same direction as that set up by the shunt coils, it is called a **cumulative compound generator**. When the series coils set up a flux in the opposite direction to that of the shunt-field winding, it is called a **differential compound generator**.

Figure 11-14 shows the schematic diagrams for the series and shunt types of generators. The two methods of connecting compound generators are illustrated in Figure 11-14(c) and (d). The short shunt compound generator has the shunt field connected in parallel with the armature only, while the long shunt has the shunt field connected in parallel with both the armature and series field.

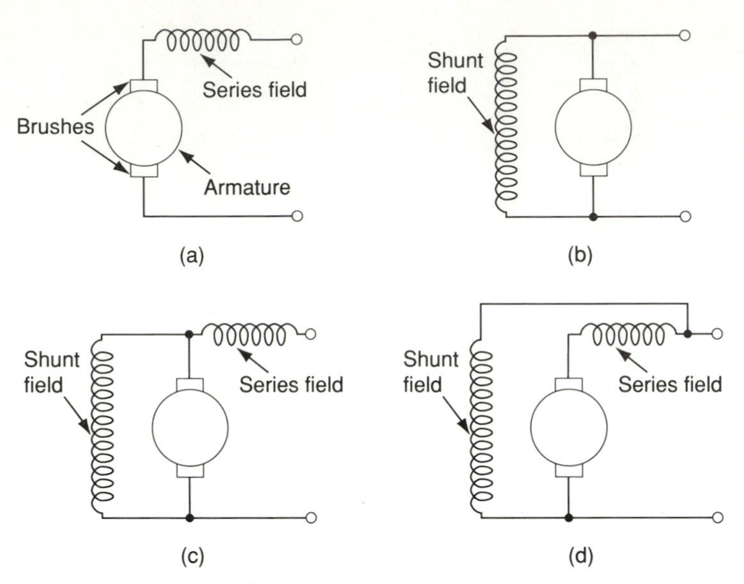

FIGURE 11-14 Field connections for (a) series, (b) shunt, (c) short-shunt compound, and (d) long-shunt compound generators.

The separately excited generator has current supplied to the field coils from an outside source, such as a storage battery or another generator. The field winding consists of a large number of turns of small wire carrying a low value of current. Therefore, its resistance is relatively high. This winding is made up of as many identical coils as there are poles, and the coils are connected in series. The two methods of connecting separately excited generators are shown in Figure 11-15.

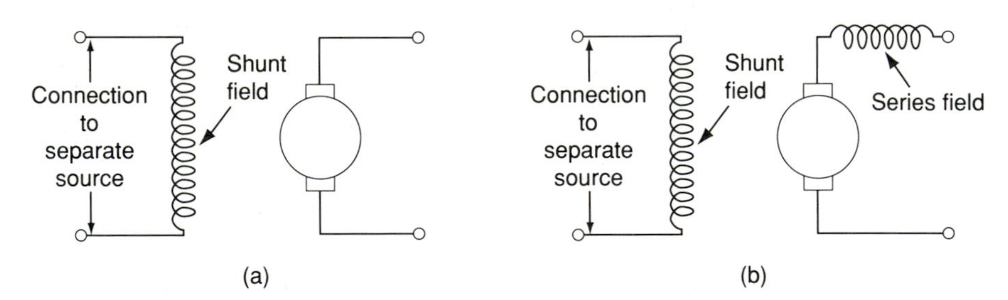

FIGURE 11-15 Separately excited generators: (a) separate excitation shunt field; (b) separate excitation shunt field compound operation.

11-7 VOLTAGE BUILDUP OF A SHUNT GENERATOR

The shunt generator excites its own field. At the instant the generator is started, there is no voltage across its terminals and no field current. As the generator comes up to speed, a small induced emf appears at its terminals, due to the cutting of the residual flux by the armature conductors. This implies that there will be a small current through the field. If the current is set up in the right direction, the flux will increase, and a higher voltage is generated. This higher voltage will force more current through the field circuit. This again increases the flux and the induced voltage, and continues to do so until the resistance of the field circuit and the bend in the magnetization curve prevent any further increase in voltage.

This voltage buildup behavior is illustrated in Figure 11-16. A saturation curve of a shunt generator is shown together with its field-resistance line, R_F, plotted to the same scale. As the generator comes up to speed, the voltage *Oa* is generated in the armature by residual magnetism. Since the field current is directly across the brushes, the current *Ob* is forced through the field. The field current *Ob* generates the voltage *Oc*, which in turn forces the current *Od* through the field. The machine continues to build up until the point where the field-resistance line crosses the saturation curve.

If the field resistance was increased by means of a field rheostat, the voltage would not build up to as high a value. Therefore, by adjusting the resistance of the field current, the value of buildup voltage can be controlled. If the resistance of the field circuit is increased to that of resistance line *T*, the machine will build up to any point between *x* and *y*, and will be unstable. If the resistance is increased even slightly above this amount, the machine will not build up much above the voltage due to residual magnetism. This is referred to as **critical field resistance.**

A shunt generator may fail to build up for any one of the following reasons:

1. *Lack of residual magnetism.* The voltage of self-excited shunt generator will not rise much above an extremely low residual value if the residual

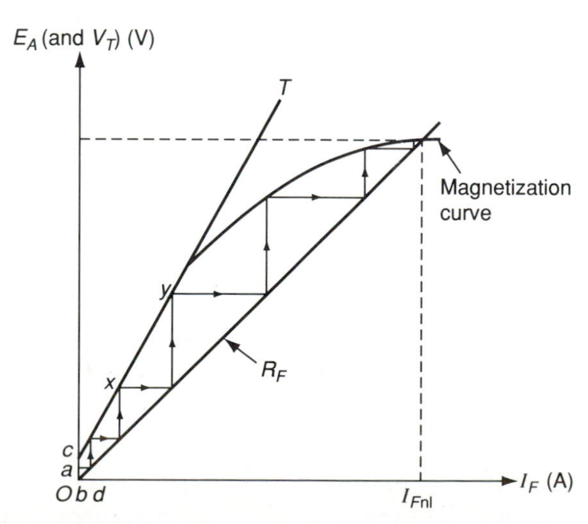

FIGURE 11-16 Buildup of self-excited shunt generator.

flux is insufficient. If the residual flux = 0, then $EA = 0$, and the voltage never builds up. A reversed field connection, at some time, may have opposed the residual magnetism and caused it to diminish to a value so small that the field current resulting from the initial induced emf is insufficient to produce an appreciable increase in the flux. Lack of residual magnetism can be determined by connecting a voltmeter across the armature terminals and bringing the generator up to speed. If the voltmeter reads practically zero, there is no residual magnetism. To solve for a lack of residual magnetism, it is necessary to "flash" the generator field. First disconnect the field from the armature circuit and connect it directly to an external dc source such as a battery. The current flow from this external dc source will leave a residual flux in the poles, which will then allow normal starting.

2. *Field connections reversed.* The direction of rotation of the generator may have been reversed, or the connections of the field may have been reversed. In either case, the residual flux produces an internal generated voltage. It is possible for a generator to build up to a satisfactory voltage and have the opposite required voltage. The easiest way to solve for this problem is to shut down the machine and reverse the external connections. This condition of a reversed field connection rarely occurs with a shunt generator that has been properly installed and is in service. However, with compound generators, a flashover, an open field in the generator, or other conditions may cause the line current to reverse and enter the armature at its positive terminal. In a situation where the internal generated voltage produces a current that results in a flux opposing the residual flux, the flux will decrease below the residual flux and it is impossible for any voltage to build up. If this problem occurs, it can be corrected by reversing the direction of rotation, reversing the field connections, or by flashing the field with opposite magnetic polarity.

3. *Field-circuit resistance too high.* If the resistance of the shunt field is greater than the critical resistance, the generator will not build up. Normally, the shunt generator would build up to a point where the magnetization curve intersects the field resistance line. In a situation where the field resistance exceeds the critical resistance, the steady-state operating voltage is essentially at the residual level and will never build up. When the field-circuit resistance is too high, the voltmeter reading after the field circuit is closed increases slightly above its initial reading and then stops increasing. The solution to this problem is to reduce the resistance of the shunt-field circuit by decreasing resistance in the field rheostat.

4. *Shunt field open-circuited.* Occasionally, the shunt-field circuit becomes open-circuited. This condition is indicated when there is no change in the voltmeter reading when the field connection to the armature terminals is

made and then broken. The solution to this problem is to test the field circuit and locate the open circuit.

11-8 ARMATURE REACTION

When the armature of a dc generator carries a load current, it becomes an electromagnet. The magnetic field produced in the armature is 90 electrical degrees from that produced by the main field. The resulting magnetic action of the armature then tends to distort and alter the direction of the uniformly distributed main field and create a slight demagnetizing effect. Since this demagnetization is detrimental to commutation, most dc generators are equipped with interpoles and compensating windings that oppose and neutralize the effects of armature reaction.

One of the most important functions of the dc generator is to commutate the armature current properly. Commutation for each individual coil element always involves a succession of extremely short periods during which it is short-circuited. Before the short-circuit occurs, current flows in one direction, while after the short-circuit, the direction of current flow is reversed. For commutation to be successful, coil elements must be short-circuited when the coil sides are cutting no flux. This implies that the brushes must be located so that the coil sides are in magnetic neutral zones during the short-circuit periods. However, it does not necessarily mean that the brushes line up on the mechanical neutral between the pole tips. The net result of this armature reaction is to shift the neutral voltage point in relation to the brush position. Because of this shifting, the commutator and brush switching function is no longer spark free. Sparking at the commutator substantially reduces the life of both the commutator and the brushes.

Before the development of interpoles and compensating windings, armature reaction was prevented by manually rotating the brush hanger mechanism until the neutral point was found. This method was quite effective for fixed current loads, but was relatively useless for loads with varying current values, as it required the brushes to be moved every time the load current changed value.

11-9 EXTERNAL CHARACTERISTIC OF SHUNT GENERATORS

If either a separately excited generator or a shunt generator is loaded after the voltage is built up, the terminal voltage will drop. There are three reasons for this voltage reduction:

1. An internal voltage drop produced by the armature circuit resistance

2. The effect of armature reaction on the air-gap flux

3. The reduction in field current caused by the preceding two reasons

Figure 11-17 shows the external characteristic of a shunt generator. Throughout the test, the speed is maintained at its rated value. An adjustable load is connected across the armature terminals. Load is then applied in steps, with simultaneous readings of the voltmeter and ammeter being taken. Because of a slight hysteresis effect, the terminal voltage at rated current may vary slightly from the value at which it was originally set.

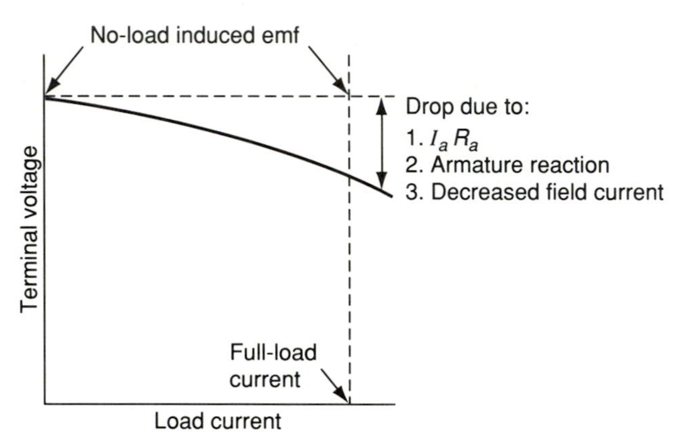

FIGURE 11-17 External characteristics of a shunt generator.

With most generators, the load can readily be carried to the breakdown point and then to a short-circuit condition at rated load current. If the load is gradually reduced, the terminal voltage will slowly increase, returning almost to its no-load value. As the load current increases, the armature current also increases, as does the armature circuit voltage drop, I_aR_a. Therefore, the terminal voltage across the shunt generator armature, V_t, decreases when load is applied. This can be expressed in equation form as

$$V_t = E_g - I_aR_a \quad (11\text{-}7)$$

where V_t = terminal voltage

E_g = generated voltage

I_a = armature circuit current

R_a = total series resistance of all the elements present in the armature circuit, including armature resistance, brush resistance, commutating field resistance, and compensating field resistance

The load circuit, field circuit, and armature circuit are all in parallel in a dc shunt generator. Because of this relationship, the three circuits can be expressed by Kirchhoff's current law. The armature circuit is the source branch of the total circuit, and its current direction is from negative to positive. The field and load circuits have a current direction that is positive to negative. In equation form,

$$I_a = I_f + I_L \tag{11-8}$$

where I_a = armature current

I_f = field current

I_L = load current

Since these three circuits are connected in parallel, their voltage relationship may be expressed using Kirchhoff's voltage law:

$$V_t = V_f = V_L$$

EXAMPLE 11-5 A 50-kW 460-V shunt generator has a total field circuit resistance of 35 Ω and a total armature circuit resistance of 0.025 Ω. Calculate (a) the load current; (b) the field circuit current; (c) the armature circuit current; (d) the generated voltage.

Solution (a) $I_L = \dfrac{\text{kW} \times 1000}{V_L} = \dfrac{50 \times 1000}{460} = 108.7 \text{ A}$

(b) $I_f = \dfrac{V_t}{R_f} = \dfrac{460}{35} = 13.14 \text{ A}$

(c) $I_a = I_f + I_L = 13.14 + 108.7 = 121.84 \text{ A}$

(d) $V_t = E_g - I_aR_a$

$$\therefore E_g = V_t + I_aR_a = 460 + (121.84)(0.025)$$
$$= 463.05 \text{ V}$$

11-10 PERCENT VOLTAGE REGULATION

To indicate the extent to which the voltage of a generator changes as the load is gradually lowered from its rated value to zero load, the term **voltage regulation** or **percent voltage regulation** is used. The American Standards Association defines percent voltage regulation as

> the final change in voltage with constant field-rheostat setting when the specified load is gradually reduced to zero, expressed as a percent of rated voltage, the speed being kept constant.

Since the speed of the prime mover and its generator usually varies somewhat between full load and no load, it is often desirable to take this into account by specifying the overall percent voltage regulation, since this includes both the generator voltage and prime-mover speed characteristics. The following equation is used for determining percent voltage regulation:

$$\%\text{ voltage regulation} = \frac{E_{nl} - E_{fl}}{E_{fl}} \times 100 \qquad (11\text{-}9)$$

where E_{nl} is the no-load (generated) voltage and E_{fl} is the full-load (rated) voltage.

EXAMPLE 11-6 The full-load voltage of a shunt generator is 240 V. What is the percent voltage regulation of the machine if the terminal emf rises to 252 V when the load is reduced to zero?

Solution

$$\%\text{ Regulation} = -\frac{E_{nl} - E_{fl}}{E_{fl}} = \frac{252 - 240}{240} \times 100$$
$$= 5\%$$

EXAMPLE 11-7 A 100-kW 240-V shunt generator has a voltage regulation of 9.6%. (a) Calculate the no-load terminal voltage. (b) Assuming that the voltage varies uniformly between no-load and full-load current, calculate the kilowatt output of the generator for a terminal voltage of 255 V.

Solution

(a)
$$9.6 = \frac{E_{nl} - 240\text{ V}}{240\text{ V}} \times 100$$
$$\therefore\ E_{nl} = \frac{9.6 \times 240\text{ V}}{100} + 240\text{ V}$$
$$= 263.04\text{ V}$$

(b)
$$I_{fl} = \frac{P}{E} = \frac{100{,}000\text{ W}}{240\text{ V}}$$
$$= 416.67\text{ A}$$

$$I_{255} = 416.67\text{ A} \times \frac{263.04\text{ V} - 255\text{ V}}{263.04\text{ V} - 240\text{ V}}$$
$$= 145.4\text{ A}$$

$$P_{255} = \frac{255\text{ V} \times 145.4\text{ A}}{1000}$$
$$= 37.08\text{ kW}$$

11-11 EXTERNAL CHARACTERISTIC OF THE SERIES GENERATOR

The series generator is self-excited and the armature, load, and field windings are in series and carry the same current. The current that is delivered to the load must also provide the necessary excitation for the series field so that a voltage is generated in the armature as well as creating a demagnetizing armature reaction effect. When the load is zero, or open-circuited, the current is zero. Under this condition, the series-field ampere-turns will be zero and the generated voltage will be the residual value, E_r. If there is a load resistance connected, a current will flow, and the series field will cause additional flux, resulting in a higher voltage being generated. At the same time, the armature will develop a demagnetizing action, and a voltage drop will occur in the armature and series–field resistances, R_a and R_{se}, respectively. Therefore, the voltage that will appear at the series-generator terminals will be stabilized at some value that is a function of the net generated voltage and the $I(R_a + R_{se})$ voltage drop. The terminal voltage, V_t, will rise with the load current as long as the overall voltage increases more rapidly than those factors that tend to reduce it. In equation form, the relation between the voltage drops and the terminal voltage may be expressed as

$$V_t = E_g - I_a(R_a + R_{se}) \tag{11-10}$$

When the terminals of a series generator are connected to an external circuit, current will flow in the field windings. If this current in the winding is in the right direction to increase the field strength, the voltage across the brushes will rise. As the resistance of the external circuit is decreased, more and more current will flow in the field windings and the voltage will continue to rise.

The terminal voltage continues to rise until a certain point is reached. If the current output is increased further, the terminal voltage, as shown in Figure 11-18, will decrease. This point depends on the following three factors:

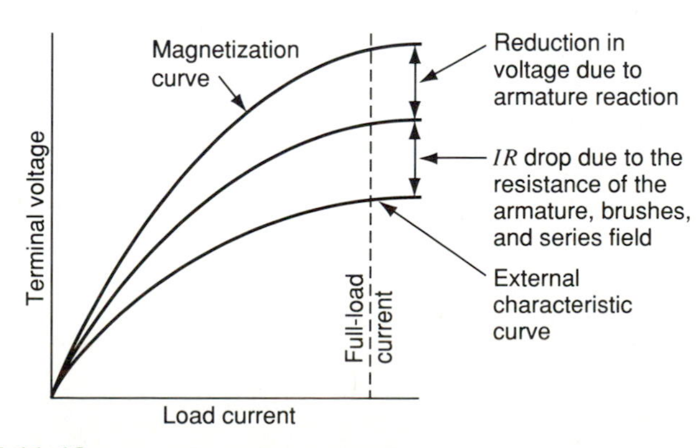

FIGURE 11-18 External characteristic of a series generator.

1. Shape of the magnetization curve
2. Armature reaction
3. Total resistance of the circuit

11-12 EXTERNAL CHARACTERISTIC OF THE COMPOUND GENERATOR

A great majority of generators in service have two sources of field excitation. One of these sources has its shunt field virtually independent of the load, while the other source, the series field, is energized by the load current. These types of machines are called compound generators. The addition of the series field, connected to aid the shunt field, is to create additional values of flux with increasing load currents so that the armature will generate greater voltages and compensate for the shunt generators inherent tendency to lose terminal voltage. It is the load current in a shunt generator that produces the armature *IR* drop, as well as creating armature reaction and a decrease in shunt-field current. By adding a few series turns to the field, these three losses can be compensated for.

If the number of additional series ampere-turns are just sufficient at full-load current to compensate for armature *IR* drop, armature reaction, and *IR* drop in the series winding, the terminal voltage at this load will be equal to the no-load voltage. Under this condition the machine is considered to be a **flat-compound generator**, and the characteristic curve will be similar to the middle curve shown in Figure 11-19. Since a flat-compound generator has a load-voltage characteristic in which the no-load and full-load voltages are equal, it is said to have zero percent voltage regulation.

If the series ampere-turns at full load are more than sufficient to compensate for the armature and series-field drop, as well as the armature reaction, the terminal voltage at this value of load will be higher than at no-load. When the rated-load voltage is higher than the no-load voltage, the generator is said

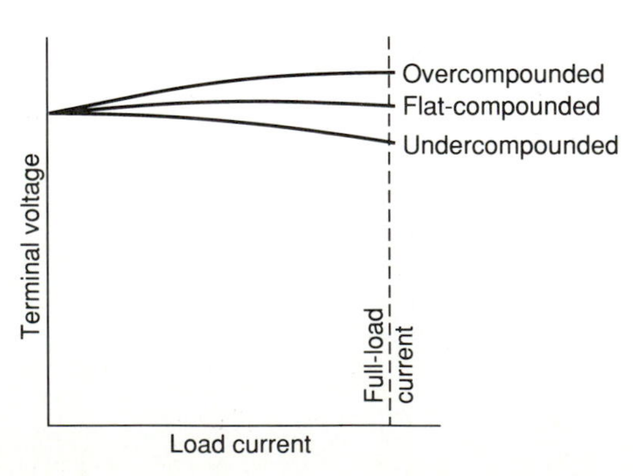

FIGURE 11-19 Compound-generator characteristics.

to be **overcompounded**. When the rated voltage is less than the no-load voltage, the generated is considered to be **undercompounded**. Very rarely would an undercompounded generator be used in industry.

The generator will be flat-compounded, overcompounded, or undercompounded, depending on the number of series-field ampere-turns with respect to the number of shunt-field ampere-turns. The **degree of compounding** is therefore determined by the design of the main field and, once established, will yield the desired no-load and full-load terminal voltages. However, the emf between these two load conditions will not, in general, change on a proportionate basis since magnetic saturation effects do not permit the flux to vary directly with the armature current. The net result is that the external characteristics for the three types of compound generator are not linear curves but are concave. For example, in the flat-compound generator, the voltages between no-load and full-load will be higher than at the terminal loads.

Usually, compound generators are manufactured to be overcompounded. This allows the degree of compounding to be controlled by diverting current away from the series field. A low-resistance shunt, or **diverter**, is connected in parallel with the series field to accomplish this control of compounding. The circuit for a series field diverter is shown in Figure 11-20. The diverter is generally made of German silver or manganin, and is located where it will have no magnetic influence. The ratio between the resistance of the diverter and the resistance of the series field will determine how much current is diverted around the series field. When the diverter resistance is extremely large, the diverted current will be small and the external characteristic will be that of an overcompounded generator. If the diverter resistance is extremely small, the diverter current will be large and the external characteristic will be that of a shunt generator.

In Figure 11-20, the line current, I_L, divides into two parts, I_{SE} and I_D, the series-field and diverter currents, respectively. Since the series-field resistance, R_{SE}, is in parallel with the diverter resistance, R_D, the total line current

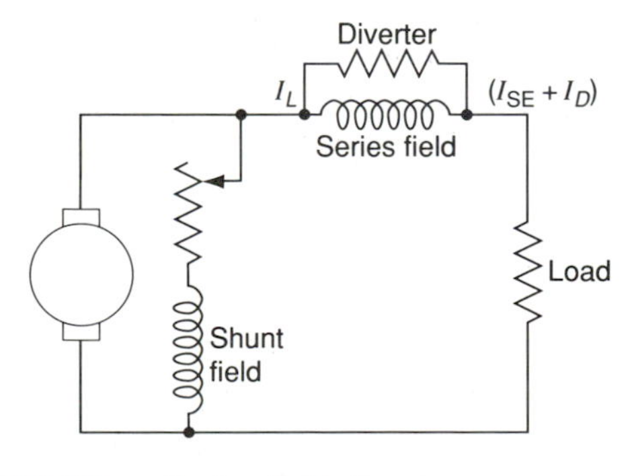

FIGURE 11-20 Series-field diverter.

will divide so that I_{SE} and I_D are related to each other by an inverse ratio of the respective resistances:

$$\frac{I_D}{I_{SE}} = \frac{R_{SE}}{R_D} \tag{11-11}$$

Since $\quad I_L = I_D + I_{SE}$, then

$$I_{SE} = I_L \frac{R_D}{R_D + R_{SE}} \tag{11-12}$$

The correct diverter resistance is generally determined experimentally. By manipulating the field rheostat, with the generator operating at proper speed, the no-load voltage is set for the desired value. Load is then applied so that rated current flows, and the terminal voltage is observed. If the terminal voltage is higher than the desired value, a diverter is connected across the series field and the full-load test is repeated. If the full-load voltage is still too high, the ohmic value of the diverter must be reduced. If the full-load voltage is too low, the diverter resistance is increased. It is necessary to use extreme caution when performing this test, since the values of current flowing through the diverter and shunt field are relatively high.

EXAMPLE 11-8 The following information is given regarding a short-shunt overcompound generator equipped with a diverter: $I_L = 1100$ A, $R_{SE} = 0.003\ \Omega$, $R_D = 0.009\ \Omega$, series-field turns/pole = 6. Calculate (a) the number of series-field ampere-turns when operated with the diverter connected; (b) the number of series-field ampere-turns when operated with the diverter disconnected.

Solution

(a) $$I_{SE} = I_L \frac{R_D}{R_D + R_{SE}} = 1100 \times \frac{0.009}{0.009 + 0.003} = 825 \text{ A}$$

$$6 \times 825 = 4950 \text{ At}$$

(b) $$6 \times 1100 = 6600 \text{ At}$$

Percentage overcompounding of a generator is defined as the voltage rise between no-load and full-load currents, calculated as a percentage of the no-load voltage. In equation form,

$$\text{percentage overcompounding} = \frac{E_{fl} - E_{nl}}{E_{nl}} \times 100 \tag{11-13}$$

11-13 LOSSES AND EFFICIENCY OF A DYNAMO

Dynamo is a term used to describe a machine that can be used as either a motor or a generator, depending on its application. In direct-current machines, losses of energy inevitably occur. These losses may be classified under two main categories:

1. *Losses due to ohmic resistance.* When current flows in a circuit the power wasted in overcoming the resistance is given as the product of the resistance and the square of the current. In dc machines, such ohmic losses occur in the armature winding, in the brushes and brush contacts, in the interpole windings, and in the shunt and series windings. Power wasted in the armature circuit is found by

$$P = I_a^2 R_a \tag{11-14}$$

   where I_a is the total current flowing through the armature circuit and R_a is the sum of the resistance of the armature, interpole windings, series windings, and brush rigging.

2. *Losses due to changing magnetization.* These losses occur in any iron that carries an alternating, or changing, flux, and in particular in the armature core plates and pole shoes. In the armature core plates, the magnetic flux alternates with a frequency that is determined by the angular speed and the number of pairs of poles. In the pole shoes, the flux changes in magnitude and direction at a frequency determined by the product of the number of slots on the armature and the speed of the armature. These variations in magnitude and direction cause the pole shoes to vibrate and cause an audible hum in the dynamo.

In a compound generator, the loss of power increases in direct proportion to the square of the current. If the terminal voltage was held constant, the loss of power in the shunt field would also be constant and would be independent of any changes occurring in the armature current. In the case of the shunt generator, to maintain a constant terminal voltage, the field current must be increased as the load increases. The losses due to ohmic resistance and changing magnetization are caused by three factors:

1. *Eddy currents.* The changing flux in the iron establishes emfs that cause current to flow through the iron. To reduce these circulating, or eddy, currents, it is necessary to construct the core out of a number of thin laminated plates. The loss may be shown to vary in proportion to the square of the thickness of the plates. In addition, the loss due to eddy currents depends

on the square of the emf induced in the plates. Consequently, it is also dependent on the product of the square of the flux density and the frequency.

2. *Hysteresis loss*. This loss occurs in the revolving armature core and is caused by the magnetic polarity of the iron changing in sequence with the changing positions of the magnetic materials under various poles. For example, when an armature-core slot is passing under a south pole, its polarity will be north, and the iron particles are oriented with their south ends pointing inward. When this same slot moves under a north pole, its polarity changes and the iron particles rapidly change orientation. This extremely fast "jerking" of the particles causes a friction in the material which results in a dissipation of heat. The hysteresis loss is magnetic in character but is a result of the mechanical movement of the armature core. For this reason, hysteresis losses are considered to be rotational losses.

3. *Friction losses*. These losses take into account the bearing and windage losses caused by the rotating armature. They may be determined by using a calibrated motor to drive the armature of the machine at the proper speed, with the brushes lifted. The losses that occur are the difference between the output of the motor and the output of the generator. Another friction loss is that caused by brush friction. This loss varies due to the type of brushes and the nature of the surface with which the brush makes contact. Brush friction loss can be found by also using a calibrated motor and noting the difference in power required for the armature turning with the brushes making contact with the commutator and with the armature turning freely.

When a dynamo is working under constant full-load conditions, the energy wasted in the windings and the iron of the armature causes its temperature to rise above the ambient temperature. The final temperature attained is such that the machine dissipates the heat as quickly as the heat is produced. Most small dc machines use a fan to dissipate this heat loss and allow a relatively large power loss without undue rise in temperature.

The output power of a dc machine is equal to the input power minus the losses that occur in the dynamo:

$$\text{power output} = \text{power input} - \text{losses} \tag{11-15}$$

The ratio of the power output to the power input is called the **efficiency**, which is usually expressed as a percentage:

$$\text{efficiency} = \frac{\text{output}}{\text{input}} \times 100 \tag{11-16}$$

11-14 PARALLELING OF DIRECT-CURRENT GENERATORS

In industrial generating stations, power is usually supplied from several smaller units instead of from one large generator, for the following reasons:

1. A generator that breaks down may be removed from the system for repairs without interrupting the power supply.
2. The efficiency of the generating station may be maintained at a high level by varying the number of generators in service. When demand is low, fewer generators can be utilized, allowing the units to operate constantly at near full-load.
3. Generators may be added to the system as power demand grows in the area to which the generating station supplies power. This provides an economical method of meeting power demand. As residential, or industrial, subdivisions expand, more generators can be brought on line.

To operate dc generators in parallel, three basic requirements must be met:

1. The generators must be connected positive terminal to positive terminal.
2. The output voltages of both generators must be approximately equal to each other.
3. The prime movers of both generators should have similar rotational speed characteristics.

When paralleling generators, the prime movers should either have constant rotational characteristics, or they should droop in speed with increasing load. A single dc generator with a drooping voltage–current characteristic is shown in Figure 11-21. Initially, load current I_{L1} is supplied by generator V_{L1}. By increasing the load on the generator to point I_{L2}, the terminal voltage falls to V_{L2}. Increasing either the speed of the generator, or the amount of field current, will cause the generated voltage to increase.

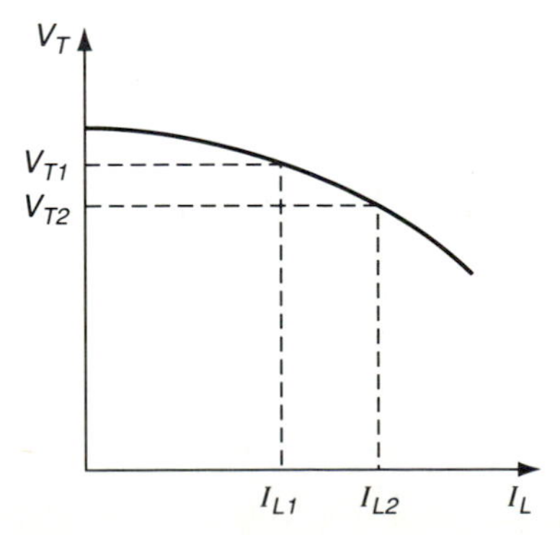

FIGURE 11-21 Single generator operating with a dropping voltage–current characteristic.

11-15 SHUNT GENERATORS IN PARALLEL

When generators are connected in parallel, they function together to supply power to a common load. Under ideal conditions, the combined rating of two or more generators is approximately equal to the total load, and each generator assumes its proportionate share of the total load. To achieve such a load division, the voltage changes of all machines must be exactly the same for equal changes in the percent change of load. Therefore, under ideal conditions, the generators must have identical external voltage versus load characteristics.

When operating in parallel, shunt generators are completely stable regardless of whether or not their external voltage versus load characteristics are identical. This is due to the drooping nature of their characteristic curves. Figure 11-22 shows two shunt generators connected in parallel through a common bus line to which the feeders of the power plant are connected. Since the bus bar, or terminal, voltage must be the same for both generators, if generator *A* attempts to take more load than generator *B*, there will be a reduction in generator *A*'s emf and a corresponding rise in the emf at generator *B*.

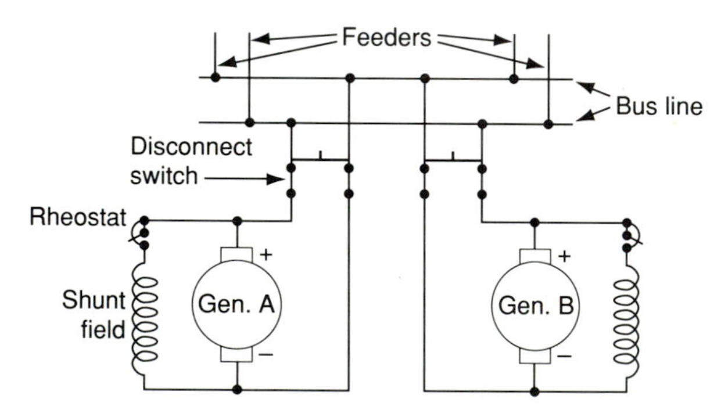

FIGURE 11-22 Shunt generators in parallel.

When two shunt generators have exactly the same characteristics, the current in each would be half the total at all times, since at the same terminal voltage the same value of current would be delivered by each. If the generators are of different capacities, their characteristics must be such that the total load will always be divided among them in the proportion of their ratings. If a 150-kW generator is connected in parallel with a 300-kW generator having the same characteristics, a load of 360 kW would divide so that the first generator supplied 120 kW and the second machine supplied 240 kW.

When the external characteristic curves are not similar, as in Figure 11-23, one generator will deliver a larger percent of its rated capacity than the other. Although both machines have different values of voltage at no-load, when full-load current flows they have the same terminal voltage.

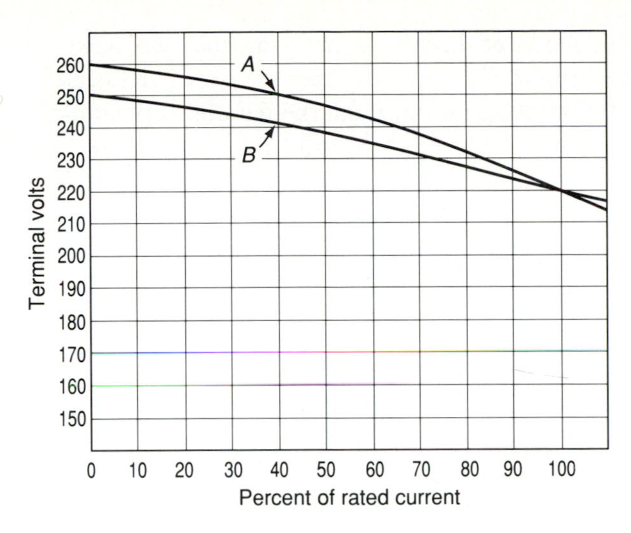

FIGURE 11-23 Characteristic curves of two shunt generators operated in parallel.

EXAMPLE 11-9 Generator A in Figure 11-23 has a rating of 200 kW, and generator B has a rating of 300 kW. Both have a rated voltage of 220 V dc. Calculate the kilowatt output of each generator as well as the total power output when the terminal voltage is 230 V.

Solution From the graph, at 230 V generator A delivers 82% of rated current and generator B delivers 70%.

$$\text{Generator A: } 0.82 \times \frac{200{,}000\ \text{W}}{220\ \text{V}} = 745.455\ \text{A}$$

$$\frac{230\ \text{V} \times 745.455\ \text{A}}{1000} = 171.455\ \text{kW}$$

$$\text{Generator B: } 0.70 \times \frac{300{,}000\ \text{W}}{220\ \text{V}} = 954.545\ \text{A}$$

$$\frac{230\ \text{V} \times 954.545\ \text{A}}{1000} = 219.545\ \text{kW}$$

$$\text{Total power output} = 171.455\ \text{kW} + 219.545\ \text{kW} = 391\ \text{kW}$$

11-16 COMPOUND GENERATORS IN PARALLEL

In Section 11-15 it was stated that shunt generators will function properly in parallel as long as they have identical characteristics. With the compound generator, it is impossible to operate in parallel even when their characteristics are identical, unless an additional feature, known as the *equalizer connection*, is provided. This is accomplished by connecting the two negative brushes together, as shown in Figure 11-24. Since the two series fields are connected in parallel, the total current divides through the parallel series fields.

$$I_T = I_{SEA} + I_{SEB} \tag{11-17}$$

The two series field currents are inversely proportional to their respective resistances.

$$\frac{I_{SEA}}{I_{SEB}} = \frac{R_{SEB}}{R_{SEA}} \tag{11-18}$$

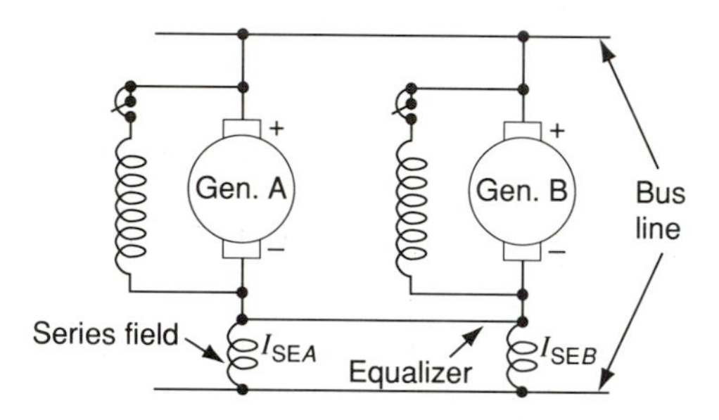

FIGURE 11-24 Compound generators in parallel.

The equalizer functions in the following manner: Assume that the generator begins to take more than its share of the load. The increase of load current will cause an increase of current not only in the series field of generator *A* but also, by means of the equalizer, in the series field of generator *B*. Therefore, both generators are affected in a similar manner, and neither is able to run away with the load.

When compound generators are not identical, they will function most satisfactorily in parallel when:

1. The external characteristics of both generators are similar.
2. The resistances of the series fields are inversely proportional to the ratings of the machines.

The first condition is accomplished by carefully adjusting the diverter resistance across the series field of either generator. The second condition is attained by placing a resistor with a low ohmic value in series with one of the series fields.

To parallel a compound generator, it is first brought up to normal speed. If an ammeter is placed in series between the armature circuit and the bus bar, it will indicate zero amperes if the voltage of the generator is the same as the voltage of the bus bar. If the voltage of the generator was a little lower than that of the bus bar, the ammeter would show a reversed deflection, indicating that current is being taken from the bus bar by the generator. If the generator voltage was somewhat greater than that of the bus bar, the ammeter would indicate in the correct direction and its reading would be the current delivered to the system by the machine. When the generator is operating at the correct voltage, the load carried by the machine is set by adjusting the shunt-field rheostat.

To take a compound generator off line, the shunt-field resistance is increased until the excitation has been sufficiently reduced and zero current flows between the armature and bus bar. The disconnect switch is then opened.

KEY TERMS

Dynamo
Electromechanical device
Generator
Motor
Yoke
Field winding
Compound field
Field pole
Shoe
Commutator
Brush
Brush rigging
Interpole
Armature
Armature winding
Plex
Simplex armature winding
Duplex armature winding
Multiplex winding
Pole pitch
Full-pitch coil
Fractional-pitch coil
Chorded winding
Lap winding
Frog-leg winding
Simplex-lap winding
Wave winding
Multiplex wave winding
Neutral plane
Frequency
Magnetization (saturation) curve
Series generator
Shunt generator
Compound generator
Cumulative compound generator
Differential compound generator
Critical field resistance
Voltage regulation
Percent voltage regulation
Flat-compound generator
Overcompounded
Undercompounded
Degree of compounding
Diverter
Percentage overcompounding
Efficiency

PROBLEMS

11-1 A duplex lap-wound armature is used in an eight-pole machine with eight brush sets each spanning two commutator bars. Determine the number of paths in the armature.

11-2 Determine the number of parallel paths in a simplex lap-wound armature having four poles.

11-3 A simplex wave-wound armature is used in a four-pole machine. How many parallel paths exist in the armature?

11-4 A triplex wave-wound armature is used in a four-pole armature. Determine the number of parallel paths.

11-5 The armature of an eight-pole generator has 6000 conductors. The armature has a duplex lap winding and the flux per pole is $120{,}000 \times 10^{-7}$ Wb. If the armature is rotating at 1800 rpm, calculate the generated voltage.

11-6 A 12-pole dc generator with a triplex lap-wound armature has a flux per pole of $80{,}000 \times 10^{-7}$ Wb. Determine the generated voltage if the armature is rotating at 1200 rpm and the total number of conductors is 4000.

11-7 Determine the frequency of the alternating voltage developed in the armature of an eight-pole generator operating at a speed of 1800 rpm.

11-8 How many poles would a generator have if it developed a frequency of 200 Hz when operating at 1200 rpm?

11-9 At what speed must the armature of a four-pole generator rotate to develop a voltage at 60 Hz?

11-10 A dc shunt generator has a generated emf of 245 V when the armature circuit current is 80 A. The resistance of the armature circuit is 0.31 Ω. Determine the terminal voltage.

11-11 A shunt generator delivers a load current of 50 A. If the armature current is 52.5 A, what is the value of the field current?

11-12 A 20-kW 230-V shunt generator has an armature circuit resistance of 0.085 Ω and a field circuit resistance of 40 Ω. Determine (a) the load current; (b) the field circuit current; (c) the armature circuit current; (d) the generated voltage.

11-13 A shunt generator has a full-load voltage of 230 V. Determine the percent regulation if the terminal voltage is 241 V when load voltage is zero.

11-14 A 200-kW 460-V shunt generator has a voltage regulation of 8.2%. Determine (a) the no-load voltage; (b) the power output for the generator with a terminal voltage of 470 V. Assume uniform variance of voltage between no-load and full-load.

11-15 A series dc generator has an armature current of 60 A and a generated emf of 250 V. The armature and brush resistance is 0.24 Ω, and the resistance of the series field is 0.09 Ω. Determine the terminal voltage.

11-16 A 100-kW 220-V dc generator has a series-field resistance of 0.025 Ω, a shunt-field resistance of 80 Ω, and an armature resistance of 0.015 Ω. Determine the generated voltage if the machine is connected (a) long shunt; (b) short shunt.

11-17 A 2500-kW 2000-V 1800-rpm separately excited dc generator has an armature circuit resistance of 0.35 Ω. Determine the induced voltage.

11-18 A 20-kW 220-V shunt generator has a winding resistance of 20 Ω and an external resistance of 12 Ω. The armature resistance is 0.05 Ω. Calculate (a) the full-load current; (b) the field current; (c) the armature current; (d) the generated voltage.

11-19 A short-shunt overcompound generator equipped with a diverter has a load current of 800 A, series-field resistance of 0.002 Ω, diverter resistance of 0.005 Ω, and a series-field turns/pole of 4. Determine (a) the number of series-field ampere-turns when operated with the diverter connected; (b) the number of series-field ampere-turns with the diverter disconnected.

11-20 A 115-V short-shunt compound generator is rated at 60 A. The machine has a series-field resistance of 0.02 Ω, an armature resistance of 0.06 Ω, and a diverter resistance of 0.035 Ω. The shunt-field current is 2.5 A. Determine the generator voltage and the total power developed by the machine.

11-21 The series-field coils of an eight-pole 100-kW 440-V compound generator each have eight turns of wire. The diverter resistance is 0.025 Ω, and the total resistance of the series field is 0.01 Ω. Determine the number of ampere-turns of each series-field coil when the machine is operating at full load.

11-22 A 20-kW 115-V long-shunt compound generator supplies a load current of 250 A at rated voltage. The machine has a rotational power loss of 1200 W and a shunt-field current of 8 A. The series-field resistance is 0.015 Ω and the armature resistance is 0.06 Ω. Determine the efficiency of the generator.

11-23 The no-load voltage of a 100-kW 220-V shunt generator is 229 V. The resistance of the shunt-field circuit is 41 Ω, and the armature resistance is 0.03 Ω. Determine at rated load: (a) the induced emf; (b) the field losses; (c) the armature losses; (d) the total power generated; (e) the voltage regulation.

11-24 A shunt generator delivers 50 kW when the terminal voltage is 220 V and the generated voltage is 232 V. If the field resistance is 45 Ω, determine the armature circuit resistance.

11-25 A 20-kW 115-V compound generator has a series-field resistance of 0.018 Ω, shunt-field resistance of 55.3 Ω, and an armature resistance of 0.15 Ω. The rotational losses of the machine are 680 W. Assuming a long-shunt connection, determine the full-load efficiency.

11-26 A 50-kW 230-V dc generator has rotational losses of 1450 W, and the shunt-field circuit draws 7.25 A. The voltage drop across the brushes is 1.8 V, and the resistance of the armature circuit is 0.022 Ω. Determine (a) the load current supplied; (b) the total losses of machine; (c) the input power; (d) the efficiency of generator at rated load.

CHAPTER

12 DC Motors

LEARNING OBJECTIVES

Upon completion of this chapter you will be able to:

- List the three quantities that affect the force acting on a current-carrying conductor.
- Explain Fleming's left-hand rule.
- Discuss the torque of a motor and the relationship between torque and power.
- Determine the armature current for given values of armature resistance, terminal voltage, and counter emf.
- Explain armature reaction in a motor.
- Discuss the relationship between motor speed and armature current.
- Explain the basic differences between shunt, series, and compound motors.
- Discuss dc servomotors and their applications in industry.
- List the two most popular types of stepper motors.
- Explain the basic operation of the brushless dc motor.
- List the three main applications of SCRs in motor control circuits.
- Define pulse width modulation (PWM) and duty cycle.
- Explain the basic operation of the chopper circuit.
- Define dynamic braking and regenerative braking.

12-1 INTRODUCTION

The electric motor is a machine that converts electrical energy into mechanical energy. The construction of a dc motor is basically the same as that of a dc generator. A dc generator may be made to function as a motor by applying a suitable source of direct voltage across the normal output electrical terminals.

When a current flows through a conductor in a magnetic field, it will tend to cause the conductor to move at right angles to the lines of force. The direction that the conductor will tend to move in will depend on the direction of the lines of force and the direction of the current in the conductor. Figure 12-1(a) shows a uniform field between the opposite poles of two magnets. The cross section of a conductor placed between two poles and carrying current toward the observer is represented in Figure 12-1(b). If the conductor of Figure 12-1(b) were placed in the magnetic field of Figure 12-1(a), the resul-

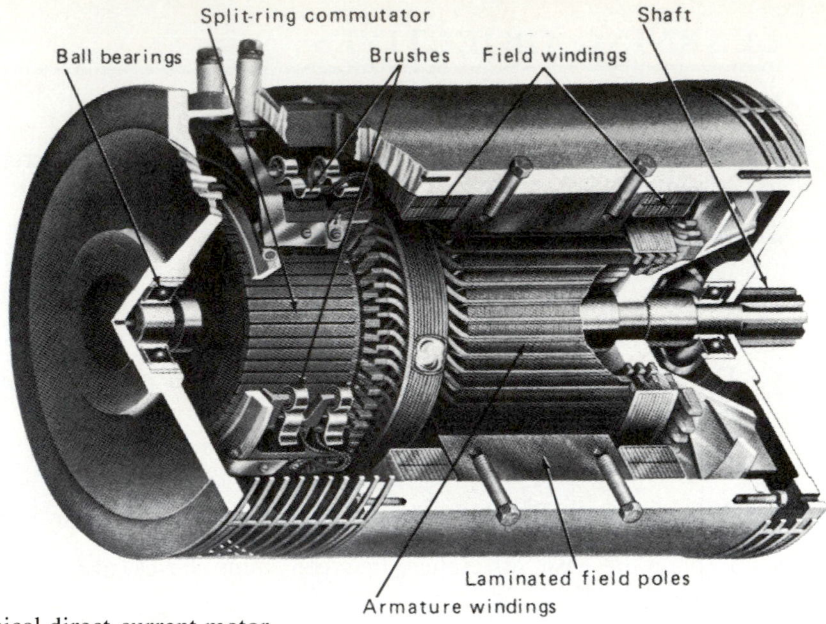

Typical direct-current motor

Source: Prentice Hall Inc. / Courtesy of General Electric Company, DC Motor and Generator Department

tant magnetic field would then be as shown in Figure 12-1(c). Above the conductor, the flux density is reduced because the two fields are opposite in direction and tend to destroy each other. Below the conductor, the flux density is increased because they are in the same direction. Magnetic lines have a tendency to straighten out, which results in the conductor in Figure 12-1(c) being forced in an upward direction. By reversing the direction of current in the conductor, as in Figure 12-1(d), the direction of motion will also be reversed. This is due to the flux density above the conductor increasing, and the flux density below the conductor decreasing.

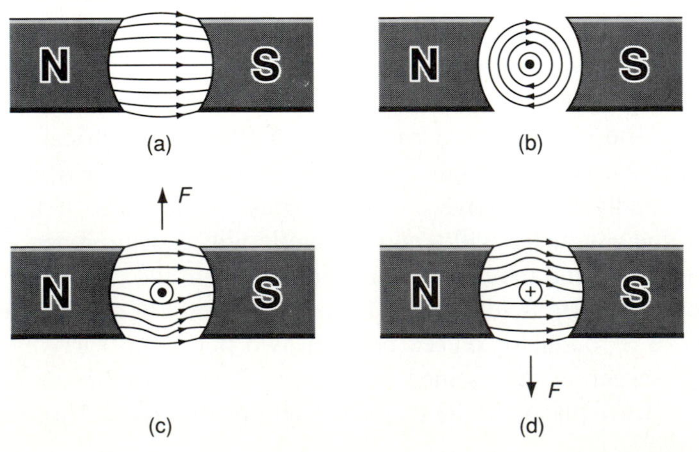

FIGURE 12-1 Force acting on conductor carrying current in a magnetic field.

12-2 FORCE ON A CONDUCTOR

The force acting on a conductor carrying current in a magnetic field is directly proportional to three quantities:

1. Strength of the field
2. Magnitude of the current
3. Length of the conductor

The following equation is used to calculate the force acting on a conductor:

$$F = BlI \qquad (12\text{-}1)$$

where F = force, in newtons

B = flux density, in webers/square meters

l = length of conductor in magnetic field, in meters

I = current, in amperes

EXAMPLE 12-1 A flat rectangular coil of 35 turns lies with its plane parallel to a magnetic field. The flux density is 0.50 Wb/m². The length of the coil transverse to the field is 22.5 cm. The current per conductor is 25 A. Determine the force in newtons that acts on each side of the coil.

Solution

$$I = \frac{\text{current}}{\text{conductor}} \times \text{number of conductors} = 25 \times 35$$

$$= 875 \text{ A}$$

$$F = BlI = (0.50 \text{ Wb/m}^2)(0.225 \text{ m})(875 \text{ A})$$

$$= 98.44 \text{ N}$$

12-3 FLEMING'S LEFT-HAND RULE

When motion in electrical machinery is discussed, it is important to realize that what is actually being implied is the **relative motion** of one thing to another. For example, the conductor may move so as to cut the lines of force of a field that remains stationary in space. In this example, as the field is stationary the relative motion of the conductor to the field is the same as its actual motion. Conversely, the conductors may remain stationary and the field moves so that its lines of force cut the conductors.

Fleming's right-hand rule relates to the direction of a magnetic field, the direction of motion of a conductor in that field, and the direction of induced emf. If a conductor is carrying current and at the same time is moving across a magnetic field, it will exert a force on its confining structure. Then force is referred to as **motor action**. The relationship between current direction, field direction, and the developed force on the conductors may be determined in a manner similar to the right-hand rule. This relation is expressed by **Fleming's left-hand rule**. The rule is illustrated in Figure 12-2 and may be stated as follows:

> Extend the thumb, the forefinger, and the middle finger at right angles to one another. Now turn the hand in such a position that the first finger points in the direction of the field or flux, and the second finger in the direction of the current in the conductor. The thumb will now point in the direction in which the conductor tends to move.

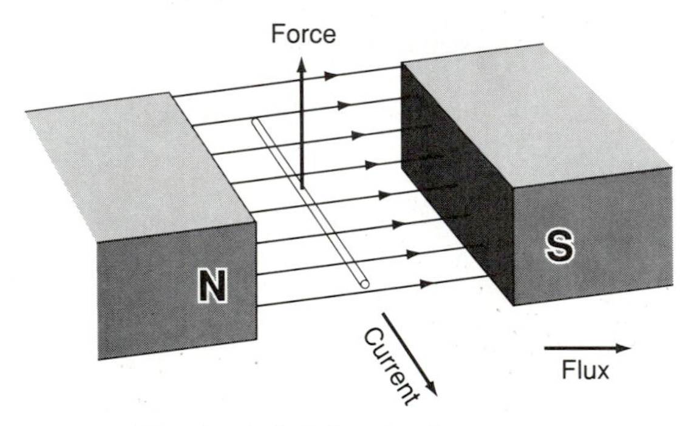

FIGURE 12-2 Fleming's left-hand rule.

12-4 TORQUE OF A MOTOR

When an armature is rotating about an axis, a tangential force is necessary to produce and maintain rotation. The total effect of the force is determined not only by its magnitude, but also by its radial distance from the axis of rotation to the line of action of the force. **Torque** is defined as the product of the force and its perpendicular distance from the axis of rotation.

Figure 21-3 shows a four-pole machine. The current is shown flowing toward the reader in the conductors under the north poles, and away from the reader in the conductors under the south poles. If the armature is rotated 90°, the direction of current in all the conductors is reversed, although the direction of force remains the same.

To develop a continuous torque in a motor, the current in each coil of the armature must be reversed, and this reversal should occur when the coil is passing through the neutral plane, or the plane of zero torque. A commutator

FIGURE 12-3 Torque developed by conductors in motor armature.

is therefore necessary to change the direction of current at the desired point. Direct current is supplied to the motor through the commutator, brushes, and coils of the machine. The commutator acts as a switch, providing the alternating current effect required to produce torque. This switching action of the commutator occurs when flux is at a minimum value. This implies that torque is dependent on flux, current, and the physical dimensions of the motor. These dimensions, such as the size of the armature and number of conductors, are of a fixed value and are represented by the constant K.

To calculate torque in a motor, the following equation may be used:

$$T = KI_a\Phi \tag{12-2}$$

where T = torque, in newton-meters

K = a constant, depending on physical dimensions of motor

I_a = armature current, in amperes

Φ = flux entering armature, in webers

EXAMPLE 12-2 When a motor is drawing 50 A from the line supply, it develops a torque of 70 N·m. Determine the torque when the field flux is reduced to 80% of its original value and current is increased to 85 A.

Solution If the armature current remained at 50 A, the value of torque due to weakening of the field alone would be $0.80 \times 70 = 56$ N·m. Since the armature current has increased to 85 A, the final value of torque will be

$$\frac{85}{50} \times 56 = 95.2 \text{ N} \cdot \text{m}$$

Electromagnetic torque developed by the armature is generally referred to as the **developed torque**. This torque is developed internally and is not able to be used entirely to drive equipment. The torque available at the shaft of the motor is less than the developed torque due to rotational losses. Some of these rotational losses include bearing friction and air movement inside the motor.

12-5 RELATION OF TORQUE AND POWER

In Section 12-4 torque was defined as a force acting at a right angle at a radial distance from a center of rotation. Since 1 newton-meter is 1 newton acting at 1 meter crank radius, this torque force when rotating must pass through 2π times its radius in one complete circle of motion.

In the English system of measurement, mechanical power is measured in horsepower, and torque is measured in foot-pounds. Horsepower is defined as the *rate of doing work* and is equal to the force multiplied by the distance divided by the time to traverse the distance. In equation form,

$$\text{hp} = \frac{TS}{5252} \qquad (12\text{-}3)$$

where T is the force, in lb, $\times$ radius of armature, in ft·lb, and S is the speed of motor, in rpm. In the SI system of measurement, mechanical power is measured in watts and is defined by the equation

$$\text{power} = T \times 2\pi r/s \qquad (12\text{-}4)$$

where T is the output torque, in newton-meters, and r/s is the number of revolutions through which the armature rotates in s seconds. To convert horsepower to watts, the following equation is used:

$$\text{power (in watts)} = \text{hp} \times 746 \qquad (12\text{-}5)$$

EXAMPLE 12-3 A dc motor has an output torque of 88 N·m when the armature is rotating at 40 rev/s. Calculate the output of the motor in horsepower.

Solution

$$\text{Power} = T \times 2\pi r/s = 88 \times 2\pi \times 40$$
$$= 22{,}116.81 \text{ W}$$

$$hp = \frac{\text{watts}}{746} = \frac{22{,}116.81\text{ W}}{746}$$
$$= 29.65\text{ hp}$$

12-6 MEASUREMENT OF TORQUE

It is often necessary to determine the efficiency of a motor at certain specific loads, and frequently over its entire range of operation. The most common method of making direct measurements of efficiency in motors is to use a **prony brake**. If the motor input, which can be measured with an ammeter and voltmeter, is known, the motor efficiency can be calculated by using the prony brake method.

A typical prony brake configuration is illustrated in Figure 12-4. Generally, the prony brake is a screw-actuated clamp brake band, which is wrapped around a rotating liquid-cooled brake drum. Any desired load may be placed on the motor by adjusting the pressure of the brake shoe blocks. This pressure, or force, which attempts to rotate the brake with the moving drum, is opposed by the brake arm. The power output of the motor is dissipated as heat which is produced as the drum turns in the brake shoes. As the brake arm is pulled by this force, a reading is displayed on the scale. The vibration filter is a mechanical filter designed to reduce brake arm vibration and make the scale easier to read with a greater degree of accuracy.

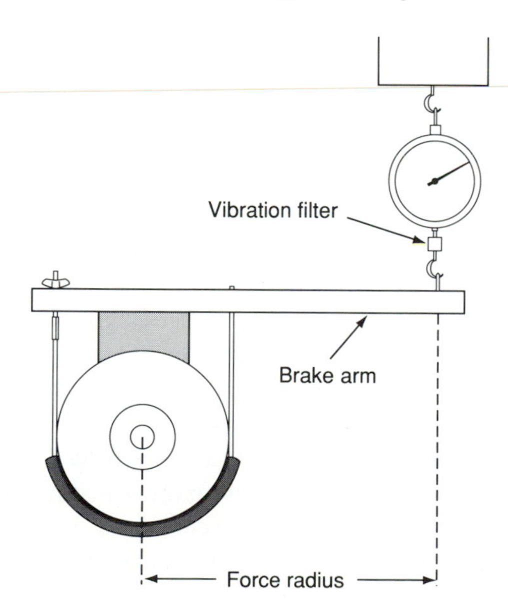

FIGURE 12-4 Prony brake for measuring the torque developed by a motor.

EXAMPLE 12-4 A dc motor is tested with a prony brake and the following data are obtained: The balance on a 1.5-m brake arm reads 32.4 N. A voltmeter and ammeter connected to the input terminals read 220 V and 42 A. The speed of the motor is found to be 28 rev/s. Determine (a) the output of the motor; (b) its efficiency at this particular load.

Solution (a) Torque, $T = 32.4\ \text{N} \times 1.5\ \text{m} = 48.6\ \text{N}\cdot\text{m}$.

$$\text{Output power} = T \times 2\pi r/s = 48.6\ \text{N}\cdot\text{m} \times 2\pi \times 28\ r/s = 8550.16\ \text{W}$$

(b) Input power $= 220 \times 42 = 9240$ W.

$$\text{Efficiency} = \frac{\text{output power}}{\text{input power}} \times 100 = \frac{8550.16\ \text{W}}{9240\ \text{W}} \times 100 = 92.5\%$$

12-7 COUNTER EMF

When the armature of a motor is rotating, it is carrying current. This action causes a magnetic field that is displaced by 90 electrical degrees from the main magnetic field. Like the main pole field, the armature field is stationary even though it results from the magnetomotive force of rotating conductors. This is because the alternating currents in the armature conductors always flow in the same direction as they pass under main-field poles of the same polarity.

When a motor armature is rotating as a result of the torque that is produced by motor action, it is also acting as a generator at the same time. If the armature is rotating due to motor action, the armature conductors continually cut through the resultant stationary magnetic field, and because of this cutting of flux, voltages are generated in the same conductors that experience the force action. Therefore, while a motor is rotating it is simultaneously functioning as a generator. The generated emf, or **counter emf**, opposes the applied voltage to the extent that the current in the armature conductors is limited to exactly the value that is required for the power requirements of the motor.

The generated voltage in the armature winding of a motor directly opposes the applied emf, and in doing so has a limiting effect on the armature current. The counter emf is the primary cause of the automatic control of the speed and torque relations of a motor as it is operated under varying load conditions.

When a voltage is applied across a conductor, as shown in Figure 12-5, current will flow away from the reader, or from point *a* to *b*. The applied emf will cause the conductor to be forced in an upward direction. As the conductor moves upward, flux is cut, resulting in an emf being induced in the opposite direction to the applied emf. The counter emf is directly proportional to the speed of the armature and to the field strength.

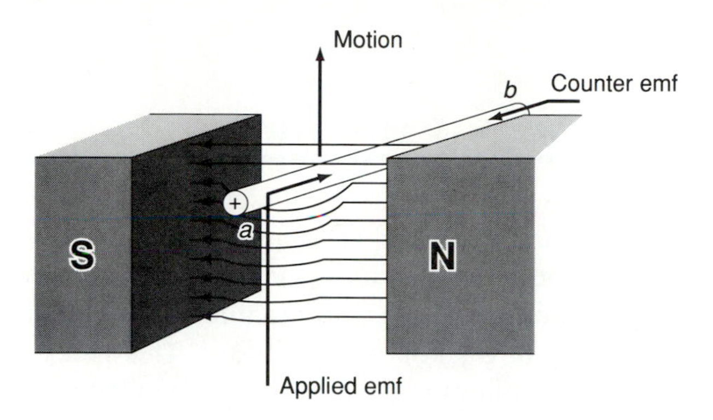

FIGURE 12-5 Relation of the direction of emfs in motor conductor.

The applied emf must be greater than the counter emf in order for current to flow in the proper direction and cause the armature to rotate. Since the applied emf must also overcome the armature resistance, the effective voltage across the armature must equal the applied voltage minus the counter emf. The current that flows through the armature is not determined by the resistance of the armature and the applied voltage; instead, it is equal to the difference between the applied voltage and the counter emf divided by the resistance of the armature. In equation form,

$$I_a = \frac{V_t - E_g}{R_a} \tag{12-6}$$

where I_a = armature current

V_t = motor terminal voltage

E_g = counter emf

R_a = armature-circuit resistance

EXAMPLE 12-5 A dc motor draws 30 A at 240 V. It has an armature circuit resistance of 0.375 Ω. Determine (a) the counter emf being developed; (b) the mechanical power developed.

Solution (a) $I_a = \dfrac{V_t - E_g}{R_a}$

$$E_g = V_t - I_aR_a = 240\text{ V} - (30\text{ A})(0.375\ \Omega)$$
$$= 228.75\text{ V}$$

$$\text{(b)}\ P = E_aI_a = 228.75\text{ V} \times 30\text{ A}$$
$$= 6.86\text{ kW}$$

12-8 ARMATURE REACTION IN A MOTOR

In all direct-current machines, the armature is an electromagnet with a north pole and a south pole. Theoretically, the armature poles should be located midway between the pole pieces, but due to reaction between the two sets of poles, the armature is caused to shift *against* the direction of rotation in a motor, and *with* the direction of rotation in a generator. This is because in a generator, the armature current flows in the direction of the induced emf, and in a motor the armature current flows in the opposite direction of the induced emf.

The current flowing in the conductors of a motor armature is shown in Figure 12-6(a). The brushes are considered to be in the geometrical neutral, and the direction of flux produced by the magnetomotive force is in an upward direction at right angles to the polar axis.

The current flowing in both the field and the armature of the motor is shown in Figure 12-6(b). The direction of flux is now in an upward diagonal direction due to the effect of the ampere-turns of both the field and armature. When current flows through the armature and field coils at the same time, the north pole of the armature will repel that of the field poles and attract the south pole. Similarly, the south pole of the armature will repel the main south pole and attract the north pole. The result of this action is that instead of the lines of force being uniformly distributed, the field will be distorted, with the lines of force being pushed away from one pole tip over onto the other.

Since the neutral plane is perpendicular to the direction of the resultant flux, it moves backward. In Figure 12-6(b) the brushes are moved backward by the angle β from the geometrical neutral. This demonstrates that in a motor it is necessary to move the brushes backward with increasing load, where the generator requires brushes to be moved forward with increasing load. If there were no counter emf, the brush axis would coincide with the neutral plane.

The amount of distortion will vary with the load on the machine. If the load is small, the current in the armature is at a low value, the field poles of the armature will be weak and have very little effect upon the main poles, and the flux will be uniformly distributed. However, if the load is increased, the magnetic field of the armature will increase, and its effect of distorting the flux from the main poles will also increase. Therefore, with a varying load,

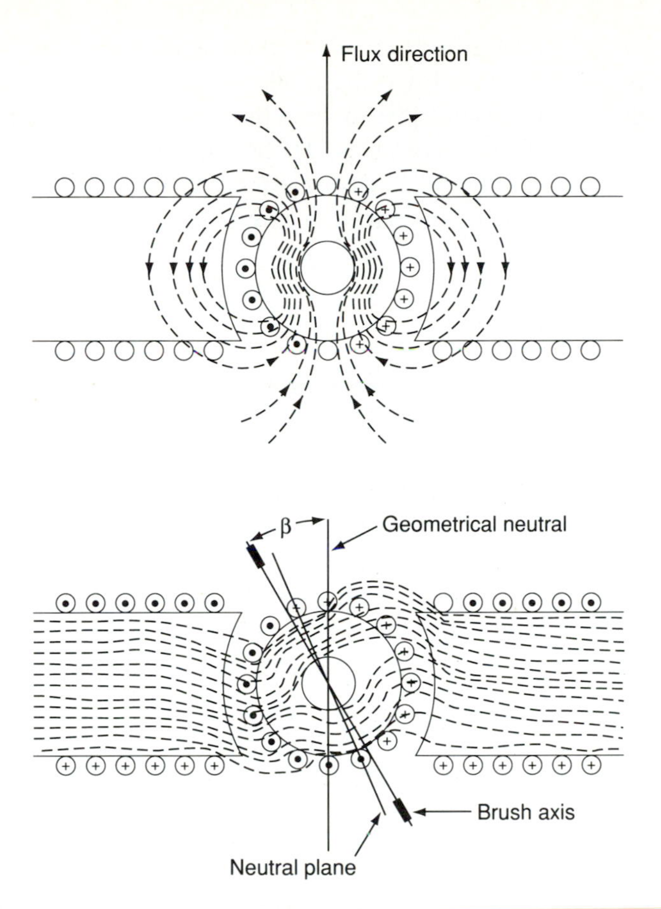

FIGURE 12-6 Armature reaction in motor.

the distribution of the field from the main poles will also vary. Consequently, the neutral point is considered to shift with the load.

As in the case of generators, motors are equipped with interpoles to eliminate the need to shift the brushes to secure good commutation. Essentially, the only difference between the interpole generator and the interpole motor is that in the generator, the interpole has the same polarity as the main pole and is ahead of the main pole in the direction of rotation. The motor also has the same polarity between the interpole and the main pole, but the interpole is located behind the main pole in the direction of rotation.

Since the interpoles have the same polarity as the armature poles, if they are both of the same magnetic strength, the interpole will neutralize the effect of the armature pole and the neutral will remain in a fixed position. Because interpoles are connected in series with the armature, the strength of the interpole's magnetic field will vary as the strength of the armature magnetic field.

At light loads the current in the armature and interpole windings will be small; consequently, the field poles of each will be weak. As the load increases, the current through the armature and interpole windings increases, and the strength of the magnetic fields increases. If the interpoles are properly designed, they will neutralize the effect of the armature pole at all variances of load, making it possible to obtain sparkless commutation with the brushes in a fixed position.

12-9 SPEED OF A MOTOR

Motors are generally designed and manufactured to deliver specific output power when they operate at certain speeds. These ratings are always indicated on the nameplate. For conventional motors, the speed will be greater than the nameplate value if the mechanical load is below the rated power output. When the speed of a motor rises or falls because of a load change, it is always accompanied by an inherent tendency on the part of the armature current to decrease or increase, respectively. If the speed of the armature is decreased while the field strength is kept constant, the emf will decrease because the total line of force passed in 1 s will be less than before the armature slowed down. Therefore, the instant the armature of a motor slows down, the emf that its conductors generate will be decreased. If the motor is idling, only a small current will be required to keep it running, just enough to overcome the friction and other internal losses. The applied voltage multiplied by the current flowing at no-load will give the power used by the motor to overcome its losses. If the motor was loaded, the loading effect would tend to retard the motion of the armature. The speed would consequently begin to drop and would continue to do so until the counter emf had been reduced to such a value as would allow sufficient current to flow to overcome the load imposed on the motor. The power requirements of the motor would now include the internal losses as well as the watts required to drive the load.

The amount of change in speed for change in load varies greatly for different types of motors. However, the speed of any motor is directly proportional to the counter emf and inversely proportional to the flux. This relationship may be defined using the equation

$$S = \frac{V_a - I_a R_a}{K\Phi} \qquad (12\text{-}7)$$

where S = speed, in rpm

V_a = armature voltage

$$I_aR_a = \text{armature voltage drop}$$

$$K = \text{a constant, depending on physical dimensions of motor}$$

$$\Phi = \text{flux entering armature, in webers}$$

EXAMPLE 12-6 A 230-V dc motor operates at full load at a speed of 1750 rpm, with 277 A flowing in the armature. The brush drop is assumed to be 3.5 V, and armature resistance was found to be 0.045 Ω. At what speed will the motor operate at no load if the armature current is then 12.3 A and the brush drop is 1.8 V? Assume that the flux remains constant.

Solution

$$\text{rpm}_1 = \frac{V_a - I_aR_a}{K\Phi} \times 1750 = \frac{(230 - 3.5) - (277 \times 0.045)}{K\Phi} = \frac{214.04}{K\Phi}$$

$$\text{rpm}_2 = \frac{V_a - I_aR_a}{K\Phi} = \frac{(230 - 1.8) - (12.3 \times 0.045)}{K\Phi} = \frac{227.65}{K\Phi}$$

$$\frac{\text{rpm}_2}{1750} = \frac{227.65}{214.04}$$

$$\therefore \text{rpm}_2 = 1750 \times \frac{227.65}{214.04} = 1861.3 \text{ rpm}$$

Speed regulation is the ability of a motor to maintain its speed when a load is applied. It is an inherent characteristic of a motor and remains the same as long as the applied voltage does not vary. The speed regulation of a motor is a comparison of its no-load speed to its full-load speed and is expressed as a percentage of full-load speed. In equation form,

$$\%\text{ speed regulation} = \frac{\text{no-load speed} - \text{full-load speed}}{\text{full-load speed}} \times 100 \quad (12\text{-}8)$$

The lower the percent speed regulation of a motor, the more constant the speed will be under varying load conditions.

EXAMPLE 12-7 The no-load speed of a dc motor is 1500 rpm. When the motor carries its rated load, the speed is reduced to 1210 rpm. Calculate the percent speed regulation of the motor.

Solution $$\% \text{ Regulation} = \frac{\text{no-load speed} - \text{full-load speed}}{\text{full-load speed}} \times 100$$
$$= \frac{1500 - 1210}{1210} \times 100$$
$$= 23.97\%$$

12-10 SHUNT MOTOR

The shunt motor shown in Figure 12-7 is connected in the same manner as a shunt generator, with its field connected directly across the line in parallel with the armature. A field rheostat is usually connected in series with the field. If such a motor has its load removed, the retarding torque is small, being due only to windage and friction. The armature will develop a counter emf that will restrict the armature current to a value that will cause the motor to develop a torque equal to the resistance torque.

Since the field winding of a shunt motor is connected across the line, the magnetic density in the field poles will be practically constant irrespective of the load. Therefore, the torque of a shunt motor varies directly as the current in the armature. If load is applied to any motor, the motor immediately tends to slow down. A motor must slow down when an external load is applied to it, because the small no-load current is just sufficient to produce a torque to overcome friction. This decrease in speed causes the generated voltage to decrease, since the field flux is constant, and more current flows through the armature. The armature current will continue to increase until a torque is developed to just equal that of the external load. Therefore, the torque of a shunt motor varies directly as the current in the armature.

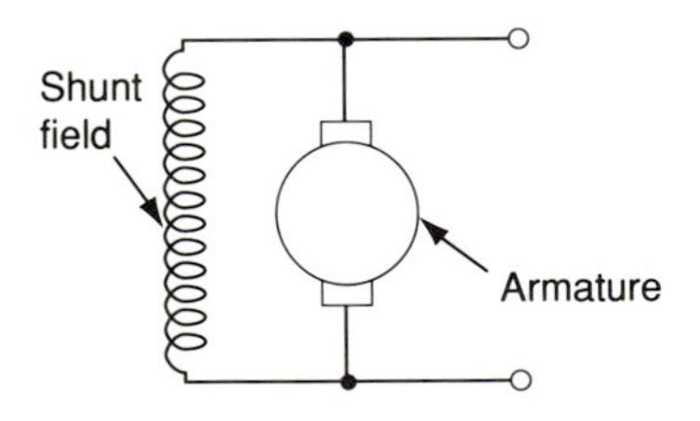

FIGURE 12-7 Dc shunt motor.

Theoretically, since the strength of the field pole is constant, the speed of the motor will be constant from no-load to full-load. In reality, the variation is only about 5 to 10% from no-load to full-load, the speed being higher at no-load than at full-load. The relationship between torque and speed in a shunt motor is illustrated in the graph of Figure 12-8. Torque is also shown plotted against armature current, and the resulting curve is practically linear. The speed of a motor varies in accordance with the equation

$$S = \frac{V_a - I_a R_a}{K\Phi}$$

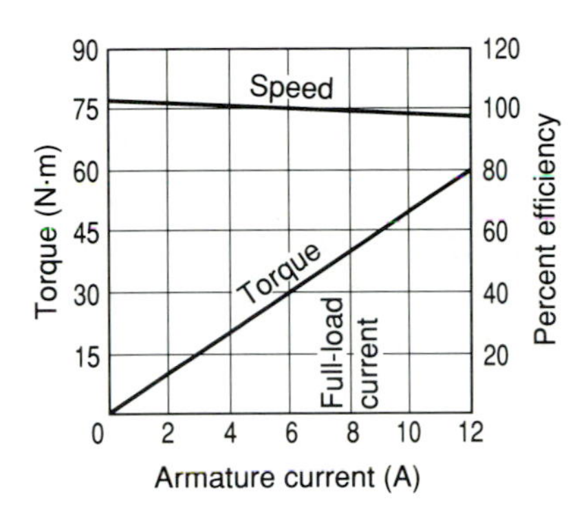

FIGURE 12-8 Torque–current characteristics of shunt motor.

This equation shows that the speed of a motor is directly proportional to the counter emf and inversely proportional to the constant K and to the field flux.

In the shunt motor, V_a, K, R_a, and Φ are virtually constant, the only variable being I_a. As the load on the motor increases, I_a increases and the numerator of equation 12-7 decreases. When the motor is carrying no-load, the value of I_a is small due to the speed and counter emf both being at maximum values. As stated earlier, the full-load speed of the shunt motor is within 95% of the no-load speed. The full-load speed is lower due to armature reaction, and the resultant field flux tends to decrease slightly with an increase of load. However, there are situations where the speed of a loaded shunt motor will either remain constant or even begin to rise above no-load speed. To stabilize the motor and prevent increases in speed, a small number of series turns are added. These windings are called **stabilization windings** and will increase the flux with an increase of load.

Shunt motors are essentially constant-speed machines and may be considered as such. Although the speed may be varied by adjusting the current through the field winding, the speed remains constant for a given value of field current. This constant-speed characteristic makes the shunt motor ideal for use in machine tools, blowers, or any other device that requires a constant-speed driving source.

If a variable resistor is inserted in either the armature or field circuit, the speed of a shunt motor can be controlled. However, adding resistance to the armature circuit will cause substantial power loss, resulting in dramatic changes in speed with a change in load. Therefore, the most common method of altering the speed of a shunt motor is by varying the resistance in the field circuit. When the resistance of the field circuit is increased, the current through the circuit is decreased, and the field flux is consequently decreased. This results in less counter emf being generated and more current flowing through the armature. Since the increase in armature current is greater than the decrease in field current, the torque increases and the armature speed increases. This creates the unusual scenario of an increase in field resistance resulting in an increase in motor speed.

By using a field rheostat for speed control, a shunt motor will run at nearly constant speed for any setting of the rheostat. This constant-speed characteristic is desirable in many industrial applications in which the speed is adjusted at a definite value, and it is important that it remain at approximately that value for varying loads.

12-11 SERIES MOTOR

The field coils of a series motor are connected in series with the armature, as shown in Figure 12-9, in the same manner as the series generator. The field has comparatively few turns of wire, and this wire must be of sufficient cross section to carry the rated armature current of the motor. Since there is only one circuit through the motor, any change in the load causes a change in the current and flux. In a series motor, the flux increases almost directly with the load. An increase in the load increases both the armature current and the flux, causing a much greater increase in the torque. Since flux is almost directly proportional to armature current, the torque is then directly proportional to the square of the armature current. In equation form,

$$T = KI_a^2 \qquad (12\text{-}9)$$

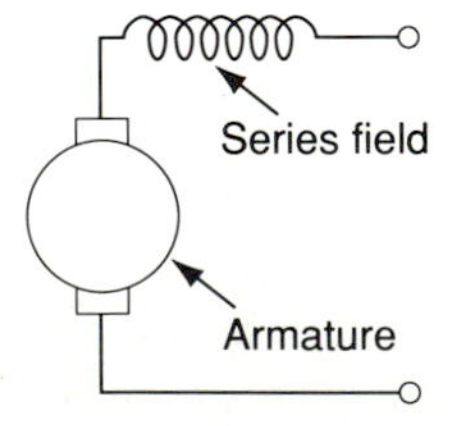

FIGURE 12-9 Dc series motor.

This relation is shown by the torque curve in Figure 12-10. If the armature current is doubled, the torque will increase four times. Since the field and armature windings are connected in series, if the current is increased, the current will be doubled not only in the armature's conductors, but also through the field coil. The latter will increase the flux from the field pole by two. Consequently, when there is twice as much current in the armature conductors acting on double the flux from the pole pieces, the result is four times the turning effort developed by the series motor.

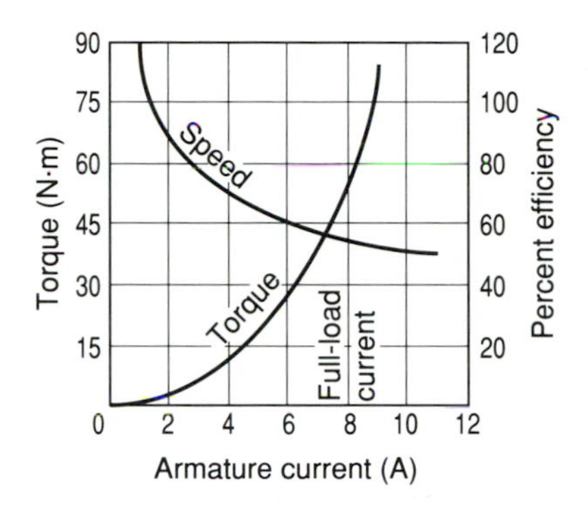

FIGURE 12-10 Torque–current characteristics of a series motor.

The speed of a series motor may be expressed mathematically as

$$S = \frac{V_a - I_a(R_a + R_{se})}{K\Phi}$$

One of the serious disadvantages of a series motor is that it will operate at a dangerously high speed if its load is removed completely. If the load is changed on the motor, the speed will vary inversely. Therefore, if the load increases, the speed decreases proportionately and the machine is incapable of operating at a constant speed. As the speed is inversely proportional to the flux, a given percentage change in produces the same percentage change in speed. When the load is decreased, the flux decreases correspondingly and the armature must speed up in order to develop the required counter emf. If the load is removed entirely, the flux becomes extremely small, causing a very high speed. For this reason the load should never be completely removed from a series motor. It is dangerous to remove the load, since the armature is almost certain to reach a speed where centrifugal force will destroy the machine. This characteristic of the torque increasing as the square of the current makes the series motors well suited for work that requires frequent starting under heavy loads, such as cranes and winches.

12-12 COMPOUND MOTORS

The construction of the compound motor is identical to that of the compound generator. When a motor is connected so that the series winding aids the shunt winding, it is said to be cumulatively compounded. If the shunt winding opposes the series winding, the motor is said to be differentially compounded. Figure 12-11 shows the two methods of connecting a compound dc motor. When the motor is connected in a differentially compounded configuration, the resulting differential mmf of the two fields will tend to reduce the total flux with increasing values of load current. At light loads, the demagnetizing action of the series field mmf is relatively small, so that it has little effect on the shunt field mmf. However, at heavy loads, the series-field current is comparatively high, which means that the total flux may be considerably less than at no-load. Because of this differential action of the two fields, the motor attempts to increase in speed with increasing values of load.

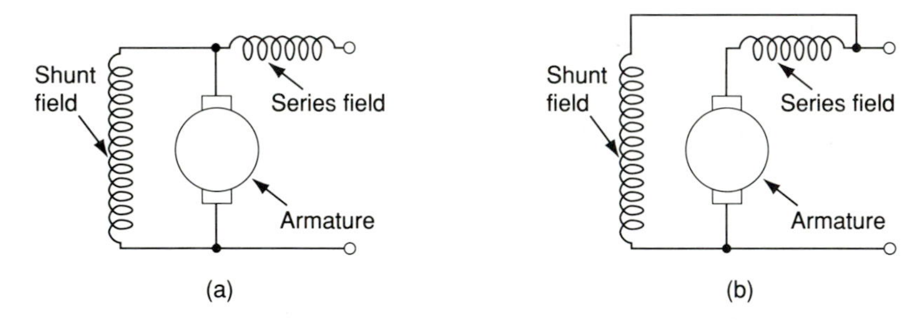

FIGURE 12-11 Dc compound motor: (a) short-shunt connection; (b) long-shunt connection.

Therefore, if the motor is designed with the series-field ampere-turns correctly proportioned with respect to shunt-field ampere-turns, it is possible to have a speed characteristic that is almost flat over most of the operating range. This condition, of zero speed regulation, will result because the armature voltage drop is mostly offset by a reduction in flux, which has an equivalent tendency to raise the speed. Consequently, differential-compound motors are very unstable when the load increases to the point where there is an increase in speed. Under this condition, the motor attempts to *run away* since every small increase in speed is accompanied by a corresponding rise in load current. This increase in load current flows through the series field and reduces the flux, resulting in a further increase in the speed of the motor.

The characteristics of the cumulative-compound motor are a combination of both the shunt and series motors. When the motor is loaded, the series windings increase the flux, and the torque increases. This increase in torque is greater than for a shunt motor; however, the increase in flux causes the speed to decrease at a more rapid rate than would occur in a shunt motor. The cumulative-compound motor will develop a high value of torque when the load is suddenly increased.

Another characteristic of this type of motor is that it does not run away when the load is removed. When load is applied rapidly, the large increase in current causes a large increase in torque. The speed of the armature is simultaneously decreasing as the torque increases, resulting in a transfer of kinetic energy from the armature to the load. This energy transfer assists the electrical system in absorbing the impact of sudden increases in power demand. The cumulative-compound motor is ideally suited for applications that require a large starting torque and a constant running speed, such as elevators. Figure 12-12 illustrates the comparative characteristics of shunt, cumulative-compound, and differential-compound motors.

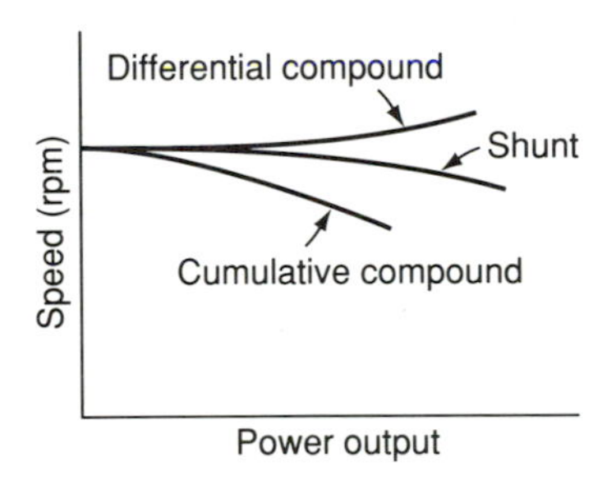

FIGURE 12-12 Characteristics speed versus power output curves for shunt, differential compound, and cumulative compound motors.

12-13 DC SERVOMOTORS

Dc servomotors are dc motors with two separate windings: the field winding and the armature winding. The speed of the servomotor is controlled by varying either the armature current or the field current. The three main types of dc servomotors are the shunt, series, and permanent-magnet motors. Dc servomotors range in size from 0.02 to 1000 hp and are well suited for a wide variety of applications, including robotics and high-performance magnetic tape drives.

Figure 12-13 shows a separately excited dc servomotor. The voltage across the field winding is controlled by an electronic amplifier. The amplifier increases the input signal to a value that can be used to drive the servomotor. The input voltage, V_{in}, is also referred to as the **error voltage**. The error voltage represents the difference between the measured signal and the desired signal. For example, assume that the servomotor is connected to a device that must be moved a certain number of degrees. If the device is a considerable distance from its desired position, the error voltage will be large and the shaft of the servomotor will turn rapidly. If the device is positioned at its desired location, the error voltage will be zero and the shaft of the motor will not turn.

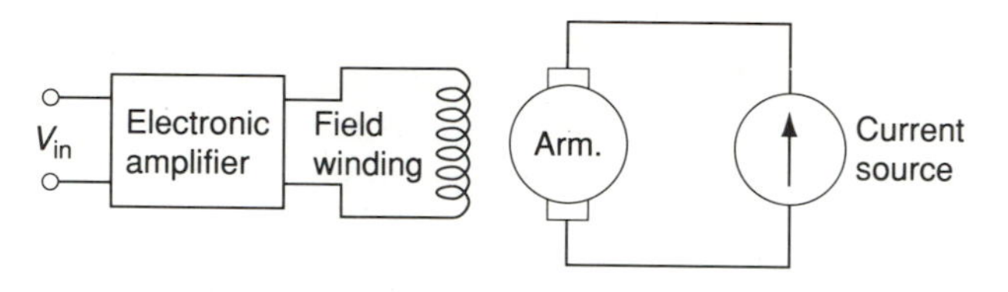

FIGURE 12-13 Field-controlled dc servomotor.

The armature is supplied by a constant-current source, which maintains a fixed value of current through the armature circuit. The equation for torque in a dc motor was given in equation 12-2 as $T = K\Phi I_a$. According to this equation, if the armature current remains constant and the field flux increases, the torque of the motor will increase. Therefore, the torque of the servomotor varies directly with the field flux and field current up to the point of saturation.

The permanent-magnet (PM) servomotor shown in Figure 12-14 uses permanent magnets for constant field excitation instead of a constant-current source. These types of motors use either Alnico or ceramic magnets. The disadvantage of Alnico magnets is that they are adversely affected by the demagnetizing effect of large armature transient currents.

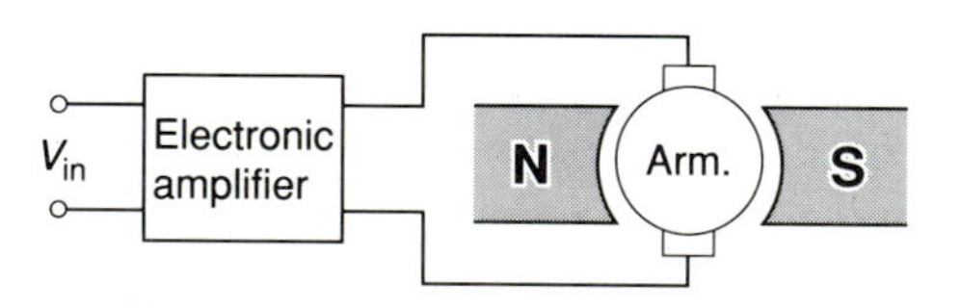

FIGURE 12-14 Permanent-magnet dc servometer.

Figure 12-15 shows the torque–speed characteristic of a dc servomotor. The highest output torque occurs when the motor is at standstill or at low operating speeds. The torque–speed curve also shows that as the device being driven by the servomotor approaches its desired position, the motor can develop a reverse torque to slow the load down and prevent **overshoot**. At this point the servomotor acts as a generator, and a countertorque is developed as a result of the armature current changing directions. This automatic slowing down of the motor is called the **damping effect.**

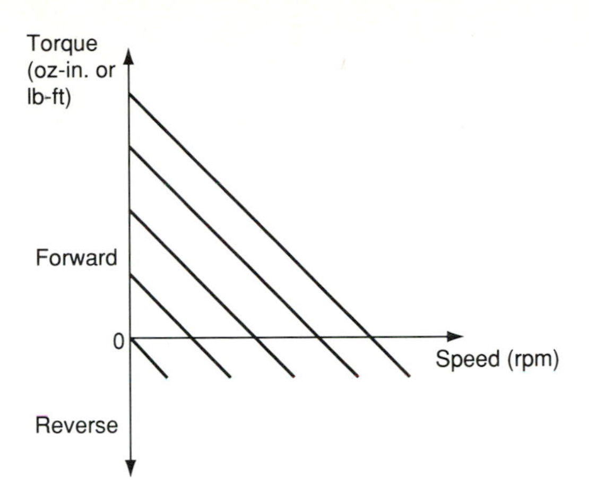

FIGURE 12-15 Torque versus speed curves for a dc servomotor.

12-14 STEPPER MOTORS

A **stepper**, or **stepping, motor** converts electric pulses into proportionate mechanical movement. Each revolution of the stepper motor's shaft is made up of a series of discrete individual **steps**. A step is defined as the angular rotation produced by the output shaft each time the motor receives a step pulse. A **step angle** is the rotation of the output shaft caused by each step, measured in degrees.

Direct-current stepping motors

Source: Prentice Hall Inc. / Superior Electric Co.

The stepper motor is one of the few motors that is essentially digital in nature, with one angular step for each input pulse supplied by power transistors. The stepper motor delivers precise incremental motions of great accuracy and is well suited for applications such as robotics, instrumentation controls, and computer equipment. Stepper motors have torques that range from 0.5 μN·m to 100 N·m, and power outputs that vary from 1 mW to 10 kW.

The most popular types of stepper motors are permanent-magnet (PM) and variable-reluctance (VR). The **permanent-magnet stepper motor** operates on the reaction between a permanent-magnet rotor and an electromagnetic field. A basic PM motor has a two-pole cylindrical rotor magnet within a four-pole slotted stator. Figure 12-16 shows the components of a permanent-magnet stepper motor. In Figure 12-16(a), one permanent magnet is shown at each end of the rotor. The stator is shown in Figure 12-16(b). Both the stator and rotor are shown as having teeth. In this motor the residual torque is developed by the magnetic force between the permanent magnet and the stator. The number of teeth on the rotor and stator determine the step angle that will occur each time the polarity of the winding is reversed. The greater the number of teeth, the smaller the step angle.

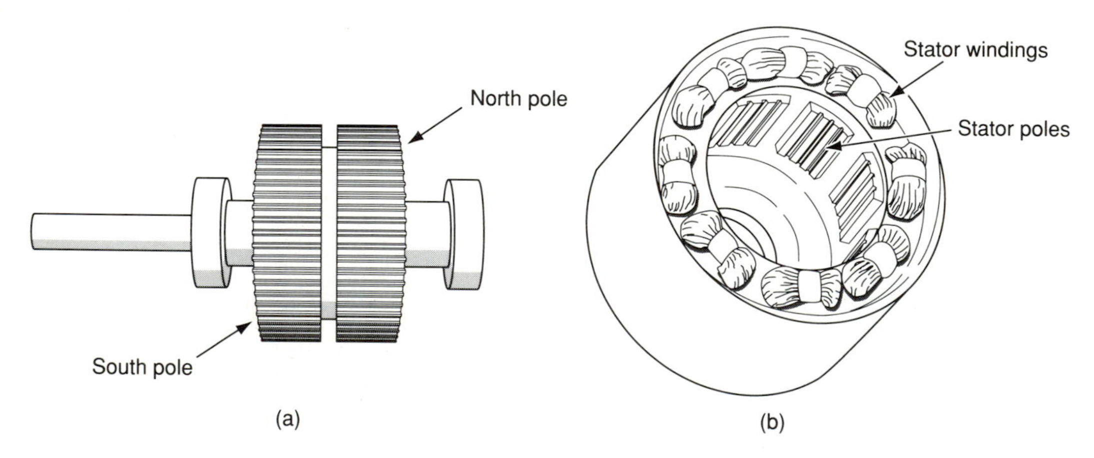

FIGURE 12-16 Components of a permanent-magnet stepper motor: (a) rotor; (b) stator.

The stator coils are usually energized with square-wave dc pulses. Exciting the stator winding by direct-current results in the formation of two magnetic stator poles, one north and one south. The torque developed by the PM stepper motor is directly proportional to the current in the stator coils. Since the rotor of a PM stepper motor is a permanent magnet, a residual, or **detent torque** exists even when the motor is not turning. The detent torque can be noticed by turning a stepper motor by hand. Whenever the rotor is moved from its position of minimum reluctance, a **restoring torque** is generated.

This torque is generally a very small percentage of the maximum torque of the motor.

The **variable-reluctance stepper motor** differs from the PM stepper motor in that it has no permanent-magnet rotor and consequently has no reluctance torque to hold the rotor at one position when turned off. This type of motor operates on the principle of minimizing the reluctance along the path of the applied magnetic field. The rotor is made of unmagnetized soft steel with teeth and slots. When a current flows through the stator windings, a torque is developed that tends to turn the rotor to a position of minimum magnetic path reluctance. By alternating the windings that are energized in the stator, the stator field changes, and the rotor is moved to a new position.

12-15 BRUSHLESS DC MOTORS

Brushless dc motors consist of stator windings and a radially magnetized permanent-magnet rotor which are commutated by means of electronic switching. The field windings are energized in sequence to produce a revolving magnetic field. Motors using electronic commutation are not affected by the arcing associated with mechanical commutation. Since arcing always exists in mechanical commutation, conventional dc motors generate radio-frequency noise. Arcing and friction also limit the brush life, resulting in the need for regular servicing of mechanically commutated motors.

The electronic commutator consists of thyristors or transistors. These semiconductor switches need firing pulses, which are dependent on the rotor position, to allow them to conduct at the correct time. In a mechanical commutator, this synchronization occurs automatically, but in an electronic system a rotor position sensor is required. The rotor position is detected in most cases by either a Hall effect magnetic sensor or a switching optical sensor.

The Hall effect sensor is a device that produces an output voltage in the presence of a magnetic field. The Hall element has a constant current passed through the sensor. A magnetic field with a flux density is applied at right angles to element. The Hall voltage is induced in a direction that is perpendicular to the current and magnetic field. The optic sensor uses a light-emitting diode (LED)–photodiode pair. At a specific time, the LED illuminates the photodiode, which in turn switches on the transistor or thyristor that controls the coil current.

Brushless dc motors have efficiencies that can exceed 75% and are capable of rotating at speeds greater than 40,000 rpm. These types of motors are very popular in the biomedical and aerospace fields. Due to the high reliability and low maintenance requirements, brushless dc motors are used as artificial heart pump motors and in cryogenic coolers. In the aerospace industry, these motors are used as gyroscopic motors. Other applications include disk drive motors, servomechanisms, video recorders, and tape transport systems.

12-16 REVERSING THE DIRECTION OF DC MOTORS

The direction of the developed torque in a motor depends on the simultaneous directions of the magnetic field and the current in the armature winding. Essentially, there are two methods for reversing the direction of a dc motor:

1. Change the direction of current flow through the armature.
2. Change the direction of current flow through the field circuit or circuits.

The most common method of reversing the current flow through the armature is to reverse the polarity of the commutator brushes. By reversing the magnetic polarity of the field circuit, the direction of current flow through the field will also be reversed.

Compound motors are reversed in virtually the same manner as shunt motors, by reversing either the current through the armature winding or the currents through both the shunt and series fields. If only one field is reversed, a differential action will result, causing instability.

12-17 ELECTRONIC CONTROL OF DC MOTORS

Semiconductor devices such as diodes and **silicon-controlled rectifiers** (SCRs) are very common in electronic control circuits for dc motors. The SCR has current ratings that range from 0.25 A to greater than 3000 A, and voltage ratings that range up to 5000 V. Silicon-controlled rectifiers can switch on in about 1 μs and turn off in about 10 to 20 μs.

The SCR is a three-terminal disk of four alternate layers of p- and n-type silicon. The layers and junctions between them are formed by precision gaseous diffusion and alloying techniques. The schematic symbol for the SCR is shown in Figure 12-17. Some of the packages used to house silicon-controlled rectifiers are shown in Figure 12-18. The larger devices are mounted in larger packages and have higher current-handling capabilities. The DO-200 package can be used for devices rated as high as 3000 A. The molded multiple-SCR package contains two or more SCRs in a single package.

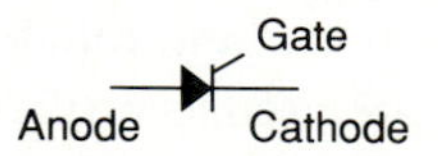

FIGURE 12-17 Electrical symbol of an SCR.

The three main applications of the SCR in motor control circuits are as follows:

1. *Rectification*. The SCR will conduct current in only one direction.
2. *Latching*. The SCR operates like an on–off switch. The SCR can be switched on by applying a small amount of control current to the gate for a brief instant of time. The SCR will remain in the conducting state until the flow of current between the anode and cathode is reduced below a specified value, or until the anode and cathode become reverse-biased.
3. *Amplification*. An SCR is capable of switching thousands of amperes of current by a control current of a few microamperes.

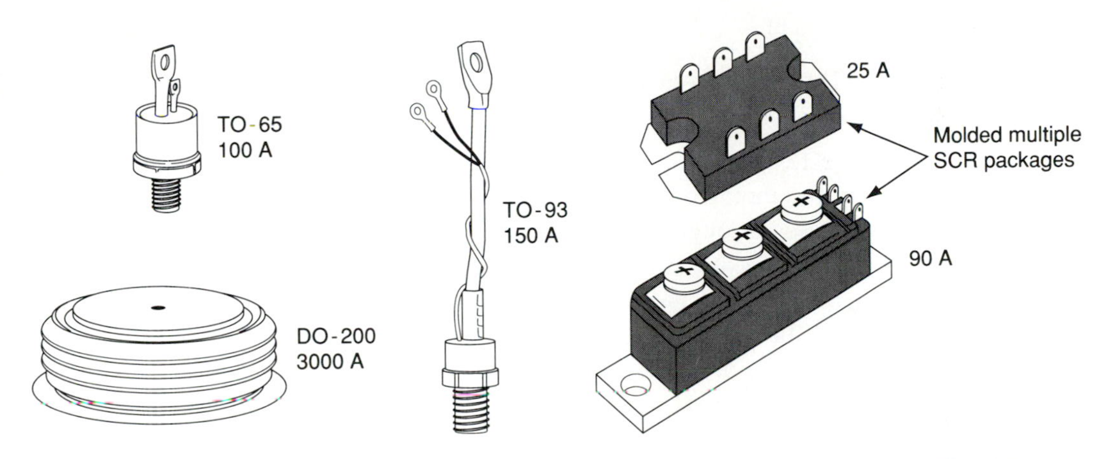

FIGURE 12-18 SCR packages.

The method of switching off an SCR is called **commutation**. SCRs are commutated by reverse-biasing the device, or by reducing the current below the manufacturer's specified value of **holding current**. The holding current is the amount of current flow through the device required to maintain conduction.

High-power solid-state controllers are an extremely efficient method of speed control for dc motors. The most common electronic controller used in industry is the **dc chopper**. A chopper is essentially an on–off switch which connects and disconnects the load to a dc voltage source. Figure 12-19(a) shows a basic chopper circuit, which is also referred to as a **pulse width modulation** (PWM) circuit. In Figure 12-19(b), the pulse width is determined by the length of time the SCR conducts. The points where the SCR turns on and off are indicated by t_{on} and t_{off}. Since the SCR cannot switch itself off once in the conducting state, a commutating circuit is required to apply a negative voltage on the SCR for a brief instant.

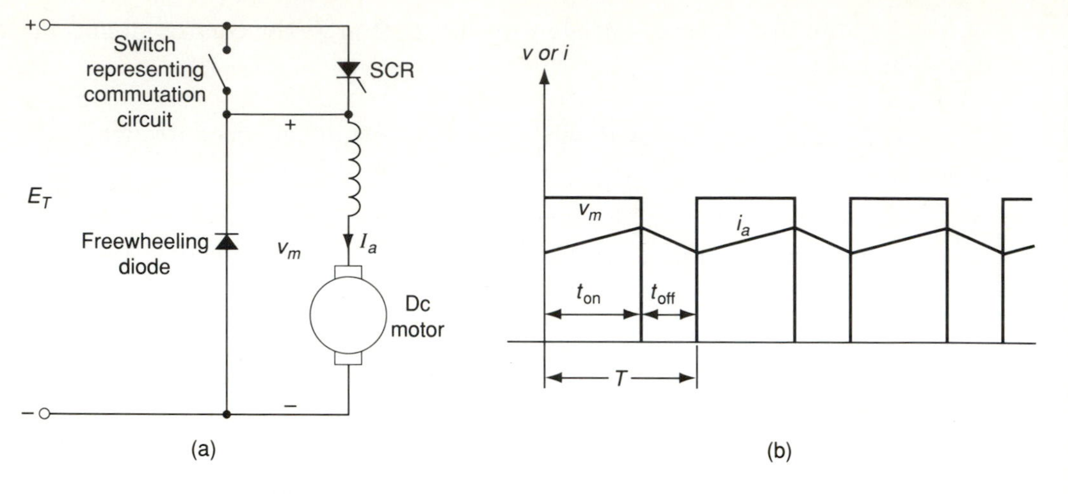

FIGURE 12-19 (a) Dc motor driven by a chopper; (b) motor current and voltage waveforms.

In Figure 12-19(b), the armature current, i_a, initially increases after t_0 and the speed of the motor picks up. When the SCR is commutated the armature current decreases through the freewheeling diode. When the SCR is triggered again, the cycle repeats. Consequently, the motor speed is controlled by varying the on–off times of the SCR. The longer the SCR is on, the greater the voltage across the motor.

The **duty cycle** of a pulse width modulation circuit is a ratio of the conducting time to the period of one cycle ($t_{on} + t_{off}$). In equation form,

$$\text{duty cycle (\%)} = \frac{t_{on}}{t_{on} + t_{off}} \times 100 \qquad (12\text{-}10)$$

Figure 12-20(a) represents a high duty cycle since the width of the positive-going pulse is a large percentage of the cycle (greater than 50%). Figure 12-20(b) shows a low duty cycle since the width of the positive-going pulse is a small percentage of the cycle (less than 50%). If the duty cycle is exactly 50%, the pulse is considered to be a **square wave**, since the positive pulse is equal to one-half cycle.

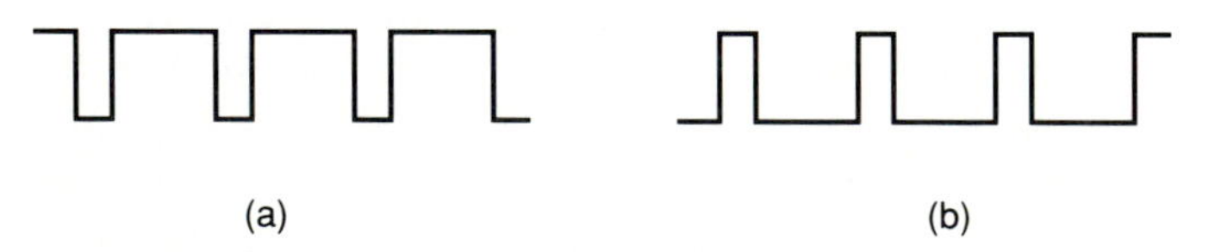

FIGURE 12-20 (a) High duty cycle; (b) low duty cycle.

EXAMPLE 12-8 A chopper is on for 20 ms and off for the next 50 ms. Determine the duty cycle.

Solution

$$\begin{aligned}\% \text{ Duty cycle} &= \frac{t_{\text{on}}}{t_{\text{on}} + t_{\text{off}}} \times 100 \\ &= \frac{20\,\text{ms}}{20\,\text{ms} + 50\,\text{ms}} \times 100 \\ &= 28.57\%\end{aligned}$$

EXAMPLE 12-9 A dc series motor is driven by a chopper circuit. The supply voltage is 100 V and the duty cycle is 25%. Determine the dc voltage applied to the motor.

Solution Duty cycle represents the ratio of the time that voltage is applied to the load. Therefore,

$$\begin{aligned}V_L = V_{\text{in}} \times \% \text{ duty cycle} &= 100 \times 0.25 \\ &= 25\text{ V}\end{aligned}$$

Another type of dc chopper circuit is shown in Figure 12-21. The configuration is called a **Jones chopper circuit**. The two SCRs are used to chop the dc voltage into a series of voltage pulses. The triggering circuit for the SCRs is not shown. The basic operation of the circuit is that SCR_1 acts like a switch. If the switch remains closed for a long period of time, the motor will reach the maximum steady-state speed. If SCR_1 is not switched on, the motor will not rotate. By alternating the triggering and switching off of SCR_1, the motor will rotate at a speed between minimum and maximum.

If we assume that the capacitor C is discharged initially, the current flows through L_1 and the dc motor circuit when SCR_1 is triggered. Since L_1 and L_2 are closely coupled, the capacitor will be charged, with the lower plate, A, becoming more positive and plate B becoming more negative. The charge on the capacitor is trapped by diode D_1. The capacitor remains charged until SCR_2 is triggered. At this point the discharge of the capacitor reverse biases SCR_1 and turns it off.

The capacitor again begins to charge, with plate A now becoming positive. SCR_2 is turned off since the flow of current between the anode and cathode is too small to maintain conduction. Each time SCR_1 is triggered the capacitor charges. The magnitude of the charge depends on the magnitude of the load current. Diode D_2 is a freewheeling diode. Its purpose is to provide a discharge path for the inductive current when SCR_1 is turned off.

The bipolar junction transistor (BJT) is also used in electronic motor control circuits. In Figure 12-21(a), four BJTs are used to control the direction of a dc servomotor. These transistors are npn type, which means that the voltage

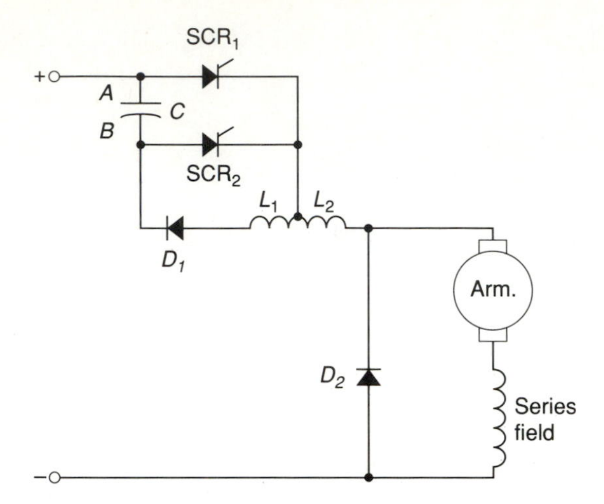

FIGURE 12-21 Dc series motor control using the Jones circuit.

at the base, *B*, must be positive in order for the transistor to conduct. When the base voltage is of a sufficient positive value, the transistor acts as a closed switch and current flows between the collector, *C*, and emitter, *E*. When the base voltage is zero, the transistor acts as an open switch and no current flows between *C* and *E*.

In Figure 12-22 the transistors are connected as a bridge circuit. The control circuitry for the base of each transistor is a voltage pulse supplied by using pulse width modulation (PWM). The length of time the transistor conducts current depends on the length of time the voltage pulse is applied at the

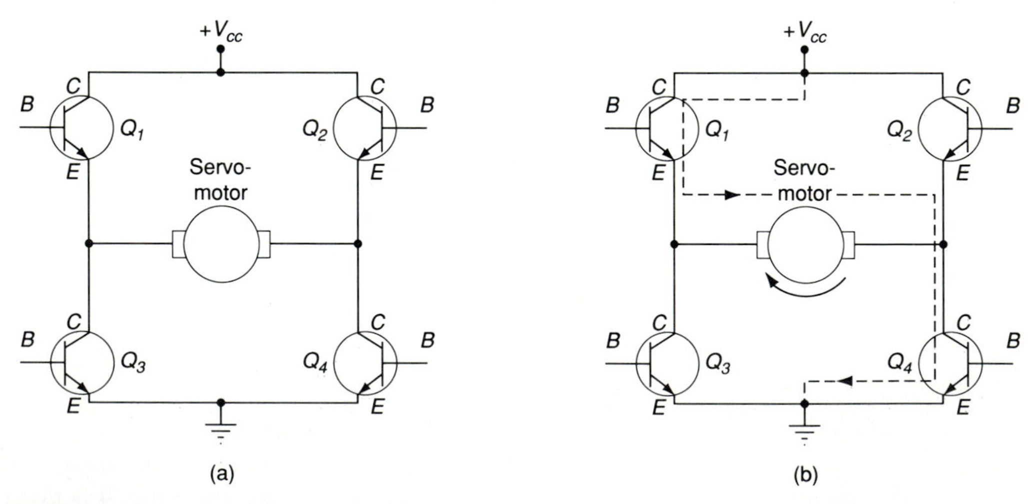

FIGURE 12-22 (a) Transistor bridge dc motor control; (b) conduction path through transistor bridge.

base. PWM is an extremely efficient method of controlling a dc servomotor. Instead of varying a dc voltage to the motor to control its torque, a series of fixed dc level pulses are applied. The torque of the motor can then be controlled by varying the width of the pulse, allowing rapid motor acceleration with a minimum of heat buildup.

In Figure 12-22(b), if transistors Q_1 and Q_4 are on and Q_2 and Q_3 are off, the motor will rotate in the direction indicated by the arrow. The current flow from source to ground is shown by the dashed line. If Q_2 and Q_3 are on and Q_1 and Q_4 are off, the direction of current flow through the motor is reversed and the motor will rotate in the opposite direction.

12-18 DC MOTOR STARTING CIRCUITS

Essentially, there are two methods for starting dc motors: manually and automatically. For many years, the standard manual starters used in industry were the three-point and four-point starters. These starters use an electromagnetic holding coil in series with the shunt field coil. When the manual starter is engaged, a large value of resistance is placed in series with the armature. As the lever on the starter is rotated, the resistance is gradually cut out of the armature circuit until the motor runs at full speed.

Automatic starters use the same principle of reduced voltage starting as that used in manual starting. The main difference between the two starters is that the automatic starter generally incorporates timing relays into the circuitry, so that the resistance is removed from the armature circuit in timed intervals.

In motor control circuits, various schematic symbols are used to represent different functions. The following terms identify and define some of the more common components that appear in motor control circuits.

Pushbutton. Pushbuttons are manually operated. They have either normally open contacts that close when the button is depressed, or normally closed contacts that are opened when the button is depressed. The schematic symbols for normally open and normally closed contacts are shown in Figure 12-23. Pushbuttons are momentary contact devices.

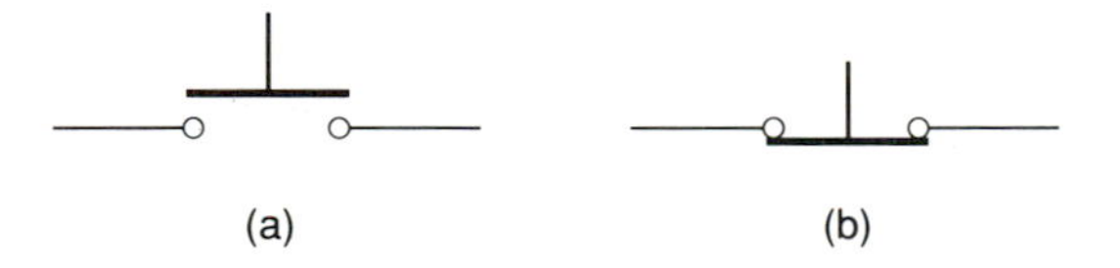

FIGURE 12-23 (a) Normally open pushbutton; (b) normally closed pushbutton.

Relay. A relay is an electromagnetic device that will open or close one or more sets of electrical contacts. This is accomplished by means of a coil, which upon energization, will either force normally open contacts to close or normally closed contacts to open. Basic single-pole relay action is shown in Figure 12-24, with Figure 12-24(a) being the nonenergized state and Figure 12-24(b) the energized state.

When the coil is energized, a north and a south pole are produced across the working gap by the solenoid winding of the coil, producing a magnetic flux. The relay is actuated, or *latched*, whenever current of sufficient intensity flows through the coil and the attractive force between the core and armature overcomes the spring tension, as shown in Figure 12-24(b). The relay remains latched as long as sufficient current flows through the coil. When the current through the coil is reduced below a specific holding value, the core becomes unmagnetized and the armature is pulled up be spring action to its unactuated position.

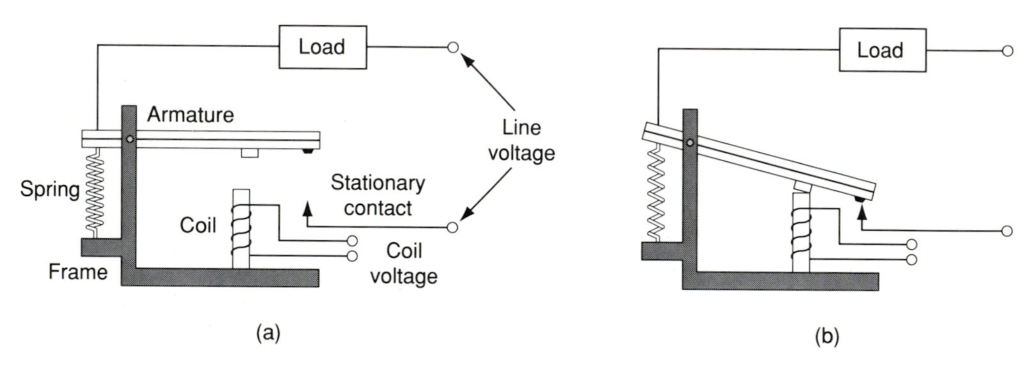

FIGURE 12-24 (a) Nonenergized relay; (b) energized relay.

The schematic symbol for a relay is shown in Figure 12-25. The relay coil is represented by a circle, as in Figure 12-25(a), and the contacts are shown as parallel lines, as in Figure 12-25(b) and (c). The current-handling capacity of the relay coil is very low compared to that of the relay contacts.

Time-delay relay. This relay functions in exactly the same way as an ordinary relay except that when the coil is energized, there is an adjustable time delay before its contacts change state. The schematic symbol for a time-delay relay is shown in Figure 12-26.

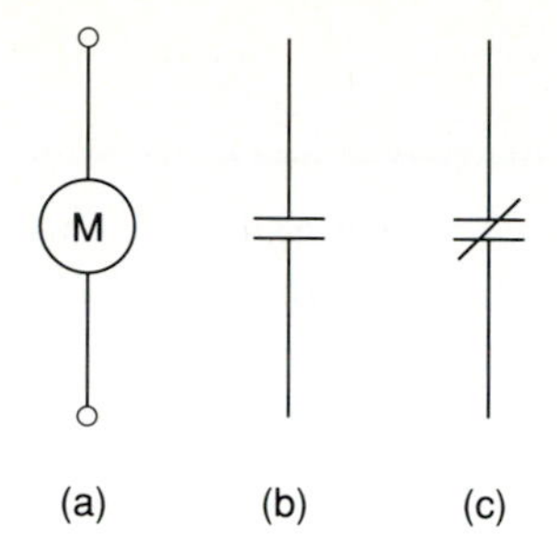

FIGURE 12-25 (a) Relay coil; (b) normally open contacts; (c) normally closed contacts.

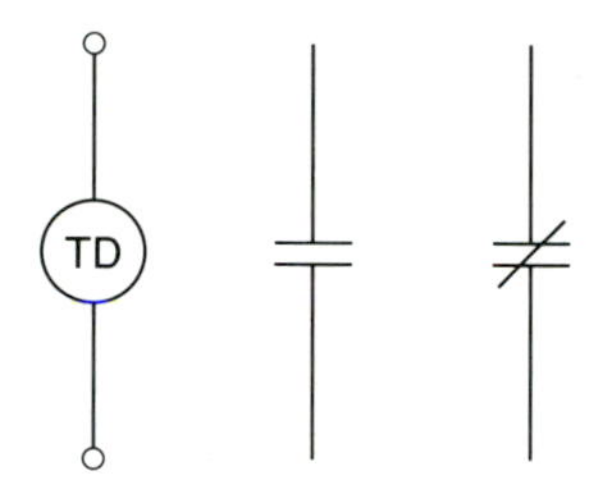

FIGURE 12-26 Time-delay relay with normally open and normally closed contacts.

Overload protection. The overload-protection device consists of a heater coil and some normally closed contacts. The heater coils are placed in series with the motor to "sense" the motor current. If the load on a motor becomes excessive, the current flowing to the motor will heat up the heater coils. The heater coils will then open and cause the normally closed contacts to open. The normally closed contacts are usually placed in series with the stop button of the control circuit. The overload protection is capable of disengaging power in both the power circuit and the control circuit. Once the heater coil has cooled down, the normally closed contacts return to their normal condition and the motor is ready to be restarted. The schematic symbols for the overload coil and contact are shown in Figure 12-27.

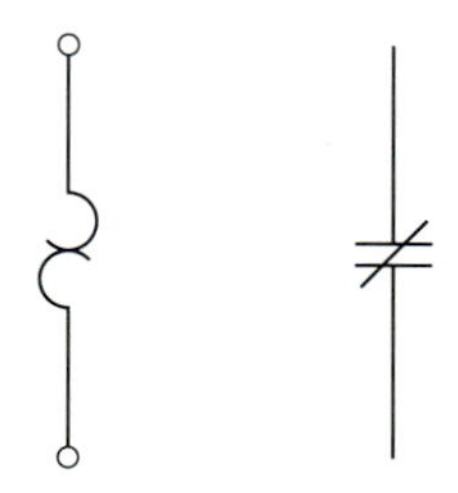

FIGURE 12-27 Overload heater coil and its normally closed contacts.

12-19 AUTOMATIC STARTERS

If a motor is at a remote location, or if a large motor is to be started, it is usually desirable to utilize an automatic starter to bring the motor up to operating speed. Manual three- and four-point starters are very effective for starting small motors, but it is quite difficult to bring a large motor up to speed with a manual starter. There are three main types of automatic starters: the counter-emf type, time-element type, and current-limit type.

Counter-emf starter

The **counter-emf starter** depends on the counter emf developed across its armature for operation. A simple diagram illustrating the principle of the counter-emf method of cutting out one step of starting resistance is shown in Figure 12-28. The instant the power is turned on, the counter emf developed by the motor is zero and the voltage across the accelerating contactor is also zero. As the motor is brought up to speed, the increasing counter emf results in a decrease in the armature current and voltage drop across the starting resistance. At the same time, the voltage across the armature increases. When this voltage reaches a predetermined value, the accelerating contactor closes and shorts out the starting resistance, which then applies full voltage across the armature. The reciprocal effect of an increase in starting current causes continued acceleration until the motor comes up to rated speed, and the armature current is again reduced to rated value. If the motor is heavily loaded, the rate of acceleration is more gradual and the rise in voltage on the acceleration contactor is slowed. This delay prevents the starting resistance from being cut out until the rise in speed and counter emf permits the contactor to close.

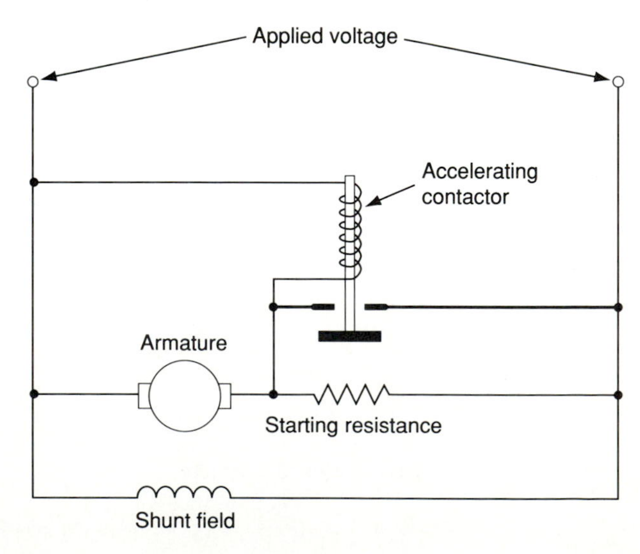

FIGURE 12-28 Counter-emf starter.

The principal disadvantage of the counter-emf starter is that any change in supply voltage may cause an erratic change in acceleration. This is because the starting contactor coils are designed to close the contactor on one value of voltage and to operate continuously on an increased voltage. This implies that the coils are sensitive to voltage change and will not function properly if fluctuations in line voltage should occur.

Time-element starter

The **time-element starter** has resistance in series with the armature, which is gradually reduced during a specified time interval. The most common method of reducing resistance in this manner is by using time-delay relays with contactors connected at different taps on a resistor, as shown in Figure 12-29. When the start button is pushed, coil M becomes energized. This instantaneously closes the four contacts marked M. The two M contacts in series with the motor are now closed, so the motor begins to accelerate with maximum resistance connected in series with the armature. The M contact in series with the coil marked TD_1 is also closed. When coil TD_1 times out, the TD_1 contacts

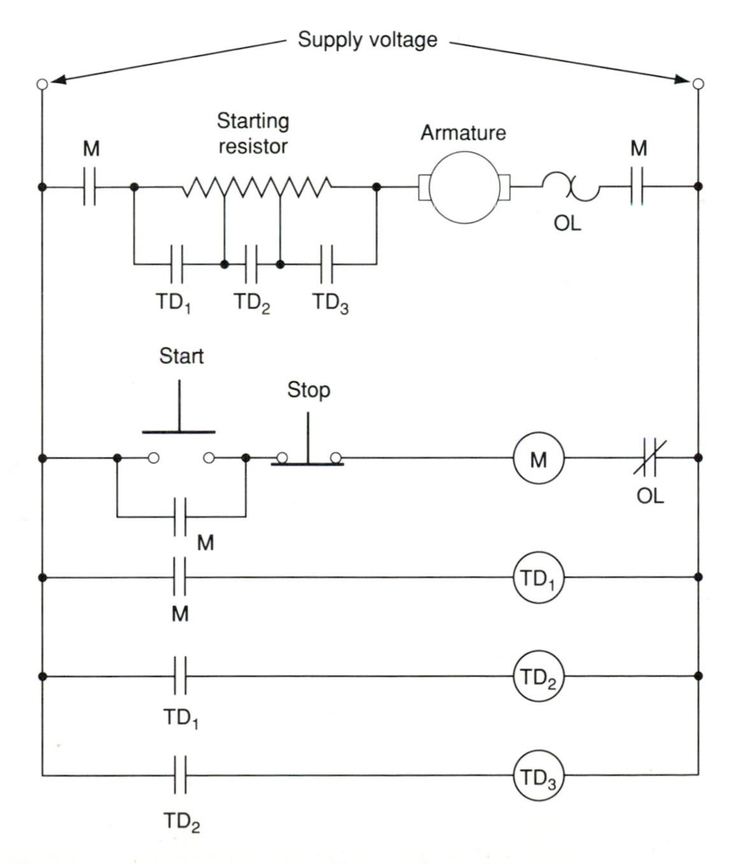

FIGURE 12-29 Motor-starting circuit using time-delay relays.

close, and a portion of the starting resistor is removed from the circuit. Coil TD_2 also begins to time out and when the TD_2 contacts close, more resistance is removed from the armature circuit. Coil TD_3 becomes energized and after a certain time interval its contacts also close. All resistance has now been removed from the circuit and the motor operates at full speed. When the stop button is pressed, coils M, TD_1, TD_2, and TD_3 become deenergized and all the contacts return to their normally open state.

The advantages of the time-element type of automatic starter are that its cost is low and it is a very reliable method of starting a motor. The disadvantage of this type of starter is that there is no method for sensing a change in load. This means that the resistance is removed at the same rate regardless of the load applied to the motor.

Current-limit starter

The **current-limit starter** uses current-limit contacts that remain open as long as a minimum value of current flows through them. If the current falls below this preset value, the contacts close. This type of contact is known as a series-

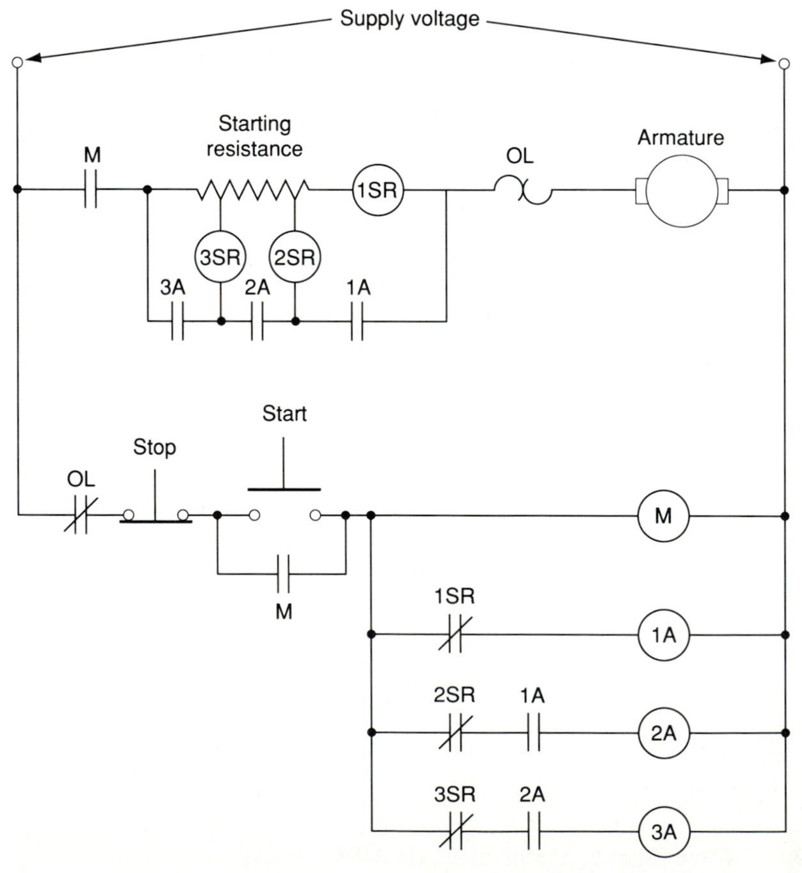

FIGURE 12-30 Current-limit acceleration starter with series relays.

lockout contact. Figure 12-30 shows a simplified schematic diagram of a current-limit starter.

When the start button is pressed, relay coil M becomes energized and this causes the two M contacts to close. The motor begins operating as current flows through the starting resistor and series relay 1SR. The 1SR coil is a fast-acting relay, so its normally closed contact is opened before coil 1A becomes energized. Therefore, the motor starts with maximum resistance in series with the armature. When the current through the armature falls back to a normal operating value, coil 1SR is no longer operational, and contact 1SR returns to its normally closed state. Relay coil 1A now becomes energized, removing a portion of the starting resistance and energizing relay 2SR. Since relay 1A's contacts are closed, when the current through coil 2SR falls below a certain value, 2SR will drop out, enabling relay 2A. The cycle then repeats until the starting resistor is removed from the circuit entirely.

12-20 CONTROLLERS

Whenever a starter is equipped with a means of governing, or regulating, the electric power delivered to a machine, it is referred to as a **controller**. A dc motor controller is capable of starting, stopping, controlling speed, reversing, and providing protection for the motor that it is controlling. A device that is commonly used for starting and controlling a series motor is the **drum controller**. Drum controllers are also used with shunt and compound motors that require frequent stopping, reversing, and starting. A drum controller derives its name from the fact that it utilizes a drum switch as the main switching element. One common type of drum controller is shown in Figure 12-31(a). It consists basically of a drum cylinder isolated from a central shaft with an operating handle attached to it. The drum has copper segments attached, which are either connected to, or isolated from, each other. A series of contact "fingers" is arranged to make contact with the segments of the drum. Although these fingers are isolated from each other, they do provide a connection between the starting resistance and the motor circuit. The contacts are held securely in place when the handle is turned by means of a notched wheel which is keyed to the shaft of the controller. When the contacts are properly positioned, a spring forces a roller into the corresponding notch and the correct position is indicated on the handle.

When the handle is turned forward by one notch, the connection points are as shown in Figure 12-31(b). The positive terminal of the dc input is connected across both the shunt and series field. The resistor connected in series with the series field provides additional resistance in the armature circuit and the motor begins turning at a reduced speed. If the handle is turned to the

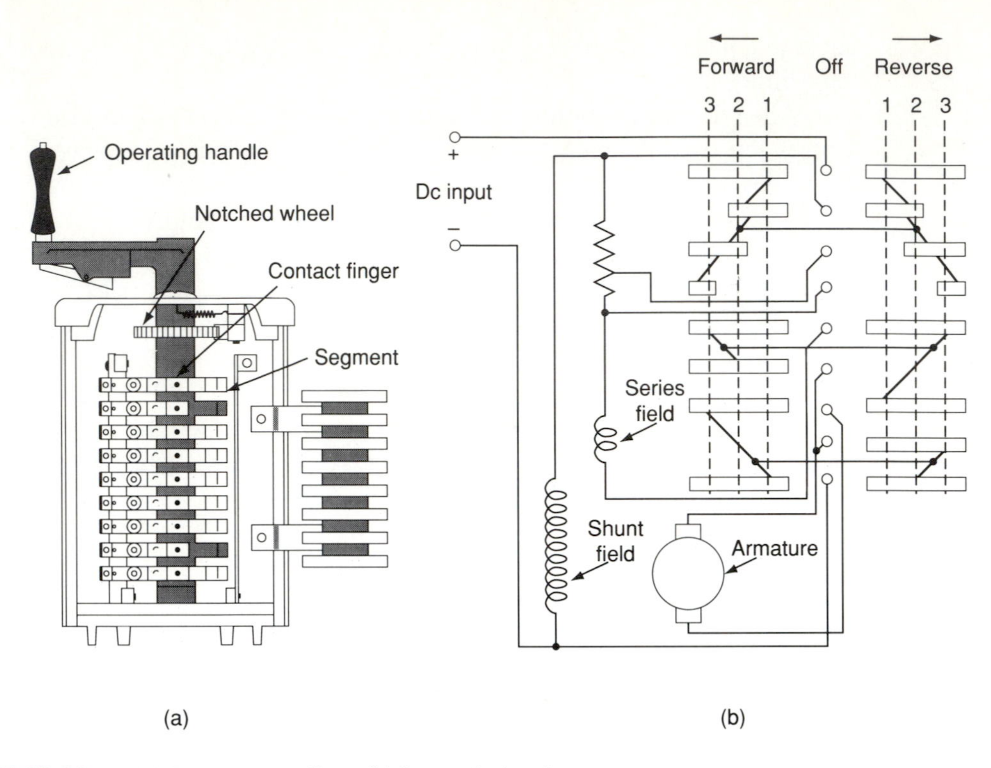

FIGURE 12-31 (a) Drum controller; (b) internal circuit.

second position of the forward setting, part of the resistance is removed from the armature circuit and the motor runs at a higher speed. When the handle is set to position 3, the resistance is shorted out of the circuit and the motor runs at full speed. Reversing the motor is accomplished by returning the handle to the off position, and then setting the handle to position 1. This changes the interconnections of the contact fingers, reversing the armature connections and consequently reversing the direction of rotation of the motor.

Since the current through motor controllers can be very high, and the controller is operated frequently, the contact fingers and corresponding contacts would burn from the "arcing and sparking" which is formed by the contacts opening. It is therefore necessary to extinguish the electric arc by means of a **magnetic blowout**. The basic principle of a magnetic blowout is illustrated in Figure 12-32. When motor current passes through the windings on either side of the contacts, the coils provide a magnetic field across the curved surface of the contacts. The contacts will remain closed as long as the coil is energized. When the contacts open, the current forms an electrical arc, which moves toward the reader. The direction of the arc is determined by Fleming's left-hand rule. As the arc is drawn out and away from the contacts, it eventually reaches a point where it breaks. The contacts are usually enclosed in a chamber called an **arc chute**. The purpose of the arc chute is to create a deionizing effect, which increases the extinguishing effect of the blowout. Since the coil of the magnetic blowout is connected in series with the line voltage, the extinguishing action is in direct proportion to the size of the electric arc formed.

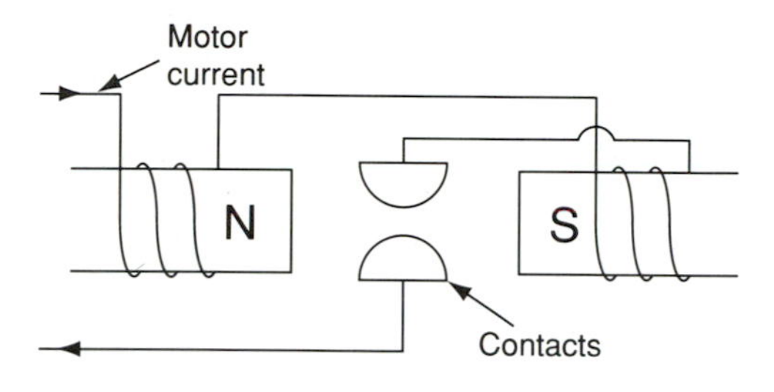

FIGURE 12-32 Magnetic blowout.

12-21 PLUGGING

Plugging is defined as a method of bringing a motor to a stop by reversing the developed torque in the armature. Plugging is used primarily when it is necessary to bring a motor to a rapid stop so that it may be reversed. One method of plugging has the field circuit left undisturbed while power is applied to the armature with reverse polarity. This action quickly causes the machine to stop and then start up with the opposite rotation. Another method involves the main switch being opened at the instant the armature comes to rest. At the instant the motor is plugged, the applied armature voltage and its counter emf are nearly equal and in the same direction. Therefore, to limit the initial rush of armature current to a reasonable value, it is necessary to insert a resistor in this circuit when the power is reversed. This added resistance will help to hold the current to a predetermined maximum. When the motor is reversed, the torque direction is reversed even though the direction of rotation remains the same.

12-22 DYNAMIC BRAKING

Dynamic braking makes use of the generator action in a motor to bring it to rest quickly. Essentially, this involves taking the energy required to drive the load and dissipating it as heat in resistors. This method saves wear and heating of the brake shoes as well as providing precise rheostatic control of speed. Dynamic braking is usually accomplished by the use of a controller which connects the field across the line while placing a resistance across the armature terminals. This produces generator action and results in a retarding action in the armature. If series motors are used, their fields must be connected across the line in series with resistance. Dynamic braking is not effective for stopping the series motor completely, since the braking action is reduced in proportion to the speed of the armature.

A connection for dynamic braking of a shunt motor is shown in Figure 12-33. The shunt field is connected directly across the line, allowing it to be constantly energized. When the start button is pressed, both relay coils are energized, and contacts *A* close while contact *B* opens. Resistor *R* is removed from the circuit while the motor is operating normally. When the stop button is pressed, contacts *A* open and contact *B* returns to its normally closed state. The armature is now removed from the power supply and is connected across resistor *R*. As the armature slows down, its rotates in a strong magnetic field and operates as a generator, establishing a high value of current through the armature circuit as well as through resistor *R*. Since the armature current is in a magnetic field as the motor slows down, a torque is established in the opposite direction to that of the armature rotation. This opposing torque quickly brings the motor to a stop.

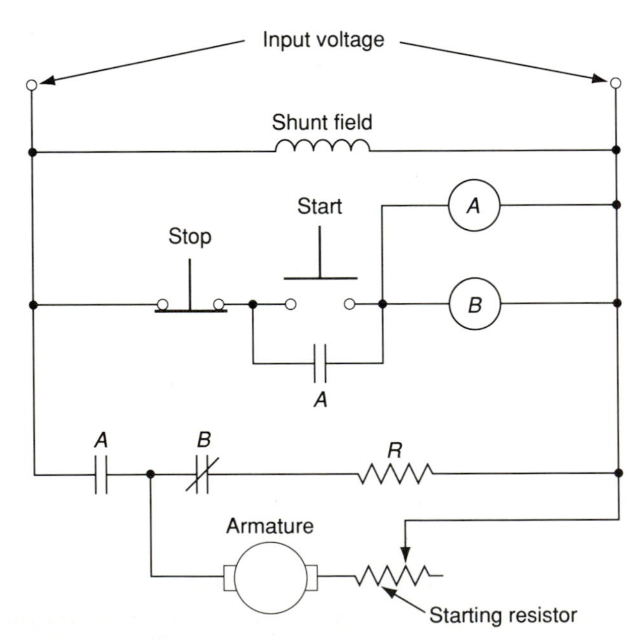

FIGURE 12-33 Dynamic braking circuit.

Dynamic braking of small-horsepower motors can also be accomplished using a transistor, as shown in Figure 12-34. In this circuit, a permanent-magnet motor is controlled by a PNP bipolar junction transistor (BJT). When the transistor is on, the resistance between the collector, *C*, and emitter, *E*, is very low. When the transistor is on, the resistance between *C* and *E* is very high. In order for the PNP transistor to conduct current, the voltage at the emitter must be more positive than the base. This condition is referred to as forward-biasing the emitter–base junction.

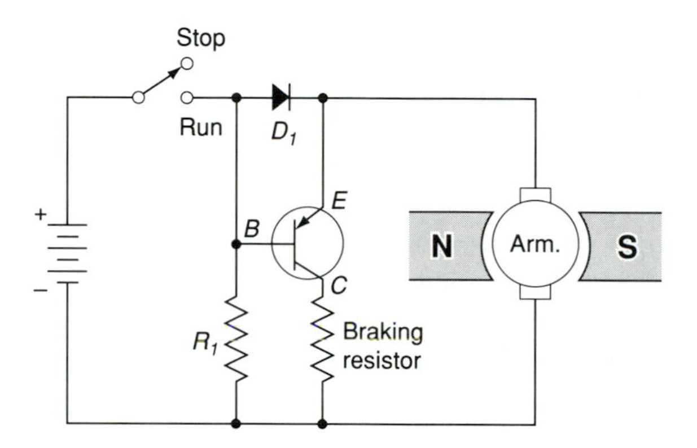

FIGURE 12-34 Transistor dynamic braking circuit.

Diode D_1 acts as a steering diode. That is, current will flow through the device in only one direction: anode to cathode. When the switch is in the RUN position, current flows through the diode to the PM motor. The transistor is not on because the emitter–base junction is reverse-biased. With the RUN switch closed, the voltage at the base is more positive than the emitter, due to the small forward voltage drop across D_1 (approximately. 0.7 V). When the switch is moved to the STOP position, the armature, which is still rotating, generates a voltage with the same polarity as the power supply, with respect to the emitter and collector of the transistor. Resistor R_1 acts as a limiting resistor for the base current when the STOP position is selected. The diode prevents any backfeed to the base of the transistor, ensuring a forward-bias condition. The transistor is now in a conducting state and acts like a closed switch, connecting the braking resistor in parallel with the armature, and dynamic braking occurs.

12-23 REGENERATIVE BRAKING

The action of **regenerative braking** is similar to that of dynamic braking. This term is applied to a system in which the load exerts a negative torque on the motor, driving it as a generator so that power is returned to the supply lines. For regenerative braking to be used, the counter emf of the motor must be greater than the line voltage. This means that the operating speed must be greater than the normal speed, such as when a hoist is being lowered. In addition to this overspeed condition, shunt motors may be made to regenerate if, at reduced speed, the field strength is increased sufficiently to cause the counter emf to exceed the line voltage. One common application of regenerative braking is in railroad locomotives, when the train is going down a long decline. Continuous braking action is used to limit the train speed as well as provide electrical energy for redistribution through the system for other equipment.

12-24 WARD–LEONARD SYSTEM OF SPEED CONTROL

The **Ward–Leonard system of speed control** uses the armature-control method of reducing speed without using series resistors. The dc motor armature is fed directly from a dc generator armature that is driven by a constant-speed prime mover. The magnitude and polarity of the dc field supply to the generator is varied by means of a rheostat and reversing switch. This implies that the motor armature is supplied by a generator that has a relatively linear voltage output from zero to full-load value. A constant-voltage source supplies both the generator field and the motor field. The motor is brought up to speed by gradually increasing the excitation of the generator field. With this type of control, as shown in Figure 12-35, the motor may be operated at any speed up to its maximum value.

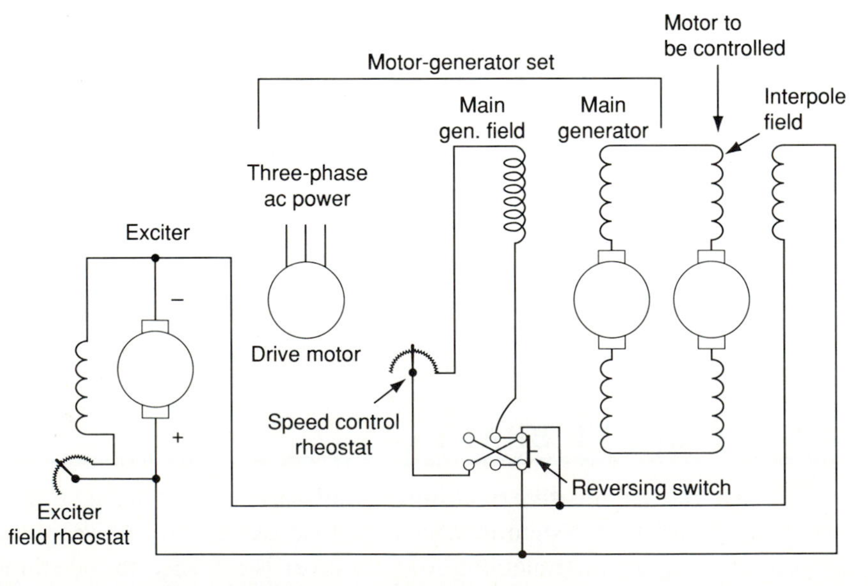

FIGURE 12-35 Ward–Leonard speed control system.

With this type of control system, the motor may be brought to a standstill quickly, merely by rapidly reducing the generator voltage. When the generator voltage is reduced below the counter emf of the motor, this counter emf sends current through the generator armature, establishing dynamic braking. The advantages of this type of speed control are that it does away with the armature rheostatic losses and instability of speed with variable loads. The disadvantage of this method of speed control is the initial expense of adding a motor–generator set, with its control equipment. The Ward–Leonard system of control is often used in connection with large motors such as elevators when a direct-current supply is not available.

KEY TERMS

Relative motion
Motor action
Fleming's left-hand rule
Torque
Developed torque
Prony brake
Counter emf (cemf)
Stabilization winding
Error voltage
Overshoot
Damping effect
Stepper (stepping) motor
Steps
Step angle
Permanent-magnet stepper motor
Detent torque
Restoring torque
Variable-reluctance stepper motor
Brushless dc motor
Silicon-controlled rectifier (SCR)
Commutation
Holding current
Dc chopper
Pulse-width modulation (PWM)
Duty cycle
Square wave
Jones chopper circuit
Counter-emf starter
Time-element starter
Current-limit starter
Controller
Drum controller
Magnetic blowout
Arc chute
Plugging
Dynamic braking
Regenerative braking
Ward–Leonard system of speed control

PROBLEMS

12-1 A current of 12 A is flowing through a conductor situated at right angles to a magnetic field with a flux density of 0.35 T. Determine the force on the conductor if the length of the conductor within the magnetic field is 15 cm.

12-2 A flat rectangular coil made of 200 turns is lying with its plane parallel to a magnetic field. The flux density is 0.022 Wb/m^2. The length of the coil transverse to the field is 18 cm. Calculate the force in newtons which acts on each side of the coil if the current per conductor is 20 A.

12-3 A conductor carries a current of 10 mA at right angles to a magnetic field where the flux density is 0.61 Wb/m^2. The length of the conductor is 25 cm. Determine the force on the conductor.

12-4 A motor that is developing a torque of 90 N·m has a sudden 30% decrease in field flux and a 15% increase in armature current. Determine the torque after these changes.

12-5 A dc motor develops a torque of 50 N·m when the armature current is 9 A. If the flux entering the armature remains constant, find the torque developed by the motor when I_a is increased to 15 A.

12-6 A motor rated at 2 hp, 120 V, will develop 2.4 hp when rotating at 1760 rpm. Determine the torque developed at 1760 rpm.

12-7 Determine the power, in watts, developed by a 15-hp motor.

12-8 When the armature of a dc motor is rotating at 2000 rpm, the motor develops an output torque of 100 N·m. Calculate the output power of the motor in watts.

12-9 A dc motor has an output torque of 75 N·m when the armature is rotating at 30 rev/s. Determine the output of the motor in horsepower.

12-10 A prony brake is used to test a dc motor and the following data are obtained: I_{in} = 75 A, V_{in} = 220 V, speed = 1800 rpm, and a 2-m brake arm balance reads 41.6 N. Determine (a) the output of motor in both watts and horsepower; (b) the efficiency at load tested.

12-11 A 37-in. prony brake arm is adjusted so that a net force of 22 lb is indicated on the scale. Determine the torque developed by the motor under test.

12-12 The armature of a 220-V dc motor draws 15 A when operating at full load and has a resistance of 2.1 Ω. Determine (a) the counter emf produced by the armature; (b) the power developed by the armature.

12-13 Find the armature current of a 240-V motor that has an armature resistance of 0.75 Ω and a counter emf of 233 V.

12-14 Assuming a constant flux, if a motor operating at 2160 rpm develops a counter emf of 220 V, what voltage is developed when the motor speed is 1800 rpm?

12-15 A 240-V dc motor has a full-load current of 150 A flowing in the armature when operating at 1800 rpm. The armature resistance is 0.08 Ω and the brush voltage drop is 2.8 V. Determine the no-load speed of the motor if the no-load armature current is 9 A and brush drop is 1.2 V. Assume a constant flux through all load conditions.

12-16 The speed of a dc motor is 1750 rpm at no-load. When full-load current flows, the motor speed is reduced to 1670 rpm. Determine the percentage of speed regulation.

12-17 The percent speed regulation of a motor is 12.5%. If the no-load speed is 1800 rpm, what is the full-load speed?

12-18 A 220-V dc shunt motor operates at 1200 rpm when the armature current is 95 A. The resistance of the armature circuit is 0.08 Ω. The motor is required to run at 900 rpm and draw an armature current of 75 A. Determine the size of resistor to be placed in series with the armature to accomplish this.

12-19 A 20-hp 240-V dc shunt motor has an armature resistance of 0.18 Ω. When the motor is operating at no-load, the speed is 1800 rpm and the armature current is 10 A. Determine (a) the size of external resistor that must be inserted in the armature circuit to reduce the speed to 1200 rpm at a rated load of 80 A; (b) the speed at rated load with no external resistance in the armature circuit. Assume a constant value of field flux.

12-20 Two dc motors, one shunt and one series, each have identical horsepower, voltage, and torque ratings. A full-load armature current of 80 A will develop 100% torque in each motor. Determine the amount of current required to develop 175% of full-load torque in each motor.

12-21 A dc motor is connected to an 80-V battery through a chopper. If the duty cycle is 45%, what is the dc voltage applied to the motor?

12-22 A chopper is on for 30 ms and off for the next 60 ms. Determine the duty cycle.

CHAPTER

Alternating Voltages and Currents

LEARNING OBJECTIVES

Upon completion of this chapter you will be able to:

- Explain the instantaneous value of a sine wave.
- Convert radians to electrical degrees, and vice versa.
- Define angular velocity.
- Discuss the average and rms values of a sine wave.
- Explain the phase relationships between alternating current and voltage.
- Discuss the differences between phasors and vectors.
- Convert polar quantities into rectangular, and vice versa.
- Explain the basic operating characteristics of ac measuring instruments such as frequency counters and oscilloscopes.
- Determine voltage and frequency values from oscilloscope displays.

13-1 INTRODUCTION

At the present time, over 95% of the electric energy used commercially is generated as alternating current. The main reason that alternating current is used is that alternating voltage may easily be raised or lowered in value. This is a tremendous advantage in electrical distribution systems, allowing ac power to be generated and distributed at a high voltage and reduced to a more practical voltage at the load. High voltage is essential to transmit large amounts of power efficiently over long distances. The voltage of direct-current systems is limited because of commutation to approximately 1500 V per generator. Alternating-current generators do not require commutators and have capacities in excess of 250,000 kVA.

When using alternating current, the voltage may be raised and lowered economically by means of transformers. This provides an efficient method of transmitting power over long distances. The weight of conductor required to transmit a given amount of power a given distance with fixed loss varies inversely as the square of the transmission voltage. For example, when the transmission voltage is doubled, the weight of conductor increases by 25%. Ac systems use transformers to step up voltage for economical power transmission, and then step down to the low voltages at which the power is utilized.

13-2 GENERATION OF SINE WAVES

In the discussion of dc generators it was explained that when a coil rotates in a uniform magnetic field, an emf is generated in that coil. This voltage is constantly varying in both direction and magnitude. Figure 13-1 shows an elementary alternating-current generator consisting of a one-turn coil, *AB*, which is free to rotate in a magnetic field. The end of the coil marked *A* is connected to slip ring *C*, while slip ring *D* is connected to the end of the coil marked *B*. The direction of coil rotation is assumed to be counterclockwise. At the instant rotation begins, conductors *A* and *B* begin to move parallel to the magnetic field. Since the conductors do not cut the magnetic field, no voltage is induced. As the coil continues to rotate the conductors begin to cut through the magnetic field and a voltage is induced in the coil. By applying Fleming's right-hand rule during the first 180° of rotation, the direction of the induced voltage is determined to be from coil *A* to *B*.

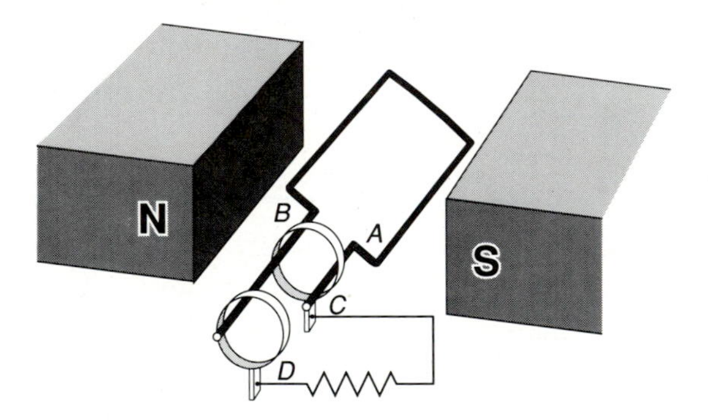

FIGURE 13-1 Simple ac generator with slip rings.

When the coil reaches 180°, the conductors once again move parallel to the flux, causing the induced voltage to fall to zero. As the counterclockwise rotation continues, conductors *A* and *B* are moving across the magnetic field in opposite directions to their original motion. The reversal of conductors results in an emf of opposite polarity being developed during the second 180° of revolution. Therefore, as the coil passes through one revolution, the induced voltage rises from zero to a maximum value in one direction, decreases through zero to a maximum value in the opposite direction, and returns to zero. This type of voltage is referred to as **alternating emf**. The value of this emf at any given time is called the **instantaneous value** and is represented by the letter e. The greatest value that e can have is the **maximum value**, E_m, or the **peak value**. The instantaneous value of current is represented by the letter i, and the peak or maximum current, by I_m. The **peak-to-peak value** of a sine wave is equal to $2E_m$ or $2I_m$.

If 360° of rotation are laid out along a horizontal axis, and the instantaneous voltages for various angular positions of the coil are plotted vertically, a wave shape similar to that of Figure 13-2 would be obtained. The actual waveform of the alternating voltage induced in a single coil is determined by the pattern of flux distribution in the air gap. In commercial ac generators, several windings are connected in series, and the alternating emf wave shape would more closely resemble the sinusoidal wave shown in Figure 13-3.

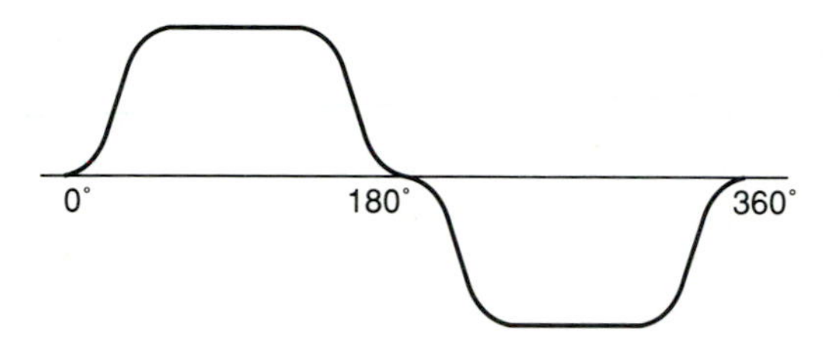

FIGURE 13-2

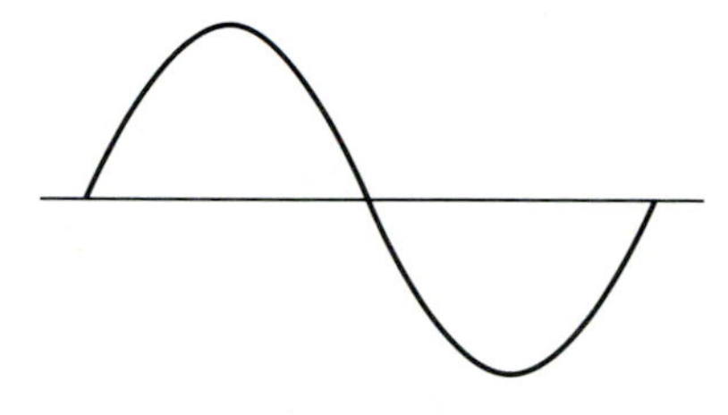

FIGURE 13-3 Ac sine wave.

To determine the instantaneous value of emf at a given instant on a sine wave, it is necessary to understand some fundamental trigonometry. Trigonometric functions are based on the relationships among the lengths of the sides of a right-angle triangle as the size of one of its angles is varied. A basic right-angle triangle is shown in Figure 13-4. The sine of the angle θ is the length of the side opposite the angle divided by the length of the hypotenuse, which is the longest side and the one opposite the right angle.

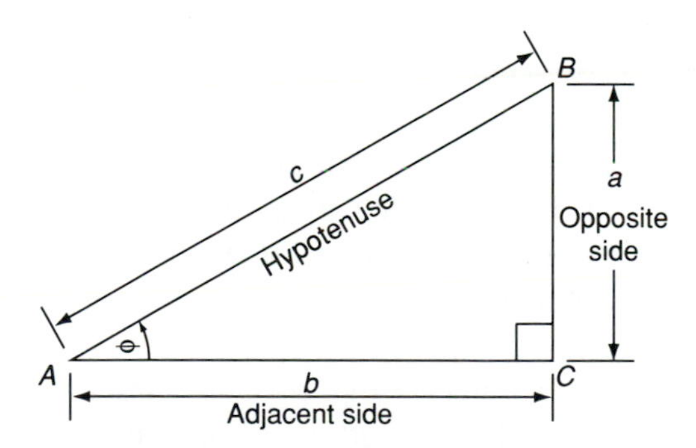

FIGURE 13-4 Definition of the sine function:

$$\text{sine of } \theta = \frac{\text{opposite side}}{\text{hypotenuse}} = \frac{a}{c}$$

The sine-wave form is the result of plotting the values of the sine against the values of angles to which each belongs. A convenient method of deriving sine values for plotting is by using a rotating **unit hypotenuse**, illustrated in

Figure 13-5. As the unit hypotenuse rotates, the distance above or below the horizontal of its endpoint A represents the sine. This value can be projected directly over to the right in the graph above the angle corresponding to theta, which is represented by the position of line OA.

Figure 13-5 shows the unit hypotenuse at 30° intervals during the first 180° of rotation. The fact that the rotating hypotenuse OA is constructed to have one unit of length makes the length of the vertical line AB equal to the sine of the angle. For example, triangle $OA \cdot B_{30}$ has a central angle 0 equal to 30°. Therefore, the length $A_{30} \cdot B_{30}$ is the sine 30° = 0.5. If a horizontal line is drawn from each A point toward the right, it intersects with a scale in the center of the figure, indicating the value of the sine.

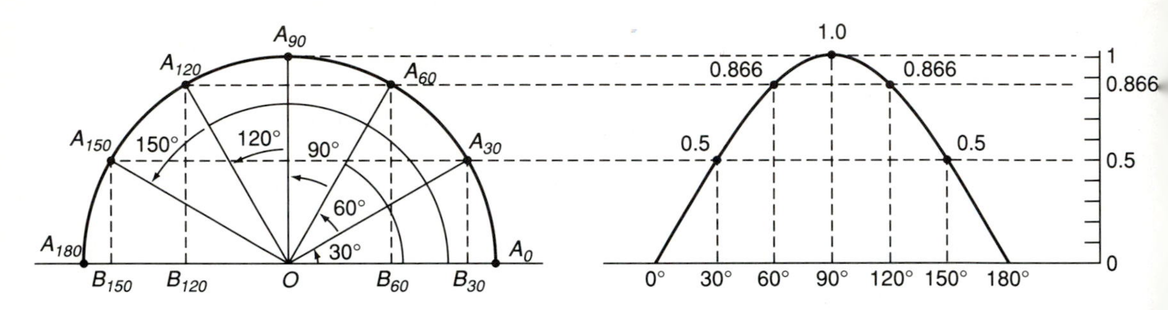

FIGURE 13-5 Relationship between sine wave and rotating armature during first 180°.

13-3 RADIAN MEASURE

In some situations it is necessary to determine the relationship between the linear velocity of a conductor and the angular velocity of the conductor. The **radian** unit is a very accurate method of calculating this relationship. In ac circuits, angles are frequently measured in radians rather than degrees. If the angle of rotation of a conductor is measured in radians, one complete revolution represents 2π radians instead of 360°.

The definition of a radian is illustrated in Figure 13-6. One radian is that angle which subtends a circular arc whose length is equal to the radius, r, of that arc. The arc subtended by 360° is a whole circle, which has a length of 2π times the radius. The central angle AOB in Figure 13-6 is equal to 1 radian (rad) because arc AB is equal to the radius OA.

The circumference around a circle equals $2\pi r$. A complete circle will therefore have 2π radians, which is subtended by 360°. In other words, 2π radians = 360°, so the number of degrees in a radian can be found by dividing 360° by 2π.

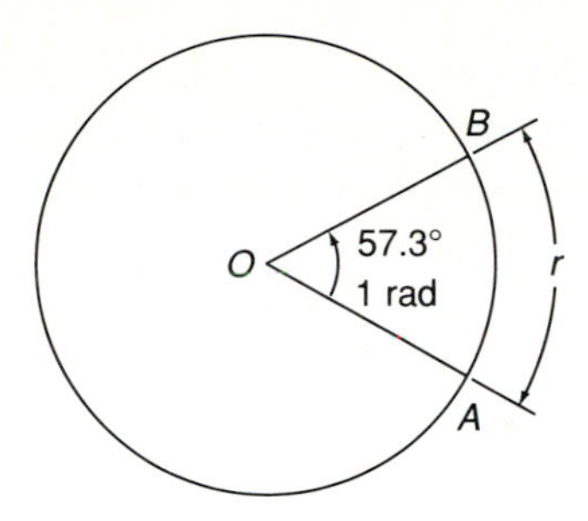

FIGURE 13-6 Definition of a radian.

$$1 \text{ radian} = \frac{360}{2\pi} = \frac{180}{\pi} = 57.3° \tag{13-1}$$

EXAMPLE 13-1 If a single loop of wire has completed three revolutions, how many radians has the coil traveled?

Solution

$$1 \text{ revolution} = 360° = 2\pi \text{ radians}$$

$$3 \text{ revolutions} = 6\pi \text{ radians} = 18.85 \text{ rad}$$

EXAMPLE 13-2 Change 2.5 rad into degrees.

Solution

$$\text{Angle (in degrees)} = \text{radians} \times \frac{180}{\pi} = \frac{2.5 \times 180}{\pi}$$

$$= 143.24°$$

EXAMPLE 13-3 A coil has rotated through 230°. What is this angle in radians?

Solution

$$\text{Radians} = \text{degrees} \times \frac{\pi}{180} = \frac{230 \times \pi}{180}$$

$$= 4.01 \text{ rad}$$

13-4 FREQUENCY

The voltage wave generated by a conductor completing the passage of one north and one south magnetic pole is called a **cycle**. Therefore, one cycle of a voltage is generated every time an armature completes 360° of rotation. The amount of time required for one cycle of change is referred to as the **time period** of the waveform. A period is then the amount of time it takes a wave to pass through a complete cycle. The time period is represented by the letter T and is measured in seconds. The number of cycles per second is defined as the **frequency** of an ac voltage or current. The unit of frequency is the **hertz**, (Hz). One hertz is equivalent to 1 cycle per second. Since the number of cycles per second is equal to the number of periods per second, the following equation may be derived:

$$f = \frac{1}{T} \tag{13-2}$$

where f is the frequency, in hertz, and T is the time, in seconds.

EXAMPLE 13-4 What is the time of one cycle when a signal has a frequency of 2650 kHz?

Solution

$$f = \frac{1}{T}$$

$$T = \frac{1}{f} = \frac{1}{2.65 \times 10^6}$$

$$= 0.377\ \mu\text{s}$$

EXAMPLE 13-5 A signal has a time period of 140 μs. Calculate the frequency.

Solution

$$f = \frac{1}{T} = \frac{1}{140 \times 10^{-6}}$$

$$= 7.14\ \text{kHz}$$

If the angle of rotation of the conductors is measured in radians instead of degrees, one complete revolution represents 2π radians. Since angular velocity is defined as degrees per second, the angle which the conductors move in 1 s may be written as

$$\text{angular velocity} = \frac{2\pi}{T} \quad \text{rad/s} \tag{13-3}$$

When determining angular velocity in radians, the symbol ω is used. Since $f = 1/T$, the equation for angular velocity may also be written as

$$\omega = 2\pi f \tag{13-4}$$

where ω is the angular velocity, in rad/s, and f is the frequency, in hertz. Since 60 Hz is the standard power-line frequency, the following relationship is quite common:

$$60\ \text{Hz} = 2\pi(60)$$

$$= 377\ \text{rad/s}$$

13-5 SINE-WAVE VALUES

In the sine wave, one complete cycle is represented by 360° or 2π radians. If the period of a sine wave is, for example, 0.3 s, each degree of the cycle would be 0.3/360 = 0.833 ms. The instantaneous value is determined from the value of $\sin\theta$ at the particular angle. At any given point, the instantaneous value, e, of the sine wave is equal to the product of the maximum value, E_m, of the sine wave and the sine of the angle corresponding to time. Expressed in equation form, the instantaneous value of voltage is

$$e = E_m \sin\theta \qquad (13\text{-}5)$$

To calculate the instantaneous value of current, the following equation may be used:

$$i = I_m \sin\theta \qquad (13\text{-}6)$$

In mathematics, the peak value is called the **amplitude** and θ is called the **argument**.

EXAMPLE 13-6 What is the instantaneous value of a sinusoidal current at 25° if $I_{max} = 2.2$ A?

Solution

$$\begin{aligned} i = I_m \sin\theta &= 2.2\text{ A} \times \sin 25° \\ &= 0.93\text{ A} \end{aligned}$$

Expressed in radian measure, the instantaneous value of voltage is calculated using the following equation:

$$e = E_m \sin\omega t \qquad (13\text{-}7)$$

where ω is the angular velocity, in rad/s, and t is the time, in seconds. Since there are 2π radians in a cycle, the instantaneous value of voltage may also be calculated as

$$e = E_m \sin 2\pi f t \qquad (13\text{-}8)$$

Because ω is expressed in radians per seconds, it is necessary to convert radians to degrees when using the sine function.

EXAMPLE 13-7 An ac waveform with a frequency of 1.3 kHz has a maximum voltage of 100 V. Calculate the instantaneous value of voltage at 58 μs. (Assume that $t = 0$ when the voltage is zero and increasing in a positive direction.)

Solution $e = E_m \sin \omega t$
$= 100 \sin[2\pi(1.3 \times 10^3)(58 \times 10^{-6})]$
$= 100 \sin 0.47 \text{ rad V}$

where 0.47 is the time angle in radians. Since 1 radian = 57.3°, e is calculated as follows:

$$e = 100 \sin(0.47 \times 57.3)$$
$$= 45.63 \text{ V}$$

EXAMPLE 13-8 An 850-Hz sine-wave signal has a maximum amplitude of 30 V. What is the instantaneous value of the signal 40 μs before it reaches its peak positive value?

Solution First, determine the period of the signal:

$$T = \frac{1}{f} = 1176.47\ \mu\text{s}$$

Next, determine the number of degrees per microsecond:

$$1176.47\ \mu\text{s} = 360°$$
$$1\ \mu\text{s} = 0.306°$$

Then

$$40\ \mu\text{s} = 40 \times 0.306°$$
$$= 12.24°$$

Since the peak maximum positive value of a sine wave occurs at 90°, 12.24° before the peak value is equal to

$$90° - 12.24° = 77.76°$$

$$e = E_m \sin \omega t = 30 \sin 77.76°$$
$$= 29.32 \text{ V}$$

13-6 AVERAGE AND RMS VALUES OF A SINE WAVE

A sinusoidal waveform has an instantaneous value that is constantly changing and a maximum value that occurs only twice in each cycle. Because of the constant fluctuating nature of a sine wave, it is often desirable to know what the **average value** of the waveform is. An average is a value obtained by adding successive, equally spaced values and then dividing the sum by the number of values. By definition, the true average, or **mean**, value of a sine wave is zero, since the positive and negative half-cycles would be identical in shape and magnitude. However, when calculating the average value of a sine wave, only one-half of the cycle is used.

The principle of average value is applied to the sine wave of Figure 13-7. The positive half-cycle of the waveform has been divided into 3° intervals. Each 3° segment represents an equal time period. Therefore, the average value of the whole sine wave is the average of all the segment-center values.

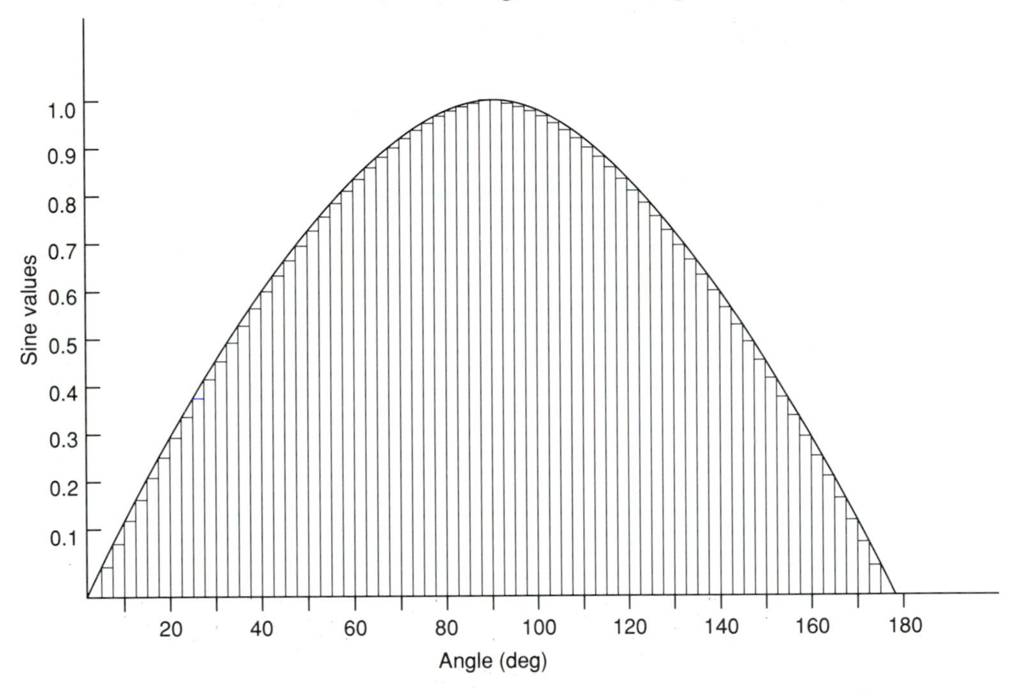

FIGURE 13-7 Method of integrating area under curve to determine average value.

If the sine values for all angles up to 180° are added and then divided by the number of values, this average equals 0.637 and the peak value is 1.

$$\text{Average value} = 0.637 \times \text{peak value}$$

$$= \frac{2}{\pi} \times E_m \tag{13-9}$$

The **effective**, or **rms**, **value** of a current or voltage is the value that will produce the same amount of power as a direct current or voltage.

Figure 13-8 shows an alternating sine wave of current having a maximum instantaneous value of $\sqrt{2}$, or 1.414, amperes. If the wave is considered over one complete cycle, the average value is zero, since there is just as much negative current as there is positive. If a dc ammeter were connected to measure this current, it would read zero. This is because a dc meter reads the average values of voltage and current.

The value of an alternating current is based on its heating effect and is defined in the following manner:

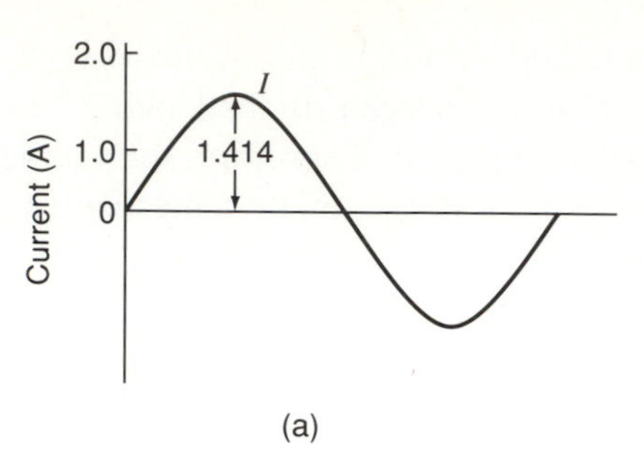

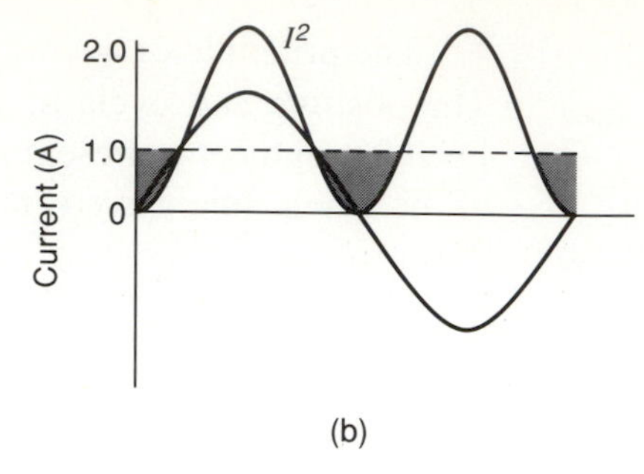

FIGURE 13-8 (a) Maximum value of sine wave; (b) current squared wave and proportion of heating effect.

> An ac ampere is the current that will produce heat in a pure resistance at the same rate as a dc ampere.

Since heating effect varies as the square of the current (I^2R), the value in amperes of the current represented by the waveform in Figure 13-8(a) must be based on the squares of the instantaneous values of current. Figure 13-8(b) shows the current wave of Figure 13-8(a) plotted together with its squared values. The maximum value of this new waveform will be 1.414^2, due to the maximum value of the original current wave being 1.414, or $\sqrt{2}$ amperes.

This squared wave of Figure 13-8(b) has a frequency which is double that of the original wave and has a horizontal axis of symmetry which is at a distance of 1.0 A above the zero axis. This squared wave has an average value of 1.0 A, which is the average of the squares of the ordinates of the current wave. If this value of current were to flow through a resistance, its heating effect over one cycle would be proportional to the area of the shaded in rectangle of Figure 13-8(b). This value of current is called the **root-mean-square**, or **rms, value** of current. Therefore, the sine wave of an ac ampere, which produces heat at the same rate as a dc ampere, has a maximum value of 1.414 A. The ratio of rms to maximum values is found by

$$E_{\text{rms}} = \frac{1}{1.414} \times E_m$$
$$= 0.707 \times E_m \quad (13\text{-}10)$$

$$I_{\text{rms}} = 0.707 \times I_m \quad (13\text{-}11)$$

When an alternating current or voltage is specified, it is always the effective, or rms, value that is implied, unless otherwise specified.

EXAMPLE 13-9 An ac voltage has an rms value of 115 V. What is the peak voltage?

Solution

$$E_m = \frac{E_{\text{rms}}}{0.707} = \frac{115\text{ V}}{0.707} = 162.66\text{ V}$$

EXAMPLE 13-10 A 240-V ac sine wave is connected to a 100-Ω resistor. Calculate the peak and rms values of current through the resistor.

Solution

$$E_m = \frac{E_{\text{rms}}}{0.707} = \frac{240\text{ V}}{0.707} = 339.46\text{ V}$$

$$I_m = \frac{E_m}{R} = \frac{339.46\text{ V}}{100\,\Omega} = 3.395\text{ A}$$

$$I_{\text{rms}} = \frac{E_{\text{rms}}}{R} = \frac{240\text{ V}}{100\,\Omega} = 2.4\text{ A}$$

EXAMPLE 13-11 Calculate the rms voltage of a signal that has a voltage of 105 V peak to peak.

Solution

$$E_{\text{p-p}} = 105\text{ V}$$

$$E_m = \tfrac{1}{2}E_{\text{p-p}} = 52.5\text{ V}$$

$$E_{\text{rms}} = 0.707 \times E_m = 0.707 \times 52.5\text{ V} = 37.12\text{ V}$$

In ac circuits, unless otherwise specified, power is calculated in terms of the effective values of current and voltage.

In some situations, it is convenient to base calculations initially using the average value of the emf over half a period, so that it becomes necessary to have some means of connecting this average value with the effective value. The correlation between these two values is called the **form factor**, and is defined as

$$\text{form factor} = \frac{\text{effective value}}{\text{average value}} \tag{13-12}$$

For sine waves the form factor is represented as

$$\frac{\text{effective value}}{\text{average value}} = \frac{\text{maximum value}}{\sqrt{2}} = \frac{\pi}{2\sqrt{2}} = 1.11 \tag{13-13}$$

13-7 PHASE RELATIONSHIPS

Whenever an alternating voltage causes a current to flow through a circuit, the current is said to be alternating at the same **fundamental frequency** as the applied voltage. Figure 13-9 shows a voltage and current waveform with different magnitudes but alternating at the same frequency. Under these conditions, when voltage and current pass through their zero values and increase to their maximum values at the same time and in the same direction, the two waves are considered to be **in phase** with each other.

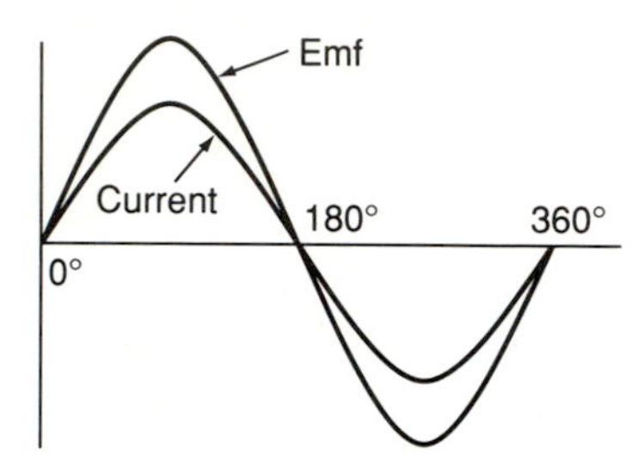

FIGURE 13-9 Currents and voltage waveforms in phase with each other.

If a sinusoidal voltage is applied to a purely resistive circuit, that is, a circuit with no capacitance or inductance, the resultant current will be as shown in Figure 13-9. However, in circuits containing combinations of resistance, capacitance, and inductance, the current and voltage waves do not pass through their maximum and minimum values at the same time. When this occurs, the current and voltage are said to be **out of phase** with each other.

Figure 13-10 shows a current and voltage waveform that has a phase displacement of 60°. Since the current waveform of Figure 13-10 is seen to start 60° later than the voltage, the current is said to **lag** the voltage by 60°. Generally, a circuit that contains a predominant amount of inductance will have a phase displacement where the current lags the voltage. A predominantly capacitive circuit will have a phase displacement where the current starts before, or **leads**, the voltage. The relationship between resistance, inductance, and capacitance is discussed in more detail in subsequent chapters.

When the current and voltage of a circuit are in phase, the difference in time between the two waves is zero, since they begin and end their cycles at the same instant. When the current and voltage are out of phase, the phase displacement is usually measured in electrical degrees and is called the **phase angle**. In Figure 13-10, the current is lagging the voltage by 60°, implying that the phase angle is equal to 60°.

Since the phase angle plays a role in determining the value of instantaneous voltage or current, it must be included in the calculation. For example, assume

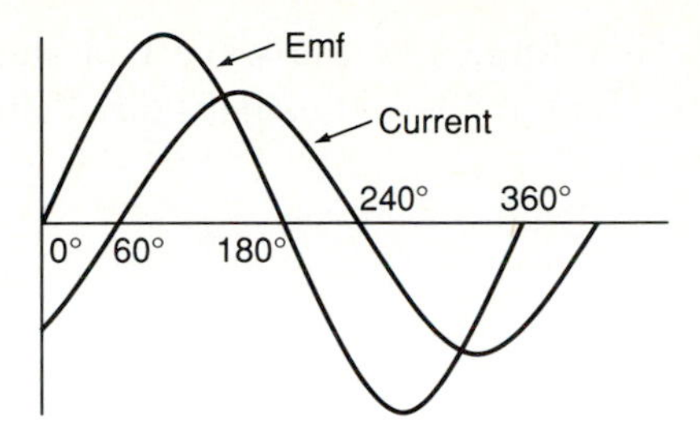

FIGURE 13-10 Voltage and current waveforms with a phase displacement of 60°.

that a circuit has an instantaneous voltage of $e = 30 \sin 100t$ and a current of $i = 3 \sin(100t + 30°)$. Since 30° has been added to the instantaneous value of current, the current wave will arrive 30° earlier than the voltage wave, so the current has a phase angle of 30° leading. If the value of current is given as $i = 3 \sin(100t - 30°)$ and the voltage is $e = 30 \sin 100t$, the current wave is delayed by 30° and is considered to be lagging.

When these extra angles are included in the general equations for e and i, they are expressed as

$$e = E_m(\sin \omega t \pm \alpha) \qquad \text{volts} \tag{13-14}$$

$$i = I_m(\sin \omega t \pm \beta) \qquad \text{amperes} \tag{13-15}$$

where α and β are the lowercase Greek letters alpha and beta, respectively.

$$\text{Phase angle, } \theta = \beta - \alpha \tag{13-16}$$

When the phase angle, theta, is negative, the phase is said to be *lagging*. When the phase angle is positive, the phase is considered to be *leading*.

EXAMPLE 13-12 The voltage $e = 120 \sin(2200t + 60°)$ V causes the current $i = 5 \sin(2200t + 25°)$ A to flow in a circuit. Determine the phase angle between e and i. Also, express this phase angle in terms of leading or lagging phase, relative to current.

Solution

$$\theta = \beta - \alpha = 25° - 60°$$
$$= -35°$$

Since the phase angle is negative, the phase angle is lagging.

The sine and cosine values which are listed in a trigonometric table show that the cosine assumes the same values as the sine, except that for the cosine, the angle is shifted for each 90°. Therefore, a cosine wave is a sine wave shifted to the left, or leading, by 90°. Figure 13-11 illustrates the relationship between a sine wave and a cosine wave.

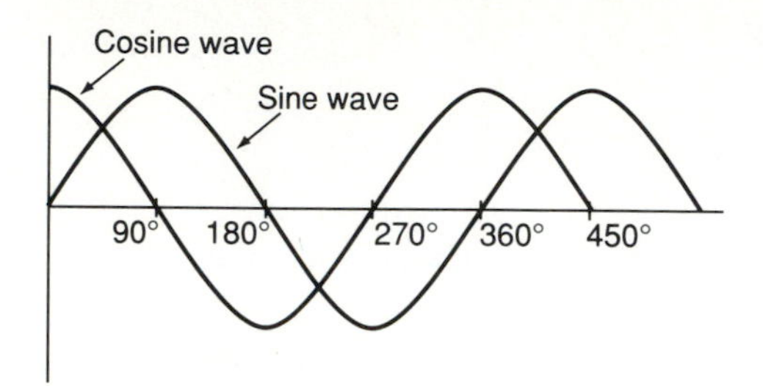

FIGURE 13-11 Displacement between sine wave and cosine wave.

To determine the phase angle between voltage and current, it is necessary to express both values in terms of either sine or cosine functions. Since a cosine wave is a sine wave shifted to the left by 90°, in order to change a cosine expression to sine form, add 90° to the angle and rewrite as a sine function.

$$\cos(\omega t + \alpha) = \sin(\omega t + \alpha + 90^\circ) \tag{13-17}$$

A sine wave is a cosine wave shifted right by 90°. To change a sine expression to cosine form, subtract 90° from the angle and rewrite as a cosine function.

$$\sin(\omega t + \alpha) = \cos(\omega t + \alpha - 90^\circ) \tag{13-18}$$

EXAMPLE 13-13 Convert $25\cos(3000t - 60^\circ)$ to sine form.

Solution

$$\begin{aligned} 25\cos(3000t - 60^\circ) &= 25\sin(3000t - 60^\circ + 90^\circ) \\ &= 25\sin(3000t + 30^\circ) \end{aligned}$$

EXAMPLE 13-14 Convert $15\sin(2500t + 35^\circ)$ to cosine form.

Solution

$$\begin{aligned} 15\sin(2500t + 35^\circ) &= 15\cos(2500t + 35^\circ - 90^\circ) \\ &= 15\cos(2500t - 55^\circ) \end{aligned}$$

The phase relationships of nonsinusoidal waveforms can be considerably different from those of sine waves. Waveforms such as the square wave, triangular wave, and sawtooth wave are called **periodic waveforms**. A periodic waveform is a repetitive waveform. By definition, *periodic* means recurring. A waveform does not have to be sinusoidal to be repetitive, although all nonsinusoidal periodic waveforms are composed of sine waves of different frequencies whose sum yields the original nonsinusoidal periodic waveform. The polarity of a periodic ac voltage reverses at regular time intervals, and the direction of a periodic ac current reverses at regular time intervals.

Complete mathematical analysis of a recurring nonsinusoidal waveform is made by applying principles that lead to the equation for the wave in a form called **Fourier series**, named after the French physicist and mathematician Baron Jean Fourier. Periodic functions of time of period T can be represented with the Fourier series as an infinite sum of sinusoids having frequencies that are multiples of the fundamental frequency $f = 1/T$. By using a Fourier series, nonsinusoidal periodic waves that can be defined over a 2π interval can be expanded into a series of sine and cosine waves of different frequencies and a constant term called the dc term, A_0. The dc term is the average value of the wave over one full cycle.

A comprehensive discussion of Fourier series is beyond the scope of this book, although it should be noted that this method of mathematical analysis is extremely useful in the design of equipment such as audio amplifiers and power transformers.

13-8 PHASORS

A **vector** quantity is defined as a quantity that has both magnitude and direction. Vector quantities are represented by an arrow, whose length represents the magnitude and whose direction represents the direction, or angle, which the force acts. Physical forces, such as those illustrated in Figure 13-12, are conveniently represented by vectors. If a force of 200 N is applied at right angles to a force of 300 N, a resultant force would be established as indicated. This force is at an angle θ with respect to the first force.

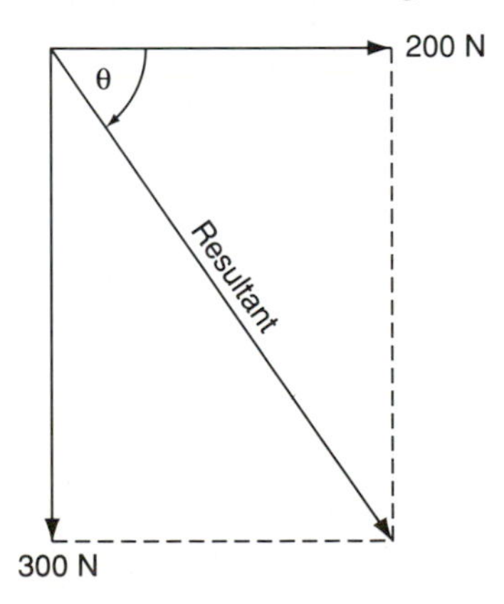

FIGURE 13-12 Representation of vector forces.

A similar technique can also be utilized in ac circuits to represent currents and voltages. Strictly speaking, ac voltages and currents are not vectors because they cannot be defined in terms of force and direction alone. These quantities also have a time relationship to each other and must be solved by using complex algebra. This type of algebra is used for solving numbers that exist in the two-dimensional complex plane. To avoid possible complications between vectors using linear algebra and ac quantities in the time domain, these current and voltage arrows have been renamed **phasors**.

Phasors do not normally represent instantaneous values, since they are of fixed length in any given situation. They are, essentially, used to represent effective, or rms, values, although they can also represent an instantaneous value for one specific instant of time. The phasor corresponds to the entire cycle of current or voltage, but is shown at only one angle, since the complete cycle is known to be a sine wave.

A **phasor diagram** consists of a number of phasors, one for each quantity represented. Each phasor is drawn from a common origin. The phasor in Figure 13-13 is stopped in one of its many possible positions. The horizontal axis is referred to as the reference axis. A phasor lying on the reference axis is considered to be the **reference phasor**. When a phasor is rotated in a counterclockwise direction from the reference axis, the resulting phase angle is taken to be positive. When the phasor is rotated in a clockwise direction, theta assumes a negative value.

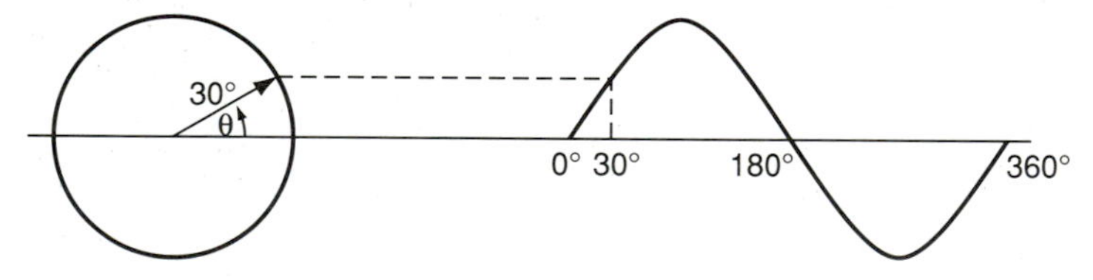

FIGURE 13-13 Instantaneous values of current from rotating phasors.

Earlier in this chapter it was mentioned that when a sinusoidal voltage is applied to a pure resistance, the resulting current is sinusoidal and is in phase with the voltage. Figure 13-14 illustrates how this information would be indicated on a phasor diagram. Note the capital letters E and I. When phasor quantities are shown as effective, or rms, values on a phasor diagram, they are usually shown in capital letters using boldface type. Instantaneous values of voltage and current are usually represented by lowercase letters. An instantaneous voltage drop is represented by the letter v, a voltage source by e, and an instantaneous current by i.

When labeling a phasor quantity, the sinusoidal expression is not used. For example,

$$e = 141.4 \sin(\omega t + 30°)$$

becomes

$$E = 100 \underline{/30°}\ V$$

where E is the rms value and $\underline{/\quad}$ denotes phase angle.

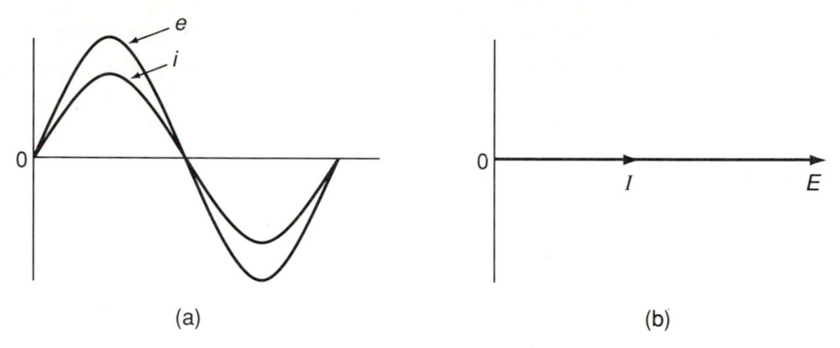

FIGURE 13-14 (a) Current and voltage relationships for a purely resistive circuit; (b) phasor diagram for a purely resistive circuit.

The sinusoidal current in a purely capacitive circuit leads the voltage by 90°. A phasor diagram for this type of circuit would be as shown in Figure 13-15(a). A phasor diagram for a purely inductive circuit would be represented by the phasor diagram shown in Figure 13-15(b). In both examples the phasor representing current is taken as the reference phasor. The choice of current or voltage as a reference phasor is arbitrary and is left up to the individual. However, when applying phasor diagrams to series circuits, it is customary to use current as a reference phasor since current is common to all parts of a series circuit. Consequently, a parallel circuit is usually represented in phasor diagrams with voltage as the reference phasor.

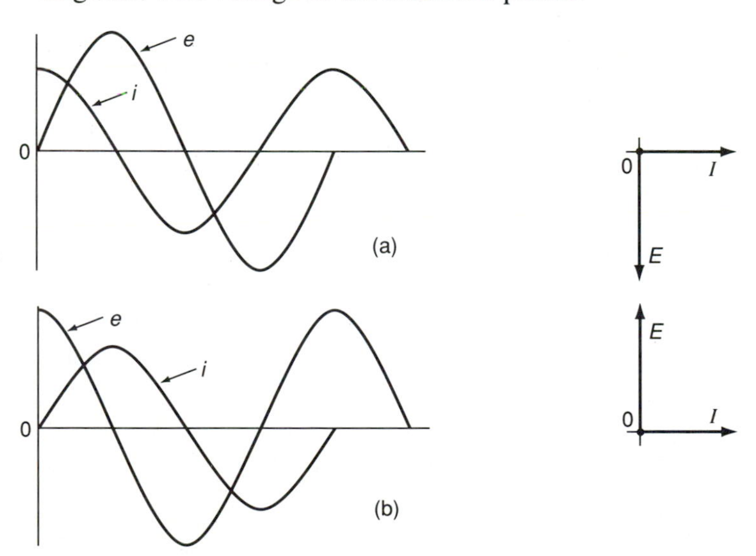

FIGURE 13-15 (a) Current and voltage relationship in a purely captive circuit. Phasor diagram shows voltage lagging current. (b) Current and voltage relationship in a purely inductive circuit. Phasor diagram shows voltage leading current.

13-9 POLAR AND RECTANGULAR FORMS OF PHASORS

There are two ways of specifying a phasor quantity. The first method, which we have been using in this chapter, is referred to as **polar form**. This method uses a phasor quantity of known magnitude lying in the coordinate plane and has a certain time quantity specified by the number of degrees between the phasor and the reference axis. The angle and magnitude of a phasor are also referred to as the **argument** and **modulus**, respectively. The polarity of a phasor quantity in polar form is determined by which quadrant the phasor lies in.

Figure 13-16 illustrates the four quadrants. The horizontal, or *X*, axis extends to the right and to the left of the origin and is designated as positive or negative. The vertical, or *Y*, axis is also given positive and negative designations. The quadrants are numbered from 1 to 4 in a counterclockwise direction from the reference axis. Any phasor lying within quadrants 1 or 2 would have a positive phase angle between 0 and 180°.

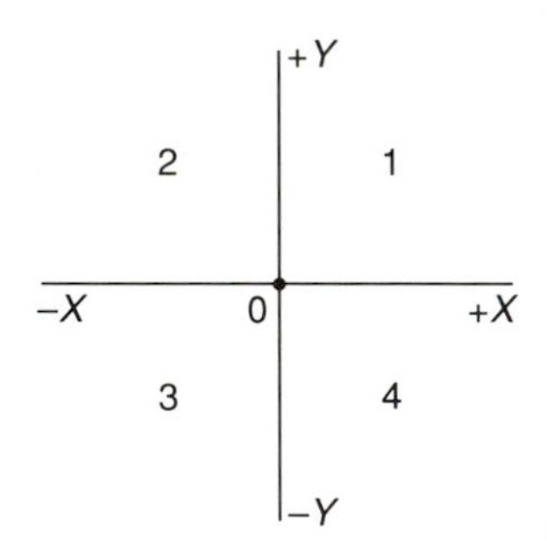

FIGURE 13-16 The four phasor quadrants.

When a phasor is rotated into the third or fourth quadrant, it may be considered as having either a positive or a negative value. Consider the phasor of Figure 13-17. As mentioned previously, angles are considered positive when rotated counterclockwise from the reference axis. When the phasor at 45° is rotated counterclockwise by 180°, the phasor is now at 225°. If the phasor is considered to be rotated clockwise by 180°, the phasor would lie at −135° from the reference axis. Either method of indicating a negative phasor is correct:

1. By placing a minus sign in front of the polar form
2. By changing the angle by 180°

The second method of specifying a phasor quantity is by **rectangular coordinates**. This method is shown in Figure 13-18, where the phasor is

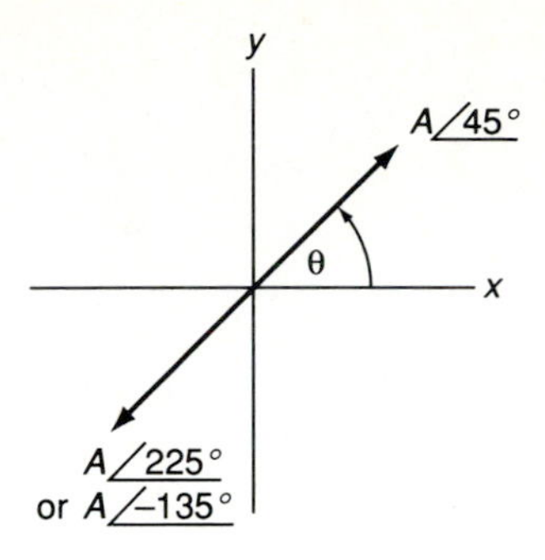

FIGURE 13-17 Reversal of a phasor.

assumed to be placed on a grid based on the axes of a graph. The axes can be either horizontal and vertical, as shown in Figure 13-17, or it can be angularly displaced. In both cases, the axes are at right angles to each other. To express a phasor in rectangular form, a sound understanding of trigonometric concepts is required. Since the horizontal component, vertical component, and phasor compose a right-angle triangle, the magnitude of the phasor is determined by using the **Pythagorean theorem**.

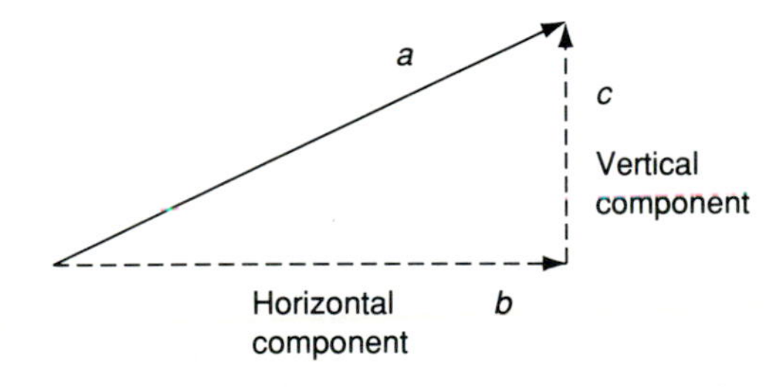

FIGURE 13-18 Rectangular form of specifying phasors.

The diagram shown in Figure 13-18 has the horizontal component identified as b, the vertical component labeled c, and the hypotenuse is identified as a. To solve for the magnitude of the phasor, the Pythagorean theorem would be

$$a = \sqrt{b^2 + c^2} \tag{13-19}$$

The sine, cosine, and tangents of phase angles are defined in the following manner:

$$\sin\theta = \frac{\text{opposite side}}{\text{hypotenuse}} = \frac{c}{a} \tag{13-20}$$

$$\cos\theta = \frac{\text{adjacent side}}{\text{hypotenuse}} = \frac{b}{a} \tag{13-21}$$

$$\tan\theta = \frac{\text{opposite side}}{\text{adjacent side}} = \frac{c}{b} \tag{13-22}$$

In order to determine the phase angle of the phasor itself, the inverse trigonometric operation of the tangent is used:

$$\theta = \tan^{-1}\frac{c}{b} \tag{13-23}$$

The horizontal component is sometimes referred to as the **real** part of the phasor. Its magnitude is determined by

$$b = a\cos\theta \tag{13-24}$$

The vertical component is occasionally referred to as the **imaginary** part of the phasor. Its magnitude is found by

$$c = a\sin\theta \tag{13-25}$$

EXAMPLE 13-15 Determine the horizontal and vertical components of the phasor $10\ \underline{/30^\circ}$.

Solution

$$\text{Horizontal component} = a\cos\theta = 10\cos 30^\circ = 8.66$$

$$\text{Vertical component} = a\sin\theta = 10\sin 30^\circ = 5$$

To check: $10 = \sqrt{8.66^2 + 5^2}$.

When phasor quantities are expressed in rectangular form, two phasor quantities with their phase differences at 90° must be shown. To do this it is necessary to distinguish between a phasor component at one basic angle and the component that has an angle of 90° with respect to the phasor at the basic angle. To distinguish between the horizontal and vertical components the *j* **operator** is used. Whenever a component is identified by having a *j* in front of it, the component is taken as being vertical. Any component given without the symbol *j* is assumed to be horizontal. Therefore, the phasor of Figure 13-18 may be described in rectangular form as

$$a = b + jc$$

The rectangular coordinates of a phasor consist of real numbers on the horizontal axis identified by + or − signs, and imaginary numbers on the vertical axis identified by $+j$ or $-j$ operators. A phasor quantity in rectangular form is considered to be a **complex number**, since it contains both real and imaginary numbers. When a complex number has its real and imaginary numbers multiplied by -1, it is referred to as the **reversal** of a complex number. The **conjugate** of a complex number multiplies only the imaginary part by -1. The four possible directions for the coordinates by the mathematical operators are shown in Figure 13-19. The j operator is a term used to indicate that an electrical quantity has gone through a 90° rotation in a counterclockwise direction from the reference axis. When a $-j$ is specified, it implies that the electrical quantity has gone through a 90° rotation in a clockwise direction from the reference axis.

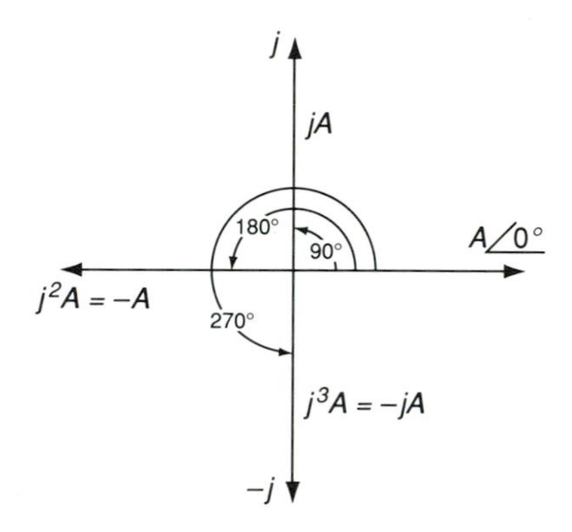

FIGURE 13-19 The j operator.

The j operator is more than just a prefix for identifying the vertical component of a phasor. Unlike most operators, j has a numerical value, $\sqrt{-1}$. In Figure 13-19, when A is multiplied by $-j$, the phasor is rotated counterclockwise by 90°. When the j operator is multiplied again, it causes a rotation of 180°. The expression is now

$$A \angle 180^\circ = j^2 A = \sqrt{-1}\sqrt{-1}\,A = -A \qquad (13\text{-}26)$$

When the effect of successive rotations through 90° is listed, Table 13-1 is developed.

TABLE 13-1 *Relation of Reference Position to j Operator*

Reference position	Operator
90° rotation	$\sqrt{-1}\,A$ or jA
180° rotation	$-1A$ or j^2A
270° rotation	$-\sqrt{-1}\,A$ or j^3A
360° rotation	$(-1)^2A$ or A or j^4A

EXAMPLE 13-16 Reverse the rectangular phasor $E = 8 - j6$.

Solution

$$E = 8 - j6$$
$$-E = -8 + j6$$

EXAMPLE 13-17 Express $48 + j30 - j^2 16$ as a rectangular phasor.

Solution Referring to Table 13-1, we have

$$-j^2 16 = +1(16)$$

Therefore,

$$48 + 16 + j30 = 64 + j30$$

13-10 CONVERSION BETWEEN POLAR AND RECTANGULAR FORMS

The Pythagorean theorem from plane geometry states that in any right-angle triangle the square of the length of the hypotenuse is equal to the sum of the squares of the lengths of the other two sides. Therefore, the sum of the square of the horizontal component and the square of the j component equals the square of the resultant. For example, consider the phasor shown in Figure 13-20. The magnitude of the phasor is 10, and since it is a right-angle triangle, theta is 45°. In Section 13-9 the equation for determining the horizontal component was given as a cos θ, and the vertical component was given as a sin θ.

To solve for line OA,

$$\begin{aligned} OA &= 10 \cos \theta \\ &= 7.07 \end{aligned}$$

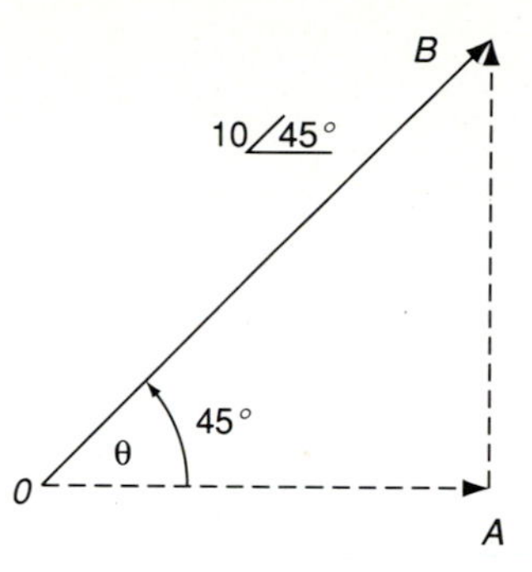

FIGURE 13-20

To solve for line AB,

$$AB = 10 \sin \theta$$
$$= j7.07$$

Therefore, the equation for changing from polar to rectangular form is

$$A\angle\theta = A \cos \theta + jA \sin \theta \qquad (13\text{-}27)$$

where A is the magnitude of the phasor.

EXAMPLE 13-18 Express the following phasors in their rectangular coordinates: (a) $220 \angle 120°$; (b) $240 \angle 1.7\pi$.

Solution

(a) $$220 \angle 120° = 220(\cos 120°) + j220(\sin 120°)$$
$$= -110 + j190.53$$

(b) $$240 \angle 1.7\pi = 240(\cos 1.7\pi) + j240(\sin 1.7\pi)$$
$$= 141.15 - j194.11$$

When converting from rectangular to polar form, the resultant and the angle it makes with the horizontal component must be found. To find the magnitude of the phasor, use of the Pythagorean theorem is required. The resultant, R, of Figure 13-21 is found by using equations 13-19 and 13-23.

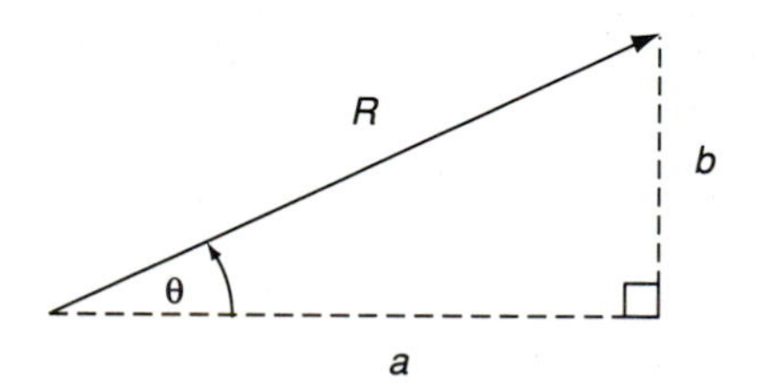

FIGURE 13-21

$$R = \sqrt{a^2 + b^2}$$

$$\theta = \tan^{-1}\frac{b}{a}$$

EXAMPLE 13-19 Convert the following from rectangular to polar form: (a) $-25 + j25$; (b) $-20 - j40$; (c) $50 - j15$.

Solution

(a) $-25 + j25 = \sqrt{(-25)^2 + 25^2}\ \underline{/\tan^{-1}(25/-25)}$
$= 35.36\ \underline{/135°}$

(b) $-20 - j40 = \sqrt{(-20)^2 + (-40)^2}\ \underline{/\tan^{-1}(-40/-20)}$
$= 44.72\ \underline{/-116.6°}$

(c) $50 - j15 = \sqrt{50^2 + (-15)^2}\ \underline{/\tan^{-1}(-15/50)}$
$= 52.2\ \underline{/-16.7°}$

13-11 PHASOR ADDITION AND SUBTRACTION

When phasors have the same phase angle, phasor addition and subtraction is a very simple procedure. For example, consider two voltages in a purely capacitive circuit, $E_1 = 35\ \underline{/-90°}$, $E_2 = 25\ \underline{/-90°}$. To find the sum of these two phasors, their magnitudes are added together.

$$E_T = E_1 + E_2 = 35\ \underline{/-90°} + 25\ \underline{/-90°}$$
$$= 60\ \underline{/-90°}\text{ V}$$

The same applies for subtraction of phasors with equal phase angles. Using the previous values, we have

$$E_T = E_1 - E_2$$
$$= 10\ \underline{/-90°}\text{ V}$$

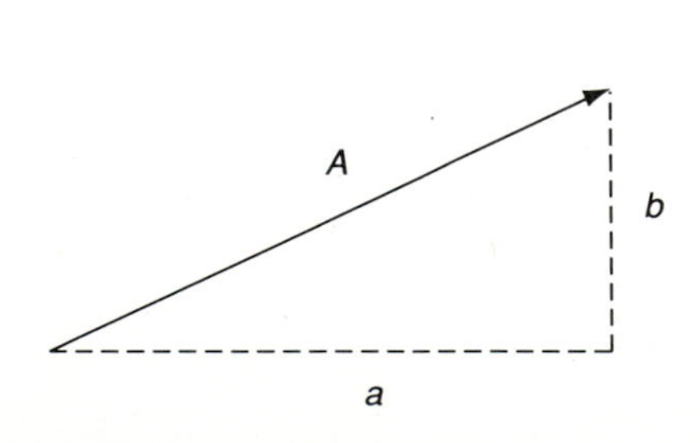

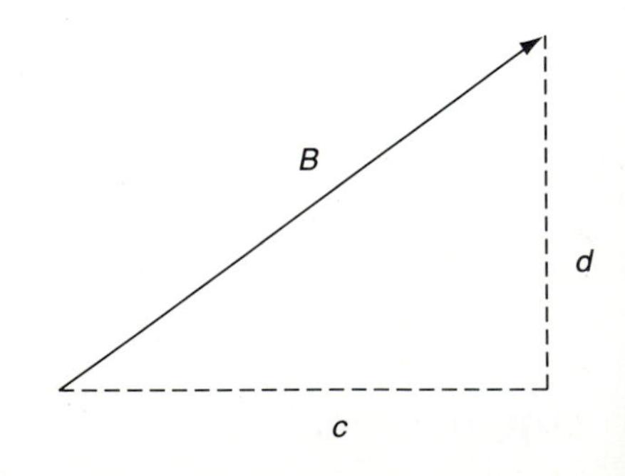

FIGURE 13-22

When adding or subtracting phasors having different phase angles, as in Figure 13-22, a different approach is required. It is necessary to express phasors in their rectangular or complex form, and by using ordinary arithmetic methods, add or subtract the real parts of each phasor, then add or subtract the imaginary parts. It is important that the imaginary and real parts be kept separate from each other. To find the sum of phasor A and phasor B in Figure 13-22, the following method may be used:

$$A = a + jb \quad \text{and} \quad B = c + jd$$

Their sum, C, is given by the equation

$$C = A + B = (a + jb) + (c + jd)$$

By combining the real and imaginary terms, the equation becomes

$$C = (a + c) + j(b + d)$$

If phasors A and B were to be subtracted, the result would be found by the following equation:

$$C = (a - c) + j(b - d)$$

Once the result has been calculated in rectangular form, the phasor is converted back to polar form.

EXAMPLE 13-20

What is the sum of $E_1 = 70 \angle 80^\circ$ V and $E_2 = 100 \angle -30^\circ$ V?

Solution

$$E_1 = 70 \cos 80^\circ + j70 \sin 80^\circ$$
$$= 12.16 + j68.94 \text{ V}$$

$$E_2 = -100 \cos 30^\circ - j100 \sin 30^\circ$$
$$= 86.6 - j50 \text{ V}$$

$$E_T = 98.76 - j18.94 \text{ V}$$

Converting to polar form, we have

$$\theta = \tan^{-1} \angle \frac{-18.94}{98.76}$$
$$= -10.86^\circ$$

$$E = \sqrt{(98.76)^2 + (18.94)^2}$$
$$= 100.56 \angle -10.86^\circ \text{ V}$$

EXAMPLE 13-21 Subtract $I_1 = 12 \angle 45^\circ\ A$ from $I_2 = 20 \angle -30^\circ\ A$.

Solution

$$I_1 = 12 \cos 45^\circ + j \sin 45^\circ$$
$$= 8.49 + j8.49 \text{ A}$$

$$I_2 = 20 \cos -30^\circ + j20 \sin -30^\circ$$
$$= 17.32 - j10$$

$$I_T = I_2 - I_1 = (17.32 - j10) - (8.49 + j8.49)$$
$$= 8.83 - j18.49$$
$$= 20.49 \angle -64.5^\circ \text{ A}$$

It is also possible to solve for phasor addition and subtraction by accurately plotting the phasors on a graph. One graphical solution technique consists of drawing the phasors to scale and adding or subtracting the phasors in a *tip-to-tail* method, as illustrated in Figure 13-23. Assume that two phasors are to be added which have magnitudes and angles of $10 \angle 61^\circ$ and $15 \angle 30^\circ$, respectively. The first step is to lay out $10 \angle 61^\circ$ with the aid of a ruler and a protractor; the scale can be 10 units of any size selected to give a convenient length. Next, start at the tip, or arrow, of the first phasor and lay out the phasor to be added, which in this case is $15 \angle 30^\circ$. This is done in the same manner as the first phasor except for the starting point, which is now the tip of the first phasor. In other words, the tail of the second phasor begins at the tip of the first phasor.

The two phasors have now been added graphically, and a phasor is now drawn from the tail of the first phasor to the tip of the second, which completes a triangle, or polygon. The length of this phasor can now be measured in the same units as those used to lay out the original phasors, and the resultant is obtained, which in this case is 24.13, as shown in Figure 13-23. The angle of the resultant can be measured with a protractor to complete the graphical addition. Although a graphical solution is always possible, the accuracy of this method is limited. For this reason, the addition and subtraction of phasors is generally accomplished by converting the phasors to rectangular form.

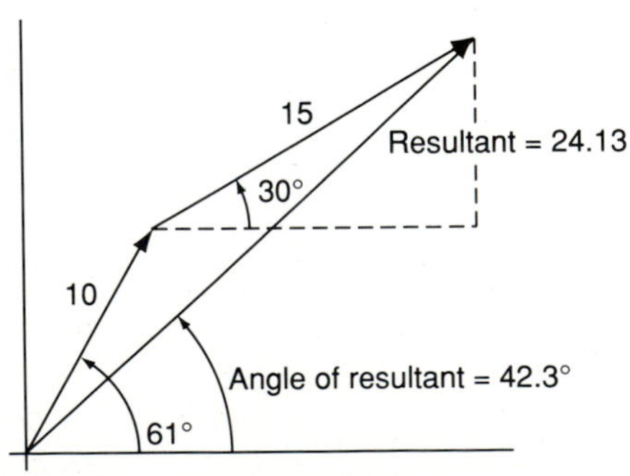

FIGURE 13-23 Addition of phasors.

13-12 PHASOR MULTIPLICATION AND DIVISION

When multiplying and dividing phasors, it is not necessary to convert them to rectangular form. To multiply phasors, multiply the magnitudes and add the angles algebraically. To divide phasors, divide the magnitudes, and subtract the angles algebraically.

EXAMPLE 13-22

Multiply $40 \underline{/25^\circ}$ by $28 \underline{/-45^\circ}$.

Solution

$$40 \times 28 = 1120$$

$$25^\circ + (-45^\circ) = -20^\circ$$

$$1120 \underline{/-20^\circ}$$

EXAMPLE 13-23

Divide $100 \underline{/60^\circ}$ by $240 \underline{/35^\circ}$.

Solution

$$\frac{100}{240} = 0.417$$

$$60^\circ - 35^\circ = 25^\circ$$

$$0.417 \underline{/25^\circ}$$

13-13 AC METERS AND INSTRUMENTS

By using a **bridge rectifier**, as shown in Figure 13-24, the D'Arsonval movement employed in dc meters can also be used to measure sinusoidal voltages and currents. A bridge rectifier converts alternating current to direct current. During the positive half-cycle, diodes D_2 and D_3 are forward-biased and the meter movement has the polarity shown in Figure 13-24. During the negative

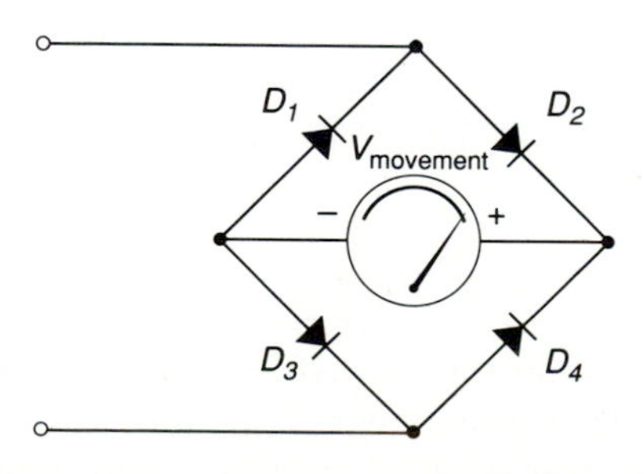

FIGURE 13-24 Full-wave bridge rectifier.

half-cycle of the ac input, diodes D_1 and D_4 are forward-biased and the current flow through the meter is the same as during the positive half-cycle. Since the direction of current flow does not change during either half-cycle, the current has been rectified.

The current through the meter movement of Figure 13-24 flows in pulses, since each alternation rises from zero to a peak value and drops back to zero again. Unless the frequency of the ac input is extremely low, the meter movement will not be able to follow the variations in the pulsating current. Instead, the meter's pointer responds to the average value of the ac sine wave, which is $0.637 \times V_{peak}$. However, the scale on the ac meter is usually calibrated in effective, or rms, values. The rms value of a sine wave is $0.707 \times V_{peak}$. Figure 13-25 shows the ac input signal to a bridge rectifier and the dc output signal.

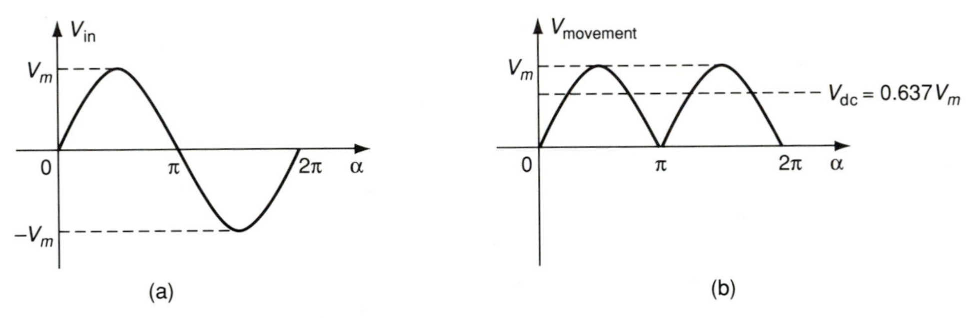

FIGURE 13-25 (a) Sinusoidal input; (b) full-wave rectified signal.

The rectifier-type D'Arsonval meter will not provide reliable readings for square waves, pulses, or irregular waveshapes. Figure 13-26 shows some typical waveforms found in electrical and electronic circuits. With the sine wave, an ac meter will read 0.707 of the peak value, which is $1.11 \times$ average. With the square wave, the meter will still indicate $1.11 \times$ average, but in this case, average is equal to peak. Therefore, the ac meter will not give an accu-

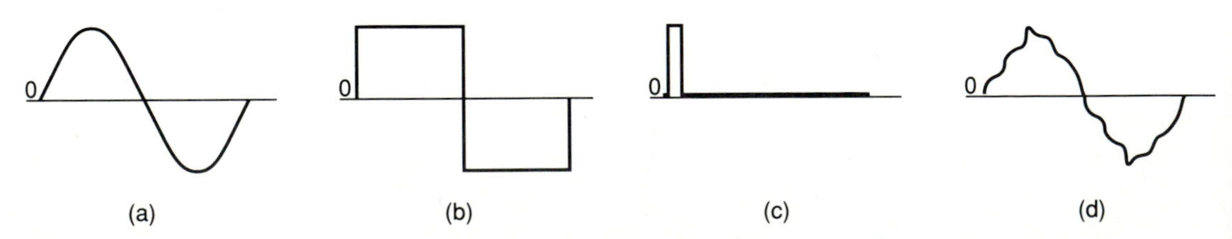

FIGURE 13-26 (a) Sine wave: average = 0.637 peak, rms = 0.707 peak, rms = 1.11 average; (b) square wave: average = peak, rms = peak, rms = average; (c) pulse: average = duty cycle × peak, rms = average; (d) irregular: cannot calculate by usual methods.

rate reading of a square wave since it will indicate that the rms value is actually 1.11 × peak. With a pulse input, the meter will still respond to the average and indicate 1.11 × average. Once again this is wrong. In the irregular waveform, the meter will respond to the nonsinusoidal wave and indicate 1.11 × average. We have no way of determining whether this value is correct or not using a bridge rectifier type of meter. In most cases, the reading would be incorrect.

The **electrodynamometer** is a meter that is capable of measuring both ac and dc voltages and currents. This type of meter is capable of very accurate measurements at low frequencies. It is similar to the D'Arsonval except that it has no permanent magnet and the magnetic field is produced by two field coils. Figure 13-27 illustrates a positive deflection of the pointer for current flowing in either direction. In Figure 13-27(a), the stationary and moving coils establish flux in such a manner that like poles are adjacent to each other. The like poles repel and cause the moving coil to be deflected clockwise. Consequently, the pointer moves upscale from left to right. When the current flow is reversed, as in Figure 13-27(b), like poles are once again adjacent to each other and the pointer moves upscale from left to right.

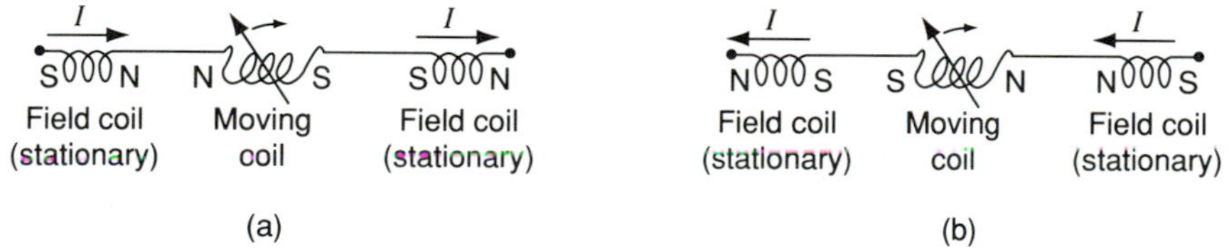

FIGURE 13-27 (a) Current flowing from left to right; (b) current flowing from right to left.

Frequency measurements can be made with the electrodynamic instrument shown in Figure 13-28. Two tuned circuits are utilized; one tuned circuit resonates just below the low end of the scale, and the other resonates just above the high end of the scale. In Figure 13-28, the center-scale frequency is 60 Hz. Each tuned circuit is connected in series with half of the fixed-coil winding. At frequencies below 60 Hz, coil F_1 has the greatest influence on deflection and the pointer moves counterclockwise. At frequencies above 60 Hz, coil F_2 has the stronger effect and the pointer moves clockwise.

A **digital electronic counter** is a sophisticated frequency meter that is capable of measuring frequencies over an extremely wide range in comparison to the electrodynamometer. Digital electronic counters are used to measure the frequency of an unknown signal, the period of any signal, and the frequency ratio of two different signals.

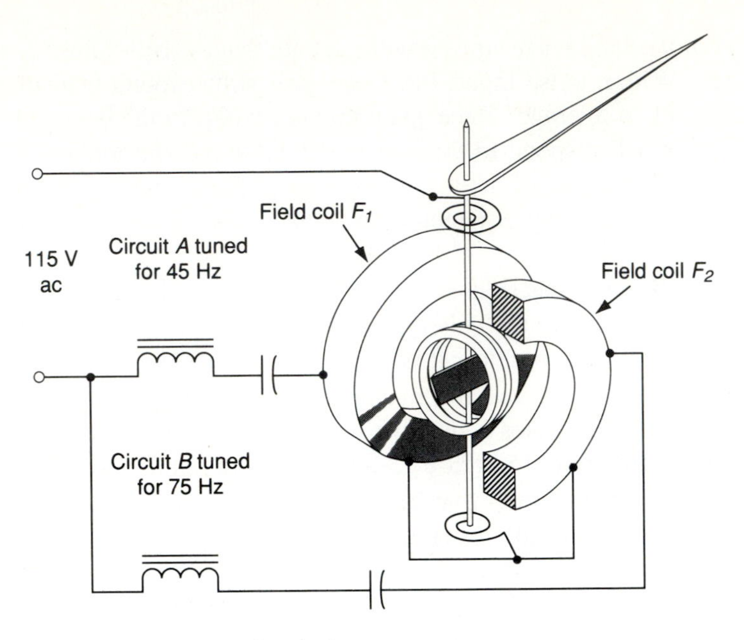

FIGURE 13-28 Resonant-circuit frequency meter.

A simplified block diagram of a digital electronic counter is shown in Figure 13-29. Input signal A may be a sine wave that is passed through an amplifier and then through a **Schmitt trigger** circuit. A Schmitt trigger processes the input signal and is of crucial importance in the operation of every

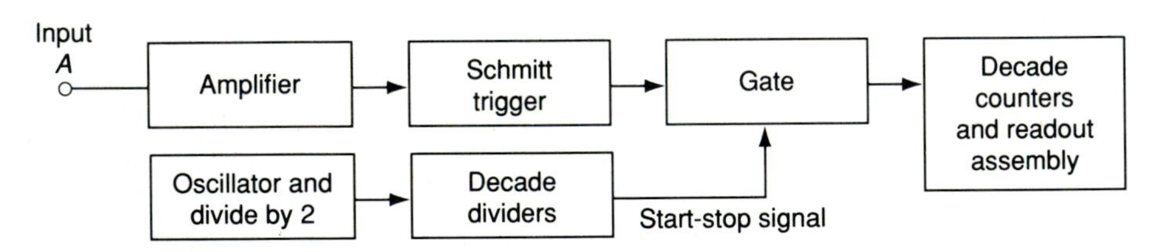

FIGURE 13-29 Block diagram of digital frequency meter.

electronic counter. The Schmitt trigger generates positive- and negative-going spikes which correspond to the sine-wave zero crossings. A typical digital electronic counter will use a 2-MHz oscillator which is divided by 2 to produce a stable 1-MHz clock. The 1-MHz clock signal is divided down in the decade dividers. The time elapsed between the start/stop signal depends on the setting of the frequency-time selector switch on the digital counter.

Figure 13-30 illustrates how the start/stop gate signal determines the actual frequency count. The length of time the start/stop gate is on determines the number of Schmitt trigger spikes allowed by the start/stop signal to pass through the gate and into the decade counters and digital readout assembly. For example, if the input signal is 1000 Hz and the frequency time selector switch is set at 1 s, the start/stop gate will be 1 s long. This means that the 1-MHz clock must have been divided down through the decade dividers by 1 million times to generate the 1 s start/stop gate. The decade counters and readout assembly receive 1000 spikes, corresponding to 1000 Hz. Therefore, the frequency displayed by a digital counter is simply the number of Schmitt trigger pulses per time period.

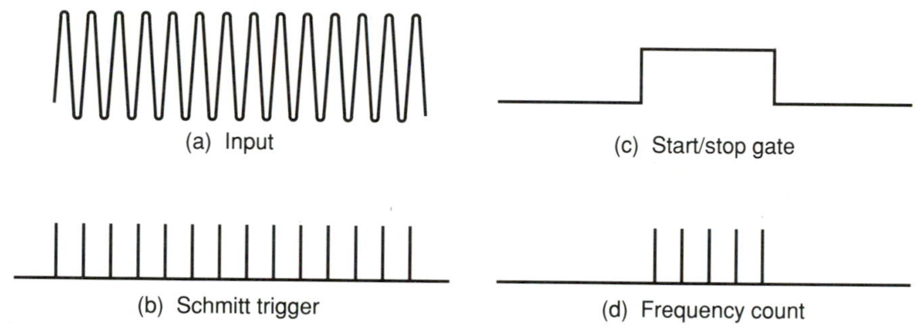

FIGURE 13-30 Waveforms for frequency measurement: (a) input; (b) Schmitt trigger; (c) start/ stop gate; (d) frequency count.

An **oscilloscope** is an electronic measuring instrument consisting of a cathode-ray tube (CRT) and various associated circuit sections. The oscilloscope is commonly used to automatically plot a particular voltage variation versus time. Almost anything can be measured on the two-dimensional graph drawn by an oscilloscope. Parameters such as phase shift, rise time, decay time, peak-to-peak voltages and currents, repetition rate, pulse duration, pulse delay time, period, and frequency can all be measured by using this versatile instrument.

The heart of the oscilloscope is the cathode-ray tube, since it performs the basic functions to convert a signal into an image. The CRT is a vacuum tube similar in shape to a TV picture tube, as illustrated in Figure 13-31. A cathode ray tube consists of an electron gun for supplying a concentrated beam of electrons, a pair of deflection plates for changing the direction of the electron beam, and a screen coated with a substance that glows when struck by the electron beam. A transparent ruled screen called a **graticule** is generally mounted in front of the fluorescent screen.

An electric current is passed through the heater of the electron gun to increase the temperature of the cathode to a point where electrons are emitted. The cathode is surrounded by a cylindrical cap that is at a negative potential. This cap, which has a small hole located along the longitudinal axis of the CRT, acts as the control grid. Since the control grid is at a negative potential, electrons are repelled away from the cylindrical walls, and consequently, stream through the hole, where they move into the electric fields of the focusing and accelerating grids.

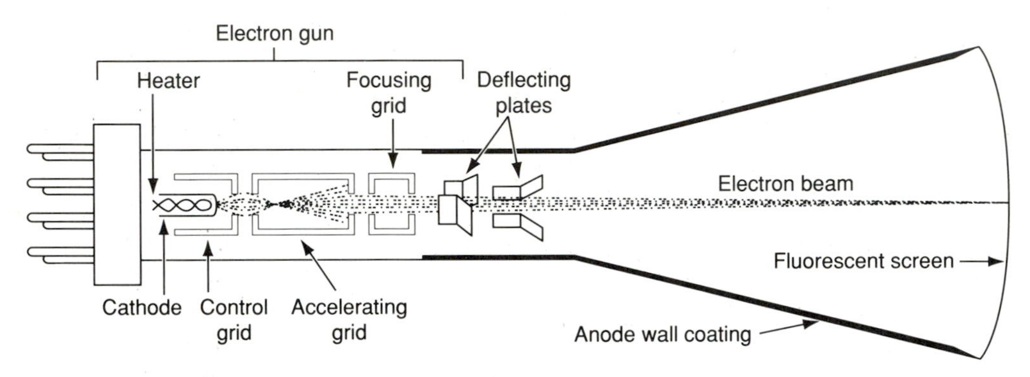

FIGURE 13-31 Basic components of a CRT.

Figure 13-32 shows a block diagram of a basic cathode-ray oscilloscope. The vertical input signal is applied to the vertical deflection plates via a multistage vertical amplifier. The vertical amplifier is the main factor in determining the sensitivity and bandwidth of an oscilloscope. The vertical sensitivity is a measure of how much the electron beam will deflect for a specified input signal.

The horizontal amplifier provides the deflection voltages required to deflect the beam across the *x*-axis of the CRT. With the switch set to **internal sync**, as it is for normal operation of the oscilloscope, the output of the vertical amplifier is applied to the **sweep generator**. The purpose of the sweep generator is to develop a voltage at the horizontal deflection plate that increases linearly with time. This linearly increasing voltage, called a **ramp voltage**, or **sawtooth waveform**, causes the beam to be deflected equal distances horizontally per unit of time.

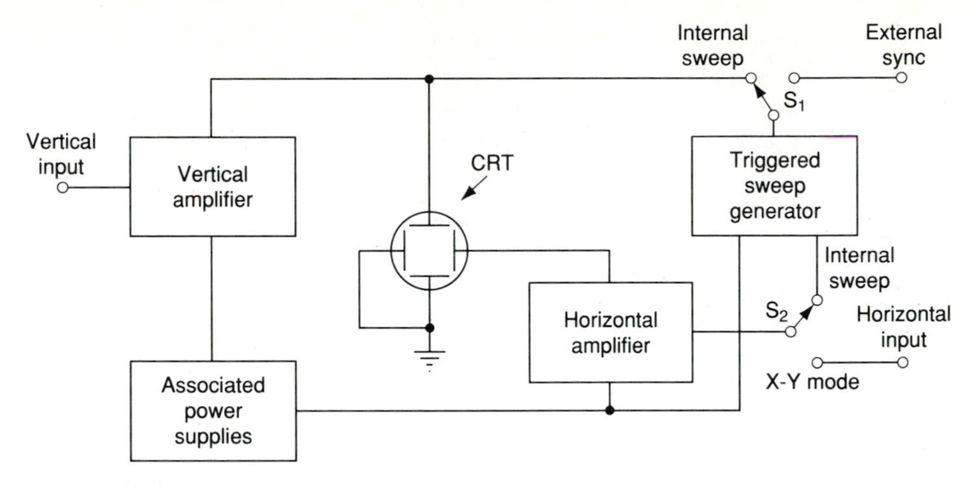

FIGURE 13-32 Block diagram of a basic oscilloscope.

Figure 13-33 illustrates a typical CRT faceplate with a graticule. Generally, graticules are laid out in an 8 × 10 pattern. Each of the 8 vertical and 10 horizontal lines divides the screen into 1-cm squares. The minor divisions shown in Figure 13-34 represent increments of 0.2 cm. The volts/div. switch and sec/div. switch on the front of an oscilloscope change the value of each major vertical and horizontal division on the graticule. For example, on the 5-V/div. setting, each of the eight major vertical divisions represents 5 V, so

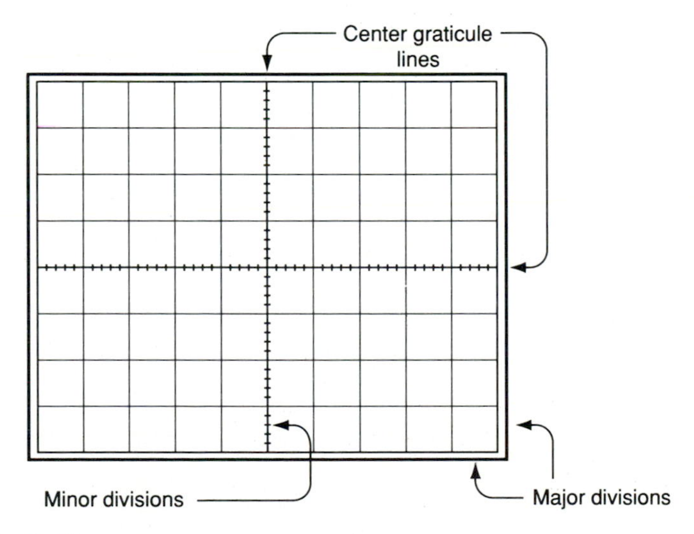

FIGURE 13-33 CRT faceplate with graticule.

the entire screen can display up to 40 V from top to bottom. If the sec/div. switch is set at 10 ms, each of the 10 major horizontal divisions represents 10 ms, and the entire screen can display up to 100 ms from left to right.

Figure 13-34 shows the front panel of a dual-trace oscilloscope. The **intensity control** varies the brightness of the displayed waveform, and the **focus control** allows the focus of the wave to be adjusted. The **beam finder** is used to locate a waveform that has been shifted off the screen. The dual-trace oscilloscope allows two waveforms to be displayed on the screen at any given time. Two coaxial-type input terminals are used to connect the oscilloscope to the input signals. The pushbuttons above the input connectors allow the oper-

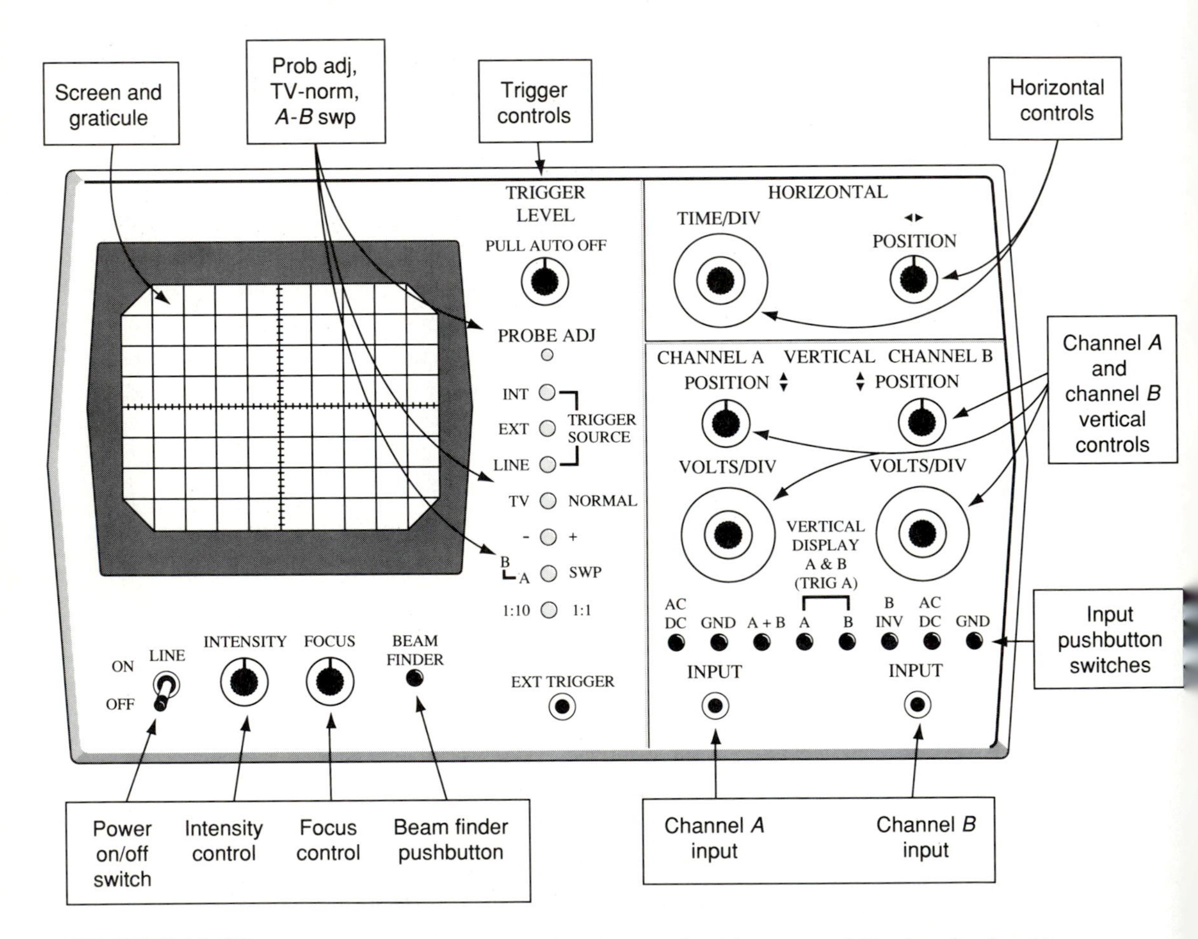

FIGURE 13-34 Oscilloscope front panel and controls. (Courtesy of Hewlett-Packard.)

ator to select ac or dc signals. The ground (GRND) pushbuttons disconnect the input signals and ground the input terminals. This allows each trace to be referenced to an appropriate position on the screen. By selecting **vertical display**, input signal A, B, or signals A and B can be displayed on the screen. If $A + B$ and the B inv. buttons are pressed, the oscilloscope displays the difference of the two inputs.

The **vertical controls** for channels A and B allows an appropriate volts/div. to be selected as well as move the waveform up or down by using the position knob. The time/div. switch selects the horizontal deflection sensitivity of the display. The horizontal position is adjusted by turning the position knob. The vernier knob in the center of both the time/div. and volts/div. must be turned fully clockwise for these controls to be calibrated.

The **trigger control** is used to provide a stable waveform display. For the waveform to be stable, the display must begin at exactly the instant that the input waveform is at its zero position. The three *trigger source* pushbuttons allow the selection of INT (internal), EXT (external), or LINE as triggering sources for the time base. When INT is selected, the time base is triggered by one of the input signals. The EXT trigger source allows the time base to be triggered from an external source connected to the EXT trigger terminal. If the LINE trigger is chosen, the time base is triggered from the line or ac power frequency. The Probe adj is used as a known signal source when calibrating the oscilloscope. In most applications, the TV-norm switch is set on the normal position. The A-B sweep button is usually set at SWP, and the time base is allowed to sweep the electron beam horizontally across the screen for waveform display. If the A-B position is selected, the time base is disconnected. In this mode, signals applied to channel B will produce vertical deflection, and signals applied to channel A produce horizontal deflection.

The most direct voltage measurement made with an oscilloscope is the peak-to-peak value. The peak-to-peak to value of voltage is calculated as follows:

$$V_{\mathrm{p-p}} = (\text{volts/div.})\,\frac{\text{no. div.}}{1} \tag{13-28}$$

The period and frequency of periodic signals are easily measured with an oscilloscope. The waveform must be displayed in such a manner that one complete cycle is displayed on the CRT screen. Accuracy is usually improved if the single cycle displayed fill as much of the horizontal distance across the screen as possible. The period is determined as

$$T = (\text{time/div.})\,\frac{\text{no. div.}}{\text{cycle}} \qquad (13\text{-}29)$$

The frequency is then calculated as the reciprocal of the period, or

$$f = \frac{1}{T}$$

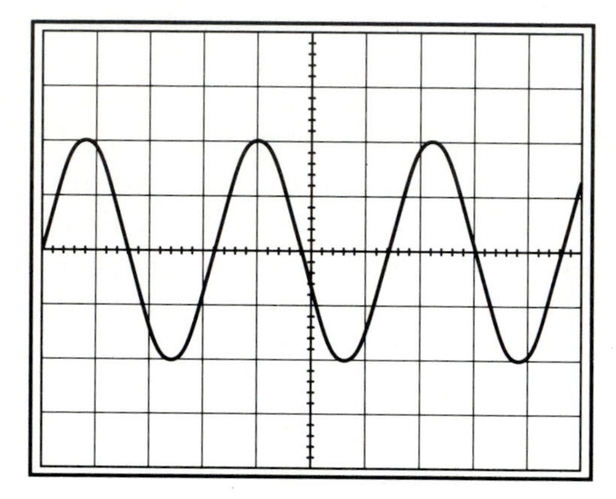

FIGURE 13-35

EXAMPLE 13-24 Calculate the peak-to-peak voltage of the sine wave shown in Figure 13-35 if the volts/div. switch is on 2 V/cm.

Solution

$$V_{p-p} = (\text{volts/div.})\,\frac{\text{no. div.}}{1} = (2\ \text{V/cm})\,\frac{4}{1}$$
$$= 8\ \text{V}$$

EXAMPLE 13-25 Determine the frequency of the sine wave shown in Figure 13-35 if the sec/cm switch is set at 1 ms/cm.

Solution

$$T = (\text{time/div.})\,\frac{\text{no. div.}}{\text{cycle}}$$
$$= (1\ \text{ms/cm})\,\frac{3.2\,\text{cm}}{\text{cycle}}$$
$$= 3.2\ \text{ms} \quad \text{or} \quad 0.0032\ \text{s}$$

$$f = \frac{1}{T} = \frac{1}{0.0032}$$
$$= 312.5\ \text{Hz}$$

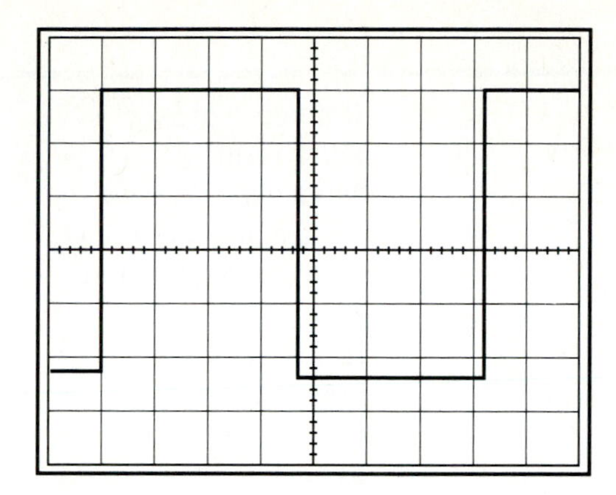

FIGURE 13-36

EXAMPLE 13-26 For the waveform shown in Figure 13-36, determine (a) the peak-to-peak voltage; (b) the frequency of the square wave. Assume that the volts/cm switch is set at 10 V/cm and the sec/cm switch is at 10 μs/cm.

Solution

(a) $$V_{p-p} = (\text{volts/div.})\,\frac{\text{no. div.}}{\text{cycle}} = (10\ \text{V/cm})\,\frac{5.4}{1} = 54\ \text{V}$$

(b) $$T = (\text{time/div.})\,\frac{\text{no. div.}}{\text{cycle}} = (10\ \mu\text{s/cm})\,\frac{7.2\ \text{cm}}{\text{cycle}} = 72\ \mu\text{s} \quad \text{or} \quad 0.000072\ \text{s}$$

$$f = \frac{1}{T} = \frac{1}{0.000072} = 13.89\ \text{kHz}$$

KEY TERMS

- Alternating emf
- Instantaneous value
- Maximum value
- Peak value
- Unit hypotenuse
- Radian
- Cycle
- Time period
- Frequency
- Hertz
- Amplitude
- Argument
- Average value
- Mean
- Effective (rms) value
- Root-mean-square (rms) value
- Form factor
- Fundamental frequency
- In phase
- Out of phase
- Lagging
- Leading
- Phase angle
- Periodic waveform
- Fourier series
- Vector
- Phasor
- Phasor diagram
- Reference phasor
- Polar form
- Modulus
- Rectangular coordinates
- Pythagorean theorem
- Real
- Imaginary
- *j* operator
- Complex number
- Reversal
- Conjugate
- Bridge rectifier
- Electrodynamometer
- Digital electronic counter
- Schmitt trigger
- Oscilloscope
- Graticule
- Internal sync
- Sweep generator
- Ramp voltage
- Sawtooth waveform
- Intensity control
- Focus control
- Beam finder
- Vertical display
- Vertical control
- Trigger control

PROBLEMS

13-1 How many radians has a single loop of wire traveled if it has completed 12 revolutions?

13-2 If a loop of wire has rotated 310°, how many radians has it traveled?

13-3 Convert 1.74 radians into degrees.

13-4 If a single loop of wire has traveled 25.13 rad, how many revolutions has the coil completed?

13-5 Calculate the time of one cycle if a signal has a frequency of 32.76 kHz.

13-6 Calculate the frequency of a signal with a time period of 15.7×10^{-6} s.

13-7 Determine the angular velocity of a phasor representing a sine wave with a frequency of 1200 Hz.

13-8 Determine the instantaneous value of a sinusoidal voltage at 40° if the maximum voltage is 140 V.

13-9 An ac waveform with a frequency of 60 Hz has a maximum current of 10 A. Determine the instantaneous value of current at 0.7×10^{-3} s. Assume that $t = 0$ when the current is zero and increasing in a positive direction.

13-10 A 1500-Hz sinusoidal waveform has a maximum amplitude of 40 V. Determine the instantaneous voltage 30 μs before the wave reaches its peak positive value.

13-11 Determine the value of a sine-wave voltage at 5 μs from the positive-going zero crossing when the peak voltage is 12 V and the frequency is 25 kHz.

13-12 An ac voltage has an rms value of 208 V. Determine the peak voltage.

13-13 A full-wave bridge rectifier is connected to a 120-V rms supply. Neglecting the voltage drop of the diodes, calculate the average output voltage of the rectifier.

13-14 A 120-V ac sine wave is connected to an 80-Ω resistor. Determine (a) the peak and (b) the rms values of current through the resistor.

13-15 If an ac waveform has a peak-to-peak voltage of 187.11 V, find the rms value.

13-16 For the following voltage and current, determine the phase angle between the two values and express this angle in terms of leading or lagging phase: $e = 120 \sin(377t + 85°)$ V, $i = 8 \sin(377t + 15°)$ A.

13-17 Convert $215 \cos(2200t - 80°)$ to sine form.

13-18 Convert $75 \sin(377t + 50°)$ to cosine form.

13-19 Express the following voltage as an rms phasor quantity: $e = 294 \sin(\omega t + 60°)$.

13-20 Determine (a) the horizontal and (b) the vertical components of the phasor $27 \angle 60°$.

13-21 Reverse the rectangular phasor $15 - j12$.

13-22 Express $120 + j80 - j^2 40$ as a rectangular phasor.

13-23 Determine (a) the conjugate and (b) the reversal of the following complex number: $9 + j14$.

13-24 Express the following phasors in their rectangular coordinates: (a) $18 \angle 200°$; (b) $120 \angle 60°$; (c) $75 \angle 1.5\pi$.

13-25 Convert the following from rectangular to polar form: (a) $50 - j71$; (b) $-3 + j15$; (c) $-25 - j40$.

13-26 Determine the phasor sum of the following two voltage phasors: $E_1 = 24 \angle 75°$ V, $E_2 = 90 \angle 40°$ V.

13-27 What is the phasor sum of $I_1 = 1.5 \angle 30°$ A and $I_2 = 2.2 \angle 15°$ A?

13-28 Subtract $E_1 = 24 \angle 75°$ V from $E_2 = 18 \angle -35°$ V.

13-29 Multiply $20.84 \angle 35°$ by $17.55 \angle -20°$.

13-30 Divide $40 \angle 110°$ by $80 \angle 60°$.

13-31 A sinusoidal signal has a peak-to-peak display of six divisions on an oscilloscope. The vertical sensitivity is set at 5 V/div. Calculate (a) E_{p-p}; (b) E_p; (c) E_{rms}.

13-32 A sinusoidal signal has a display of eight horizontal divisions on an oscilloscope for one cycle. The sec/div. switch is set at 50 μs/div. Calculate the frequency of the signal.

13-33 Calculate the phase angle, or the angle of lag, of wave B in Figure 13-37.

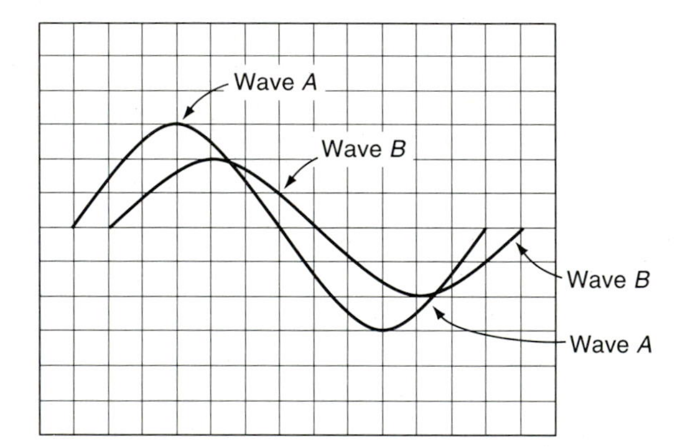

FIGURE 13-37

13-34 For the nonsinusoidal waveform shown in Figure 13-38, calculate (a) the positive and negative amplitudes of the signal; (b) the period and frequency of the signal. Assume that the vertical and time-base controls are set for 10 V/cm and 5 ms/cm, respectively.

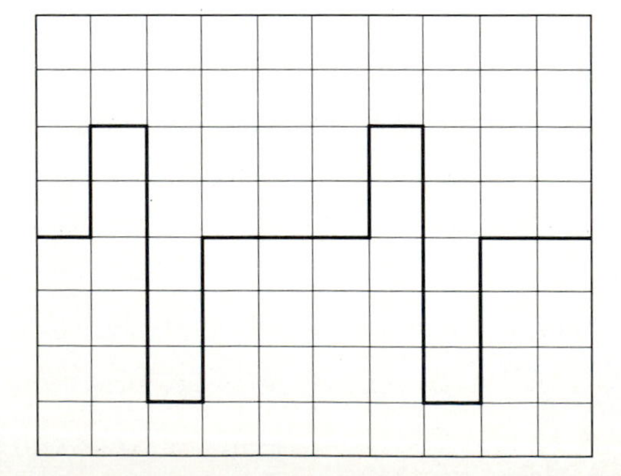

FIGURE 13-38

CHAPTER

Single-Phase Alternating-Current Circuits

LEARNING OBJECTIVES

Upon completion of this chapter you will be able to:

- Explain the effects of inductive reactance and capacitive reactance on an ac circuit.
- Define impedance.
- Utilize the voltage divider rule in ac circuit calculations.
- Explain admittance and susceptance in ac circuits.
- Determine equivalent circuits for ac calculations.
- Discuss power in ac circuits and the differences between apparent power, real power, and reactive power.
- Utilize the power triangle for circuit calculations.
- Explain the principles and applications of power factor correction.
- Discuss resonance in ac circuits.
- Explain the Q factor of a series circuit.
- Discuss bandwidth of resonant circuits.
- Describe basic filter circuits.
- Explain the difference between active and passive filters.

14-1 INTRODUCTION

An electric circuit energized by a single alternating voltage is referred to as a **single-phase circuit**. In Chapter 13 it was mentioned that in an ac circuit which contains only resistance, the current and voltage are said to be in phase with each other. If the single-phase circuit contains only inductance, the current lags the voltage by 90°. In this type of circuit the ratio of voltage to current is equal to the inductive reactance of the circuit in ohms. When an ac circuit contains only capacitance, the current leads the applied voltage by 90°, and the ratio of voltage to current is equal to the capacitive reactance of the circuit in ohms. These three types of circuits are easily solved for since Ohm's law may be directly applied to solve for current, voltage, resistance, or reactance.

In a single-phase circuit that contains resistance and reactance, the opposition to current flow is the combined phasor sum of these quantities. This total opposition to the flow of alternating current is called impedance. The impedance of a circuit is expressed in ohms and its symbol is the letter Z.

14-2 RESISTANCE IN AC CIRCUITS

The simplest ac circuit is one that contains resistance only. The relationship between voltage and current is given by Ohm's law ($E = IR$). In such a circuit, the only voltages involved are the applied emf and the voltage drop across the resistor. The voltage and current magnitudes have a constant ratio that is equal to the resistance and is not a function of time.

A sinusoidal voltage is applied to a resistance as shown in Figure 14-1. The equation for solving for the instantaneous current is given as

$$i = I_m \sin \omega t = I_m \sin 2\pi ft \tag{14-1}$$

where f is the frequency, in cycles per second. The resulting time variations in the current and voltage waveforms are shown in Figure 14-2. Since the applied emf must be entirely utilized in forcing the current through the resistor, R, the applied voltage must be equal to the voltage drop across the resistor. Therefore, the circuit voltage, e, at any instant is equal to the IR drop.

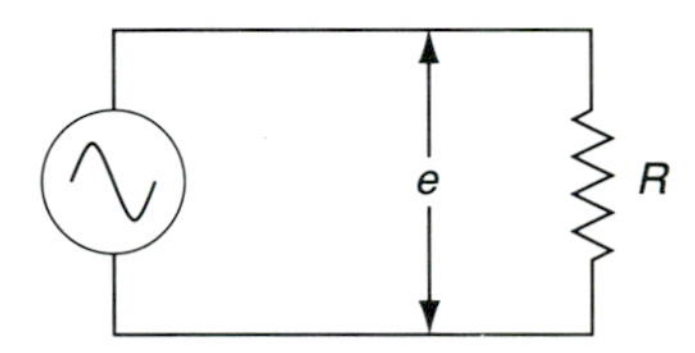

FIGURE 14-1 Resistive load connected across an ac source.

$$\begin{aligned} e &= IR \\ &= I_m \times R \sin \omega t \end{aligned} \tag{14-2}$$

From the waveforms shown in Figure 14-2, it is apparent that the voltage wave and the current wave are both sinusoidal and have the same frequency. When $\omega t = \pi/2$, or 90°, $\sin \omega t = 1$, and both the current and the voltage are simultaneously at their maximum positive values.

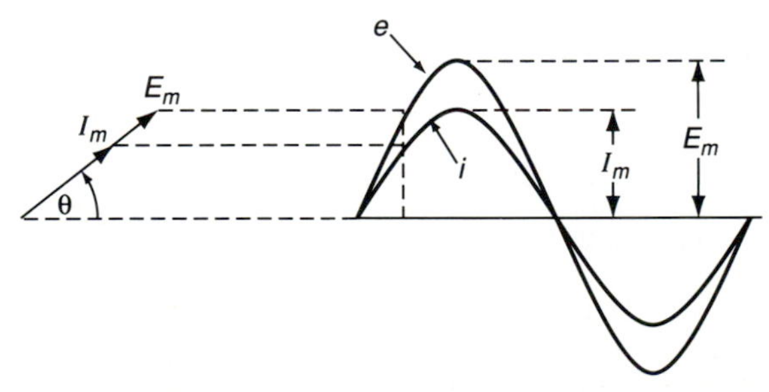

FIGURE 14-2 Instantaneous and phasor values of current and voltage in a purely resistive circuit.

Since both current and voltage waveforms are in phase, the radius phasors that determine their waves must also be in phase with each other. In Figure 14-3(a) the current phasor and voltage phasor are shown at their maximum values. In Figure 14-3(b) the phasors are shown at their effective, or rms, values.

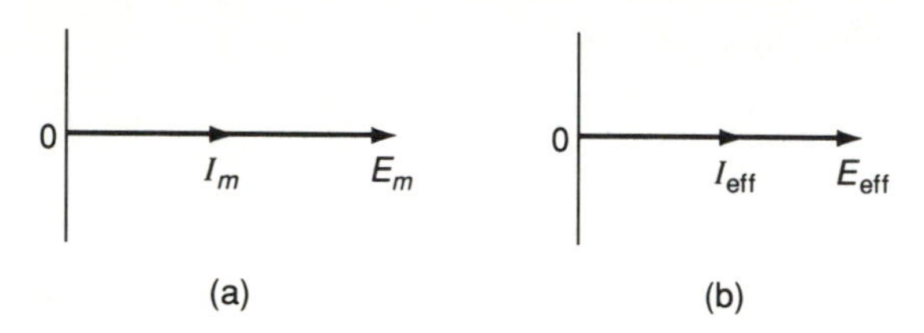

FIGURE 14-3 Phasor diagram of (a) maximum and (b) effective values of voltage and current.

EXAMPLE 14-1 The applied emf of the circuit shown in Figure 14-1 is $e = 200\sin(377t)$. Calculate (a) the instantaneous value of voltage at $t = 0.002$ s; (b) the effective value of current for a resistance of 1200 Ω.

Solution

(a) $$e = 200\sin[(377)(0.002)(57.3)] = 136.92\text{ V}$$

(b) $$I_{\text{rms}} = \frac{E_{\text{rms}}}{R} = \frac{(0.707)(200\text{ V})}{1200\,\Omega} = 117.83\text{ mA}$$

14-3 INDUCTIVE REACTANCE

When inductance was discussed in dc circuits, the opposition that an inductance offers to a changing current was called self-induced voltage, or counter emf, and was measured in volts. Since a coil reacts to a current change by generating a counter emf, a coil is said to be reactive. Therefore, the opposition to current flow offered by a coil is called **inductive reactance**, which is measured in ohms and represented by the symbol X_L. The equation for determining inductive reactance is given as

$$X_L = 2\pi fL \qquad (14\text{-}3)$$

where X_L = inductive reactance, in ohms

f = frequency, in hertz

L = inductance, in henrys

EXAMPLE 14-2 Calculate the inductive reactance of a 12-H coil at 60 Hz.

Solution
$$X_L = 2\pi fL = 2\pi(60 \text{ Hz})(12 \text{ H})$$
$$= 4523.89 \ \Omega$$

EXAMPLE 14-3 The effective voltage across an inductor is 50 V when the effective current is 200 mA. The frequency of the circuit is 60 Hz. Determine the inductance.

Solution
$$X_L = \frac{E_{\text{rms}}}{I_{\text{rms}}} = \frac{50 \text{ V}}{0.2 \text{ A}}$$
$$= 250 \ \Omega$$
$$L = \frac{X_L}{2\pi f} = \frac{250 \ \Omega}{2\pi(60)}$$
$$= 0.66 \text{ H}$$

EXAMPLE 14-4 The instantaneous voltage across a 5-H inductor is $e = 30 \sin 150t$. Calculate the instantaneous current when $t = 25$ ms.

Solution
$$X_L = 2\pi fL = \omega L = 150(5) = 750 \ \Omega$$
$$I_m = \frac{E_m}{X_L} = \frac{30 \text{ V}}{750 \ \Omega} = 0.04 \text{ A}$$

In an inductive circuit, i lags e by 90°.

$$i = I_m \sin(\omega t - 90°) = 0.04 \sin[150(0.025)(57.3) - 90°]$$
$$= 32.82 \text{ mA}$$

To calculate series and parallel combinations of inductive reactance, the same method is used as for solving resistance combinations. To find the total reactance of two or more series-connected inductors,

$$X_{LT} = X_{L1} + X_{L2} + \cdots + X_{LN} \qquad (14\text{-}4)$$

For two parallel-connected inductors, the product-over-sum equation may be used:

$$X_{LT} = \frac{X_{L1} \times X_{L2}}{X_{L1} + X_{L2}} \qquad (14\text{-}5)$$

To solve for two or more parallel-connected inductors, we have

$$\frac{1}{X_{LT}} = \frac{1}{X_{L1}} + \frac{1}{X_{L2}} + \frac{1}{X_{L3}} + \cdots + \frac{1}{X_{LN}} \qquad (14\text{-}6)$$

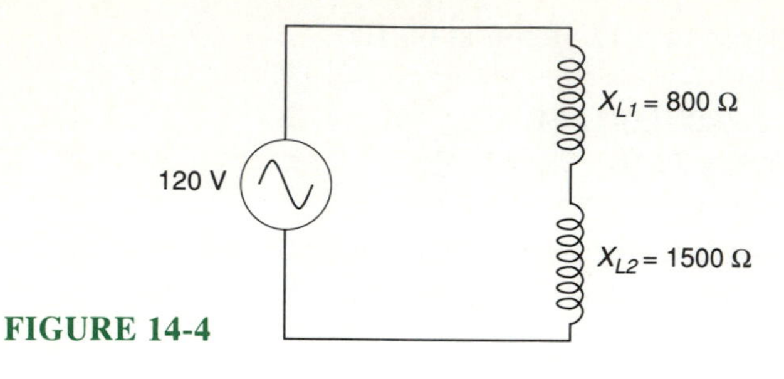

FIGURE 14-4

EXAMPLE 14-5 For the circuit shown in Figure 14-4, calculate the values of X_{LT}, I_T, V_{L1}, and V_{L2}.

Solution

$$X_{LT} = X_{L1} + X_{L2} = 800\,\Omega + 1500\,\Omega = 2300\,\Omega$$

$$I_T = \frac{E_T}{X_{LT}} = \frac{120\,\text{V}}{2300\,\Omega} = 52.17\,\text{mA}$$

$$V_{L1} = I_T X_{L1} = 52.17\,\text{mA} \times 800\,\Omega = 41.74\,\text{V}$$

$$V_{L2} = I_T X_{L2} = 52.17\,\text{mA} \times 1500\,\Omega = 78.26\,\text{V}$$

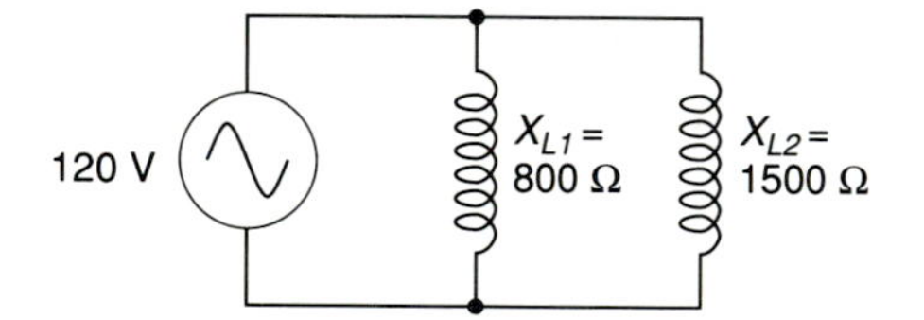

FIGURE 14-5

EXAMPLE 14-6 For the circuit shown in Figure 14-5, calculate the values of X_{LT}, I_{L1}, I_{L2}, and I_T.

Solution

$$X_{LT} = \frac{X_{L1} \times X_{L2}}{X_{L1} + X_{L2}} = \frac{800\,\Omega \times 1500\,\Omega}{800\,\Omega + 1500\,\Omega} = 521.74\,\Omega$$

$$I_{L1} = \frac{120\,\text{V}}{800\,\Omega} = 150\,\text{mA}$$

$$I_{L2} = \frac{120\,\text{V}}{1500\,\Omega} = 80\,\text{mA}$$

$$I_T = I_{L1} + I_{L2} = 150\,\text{mA} + 80\,\text{mA} = 230\,\text{mA}$$

14-4 CAPACITIVE REACTANCE

Capacitance was defined previously as that quality of a circuit that enables energy to be stored in an electric field. A capacitor is a device that possesses the quality of capacitance. In the discussion of dc circuits, it was shown that a capacitor is charged when a voltage is applied to its terminals and discharged when it is short-circuited. If an alternating voltage is applied to a capacitor, it will be alternately charged and discharged, and an alternating current of the same frequency will flow in the circuit.

When an alternating emf is applied to the capacitor shown in Figure 14-6(a), the charge on the plate at any instant is proportional to the voltage E and will therefore be in phase with this voltage. Since the voltage is alternating, the capacitor is charged first in one direction and then in the other, and an alternating current will flow. During the first 90° of the sine wave, the voltage is increasing as the capacitor is being charged.
At 90° of the sine wave, the rate of change in voltage is zero, and zero current flows, even though the charge in the capacitor is at a maximum. Between 90 and 180° the voltage is decreasing, and consequently the charge on the capacitor is decreasing. This means that the capacitor must also be discharging at this point in the cycle. For the capacitor to discharge, the current must flow in the opposite direction to the applied voltage. This reverse polarity of current is shown below the horizontal line of Figure 14-6(b). Between 180 and 270°, the capacitor is being charged in the opposite direction, but the current is once again considered to have the same direction as the applied emf. The final 90° of the cycle show the current in the opposite direction to the voltage as the capacitor discharges. For the purely capacitive circuit of Figure 14-6(a), the current leads the voltage by 90°. This relationship is shown in the phasor diagram of Figure 14-6(c).

As in the case of a pure inductor, a pure capacitor has no resistance component. The ratio of effective voltage across a capacitor to the effective current, called the **capacitive reactance**, is represented by the symbol X_C.

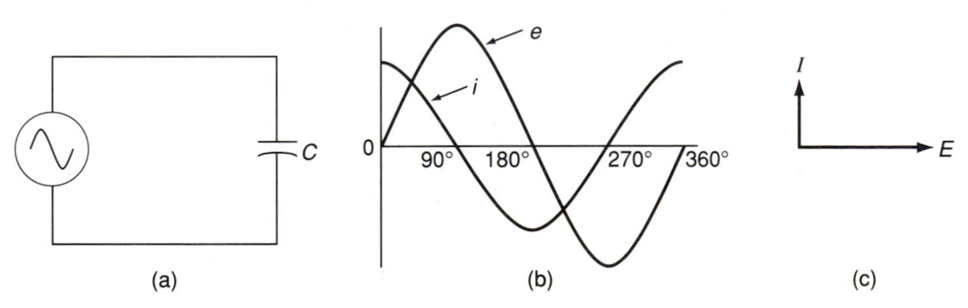

FIGURE 14-6 Effect of capacitance in an ac circuit.

Capacitive reactance is measured in ohms and is defined as the opposition offered by a capacitor or by any capacitive circuit to the flow of current. The reactance of a capacitor varies inversely with its capacitance. The rate by which the voltage changes is determined by the frequency of the applied voltage. Since the amount of current flowing in a capacitive circuit is determined by the rate at which the voltage is changing, the lower the frequency, the slower the rate of change, and the less current will flow. The rate of change of voltage in an ac circuit is determined by the **angular velocity** of the applied voltage. The reactance of a capacitor varies inversely with capacitance and with angular velocity and is determined by the following equation:

$$X_C = \frac{1}{2\pi f C} \quad (14\text{-}7)$$

where X_C = capacitive reactance, in ohms

f = frequency, in hertz

C = capacitance, in farads

EXAMPLE 14-7 A 100-μF capacitor is supplied from a 120-V ac supply with a frequency of 60 Hz. Determine (a) the capacitive reactance; (b) the current that flows in the circuit.

Solution

(a) $$X_C = \frac{1}{2\pi f C} = \frac{1}{2\pi(60\text{ Hz})(100 \times 10^{-6}\text{ F})} = 26.53\ \Omega$$

(b) $$I_{\text{rms}} = \frac{E_{\text{rms}}}{X_C} = \frac{120\text{ V}}{26.53\ \Omega} = 4.52\text{ A}$$

EXAMPLE 14-8 A capacitor has a reactance of 250 Ω at 550 Hz. What is its reactance at 60 Hz?

Solution If the reactance at one frequency is known, the reactance at another frequency can be found by multiplying the known reactance by the ratio of the two frequencies.

$$\frac{X_{C1}}{X_{C2}} = \frac{f_1}{f_2}$$

Therefore,

$$X_{C2} = X_{C1}\frac{f_1}{f_2} = 250\ \Omega \times \frac{550\ \text{Hz}}{60\ \text{Hz}}$$
$$= 2291.67\ \Omega$$

To calculate series and parallel combinations of capacitive reactance, the same method is used as for solving inductive reactance combinations. To find the total reactance of two or more series-connected capacitors, we use

$$X_{CT} = X_{C1} + X_{C2} + X_{C3} + \cdots + X_{CN} \tag{14-8}$$

To solve for two or more parallel-connected capacitors,

$$\frac{1}{X_{CT}} = \frac{1}{X_{C1}} + \frac{1}{X_{C2}} + \frac{1}{X_{C3}} + \cdots + \frac{1}{X_{CN}} \tag{14-9}$$

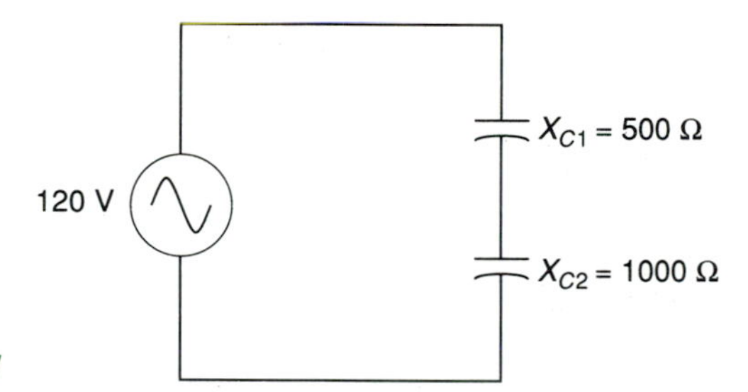

FIGURE 14-7

EXAMPLE 14-9 For the circuit shown in Figure 14-7 determine the values of X_{CT}, I_T, V_{C1}, and V_{C2}.

Solution

$$X_{CT} = X_{C1} + X_{C2} = 500\ \Omega + 1000\ \Omega = 1500\ \Omega$$

$$I_T = \frac{E}{X_{CT}} = \frac{120\ \text{V}}{1500\ \Omega} = 80\ \text{mA}$$

$$V_{C1} = I_T X_{C1} = 80\ \text{mA} \times 500\ \Omega = 40\ \text{V}$$

$$V_{C2} = I_T X_{C2} = 80\ \text{mA} \times 1000\ \Omega = 80\ \text{V}$$

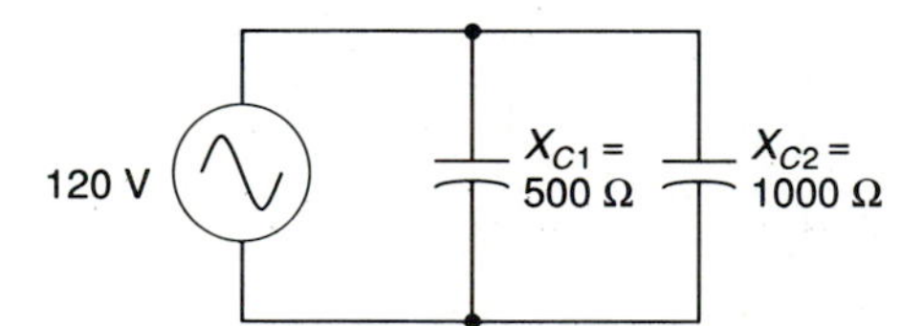

FIGURE 14-8

EXAMPLE 14-10 For the circuit shown in Figure 14-8, calculate the values of X_{CT}, I_{C1}, I_{C2}, and I_T.

Solution Since only two capacitors are given, the product-over-sum equation may be used.

$$X_{CT} = \frac{X_{C1} \times X_{C2}}{X_{C1} + X_{C2}} = \frac{500\,\Omega \times 1000\,\Omega}{500\,\Omega + 1000\,\Omega} = 333.33\,\Omega$$

$$I_{C1} = \frac{120\text{ V}}{500\,\Omega} = 0.24\text{ A}$$

$$I_{C2} = \frac{120\text{ V}}{1000\,\Omega} = 0.12\text{ A}$$

$$I_T = I_{C1} + I_{C2} = 0.24\text{ A} + 0.12\text{ A} = 0.36\text{ A}$$

14-5 IMPEDANCE

Impedance is defined as the total opposition to current flow in an ac circuit. Impedance may be pure resistance or pure reactance, but usually it is a combination of resistance and reactance. The symbol for impedance is the letter Z, and it is measured in ohms. The current in an ac circuit is directly proportional to the voltage across the circuit and inversely proportionate to the impedance of the circuit. In equation form,

$$Z = \frac{E}{I} \tag{14-10}$$

Since the amount of impedance is the ratio between voltage and current, the impedance angle is the difference in phase angle between the voltage and current.

EXAMPLE 14-11 The voltage and current for an ac load are $120\,\angle 45^\circ$ V and $2.5\,\angle 30^\circ$ A, respectively. Determine the impedance of the circuit.

Solution

$$Z = \frac{E}{I} = \frac{120\,\angle 45^\circ\text{ V}}{2.5\,\angle 30^\circ\text{ A}} = 48\,\angle 15^\circ\,\Omega$$

EXAMPLE 14-12 The instantaneous values for an ac load are $e = 3\sin(\omega t + 60^\circ)$ V and $i = 15\sin(\omega t + 20^\circ)$ mA. Determine the impedance of the circuit.

Solution Solve for the impedance by using the maximum values for the current and voltage.

$$Z = \frac{E_{\text{rms}}}{I_{\text{rms}}} = \frac{E_m}{I_m} = \frac{3 \angle 60^\circ \text{ V}}{0.015 \angle 20^\circ \text{ A}}$$
$$= 200 \angle 40^\circ \ \Omega$$

In ac circuits containing both resistance and reactance, the rectangular form of impedance is

$$Z = R \pm jX \tag{14-11}$$

When an ac circuit contains resistance and reactance, the total voltage drop, IZ, is equal to the vectorial sum of the resistance, IR, and the reactance, IX, drops. By the Pythagorean theorem,

$$IZ = \sqrt{(IR)^2 + (IX)^2} \tag{14-12}$$

If the current, I, is canceled, the equation becomes

$$Z = \sqrt{R^2 + X^2} \tag{14-13}$$

Since impedance, resistance, and reactance are fixed quantities which do not vary with time, they may be represented by a **vector diagram**. These types of diagrams are usually drawn in rectangular form to distinguish them from phasor diagrams. Vector diagrams containing R, X, and Z are referred to as **impedance triangles** and can prove to be very useful in problem solving. Two examples of impedance triangles are illustrated in Figure 14-9. The circuit shown in Figure 14-9(a) represents an inductance and resistance connected in series, while the impedance triangle of Figure 14-9(b) represents a capacitor and resistor connected in series. The reactive component in both circuits is on the vertical axis, so if a circuit contained both capacitance and inductance, the two values would be subtracted from each other.

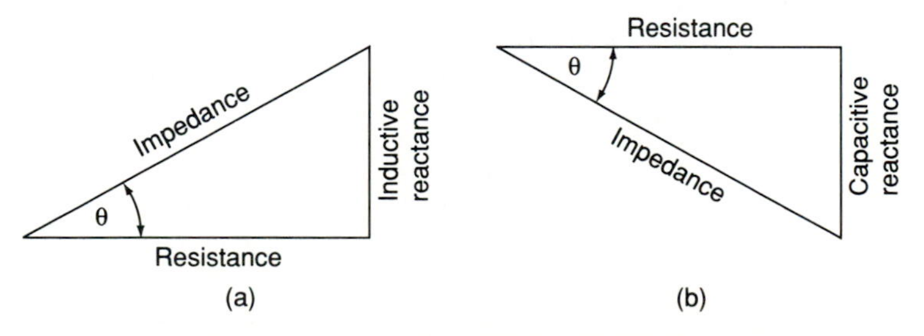

FIGURE 14-9 Impedance triangle: (a) resistance and inductive reactance in series; (b) resistance and capacitive reactance in series.

The phase angle theta is solved for in the same manner as for current and voltage relationships:

$$\theta = \tan^{-1}\frac{X}{R} \tag{14-14}$$

EXAMPLE 14-13 The impedance of a circuit is given as $Z = 24 \angle -34.5^\circ\ \Omega$. Determine the amount of resistance and reactance in this circuit.

Solution Convert to rectangular form.

$$\begin{aligned} Z = 24 \angle -34.5^\circ\ \Omega &= 24\cos(-34.5^\circ) + j24\sin(-34.5^\circ) \\ &= 19.78 - j13.59\ \Omega \end{aligned}$$

The negative j component indicates that the reactance is capacitive. The impedance triangle would be as shown in Figure 14-10.

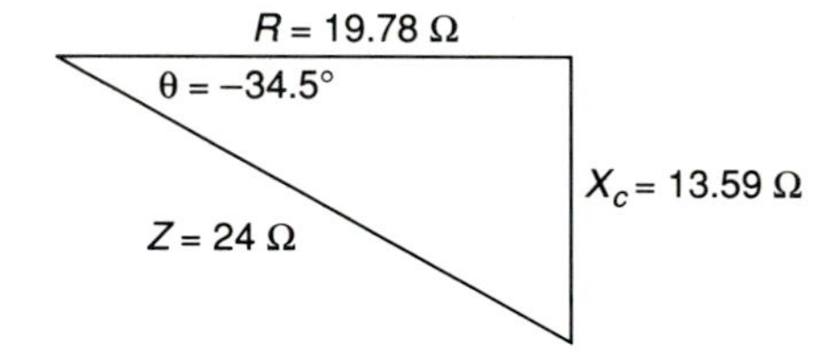

FIGURE 14-10

14-6 SERIES *RL* CIRCUIT

A series circuit containing resistance, R, and inductance, L, is shown in Figure 14-11(a). The phasor diagram for this circuit is illustrated in Figure 14-11(b). Since current is the same through all parts of a series circuit, the current phasor is drawn as the reference phasor. The voltage, V_R, across resistance, R, is always in phase with the current through the resistance. Because the voltage across an inductance leads the current by 90°, the phasor V_L is drawn 90° ahead of the reference phasor. The applied voltage, E, is the resultant of the two component voltages V_R and V_L.

The impedance triangle illustrated in Figure 14-11(c) is obtained by the rectangular coordinates for impedance.

$$Z = R + jX \tag{14-15}$$

which expressed in polar coordinates is

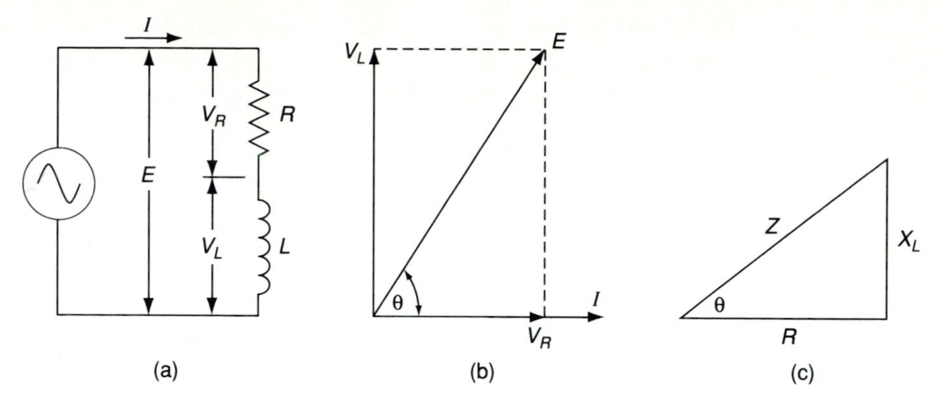

FIGURE 14-11 (a) Series *RL* circuit; (b) phasor diagram; (c) impedance triangle.

$$Z = \sqrt{R^2 + X_L^2} \quad \angle \tan^{-1} \frac{X_L}{R} \tag{14-16}$$

The triangle formed by Z, R, and X_L is similar to that formed by E, V_L, and V_R in the phasor diagram of Figure 14-11(b). It is evident that the impedance of this type of circuit equals the vector sum of the resistance and the inductive reactance. Also, it is evident from the impedance triangle that

$$\theta = \sin^{-1} \frac{X_L}{Z} = \cos^{-1} \frac{R}{Z} = \tan^{-1} \frac{X_L}{R} \tag{14-17}$$

$$X_L = Z \sin \theta \tag{14-18}$$

$$R = Z \cos \theta \tag{14-19}$$

$$X_L = R \tan \theta \tag{14-20}$$

To determine the resultant, E, rectangular form may be used:

$$E = V_R + jV_L \tag{14-21}$$

or polar form:

$$E = \sqrt{V_R^2 + V_L^2} \quad \angle \tan^{-1} \frac{V_L}{V_R}$$

EXAMPLE 14-14 A series RL circuit contains a 100-Ω resistor and a 0.35-H coil connected to a 120-V 60-Hz supply. Determine the voltages across the resistor and inductor. Also, draw the voltage phasor and impedance triangle for the circuit.

Solution

$$X_L = 2\pi fL = 2\pi(60\text{ Hz})(0.35\text{ H}) = 131.95\ \Omega$$

$$Z = 100 + j131.95 = \sqrt{100^2 + 131.95^2}\ \angle \tan^{-1}\frac{131.95}{100}$$

$$= 165.56\ \angle 52.8^\circ\ \Omega$$

The impedance angle of 52.8° is the same angle by which the voltage leads the current.

$$V_R = 120\cos 52.8^\circ = 72.55\text{ V}$$

$$V_L = 120\sin 52.8^\circ = 95.58\text{ V}$$

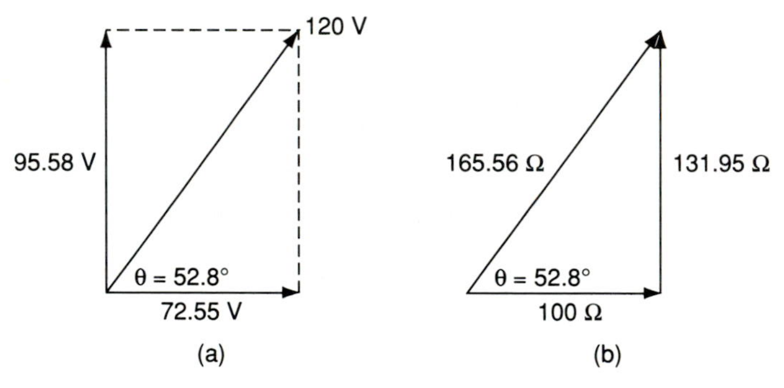

FIGURE 14-12 (a) Voltage phasor and (b) impedance triangle for Example 14-14.

EXAMPLE 14-15 A 2-H coil and a 500-Ω resistor are connected to a 240-V 60-Hz supply. Determine the magnitude and phase of the current with respect to the applied voltage of the circuit.

Solution

$$X_L = 2\pi fL = 2\pi(60\text{ Hz})(2\text{ H}) = 754\ \Omega$$

$$\theta = \tan^{-1}\frac{X_L}{R} = \tan^{-1}\frac{754\ \Omega}{500\ \Omega} = 56.5^\circ$$

$$Z = \sqrt{R^2 + X^2} = \sqrt{500^2 + 754^2} = 904.71\ \Omega$$

$$I = \frac{E}{Z} = \frac{240\text{ V}}{904.71\ \Omega} = 265.28\text{ mA}$$

Since this circuit contains inductance, the current **lags** the voltage by 56.5°.

$$I = 265.28\ \angle -56.5^\circ\text{ mA}$$

EXAMPLE 14-16 An *RL* series circuit has an applied voltage of $120 \angle 0°$ V and the current through the circuit is $15 - j20$ A. The frequency of the circuit is 60 Hz. Calculate the value of resistance and inductance.

Solution

$$I = 15 - j20 \text{ A}$$
$$= 25 \angle -53.1° \text{ A}$$

$$Z = \frac{E}{I} = \frac{120 \angle 0° \text{ V}}{25 \angle -53.1° \text{ A}} = 4.8 \angle 53.1° \ \Omega$$
$$= 2.88 + j3.84$$

Therefore,

$$R = 2.88 \ \Omega$$

$$X_L = 2\pi f L = 3.84 \ \Omega$$

$$L = \frac{X_L}{2\pi f} = \frac{3.84 \ \Omega}{2\pi(60 \text{ Hz})} = 10.19 \text{ mH}$$

14-7 SERIES *RC* CIRCUIT

In an ac series circuit that contains resistance and capacitance, the applied voltage can be resolved into two components: V_R, which is in phase with the current, and V_C, which lags the current by 90°. The phasor diagram of Figure 14-13(b) shows this relationship. Current is used as a reference and is drawn on the horizontal axis. The phasor sum of V_R and V_C gives the applied voltage E. The value of the phase angle between the applied voltage and the current depends on the ratio of V_C to V_R. Expressed in rectangular coordinates,

$$E = V_R - jV_C \tag{14-22}$$

In polar form,

$$E = \sqrt{V_R^2 + V_C^2} \ \angle \tan^{-1} \frac{V_C}{V_R} \tag{14-23}$$

The impedance triangle of Figure 14-13(c) is similar to the triangle formed by E, V_R, and V_C in the phasor diagram of Figure 14-13(b).

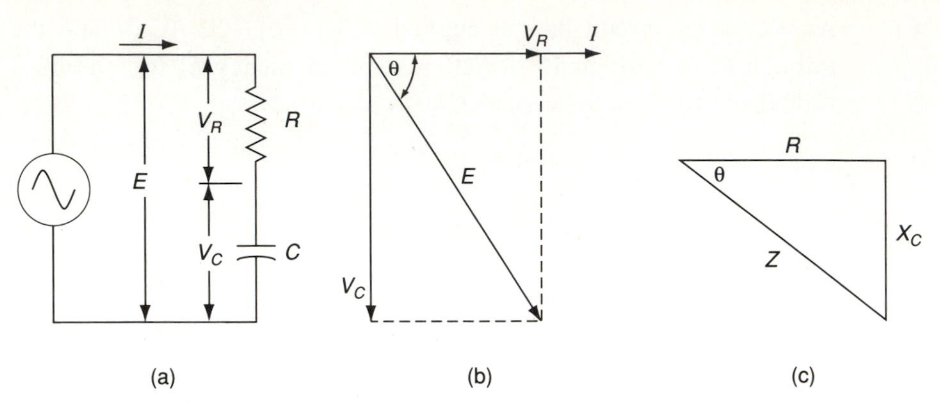

FIGURE 14-13 (a) Series *RC* circuit; (b) phasor diagram; (c) impedance triangle.

The rectangular coordinates for impedance in an *RC* series circuit are

$$Z = R - jX_C \tag{14-24}$$

In polar form,

$$Z = \sqrt{R^2 + X_C^2} \ \angle \tan^{-1} \frac{-X_C}{R}$$

It is evident from the impedance triangle that the impedance of this type of circuit equals the vector sum of the resistance and the capacitive reactance. Also, it is evident from the impedance triangle that

$$\theta = \sin^{-1} \frac{X_C}{Z} = \cos^{-1} \frac{R}{Z} = \tan^{-1} \frac{X_C}{R} \tag{14-25}$$

$$X_C = Z \sin \theta \tag{14-26}$$

$$R = Z \cos \theta \tag{14-27}$$

$$X_C = R \tan \theta \tag{14-28}$$

EXAMPLE 14-17 A resistance of 60 Ω is connected in series with a 50-μF capacitance across a 220-V 60-Hz supply. Determine (a) the impedance; (b) the current.

Solution

(a) $$X_C = \frac{1}{2\pi f C} = \frac{1}{2\pi(60\text{ Hz})(50 \times 10^{-6}\text{F})} = 53.05\ \Omega$$

$$Z = 60 - j53.05\ \Omega = \sqrt{60^2 + (-53.05)^2} \ \angle \tan^{-1}\frac{-53.05}{60}$$

$$= 80.09\ \angle -41.5\ \Omega$$

(b) $$I = \frac{E}{Z} = \frac{220\ \text{V}}{80.09\ \angle -41.5^\circ\ \Omega}$$

$$= 2.75\ \angle 41.5^\circ\ \text{A}$$

EXAMPLE 14-18 Determine the value of capacitance required to be connected in series with a 1200-Ω resistor to limit the current in the circuit to 25 mA. The supply voltage is 120 V, 60 Hz.

Solution

$$Z = \frac{E}{I} = \frac{120\ \text{V}}{25\ \text{mA}} = 4800\ \Omega$$

$$X_C = \sqrt{Z^2 - R^2} = \sqrt{4800^2 - 1200^2} = 4647.58\ \Omega$$

$$X_C = \frac{1}{2\pi fC}$$

Therefore,

$$C = \frac{1}{2\pi fX_C} = \frac{1}{2\pi(60\ \text{Hz})(4647.58\ \Omega)}$$

$$= 0.57\ \mu\text{F}$$

EXAMPLE 14-19 A 20-Ω resistor and a 10-μF capacitor are connected in series to a circuit with an instantaneous value of current of $i = 3\sin(3500t + 90^\circ)$ A. Calculate the values of V_R, V_C, and E. Also, draw the phasor diagram for these values.

Solution

$$i = 3\sin(3500t + 90^\circ)\ \text{A}$$

$$I = \frac{3}{\sqrt{2}\ \angle 90^\circ} = 2.12\ \angle 90^\circ\ \text{A}$$

$$X_c = \frac{1}{\omega C} = \frac{1}{3500 \times (10 \times 10^{-6}\ \text{F})} = 28.57\ \Omega$$

$$V_R = IR = 2.12\ \angle 90^\circ\ \text{A} \times 20\ \angle 0^\circ\ \Omega = 42.4\ \angle 90^\circ\ \text{V}$$

$$V_C = IX_C = 2.12\ \angle 90^\circ \times 28.57\ \angle -90^\circ = 60.57\ \angle 0^\circ\ \text{V}$$

$$E = V_R + V_C = 42.4\ \angle 90^\circ + 60.57\ \angle 0^\circ = 73.94\ \angle 35^\circ\ \text{V}$$

The phasor diagram is shown in Figure 14-14.

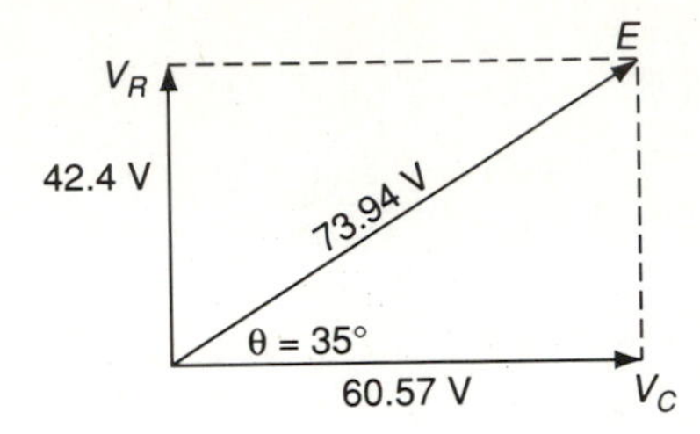

FIGURE 14-14

14-8 SERIES *RLC* CIRCUIT

When an ac circuit consists of resistance, inductance, and capacitance connected in series, as shown in Figure 14-15(a), the components of the applied voltage E are the voltage drop across the resistor, V_R, and the total reactance voltage, V_X. As mentioned earlier, the total reactance is the difference between the capacitive reactance and the inductive reactance. In equation form, the reactance voltage is expressed as

$$jV_X = j(V_L - V_C) \tag{14-29}$$

The applied voltage is expressed in rectangular coordinates by the equation

$$E = V_R + j(V_L - V_C) \tag{14-30}$$

and is expressed in polar coordinates as

$$E = \sqrt{V_R^2 + (V_L - V_C)^2} \,\angle \tan^{-1}\frac{V_L - V_C}{V_R}$$

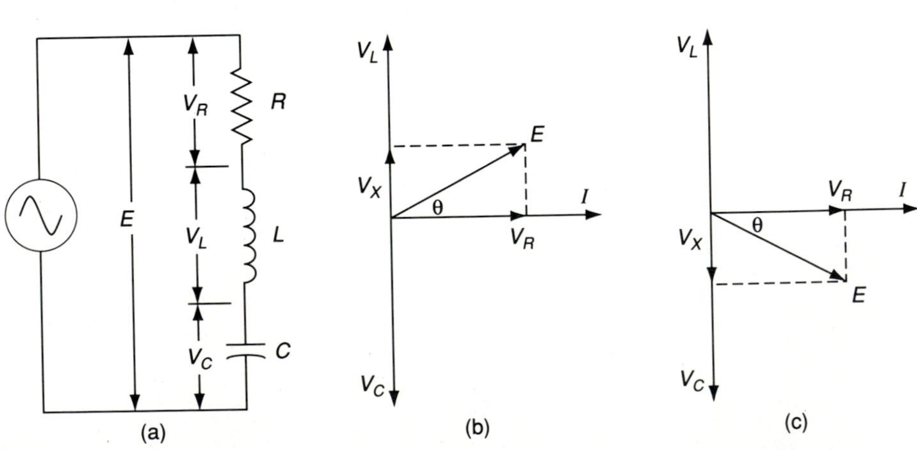

FIGURE 14-15 (a) Series *RLC* circuit; (b) phasor diagram for $X_L > X_C$; (c) phasor diagram for $X_C > X_L$.

The phasor diagram of Figure 14-15(b) is for a circuit in which V_L is greater than V_C. The current I is plotted on the reference axis, V_R is in phase with I, and V_X is plotted 90° ahead of I. Since V_L is greater than V_C, the total reactive voltage V_X is considered as an inductive reactance voltage.

Figure 14-15(c) is the phasor diagram for a circuit in which V_L is less than V_C. When this condition exists, V_X lags 90° behind I and is treated as a capacitive reactance voltage. The impedance triangle of Figure 14-16(a) is for a series circuit in which $X_L > X_C$. Figure 14-16(b) is the impedance triangle for a series circuit in which $X_L < X_C$.

The total, or equivalent, reactance in rectangular form expressed by the equation

$$jX = j(X_L - X_C) \tag{14-31}$$

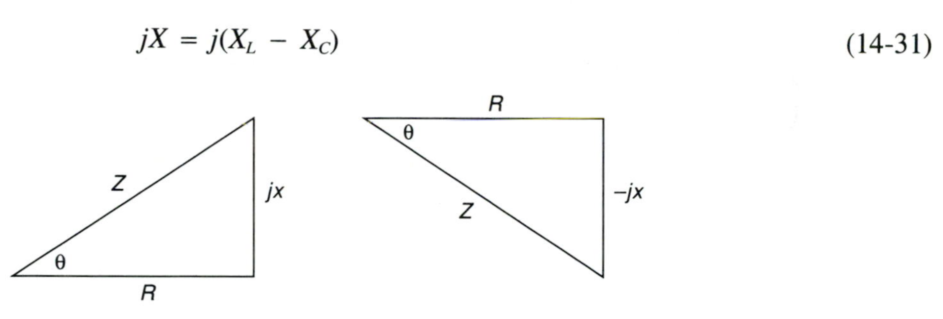

FIGURE 14-16 Series *RLC* impedance triangles.

It is obvious that the impedance is the vector sum of R and X regardless of whether X is inductive or capacitive in its effect. Expressed in rectangular form,

$$Z = R + jX \tag{14-32}$$

In polar coordinates,

$$Z = \sqrt{R^2 + X^2} \angle \tan^{-1}\frac{X}{R} \tag{14-33}$$

EXAMPLE 14-20 An ac series circuit has a resistance of 15 Ω, an inductive reactance of 40 Ω, and a capacitive reactance of 50 Ω when connected across a 120-V 60-Hz supply. Determine (a) the impedance of the circuit; (b) the current through the circuit; (c) the values of V_R, V_L, and V_C.

Solution (a) $Z = R + j(X_L - X_C)$

$$= 15 + j(40 - 50) = \sqrt{15^2 + (-10)^2} \angle \tan^{-1}\frac{-10}{15}$$

$$= 18.03 \angle -33.7^\circ\ \Omega$$

$$\text{(b)}\ I = \frac{E}{Z} = \frac{120 \angle 0^\circ\ \text{V}}{18.03 \angle -33.7^\circ\ \Omega} = 6.66 \angle 33.7^\circ\ \text{A}$$

$$\text{(c)}\ V_R = IR = (6.66 \angle 33.7^\circ\ \text{A})(15 \angle 0^\circ\ \Omega) = 99.9 \angle 33.7^\circ\ \text{V}$$

$$V_L = IX = (6.66 \angle 33.7^\circ\ \text{A})(40 \angle 90^\circ\ \Omega) = 266.4 \angle 123.7^\circ\ \text{V}$$

$$V_C = IX = (6.66 \angle 33.7^\circ\ \text{A})(50 \angle -90^\circ\ \Omega) = 333 \angle -56.3^\circ\ \text{V}$$

EXAMPLE 14-21 Calculate the total impedance of a 500-Hz series circuit containing a 0.2-H coil, 0.5-μF capacitor, and a 10-Ω resistor.

Solution

$$X_L = 2\pi fL = 2\pi(500\ \text{Hz})(0.2\ \text{H}) = 628.32\ \Omega$$

$$X_C = \frac{1}{2\pi fC} = \frac{1}{2\pi(500\ \text{Hz})(0.5 \times 10^{-6}\ \text{F})} = 636.62\ \Omega$$

$$X = X_L - X_C = 628.32\ \Omega - 636.62\ \Omega = -8.3\ \Omega$$

$$Z = \sqrt{R^2 + X^2}\ \angle \tan^{-1}\frac{X}{R} = \sqrt{10^2 + (-8.3)^2}\ \angle \tan^{-1}\frac{-8.3}{10}$$
$$= 13 \angle -39.7\ \Omega$$

EXAMPLE 14-22 A series circuit contains a 20-Ω resistance, 0.02-H inductance, and 5-μF capacitance. The instantaneous voltage for the circuit is $e = 45.5 \sin(7500t)$ V. Calculate (a) the total impedance; (b) the instantaneous current.

Solution

$$\text{(a)}\ Z = R + jX_L - jX_C = R + j(\omega L - \frac{1}{\omega C})$$
$$= 20 + j[(7500)(0.02) - \frac{1}{(7500)(5 \times 10^{-6})}]$$
$$= 20 + j123.33$$
$$= 124.94 \angle 80.8^\circ\ \Omega$$

$$\text{(b)}\ E = 0.707 \times 45.5 = 32.17 \angle 0^\circ\ \text{V}$$
$$I = \frac{E}{Z} = \frac{32.17 \angle 0^\circ\ \text{V}}{124.94 \angle 80.8^\circ\ \Omega}$$
$$= 257.48 \angle -80.8^\circ\ \text{mA}$$

Since 257.48 mA is the effective value of current, it is necessary to use the $\sqrt{2}$ multiplier to determine the instantaneous value.

$$i = (257.48 \times 10^{-3})\sqrt{2}\sin(7500t - 80.8^\circ)\ \text{A}$$
$$= 364.14 \sin(7500t - 80.8^\circ)\ \text{mA}$$

14-9 VOLTAGE DIVIDER RULE

The format used for the voltage divider rule in ac circuits is exactly the same as that used for dc circuits. That is,

$$V_X = \frac{(Z_x)(E)}{Z_T} \qquad (14\text{-}34)$$

where V_X = voltage across one or more series-connected components

Z_X = total impedance of component, or components

Z_T = total impedance of series circuit

E = total applied voltage of circuit

EXAMPLE 14-23 For the circuit shown in Figure 14-17, calculate the voltages V_R and V_L, using the voltage divider rule.

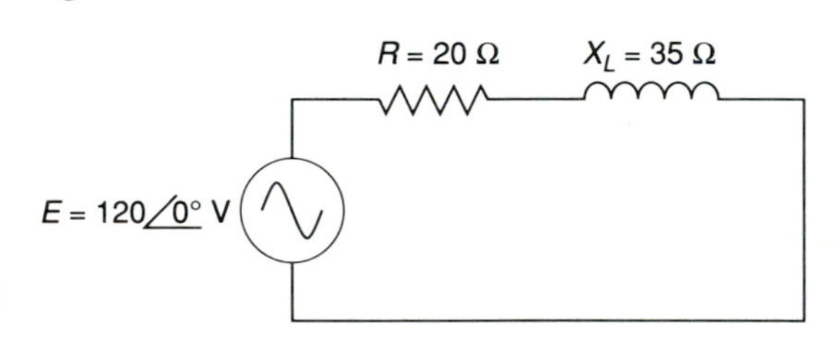

FIGURE 14-17

Solution

$$V_R = \frac{R \times E}{R + X_L} = \frac{(20\,\underline{/0^\circ}\,\Omega)(120\,\underline{/0^\circ}\,\Omega)}{20\,\underline{/0^\circ}\,\Omega + 35\,\underline{/90^\circ}} = 59.54\,\underline{/-60.3^\circ}\text{ V}$$

$$V_L = \frac{X_L \times E}{R + X_L} = \frac{(35\,\underline{/90^\circ}\,\Omega)(120\,\underline{/0^\circ}\text{ V})}{40.31\,\underline{/60.3^\circ}\,\Omega} = 104.19\,\underline{/29.7^\circ}\text{ V}$$

EXAMPLE 14-24 A 60-Hz ac series circuit contains a 1200-Ω resistor, a 3-H inductor, and a 5-μF capacitor. The applied voltage is $E = 100\,\underline{/30^\circ}$ V. Calculate V_R, V_L, and V_C using the voltage divider rule.

Solution

$$X_L = 2\pi fL = 2\pi(60\text{ Hz})(3\text{ H}) = 1130.97\ \Omega$$

$$X_C = \frac{1}{2\pi fC} = \frac{1}{2\pi(60\text{ Hz})(5 \times 10^{-6}\text{ F})} = 530.52\ \Omega$$

$$V_R = \frac{R \times E}{R + X_L + X_C}$$

$$= \frac{(1200\,\underline{/0^\circ}\,\Omega)(100\,\underline{/30^\circ}\text{ V})}{1200\,\underline{/0^\circ}\,\Omega + 1130.97\,\underline{/90^\circ}\,\Omega + 530.52\,\underline{/-90^\circ}\,\Omega}$$

$$= \frac{120 \times 10^3 \angle 30^\circ}{1342.29 \angle 26.6^\circ \ \Omega}$$
$$= 89.4 \angle 3.4^\circ \text{ V}$$

$$V_L = \frac{X_L \times E}{R + X_L + X_C} = \frac{(1130.97 \angle 90^\circ \ \Omega)(100 \angle 30^\circ \text{ V})}{1342.29 \angle 26.6^\circ \ \Omega}$$
$$= 84.27 \angle 93.4^\circ \text{ V}$$

$$V_C = \frac{X_C \times E}{R + X_L + X_C} = \frac{(530.52 \angle -90^\circ \ \Omega)(100 \angle 30^\circ \text{ V})}{1342.29 \angle 26.6^\circ \ \Omega}$$
$$= 39.52 \angle -86.6^\circ \text{ V}$$

14-10 ADMITTANCE AND SUSCEPTANCE

In the discussion of resistance in dc circuits, it was demonstrated how the reciprocal of resistance, or conductance, was utilized in problem solving. In equation form, conductance was expressed as

$$G = \frac{1}{R} \tag{14-35}$$

where G is the conductance, in siemens, and R is the resistance, in ohms.

In ac circuit applications, the ease with which current flows in a given component, or circuit, is called **admittance**, which is measured in siemens and represented by the letter Y. Admittance is defined as the reciprocal of impedance and is considered to be a vectorial quantity.

$$Y = \frac{1}{Z} \tag{14-36}$$

where Y is the admittance, in siemens, and Z is the impedance, in ohms.

Since admittance is a vectorial quantity, it must have two rectangular components. For impedance circuits these components are resistance and reactance. For admittance circuits these components are **conductance** and **susceptance**. Susceptance is the reciprocal of reactance, and is represented by the letter B.

$$Y = G \pm jB \tag{14-37}$$

where Y = admittance, in siemens

G = conductance, in siemens

B = susceptance, in siemens

The sum of conductance and susceptance in a circuit is found by phasor addition. The "$\pm j$" in equation 14-37 is used to stipulate the type of susceptance in the circuit. The "$+j$" is used for capacitive susceptance, B_C, while a "$-j$" represents inductive susceptance, B_L.

When resistors are connected in parallel, the total conductance is calculated by the equation

$$G_T = G_1 + G_2 + G_3 + \cdots + G_N \tag{14-38}$$

The total susceptance in a parallel circuit is determined by

$$B_T = B_1 + B_2 + B_3 + \cdots + B_N \tag{14-39}$$

When the admittance of each parallel branch is known, the total admittance is found by the equation

$$Y_T = Y_1 + Y_2 + Y_3 + \cdots + Y_N \tag{14-40}$$

To solve ac circuit problems which contain impedance, Ohm's law can be used to find the current, $I = E/Z$. The equivalent equation for calculating current in a circuit containing admittance is

$$I = E \times Y \tag{14-41}$$

The parallel circuit of Figure 14-18 consists of two branch circuits with resistance and inductance in one branch, and resistance and capacitance in the other. If the applied voltage is considered to have a phase angle of 0° and lie on the horizontal axis, the current I_1 can be calculated as follows:

$$\begin{aligned} I_1 &= \frac{1}{R_1 + jX_1} = \frac{1}{R_1 + jX_l} \times \frac{R_1 - jX_1}{R_1 - jX_1} \\ &= \frac{R_1 - jX_1}{R_1^2 + X_1^2} = \frac{R_1}{R_1^2 + X_1^2} - j\frac{X_1}{R_1^2 + X_1^2} \end{aligned} \tag{14-42}$$

The admittance of an ac circuit is defined as the ratio of I/V, or the **current per volt**. Therefore, Y could be substituted for I_1 in equation 14-42 and the admittance of the first branch becomes

$$Y_1 = \frac{R_1}{R_1^2 + X_1^2} - j\frac{X_1}{R_1^2 + X_1^2} \tag{14-43}$$

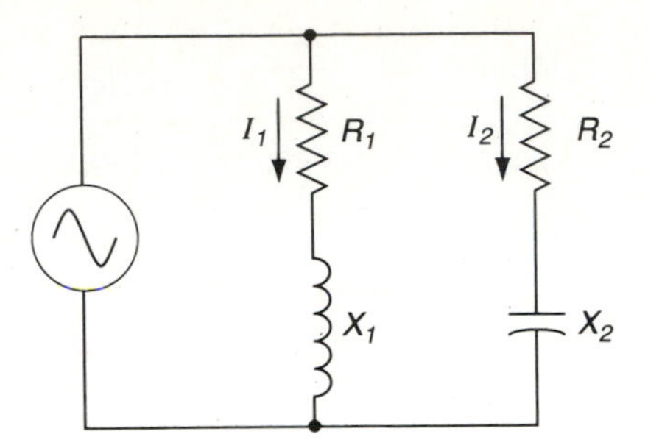

FIGURE 14-18

The horizontal component of the current per volt is the conductance, and the vertical component is the susceptance. These two components are determined by the following equations:

$$G_1 = \frac{R_1}{R_1^2 + X_1^2} \tag{14-44}$$

$$B_1 = \frac{X_1}{R_1^2 + X_1^2} \tag{14-45}$$

Since the admittance is complex in form and contains both real and imaginary parts, it can also be expressed in phasor notation using the Pythagorean theorem:

$$Y = \sqrt{G^2 + B^2} \;\angle \pm\tan^{-1}\frac{B}{G} \tag{14-46}$$

When solving parallel ac circuit problems it is possible to utilize an **admittance triangle**, which is based on the same principles as its counterpart, the impedance triangle. Figure 14-19(a) shows an admittance triangle containing conductance and capacitive susceptance. The admittance triangle shown in Figure 14-19(b) contains conductance and inductive susceptance.

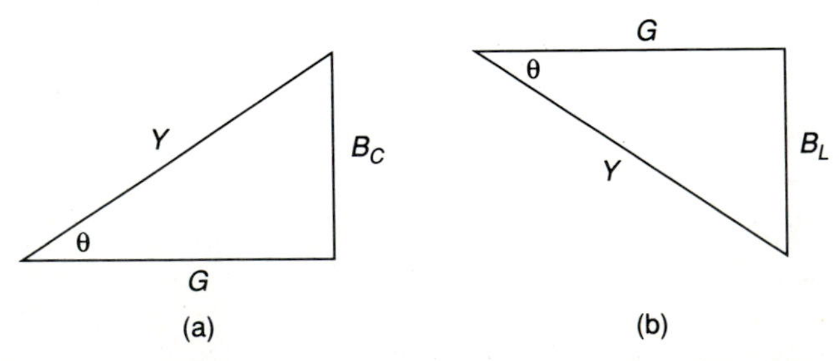

FIGURE 14-19 Admittance triangles: (a) capacitive susceptance; (b) inductive susceptance.

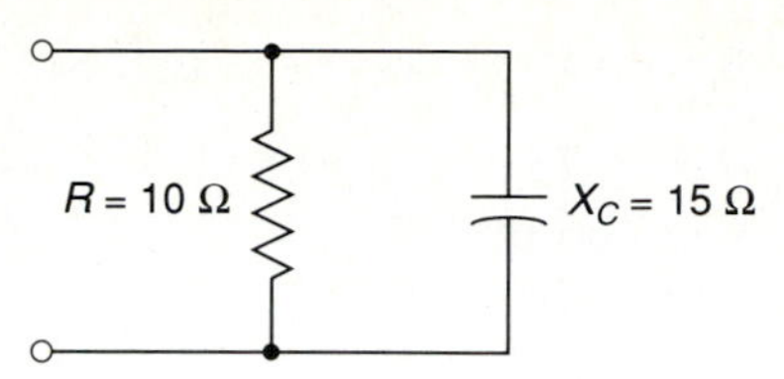

FIGURE 14-20

EXAMPLE 14-25 Determine the total admittance and impedance for the parallel circuit of Figure 14-20.

Solution

$$G = \frac{1}{R} = \frac{1}{10} = 0.1 \text{ S}$$

$$B_C = \frac{1}{X_C} = \frac{1}{15} = 0.067 \text{ S}$$

$$Y_T = \sqrt{G^2 + B^2} \angle \tan^{-1}\frac{B}{G}$$

$$= \sqrt{(0.1)^2 + (0.067)^2} \angle \tan^{-1}\frac{0.067}{0.1}$$

$$= 0.12 \angle 33.82^\circ \text{ S}$$

$$Z_T = \frac{1}{Y_T \angle -\theta_T} = \frac{1}{0.12 \angle -33.82^\circ}$$

$$= 8.33 \angle -33.82^\circ \ \Omega$$

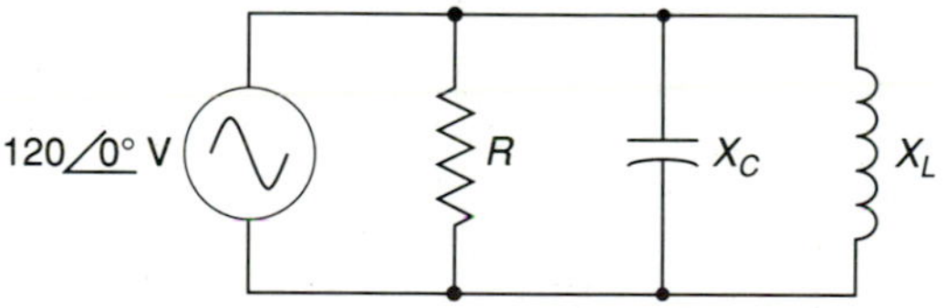

FIGURE 14-21

EXAMPLE 14-26 The circuit shown in Figure 14-21 contains the following values: $R = 800\ \Omega$, $X_C = 1200\ \Omega$, and $X_L = 1500\ \Omega$. Determine (a) the susceptance; (b) the conductance; (c) the admittance; and (d) the line current.

Solution

(a) $$G = \frac{1}{R} = \frac{1}{800} = 1.25 \text{ mS}$$

(b) $$B_C = \frac{1}{X_C} = \frac{1}{1200} = 0.833 \text{ mS}$$

$$B_L = \frac{1}{X_L} = \frac{1}{1500} = 0.667 \text{ mS}$$

$$B_T = B_C - B_L = 0.833 - 0.667 = 0.166\,\text{mS}$$

$$\text{(c)}\ Y = G + jB = 1.25\ \text{mS} + j0.833 - j0.667$$
$$= 1.25 + j0.166\ \text{mS}$$
$$= 1.26\ \underline{/7.56^\circ}\ \text{mS}$$

$$\text{(d)}\ I = E \times Y = (120\ \underline{/0^\circ}\ \text{V})(1.26\ \underline{/7.56^\circ}\ \text{mS})$$
$$= 151.2\ \underline{/7.6^\circ}\ \text{mA}$$

EXAMPLE 14-27 An ac parallel circuit contains a 60-Ω resistance and an 80-Ω inductive reactance in branch 1. Branch 2 consists of a 100-Ω resistance, 150-Ω inductive reactance, and 250-Ω capacitive reactance connected in series. Determine the total (a) conductance; (b) susceptance; (c) admittance.

Solution

$$\text{(a)}\ Y_1 = \frac{1}{R_1 + jX_1} = \frac{1}{60 + j80} = 6 - j8\ \text{mS}$$
$$= 0.01\ \underline{/-53.13^\circ}\ \text{S}$$

$$Y_2 = \frac{1}{R_2 + j(X_L - X_C)} = \frac{1}{100 - j100} = 5 + j5\ \text{mS}$$
$$= 7.07\ \underline{/45^\circ}\ \text{mS}$$

$$G = G_1 + G_2 = 6\ \text{mS} + 7.07\ \text{mS} = 13.07\ \text{mS}$$

$$\text{(b)}\ jB = -j8\ \text{mS} + j5\ \text{mS} = -j3\ \text{mS}$$

$$\text{(c)}\ Y_T = Y_1 + Y_2 = (6 - j8\ \text{mS}) + (5 + j5\ \text{mS})$$
$$= 11 - j3\ \text{mS}$$
$$= 11.4\ \underline{/-15.3^\circ}\ \text{mS}$$

14-11 THE PARALLEL *RLC* CIRCUIT

The parallel circuit shown in Figure 14-22(a) has three load branches, and the supply voltage E is common to all components. The phasor diagram may be determined in the same manner as for a series *RLC* circuit except that the diagram now consists of current phasors. The phasor diagram of Figure 14-22(b) shows the relationship between the currents in the three components of the circuit. The reference phasor, E, is drawn with the current through the pure resistor, I_R, in phase with it. Since the inductive current lags the applied voltage, the I_L phasor is drawn 90° behind E. The capacitive current leads E by 90°, so it is drawn 90° ahead of the reference phasor.

The various factors of a parallel circuit can be determined with the least difficulty by observing the following order:

1. Find the impedance of each branch.
2. Find the current of each branch.
3. Draw a phasor diagram of the currents.
4. Find the total in-phase, or resistance, current.
5. Find the total reactance current.
6. Determine the line current. Once the individual branch currents and their phase angles have been found, the total line current is obtained by phasor addition.
7. Find the impedance of the circuit. The total impedance is obtained by applying Ohm's law to the total current and the applied voltage of the circuit.

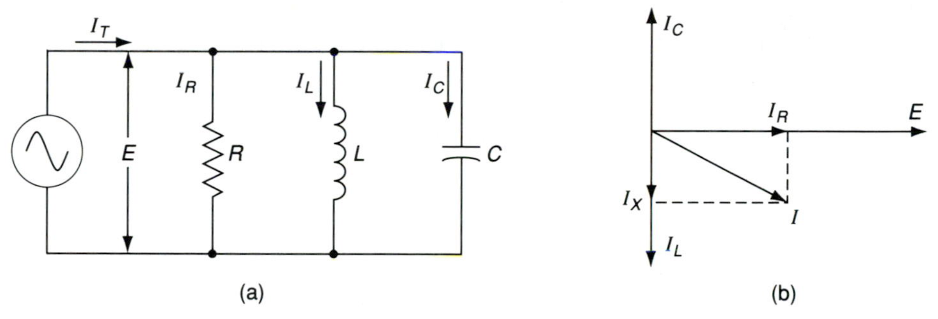

FIGURE 14-22 (a) Parallel RLC circuit; (b) phasor diagram in which $I_L > I_C$.

The total impedance in a parallel ac circuit is obtained in the same manner as the total resistance in a parallel dc circuit. That is,

$$\frac{1}{Z_T} = \frac{1}{Z_1} + \frac{1}{Z_2} + \frac{1}{Z_3} + \cdots + \frac{1}{Z_N} \tag{14-47}$$

or

$$Z_T = \frac{Z_1 \times Z_2}{Z_1 + Z_2} \tag{14-48}$$

To determine the total current in a circuit when the individual branch currents are known, the following equation may be used:

$$I_T = I_R + jI_X \tag{14-49}$$

or, in polar form,

$$I_T = \sqrt{I_R^2 + I_X^2}\ \angle \tan^{-1}\frac{I_X}{I_R} \tag{14-50}$$

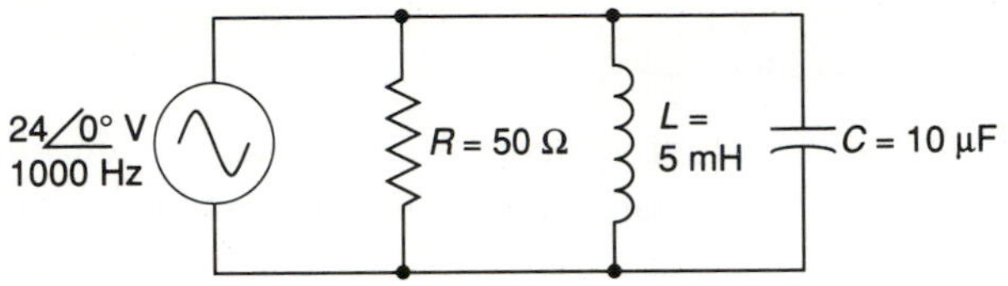

FIGURE 14-23

EXAMPLE 14-28 For the circuit shown in Figure 14-23, calculate I_R, I_L, I_C, I_T, and their respective phase angles.

Solution

$$X_L = 2\pi fL = 2\pi(1000\text{ Hz})(5 \times 10^{-3}\text{ H}) = 31.42\ \Omega$$

$$X_C = \frac{1}{2\pi fC} = \frac{1}{2\pi(1000\text{ Hz})(10 \times 10^{-6}\text{ F})} = 15.92\ \Omega$$

$$I_R = \frac{E}{R} = \frac{24\angle 0^\circ\text{ V}}{50\angle 0^\circ\ \Omega} = 0.48\angle 0^\circ\text{ A}$$

$$I_L = \frac{E}{X_L} = \frac{24\angle 0^\circ\text{ V}}{31.42\angle 90^\circ\ \Omega} = 0.76\angle -90^\circ\text{ A}$$

$$I_C = \frac{E}{X_C} = \frac{24\angle 0^\circ\text{ V}}{15.92\angle -90^\circ\ \Omega} = 1.51\angle 90^\circ\text{ A}$$

$$\begin{aligned} I_T &= I_R + I_L + I_C \\ &= 0.48\angle 0^\circ\text{ A} + 0.76\angle -90^\circ\text{ A} + 1.51\angle 90^\circ\text{ A} \\ &= 0.89\angle 57^\circ\text{ A} \end{aligned}$$

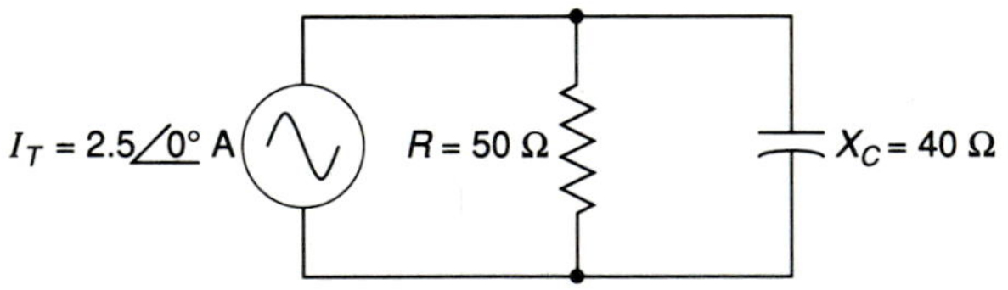

FIGURE 14-24

EXAMPLE 14-29 For the circuit shown in Figure 14-24, determine the applied voltage.

Solution

$$\begin{aligned} Z_T &= \frac{R \times X_C}{R + X_C} = \frac{(30\angle 0^\circ\ \Omega)(40\angle -90^\circ\ \Omega)}{30\angle 0^\circ + 40\angle -90^\circ\ \Omega} \\ &= \frac{1200\angle -90^\circ\ \Omega}{50\angle -53^\circ\ \Omega} \end{aligned}$$

$$= 24 \angle -36.9^\circ \ \Omega$$

$$E = I_T Z_T = (2.5 \angle 0^\circ \text{ A})(24 \angle -36.9^\circ \ \Omega)$$
$$= 60 \angle -36.9^\circ \text{ V}$$

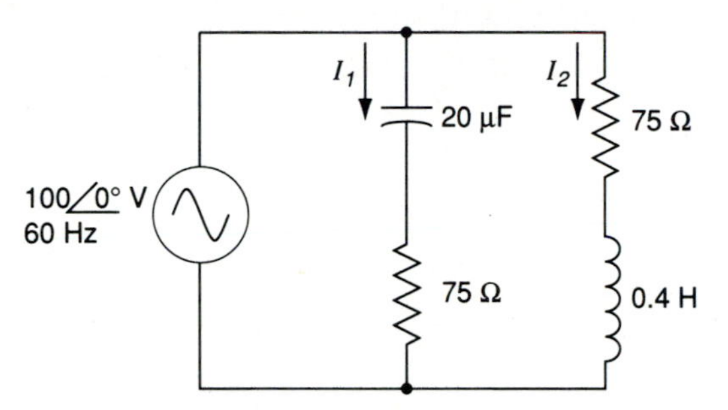

FIGURE 14-25

EXAMPLE 14-30 For the circuit shown in Figure 14-25, calculate the current in each branch and the total impedance.

Solution

$$X_L = 2\pi f L = 2\pi(60\text{ Hz})(0.4\text{ H}) = 150.8\ \Omega$$

$$X_C = \frac{1}{2\pi f C} = \frac{1}{2\pi(60\text{ Hz})(20 \times 10^{-6}\text{ F})} = 132.63\ \Omega$$

$$Z_1 = \sqrt{R^2 + X_C^2} \ \angle \tan^{-1}\frac{X_C}{R}$$
$$= \sqrt{75^2 + 132.63^2} \ \angle \tan^{-1}\frac{-132.63}{75}$$
$$= 152.37 \angle -60.5^\circ \ \Omega$$

$$Z_2 = \sqrt{R^2 + X_L^2} \ \angle \tan^{-1}\frac{X_L}{R}$$
$$= \sqrt{75^2 + 150.8^2} \ \angle \tan^{-1}\frac{150.8}{75}$$
$$= 168.42 \angle 63.6^\circ \ \Omega$$

$$I_1 = \frac{E}{Z_1} = \frac{100 \angle 0^\circ}{152.37 \angle -60.5^\circ} = 0.66 \angle 60.5^\circ \text{ A}$$

$$I_2 = \frac{E}{Z_2} = \frac{100 \angle 0^\circ}{168.42 \angle 63.6^\circ} = 0.59 \angle -63.6^\circ \text{ A}$$

$$Z_T = \frac{Z_1 \times Z_2}{Z_1 + Z_2} = \frac{(152.37 \angle -60.5^\circ \ \Omega)(168.42 \angle 63.6^\circ \ \Omega)}{152.37 \angle -60.5^\circ \ \Omega + 168.42 \angle 63.6^\circ \ \Omega}$$
$$= \frac{25{,}662.16 \angle 3.1^\circ}{151.02 \angle 6.94^\circ}$$
$$= 169.93 \angle -3.84^\circ \ \Omega$$

14-12 CURRENT DIVIDER RULE

The current divider rule for ac circuits follows the same format as that for dc circuits. Figure 14-26 shows two parallel impedances; to calculate the branch currents, the following equations are used:

$$I_1 = \frac{Z_2 \times I_T}{Z_1 + Z_2} \quad \text{or} \quad I_2 = \frac{Z_1 \times I_T}{Z_1 + Z_2} \tag{14-51}$$

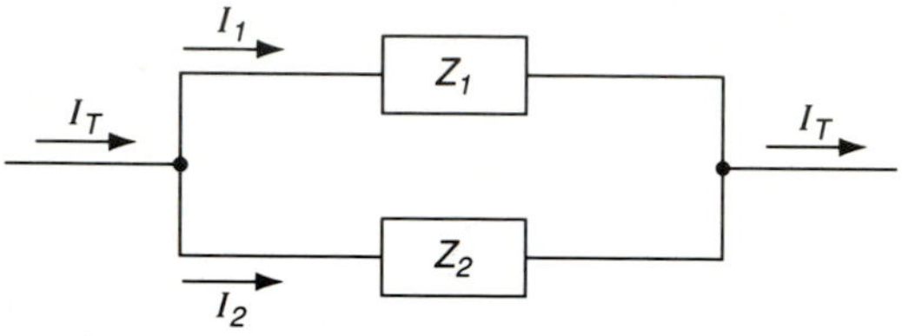

FIGURE 14-26 Parallel impedances.

EXAMPLE 14-31 For the circuit shown in Figure 14-27, calculate the current through each branch using the current divider rule.

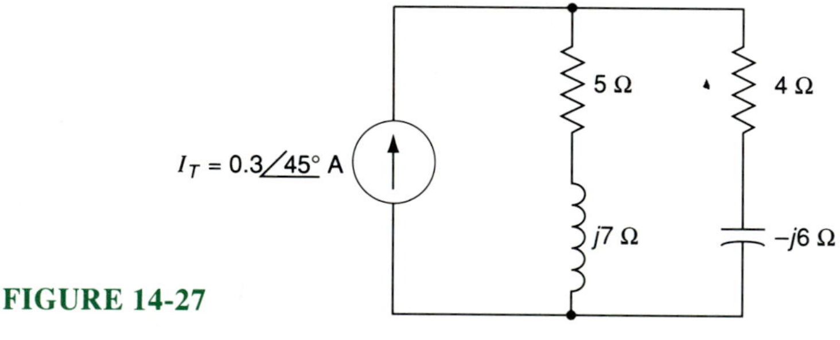

FIGURE 14-27

Solution

$$Z_1 = \sqrt{5^2 + 7^2} \angle \tan^{-1}\frac{7}{5}$$
$$= 8.6 \angle 54.5^\circ \ \Omega$$

$$Z_2 = \sqrt{4^2 + 6^2} \angle \tan^{-1}\frac{-6}{4}$$
$$= 7.21 \angle -56.3^\circ \ \Omega$$

$$I_1 = \frac{Z_2 \times I_T}{Z_1 + Z_2} = \frac{(7.21 \angle -56.3^\circ \ \Omega)(0.3 \angle 45^\circ \ \text{A})}{8.6 \angle 54.5^\circ \ \Omega + 7.21 \angle -56.3^\circ \ \Omega}$$
$$= 239.01 \angle -17.7^\circ \ \text{mA}$$

$$I_2 = \frac{Z_1 \times I_T}{Z_1 + Z_2} = \frac{(8.6 \angle 54.5^\circ \ \Omega)(0.3 \angle 45^\circ \ \text{A})}{9.05 \angle 6.34^\circ \ \Omega}$$
$$= 285.08 \angle 93.1^\circ \ \text{mA}$$

14-13 SERIES-PARALLEL CIRCUITS

Circuits that contain combinations of series and parallel impedances are referred to as series–parallel circuits. The easiest method for solving series–parallel impedances is to convert them to phasor quantities and use the same methods as for solving resistor problems in dc circuits.

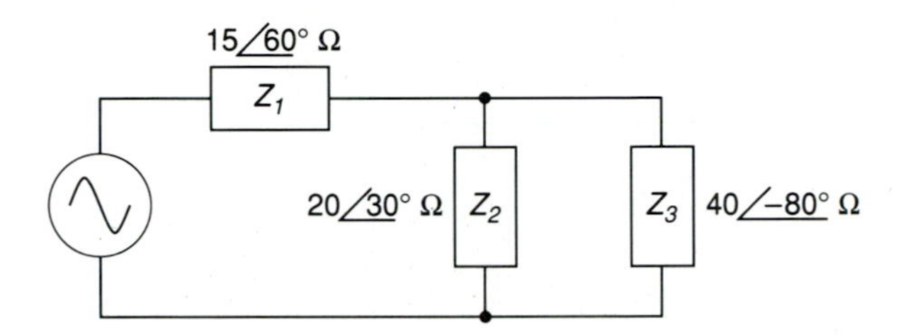

FIGURE 14-28

EXAMPLE 14-32 Determine the total impedance of the circuit shown in Figure 14-28.

Solution By finding the impedance of Z_2 and Z_3, the circuit will reduce to a series combination.

$$Z_{2\|3} = \frac{Z_2 \times Z_3}{Z_2 + Z_3} = \frac{(20\,\angle 30^\circ\ \Omega)(40\,\angle -80^\circ\ \Omega)}{20\,\angle 30^\circ\ \Omega + 40\,\angle -80^\circ}$$

$$= \frac{800\,\angle -50^\circ\ \Omega}{38.11\,\angle -50^\circ\ \Omega}$$

$$= 21\,\angle 0^\circ\ \Omega$$

$$Z_T = Z_1 + Z_{2\|3}$$

$$= 15\,\angle 60^\circ\ \Omega + 21\,\angle 0^\circ\ \Omega$$

$$= 31.32\,\angle 24.5^\circ\ \Omega$$

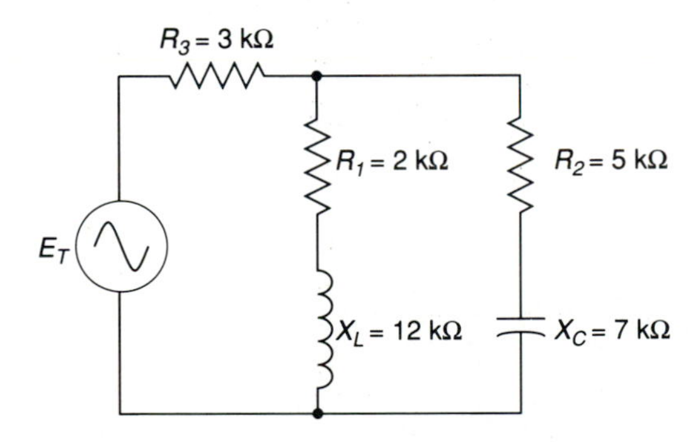

FIGURE 14-29

EXAMPLE 14-33 For the circuit shown in Figure 14-29, determine E_{XL} and E_{XC}.

Solution

$$Z_1 = \sqrt{R_1^2 + X_L^2} \angle \tan^{-1}\frac{X_L}{R_1}$$
$$= \sqrt{2000^2 + 12{,}000^2} \angle \tan^{-1}\frac{12\text{ k}\Omega}{2\text{ k}\Omega}$$
$$= 12.17 \angle 80.5^\circ \text{ k}\Omega$$

$$Z_2 = \sqrt{R_2^2 + X_C^2} \angle \tan^{-1}\frac{-X_C}{R_2}$$
$$= \sqrt{5000^2 + 7000^2} \angle \tan^{-1}\frac{-7\text{ k}\Omega}{5\text{ k}\Omega}$$
$$= 8.6 \angle -54.5^\circ \text{ k}\Omega$$

$$Z_{1\|2} = \frac{Z_1 \times Z_2}{Z_1 + Z_2} = \frac{(12.17 \angle 80.5^\circ \text{ k}\Omega)(8.6 \angle -54.5^\circ \text{ k}\Omega)}{12.17 \angle 80.5^\circ \text{ k}\Omega + 8.6 \angle -54.5^\circ \text{ k}\Omega}$$
$$= 12.17 \angle -9.5^\circ \text{ k}\Omega$$

$$Z_T = Z_{1\|2} + R_3 = 12.17 \angle -9.5^\circ \text{ k}\Omega + 3 \angle 0^\circ \text{ k}\Omega$$
$$= 15.14 \angle -7.63^\circ \text{ k}\Omega$$

$$I_T = \frac{E_T}{Z_T} = \frac{120 \angle 0^\circ \text{ V}}{15.14 \angle -7.63^\circ \text{ k}\Omega}$$
$$= 7.93 \angle 7.63^\circ \text{ mA}$$

$$E_{XL} = jX_L \times I_1$$

$$I_1 = \frac{E_1}{Z_1} = \frac{I_T(Z_{1\|2})}{Z_1}$$

$$E_{XL} = \frac{(X_L)(I_T)(Z_{1\|2})}{Z_1}$$
$$= \frac{(12{,}000 \angle 90^\circ \Omega)(0.00793 \angle 7.63^\circ \text{ A})(12{,}170 \angle -9.5^\circ \Omega)}{12{,}170 \angle 80.5^\circ \Omega}$$
$$= 95.16 \angle 7.6^\circ \text{ V}$$

$$E_{XC} = -jX_C \times I_2$$

$$I_2 = \frac{E_2}{Z_2} = \frac{I_T(Z_{1\|2})}{Z_2}$$

$$\therefore \quad E_{XC} = \frac{(X_C)(I_T)(Z_{1\|2})}{Z_2}$$
$$= \frac{(7000 \angle -90^\circ \Omega)(0.00793 \angle 7.63^\circ \text{ A})(12{,}170 \angle -9.5^\circ \Omega)}{8600 \angle -54.5^\circ}$$
$$= 78.55 \angle -37.4^\circ \text{ V}$$

14-14 EQUIVALENT CIRCUITS

When solving ac circuit problems, it is sometimes desirable to substitute an equivalent series circuit for a parallel combination, or an equivalent parallel circuit for a series combination. A series impedance that draws current of the same magnitude and phase angle as a parallel circuit when connected across the same applied voltage is called the **equivalent series circuit** of the parallel combination. Figure 14-30(a) shows a resistance and reactance connected in parallel. When these components are connected in series, as in Figure 14-30(b), a certain resistance value R_S and reactance value X_S will produce the same effect as the original resistance and reactance connected in parallel. In equation form,

$$R_S = \frac{R_p \times X_p^2}{R_p^2 + X_p^2} \qquad (14\text{-}52)$$

$$X_S = \frac{X_p \times R_p^2}{R_p^2 + X_p^2} \qquad (14\text{-}53)$$

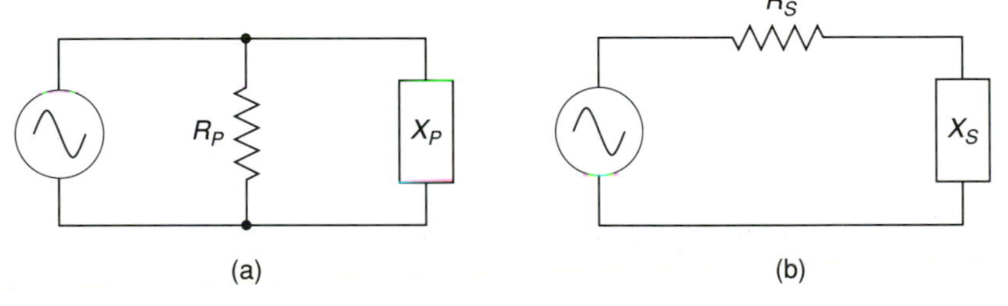

FIGURE 14-30 (a) Equivalent parallel circuit; (b) equivalent series circuit.

EXAMPLE 14-34 Calculate the resistance and reactance values of a series equivalent circuit that may be substituted for a parallel circuit containing a resistance of 60 Ω and a reactance of 80 Ω.

Solution

$$R_S = \frac{R_p \times X_p^2}{R_p^2 + X_p^2} = \frac{(60) \times (80)^2}{(60)^2 + (80)^2} = 38.4\,\Omega$$

$$X_S = \frac{X_p \times R_p^2}{R_p^2 + X_p^2} = \frac{(80) \times (60)^2}{(60)^2 + (80)^2} = 28.8\,\Omega$$

In order to match line current and phase angle between a parallel circuit and an equivalent series circuit, the total impedance Z and the circuit phase angle θ must be the same for both circuits. Therefore,

$$R_S = Z_T \cos\theta \tag{14-54}$$

$$X_S = Z_T \sin\theta \tag{14-55}$$

If θ is positive, the following equivalent series inductance is used:

$$L = \frac{X_s}{2\pi f} \tag{14-56}$$

If θ is negative, the following equivalent series capacitance is used:

$$C = \frac{1}{2\pi f X_s} \tag{14-57}$$

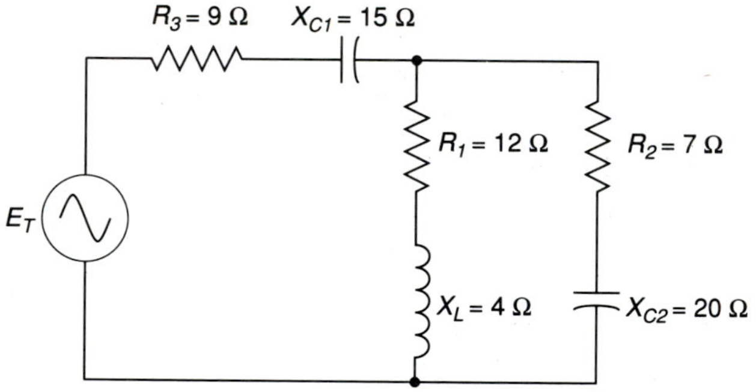

FIGURE 14-31

EXAMPLE 14-35 Determine the equivalent series circuit for the network shown in Figure 14-31.

Solution

$$\begin{aligned}
Z_1 &= \sqrt{R_1^2 + X_L^2}\ \angle \tan^{-1}\frac{X_L}{R_1} \\
&= \sqrt{12^2 + 4^2}\ \angle \tan^{-1}\frac{4}{12} \\
&= 12.65\ \angle 18.44^\circ\ \Omega \\
Z_2 &= \sqrt{R_2^2 + X_C^2}\ \angle \tan^{-1}\frac{-X_C^2}{R_2} \\
&= \sqrt{7^2 + 20^2}\ \angle \tan^{-1}\frac{-20}{7} \\
&= 21.19\ \angle -70.71^\circ\ \Omega
\end{aligned}$$

$$\begin{aligned}
Z_{1\|2} &= \frac{Z_1 \times Z_2}{Z_1 + Z_2} \\
&= \frac{(12.65\ \angle 18.44^\circ\ \Omega)(21.19\ \angle -70.71^\circ\ \Omega)}{12.65\ \angle 18.44^\circ\ \Omega + 21.19\ \angle -70.71^\circ\ \Omega} \\
&= 10.79\ \angle -12.2^\circ\ \Omega
\end{aligned}$$

In rectangular form,

$$Z_{1\|2} = 10.79(\cos -12.2° + j \sin -12.2°)$$
$$= 10.55 - j2.28\ \Omega$$

$$Z_T = R - jX_C + Z_{1\|2}$$
$$= 9 - j15 + 10.55 - j2.28$$
$$= 19.55 - j17.28\ \Omega$$

The series equivalent circuit is a resistor of 19.6 Ω in series with a capacitive reactance of 17.3 Ω.

When a circuit has resistance and reactance connected in series, a certain resistance value, R_P, and reactance value, X_P, will produce the equivalent parallel combination as the original series- connected circuit. This circuit is known as the **equivalent parallel circuit** and is calculated in the following manner:

$$R_P = \frac{R_S^2 + X_S^2}{R_S} \tag{14-58}$$

$$X_P = \frac{R_S^2 + X_S^2}{X_S} \tag{14-59}$$

14-15 POWER IN AC CIRCUITS

In the study of dc circuits, power was defined as the product of voltage and current. In ac circuits this will be true only when the current is in phase with the voltage, such as circuits containing only resistance. In cases where the current is not in phase with the voltage, the product of the voltage and current will not be equal to the power actually consumed by the circuit. In an ac circuit that contains reactance, the current through the circuit either leads or lags the applied voltage. This implies that there is a phase angle between the two components which must be taken into account when calculating the power consumed by the circuit.

To find the corresponding value of power between resistive and reactive circuits, the product of voltage and current is multiplied by the cosine of the phase angle. The **average power**, or **real power**, as it is sometimes called, is the power delivered to, and dissipated by, the load. This is the power indicated when measured by a wattmeter.

The equation for determining average power in an ac circuit is given as

$$P = E_{\text{rms}} \times I_{\text{rms}} \times \cos\theta \quad \text{or} \quad \frac{E_m \times I_m}{2} \times \cos\theta \tag{14-60}$$

Figure 14-32(a) shows a resistor connected across an ac source. Figure 14-32(b) indicates how voltage, current, and the power dissipated in the resistor change with time. The power varies between zero and maximum and has a frequency double that of the source. The dashed horizontal line represents the average power, which is half of the maximum instantaneous power. The average power is considered to be the direct-current equivalent of the ac power supplied by the source.

In a purely resistive circuit, since voltage and current are in phase, $\theta = 0°$. Therefore,

$$\begin{aligned} P &= E_{\text{rms}} \times I_{\text{rms}} \times \cos 0° \\ &= E \times I \times 1 \end{aligned} \tag{14-61}$$

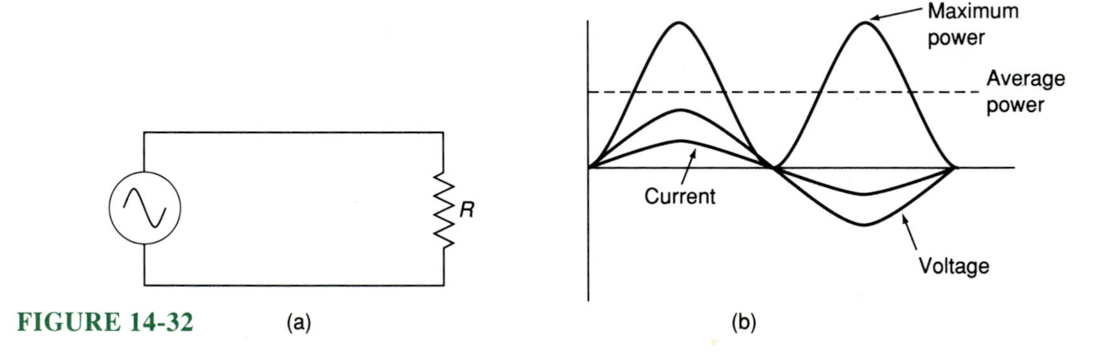

FIGURE 14-32 (a) (b)

Figure 14-33(a) shows an ac voltage source applied across a pure inductor. In Figure 14-33(b), the product of the voltage and current crosses the zero axis. For one-half of the cycle, energy is being supplied to the inductor. During the other half of the cycle, energy is being returned by the inductor to the source. Consequently, the total power dissipated in a cycle by an ideal inductor is zero.

A practical inductor will contain a small amount of internal resistance, so there will be some heat dissipation during the cycle. An inductor is often referred to as a **storage element**, which receives electrical energy, stores it in the form of a magnetic field, and returns the energy to the circuit at a later time.

In a purely inductive circuit, voltage leads current by 90°. The average power is calculated as

$$P = E_{rms} \times I_{rms} \times \cos 90^\circ$$
$$= E \times I \times 0$$
$$= 0 \tag{14-62}$$

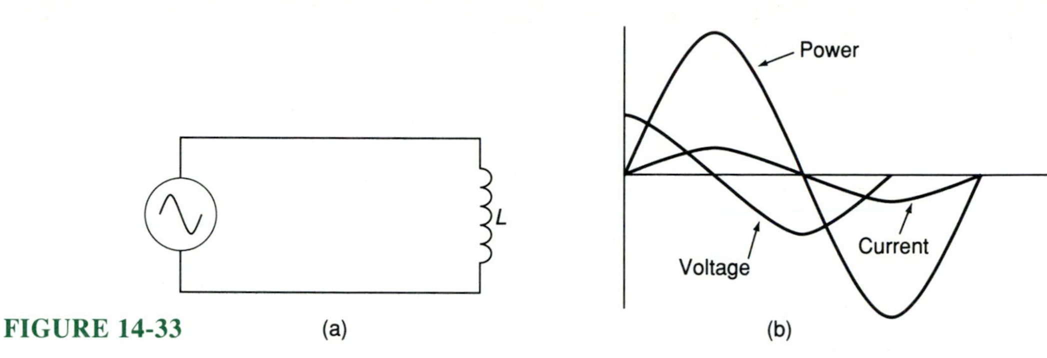

FIGURE 14-33

A capacitor is shown connected across an ac voltage source in the circuit of Figure 14-34(a). Figure 14-34(b) shows the wave shapes for the current, voltage, and instantaneous power for the circuit of Figure 14-34(a). A capacitor, like an inductor, is an energy storage element that stores energy during one half-cycle and returns energy during the other half-cycle.

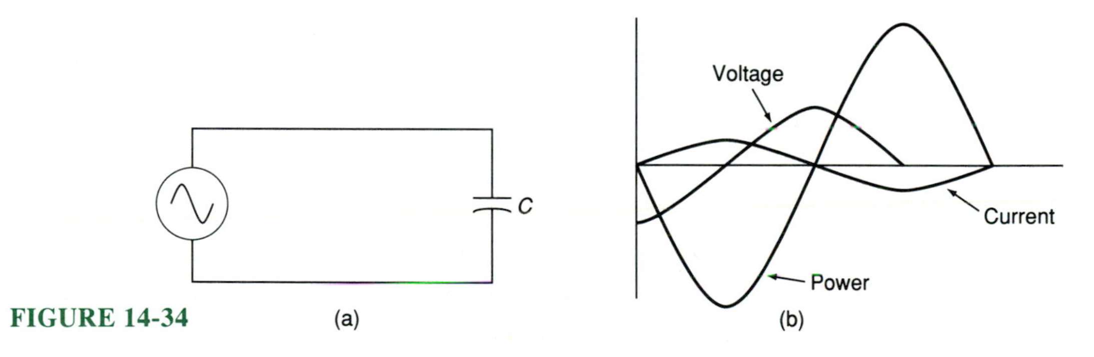

FIGURE 14-34

In a purely capacitive circuit, voltage lags current by 90°. The average power is determined by the following equation:

$$P = E_{rms} \times I_{rms} \times \cos -90^\circ$$
$$= E \times I \times 0$$
$$= 0 \tag{14-63}$$

The average power in a purely capacitive circuit is also zero.

EXAMPLE 14-36 Calculate the average power dissipated by the circuit shown in Figure 14-35.

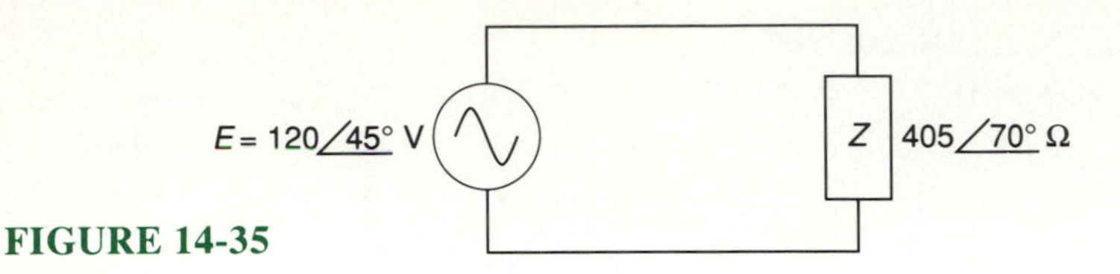

FIGURE 14-35

Solution

$$I = \frac{E}{Z} = \frac{120\angle 45^\circ\ \text{V}}{405\angle 70^\circ\ \Omega}$$
$$= 0.296\angle -25^\circ\ \text{A}$$

The phasor diagram of Figure 14-36 shows the angular displacement between the applied voltage and the current flowing through the circuit.

$$\theta = \beta - \alpha = 45^\circ - (-25^\circ)$$
$$= 70^\circ$$

$$P = EI\cos\theta = (120\ \text{V})(0.296\ \text{A})\cos 70^\circ$$
$$= 12.15\ \text{W}$$

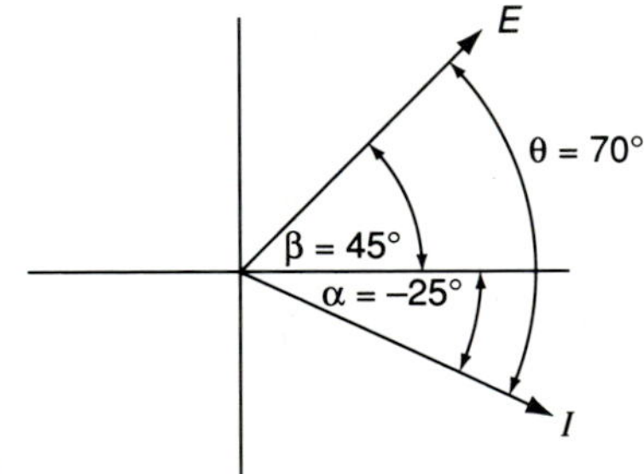

FIGURE 14-36

EXAMPLE 14-37 Calculate the average power dissipated by a circuit having an instantaneous voltage of $e = 70\sin(\omega t + 60^\circ)$ V, and an instantaneous current of $i = 15\sin(\omega t + 80^\circ)$ A.

Solution

$$E_m = 70\angle 60^\circ\ \text{V}$$

$$I_m = 15\angle 80^\circ\ \text{A}$$

$$\theta = \beta - \alpha = 60^\circ - 80^\circ$$
$$= -20^\circ$$

$$P = \frac{E_m \times I_m}{2} \times \cos\theta$$
$$= \frac{70\ \text{V} \times 15\ \text{A}}{2} \times \cos(-20^\circ)$$
$$= 493.34\ \text{W}$$

The product of voltage and current is sometimes referred to as the **apparent power**. Since it is not actually a true power measurement, it is more commonly known as **volt-amperes** and is represented by the letter S. In equation form,

$$S = E \times I \tag{14-64}$$

EXAMPLE 14-38 An ac circuit has an impedance of $55 \angle -30^\circ \ \Omega$, and an effective voltage of $110 \angle 45^\circ$ V. Calculate (a) the true power circuit; (b) the apparent power circuit.

Solution

$$I = \frac{E}{Z} = \frac{110 \angle 45^\circ \text{ V}}{55 \angle -30^\circ \ \Omega} = 2 \angle 75^\circ \text{ A}$$

$$\theta = 45^\circ - 75^\circ = -30^\circ$$

(a) $P = EI \cos\theta = (110 \text{ V})(2 \text{ A}) \cos(-30^\circ) = 190.5 \text{ W}$

(b) $S = EI = (110 \text{ V})(2 \text{ A}) = 220 \text{ VA}$

The ratio of true power to apparent power is called the **power factor** (PF) of the circuit.

$$\text{PF} = \frac{\text{true power}}{\text{apparent power}} = \frac{E_{\text{rms}} \times I_{\text{rms}} \cos\theta}{E_{\text{rms}} \times I_{\text{rms}}} = \cos\theta \tag{14-65}$$

Equation 14-65 shows that the power factor is equal to $\cos\theta$. For this reason, the phase angle is often referred to as the **power factor angle**. In an inductive circuit, where the current lags the voltage, the power factor is said to be a **lagging power factor**. If the current leads the voltage, the power factor is called a **leading power factor**. The value of the power factor depends on how much the current and voltage are out of phase with each other. Therefore, the power factor of a circuit is determined by the relative amounts of resistance and reactance in the circuit. If the impedance is known, the power factor is also known, since it is also the ratio of resistance to impedance.

$$\text{PF} = \frac{R}{Z} \tag{14-66}$$

In a purely reactive circuit, the voltage and current are 90° out of phase with each other, and the power dissipated is zero. However, reactance does draw current from the supply and the product of this current, and the applied voltage has units of joules/second, which is the same unit as the watt. Since capacitance and inductance cannot dissipate power, the reactive component alternately stores and releases power as its magnetic field alternately builds up and collapses. The product of the voltage and current of the reactive component is directly proportional to the amount of energy stored and returned to the reactive component each time the current changes direction. This product is referred to as **reactive power** and is represented by the letter Q.

Reactive power is received from the supply during the first 180° and returned to the supply during the second 180° of each cycle. For this reason, reactive power is also called **wattless power**. To distinguish between true power and reactive power, true power is measured in watts, and reactive power is measured in **volt-ampere-reactive** (VAR). The term "VAR" is used mostly in high-power applications and by electric utilities. For a pure inductance or capacitance, the reactive power is calculated using the following equation:

$$Q = E_L I_L = I^2 X_L = \frac{E_L^2}{X_L} \tag{14-67}$$

or

$$Q = E_C I_C = I^2 X_C = \frac{E_C^2}{X_C} \tag{14-68}$$

The reactive power can also be calculated as

$$Q = E_{\text{rms}} \times I_{\text{rms}} \sin \theta \tag{14-69}$$

EXAMPLE 14-39 Calculate the reactive power of a 15-μF capacitor drawing 3.51 A from a 400-Hz supply.

Solution

$$X_C = \frac{1}{2\pi f C} = \frac{1}{2\pi(400\text{ Hz})(15 \times 10^{-6}\text{ F})} = 26.53\ \Omega$$

$$Q = I^2 X_C = (3.51\text{ A})^2(26.53\ \Omega) = 326.85\text{ VAR}$$

EXAMPLE 14-40 The instantaneous voltage across a 5-H inductor is $e = 30 \sin 100t$ V. Determine the reactive power of the inductor.

Solution

$$e = 30 \sin 100t$$

$$X_L = \omega L = (100)(5) = 500\ \Omega$$

$$I_m = \frac{E_m}{X_L} = \frac{30\text{ V}}{500\ \Omega} = 60\text{ mA}$$

$$E_{\text{rms}} = 0.707 \times E_m = 0.707 \times 30\,\text{V} = 21.21\,\text{V}$$

$$I_{\text{rms}} = 0.707 \times I_m = (0.707)(60 \times 10^{-3}\,\text{A}) = 42.42\,\text{mA}$$

$$Q = E_L I_L = (21.21\,\text{V})(42.4 \times 10^{-3}\,\text{A}) = 0.9\,\text{VAR}$$

14-16 THE POWER TRIANGLE

The ratio of real power to apparent power can be drawn as the sides of a right-angle triangle, where real power is the side adjacent to angle θ and apparent power is the hypotenuse. Since real power is an in-phase, or resistive, component, it would lie on the horizontal axis. Reactive power is drawn on the vertical axis. This type of triangle, called a **power triangle,** is shown in Figure 14-37. The two power triangles illustrated are based on current as the reference phasor, so they are assumed to be the triangles of a series circuit. Power triangles based on parallel circuits with the voltage as the reference phasor would have their reactive components drawn in the opposite direction to that shown in Figure 14-37. The power triangle of a series circuit corresponds to an impedance triangle, whereas the power triangle of a parallel circuit corresponds to an admittance triangle.

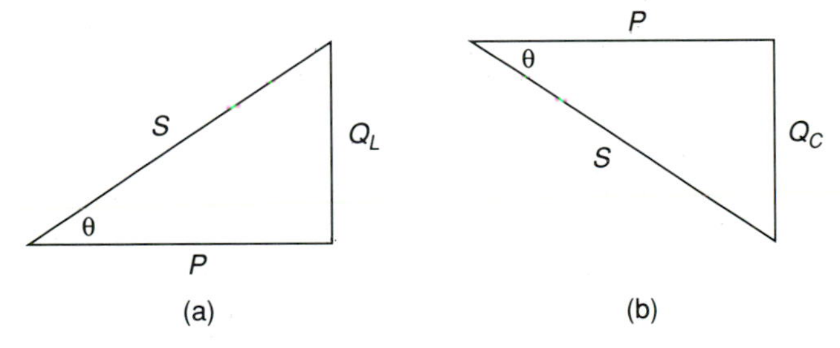

FIGURE 14-37 Power triangles: (a) resistor and inductor in series; (b) resistor and capacitor in series.

The power triangle for a series-connected resistor and inductor is shown in Figure 14-37(a), and the triangle for a resistor and capacitor is shown in Figure 14-37(b). If a circuit contains both inductive and capacitive power, the reactive component of the power triangle will be determined by the difference between the reactive power delivered to each. In the right-angle triangle of Figure 14-37, the Pythagorean theorem proves that

$$S = \sqrt{P^2 + Q^2} \qquad (14\text{-}70)$$

where S = apparent power, in volt-amperes

P = true power, in watts

Q = reactive power, in VAR

Since θ represents the power factor of the circuit, power triangles are very useful in solving problems related to power factor.

The total apparent power of an ac circuit must be determined by using the Pythagorean theorem for the total real and reactive powers. The total apparent power cannot be determined from the apparent power of each individual branch. Also, when solving for series or parallel circuits, the method of calculating total real, reactive, or apparent power is done simply by finding the sum of each branch. For these types of circuits, the total real and reactive power is found by taking the rectangular form of the apparent power of each branch and adding each branch to find the sum of the real and imaginary values.

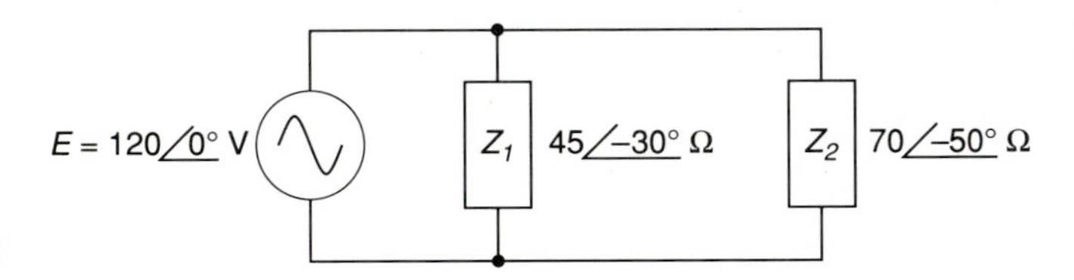

FIGURE 14-38

EXAMPLE 14-41 For the circuit shown in Figure 14-38, determine the total (a) real power; (b) reactive power; (c) apparent power, (d) power factor; (e) power triangle.

Solution

$$I_1 = \frac{E}{Z_1} = \frac{120\,\angle 0^\circ}{45\,\angle -30^\circ} = 2.67\,\angle 30^\circ \text{ A}$$

$$I_2 = \frac{E}{Z_2} = \frac{120\,\angle 0^\circ}{70\,\angle -50^\circ} = 1.71\,\angle 50^\circ \text{ A}$$

$$\begin{aligned} S_1 = EI_1 &= (120\,\angle 0^\circ)(2.67\,\angle 30^\circ) \\ &= 320.4\,\angle 30^\circ \\ &= 277.5 + j160.2 \end{aligned}$$

$$\begin{aligned} S_2 = EI_2 &= (120\,\angle 0^\circ)(1.71\,\angle 50^\circ) \\ &= 205.2\,\angle 50^\circ \\ &= 131.9 + j157.2 \end{aligned}$$

(a) $P_T = P_1 + P_2 = 277.5 + 131.9 = 409.4$ W

(b) $Q_T = Q_1 + Q_2 = j160.2 + j157.2 = 317.4$ VAR

(c) $S_T = S_1 + S_2$
$= (277.5 + j160.2) + (131.9 + j157.2)$
$= 518 \text{ VA}$

(d) $\text{PF} = \dfrac{P}{S} = \dfrac{409.4}{518} = 0.79 \quad \text{or} \quad 79\%$

Since the impedances of the branches contain negative imaginary values, they can be assumed to be capacitive. Therefore, the power factor of the circuit is leading, and the power triangle would be as shown in Figure 14-39.

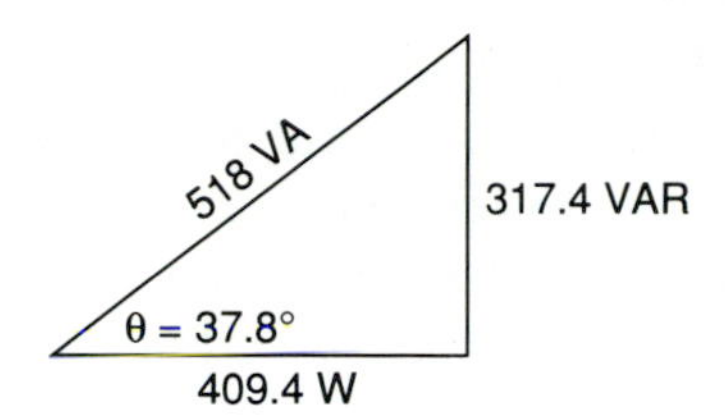

FIGURE 14-39

EXAMPLE 14-42 For the circuit shown in Figure 14-40, calculate the total (a) true power; (b) reactive power; (c) apparent power; (d) power factor.

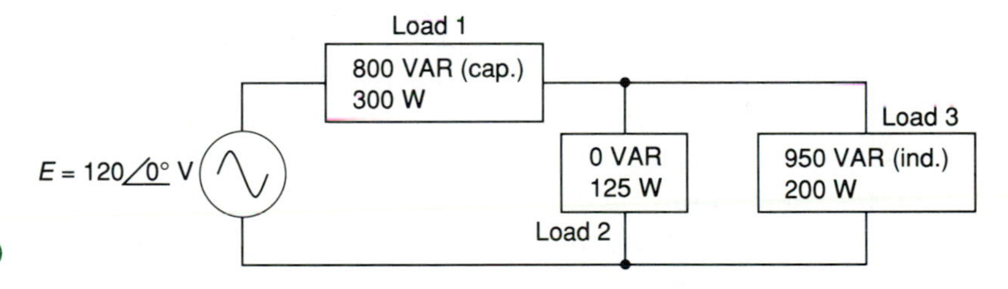

FIGURE 14-40

Solution

(a) $P_T = P_1 + P_2 + P_3 = 300 + 125 + 200 = 625 \text{ W}$

(b) $Q_T = Q_1 + Q_2 + Q_3 = (-800) + 0 + 950 = 150 \text{ VAR}$

(c) $S_T = \sqrt{P_T^2 + Q_T^2} = \sqrt{(625)^2 + (150)^2} = 642.75 \text{ VA}$

(d) $\text{PF} = \dfrac{P}{S} = \dfrac{625}{642.75} = 0.97 \text{ lag}$

Since the circuit shown in Figure 14-40 is predominately inductive, the current lags the voltage, producing a lagging power factor.

14-17 POWER FACTOR CORRECTION

Industrial loads such as welding machines, motors, and transformers are usually inductive. This results in most factories having a lagging power factor. The lagging power factor requires a greater amount of apparent power to be supplied to the user. Under ideal conditions, the power factor of a system is unity, or 1. As the power factor decreases, the current required to supply a constant wattage load increases, requiring generators, transformers, switching equipment, and line conductors to be larger to supply these low-power-factor loads. This increases the cost per kilowatthour of the equipment and results in an inefficient use of electric power.

Since the load current $= P/E \cos \theta$, the power factor of the circuit must be increased to reduce the line current. A transmission line may be operating above rated current due to the low power factor of the load it supplies. If the power factor of the load is improved, the line current required will decrease, the I^2R losses will also decrease, and the overall efficiency of the system will improve.

It is very common for the electric utility company to base its rates on the maximum current drawn by the load. This value of current is what is used to calculate the size of feeder conductors and distribution transformers on the utility's power grid. Since a utility customer is billed in terms of the volt-amperes used, it is imperative that the wasted, or reactive, power be kept to an absolute minimum value. If the number of VAR is extremely low, the apparent power and the true power are as close to being equal as possible. Equipment that is installed to improve the power factor is referred to as **power factor correction**.

Power factor correction for an inductive load is accomplished by connecting capacitance in parallel with the load. The leading current of the capacitor branch supplies a lagging component of current to the inductive load and tends to reduce the line current accordingly. As mentioned previously, this action improves the efficiency of transmission by reducing the line current and I^2R losses. To improve the power factor of loads such as induction motors, it is possible to connect capacitors directly across the terminals of the motor. However, most industrial plants will have power factor correction equipment installed at the junction points of main feeders. In large factories it is not uncommon for banks of capacitors to be installed in separate rooms to compensate for inductive loads.

To illustrate the effect of power factor correction, consider the circuit of Figure 14-41(a). The inductive loads is rated at 15 A, 1000 V, with a power factor of 0.5 lagging. The true power absorbed by this load is calculated as

$$P = EI \cos \theta = 1000 \times 15 \times 0.5 = 7500 \text{ W}$$

The apparent power is calculated as

$$S = EI = 1000 \times 15 = 15{,}000 \text{ VA}$$

and the reactive power is

$$Q = P \tan \theta = 7500 \tan 60° = 12{,}990 \text{ VAR}$$

The power triangle for this circuit is shown in Figure 14-41(b).

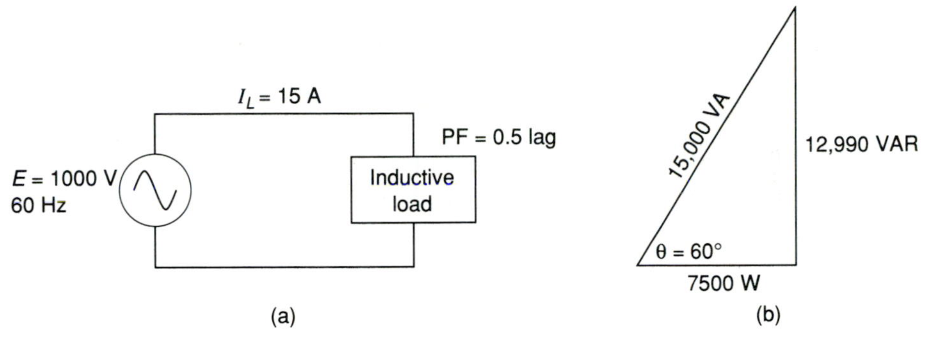

FIGURE 14-41 (a) Inductive circuit; (b) power triangle.

To reduce the power factor of Figure 14-41 to unity, a capacitor with a value of 12,990 VAR is placed in parallel with the inductive load. The amount of capacitance in farads is calculated as follows:

$$X_C = \frac{E^2}{Q} = \frac{1000^2}{12{,}990} = 76.98\ \Omega$$

$$C = \frac{1}{2\pi f X_C} = \frac{1}{2\pi(60\text{ Hz})(76.98\ \Omega)} = 34.46\ \mu\text{F}$$

Therefore, a 34.46-μF capacitor would be required to be placed in parallel with the inductive load of Figure 14-41(a) to reduce the PF to unity.

EXAMPLE 14-43 A load with an 80% lagging power factor draws 300 W from a 120-V 60-Hz supply. Calculate the size of capacitor required to improve the power factor to 0.9 lagging.

Solution $\theta = \cos^{-1} \text{PF} = \cos^{-1} 0.8 = 36.87°$

The original VAR value is found by

$$Q_1 = P \tan \theta = 300 \tan 36.87° = 225 \text{ VAR}$$

Required PF = 0.9

$$\theta = \cos^{-1}\text{PF} = \cos^{-1}0.9 = 25.84°$$

$$Q_2 = P\tan\theta = 300\tan 25.84° = 145.28\text{ VAR}$$

To find the required VAR, subtract the original VAR value, Q_1, from the required VAR value, Q_2.

$$Q_C = Q_1 - Q_2 = 225\text{ VAR} - 145.28\text{ VAR} = 79.72\text{ VAR}$$

$$X_C = \frac{E^2}{Q_C} = \frac{120^2}{79.72} = 180.63\ \Omega$$

$$C = \frac{1}{2\pi f X_C} = \frac{1}{2\pi(60\text{ Hz})(180.63\ \Omega)} = 14.69\ \mu\text{F}$$

EXAMPLE 14-44 For the circuit shown in Figure 14-42, calculate (a) the apparent power and power factor of the combined load; (b) the value of the reactive component required to correct the power factor to unity.

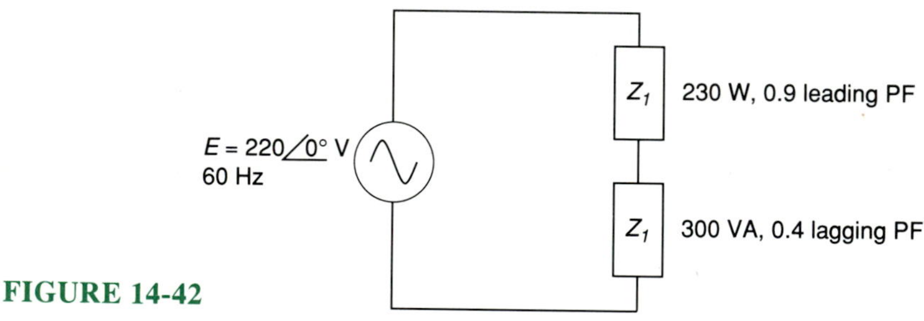

FIGURE 14-42

Solution (a) For Z_1:

$$\cos\theta = \cos^{-1}\text{PF} = \cos^{-1}0.9 = -25.84°$$

$$P_1 = 230\text{ W}$$

$$S_1 = \frac{P_1}{\text{PF}} = \frac{230}{0.9} = 255.56\ \underline{/-25.84°}\text{ VA}$$
$$= 230 - j111.39\text{ VA}$$

For Z_2:

$$\cos\theta = \cos^{-1}\text{PF} = \cos^{-1}0.4 = 66.42°$$

$$S_2 = 300\ \underline{/66.42°}\text{ VA} = 120 + j274.95$$

$$S_T = S_1 + S_2 = (230 - j111.39) + (120 + j274.95)$$
$$= 350 + j163.56$$
$$= 386.33\ \underline{/25.1°}\text{ VA}$$

(b) $P_T = S_T \cos\theta = 386.33 \cos 25.1° = 349.85$ W. The required power factor is unity; therefore,

$$S(\text{required}) = \frac{P}{\cos\theta_{\text{req}}} = \frac{349.85\ \text{W}}{1} = 349.85\ \underline{/0°}\ \text{VA}$$
$$= 349.85 + j0\ \text{VA}$$

The apparent power of the reactive component required to correct the PF to unity is calculated by

$$S_X = S_{\text{req}} - S_T = (349.85 + j0) - (350 + j163.56)$$
$$= -j163.56$$

Since the imaginary part of the reactive component is negative, the component is considered to be capacitive. The value of capacitance is found by

$$X_C = \frac{E^2}{Q_C} = \frac{220^2}{163.56} = 295.92\ \Omega$$

$$C = \frac{1}{2\pi f X_C} = \frac{1}{2\pi(60\ \text{Hz})(295.92\ \Omega)} = 8.97\ \mu\text{F}$$

The amount of capacitance required to correct the overall power factor of the circuit to unity is 8.97 μF.

14-18 EFFECTIVE RESISTANCE

When alternating current flows in a conductor, there are a larger number of flux linkages affecting the portion near the center of the conductor than near the surface. This phenomenon, known as the **skin effect**, causes a conductor to offer a higher resistance to an alternating current than it offers to a direct current. Skin effect is one factor that contributes to the **ac resistance**, or **effective resistance** of a conductor. The effective resistance of an ac circuit is directly proportional to the frequency. Consequently, high-frequency electronic circuits are especially susceptible to the skin effect. At frequencies above 10,000 MHz, a plating of gold or silver on the surface of a thin-walled hollow conductor is often utilized. Power transmission companies generally use aluminum cable with a steel core for strength. The current in the steel core is negligible in comparison with the aluminum due to the skin effect and the higher conductivity of aluminum.

The following factors comprise the effective resistance of an electrical or electronic circuit:

1. Dielectric hysteresis loss.
2. Dielectric leakage loss.
3. Eddy current loss.
4. Magnetic hysteresis loss.
5. Ohmic (dc) resistance.
6. Radiation loss.
7. Skin effect.
8. Temperature coefficient.

Losses such as eddy current and hysteresis also increase as frequency increases. This is why inductors with laminated iron cores are rarely used for radio-frequency applications. **Radiation loss** is a phenomenon that is quite pronounced in antenna circuits. Antennas use power by radiating it away in the form of a radio wave. The resistance of an antenna has two components: radiation resistance and ohmic (dc) resistance. Radiation resistance is defined as the ratio of the power radiated by the antenna to the square of the current at the feed point. The radiation resistance of an antenna may be several hundred ohms, while the dc resistance is less than 0.1 Ω.

Effective resistance in high-frequency circuits can be reduced by using fine multistrand conductors, called *Litz wire*. Since radio-frequency currents tend to flow near the surface of a conductor, Litz wire reduces ac resistance by insulating each strand of wire, allowing more current flow per circular mil.

14-19 RESONANCE IN SERIES AC CIRCUITS

An ac series circuit is considered to be in **resonance** when the inductive reactance, X_L, equals the capacitive reactance, X_C. Under this condition, the total reactance, X, equals zero and the impedance is equal to the resistance of the circuit. The phase angle between the current and voltage in a resonant circuit is zero. Therefore, the current in an ac series circuit reaches its greatest magnitude when the network is in resonance. The condition of resonance exists for a given circuit *at only one frequency*.

The *RLC* circuit shown in Figure 14-43 has an impedance defined by the equation

$$Z = R + j(X_L - X_C)$$

The impedance of an ac circuit varies with frequency. This is because inductive reactance varies directly with frequency, while capacitance varies inversely with frequency. The relationship between frequency and reactance are expressed by the equations

$$X_L = \omega L = 2\pi fL$$

$$X_C = \frac{1}{\omega C} = \frac{1}{2\pi fC}$$

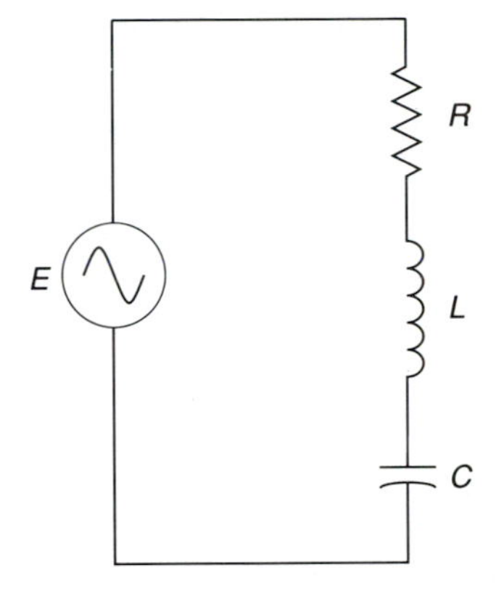

FIGURE 14-43 Series *RLC* circuit.

Since X_L has a linear relationship with frequency, its graph is shown in Figure 14-44 as having a straight line. As X_C is inversely related to frequency, the resulting graph of the quantity is curved. The total reactance is shown as a dashed line. The point where the total reactance intersects the reference axis is the **resonant frequency**, f_r. At resonant frequency, $X_L = X_C$. By using the reactance equations, the following formula is derived:

$$2\pi f_r L = \frac{1}{2\pi f_r C}$$

$$f_r = \frac{1}{(2\pi)^2 LC}$$

$$f_r = \frac{1}{2\pi \sqrt{\mathrm{LC}}} \qquad (14\text{-}71)$$

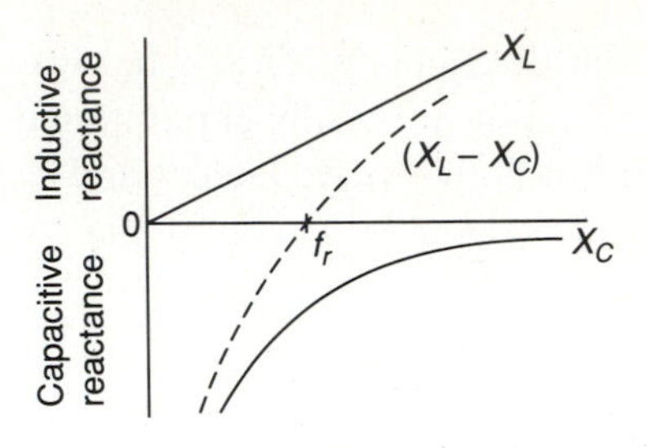

FIGURE 14-44 Resonant frequency curve.

The graphs shown in Figure 14-45 are referred to as **response curves**. Figure 14-45(a) is the amplitude–frequency response curve for a series circuit, which illustrates the rise of current amplitude to a maximum value at the resonant frequency. Below f_r the impedance of the circuit is mainly capacitive, above f_r the impedance is essentially inductive. As the frequency increases, the circuit impedance decreases until $Z = R$. Since current is the reciprocal of impedance, when the impedance is at its minimum, the maximum amount of current flows in the circuit.

Figure 14-45(b) is a phase–frequency response curve for a series circuit. At frequencies below f_r, the current is *leading* the supply voltage. At frequencies above f_r, the current is *lagging* the supply voltage. At the resonant frequency the circuit has resistance but no effective reactance, so the current is in phase with the applied voltage, and the phase angle is zero. Since maximum current flows when $Z = R$, the resonant current is calculated by Ohm's law.

$$I_r = \frac{E}{R}$$

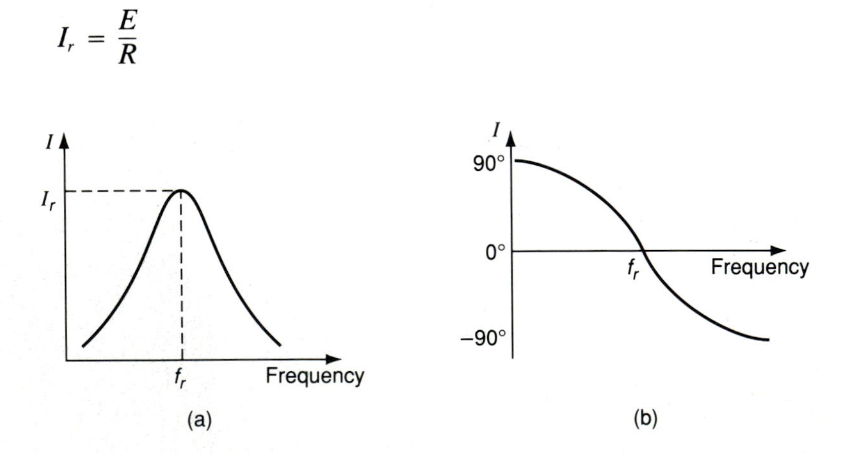

FIGURE 14-45

EXAMPLE 14-45 A 100-V series *RLC* circuit has a resistance of 1200 Ω, an inductance of 0.2 H, and a capacitance of 0.057 μF. Calculate the frequency at which the circuit resonates.

Solution $f_r = \dfrac{1}{2\pi\sqrt{LC}}$

$$= \frac{1}{2\pi\sqrt{(0.2\text{ H})(0.057 \times 10^{-6}\text{ F})}}$$
$$= 1490.62\text{ Hz}$$

EXAMPLE 14-46 A coil with an inductance of 10 mH and a resistance of 2.5 Ω is connected in series with a capacitor and a 24-V 400-Hz supply. Determine (a) the value of capacitance that will cause the circuit to be in resonance; (b) the current at resonant frequency.

Solution (a) $f_r = \dfrac{1}{2\pi\sqrt{LC}}$

$$C = \frac{1}{4\pi^2 f_r^2 L}$$
$$\therefore \quad = \frac{1}{4\pi^2(400)^2(0.01)}$$
$$= 15.83\ \mu\text{F}$$

(b) $I_r = \dfrac{E}{R} = \dfrac{24}{2.5} = 9.6\text{ A}$

The energy stored in a resonant circuit is constant; although the energy stored in the electric field of a capacitor varies from zero to a maximum and back to zero each half-cycle, and the energy stored in the magnetic field of a coil varies in a similar manner, the total energy stored is not changing with time. The resonant frequency is the frequency at which the coil releases energy at the same rate as the capacitor requires it during one quarter-cycle, and absorbs energy at the same rate the capacitor releases it during the next quarter-cycle. Consequently, the external circuit is not required to supply energy to either the coil or the capacitor. Instead, the external circuit supplies the losses that result from resistance in the resonant circuit.

14-20 *Q* OF A SERIES CIRCUIT

The ratio between the reactive power of the capacitance or inductance at resonance and the real power of a resonant circuit is referred to as the ***Q* factor**, Q_s. The Q factor is a measure of the quality of a resonance circuit. A resonance circuit is considered to be of good quality if it is capable of storing a required amount of energy with a minimum of energy loss. The Q factor indicates the amount of energy stored in a circuit compared to the amount of energy dissipated by a circuit. The smaller the level of dissipation, the higher the Q factor.

$$Q_s = 2\pi \frac{\text{maximum energy stored}}{\text{energy dissipated per cycle}}$$

The Q factor of a series circuit is also known as the *voltage magnification factor*. The voltage across the entire circuit is magnified so that it appears across each reactance multiplied by the value of Q. Whenever Q_s is greater than 1, the voltage across the coil and/or the capacitor in the circuit exceeds the supply voltage.

Figure 14-46 shows the voltage drops across a series-connected capacitor, inductor, and resistor. As the supply frequency varies, the voltages V_C, V_L, and V_R also change. At resonant frequency V_R is at its maximum value and is equal to the supply voltage. V_L and V_C are greater than V_R at resonance but are not at their maximum values. This condition does not violate Kirchhoff's voltage law since V_L and V_C are 180° out of phase with each other. In this example, V_C is at its maximum value at a frequency below f_r and V_L is at its maximum at a value above f_r.

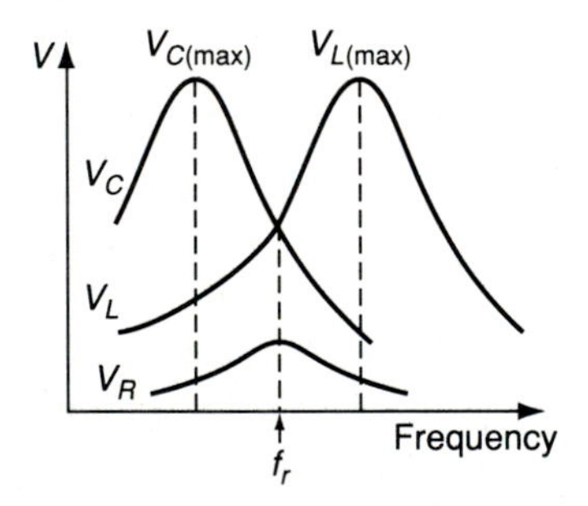

FIGURE 14-46 Voltage relations in a series *RLC* resonant circuit.

In a series resonant circuit containing inductance, the Q factor is calculated using the equation

$$Q_s = \frac{X_L}{R_s} = \frac{\omega_s L}{R_s} \tag{14-72}$$

where Q_s = figure of merit

X_L = inductive reactance at resonant frequency

R_s = resistance in series with X_L

Q is a numerical factor that is not expressed in units. The Q point for a series resonant circuit containing capacitance is determined using the following equation:

$$Q_s = \frac{X_c}{R_s} = \frac{1}{\omega_s C R_s} \tag{14-73}$$

The Q factor in a series resonant circuit containing resistance, inductance, and capacitance is found by the following equation:

$$Q_s = \frac{1}{R}\sqrt{\frac{L}{C}} \tag{14-74}$$

EXAMPLE 14-47 A series resonant circuit consists of a 750-μH inductance with an internal resistance of 9 Ω and a 0.035-μF capacitor. Determine the Q factor of the circuit.

Solution

$$\begin{aligned} Q_s &= \frac{1}{R}\sqrt{\frac{L}{C}} \\ &= \frac{1}{9}\sqrt{\frac{750 \times 10^{-6}}{0.035 \times 10^{-6}}} \\ &= 16.3 \end{aligned}$$

Series resonant circuits in portable radio receivers are tuned by means of a variable capacitor. Automobile radios typically use variable inductors for tuning purposes. In a television receiver, the inductance is varied as a coarse adjustment, and the circuit is fine tuned by varying the capacitance of the resonant circuit. To tune in a specific station using a portable radio receiver it is necessary to vary the capacitance until its reactance at that frequency becomes equal in magnitude to the inductive reactance of the coil in series with it. The desired signal will then send current at its own frequency through the circuit, and it will be considerably larger than currents at any other frequency.

Other signals from broadcast stations which operate at different frequencies on adjacent channels will also appear as voltages applied to the same series resonant circuit. If the circuit has a relatively large Q, that is, if the circuit is selective, the current versus frequency-response curve will be quite steep on both sides of the peak value. Figure 14-47 shows three different values of Q. From this response curve it can be seen that the higher the Q, the higher the current peak at resonance and the steeper the sides of the curve. In a radio circuit, the high Q value results in unwanted currents from other stations being too small to cause interference with reception of the resonant-frequency signal.

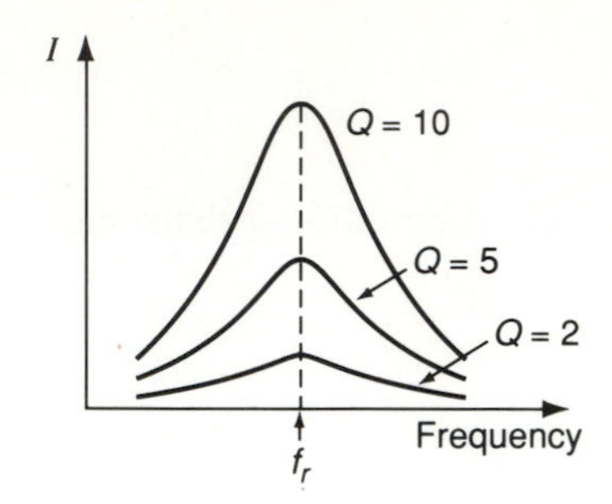

FIGURE 14-47

Figure 14-48 shows a simplified schematic of the first radio-frequency (RF) stage of a radio receiver. Radio signals are broadcast from a transmitter via electromagnetic waves which multiply as they travel through the atmosphere. The action of the electromagnetic waves cutting across the receiving antenna induce small voltages. When a voltage is induced into the antenna it causes a current to flow in the primary winding of the transformer. As a result, a voltage is also induced in the secondary winding of the transformer. By adjusting the variable, or tuning capacitor, the circuit may be made to resonate at any one of a wide variety of radio frequencies.

When a radio signal voltage is induced in the coil of a series circuit tuned to the frequency of the signal voltage, the voltage across the variable capacitor is ''picked off'' and applied to the input of a voltage amplifier. The resonance frequency causes a resonance rise in voltage to appear across the tuning capacitor. This capacitor voltage, which is Q times the induced voltage of the inductor, has the same frequency. Consequently, a very small signal of a few microvolts can be converted to a much larger signal before being amplified by the radio's electronic circuitry.

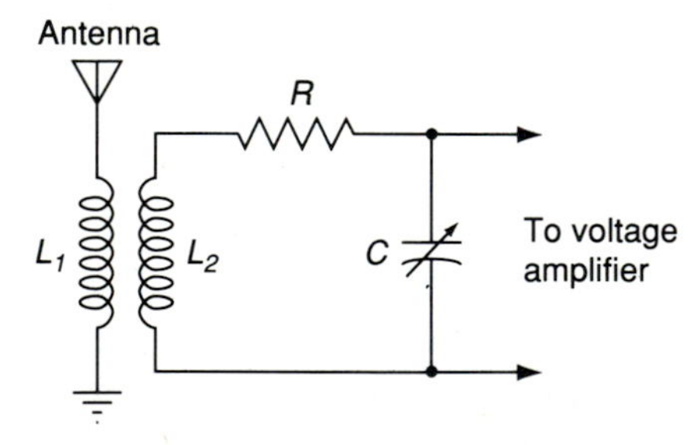

FIGURE 14-48 Tuned series resonant circuit.

EXAMPLE 14-48 In the circuit of Figure 14-48, $L_2 = 100\ \mu H$. Determine the range of adjustment of the tuning capacitor which will allow the circuit to resonate at frequencies ranging from 535 to 1605 kHz.

Solution At $f_r = 535$ kHz:

$$C = \frac{1}{4\pi^2(535 \times 10^3)^2(100 \times 10^{-6})}$$
$$= 884.98 \text{ pF}$$

At $f_r = 1605$ kHz:

$$C = \frac{1}{4\pi^2(1605 \times 10^3)^2(100 \times 10^{-6})}$$
$$= 98.33 \text{ pF}$$

14-21 BANDWIDTH OF RESONANT CIRCUITS

The **bandwidth** (BW) of a series resonant circuit is defined as the total number of cycles below and above the resonant frequency for which the current is equal to or greater than 70.7% of its resonant value. The width of this band of frequencies is also referred to as the **bandpass** of the circuit. The frequency-response curve of the series resonant circuit is shown in Figure 14-49. This curve has a maximum current at resonance. When the current drops to 0.707 of the peak value, the power in the circuit is reduced to 50% of the maximum value. The two frequencies in the curve which are at 0.707 of the maximum current are called **band**, or **half-power**, **frequencies**. These frequencies are identified on the curve as f_1 and f_2, and are often referred to as the **critical frequencies**, or **cutoff frequencies**, of a resonant circuit. f_1 is considered to be the lower half-power frequency and f_2 is taken as the upper half-power frequency. The half-power relationship is derived in the following manner:

$$P_m = (I_m)^2R \tag{14-75}$$

$$P_{f1} = (0.707 \times I_m)^2R$$
$$= 0.5(I_m)^2R$$
$$= 0.5 \times P_m$$

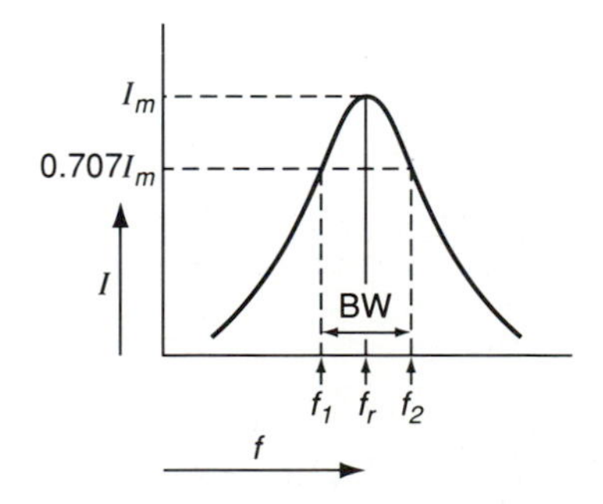

Bandwidth of a series resonant circuit.

The resonant frequency can be determined from the critical frequencies by the equation

$$f_r = \sqrt{f_1 f_2} \tag{14-76}$$

Since bandwidth is considered to be the difference between the two conditions where the impedance of the circuit is minimal, bandwidth can be expressed mathematically as

$$\text{BW} = f_2 - f_1 \tag{14-77}$$

The bandwidth can also be given in terms of resonant frequency and quality factor using the formula

$$\text{BW} = \frac{f_r}{Q} \tag{14-78}$$

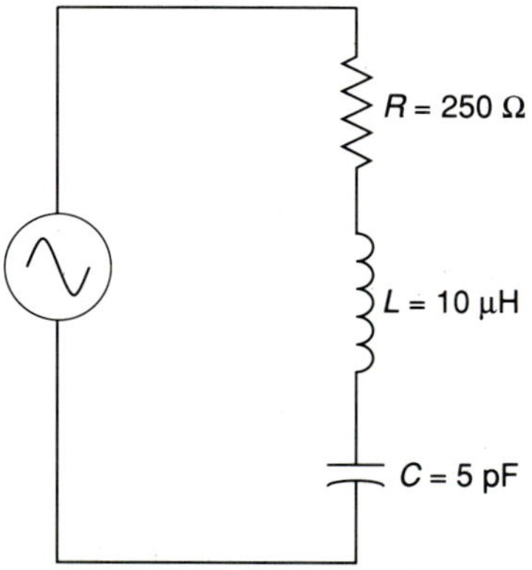

FIGURE 14-50

EXAMPLE 14-49 For the circuit shown in Figure 14-50, calculate the bandwidth at resonant frequency.

Solution

$$f_r = \frac{1}{2\pi\sqrt{LC}} = \frac{1}{2\pi\sqrt{(10 \times 10^{-6}\,\text{H})(5 \times 10^{-12}\,\text{F})}} = 22.51\,\text{MHz}$$

$$X_L = 2\pi f_r L = 2\pi(22.51 \times 10^6\,\text{Hz})(10 \times 10^{-6}\,\text{H}) = 1414.35\,\Omega$$

$$Q_s = \frac{X_L}{R} = \frac{1414.35\,\Omega}{250\,\Omega} = 5.66$$

$$\text{BW} = \frac{f_r}{Q} = \frac{22.51 \times 10^6\,\text{Hz}}{5.66} = 3.98\,\text{MHz}$$

EXAMPLE 14-50 If the applied voltage for the circuit of Figure 14-50 is $24 \angle 0^\circ$ V, determine the power dissipated in the circuit at the half-power frequencies.

Solution

$$Z_T = R = 250\ \Omega$$

$$I = \frac{E}{Z_T} = \frac{24 \angle 0^\circ}{250 \angle 0^\circ} = 0.096 \angle 0^\circ \text{ A}$$

$$P_{f1} = \tfrac{1}{2}P_{\text{max}} = \tfrac{1}{2}I^2_{\text{max}}R = (\tfrac{1}{2})(0.096)^2(250)$$
$$= 1.15 \text{ W}$$

EXAMPLE 14-51 A series resonant circuit has a resistance of 2 kΩ and half-power frequencies of 15 kHz and 125 kHz. Determine (a) the bandwidth; (b) the resonant frequency.

Solution

(a) $\text{BW} = f_2 - f_1 = 125 - 15 = 110 \text{ kHz}$

(b) $f_r = \sqrt{f_1 f_2} = \sqrt{(125)(15)} = 43.3 \text{ kHz}$

A circuit with a frequency-response curve similar to that shown in Figure 14-49 is said to favor signals at, or near, the resonant frequency. The ability of a resonant circuit to select one particular frequency and reject other frequencies is called the **selectivity** of the circuit. The smaller the bandwidth, the higher the selectivity. Generally, a value of Q that is less than 10 is considered to be a wideband frequency, while a Q value greater than 10 is said to be a narrowband frequency.

14-22 RESONANCE IN PARALLEL AC CIRCUITS

When the inductance and capacitance of a parallel circuit have equal values of reactance, the circuit is called a **parallel resonant circuit**. The parallel resonant circuit is usually connected as shown in Figure 14-51. This circuit is often referred to as a **tank circuit,** due to the storage of energy in the inductor and capacitor. In a tank circuit, the resonance frequency is influenced by how much resistance there is in the circuit. Parallel-resonant circuits are designed so that they have as little loss as possible, which means that a Q of a very high value is preferable. When Q is greater than 10, the input impedance of a two-branch, parallel resonant circuit is equal to Q times the reactance of either branch, or it is also equal to Q times the resistance of the loop.

If the resistor shown in Figure 14-51 were removed, there would be a pure inductor connected in parallel with a pure capacitor. In this situation, the line current at resonance would be zero, since the current flowing through the two reactances would be of equal and opposite value. The series equivalent impedance of this circuit becomes infinite at resonance, due to the line current being zero and $Z = E/I$. The equation for the impedance of a parallel resonance circuit is expressed as follows:

$$Z_r = \frac{X_L \times X_C}{R} = \frac{L}{RC} \tag{14-79}$$

where Z_r is the impedance at resonance.

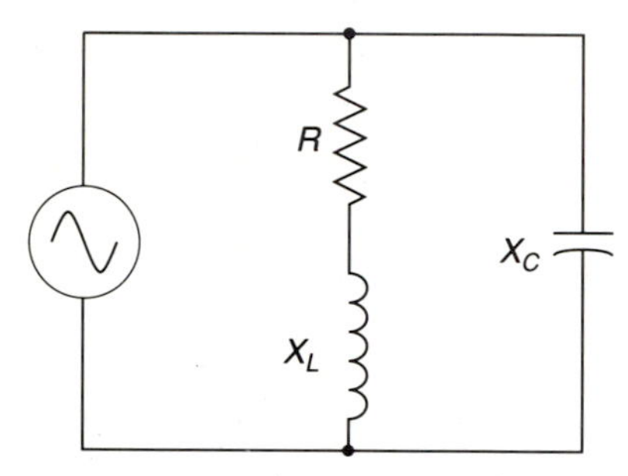

FIGURE 14-51 Parallel resonant circuit.

If the impedance of the tank circuit shown in Figure 14-51 is plotted to a logarithmic frequency base, the impedance will peak at the resonance frequency. Since the condition for parallel resonance is that $I_L = I_C$, which implies that $X_L = X_C$, the resonant frequency for a parallel resonant circuit is calculated in the same manner as for a series resonant circuit. That is,

$$f_r = \frac{1}{2\pi\sqrt{LC}}$$

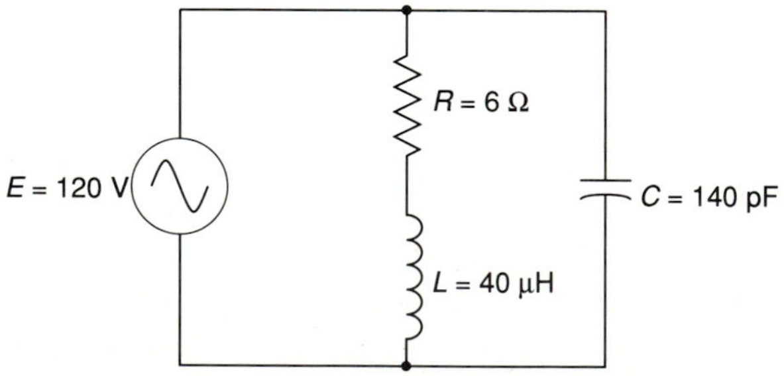

E = 120 V
R = 6 Ω
L = 40 μH
C = 140 pF

FIGURE 14-52

EXAMPLE 14-52 For the circuit shown in Figure 14-52, calculate (a) the resonant frequency; (b) the total impedance; (c) the line current.

Solution (a) $f_r = \dfrac{1}{2\pi\sqrt{LC}} = \dfrac{1}{2\pi\sqrt{(40 \times 10^{-6}\,\text{H})(140 \times 10^{-12}\,\text{F})}}$
$= 2.13\ \text{MHz}$

(b) $Z_r = \dfrac{L}{RC} = \dfrac{40 \times 10^{-6}\,\text{H}}{(6\,\Omega)(140 \times 10^{-12}\,\text{F})} = 47.62\ \text{k}\Omega$

(c) $I = \dfrac{E}{Z} = \dfrac{120\ \text{V}}{47.62 \times 10^3} = 2.52\ \text{mA}$

The frequency-response curves for a parallel *RLC* circuit are shown in Figure 14-53. In Figure 14-53(a) it can be seen that the impedance is at its maximum value, Z_r, at resonance. At frequencies below resonance, the parallel circuit is inductive. At frequencies above resonance, the parallel circuit is capacitive. In Figure 14-53(b), the current is at its minimum value, I_r, at resonance. At very high frequencies, the capacitor acts as a short circuit and the current becomes very large.

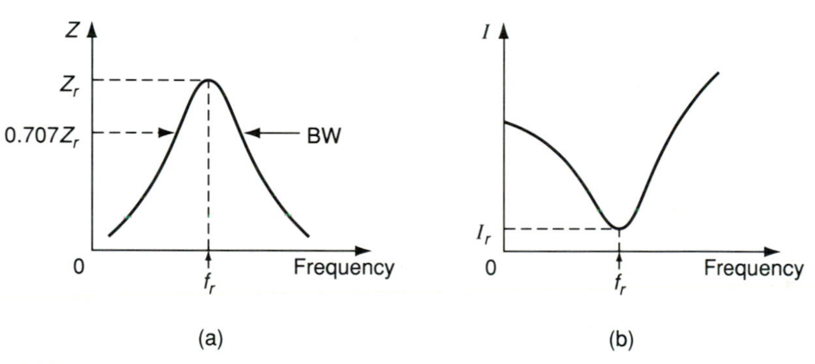

FIGURE 14-53 Parallel resonant circuit frequency-response curves: (a) impedance; (b) current.

At resonance, the current in an *RLC* circuit is determined entirely by the resistance of the circuit. The capacitive and inductive currents can be of any magnitude and their net value will be zero due to phase displacement. Therefore, in a parallel resonant circuit the capacitive and inductive currents can be considerably larger than the total circuit current and still not violate Kirchhoff's current law. In a parallel resonant circuit, the quality factor is referred to as the **current magnification factor**, Q_p. The total circuit current is magnified so that it is Q_p times as large in either reactive component.

The Q factor of a parallel resonant circuit will also have an effect on the bandwidth and selectivity of a circuit, as shown in Figure 14-54. The high-resistance response curve shows that the bandwidth is narrower and the selectivity is greater than for a low-resistance *RLC* circuit. The points f_1 and f_2 represent the critical frequencies of the high-resistance circuit, while f_1, and

f_2, indicate the critical frequencies of the low-resistance response curve.

Since resonant frequency depends entirely on the reactive components in the circuit, by varying R, the Q of the circuit can be adjusted for different applications. For example, in audio equipment that is sensitive to radio-frequency interference the circuit Q can be reduced by decreasing the resistance of the parallel resonant circuit. This effect is known as **damping**, and variable resistors in parallel resonant circuits are often referred to as **damping resistors.**

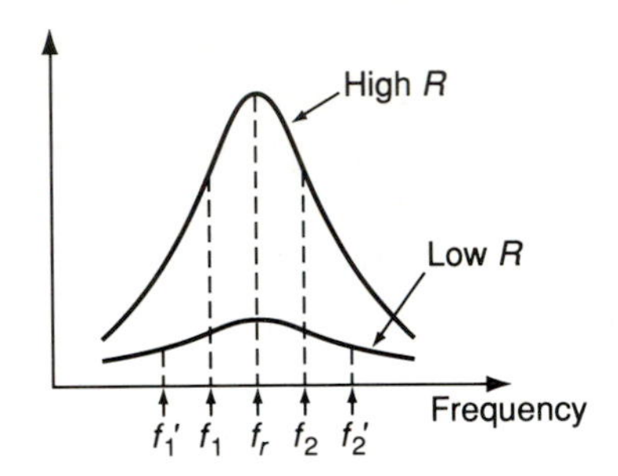

FIGURE 14-54 Effect of resistance on bandwidth and selectivity.

14-23 THE DECIBEL

The logarithm of the ratio of two voltages, currents, or power levels is typically measured in a unit known as the **bel**. Named after Alexander Graham Bell, this unit of measure was originally used to provide comparisons between the power-loss properties of various telephone lines. In most electronic circuits, the bel represents too large a unit for practical applications. For this reason, the **decibel** (dB), which is one-tenth of a bel, is used.

$$\text{dB} = 10 \log_{10} \frac{P_{\text{out}}}{P_{\text{in}}} \tag{14-80}$$

When the output power and the input power are equal, the power ratio is 1 and the decibel level is 0 dB. In situations where the input power is greater than the output power, the power ratio is less than 1 and the decibel level is negative. The reduction in power level of a signal is referred to as **attenuation**. When the power level is greater than 1, the increase in power level of the signal is called **gain**.

EXAMPLE 14-53 The input to a power amplifier is 0.1 W and the output is 20 W. What is the gain of the amplifier in dB?

Solution

$$\begin{aligned} \text{dB} &= 10 \log_{10} \frac{P_{\text{out}}}{P_{\text{in}}} \\ &= 10 \log_{10} \frac{20}{0.1} \\ &= 23.01 \end{aligned}$$

Since decibels are logarithmic, the gains expressed in decibels can be added instead of multiplied. The following example illustrates this procedure.

EXAMPLE 14-54 A directional antenna has a gain of 4 dB and the lead-in coaxial cable has a loss of 1.5 dB. The cable is fed to a tuner-preamplifier with a gain of 54 dB, and then to an amplifier with a 10-dB gain. The system is connected to a speaker with an 8-dB loss. Determine the total system gain.

Solution

Antenna gain = +4.0 dB
Coaxial cable = −1.5 dB
Tuner-preamp = +54.0 dB
Amplifier = +10.0 dB
Speaker = −8.0 dB
Total gain = 58.5

Although decibels are used primarily in expressing power gain and loss, you will also see them used on many occasions to express the voltage or current gain or loss in a circuit. The gain of an amplifier is often expressed as a voltage ratio rather than a power ratio. The formulas for expressing the ratio of two voltages or two currents in decibels are

$$\text{dB} = 20 \log_{10} \frac{V_{\text{out}}}{V_{\text{in}}} \tag{14-81}$$

$$\text{dB} = 20 \log_{10} \frac{I_{\text{out}}}{I_{\text{in}}} \tag{14-82}$$

The most important thing to remember about using current and voltage values is that the decibel figure is meaningless unless the input and output impedances are equal.

EXAMPLE 14-55 At a particular frequency the input voltage to the four-terminal network shown in Figure 14-55 is 5 V and the output voltage is 18 V. If the input and output impedances are equal, determine the voltage ratio in decibels.

Solution

$$\text{dB} = 20 \log_{10} \frac{V_{\text{out}}}{V_{\text{in}}}$$

$$= 20 \log_{10} \frac{18}{5}$$

$$= 11.13 \text{ dB}$$

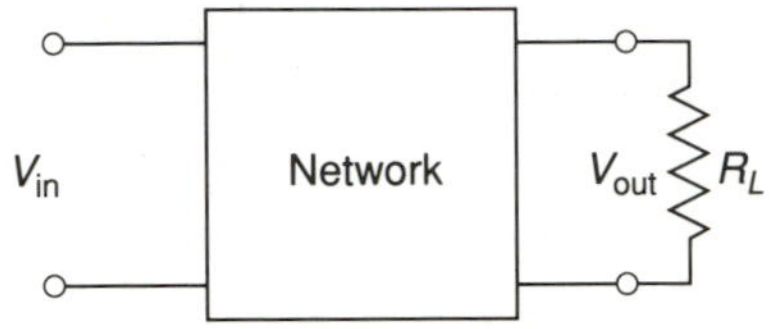

FIGURE 14-55 Four-terminal network.

14-24 FILTER CIRCUITS

Filter circuits, called **filters**, are used to block or pass a specific range of frequencies. Filters are either passive or active. **Passive filters** consist of resistors, capacitors, and/or inductors, and do not contain an amplifying device. An **active filter** uses some type of amplifying device, such as a transistor and/or operational amplifier, in combination with resistors and capacitors to obtain the desired filtering effect.

Filters can be classified into four main types depending on which frequency components of the input signal are passed on to the output signal. The four types of filters are low-pass, high-pass, band-pass, and band-stop.

Low-pass filters

A low-pass filter is a circuit that has a constant output voltage up to a cutoff, or critical, frequency, f_c. The frequencies above f_c are effectively shorted to ground and are described as being in the attenuation band, or **stopband**. The frequencies below f_c are said to be in the **passband**. Figure 14-56 shows a basic low-pass filter using an *RC* network. At low frequencies, X_c is greater than R, and most of the input signal will appear at the output. The cutoff frequency occurs when $X_c = R$. At this frequency, $V_{\text{out}} = 0.707V_{\text{in}}$.

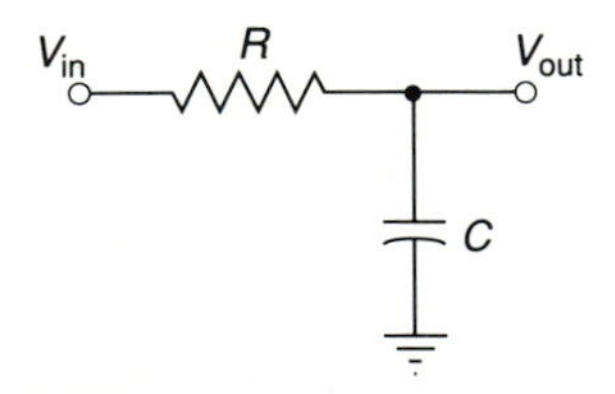

FIGURE 14-56 *RC* low-pass filter.

The response curve for a low-pass filter is shown in Figure 14-57. When the output voltage is 0.707 of the input, the power output is 50% of the input power. This condition exists at f_c, and for this reason the cutoff frequency is also referred to as the **half-power point**. If $P_{out} = 1/2\ P_{in}$, a 3-dB loss of signal has occurred.

The cutoff frequency of a low-pass filter is found by the equation

$$f_c = \frac{1}{2\pi RC}\ \text{Hz} \tag{14-83}$$

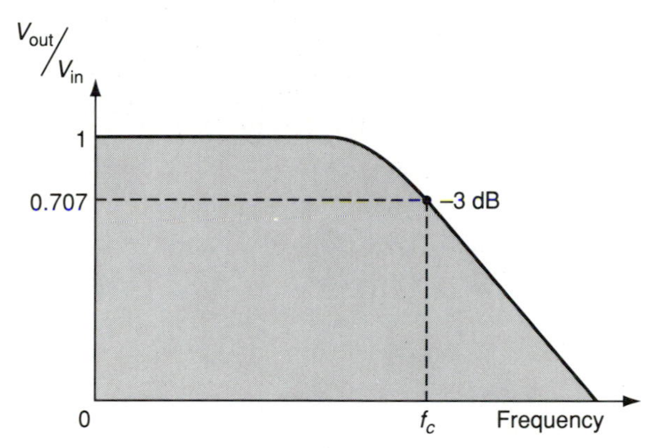

FIGURE 14-57 Response curve for low-pass filter.

EXAMPLE 14-56 Determine the cutoff frequency of the circuit of Figure 14-56 when $C = 0.01\ \mu\text{F}$ and $R = 2.2\ \text{k}\Omega$.

Solution

$$\begin{aligned} f_c &= \frac{1}{2\pi RC} \\ &= \frac{1}{2\pi(2200\ \Omega)(0.01 \times 10^{-6}\ \text{F})} \\ &= 7.23\ \text{kHz} \end{aligned}$$

High-pass filters

A high-pass filter allows frequencies above the cutoff frequency to appear at the output, but blocks or shorts to ground all those frequencies below the −3-dB point on the curve. By interchanging the two circuit elements of the low-pass *RC* filter we obtain the equivalent high-pass filter shown in Figure 14-58. Once again the capacitive reactance is large at low frequencies and small at high frequencies. The capacitor in Figure 14-58 is occasionally referred to as a **coupling capacitor**. The cutoff frequency for an *RC* high-pass filter is determined using the same equation as for the *RC* low-pass filter. That is,

$$f_c = \frac{1}{2\pi RC}$$

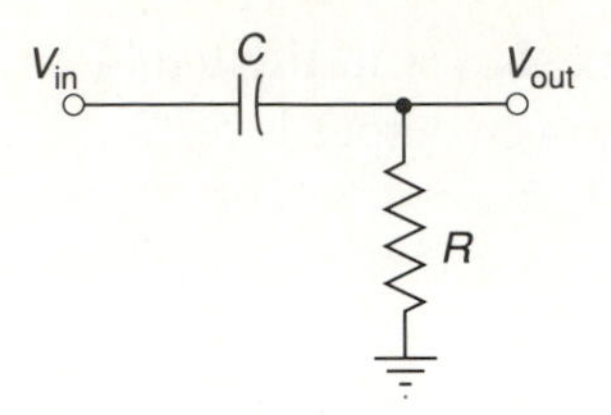

FIGURE 14-58 *RC* high-pass filter.

An example of an *RL* high-pass filter is shown in Figure 14-59. In this circuit there is a high value of inductive reactance at high frequencies. The result is that the high frequencies produce a large output voltage across the inductor. At low frequencies, X_L will be small, and the low frequencies will be effectively shorted to ground. It should be noted that *RL* filters are not used in active circuits, due to their size, weight, and losses.

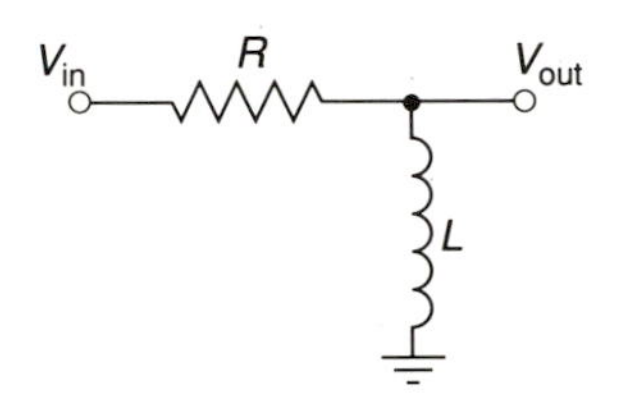

FIGURE 14-59 *RL* high-pass filter.

The equation for determining f_c is the same for both the high-pass and low-pass *RL* circuits. That is,

$$f_c = \frac{R}{2\pi L} \tag{14-84}$$

Figure 14-60 shows the response curve for a high-pass filter circuit. The cutoff frequency on the high-pass response curve is called the low cutoff frequency.

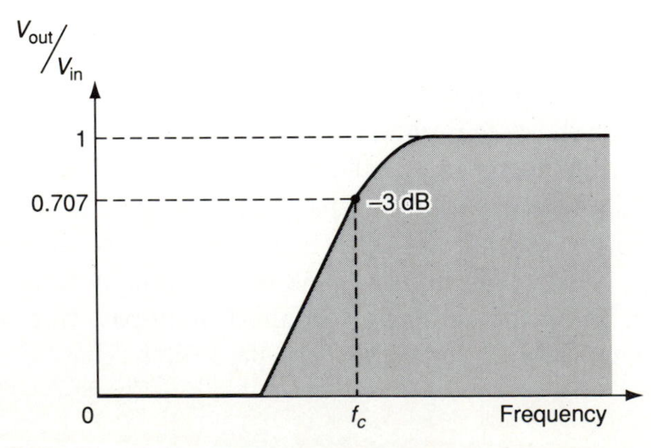

FIGURE 14-60 High-pass filter response curve.

EXAMPLE 14-57 Design an *RL* high-pass filter that will have a cutoff frequency of 1500 Hz.

Solution Assume that $R = 1\ k\Omega$.

$$f_c = \frac{R}{2\pi L}$$

Therefore,

$$L = \frac{R}{2\pi f_c} = \frac{1000\ \Omega}{2\pi(1500\ Hz)} = 106.1\ mH$$

Band-pass filters

A band-pass filter is a circuit that allows a certain range or band of frequencies to pass through it relatively unattenuated. A band-pass filter is equivalent to combining a low-pass filter and a high-pass filter. The resulting response curve would appear as shown in Figure 14-61. The cutoff frequency of the high-pass section becomes the lower-frequency limit in the passband, f_1. The upper frequency in the passband, f_2, is due to the cutoff frequency in the low-pass section. The passband, or bandwidth, is the difference between f_2 and f_1. At resonant frequency, f_r, $X_L = X_c$ and the output voltage is equal to the input signal.

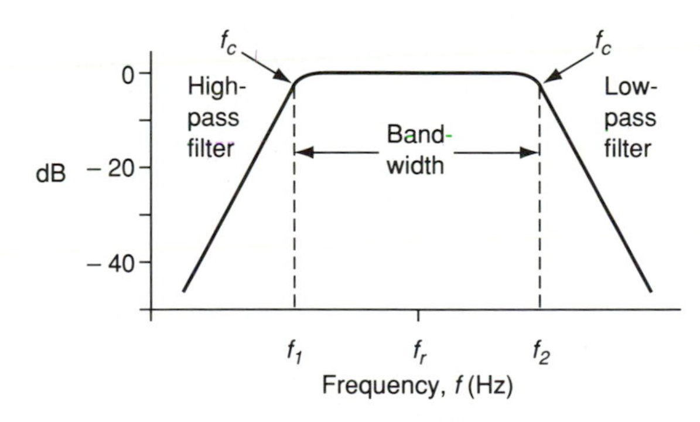

FIGURE 14-61 Response curve of a band-pass filter.

Depending on the design and purpose of the filter, the bandwidth may be very wide or very narrow. Some of the common uses of bandpass filters are as follows:

1. Audio: to equalize sound levels.
2. Communications: to select a narrow band of radio frequencies.
3. Audio: speaker crossover networks.

Band-stop filters

A band-stop filter rejects signals at frequencies within a specified band and passes signals at all other frequencies. This type of filter is also referred to as a band-reject, notch filter, wavetrap, band-elimination, or band-suppression filter. Like the band-pass filter, the band-stop filter may also be formed by combining a low-pass and a high-pass filter. However, the band-stop filter is designed so that the cutoff frequency of the low-pass filter is below that of the high-pass filter, as shown in Figure 14-62. In this response curve, the band-pass filter rejects frequencies between 100 Hz and 10 kHz.

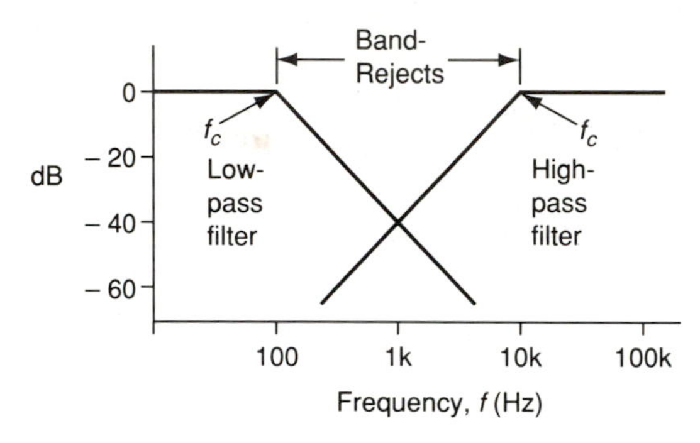

FIGURE 14-62 Response curve of band-stop filter.

Figure 14-62 represents a straight-line approximation of the magnitude and phase angle response to frequency. These approximations of actual curves are called **Bode plots**, named after Hendrik W. Bode. The sharp corners, which occur at cutoff frequency, are called **breakpoints**. The breakpoints are the points at which the gain drops to 3 dB below the midfrequency gain. The midfrequency gain of the plot is the center section, which has a fairly constant gain.

To draw a Bode plot properly, you must use **semilog** graph paper. The semilog paper has linear graduations in the vertical axis and logarithmic graduations in the horizontal axis. The linear axis is labeled in units of decibels and the log axis is labeled in units of frequency, radians per second, or angular velocity. The Bode diagram is plotted by drawing the midfrequency gain from breakpoint to breakpoint. From each breakpoint straight lines are then drawn with slopes of −20 dB per decade. A **decade** represents a tenfold change in frequency. A slope of −20 dB/decade is also equal to a change of −6 dB/octave. An **octave** is a doubling of frequency. For example, if the frequency changes from 1000 Hz to 2000 Hz, it has changed one octave.

14-25 CONSTANT-*k* FILTERS

Constant-*k* filters are *LC* circuits in which the product of the reactances at any frequency is constant. These types of filters, consisting of capacitors with negligible leakage and inductors with negligible leakage, are used extensively in communication systems. The constant-*k* filter permits maximum power transfer to a constant load within the passband.

Figure 14-63 shows the attenuation curves for two different types of filters. Curve A represents the response curve for a low-pass *RC* or *RL* filter, and curve B represents a low-pass constant-*k* filter. The *RC* and *RL* filters discussed in Section 14-23 have a wider attenuation curve and produce more insertion loss than do constant-*k* filters. **Insertion loss** is the difference between the power received at the load before insertion (installation) of a filter and the power after insertion. This loss is stated as the log of a ratio of power output to power input and is the result of power loss due to resistance in the circuit. In a constant-*k* filter, the resistance is negligible and consequently there is no significant power loss.

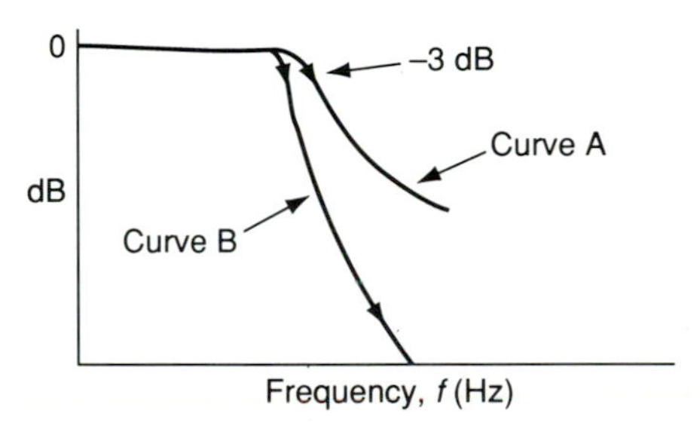

FIGURE 14-63 Attenuation curves. Curve A represents a low-pass *RC* or *RL* filter. Curve B is the attenuation curve for constant-*k* reactive filter.

There are three main types of constant-*k* filters: L-type, T-type, and π-type. Figure 14-64 shows an L-type, high-pass, and low-pass constant-*k* filter. A common application of this type of filter is the crossover network to the woofer and tweeter speakers in an audio system.

The T-type constant-*k* filter increases the sharpness of attenuation by dividing the inductance of capacitance values into two parts. The T-type filter is generally used when the filter is to be connected to a low-impedance source. Figure 14-65 shows a T-type, constant-*k*, low-pass and high-pass filter.

A Π-type constant-*k* filter is generally used when connecting to a high-impedance source. Figure 14-66 shows a Π-type, constant-*k*, low-pass and high-pass filter.

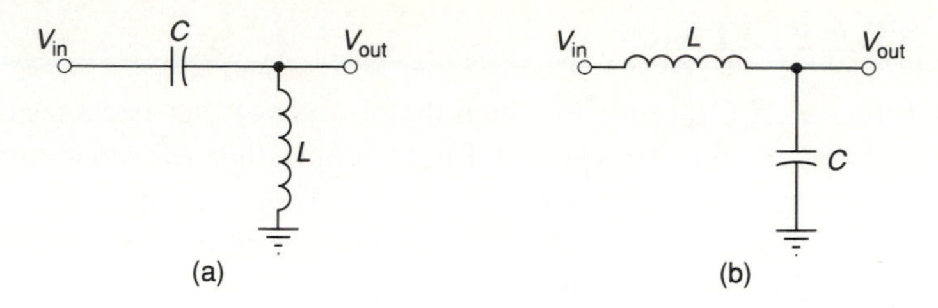

FIGURE 14-64 L-type constant-k filter: (a) high-pass; (b) low-pass.

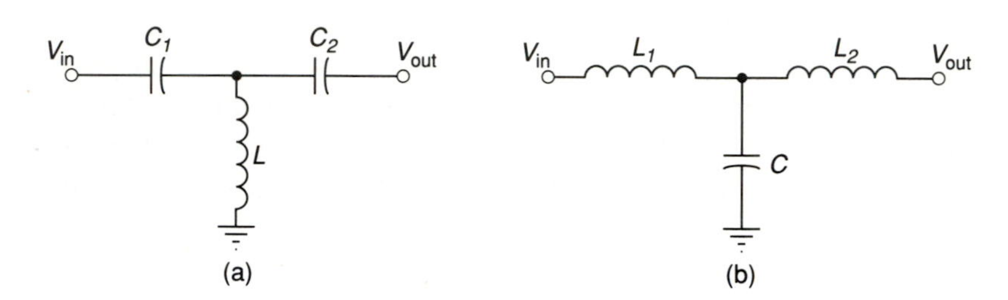

FIGURE 14-65 T-type constant-k filter: (a) high-pass; (b) low-pass.

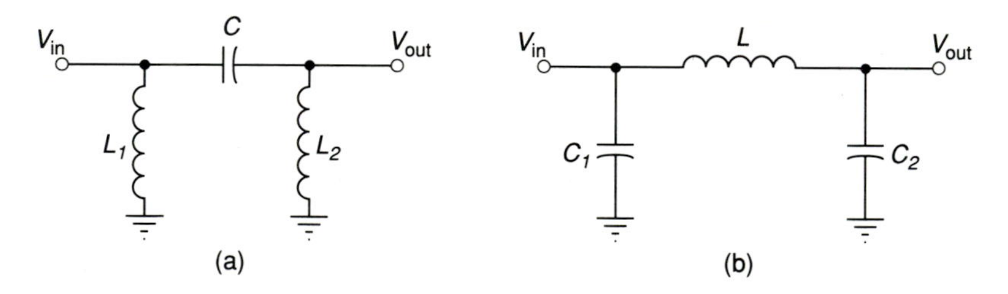

FIGURE 14-66 Π-type constant-k filter: (a) high-pass; (b) low-pass.

The values of inductance or capacitance to be used in a low-pass L-type constant-k filter to produce a specific value of cutoff frequency for a constant load (R_L) are found by

$$L = \frac{R_L}{\pi f_c} \tag{14-85}$$

$$C = \frac{1}{\pi R_L f_c} \tag{14-86}$$

The equations to determine the value of inductance or capacitance to be used in a high-pass L-type constant-k filter to produce a specific cutoff frequency for a constant load (R_L) are

$$L = \frac{R_L}{4\pi f_c} \tag{14-87}$$

$$C = \frac{1}{4\pi f_c R_L} \tag{14-88}$$

Equations 14-85 through 14-88 must be slightly modified to be applied to T-type and Π-type filters. The following modifications are required:

High-pass filter:

1. The value of the series components in the T-type filter must be multiplied by 2.
2. The value of the shunt components in the Π-type filter must be multiplied by 2.

Low-pass filter:

1. The value of the series components in the T-type filter must be divided by 2.
2. The value of shunt components in the Π-type filter must be divided by 2.

EXAMPLE 14-58 Design a low-pass constant-*k* L-type filter that is to be connected to an 8-Ω load with a cutoff frequency of 2.5 kHz.

Solution

$$L = \frac{R_L}{\pi f_c} = \frac{8\,\Omega}{\pi(2500\text{ Hz})} = 1.02\text{ mH}$$

$$C = \frac{1}{\pi R_L f_c} = \frac{1}{\pi(8\,\Omega)(2500\text{ Hz})} = 15.92\,\mu\text{F}$$

14-26 ACTIVE FILTERS

Active filters are rapidly replacing *RC*, *RL*, and constant-*k* passive filters in almost every application. Although passive *LC* filters are capable of producing sharp attenuation curves, the inductors are bulky and expensive. An increasingly feasible alternative is the active filter, which uses an operational amplifier (op-amp). The introduction of the op-amp has completely eliminated the need for inductors in filter circuits, resulting in a large reduction in space and cost. The design of active filters to fit several specifications has been reduced to a matter of looking up circuit component values in tables.

An op-amp consists of many direct-coupled transistors built into one integrated circuit (IC) to produce a very high voltage gain. Figure 14-67 shows the schematic symbol for an op-amp. The two inputs for the op-amp are identified as inverting (−) and noninverting (+). The following properties are associated with an ideal op-amp.

1. Infinite voltage gain ($A_v \rightarrow \infty$)
2. Infinite input impedance ($Z_{\text{in}} \rightarrow \infty$)
3. Zero output impedance ($Z_{\text{out}} \rightarrow 0$)
4. Infinite bandwidth

In practice, none of these properties are achievable. However, they can be approximated closely for many filtering applications.

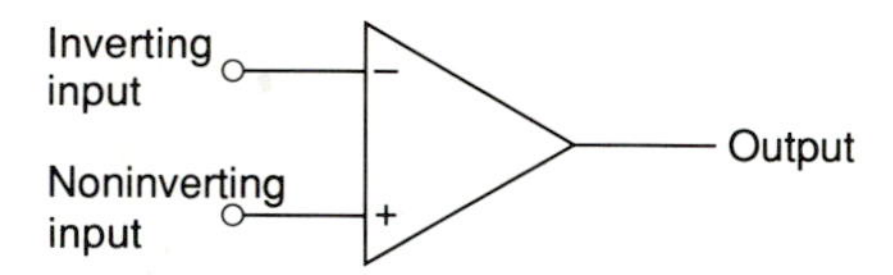

FIGURE 14-67 Op-amp terminals.

Two basic active filter circuits are shown in Figure 14-68. For these circuits, the gain is found by Z_2/Z_1. In Figure 14-68(a), $Z_2 = C$ and R_2, while $Z_1 = R_1$. This circuit is a low-pass active filter. At low frequencies, the gain approaches R_2/R_1 because the capacitive reactance is large enough that it has a negligible effect in parallel with R_2. At high frequencies, the reactance is small enough that Z_2 is dominated by the reactive component and the gain approaches zero. The cutoff frequency for the low-pass active filter is found by

$$f_c = \frac{1}{2\pi R_2 C}$$

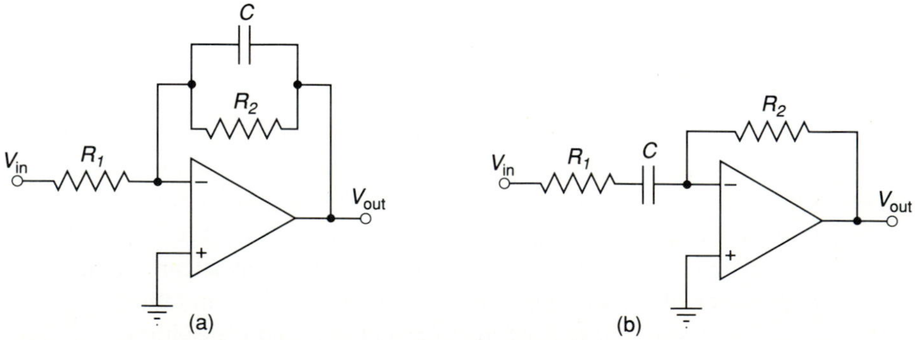

FIGURE 14-68 Basic active filters: (a) low-pass; (b) high-pass.

A high-pass active filter is shown in Figure 14-68(b). In this circuit, the gain approaches zero at very low frequencies and approaches R_2/R_1 at high frequencies. The cutoff frequency for this circuit is

$$f_c = \frac{1}{2\pi R_1 C}$$

A filter with one *RC* network that produces a -20-dB/decade rate of attenuation is referred to as **single-pole**, or **first-order**, **filter**. A **two-pole filter** uses two *RC* networks to produce a roll-off rate of -40-db/decade. Figure 14-69 shows an example of a two-pole active low-pass filter.

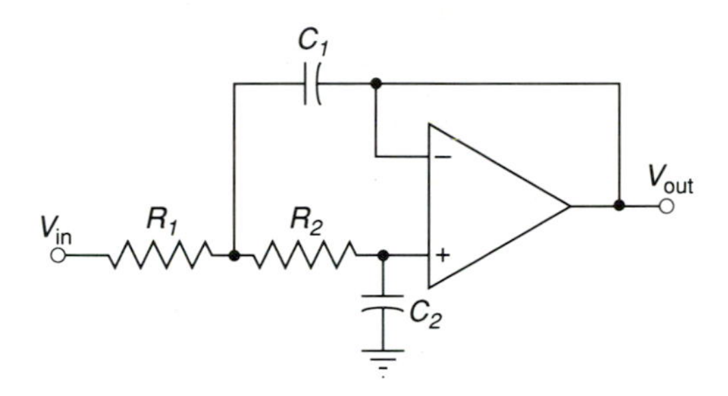

FIGURE 14-69 Two-pole active low-pass filter.

One of the *RC* networks in Figure 14-69 is made up of R_1 and C_1, and the other *RC* network is formed by R_2 and C_2. The cutoff frequency of this filter is found using the following equation:

$$f_c = \frac{1}{2\pi\sqrt{R_1 R_2 C_1 C_2}} \tag{14-89}$$

KEY TERMS

Single-phase circuit
Inductive reactance
Capacitive reactance
Angular velocity
Impedance
Vector diagram
Impedance triangle
Admittance
Conductance
Susceptance
Current per volt
Admittance triangle
Equivalent series circuit
Equivalent parallel circuit
Average (real) power
Storage element
Apparent power
Volt-ampere
Power factor
Power factor angle
Lagging power factor
Leading power factor
Reactive power
Wattless power
Volt-ampere-reactive
Power triangle
Power factor correction
Skin effect
Ac resistance
Effective resistance
Radiation loss
Resonance
Resonant frequency
Response curve
Q factor
Bandwidth
Bandpass
Band (half-power) frequency
Critical frequency
Selectivity
Parallel resonant circuit
Tank circuit
Current magnification factor
Damping
Damping resistor
Bel
Decibel
Attenuation
Gain
Filter
Passive filter
Active filter
Stopband
Passband
Half-power point
Coupling capacitor
Bode plot
Breakpoint
Semilog
Decade
Octave
Insertion loss
Single-pole (first order) filter
Two-pole filter

PROBLEMS

14-1 A 2.2-kΩ resistor is connected to an ac source. The applied emf of the circuit is $e = 170 \sin(377t)$. Determine (a) the rms value of current through the resistor; (b) the instantaneous value of voltage at $t = 0.0025$ s.

14-2 If a 2.2-kΩ resistor is connected to an applied emf with $e = 100 \sin 100\pi t$, determine the instantaneous voltage at $t = 0.008$ s.

14-3 Determine the inductive reactance of a 5.5-mH choke coil at 25 kHz.

14-4 The reactance of a coil at 60 Hz is 3350.1 Ω. Determine the inductance.

14-5 When an inductor is connected to a 24-V 60-Hz supply, an rms current of 40 mA flows. Determine the inductance of the coil.

14-6 A voltage of $120 \angle 45°$ V is applied across a purely inductive load with a reactance of 8 Ω. Determine the resulting current.

14-7 The instantaneous voltage across a 25-mH choke is $e = 22 \sin(377t)$. Determine the instantaneous current when $t = 10$ ms.

14-8 The equation for a sinusoidal voltage is $e = 200 \sin(5250t + 90°)$. Determine the instantaneous voltage at $t = 100$ μs.

14-9 Two inductors, $X_{L1} = 75\ \Omega$ and $X_{L2} = 100\ \Omega$, are connected in series to a 24-V supply. Determine (a) the total inductive reactance; (b) the total current; (c) the voltage drop across L_1; (d) the voltage drop across L_2.

14-10 Two parallel-connected inductors, $X_{L1} = 350\ \Omega$ and $X_{L2} = 500\ \Omega$, are connected to a 208-V supply. Calculate (a) the total inductive reactance; (b) the current through L_1; (c) the current through L_2; (d) the total current in circuit.

14-11 Determine the capacitive reactance of a 0.68-μF capacitor when connected to a 1200-Hz supply.

14-12 A capacitor connected to a 60-Hz supply has a capacitive reactance of 2285 Ω. Determine its capacitance.

14-13 A capacitor has a reactance of 600 Ω at 1 kHz. Find its capacitance when connected to a 60-Hz supply.

14-14 Two capacitors are connected in series to a 24-V 60-Hz supply. C_1 has a capacitance of 10 μF and $C_2 = 15\ \mu$F. Determine (a) the total capacitive reactance of circuit; (b) the total current; (c) the voltage drop across C_1 and C_2.

14-15 Two capacitors, $C_1 = 220\ \mu$F and $C_2 = 300\ \mu$F, are connected in parallel across a 120-V 60-Hz source. Find (a) the total capacitive reactance; (b) the current through each parallel path; (c) the total current.

14-16 Determine the impedance of an ac load having a load voltage of $12 \angle 60°$ V and a load current of $30 \angle 25°$ mA.

14-17 Calculate the impedance of an ac load having an instantaneous load voltage of $e = 64 \sin(\omega t + 45°)$ V and an instantaneous current of $i = 85 \sin(\omega t + 30°)$ mA.

14-18 The impedance of a circuit is given as $Z = 600 \angle -75°\ \Omega$. Determine the amount of resistance and reactance in the circuit.

14-19 A series *RL* circuit consists of a 1200-Ω resistor and a 5-H coil connected to a 208-V 60-Hz supply. Determine (a) the circuit impedance; (b) the voltage drop across the resistor and inductor.

14-20 A 25-mH coil and a 200-Ω resistor are connected in series to a 24-V 1-kHz supply. Calculate the magnitude and phase angle of the current with respect to the voltage of the circuit.

14-21 A *RL* series circuit consists of an applied voltage of $120 \angle 0^\circ$ V, 60 Hz, and the current through the circuit is $2.7 - j3.4$ A. Determine the value of resistance and reactance of the circuit.

14-22 A series *RC* circuit consisting of a 1.2-kΩ resistor and a 6-μF capacitor is connected to a $12 \angle 0^\circ$ V, 60-Hz power supply. Determine (a) the impedance; (b) the current that will flow through the circuit.

14-23 What size capacitor must be connected to a 200-Ω resistor to cause a current of 100 mA to flow in a series *RC* circuit? The supply voltage is 120 V, 60 Hz.

14-24 An ac *RLC* series circuit has the following values: $R = 10\ \Omega$, $X_C = 6\ \Omega$, and $X_L = 12\ \Omega$. The supply voltage is 24 V, 60 Hz. Determine (a) the circuit impedance; (b) the circuit current; (c) the value of V_R, V_L, and V_C.

14-25 Determine the total impedance of a 1-kHz *RLC* series circuit with the following components: $R = 2.2\ \text{k}\Omega$, $L = 0.1$ H, and $C = 0.68\ \mu$F.

14-26 A series *RLC* circuit has the following values: $R = 8\ \Omega$, $C = 220\ \mu$F, and $L = 25$ mH. If the circuit has an instantaneous voltage of $e = 17 \sin(377t)$ V, determine the instantaneous current.

14-27 A 300-Ω resistor and an inductor with a reactance of 250 Ω are connected in series across a $120 \angle 0^\circ$ V supply. Using the voltage divider rule, calculate the voltage drop across each component.

14-28 A 12-Ω resistor is connected in parallel with a 100-μF capacitor. If the applied voltage is $120 \angle 0^\circ$ V, 60 Hz, determine the total current.

14-29 For the circuit shown in Figure 14-70, calculate the current flowing through each branch and the total impedance of the circuit.

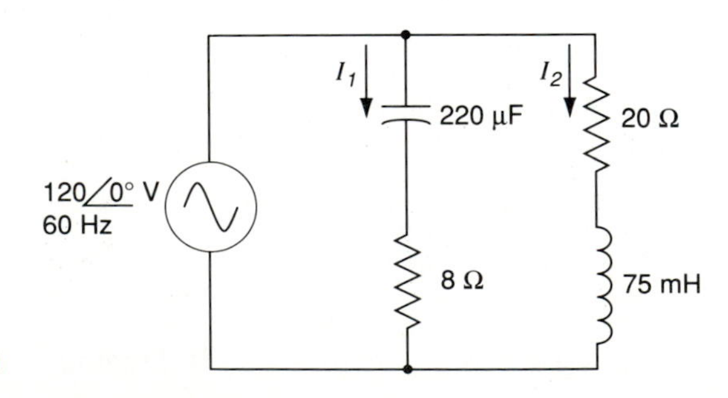

FIGURE 14-70

14-30 Determine the total impedance of the circuit shown in Figure 14-71.

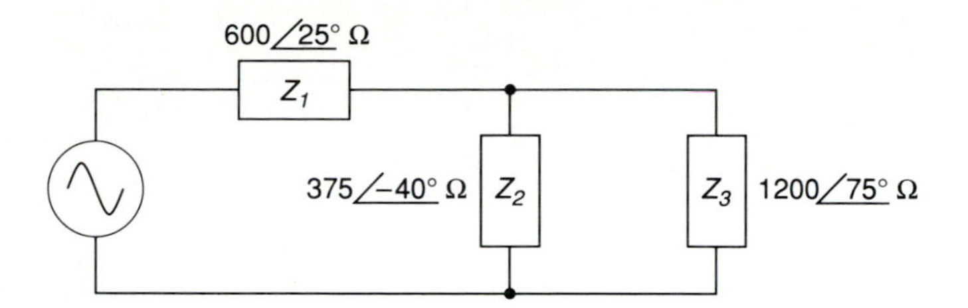

FIGURE 14-71

14-31 Calculate the resistance and reactance values of a series equivalent circuit that may be substituted for a parallel circuit containing a resistance of 12 Ω and reactance of 18 Ω.

14-32 The circuit shown in Figure 14-72 contains the following values: $R = 20\ \Omega$, $X_C = 50\ \Omega$, and $X_L = 80\ \Omega$. Determine (a) the susceptance; (b) the conductance; (c) the admittance; (d) the total current of circuit.

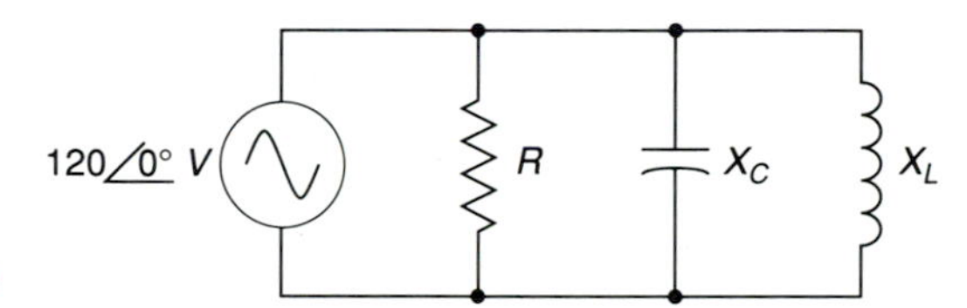

FIGURE 14-72

14-33 Determine the average power dissipated by the circuit shown in Figure 14-73.

FIGURE 14-73

14-34 Calculate the average power dissipated by a circuit having an instantaneous voltage of $e = 26 \sin(\omega t + 45°)$ V and an instantaneous current of $i = 0.5 \sin(\omega t - 15°)$ A.

14-35 An ac circuit has an impedance of $200\angle 60°\ \Omega$, and an rms voltage of $50\angle -25°$ V. Determine (a) the true power; (b) the apparent power of the circuit.

14-36 Determine the reactive power of a 2-H inductor drawing 400 mA from a 60-Hz power supply.

14-37 For the circuit shown in Figure 14-74, determine the total (a) the true power; (b) the reactive power; (c) the apparent power; (d) the power factor.

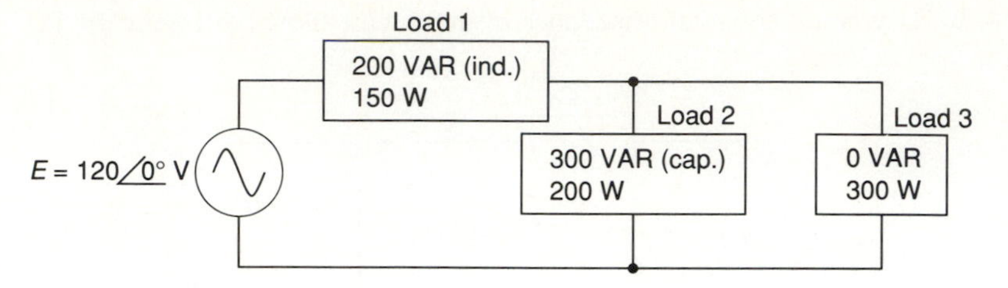

FIGURE 14-74

14-38 A load with an 85% lagging power factor draws 800 W from a 120-V 60-Hz source. Determine the size of capacitor required to improve the power factor to unity.

14-39 A 24-V series *RLC* ac circuit has the following values: $R = 8\ \Omega$, $L = 5$ H, and $C = 0.068\ \mu$F. Determine the frequency at which the circuit resonates.

14-40 A coil with an inductance of 5 mH and a resistance of 1.2 Ω is connected in series with a capacitor and a 12-V 1-kHz supply. Determine (a) the value of capacitance that will cause the circuit to be in resonance; (b) the current at resonant frequency.

14-41 Determine the Q factor of an ac circuit with a 20-mH inductor having an internal resistance of 5 Ω connected in series with a 22-μF capacitor.

14-42 An ac series *RLC* circuit consists of the following components: $R = 80\ \Omega$, $L = 0.2$ mH, and $C = 100$ pF. Calculate the bandwidth at resonant frequency.

14-43 A series circuit has a resonant frequency of 100 kHz and a bandwidth of 50 kHz. Calculate the half-power frequencies.

14-44 For the circuit shown in Figure 14-75, determine: (a) the resonant frequency; (b) the total impedance; (c) the line current.

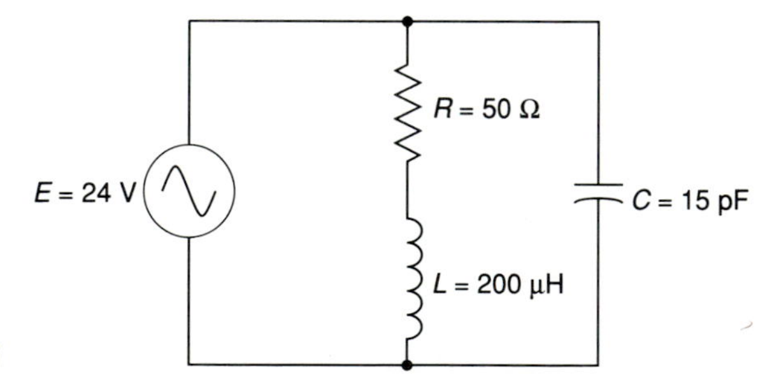

FIGURE 14-75

14-45 The output of a high-pass filter increases from 25 mV to 200 mV as the frequency increases from 5 kHz to 25 kHz. If the input level remains constant, calculate the dB increase.

14-46 Determine the power gain in dB of an amplifier in which the input power is 20 mW and the output power is 100 W.

14-47 The input power to a communications cable is 1 W and the output is 200 mW. Determine the cable loss in decibels.

14-48 How much output power will be produced by an amplifier capable of 30 dB gain if fed an input of 20 mW?

14-49 A power amplifier increases a signal level by 12 dB. If the output signal level is 40 W, calculate the input level.

14-50 Determine the cutoff frequency of the circuit of Figure 14-76 when $C = 0.68\ \mu F$ and $R = 1\ k\Omega$.

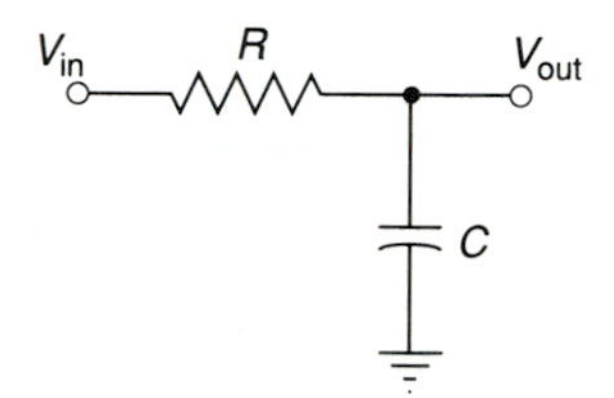

FIGURE 14-76

14-51 An *RL* high-pass filter consists of a resistor $R = 470\ \Omega$ and an inductor $L = 50$ mH. Calculate the cutoff frequency.

14-52 A band-pass filter is to be constructed using a high-pass *RC* filter and a low-pass *RC* filter. The high-pass filter has a resistor $R = 2\ k\Omega$ and capacitor $C = 10$ nF. The low-pass filter has $R = 3\ k\Omega$ and $C = 40$ nF. Determine the bandwidth of the circuit.

14-53 Design a high-pass constant-*k* L-type filter that is to be connected to a 4-Ω load with a cutoff frequency of 1200 Hz.

14-54 For the circuit shown in Figure 14-77, calculate the cutoff frequency.

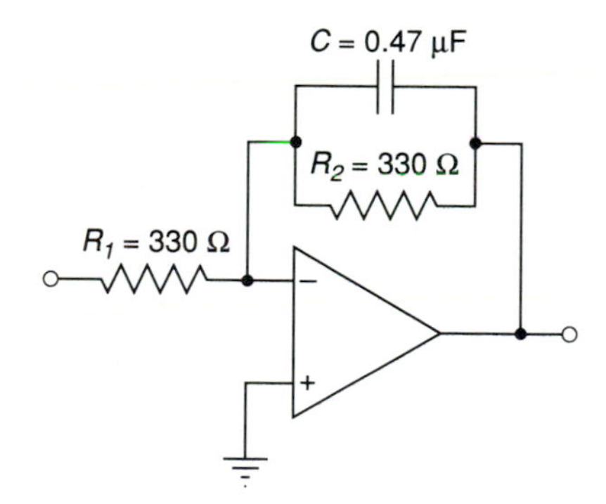

FIGURE 14-77

14-55 What value of capacitance is required to produce a cutoff frequency of 3.6 kHz in the circuit of Figure 14-78?

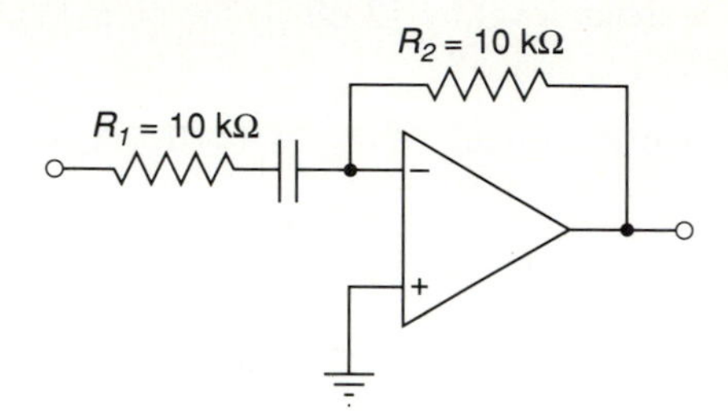

FIGURE 14-78

14-56 For the circuit shown in Figure 14-79, $R_1 = R_2 = 1$ kΩ. Calculate the capacitance values to produce a cutoff frequency of 2.5 kHz.

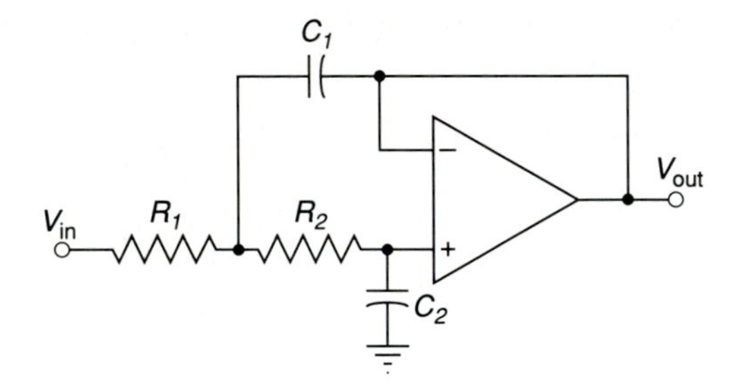

FIGURE 14-79

CHAPTER

15 Alternating-Current Circuit Analysis

LEARNING OBJECTIVES

Upon completion of this chapter you will be able to:

- Convert ac voltage sources to ac current sources, and vice versa.
- Apply loop analysis to ac circuits.
- Understand Norton's theorem and its application to circuit analysis.
- Apply superposition to a circuit with more than one voltage or current source.
- Understand the application of determinants in ac circuit analysis.
- Apply Norton's theorem to ac circuit problems.
- Define Thévenin's theorem and understand how to use it to reduce an ac circuit to a simple equivalent circuit.
- Apply Norton's theorem to ac circuits.
- Use the reciprocity theorem in ac circuit analysis.
- Discuss the application of ac bridges to solve for unknown resistances and reactances.
- Define the maximum power transfer theorem.
- Convert wye connections into delta connections, and vice versa.

15-1 INTRODUCTION

The most important network theorems used in elementary circuit analysis were introduced and described in Chapter 5 during our study of direct-current circuits. The main difference in applying the network theorems of Chapter 5 to ac circuits is that complex impedances are used instead of simple resistances. Also, all mathematical operations such as addition, subtraction, multiplication, and division are done in complex form.

Many calculators in use today are capable of directly performing mathematical functions on complex numbers, which greatly reduces the time required to solve such problems. The examples in this chapter are solved without a calculator using traditional mathematical computations. In addition to the theorems examined in dc circuit analysis, ac bridges and the reciprocity theorem are also discussed in this chapter. The principle of reciprocity has many applications in the electrical and electronic fields, particularly antennas. Ac bridge circuits are used extensively in commercially produced instruments for measuring the characteristics of electrical and electronic components.

15-2 SOURCE CONVERSIONS

When applying some of the circuit analysis techniques discussed in this chapter, it may be necessary to convert a voltage source to a current source, or vice versa. A voltage source that is connected in series with an impedance can be replaced by an equivalent circuit consisting of a current source connected in parallel with the original impedance simply by applying Ohm's law. The circuit shown in Figure 15-1(a) consists of a voltage source and a series-connected impedance. To determine the equivalent current source, the voltage source is short circuited, and Ohm's law is applied. $I = E/Z$, which is the value of the current source. The value of impedance remains the same in both the voltage source and the equivalent current source, as shown in Fig. 15-1(b). To convert from a current source to an equivalent voltage source, the procedure is reversed.

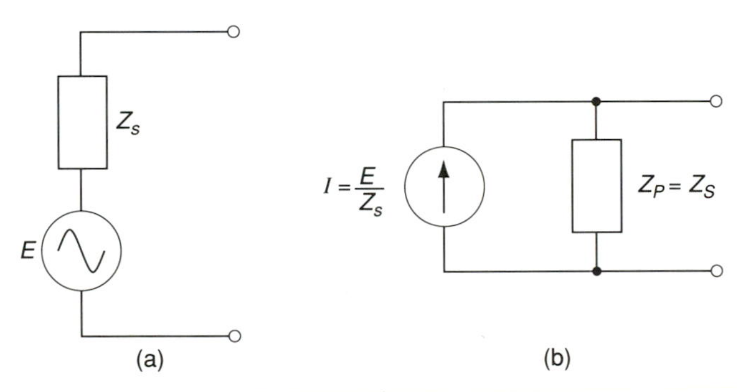

FIGURE 15-1 (a) Voltage source and series impedance; (b) equivalent current course.

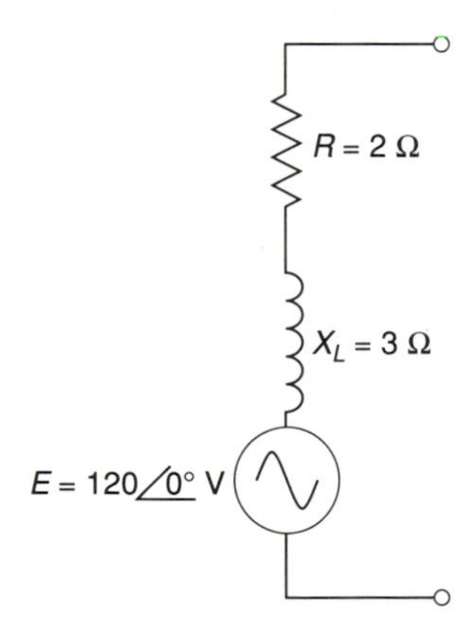

FIGURE 15-2

XAMPLE 15-1 Convert the voltage source of Figure 15-2 to a current source.

Solution $$I = \frac{E}{Z} = \frac{120\angle 0°}{2 + j3} = 33.28\angle -56.3°\text{ A}$$

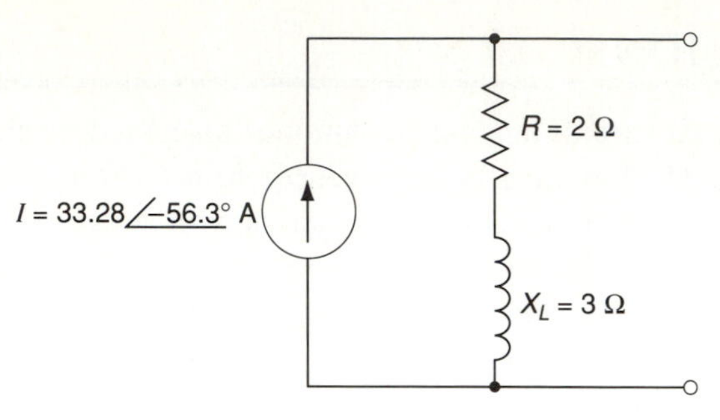

FIGURE 15-3 Equivalent current source for Example 15-1.

EXAMPLE 15-2 Convert the current source of Figure 15-4 to a voltage source.

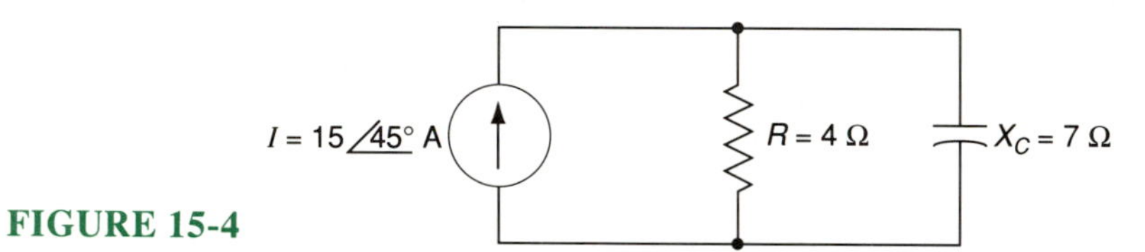

FIGURE 15-4

Solution

$$Z_p = Z_s = \frac{(4\angle{0^\circ}\ \Omega)(7\angle{-90^\circ}\ \Omega)}{4\angle{0^\circ}\ \Omega + 7\angle{-90^\circ}\ \Omega} = 3.47\angle{-29.7^\circ}\ \Omega$$

$$E = I \times Z_s = (15\angle{45^\circ}\ \text{A})(3.47\angle{-29.7^\circ}\ \Omega) = 52.1\angle{15.3^\circ}\ \text{V}$$

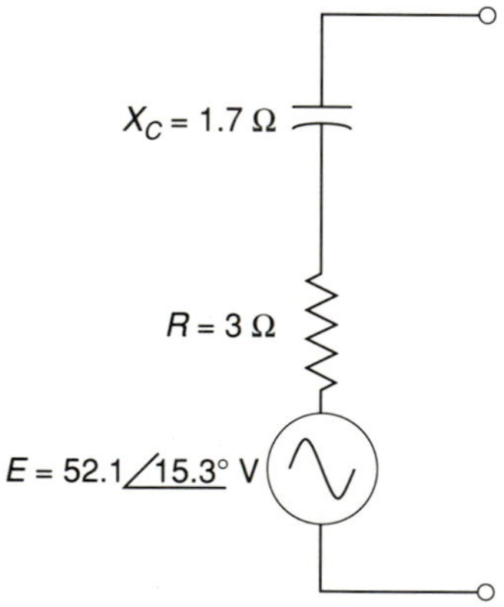

FIGURE 15-5 Equivalent voltage source for Example 15-2.

15-3 LOOP ANALYSIS OF AC CIRCUITS

The method of solving ac circuit problems using loop analysis is the same as the method used for dc circuits, except that impedance and phasor quantities are used. Loop currents are always drawn in a clockwise direction in ac circuits. Before drawing the loop currents, it is good practice to combine all the components in a branch into a single branch impedance. Since loop equations are based on Kirchhoff's voltage law, this law may be stated as follows for an ac circuit:

> In any closed ac circuit, the phasor sum of the voltage drops must equal the phasor sum of the applied voltages.

Very often a particular problem can be solved by determining the current in a single element or branch of a network. Under these conditions, the loops should be chosen so that only one unknown current flows through the element in question. In this way the simultaneous equations need only be solved once.

The steps for applying loop analysis of ac circuits are based on the steps taken for solving a dc circuit problem:

1. Draw all loop currents in a clockwise direction.
2. Indicate the polarities within each loop for each impedance. Show the polarity of the voltage drop with respect to the direction of the loop current. An impedance with two loop currents in opposite directions will have opposite voltage polarities with respect to the two loop currents.
3. Label all voltage sources with their correct polarity. Voltage sources are assumed to be positive if the current leaves the source in a clockwise direction.
4. Apply Kirchhoff's voltage law around each closed loop. When writing loop equations, assume current flow from the negative terminal to the positive terminal of the component to be a voltage rise which would have a positive complex quantity. When current flow is from the positive terminal to the negative, the component is considered to be a voltage drop and would have a negative complex quantity.
5. Use determinants to solve the resulting linear equations.

EXAMPLE 15-3 Calculate the current through the resistor in the circuit shown in Figure 15-6.

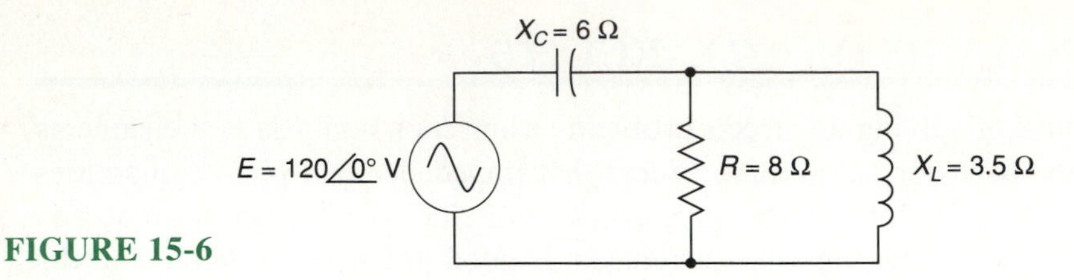

FIGURE 15-6

Solution Draw two loops (Figure 15-7) and write the resulting equations.

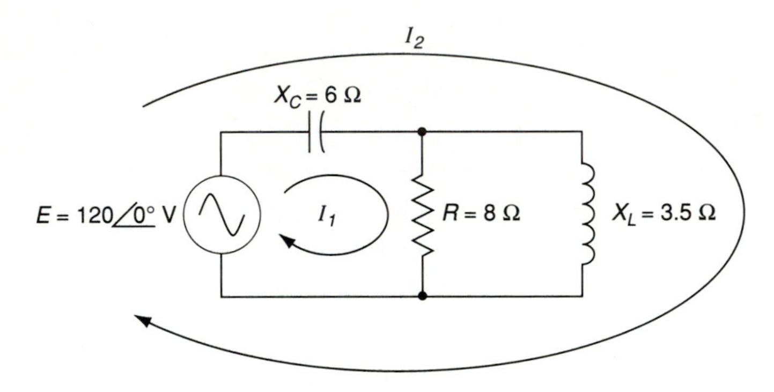

FIGURE 15-7 Loops I_1 and I_2 for Example 15-3.

For loop 1: $E = I_1(R - jX_C) - jX_C I_2$

$$120 = I_1(8 - j6) - j6I_2$$

For loop 2: $E = -jX_C I_1 + I_2(-jX_C + jX_L)$

$$120 = -j6I_1 + I_2(-j6 + j3.5)$$

The two equations can now be solved by using Cramer's rule. The determinant of the denominator is made up of the coefficients of X.

$$X = \frac{\begin{vmatrix} c & b \\ c & b \end{vmatrix}}{\begin{vmatrix} a & b \\ a & b \end{vmatrix}}$$

By using the four coefficients of the two equations, the denominator becomes

$$\begin{vmatrix} 8 - j6 & -j6 \\ -j6 & -j2.5 \end{vmatrix}$$

To find I_1, the determinant in the numerator is obtained from the determinant of the denominator by replacing the first column with the applied voltages on the left side of each equation.

$$\begin{vmatrix} 120 & -j6 \\ 120 & -j2.5 \end{vmatrix}$$

The solution for I_1 is now set up by using the foregoing determinants:

$$I_1 = \frac{\begin{vmatrix} 120 & -j6 \\ 120 & -j2.5 \end{vmatrix}}{\begin{vmatrix} 8 - j6 & -j6 \\ -j6 & -j2.5 \end{vmatrix}}$$

$$= \frac{120(-j2.5) - 120(-j6)}{-j2.5(8 - j6) - (j6)^2}$$

$$= \frac{-j300 + j720}{-j20 - 15 + 36}$$

$$= \frac{420\,\underline{/90^\circ}}{29\,\underline{/-43.6^\circ}}$$

$$= 14.48\,\underline{/133.6^\circ}\text{ A}$$

EXAMPLE 15-4 Using loop analysis, find the currents I_1 and I_2 for the circuit shown in Figure 15-8.

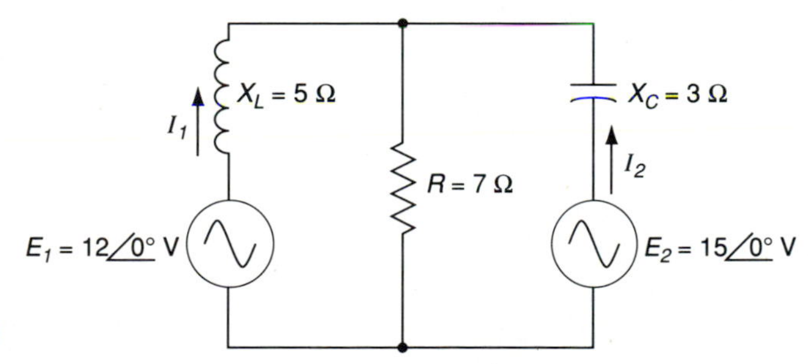

FIGURE 15-8

Solution Draw the loops currents and label the voltage rises and drops, as shown in Figure 15-9.

Loop 1: $E_1 = X_L I_1 + R(I_1 - I_2)$

$$12 = +j5I_1 + 7(I_1 - I_2)$$

$$12 = I_1(j5 + 7) - 7I_2$$

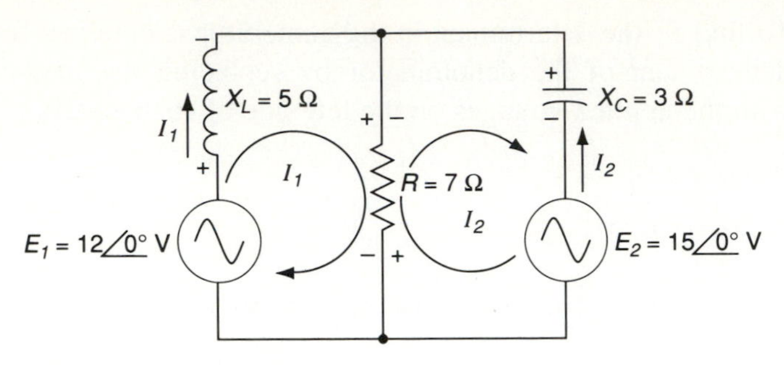

FIGURE 15-9 Loop currents and load polarities for Example 15-4.

Since the direction of current in loop 2 is chosen as counterclockwise, the voltage source is given a negative value.

$$\text{Loop 2: } -E_2 = X_C I_2 + R(I_2 - I_1)$$

$$-15 = -j3I_2 + 7(I_2 - I_1)$$

$$-15 = -7I_1 + (-j3 + 7)I_2$$

Using Cramer's determinant rule yields

$$X = \frac{\begin{vmatrix} c_1 & b_1 \\ c_2 & b_2 \end{vmatrix}}{\begin{vmatrix} a_1 & b_1 \\ a_2 & b_2 \end{vmatrix}} \qquad Y = \frac{\begin{vmatrix} a_1 & c_1 \\ a_2 & c_2 \end{vmatrix}}{\begin{vmatrix} a_1 & b_1 \\ a_2 & b_2 \end{vmatrix}}$$

$$I_1 = \frac{\begin{vmatrix} 12 & -7 \\ -15 & 7 - j3 \end{vmatrix}}{\begin{vmatrix} j5 + 7 & -7 \\ -7 & 7 - j3 \end{vmatrix}}$$

$$= \frac{12(7 - j3) - (-15)(-7)}{(j5 + 7)(7 - j3) - (-7)(-7)}$$

$$= \frac{41.68\,\angle -120.2°}{20.4\,\angle 43°}$$

$$= 2.04\,\angle -163.2° \text{ A}$$

$$I_2 = \frac{\begin{vmatrix} j5 + 7 & 12 \\ -j3 & -15 \end{vmatrix}}{\begin{vmatrix} j5 + 7 & -7 \\ -j3 & 7 - j3 \end{vmatrix}}$$

$$= \frac{-15(j5 + 7) - (-7)(12)}{(j5 + 7)(7 - j3) - (-7)(-7)}$$

$$= \frac{77.88 \angle -105.64°}{20.4 \angle 43°}$$

$$= 3.82 \angle -148.6° \text{ A}$$

EXAMPLE 15-5 Given the choice of loop currents shown in Figure 15-10, solve for I_1, I_2, and I_3.

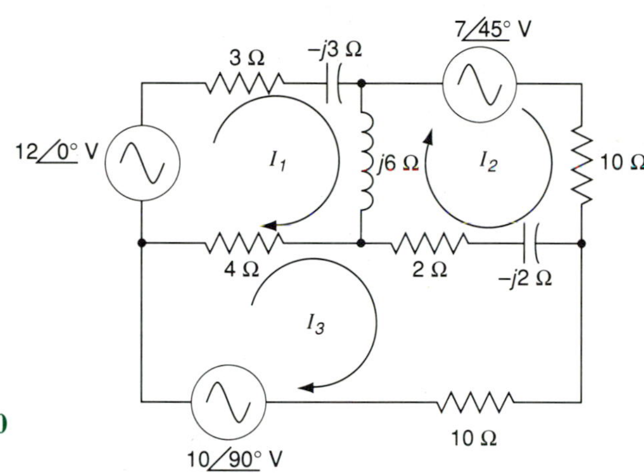

FIGURE 15-10

Solution Determine the three simultaneous equations.

Loop 1: $12 \angle 0° = I_1(3 - j3) + j6(I_1 - I_2) + 4(I_1 - I_3)$

Loop 2: $-(7 \angle 45°) = I_2(10) + (2 - j2)(I_2 - I_3) + j6(I_2 - I_1)$

Loop 3: $-(10 \angle 90°) = I_3(10) + 4(I_3 - I_1) + (2 - j2)(I_3 - I_2)$

The gathering of like terms results in the following equations:

$$(7 + j3)I_1 - (j6)I_2 - (4)I_3 = 12 \angle 0°$$

$$-(j6)I_1 + (12 + j4)I_2 - (2 - j2)I_3 = -(7 \angle 45°)$$

$$-(4)I_1 - (2 - j2)I_2 + (16 - j2)I_3 = -(10 \angle 90°)$$

As was the case with dc circuits, third-order determinants are used to solve for three unknowns in ac circuits.

$$X = \frac{\begin{vmatrix} d_1 & b_1 & c_1 \\ d_2 & b_2 & c_2 \\ d_3 & b_3 & c_3 \end{vmatrix}}{\begin{vmatrix} a_1 & b_1 & c_1 \\ a_2 & b_2 & c_2 \\ a_3 & b_3 & c_3 \end{vmatrix}} \qquad Y = \frac{\begin{vmatrix} a_1 & d_1 & c_1 \\ a_2 & d_2 & c_2 \\ a_3 & d_3 & c_3 \end{vmatrix}}{\begin{vmatrix} a_1 & b_1 & c_1 \\ a_2 & b_2 & c_2 \\ a_3 & b_3 & c_3 \end{vmatrix}} \qquad Z = \frac{\begin{vmatrix} a_1 & b_1 & d_1 \\ a_2 & b_2 & d_2 \\ a_3 & b_3 & d_3 \end{vmatrix}}{\begin{vmatrix} a_1 & b_1 & c_1 \\ a_2 & b_2 & c_2 \\ a_3 & b_3 & c_3 \end{vmatrix}}$$

Once the simultaneous equations have been arranged in the form above where X, Y, and Z represent I_1, I_2, and I_3, respectively, the first and second columns are rewritten to the right of each determinant, and the product of the diagonals are added or subtracted according to their position.

Loop 1:

$$I_1 = \frac{d_1b_2c_3 + d_3b_1c_2 + d_2b_3c_1 - d_3b_2c_1 - d_1b_3c_2 - d_2b_1c_3}{a_1b_2c_3 + a_3b_1c_2 + a_2b_3c_1 - a_3b_2c_1 - a_1b_3c_2 - a_2b_1c_3}$$
$$= \frac{3002.8 - j558.6}{1536.1 + j698.6}$$
$$= 1.81 \angle -35^\circ \text{ A}$$

Loop 2:

$$I_2 = \frac{a_1d_2c_3 + a_3d_1c_2 + a_2d_3c_1 - a_3d_2c_1 - a_1d_3c_2 - a_2d_1c_3}{a_1b_2c_3 + a_3b_1c_2 + a_2b_3c_1 - a_3b_2c_1 - a_1b_3c_2 - a_2b_1c_3}$$
$$= \frac{61.2 + j187.8}{1536.1 + j698.6}$$
$$= 0.12 \angle 47.5^\circ \text{ A}$$

Loop 3:

$$I_3 = \frac{a_1b_2d_3 + a_3b_1d_2 + a_2b_3d_1 - a_3b_2d_1 - a_1b_3d_2 - a_2b_1d_3}{a_1b_2c_3 + a_3b_1c_2 + a_2b_3c_1 - a_3b_2c_1 - a_1b_3c_2 - a_2b_1c_3}$$
$$= \frac{1334 - j920}{1536.1 + j698.6}$$
$$= 0.96 \angle -59^\circ \text{ A}$$

15-4 NODAL ANALYSIS OF AC CIRCUITS

A **node** was defined previously as a point in a circuit that is common to two or more branches. If three or more branches are joined at one node, it is referred to as a **principal node**. In ac circuits, loop analysis relates to voltage sources and impedance, while **nodal analysis** deals with current sources and admittance. Nodal analysis consists of writing Kirchhoff's current law equations for each node in a circuit. This results in a simultaneous equation being written for each unknown node voltage in the circuit.

A node voltage is the potential difference between a given node and a specific node referred to as the **reference node.** The reference node is generally the common point for all current sources in a network and is illustrated by a ground symbol. The reference node is not necessarily physically connected to ground, although it is assumed to be at zero volts with respect to ground. All node voltages are considered positive with respect to the reference node. The procedure for nodal analysis of ac circuits is as follows:

1. Convert all voltage sources to current sources.
2. Choose the reference node and assign it a subscripted voltage label of N_0.
3. Label the voltage at each of the remaining nodes. The first node is marked N_1, the second N_2, and so on.
4. Write the nodal equations. Draw and label the currents flowing into and out of each node where the voltage is unknown. If a source supplies current to a node, it is identified with a positive sign; if current flows toward the source from a node, it is given a negative symbol.
5. Solve the resulting simultaneous equations for the unknown node voltages.

The circuit shown in Figure 15-11 contains two principal nodes. The lower principal node is taken as the reference since it is common to both voltage sources. Before nodal analysis can be applied to voltage sources which have an internal impedance, these voltage sources must be converted to equivalent current sources. The method of converting a voltage source to a current source was demonstrated earlier in this chapter.

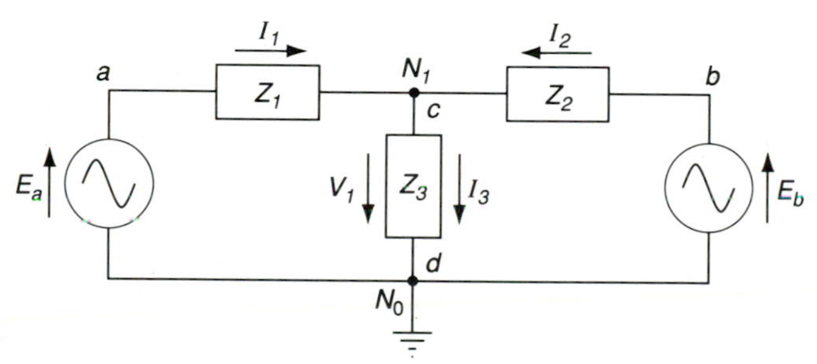

FIGURE 15-11

To convert the voltage source E_a to a current source, assume that Z_1 is the internal impedance of the source and V_1 is the potential difference of node N_1 with respect to the reference node. By applying Kirchhoff's voltage law, the source E_a is expressed as

$$E_a = I_1 Z_1 + V_1$$

or

$$I_1 = \frac{E_a}{Z_1} - \frac{V_1}{Z_1}$$

If the values of E_a and Z_1 are known, the equation above states that the current flowing into the principal node, N_1, is equal to a specified current (E_a/Z_1)

minus the current V_1/Z_1. The specified current (E_a/Z_1) may be considered as a current source across the principal node N_1 and the reference node N_0. I_1 is considered to be the value of the current source if impedance Z_1 is placed in parallel with the current source. Therefore, the voltage sources of Figure 15-11 may be replaced with the current sources shown in Figure 15-12.

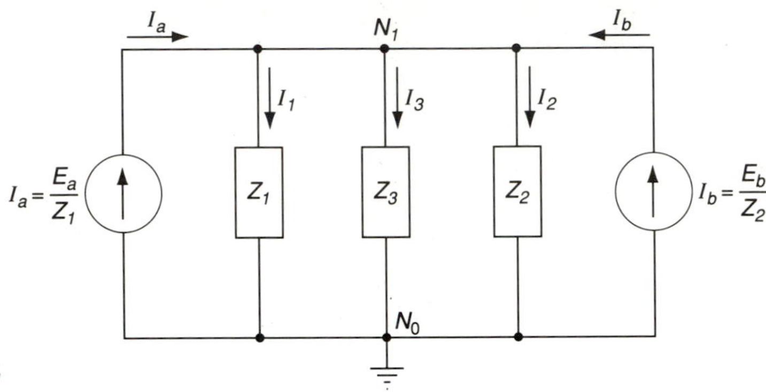

FIGURE 15-12

The current equations for the principal node N_1 can be written in terms of voltage drops and admittances as follows:

$$\text{current leaving node } N_1 = \text{current entering node } N_1$$

$$Y_1V_1 + Y_3V_1 + Y_2V_1 = Y_1E_a + Y_2E_b$$

Figure 15-12 can then be redrawn to show the currents entering and leaving node N_1, as shown in Figure 15-13.

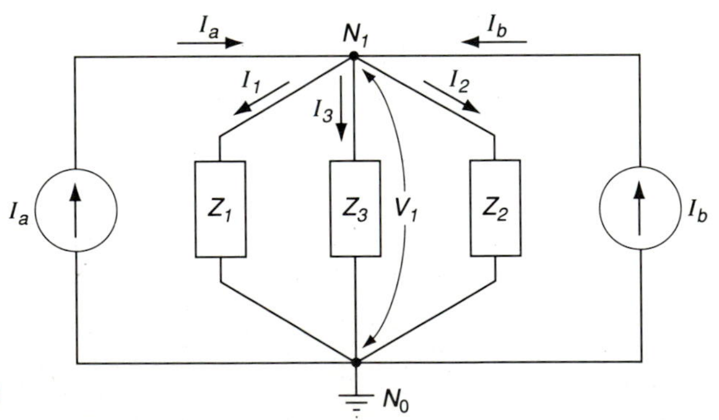

FIGURE 15-13

The nodal method of circuit analysis is generally less time consuming than loop analysis, provided that the number of nodes does not exceed the number of meshes, or loops. If N represents the number of nodes in a circuit, then $N - 1$ node equations are required, which are derived by applying Kirchhoff's current law.

EXAMPLE 15-6 For the circuit shown in Figure 15-14, calculate the voltage V_1 and the currents I_1, I_2, and I_3.

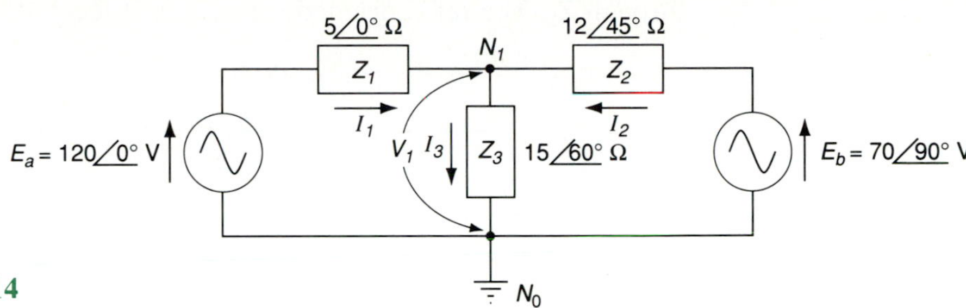

FIGURE 15-14

Solution

$$V_1(Y_1 + Y_2 + Y_3) = Y_1E_a + Y_2E_b$$

$$V_1\left(\frac{1}{5\angle 0^\circ} + \frac{1}{12\angle 45^\circ} + \frac{1}{15\angle 60^\circ}\right) = \frac{120\angle 0^\circ}{5\angle 0^\circ} + \frac{70\angle 90^\circ}{12\angle 45^\circ}$$

$$V_1 = \frac{28.425\angle 8.34^\circ}{0.3147\angle -21.77^\circ}$$
$$= 90.324\angle 30.11^\circ \text{ V}$$

$$I_3 = V_1Y_3$$
$$= (90.324\angle 30.11^\circ)(0.0667\angle -60^\circ \text{ S})$$
$$= 6.02\angle -29.9^\circ \text{ A}$$

$$I_1' = E_aY_1 - V_1Y_1$$
$$= (120\angle 0^\circ \text{ V})(0.2\angle 0^\circ \text{ S}) - (90.324\angle 30.11^\circ \text{ V})(0.2\angle 0^\circ \text{ S})$$
$$= 12.34\angle -47.3^\circ \text{ A}$$

$$I_2 = E_bY_2 - V_1Y_2$$
$$= (70\angle 90^\circ \text{ V})(0.0833\angle -45^\circ \text{ S}) - (90.324\angle 30.11^\circ \text{ V})(0.0833\angle -45^\circ \text{ S})$$
$$= 6.83\angle 117.5^\circ \text{ A}$$

EXAMPLE 15-7 For the circuit shown in Figure 15-15, calculate the current through the series connected coil and resistor.

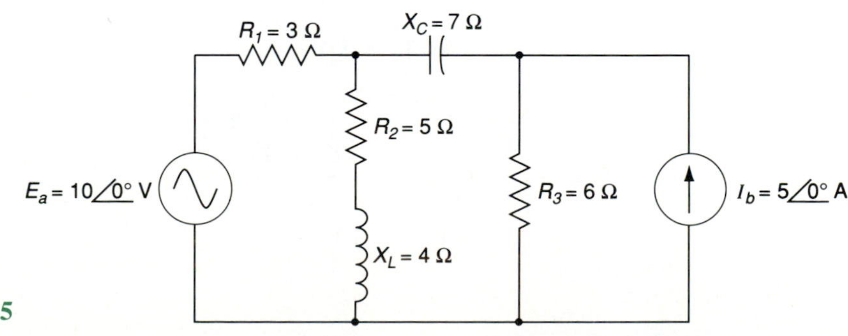

FIGURE 15-15

Solution Redraw the circuit so that it contains impedance values only, as shown in Figure 15-16. The current to be solved for will flow through Z_2. The reference node chosen is at the bottom of Z_2.

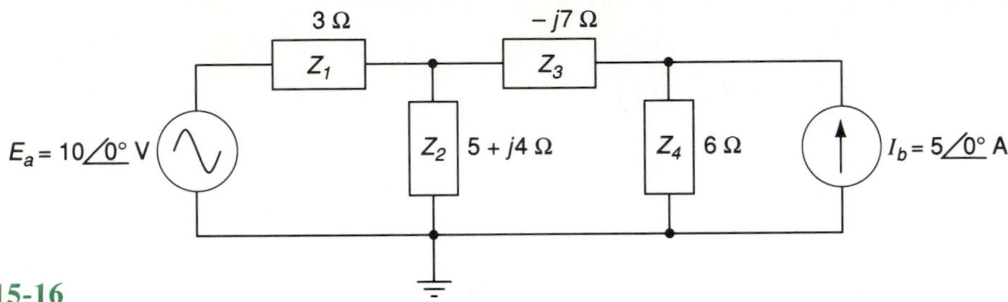

FIGURE 15-16

Solution Convert the voltage source to a current source by placing Z_1 in parallel with E_a, as shown in Figure 15-17.

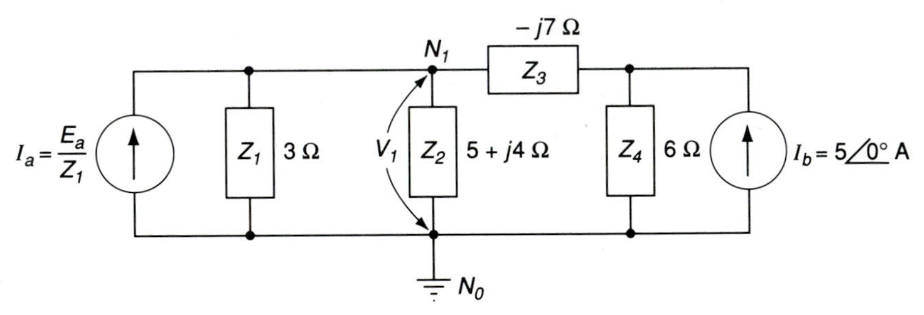

FIGURE 15-17

The nodal equations can now be written:

$$V_1(Y_1 + Y_2 + Y_3) - V_2(Y_3) = I_b$$

$$V_2(Y_3 + Y_4) - V_1(Y_3) = I_a$$

By using Cramer's rule for determinants, the voltage across Z_2 can be found.

$$V_1(Y_1 + Y_2 + Y_3) - V_2(Y_3) = I_b$$

$$-V_1(Y_3) + V_2(Y_3 + Y_4) = I_a$$

$$V_1 = \frac{\begin{vmatrix} I_b & -Y_3 \\ I_a & Y_3 + Y_4 \end{vmatrix}}{\begin{vmatrix} Y_1 + Y_2 + Y_3 & -Y_3 \\ -Y_3 & Y_3 + Y_4 \end{vmatrix}}$$

$$= \frac{I_b(Y_3 + Y_4) + I_a(Y_3)}{(Y_1 + Y_2 + Y_3)(Y_3 + Y_4) - Y_3^2}$$

$$= \frac{I_b(Y_3 + Y_4) + I_a(Y_3)}{Y_1Y_3 + Y_1Y_4 + Y_2Y_3 + Y_2Y_4 + Y_3^2 + Y_3Y_4 - Y_3^2}$$

$$= \frac{4.22 \angle 78.4^\circ}{0.1 \angle 40^\circ}$$

$$= 42.2 \angle 38.4^\circ \text{ V}$$

The current through Z_2 is now found by applying Ohm's law.

$$I_2 = \frac{V_1}{Z_2} = \frac{42.2 \angle 38.4^\circ}{6.4 \angle 38.7^\circ} = 6.6 \angle 0^\circ \text{ A}$$

15-5 SUPERPOSITION THEOREM

When analyzing circuits containing only two loops or nodes, simultaneous equations and determinants are very effective. However, when dealing with circuit problems containing three or more simultaneous equations, these two methods are cumbersome and time consuming. One method that saves time and substantially reduces the possibility of error is the **superposition theorem**. The disadvantage of this method is that it can be applied only to circuits containing linear components. The superposition theorem is extremely useful for analyzing circuits containing sources that have different operating frequencies. Loop and nodal analysis techniques cannot be applied to these types of circuits since those two methods involve phasors and impedances which are applicable at only a single operating frequency. The superposition theorem is defined as follows:

> The branch voltage or current produced by several sources supplying the same circuit is the sum of the voltages or currents produced by each source acting independently, with the other sources replaced by their internal impedances.

The method of circuit analysis using the superposition theorem consists of the following steps:

1. Replace all but one of the supply sources with their own internal impedance. If the internal impedance is negligible, short-circuit the source.
2. Solve for the current distribution around the circuit by using Ohm's law.
3. Substitute the remaining source for one of the sources that had been replaced by an internal impedance, and repeat the procedure of finding the

currents around the network. Substitute each voltage source until all values of current have been determined for each voltage source.

4. After each voltage source has been applied independently, find the phasor sum of the currents in each branch of the circuit.

All the voltages except the one for which currents are being calculated are assumed to be zero. Any impedances associated with the voltage source must be left connected in the circuit regardless of whether the voltage is assumed to be zero or it is the one considered to be acting as the independent source. When a voltage source is not specified as having an internal impedance, it is assumed to be an ideal source and is simply short-circuited when calculating the effect of another source on the circuit. A current source that has no internal impedance is replaced by an open circuit, to represent infinite impedance.

EXAMPLE 15-8 Using the superposition theorem, find the voltage across R_2 for the circuit shown in Figure 15-18.

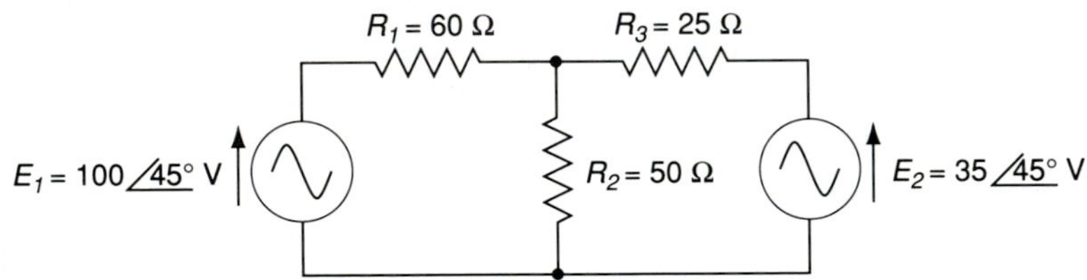

FIGURE 15-18

Solution Since E_2 has no internal impedance, it is replaced by a short circuit, as shown in Figure 15-19.

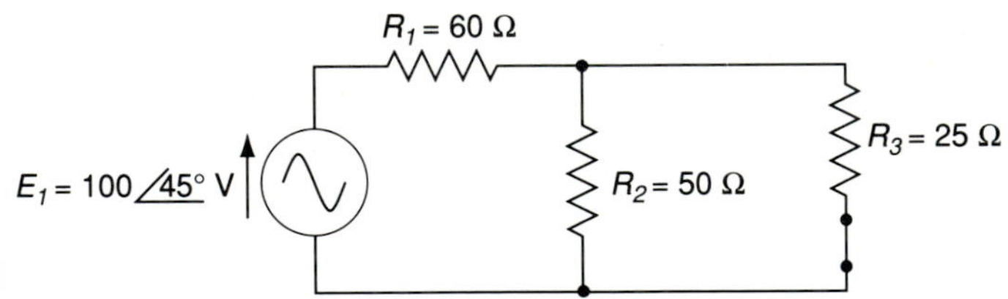

FIGURE 15-19

$$R_{2\|3} = \frac{R_2 \times R_3}{R_2 + R_3} = \frac{(50\,\Omega)(25\,\Omega)}{50\,\Omega + 25\,\Omega} = 16.67\,\Omega$$

$$R_T = R_1 + R_{2\|3} = 60\,\Omega + 16.67\,\Omega = 76.67\,\Omega$$

The voltage across R_2 and R_3 can now be found by the voltage divider rule.

$$V_{R2\|3} = \frac{R_{2\|3} \times E_1}{R_T} = \frac{16.67\,\Omega}{76.67\,\Omega} \times 100\ \angle 45^\circ\ \text{V}$$
$$= 21.74\ \angle 45^\circ\ \text{V}$$

The next step, as shown in Figure 15-20, involves replacing E_1 by its source impedance and repeating the procedure.

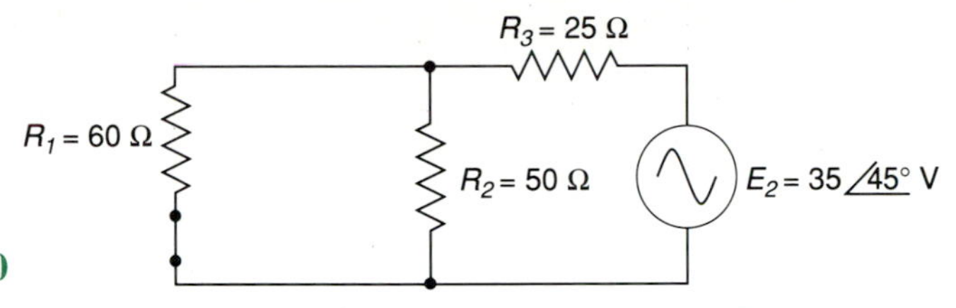

FIGURE 15-20

$$R_{1\|2} = \frac{(60\ \Omega)(50\ \Omega)}{60\ \Omega + 50\ \Omega} = 27.27\ \Omega$$

$$R_T = R_3 + R_{1\|2} = 25\ \Omega + 27.27\ \Omega = 52.27\ \Omega$$

$$V_{R1\|2} = \frac{R_{1\|2}}{R_T} \times E_2 = \frac{27.27\ \Omega}{52.27\ \Omega} \times 35\ \angle 45°\ \text{V}$$
$$= 18.26\ \angle 45°\ \text{V}$$

The voltage across R_2 can now be determined by finding the phasor sum of $V_{R1\|2}$ and $V_{R2\|3}$.

$$V_{R2} = V_{R1\|2} + V_{R2\|3} = 18.26\ \angle 45°\ \text{V} + 21.74\ \angle 45°\ \text{V}$$
$$= 40\ \angle 45°\ \text{V}$$

EXAMPLE 15-9 Calculate the current in branch *cd* for the circuit shown in Figure 15-21.

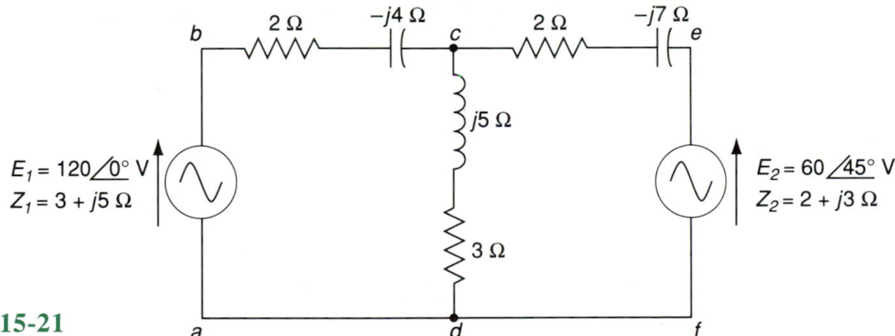

FIGURE 15-21

Solution Replace E_2 with its internal impedance, as shown in Figure 15-22.

The voltage across Z_{cd} due to E_1 is found by using the voltage divider rule. V' indicates that this is the voltage due to E_1 alone.

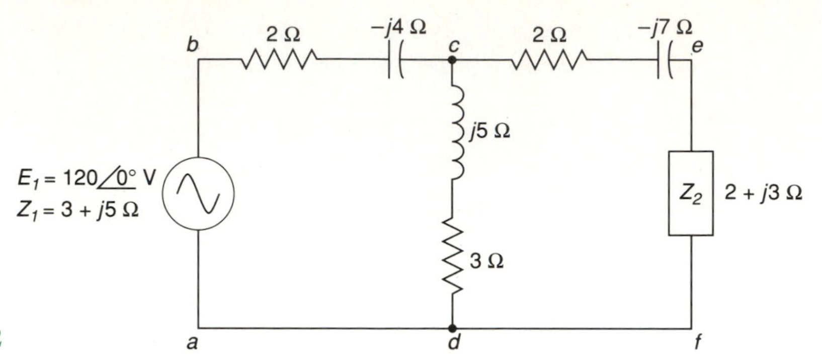

FIGURE 15-22

$$
\begin{aligned}
V'_{cd} &= \frac{Z_{cef} \| Z_{cd}}{Z_{cba} + Z_{cef} \| Z_{cd}} \times E_1 \\
&= \frac{4.64 + j0.48}{5 + j1 + 4.64 + j0.48} \times 120 \underline{/0^\circ} \text{ V} \\
&= 57.36 \underline{/-2.82^\circ} \text{ V}
\end{aligned}
$$

The next step is to replace E_1 with its internal impedance, as shown in Figure 15-23.

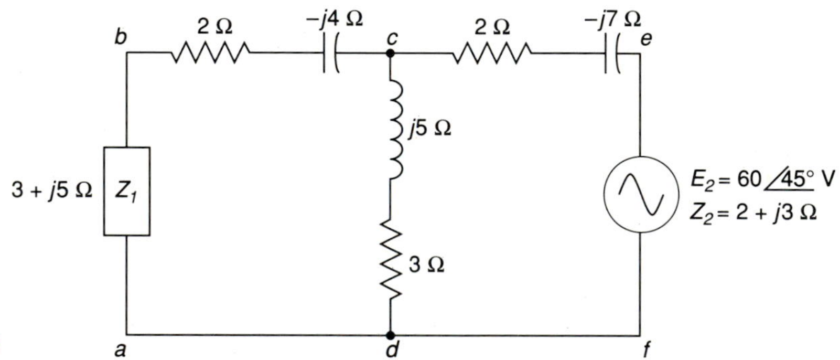

FIGURE 15-23

The voltage across Z_{cd} due to E_2 is also found by the voltage divider rule.

$$
\begin{aligned}
V''_{cd} &= \frac{Z_{cba} \| Z_{cd}}{Z_{cef} + Z_{abc} \| Z_{cd}} \times E_2 \\
&= \frac{2.48 + j1.64}{4 - j4 + 2.48 + j1.64} \times 60 \underline{/45^\circ} \text{ V} \\
&= 25.86 \underline{/98.5^\circ} \text{ V}
\end{aligned}
$$

V_{cd} is now found by adding V'_{cd} and V''_{cd}.

$$V_{cd} = V'_{cd} + V''_{cd}$$
$$= 57.36 \angle -2.82^\circ \text{ V} + 25.86 \angle 98.5^\circ \text{ V}$$
$$= 58.11 \angle 23.05^\circ \text{ V}$$

I_{cd} can now be determined by Ohm's law.

$$I_{cd} = \frac{V_{cd}}{Z_{cd}} = \frac{58.11 \angle 23.05^\circ \text{ V}}{5.83 \angle 59^\circ \, \Omega}$$
$$= 9.97 \angle -36^\circ \text{ A}$$

EXAMPLE 15-10 Use the superposition theorem to find the current through the capacitor in the circuit shown in Figure 15-24.

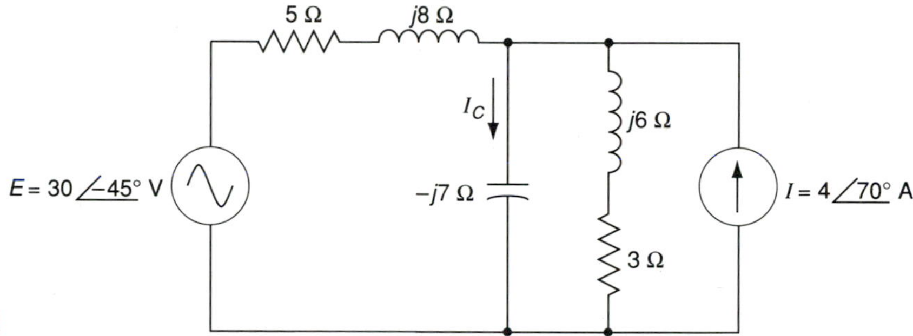

FIGURE 15-24

Solution Redraw the circuit, as shown in Figure 15-25, with the current source open-circuited.

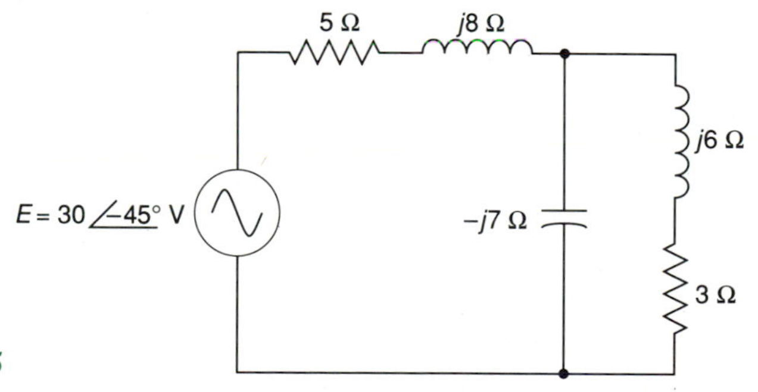

FIGURE 15-25

$$Z_T = \frac{(-j7)(3 + j6)}{-j7 + 3 + j6} + 5 + j8$$
$$= 14.7 - j2.11 + 5 + j8$$
$$= 19.7 + j5.9$$
$$= 20.56 \angle 16.7^\circ \, \Omega$$

$$I_1 = \frac{E}{Z_T} = \frac{30\angle -45^\circ \text{ V}}{20.56\angle 16.7^\circ\ \Omega} = 1.46\angle -61.8^\circ \text{ A}$$

To find the current through the capacitor with the current source open-circuited, the current divider rule is used.

$$I' = I_1 \times \frac{3 + j6}{3 + j6 - j7} = (1.46\angle -61.8^\circ \text{ A})(2.12\angle 81.9^\circ\ \Omega)$$
$$= 3.07\angle 20.1^\circ \text{ A}$$

The circuit is now redrawn with the voltage source short-circuited, as shown in Figure 15-26. Next, the parallel combination of the resistors and inductors is determined. Ignore the capacitor in this calculation.

$$Z' = \frac{(5 + j8)(3 + j6)}{5 + j8 + 3 + j6} = 1.896 + j3.44$$
$$= 3.93\angle 61.14^\circ\ \Omega$$

The current divider rule is applied to find the current I''.

$$I'' = I\frac{Z'}{Z' + j7}$$
$$= 4\angle 70^\circ \text{ A} \times \frac{3.93\angle 61.14^\circ\ \Omega}{3.93\angle 61.14^\circ\ \Omega + 7\angle -90^\circ\ \Omega}$$
$$= 3.9\angle 193.12^\circ \text{ A}$$

The current through the capacitor can now be determined.

$$I_C = I' + I'' = 3.07\angle 20.1^\circ \text{ A} + 3.9\angle 193.12^\circ \text{ A}$$
$$= 0.9\angle 168.5^\circ \text{ A}$$

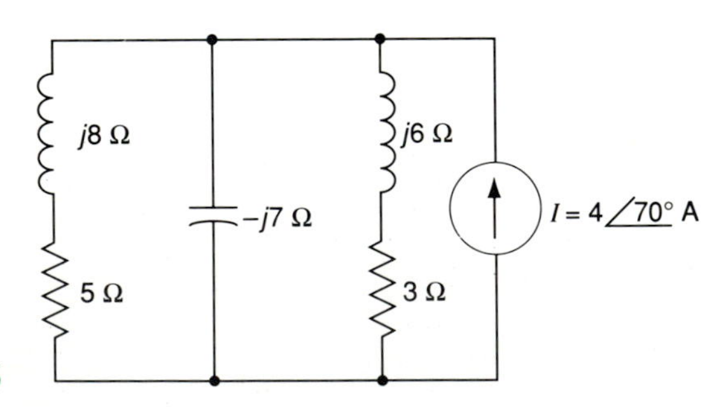

FIGURE 15-26

15-6 THÉVENIN'S THEOREM

Thévenin's theorem for ac circuits is the same as for dc circuits, except impedance is used instead of resistance. Thévenin's theorem for ac circuits is stated as follows:

> Any two-terminal circuit containing voltage and/or current sources may be replaced by a single voltage source in series with a single impedance. This equivalent series circuit is capable of providing the same current through the load as the original circuit.

This means that if the current or voltage of a single component or branch of a circuit is required, no matter how complicated the rest of the circuit may be, it can be reduced to a voltage source in series with an impedance. The reduced circuit is generally referred to as a **Thévenin equivalent circuit.** The Thévenin equivalent circuit of an ac network is applicable at one frequency only.

The steps taken in applying Thévenin's theorem to ac circuits are as follows:

1. Represent each source in its phasor form and each element by its impedance.
2. Remove the portion of the circuit where the Thévenin equivalent circuit is to be determined.
3. Determine the Thévenin equivalent impedance, Z_{TH}, by short- circuiting all voltage sources and open-circuiting all current sources. The series impedance of the Thévenin equivalent circuit is equal to the total impedance of the branch.
4. Determine the Thévenin equivalent voltage, E_{TH}, across the portion of the circuit removed in step 2. The value of Thévenin voltage for the equivalent circuit is found by solving for the simultaneous equations derived in the original circuit. Return all sources to their original position.
5. Draw the Thévenin equivalent circuit. Place the impedance removed in step 2 across the terminals of the Thévenin equivalent circuit.

EXAMPLE 15-11 For the circuit shown in Figure 15-27, calculate the current through Z_L using Thévenin's theorem.

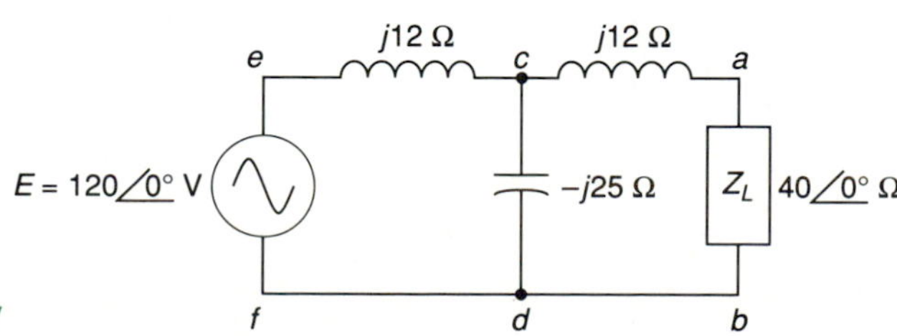

FIGURE 15-27

Solution

$$I_{fecd} = \frac{120\,\underline{/0°}\text{ V}}{12\,\underline{/90°}\,\Omega + 25\,\underline{/-90°}\,\Omega} = 9.231\,\underline{/90°}\text{ A}$$

$$\begin{aligned} V_{cd} = E_{\text{TH}} &= (I_{fecd})(Z_{cd}) \\ &= (9.231\,\underline{/90°}\text{ A})(25\,\underline{/-90°}\,\Omega) \\ &= 230.77\,\underline{/0°}\text{ V} \end{aligned}$$

According to Thévenin's theorem, the current through Z_L is equal to E_{TH} divided by the sum of the load impedance and the impedance of the rest of the circuit with the load impedance disconnected. Therefore, the impedance of the network with Z_L disconnected and when the emf in the branch ef is assumed to be zero is determined as

$$\begin{aligned} Z_{\text{TH}} &= Z_{ac} + \frac{(Z_{ce})(Z_{cd})}{Z_{ce} + Z_{cd}} \\ &= 12\,\underline{/90°}\,\Omega + \frac{(12\,\underline{/90°}\,\Omega)(25\,\underline{/-90°}\,\Omega)}{12\,\underline{/90°}\,\Omega + 25\,\underline{/-90°}\,\Omega} \\ &= 35.08\,\underline{/90°}\,\Omega \end{aligned}$$

The current through Z_L can now be determined by applying Ohm's law.

$$I_{\text{load}} = \frac{E}{Z_{\text{TH}} + Z_L} = \frac{230.77\,\underline{/0°}\text{ V}}{35.08\,\underline{/90°}\,\Omega + 40\,\underline{/0°}\,\Omega} = 4.33\,\underline{/-41.2°}\text{ A}$$

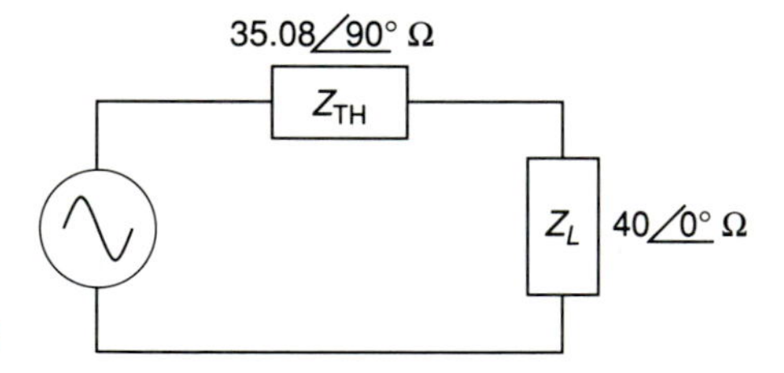

FIGURE 15-28

EXAMPLE 15-12 Using Thévenin's theorem, find the current through Z_L for the circuit shown in Figure 15-29.

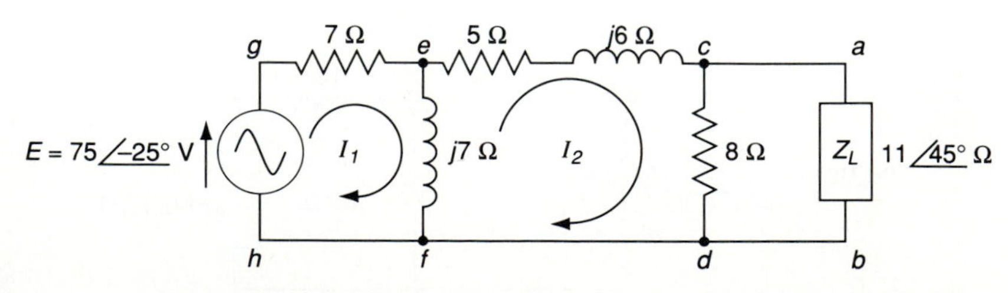

FIGURE 15-29

Solution Solve first for E_{TH}:

$$Z_{ecd} = 13 + j6 = 14.318 \angle 24.78^\circ\ \Omega$$

$$Z_{ef} = (13 + j6)\|j7 = \frac{14.318 \angle 24.78^\circ \times 7 \angle 90^\circ}{18.385 \angle 45^\circ}$$
$$= 5.45 \angle 69.78^\circ\ \Omega$$

$$V_{ef} = \frac{5.45 \angle 69.78^\circ}{10.25 \angle 29.93^\circ} \times 75 \angle -25^\circ$$
$$= 39.88 \angle 14.85^\circ\ \text{V}$$

$$E_{TH} = V_{cd} = \frac{8 \angle 0^\circ\ \Omega}{14.318 \angle 24.78^\circ\ \Omega} \times 39.88 \angle 14.85^\circ\ \text{V}$$
$$= 22.28 \angle -9.93^\circ\ \text{V}$$

$$Z_{TH} = [(7 \parallel j7) + (5 + j6)]\|8$$
$$= [4.95 \angle 45^\circ + (5 + j6)]\|8$$
$$= (3.5 + j3.5 + 5 + j6)\|8 = (8.5 + j9.5)\|8$$
$$= \frac{12.75 \angle 48.18^\circ\ \Omega \times 8 \angle 0^\circ\ \Omega}{19.04 \angle 29.93^\circ\ \Omega}$$
$$= 5.36 \angle 18.25^\circ\ \Omega$$

The circuit is now redrawn, as shown in Figure 15-30, with the Thévenin equivalent voltage and impedance in series with the load. By applying Ohm's law, the current through Z_L is determined.

$$I_L = \frac{E_{TH}}{Z_{TH} + Z_L} = \frac{22.28 \angle -9.93^\circ\ \text{V}}{5.36 \angle 18.25^\circ\ \Omega + 11 \angle 45^\circ\ \Omega}$$
$$= 1.39 \angle -46.24^\circ\ \text{A}$$

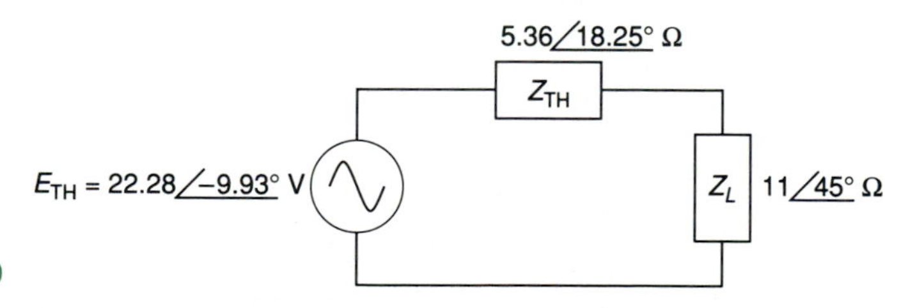

FIGURE 15-30

EXAMPLE 15-13 The unbalanced bridge network shown in Figure 15-31 is supplied with power from a 120-V source. Calculate the current in the 17-Ω detector connected across *b–d*.

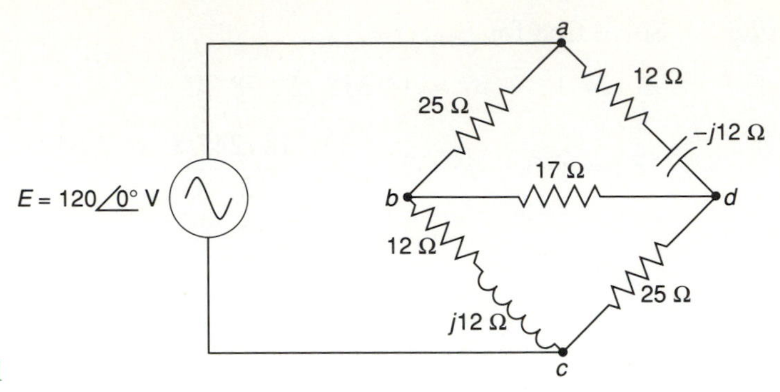

FIGURE 15-31

Solution The open-circuit voltage may be determined by considering the branches *abc* and *adc* as voltage dividers.

$$E_{TH} = V_{bc} - V_{dc} = E_{ac}\frac{Z_{bc}}{Z_{bcd}} - E_{ac}\frac{Z_{cd}}{Z_{adc}}$$
$$= 120\ \underline{/0^\circ} \times \frac{12 + j12}{37 + j12} - 120\ \underline{/0^\circ} \times \frac{25}{37 - j12}$$
$$= 26.73\ \underline{/180^\circ}\ \text{V}$$

By removing impedance *b–d* and assuming that voltage E_{ac} is reduced to zero, the equivalent impedance may be calculated.

$$Z_{TH} = \frac{Z_{abc}}{Z_{ab} + Z_{bc}} + \frac{Z_{adc}}{Z_{ad} + Z_{dc}}$$
$$= \frac{25(12 + j12)}{25 + 12 + j12} + \frac{25(12 - j12)}{25 + 12 - j12}$$
$$= (9.72 + j4.96) + (9.72 - j4.96)$$
$$= 19.44\ \underline{/0^\circ}\ \Omega$$

The total equivalent circuit is obtained by adding the impedance of branch *bd* to the Thévenin's equivalent as shown in Figure 15-32. The current through the detector can now be calculated:

$$I_L = \frac{V_{bd}}{Z_{TH} + Z_L} = \frac{26.73\ \underline{/180^\circ}\ \text{V}}{19.44\ \underline{/0^\circ}\ \Omega + 17\ \underline{/0^\circ}\ \Omega}$$
$$= 1.375\ \underline{/180^\circ}\ \text{A}$$

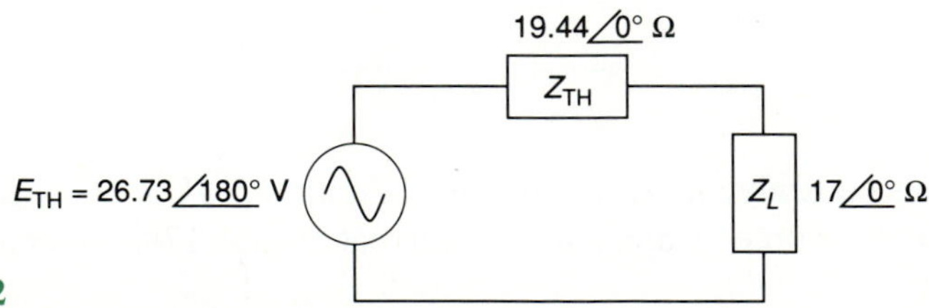

FIGURE 15-32

EXAMPLE 15-14 Use Thévenin's theorem to determine the current flowing between points c–d in the circuit shown in Figure 15-33.

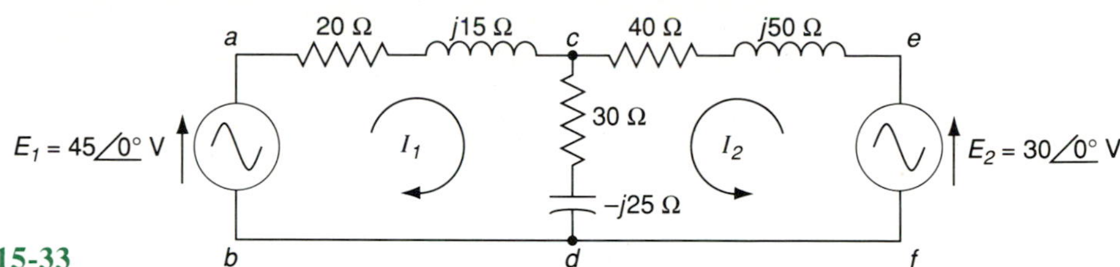

FIGURE 15-33

Solution Remove branch cd from the circuit, as shown in Figure 15-34(a).

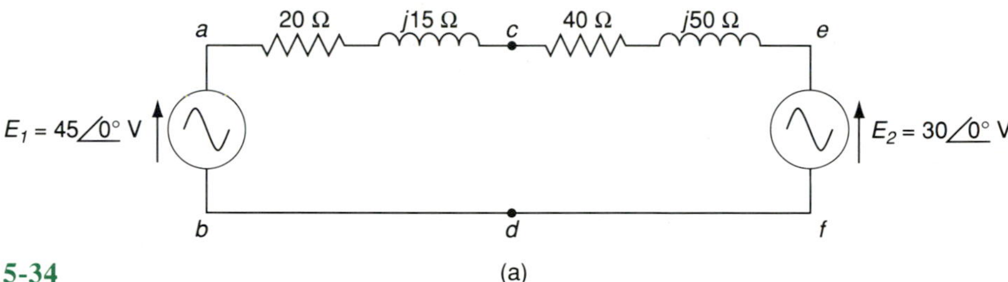

FIGURE 15-34 (a)

Apply Kirchhoff's voltage law around the circuit.

$$45 \angle 0^\circ - I(20 + j15) - I(40 + j50) - 30 \angle 0^\circ = 0$$

which reduces to

$$I = 0.17 \angle -47.29^\circ \text{ A}$$

The Thévenin voltage is now found.

$$\begin{aligned} E_{TH} &= 45 \angle 0^\circ - (0.17 \angle -47.29^\circ)(20 + j15) \\ &= 40.84 \angle 1.08^\circ \text{ V} \end{aligned}$$

Next, short-circuit the voltage sources, as shown in Figure 15-34(b), and determine the Thévenin equivalent impedance.

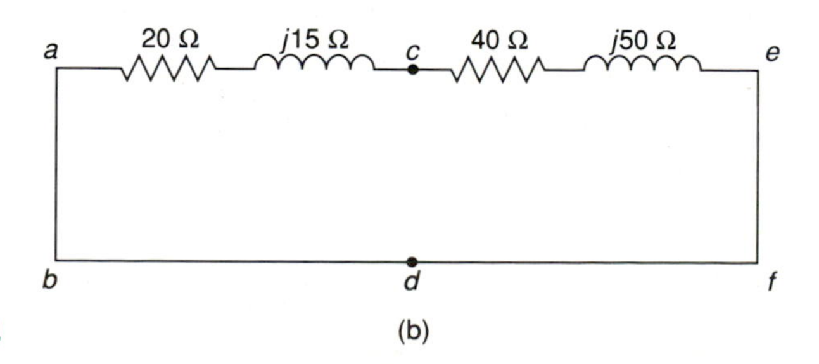

FIGURE 15-34 (b)

$$\frac{1}{Z_{TH}} = \frac{1}{20 + j15} + \frac{1}{40 + j50}$$

$$Z_{TH} = 18.096 \angle 40.92^\circ \ \Omega$$

The Thévenin equivalent circuit is now drawn with the impedance between points c–d, as shown in Figure 15-34(c). The current flowing between points c–d is calculated by Ohm's law.

$$I_{cd} = \frac{E_{TH}}{Z_{TH} + Z_{cd}} = \frac{40.84 \angle -1.08^\circ}{18.096 \angle 40.92^\circ + 30 - j25}$$

$$= 0.895 \angle 17.83^\circ \ \text{A}$$

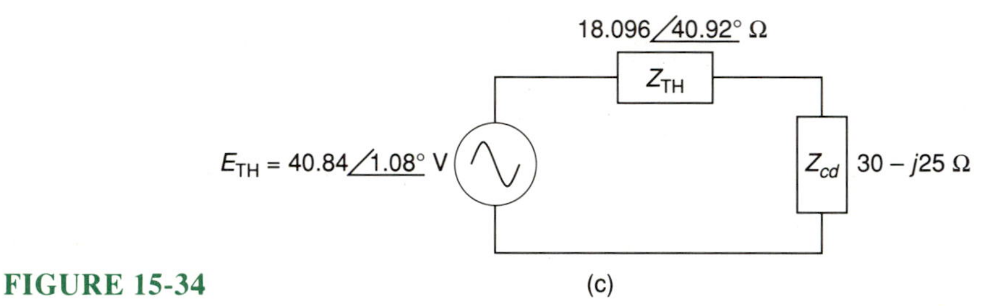

FIGURE 15-34

When analyzing electronic circuits, such as transistor models, quite often a network will contain both an independent source and a dependent source. When applying Thévenin's theorem to these types of circuits, a slightly different approach must be taken. An independent source, such as a voltage source, may be disabled for network analysis simply by shorting it out. However, networks containing dependent sources cannot be disabled. To obtain accurate results from the circuit under consideration, it is necessary to use both open-circuit and short-circuit test data at the terminals being analyzed. The following example illustrates a method for obtaining a Thévenin equivalent circuit for a network containing both an independent source and a dependent source.

EXAMPLE 15-15 For the network of Figure 15-35, calculate the Thévenin equivalent circuit with respect to terminals a–b.

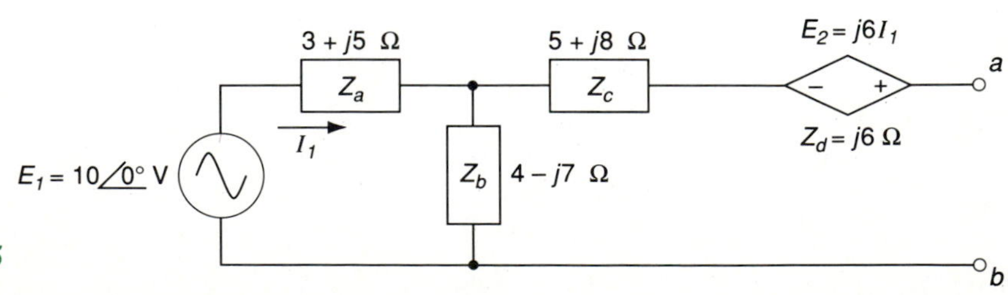

FIGURE 15-35

Solution Calculate the open-circuit voltage, V_o.

$$V_o = E_{TH} = \frac{E_1 Z_b}{Z_a + Z_b} + Z_d I_1$$

$$I_1 = \frac{E_1}{Z_a + Z_b}$$

$$\begin{aligned} E_{TH} &= \frac{E_1(Z_b + Z_d)}{Z_a + Z_b} \\ &= \frac{10\angle 0^\circ(4 - j7 + j6)}{3 + j5 + 4 - j7} \\ &= \frac{41.23\angle -14.04^\circ}{7.28\angle -15.95^\circ} \\ &= 5.66\angle 1.91^\circ \text{ V} \end{aligned}$$

Next, calculate the current that will flow in the circuit with terminals $a - b$ short-circuited, as shown in Figure 15-36(a). Write loop equations for the circuit of the figure.

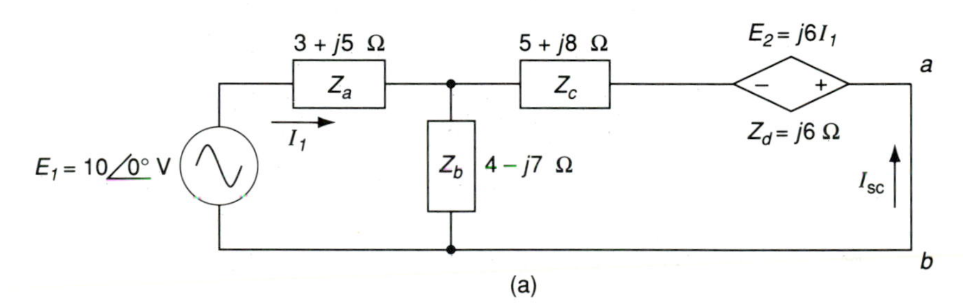

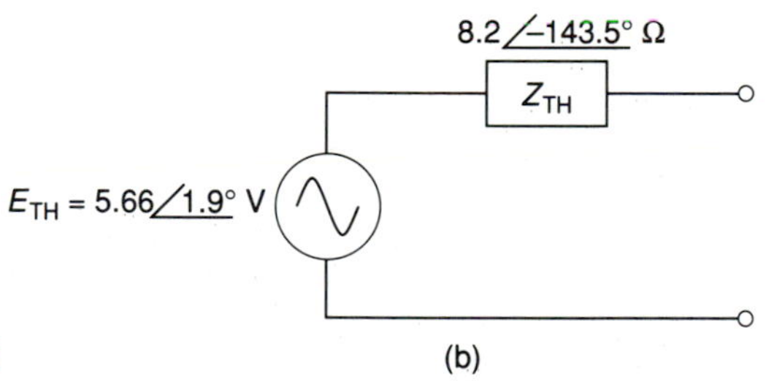

FIGURE 15-36

$$E_1 = (Z_a + Z_b)I_1 + Z_b I_{sc}$$

$$0 = (Z_b + Z_d)I_1 + (Z_b + Z_c)I_{sc}$$

By combining the two loop equations, the value of I_{sc} is determined.

$$I_{sc} = \frac{E_1(Z_b + Z_d)}{(Z_b + Z_d)Z_b - (Z_a + Z_b)(Z_b + Z_c)}$$
$$= \frac{(10\angle 0^\circ)(4 - j7 + j6)}{(4 - j)(4 - j7) - (7 - j2)(9 + j)}$$
$$= 0.69\angle 145.42^\circ \text{ A}$$

The Thévenin impedance is now determined by Ohm's law.

$$Z_{TH} = \frac{E_{TH}}{I_{sc}} = \frac{5.66\angle 1.9^\circ \text{ V}}{0.69\angle 145.42^\circ \text{ A}}$$
$$= 8.2\angle -143.52^\circ\ \Omega$$

The Thévenin equivalent circuit is shown in Figure 15-36(b).

15-7 NORTON'S THEOREM

The dual of the Thévenin's equivalent circuit is the Norton's equivalent circuit. The Norton equivalent circuit consists of a current source in parallel with an equivalent impedance. **Norton's theorem** states:

> Any two-terminal circuit that contains impedances, voltage sources, and/or current sources can be replaced by an equivalent circuit consisting of a single current source in parallel with a single impedance. The value of the current source is equal to the short-circuit current from the original circuit.

The steps for applying Norton's theorem are as follows:

1. Determine the short-circuit current and the circuit terminals.
2. Remove all voltage and current sources from the circuit. Short-circuit the voltage sources and open-circuit the current sources. Any internal resistance associated with the source must be included in the circuit.
3. The equivalent current I_N is found by replacing the voltage and current sources, and by short-circuiting the desired impedance.
4. Insert the equivalent impedance, Z_N, in parallel with the equivalent current, I_N, to form the Norton equivalent circuit.

EXAMPLE 15-16 For the circuit shown in Figure 15-37, use Norton's theorem to determine the current through R_L.

Solution Disconnect R_L, and determine the equivalent impedance, Z_N. Since Z_N is the equivalent impedance at point *ab*, Z_N is considered to be in parallel with point *cd*. The voltage source is short-circuited, as shown in Figure 15-38(a).

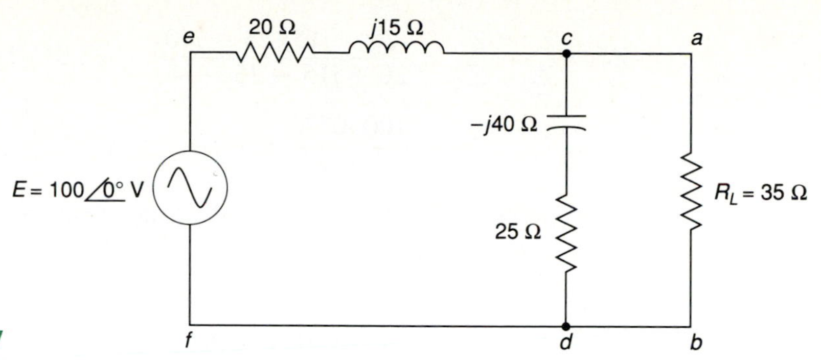

FIGURE 15-37

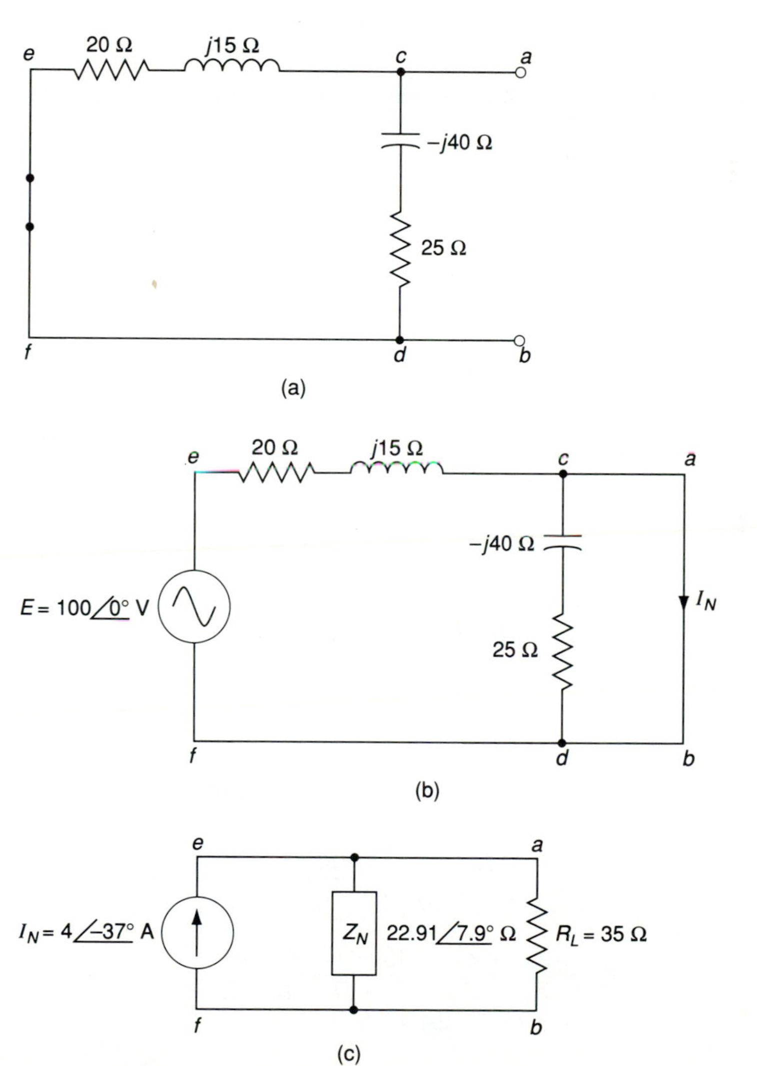

FIGURE 15-38 (a, b) Redrawn circuits and (c) Norton equivalent circuit for Example 15-16.

$$Z_N = \frac{Z_{ec} \times Z_{cd}}{Z_{ec} + Z_{cd}} = \frac{(20 + j15)(25 - j40)}{20 + j15 + 25 - j40} = 22.91 \underline{/7.9°} \ \Omega$$

$$I_N = I_{ec} = \frac{E}{Z_{ec}} = \frac{100 \underline{/0°} \text{ V}}{20 + j15} = 4 \underline{/-37°} \text{ A}$$

The current through R_L can now be determined by using the current divider rule.

$$I_{ab} = \frac{I_N \times Z_N}{Z_{ab} + Z_N} = \frac{(4 \underline{/-37°} \text{ A})(22.91 \underline{/7.9°} \ \Omega)}{35 \underline{/0°} \ \Omega + 22.91 \underline{/7.9°} \ \Omega}$$
$$= 1.59 \underline{/-32.1°} \text{ A}$$

EXAMPLE 15-17 Using Norton's theorem, find the current I_{cd} for the circuit shown in Figure 15-39.

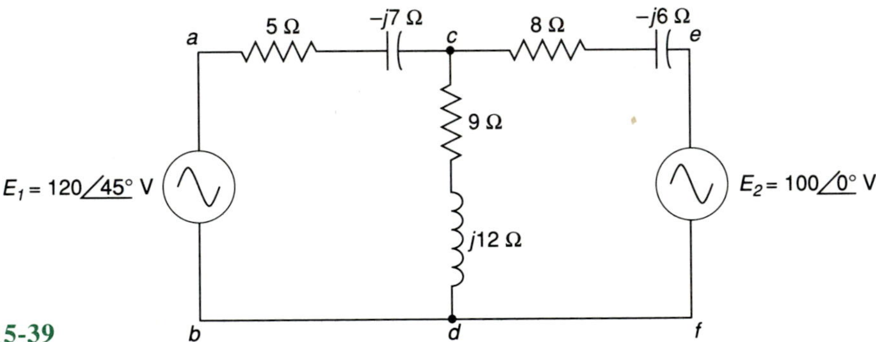

FIGURE 15-39

Solution Redraw the circuit with Z_{cd} short-circuited, as shown in Figure 15-40(a). Convert the voltage sources to current sources, as shown in Figure 15-40(b).

$$I_1 = \frac{E_1}{Z_{ac}} = \frac{120 \underline{/45°} \text{ V}}{5 - j7} = 13.95 \underline{/99.5°} \text{ A}$$

$$I_2 = \frac{E_2}{Z_{ce}} = \frac{100 \underline{/0°} \text{ V}}{8 - j6} = 10 \underline{/36.9°} \text{ A}$$

$$I_N = I_1 + I_2 = 13.95 \underline{/99.5°} \text{ A} + 10 \underline{/36.9°} \text{ A}$$
$$= 20.57 \underline{/73.9°} \text{ A}$$

$$Z_N = \frac{Z_{ac} \times Z_{ce}}{Z_{ac} + Z_{ce}} = \frac{(5 - j7)(8 - j6)}{5 - j7 + 8 - j6}$$
$$= 4.68 \underline{/-46.3°} \ \Omega$$

$$I_{cd} = \frac{I_N \times Z_N}{Z_{cd} + Z_N} = \frac{(20.57 \underline{/73.9°} \text{ A})(4.68 \underline{/-46.3°} \ \Omega)}{9 + j12 + 4.68 \underline{/-46.3°} \ \Omega}$$
$$= 6.44 \underline{/-7.6°} \text{ A}$$

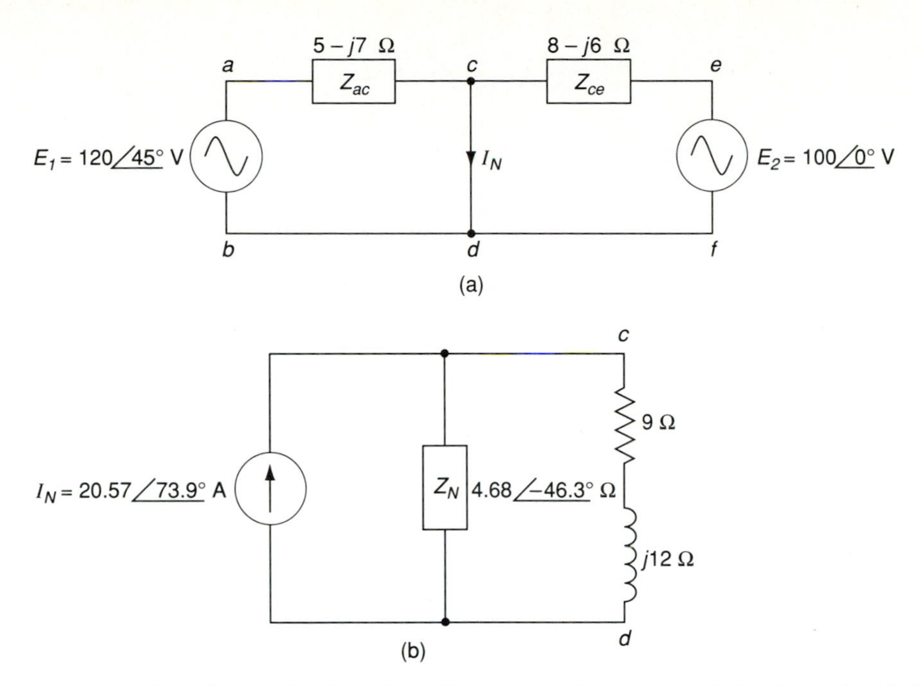

(a) Redrawn circuit and (b) Norton's equivalent circuit for Example 15-17.

15-8 RECIPROCITY THEOREM

The **reciprocity theorem** is stated in the following manner:

> If any voltage source located at one point in a circuit produces a current at a second point in the circuit, the same source of voltage acting at the second point will produce the same current at the first point.

The reciprocity theorem is illustrated in the following example.

EXAMPLE 15-18 According to the reciprocity theorem, if the voltage source, E, shown in Figure 15-41, is inserted in branch *ef* and branch *ab* is shorted, the current that flows in *ab* will be identical to the current flowing in branch *ef* when the voltage source was originally applied to the circuit. To verify this theorem, calculate the current in branch *ef* with the voltage source applied, and branch *ab* with the source short-circuited.

Solution

$$Z_{cf} = \frac{Z_{cef} \times Z_{cd}}{Z_{cef} + Z_{cd}} = \frac{(7 - j2)(-j8)}{7 - j2 - j8}$$
$$= 3 - j3.71$$

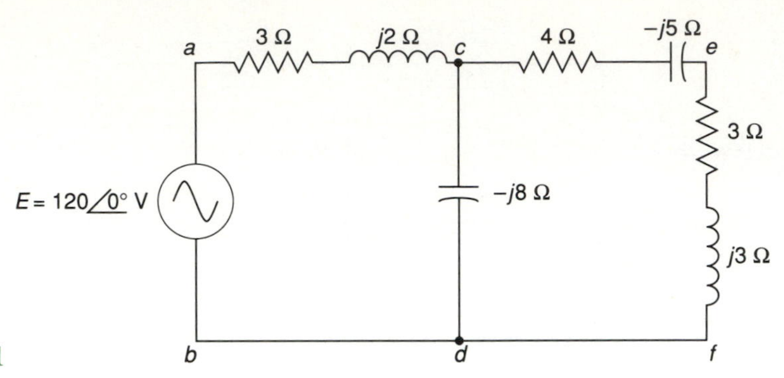

FIGURE 15-41

$$Z_{af} = Z_{ac} + Z_{cf} = 3 + j2 + 3 - j3.71$$
$$= 6 - j1.71$$

$$I_{af} = \frac{E}{Z_{af}} = \frac{120 + j0}{6 - j1.71} = 19.23 \underline{/15.9^\circ} \text{ A}$$

$$V_{cf} = I_{af} \times Z_{cf} = (19.23 \underline{/15.9^\circ} \text{ A})(4.77 \underline{/-51.04^\circ} \ \Omega)$$
$$= 91.73 \underline{/-35.14^\circ} \text{ V}$$

$$I_{ef} = \frac{V_{cf}}{Z_{cef}} = \frac{91.73 \underline{/-35.14^\circ} \text{ V}}{7 - j2} = 12.6 \underline{/-19.2^\circ} \text{ A}$$

To verify that the currents in branches *ab* and *ef* are the same when the voltage source is substituted, assume that the 120-V source is inserted in branch *ef* and that *ab* is shorted. The current in *ab* will be calculated in a similar manner to that shown above.

$$Z_{cb} = \frac{Z_{cab} \times Z_{cd}}{Z_{cab} + Z_{cd}} = \frac{(3 + j2)(-j8)}{3 + j2 - j8} = 4.27 + j0.53$$
$$= 4.3 \underline{/7.1^\circ} \ \Omega$$

$$Z_{fcb} = Z_{fc} + Z_{cb} = 7 - j2 + 4.27 + j0.53$$
$$= 11.27 - j1.47 \ \Omega$$

$$I_{fcb} = \frac{E}{Z_{fcb}} = \frac{120 + j0}{11.27 - j1.47} = 10.56 \underline{/7.43^\circ} \text{ A}$$

$$V_{cb} = I_{fcb} \times Z_{cb} = (10.56 \underline{/7.43^\circ} \text{ A})(4.3 \underline{/7.1^\circ} \ \Omega)$$
$$= 45.41 \underline{/14.53^\circ} \text{ V}$$

$$I_{ab} = \frac{V_{cb}}{Z_{cab}} = \frac{45.41 \underline{/14.53^\circ}}{3 + j2} = 12.6 \underline{/-19.2^\circ} \text{ A}$$

Therefore,

$$I_{ab} = I_{ef}$$

According to the reciprocity theorem, the ratio of the emf in branch 1 of a circuit to the current it causes in branch 2 is the same as the ratio of a voltage placed in branch 2 to the current it would cause in branch 1. This ratio of the voltage in one branch to the current in another branch is called the **transfer impedance.** Although the reciprocity theorem is useful in some situations, its applications are limited due to the fact that the theorem may only be applied to circuits having a single voltage source. Therefore, multisource ac networks cannot be solved using the reciprocity theorem.

15-9 AC BRIDGES

One method that is used to measure the separate resistive and reactive components of an impedance is the **ac bridge**. A variation of the Wheatstone bridge is shown in Figure 15-42. This ac bridge can be used to measure an unknown impedance by balancing the bridge. In addition to measuring impedances, the ac bridge can also be used for phase shifting, filtering out undesired signals, and frequency measurement of audio signals.

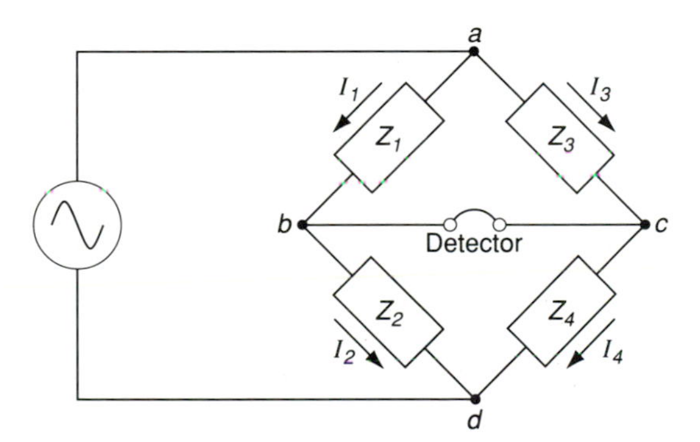

FIGURE 15-42 Ac bridge circuit.

The operation of the ac bridge is dependent on the fact that when certain circuit conditions exist, the detector current will be zero. When this occurs, the detector is said to be in a null or balanced condition. Under these conditions, the voltage from a to b equals the voltage from a to c. Therefore,

$$I_1Z_1 = I_3Z_3 \tag{15-1}$$

$$I_2Z_2 = I_4Z_4 \tag{15-2}$$

Under balanced conditions, the relationship between the four impedances can also be expressed as

$$\frac{Z_1}{Z_2} = \frac{Z_3}{Z_4} \tag{15-3}$$

which is also

$$Z_1Z_4 = Z_2Z_3 \tag{15-4}$$

Equation 15-4, known as the **general bridge equation**, applies to any balanced four-arm bridge circuit. To be balanced, at least one of the known impedances must contain a resistive and a reactive component. When dealing with an unbalanced bridge, equation 15-4 cannot be used, which means that the conventional analysis techniques discussed earlier in this chapter must be applied to solve for the unknown voltages and currents in the network.

The ac bridge shown in Figure 15-43 is used to measure the impedance of a capacitive circuit. This network is known as a **capacitance comparison bridge**. The impedance of the arms of the capacitance comparison bridge can be written as

$$Z_1 = R_1 \qquad Z_2 = R_2 - jX_{C2}$$

$$Z_3 = R_3 \qquad Z_4 = R_X - jX_{CX}$$

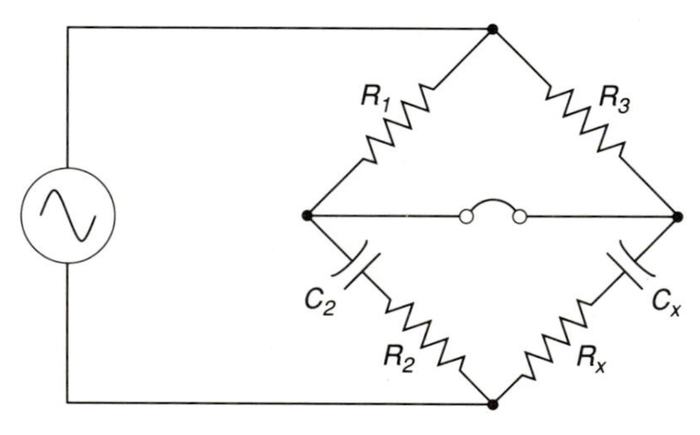

FIGURE 15-43 Capacitance bridge.

By substituting these values into the general bridge equation, the resulting balance equation is expressed as

$$R_1(R_X - jX_{CX}) = (R_2 - jX_{C2})R_3$$

This equation can be simplified by multiplying and grouping the real and imaginary terms, which results in

$$R_1R_X - jR_1X_{CX} = R_3R_2 - jR_3X_{C2}$$

Under balanced conditions, the unknown capacitance, C_X, and series resistance, R_X, can be determined by further reducing the equation above.

$$R_X = R_2 \frac{R_3}{R_1} \qquad (15\text{-}5)$$

$$C_X = C_2 \frac{R_1}{R_3} \qquad (15\text{-}6)$$

The capacitance comparison bridge is shown as containing a series combination of resistance and capacitance. By representing the unknown impedance in this manner, it is implied that the impedance is an unknown series *RC* circuit and is generally referred to as an equivalent series circuit.

EXAMPLE 15-19 For the capacitance bridge shown in Figure 15-44, determine the equivalent series circuit.

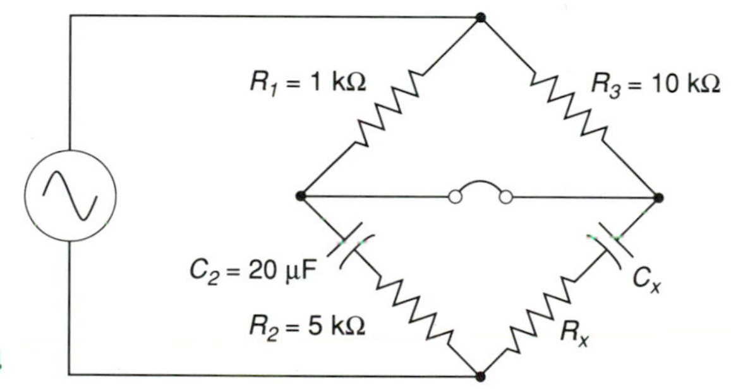

FIGURE 15-44

Solution

$$R_X = R_2 \frac{R_3}{R_1} = 5 \times 10^3 \frac{10 \times 10^3}{1 \times 10^3}$$
$$= 50 \text{ k}\Omega$$

$$C_X = C_2 \frac{R_1}{R_3} = 20 \times 10^{-6} \frac{1 \times 10^3}{10 \times 10^3}$$
$$= 2 \ \mu\text{F}$$

Although it is possible to measure the impedance of an inductive circuit by constructing an inductance comparison bridge similar to the capacitance bridge, it is not very practical. Standard inductors are quite large physically, and also quite expensive. For this reason, the **Maxwell–Wein bridge** is more commonly used for measuring inductance. The Maxwell–Wein bridge is shown in Figure 15-45. The values for an equivalent series circuit containing resistance and inductance are determined using the following equations:

$$R_X = \frac{R_2 R_3}{R_1} \tag{15-7}$$

$$L_X = C_S R_2 R_3 \tag{15-8}$$

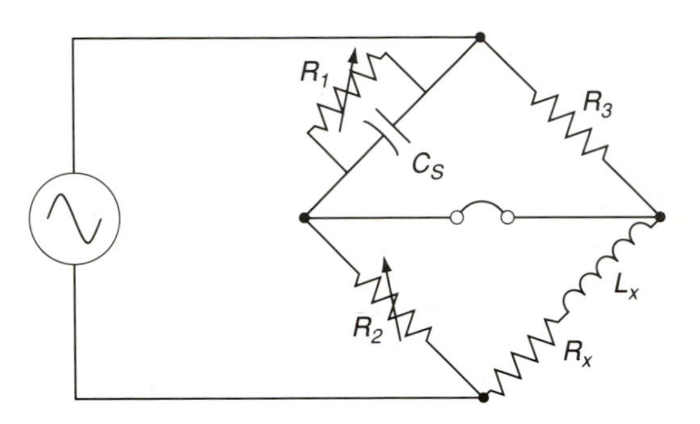

FIGURE 15-45 Maxwell bridge.

EXAMPLE 15-20 For the Maxwell–Wein bridge shown in Figure 15-46, calculate the equivalent series circuit required to give zero deflection of the null detector.

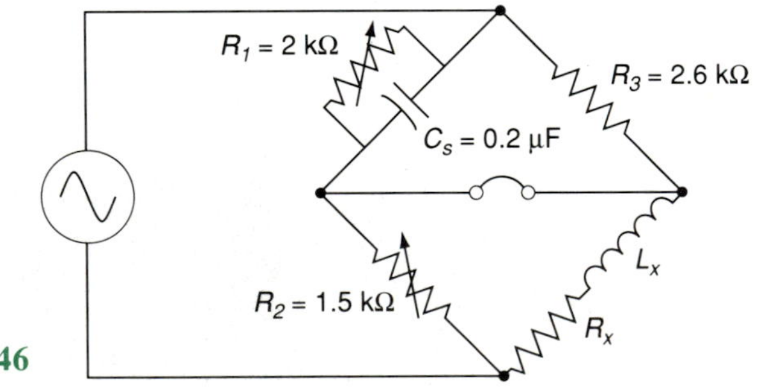

FIGURE 15-46

Solution

$$R_X = \frac{R_2 R_3}{R_1} = \frac{(1500\ \Omega)(2600\ \Omega)}{2000\ \Omega} = 1950\ \Omega$$

$$L_X = C_S R_2 R_3 = (0.2 \times 10^{-6}\ \text{F})(1.5 \times 10^{3}\ \Omega)(2.6 \times 10^{3}\ \Omega) = 0.78\ \text{H}$$

In situations where L_X is very large compared to R_X, the Maxwell–Wein bridge is not particularly effective. When the components of an impedance that has a very low power factor are to be measured, it is more practical to use a **Hay bridge**, as shown in Figure 15-47. Although the method of determining unknown impedance is more complicated using the Hay bridge than using the Maxwell–Wein bridge, the Hay bridge calculation is considerably more accurate. The following formulas for the unknowns R_X and L_X are found to specify the balance conditions:

$$R_X = \frac{R_1 R_2 R_3 \omega^2 C^2}{1 + R_1^2 \omega^2 C^2} \tag{15-9}$$

$$L_X = \frac{R_2 R_3 C}{1 + R_1^2 \omega^2 C^2} \tag{15-10}$$

EXAMPLE 15-21 The Hay bridge circuit of Figure 15-47 has the following values: R_1 = 14.7 kΩ, C = 500 pF, R_2 = 10 kΩ, R_3 = 10 kΩ, and f = 5 kHz. Determine the values of L_X and R_X for this circuit.

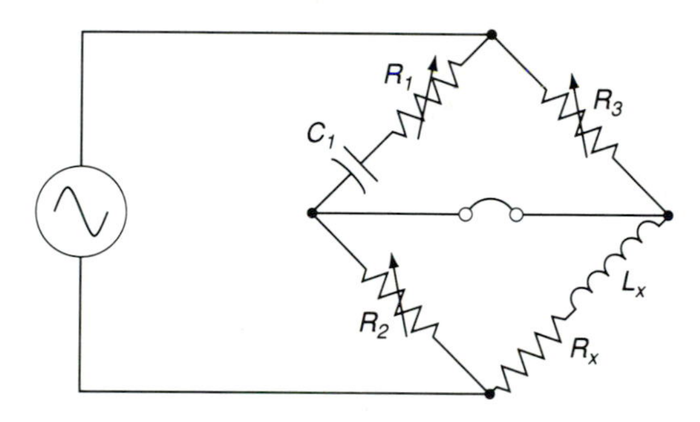

FIGURE 15-47 Hay bridge for measuring inductance.

Solution

$$\begin{aligned} R_X &= \frac{R_1 R_2 R_3 \omega^2 C^2}{1 + R_1^2 \omega^2 C^2} \\ &= \frac{(14.7\ \text{k}\Omega)(10\ \text{k}\Omega)(10\ \text{k}\Omega)(2\pi \times 5\ \text{kHz})^2(500 \times 10^{-12}\ \text{F})^2}{1 + (14.7\ \text{k}\Omega)^2(2\pi \times 5\ \text{kHz})^2(500 \times 10^{-12}\ \text{F})^2} \\ &= 344.35\ \Omega \end{aligned}$$

$$\begin{aligned} L_X &= \frac{R_2 R_3 C}{1 + R_1^2 \omega^2 C^2} \\ &= \frac{(10\ \text{k}\Omega)(10\ \text{k}\Omega)(500 \times 10^{-12}\ \text{F})}{1 + (14.7\ \text{k}\Omega)^2(2\pi \times 5\ \text{kHz})^2(500 \times 10^{-12})^2} \\ &= 47.47\ \text{mH} \end{aligned}$$

15-10 MAXIMUM POWER TRANSFER THEOREM

When considering a purely resistive circuit, in order for a voltage source or current source to deliver maximum power to its load, the load resistance must equal the source resistance. However, when applied to ac circuits containing reactive components, the **maximum power transfer theorem** states:

> The maximum power will be supplied to the load when the load impedance is the complex conjugate of the Thévenin impedance of the source.

The load impedance is the conjugate of the internal source impedance when the load resistance equals the internal resistance of the source. Therefore, the reactance of the load is equal in magnitude but opposite in direction to the reactance of the source. In equation form,

$$Z_L = Z_{TH} \qquad \theta_L = -\theta_{TH} \tag{15-11}$$

If Z_{TH} is a completely arbitrary impedance,

$$Z_{TH} = R \pm jX = A \angle \theta^\circ \tag{15-12}$$

The load impedance for the maximum power transfer is determined by the equation

$$Z_L = R \pm jX = A \angle -\theta^\circ \tag{15-13}$$

Consider the circuit shown in Figure 15-48. The implications in the equations above is that maximum power output occurs when the reactive components of the source and load are equal and opposite in direction, and when the resistive components of the source and load are equal. Therefore, when an L_R load is connected to an ac source, the source must contain an equal amount of resistance and capacitance to compensate for deficiencies caused by the load. For the circuit shown in Figure 15-48, the total current, I, is calculated as

$$I = \frac{E}{Z_{TH} + Z_{load}} = \frac{E}{(R_{TH} + R_{load}) + j(X_{TH} + X_{load})}$$

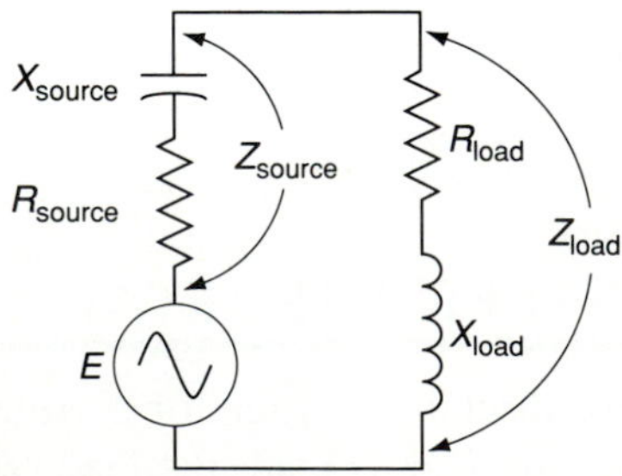

FIGURE 15-48

Since load power is dissipated only in the resistive component of the load, the load power is calculated in the following manner:

$$P_{\text{load}} = I^2R_{\text{load}} = \frac{E^2R_{\text{load}}}{(R_{\text{TH}} + R_{\text{load}})^2 + (X_{\text{TH}} + X_{\text{load}})^2}$$

EXAMPLE 15-22 The Thévenin equivalent circuit shown in Figure 15-49 consists of $Z_{\text{TH}} = 15 + j25$, $E = 100 \angle 0°$ V. Z_L is a purely resistive load. Determine the value of Z_L for which the source delivers maximum power to the load. Also, find the value of the maximum power.

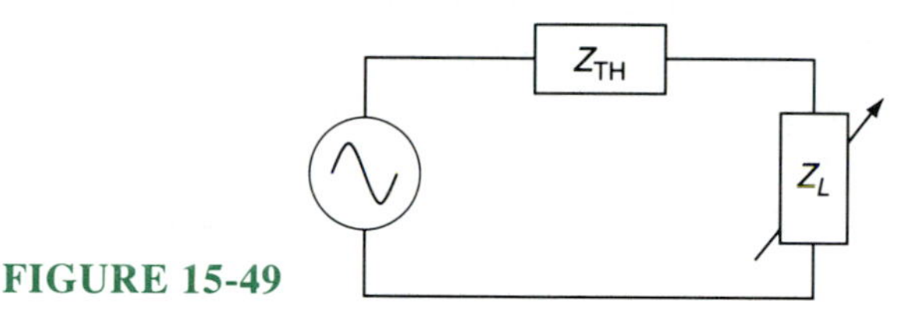

FIGURE 15-49

Solution Maximum power transfer occurs when

$$Z_{\text{TH}} = 15 + j25$$

Therefore,

$$Z_L = 15 - j25$$

$$Z_T = 15 + j25 + 15 - j25 = 30\ \Omega$$

$$I = \frac{100 \angle 0°\ \text{V}}{30\ \Omega} = 3.33\ \text{A}$$

$$P_{\text{load}} = I^2Z_{\text{load}} = (3.33)^2 30 = 332.57\ \text{W}$$

EXAMPLE 15-23 For the circuit shown in Figure 15-50, determine the value of load impedance for maximum load power.

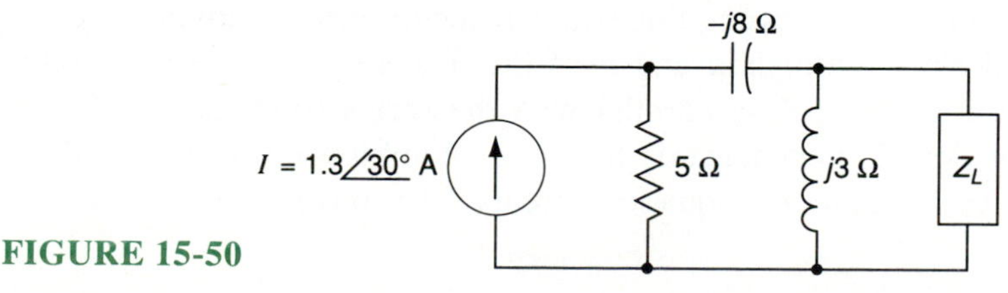

FIGURE 15-50

Solution $Z_{\text{TH}} = \dfrac{(5 - j8)(j3)}{5 - j8 + j3} = 4 \underline{/77^\circ}\ \Omega$

Since $Z_{\text{TH}} = Z_{\text{L}}$ and $\theta_{\text{TH}} = -\theta_L$, the value of load impedance for maximum load power is

$Z_L = 4 \underline{/-77^\circ}\ \Omega$

15-11 DELTA–WYE AND WYE–DELTA CONVERSIONS

Many electrical connections take the shape of a **delta** (Δ), which is also known as a **pi** (Π) **network**. These networks are illustrated in Figure 15-51(a) and (b). Another method of connecting three loads is shown in Figure 15-51(c) and (d). This method is referred to as a **wye** (Y) or **tee** (T) **network**. The equivalent circuit method of network analysis relies on the fact that it is possible to reduce complex networks to simpler versions of the original circuit. There are, however, certain network configurations which cannot be resolved by series–parallel combinations alone. These configurations may often be solved by the use of a Y–Δ conversion. This conversion results in the three Y-connected impedances being converted into a configuration, or vice versa.

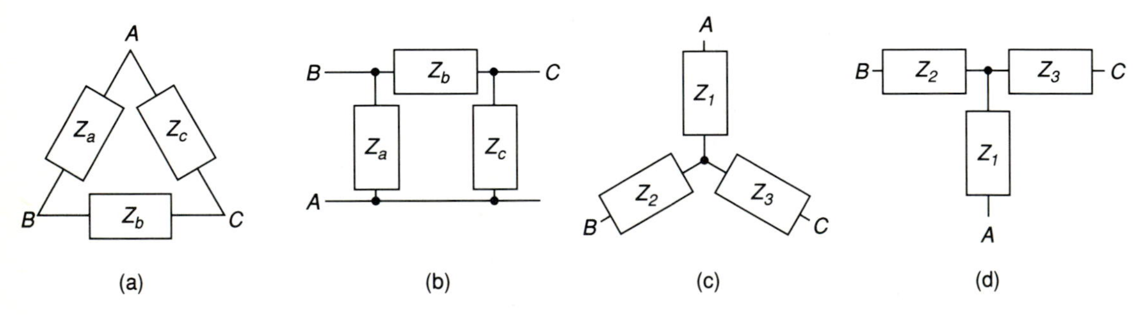

FIGURE 15-51 (a) Δ Network; (b) Π network; (c) Y network; (d) T network.

The circuits shown in Figure 15-51(a) and (c) are Δ and Y networks, respectively. If these circuits are to be equivalent, the impedance between any pair of terminals must be the same in the Y as in the Δ. Three simultaneous equations may be written expressing this equivalence of terminal impedances. Therefore, considering terminals *a* and *b* of the Y network, the equivalent impedance is the impedance Z_c in parallel with the series combination of Z_b and Z_a. The equivalent Y impedance is the series combination of Z_1 and Z_3. The following three simultaneous equations are used to convert Δ impedances to Y impedances:

$$Z_1 = \frac{Z_a Z_c}{Z_a + Z_b + Z_c} \tag{15-14}$$

$$Z_2 = \frac{Z_a Z_b}{Z_a + Z_b + Z_c} \tag{15-15}$$

$$Z_3 = \frac{Z_b Z_c}{Z_a + Z_b + Z_c} \tag{15-16}$$

To convert from Y to Δ:

$$Z_a = \frac{Z_1 Z_2 + Z_2 Z_3 + Z_3 Z_1}{Z_3} \tag{15-17}$$

$$Z_b = \frac{Z_1 Z_2 + Z_2 Z_3 + Z_3 Z_1}{Z_1} \tag{15-18}$$

$$Z_c = \frac{Z_1 Z_2 + Z_2 Z_3 + Z_3 Z_1}{Z_2} \tag{15-19}$$

One method that helps in the process of network conversion is to draw the wye network inside the delta, as shown in Figure 15-52. Each impedance of the delta is referred to as a *side* of the triangle formed by the delta. For Δ–Y conversions, each impedance of the wye is given as the product of the two *adjacent* sides of the delta divided by the sum of the delta impedances. For Y–Δ conversions, the sum of products is determined by $N = Z_1 Z_2 + Z_2 Z_3 + Z_3 Z_1$, and then each impedance of a side of the delta is equal to N divided by the *opposite* impedance of the wye.

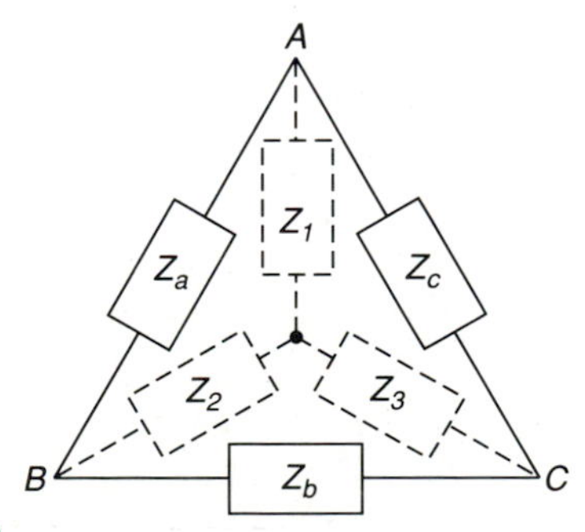

FIGURE 15-52

EXAMPLE 15-24 Determine the single equivalent impedance that will replace the network of Figure 15-53 between terminals a and d.

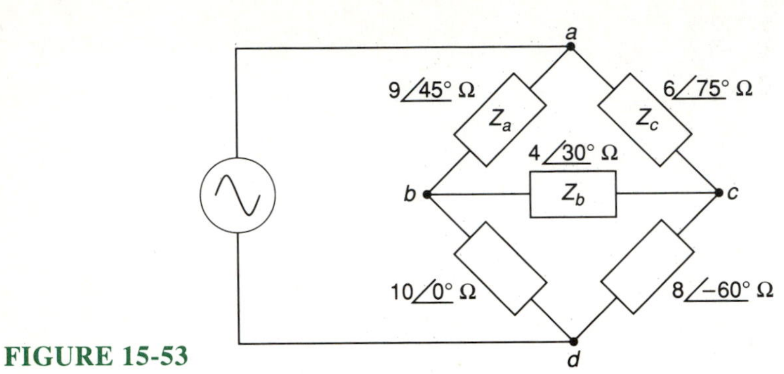

FIGURE 15-53

Solution In the circuit, no impedances are directly in parallel or directly in series. However, sections *acb* and *dcb* both form Δ networks. Either of these may be converted to an equivalent Y network. In this solution, *acb* is chosen to be converted to a Y, as shown in Figure 15-54(a).

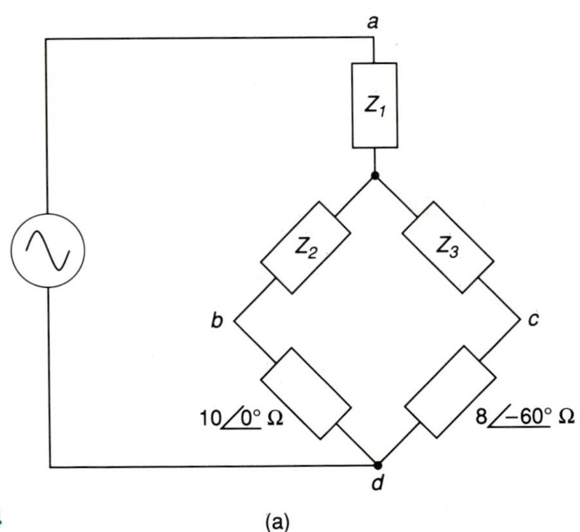

FIGURE 15-54 (a)

$$\begin{aligned} Z_1 &= \frac{Z_a Z_c}{Z_a + Z_b + Z_c} \\ &= \frac{(9\angle 45^\circ\ \Omega)(6\angle 75^\circ\ \Omega)}{(9\angle 45^\circ\ \Omega) + (4\angle 30^\circ\ \Omega) + (6\angle 75^\circ\ \Omega)} \\ &= 2.97\angle 68.8^\circ\ \Omega \end{aligned}$$

$$Z_2 = \frac{Z_a Z_b}{Z_a + Z_b + Z_c}$$

$$= \frac{(9\,\underline{/45^\circ}\ \Omega)(4\,\underline{/30^\circ}\ \Omega)}{18.167\,\underline{/51.21^\circ}\ \Omega}$$

$$= 1.98\,\underline{/23.8^\circ}\ \Omega$$

$$Z_3 = \frac{Z_b Z_c}{Z_a + Z_b + Z_c}$$

$$= \frac{(4\,\underline{/30^\circ}\ \Omega)(6\,\underline{/75^\circ}\ \Omega)}{18.167\,\underline{/51.21^\circ}\ \Omega}$$

$$= 1.32\,\underline{/53.8^\circ}\ \Omega$$

The circuit that results from replacing the *ACB* network is shown in Figure 15-54(a). In this circuit, Z_2 and the $10\,\underline{/0^\circ}\ \Omega$ impedance Z_{bd} are connected in series, similar to impedances Z_3 and Z_{cd}. Therefore, the circuit is now solved as a series–parallel circuit, as shown in Figure 15-54(b).

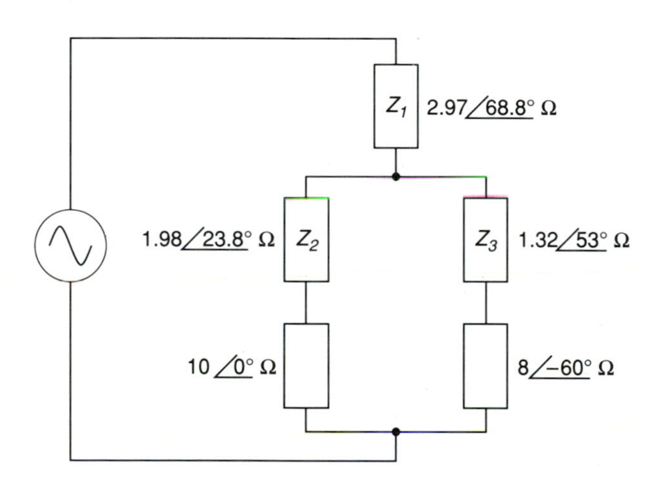

FIGURE 15-54 (b)

$$Z_{eq} = Z_1 + \frac{(Z_2 + Z_{bd})(Z_3 + Z_{cd})}{Z_2 + Z_{bd} + Z_3 + Z_{cd}}$$

$$= 2.97\,\underline{/68.8^\circ}\ \Omega$$

$$+ \frac{(1.98\,\underline{/23.8^\circ}\ \Omega + 10\,\underline{/0^\circ}\ \Omega)(1.32\,\underline{/53.8^\circ}\ \Omega + 8\,\underline{/-60^\circ}\ \Omega)}{1.98\,\underline{/23.8^\circ}\ \Omega + 10\,\underline{/0^\circ} + 1.32\,\underline{/53.8^\circ}\ \Omega + 8\,\underline{/-60^\circ}\ \Omega}$$

$$= 5.55\,\underline{/2^\circ}\ \Omega$$

EXAMPLE 15-25 Use network reduction to determine the voltage across points *b–e* for the circuit shown in Figure 15-55.

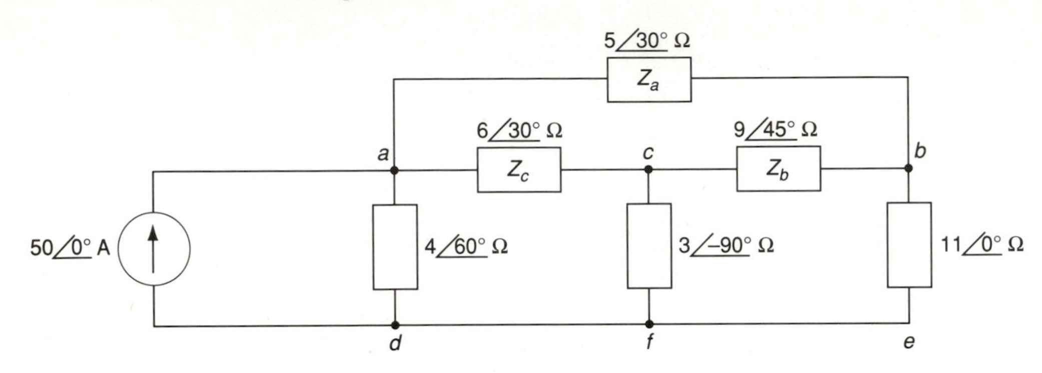

FIGURE 15-55 Circuit for Example 15-24.

Solution The three impedance segments formed by junctions *a*, *b*, and *c* is a Π(Δ) network. The equivalent wye network is found in the same manner as that shown in Example 15-23.

$$\begin{aligned} Z_1 &= \frac{Z_a Z_c}{Z_a + Z_b + Z_c} \\ &= \frac{(5\,\underline{/30^\circ}\,\Omega)(6\,\underline{/30^\circ}\,\Omega)}{5\,\underline{/30^\circ}\,\Omega + 6\,\underline{/30^\circ}\,\Omega + 9\,\underline{/45^\circ}\,\Omega} \\ &= 1.5\,\underline{/22.6^\circ}\,\Omega \end{aligned}$$

$$\begin{aligned} Z_2 &= \frac{Z_a Z_b}{Z_a + Z_b + Z_c} \\ &= \frac{(5\,\underline{/30^\circ}\,\Omega)(9\,\underline{/45^\circ}\,\Omega)}{19.993\,\underline{/37.4^\circ}\,\Omega} \\ &= 2.25\,\underline{/37.6^\circ}\,\Omega \end{aligned}$$

$$\begin{aligned} Z_3 &= \frac{Z_b Z_c}{Z_a + Z_b + Z_c} \\ &= \frac{(6\,\underline{/30^\circ}\,\Omega)(9\,\underline{/45^\circ}\,\Omega)}{19.993\,\underline{/37.4^\circ}\,\Omega} \\ &= 2.7\,\underline{/37.6^\circ}\,\Omega \end{aligned}$$

The replacement of the Δ with the equivalent Y, and the conversion of the $50\,\underline{/0^\circ}$ A, $4\underline{/60^\circ}$ Ω source is shown in Figure 15-56(a). The two series combinations are then simplified and the resulting $200\,\underline{/60^\circ}$ V, $5.3\,\underline{/50^\circ}$ Ω source is converted to a current source. The circuit now appears as in Figure 15-56(b). The circuit can now be converted back to a voltage source, as shown in Figure 15-56(c), and the voltage across points b–e is found by the voltage divider rule.

$$V_{be} = 98\ \angle 67.3^\circ\ \text{V} \times \frac{11\ \angle 0^\circ\ \Omega}{11\ \angle 0^\circ\ \Omega + 2.65\ \angle 58.5^\circ\ \Omega + 2.25\ \angle 37.6^\circ\ \Omega}$$

$$= 73.7\ \angle 52.9^\circ\ \text{V}$$

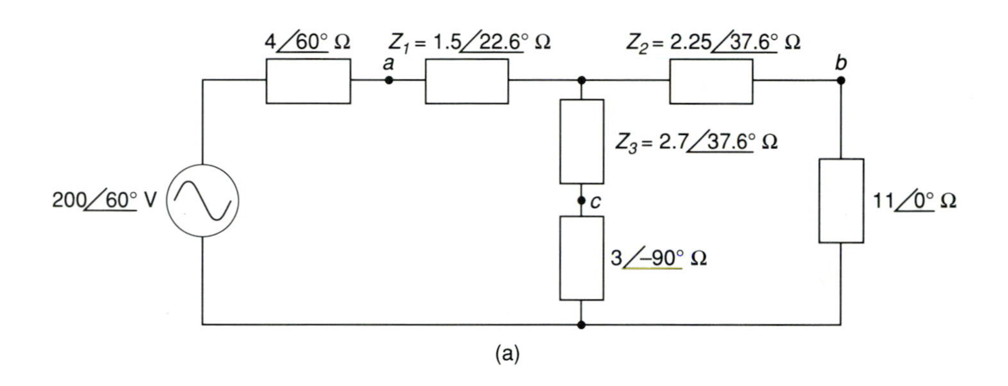

(a)

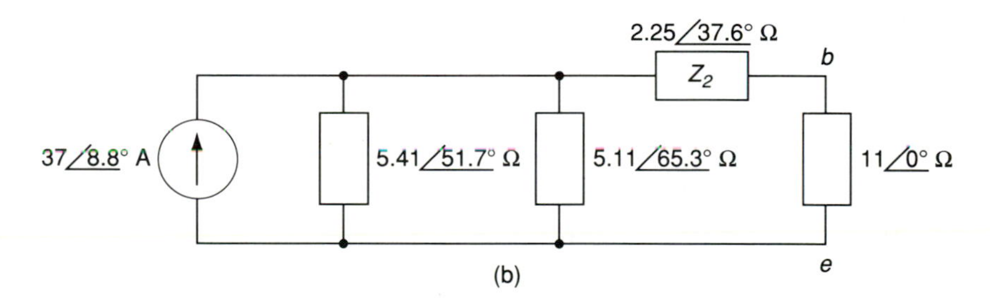

(b)

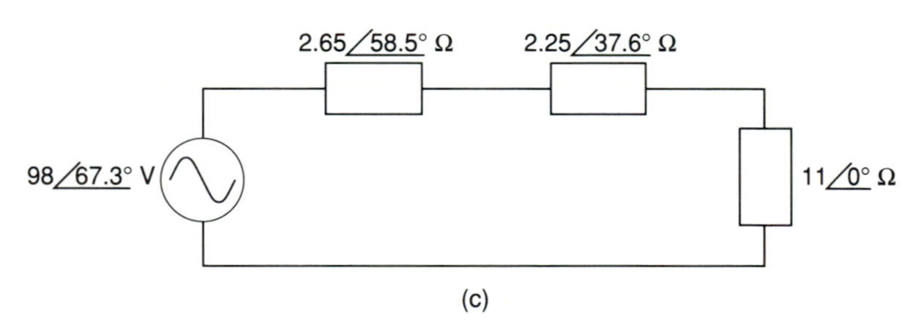

(c)

FIGURE 15-56

EXAMPLE 15-26 Find the delta that will replace the wye system shown in Figure 15-57.

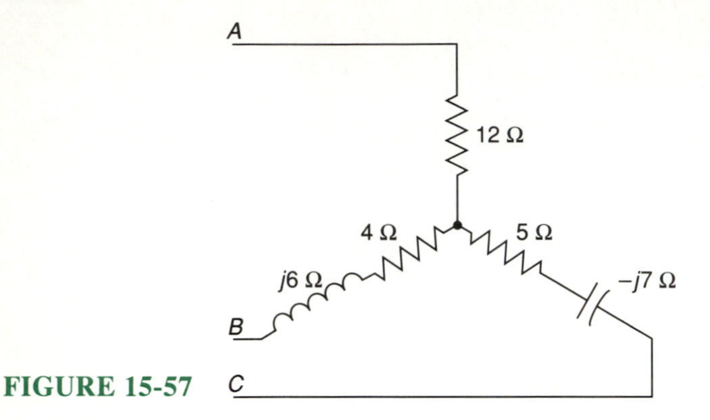

FIGURE 15-57

Solution

$$Z_a = \frac{Z_1Z_2 + Z_2Z_3 + Z_3Z_1}{Z_3}$$

$$= \frac{(12)(4 + j6) + (4 + j6)(5 - j7) + (5 - j7)(12)}{5 - j7}$$

$$= (12.43 + j15.39)\ \Omega$$

$$Z_b = \frac{Z_1Z_2 + Z_2Z_3 + Z_3Z_1}{Z_1}$$

$$= \frac{169.926 - j10.066}{12}$$

$$= (14.16 + j0.84)\ \Omega$$

$$Z_c = \frac{Z_1Z_2 + Z_2Z_3 + Z_3Z_1}{Z_2}$$

$$= \frac{169.926 - j10.066}{4 + j6}$$

$$= (11.91 - j20.38)\ \Omega$$

The equivalent delta circuit is shown in Figure 15-58.

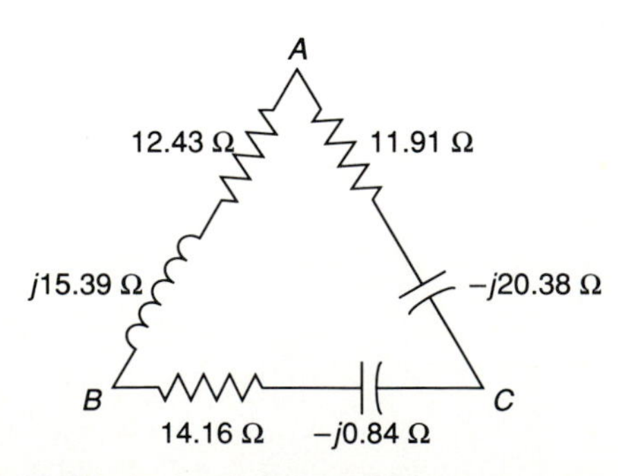

FIGURE 15-58 Equivalent delta circuit for Example 15-25.

KEY TERMS

Node
Principal node
Nodal analysis
Reference node
Superposition theorem
Thévenin's theorem
Thévenin equivalent circuit
Norton's theorem
Reciprocity theorem
Transfer impedance
Ac bridge
General bridge equation
Capacitance comparison bridge
Maxwell–Wein bridge
Hay bridge
Maximum power transfer theorem
Delta (pi) network
Wye (tee) network

PROBLEMS

15-1 Convert the voltage source of Figure 15-59 to a current source.

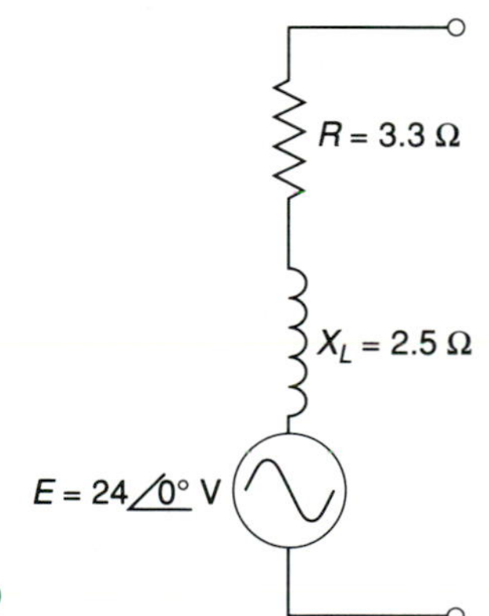

FIGURE 15-59

5-2 Convert the current source of Figure 15-60 to a voltage source.

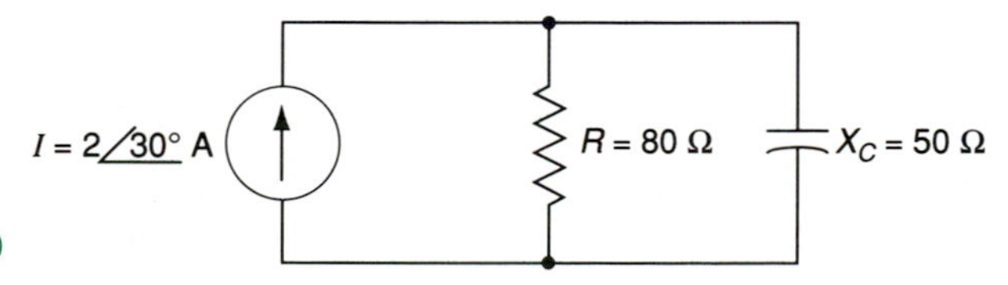

FIGURE 15-60

15-3 Using loop analysis, find currents I_1 and I_2 for the circuit shown in Figure 15-61.

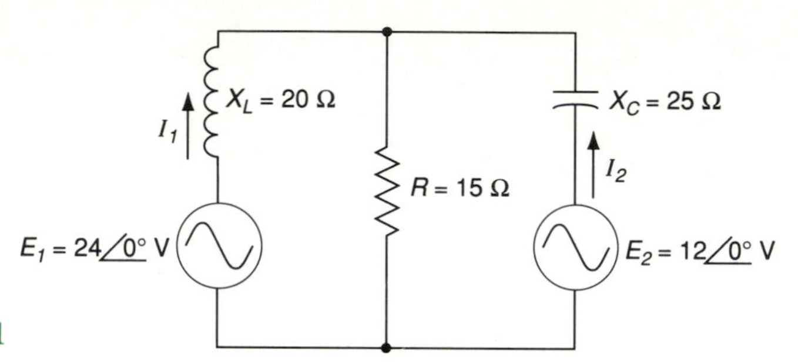

FIGURE 15-61

15-4 For the circuit shown in Figure 15-62, use loop analysis to determine currents I_1, I_2, and I_3. All impedances are $8\ \angle 60°\ \Omega$.

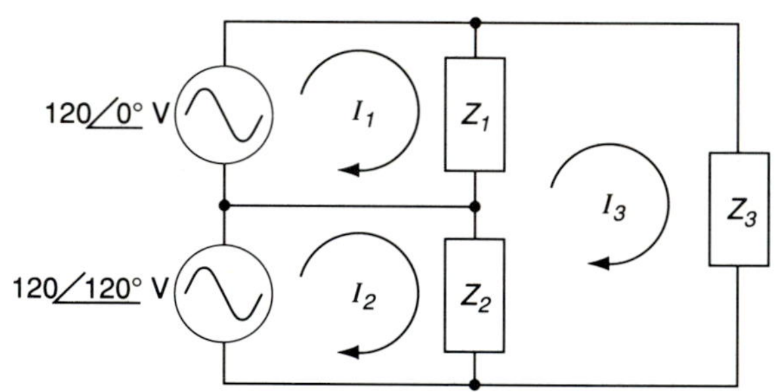

FIGURE 15-62

15-5 For the circuit shown in Figure 15-63, use nodal analysis to determine the currents I_1, I_2, and I_3.

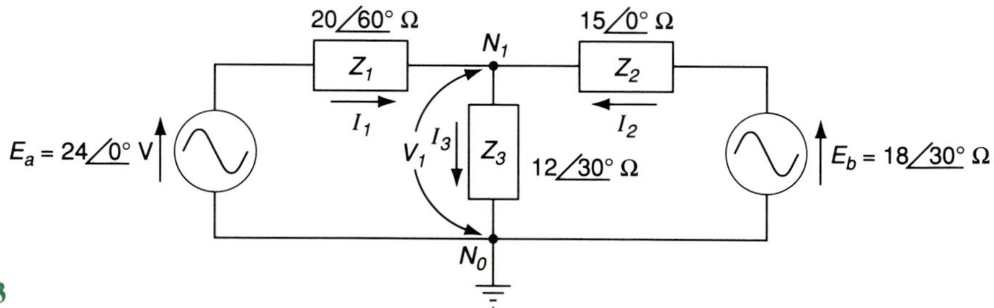

FIGURE 15-63

15-6 For the circuit of Figure 15-64, use nodal analysis to determine (a) the current through the series-connected resistor and capacitor; (b) the current through resistor R_3.

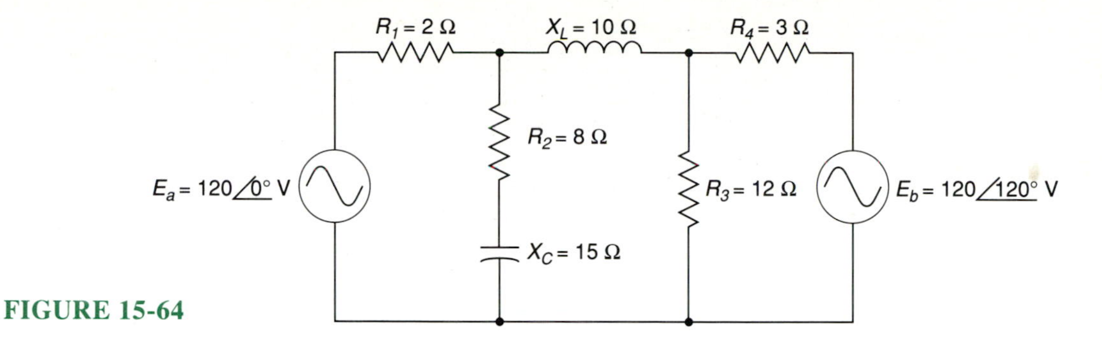

FIGURE 15-64

15-7 Use the superposition theorem to determine the voltage dropped across resistor R_2 in the circuit of Figure 15-65.

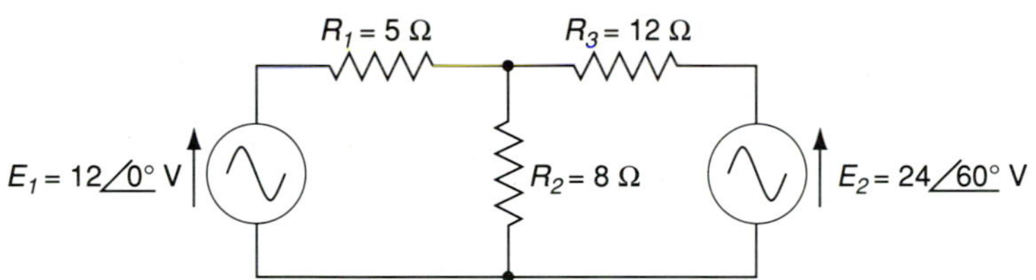

FIGURE 15-65

15-8 For the circuit shown in Figure 15-66, determine the value of current flowing through impedance Z_3 by use of the superposition theorem.

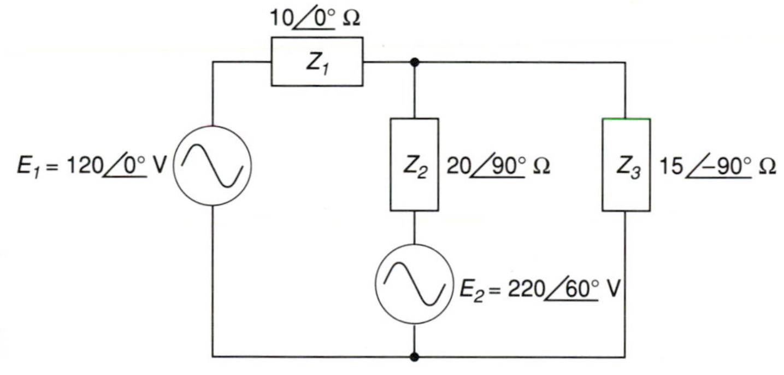

FIGURE 15-66

15-9 Repeat Problem 15-8 for an impedance of $Z_3 = 25\angle 40°\ \Omega$.

15-10 For the circuit shown in Figure 15-67, use the superposition theorem to find the current through the 20-Ω resistor.

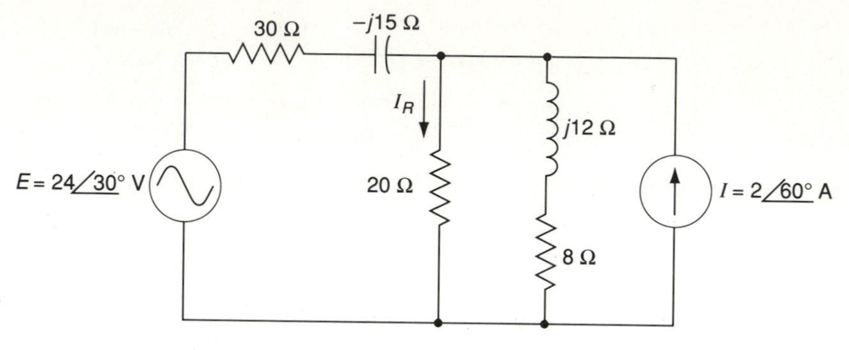

FIGURE 15-67

15-11 Use Thévenin's theorem to determine the current through Z_L in Figure 15-68.

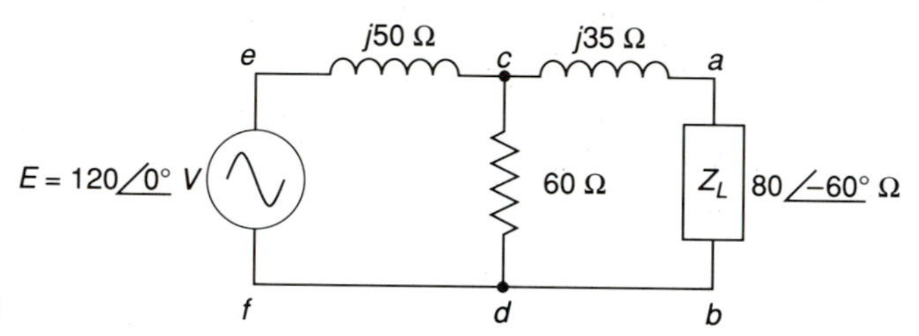

FIGURE 15-68

15-12 For the circuit of Figure 15-69, determine the current through the 15-Ω resistor using Thévenin's theorem.

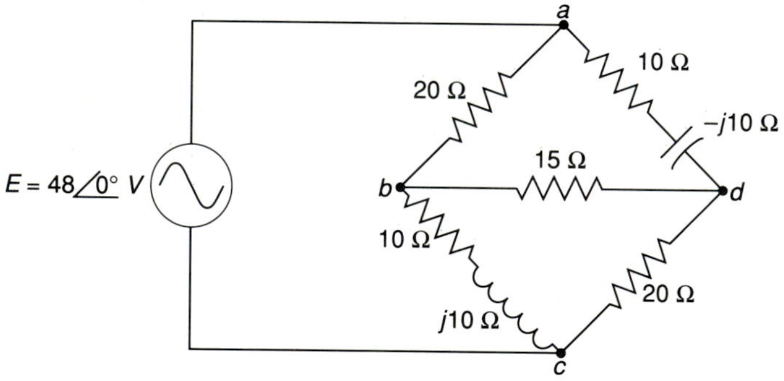

FIGURE 15-69

15-13 For Figure 15-70, find the Thévenin voltage for the circuit external to R_L.

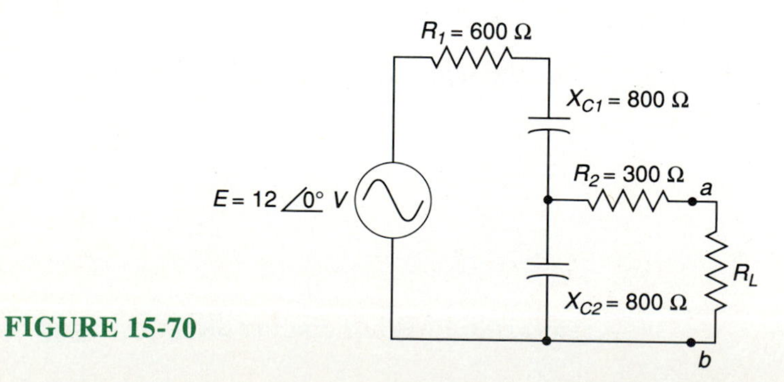

FIGURE 15-70

15-14 For the circuit shown in Figure 15-71, use Thévenin's theorem to solve for the current through the 5-Ω resistor between points *a* and *b*.

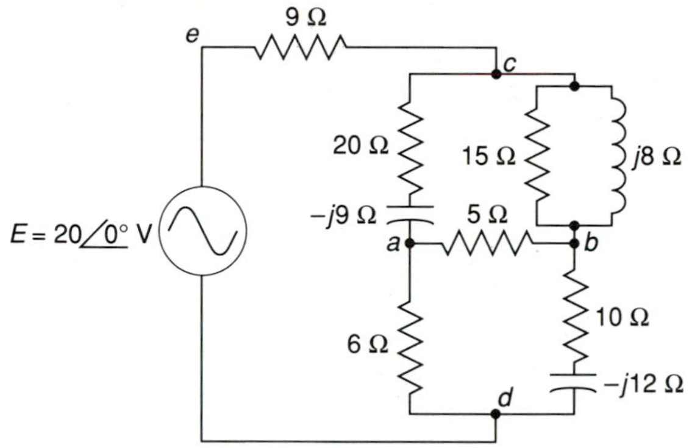

FIGURE 15-71

15-15 If the circuit shown in Figure 15-71 has a source voltage of $E = 120 \angle 60^\circ$ V and the resistance between points *a* and *b* is 12 Ω, determine the equivalent Thévenin voltage and impedance for the circuit.

15-16 For the circuit shown in Figure 15-72, use Norton's theorem to determine the current through R_L.

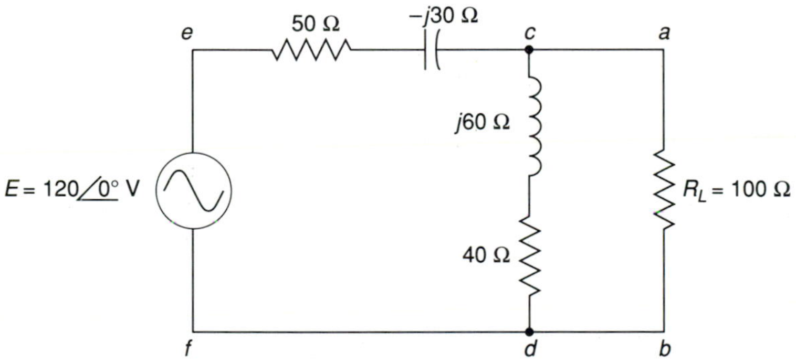

FIGURE 15-72

15-17 If the circuit of Figure 15-72 has an applied voltage of $50 \angle 30^\circ$ V and a load resistance of 75 Ω, what would be the load current? Solve by using Norton's theorem.

15-18 Use Norton's theorem to find the current through branch *cd* in the circuit of Figure 15-73.

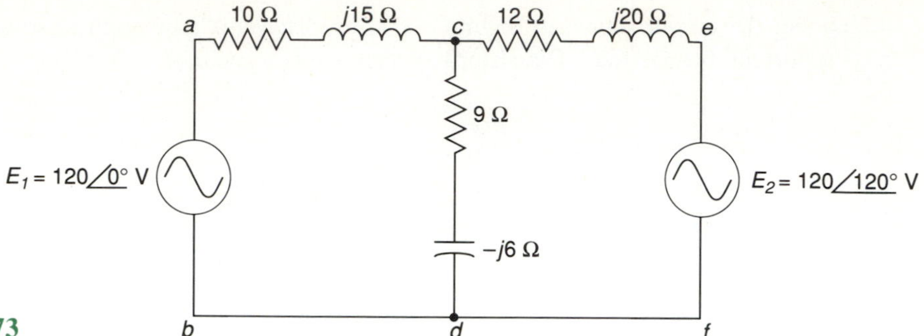

FIGURE 15-73

15-19 Apply the reciprocity theorem to the circuit shown in Figure 15-74 to find the current through Z_L. Insert the source voltage between points *a* and *b* to prove the theorem.

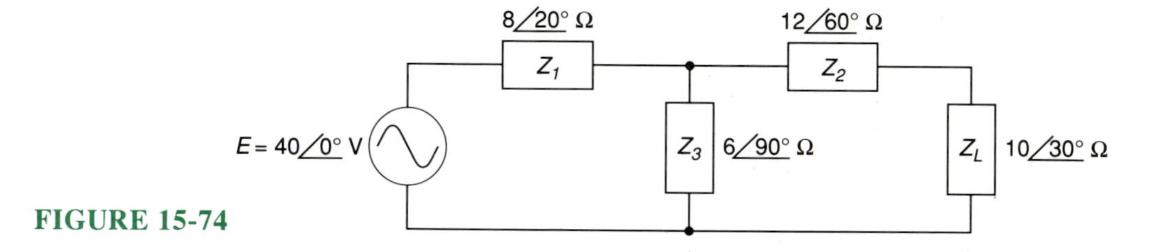

FIGURE 15-74

15-20 If the applied voltage of the circuit in Figure 15-74 changes to 120 $\angle 30°$ V and the load impedance changes to 25 $\angle 60°$ Ω, determine the value of current that will flow through R_L.

15-21 For the capacitance bridge shown in Figure 15-75, determine the equivalent series circuit.

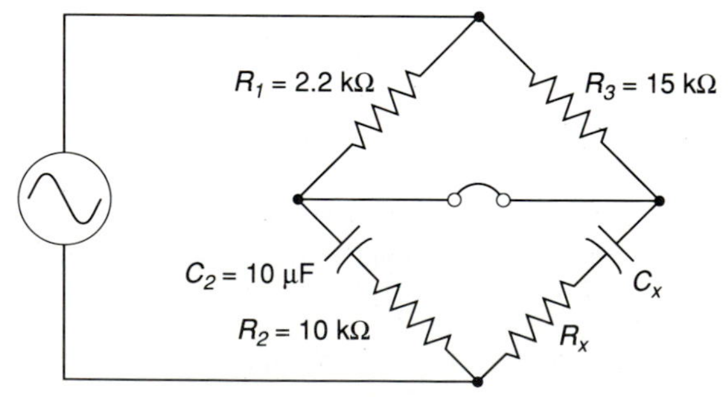

FIGURE 15-75

15-22 A capacitance comparison bridge is used to measure the impedance of a capacitive circuit. The bridge constants are as follows: $R_3 = 50$ kΩ, $R_1 = 5$ kΩ, $R_2 = 20$ kΩ, $C_2 = 40$ μF. Determine the equivalent series circuit of the unknown impedance.

15-23 For the Maxwell–Wein bridge shown in Figure 15-76, determine the equivalent series circuit required to produce zero deflection of the null detector.

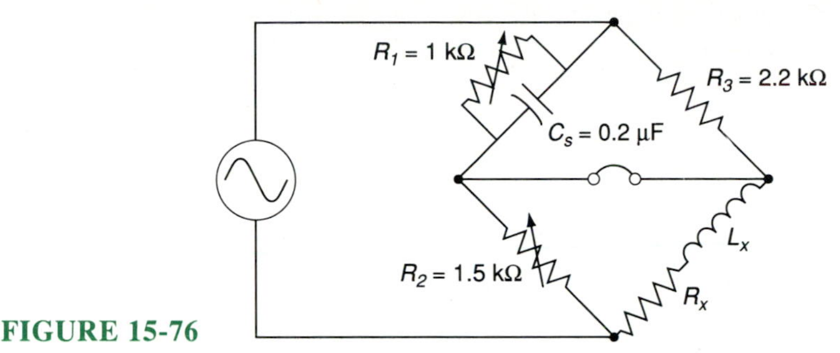

FIGURE 15-76

15-24 The components of an unknown impedance were measured by means of a Maxwell–Wein bridge. The values of resistance and capacitance were found, at balance, to be $R_1 = 600\ \Omega$, $R_2 = 200\ \Omega$, $R_3 = 850\ \Omega$, and $C_s = 220$ pF. Determine the inductance and unknown resistance of the circuit.

15-25 The Thévenin equivalent circuit shown in Fig. 15-77 consists of $Z_{TH} = 27 + j40\ \Omega$, and $E = 120\ \angle 0°$ V. Determine (a) the required load impedance so that maximum power transfer occurs; (b) the load current; (c) the apparent power, active power, and reactive power drawn by the load.

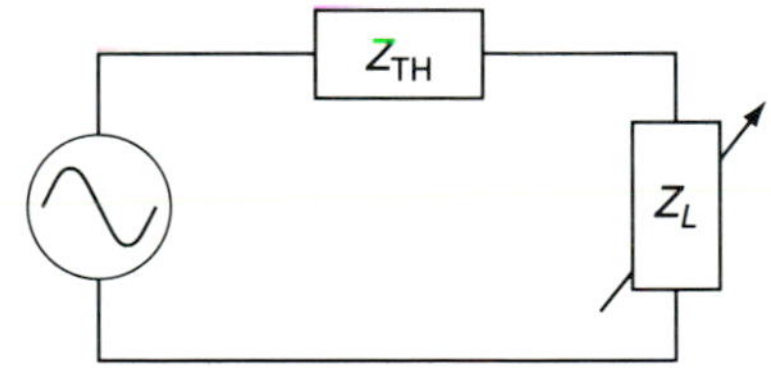

FIGURE 15-77

15-26 Repeat Problem 15-25 using a source voltage of $E = 24\ \angle 30°$ V and $Z_{TH} = 3.0 - j5.0\ \Omega$.

15-27 The circuit shown in Figure 15-78 is the Thévenin equivalent of a more complex network. A load impedance consisting of a 20-mH coil in series with a 20-Ω resistor is to be connected between points *a* and *b*. If the source frequency is 20 kHz, what value of power is delivered to the load?

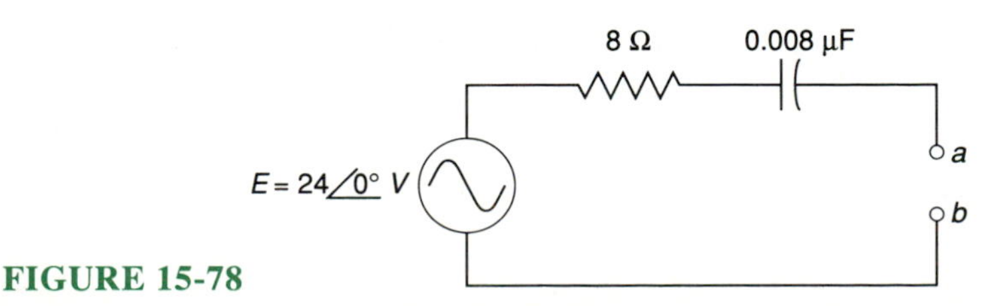

FIGURE 15-78

15-28 If the frequency of the circuit in Problem 15-27 changed to 100 kHz, determine the power delivered to the load.

15-29 Calculate the single equivalent impedance that will replace the circuit of Figure 15-79 between points *a* and *d*.

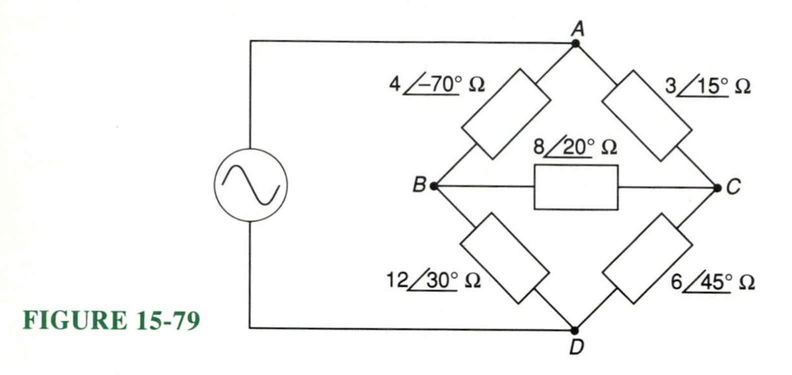

FIGURE 15-79

15-30 Determine the single equivalent impedance for the circuit shown in Figure 15-80.

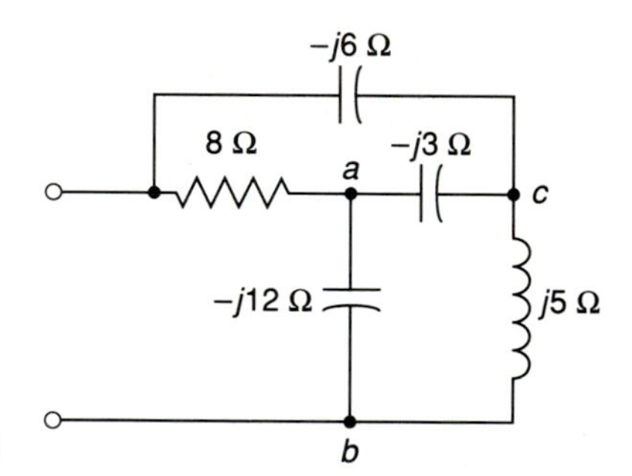

FIGURE 15-80

15-31 Find the equivalent delta of the wye circuit shown in Figure 15-81.

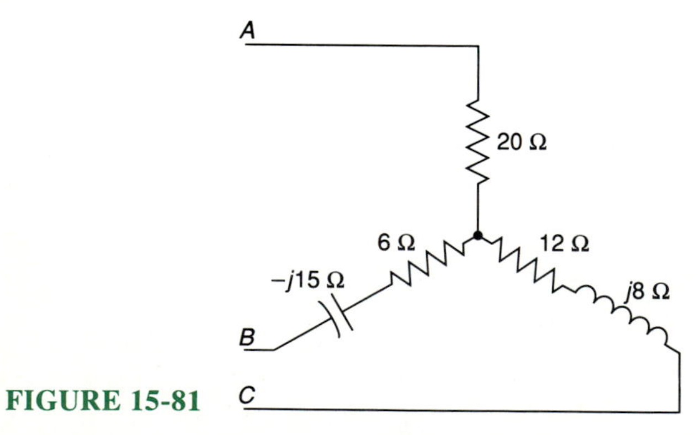

FIGURE 15-81

CHAPTER

16 Three-Phase Circuits

LEARNING OBJECTIVES

Upon completion of this chapter you will be able to:

- Understand the application of double-subscript notation in three-phase circuits.
- Explain the principles of polyphase voltage generation.
- Determine the phase relationships of balanced three-phase systems.
- Define wye connections and delta connections for three-phase circuits.
- Understand the three-phase connections required for power generation and distribution.
- Determine power in balanced and unbalanced three-phase systems.
- Explain the two-wattmeter and three-wattmeter methods of power measurement.
- Determine one-line equivalent circuits for balanced loads.
- Calculate values of current and voltage in unbalanced three-phase systems.
- Determine power factor in unbalanced three-phase loads.
- Explain phase sequence effects and measurement.
- Describe harmonics in three-phase systems.

16-1 INTRODUCTION

A **polyphase system** consists of two or more equal voltages with fixed phase differences that supply power to various ac loads. The two-phase system has two equal voltages which are 90° out of phase with each other. The three-phase system has three voltages which have a phase difference of 120°. In many industrial applications of alternating current, there are certain disadvantages to using single-phase power. In a single-phase circuit, the power delivered is pulsating. When the power factor of a single-phase system is unity, the power is zero twice in each cycle, or 120 times per second in a conventional system. If the power factor is less than unity, the power is zero four times in each cycle. In a three-phase system, although the power of any individual phase is pulsating, the total power is constant if the loads are balanced.

The power rating of a given motor or generator is greater when operated as a three-phase machine than when operated single-phase. If the single-phase rating of an ac machine is taken as 100%, the three-phase rating of the same machine will be about 50% greater. Armature reaction in single-phase synchronous generators and motors produce flux pulsations in their respective magnetic circuits, which cause eddy-current and hysteresis losses. These effects do not occur in polyphase machines with balanced loads. Single-phase induction motors inherently have zero starting torque and require special windings or other means to start them, but polyphase induction motors develop a positive starting torque that can be quite large. These characteristics are discussed in more detail in the chapters dealing with ac machines.

The transmission of power by a three-phase system requires conductors that are 25% smaller than single-phase systems, resulting in considerable savings in power distribution costs. Also, to supply power to a single-phase load from a polyphase system, all that is required is to connect one of the single-phase voltages of the system to the load. However, it is not possible to energize a polyphase load from a single-phase system without first converting the single-phase system into a polyphase system. The equipment required for a conversion of this nature is quite expensive.

16-2 DOUBLE-SUBSCRIPT NOTATION

The use of double subscripts to specify a voltage is a powerful aid in analyzing three-phase circuits. The directions of applied voltages and voltage drops are clearly indicated by their polarities in dc circuits. Alternating currents and voltages do not have definite polarities, such as positive and negative, since their polarities are reversing every cycle. The only time an ac sine wave is considered to be either positive or negative is when it is referring to an instantaneous value of voltage or current. Therefore, when combining voltages and currents in a network, it is good practice to utilize a method that designates polarity and can be referenced to phasor quantities. One such method of ac circuit and phasor notation is called the **double-subscript method**.

To illustrate the use of double-subscript notation, consider the voltages of the two coils shown in Figure 16-1(a). If these coils are displaced by 120 electrical degrees, and assuming that the emf as traced from point *C* to *D* lags the emf as traced from *A* to *B* by 120°, the end of coil *B* could be connected to the end of coil *C* to give the correct angular relationship for these two coils. When analyzing this type of circuit, it is important to know the instantaneous polarity of the voltage and the direction of flow of the current. The first subscript of double-subscript notation is used to indicate the starting point of the coil, and the second subscript is the endpoint.

The voltage E_{AD} is determined by the following phasor relationship:

$$E_{AD} = E_{AB} + E_{CD}$$

The phasor diagram of Figure 16-1(b) shows the relationship between the two coil voltages and their sum. If the voltages of the two coils in Figure 16-1(a) are traced from D to A, the resultant would be

$$E_{DA} = E_{DC} + E_{BA}$$

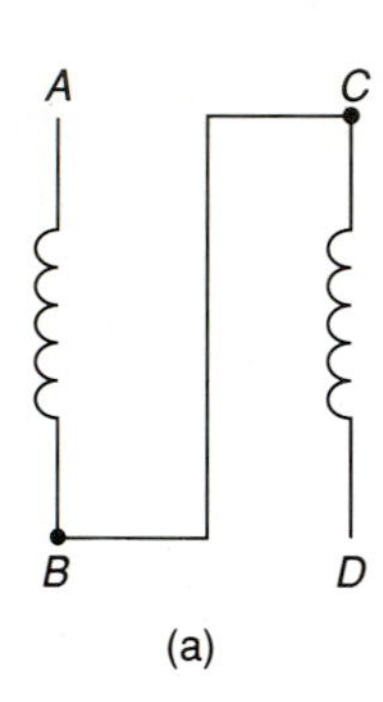

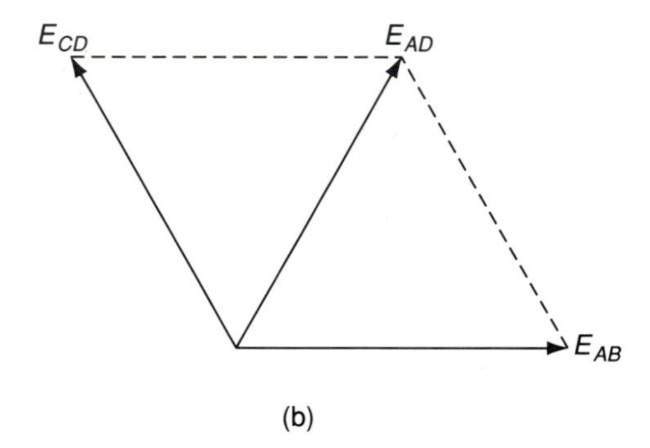

FIGURE 16-1 (a) Two coils displaced by 120 electrical degrees; (b) resulting phase relationship of coils.

Since 180° displacement of a phasor may be indicated either by interchanging the first and second subscripts or by use of the negative sign, the three phasors of Figure 16-1 may be represented in the following manner:

$$E_{AB} = -E_{BA}, \quad E_{CD} = -E_{DC}, \quad E_{AD} = -E_{DA}$$

Double-subscript notation is also well suited for currents, particularly in the application of Kirchhoff's current law. Consider the circuit of Figure 16-2, which consists of three wye-connected coils. At the junction, n, the currents combine to flow in coil nc. These currents can, therefore, be designated as I_{an}, I_{bn}, and I_{nc}. This relationship can be expressed by Kirchhoff's current law in the following manner:

$$I_{an} + I_{bn} = I_{nc}$$

As with voltages, reversing the subscripts of a current reverses its sign.

$$I_{an} + I_{bn} = -I_{cn}$$

By moving $-I_{cn}$ to the left side, the equation becomes

$$I_{an} + I_{bn} + I_{cn} = 0$$

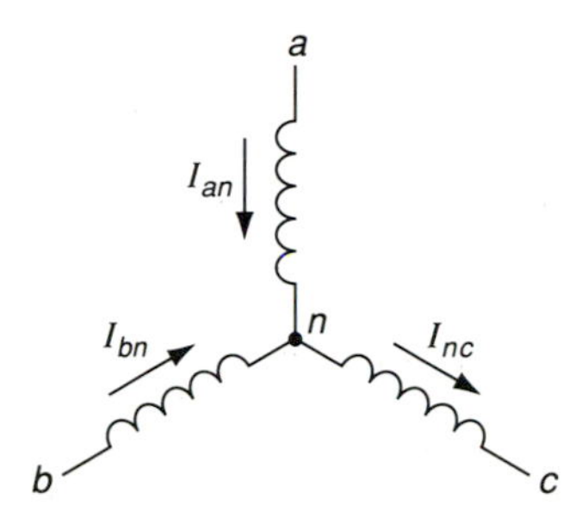

FIGURE 16-2 Double-subscript notation indicating current direction.

Polyphase circuit problems usually involve several voltages and currents and often appear to be quite complex in nature. These problems may be greatly simplified by first drawing and labeling the circuit diagram, and then constructing the phasor diagram for the circuit with all currents and voltages indicated by double-subscript notation. The advantages of using this subscript notation when analyzing polyphase circuits will become evident in the following sections.

16-3 TWO-PHASE VOLTAGE GENERATION

Polyphase ac generators have two or more single-phase windings symmetrically spaced around the stator. In a two-phase ac generator, there are two single-phase windings physically spaced so that the ac voltage induced in one winding is 90° out of phase with the voltage induced in the other winding.

An elementary two-phase generator is shown in Figure 16-3(a). The two coils, *A* and *B*, are placed perpendicular to each other on the rotor. When the coils are rotated in a constant magnetic field, each coil generates a sinusoidal waveform with a fixed 90° phase difference. With each coil having an equal number of turns, the phasor and instantaneous voltages have equal magnitudes, as shown in Figure 16-3(b) and (c). If one terminal of each coil is joined together, the node that is formed is referred to as the **neutral**, *N*. Conductors that are connected to the other ends of the coils are called **line wires**, or conductors. The voltage from each line to the neutral is the **phase voltage**. Voltage between two lines is simply referred to as **line voltage**. The value of line voltage is determined as the potential difference between two individual lines.

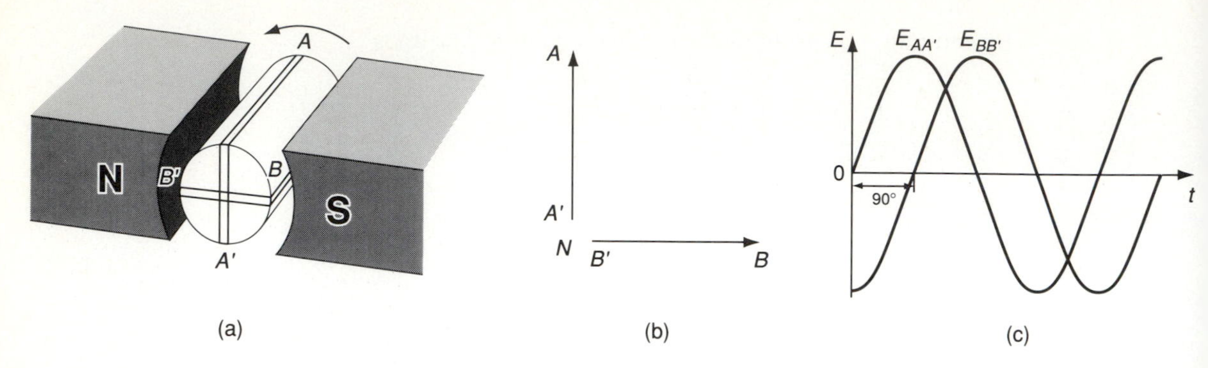

FIGURE 16-3 (a) Two-phase generator; (b) phasor diagram; (c) voltage $E_{BB'}$ shown lagging $E_{AA'}$ by 90°.

The phasor diagram shown in Figure 16-3(b) has the following phase voltages:

$$E_{BN} = E_{BB'} = E\angle 0°$$

$$E_{AN} = E_{AA'} = E\angle 90°$$

If the coil ends A' and B' are connected to form a neutral, the line voltage, E_{AB}, is the phasor sum of the two phase voltages.

$$E_{AB} = E_{AN} + E_{NB} = E\angle 90° + E\angle 180° = \sqrt{2}E\angle 135°$$

16-4 GENERATION OF THREE-PHASE VOLTAGES

The three-phase ac generator has three windings spaced so that the voltage induced in each winding is 120° out of phase with the voltages in the other two windings. Figure 16-4(a) shows three identical coils, AA', BB', and CC', which are fastened together and mounted on a rotor. This combination of coils and axis constitutes an elementary three-phase armature. When the armature is in the position shown in Figure 16-4(a), the coil AA' is in the magnetic neutral plane and its induced emf is zero. This is illustrated in Figure 16-4(c), where the induced emf E starts at its zero value and increases in a positive direction when the rotor is turned in a counterclockwise direction. When the armature has rotated through 120 electrical degrees, the emf in coil BB' will be zero and increasing in a positive direction.

When the armature has rotated through 240 electrical degrees, the emf in coil CC' will be zero and increasing in a positive direction. The sequence of phase rotation, as shown in Figure 16-4(b), is AA', BB', and CC', and the phasors would pass a fixed point in the order A–B–C–A–B–C . . . , as shown in voltage plot of Figure 16-4(c).

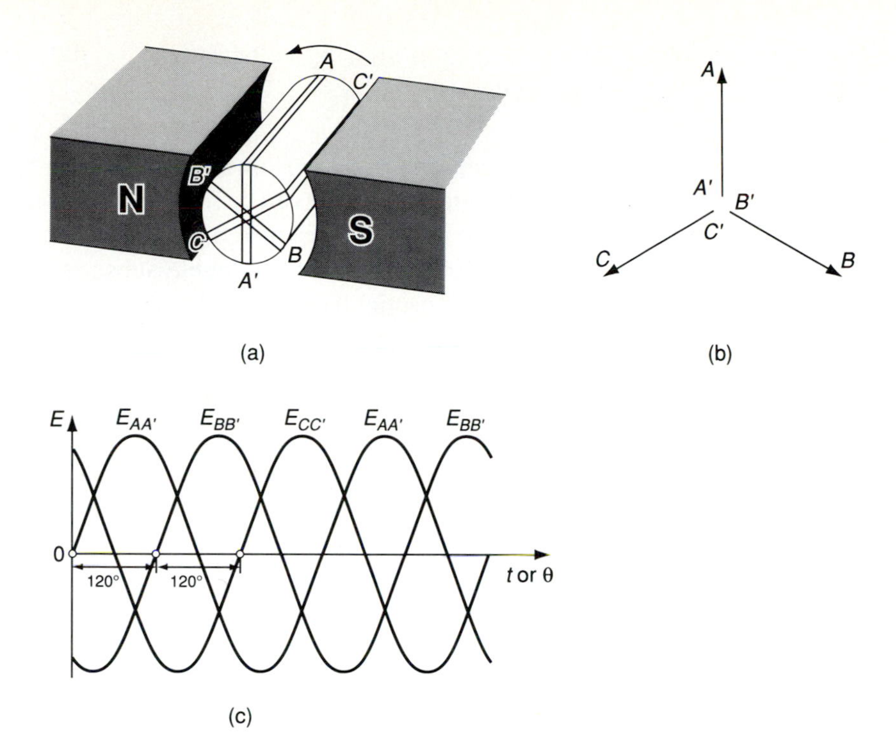

FIGURE 16-4 (a) Three-phase generator; (b) phasor diagram; (c) three-phase voltages displaced by 120°.

The order in which the phase voltages of a polyphase system pass through their positive maximum and other corresponding instantaneous values is called the **phase sequence** of the system. If the rotor of the elementary generator of Figure 16-4(a) is rotated clockwise, the rotation remains counterclockwise, and coils *B* and *C* are reversed, the phase sequence is now *A–C–B–A–C–B* . . . , as shown in Figure 16-5.

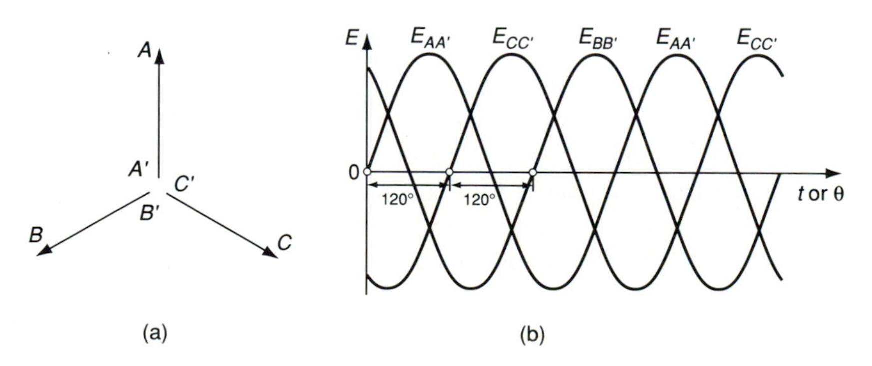

FIGURE 16-5 (a) Phase sequence shown as *ACB*; (b) resulting phase displacement.

EXAMPLE 16-1 Determine the phase sequence of the following voltages: $E_{AB} = 100 \angle 45^\circ$ V, $E_{BC} = 100 \angle -75^\circ$ V, $E_{CA} = 100 \angle 165^\circ$ V.

Solution By plotting the three phasors, as shown in Figure 16-6, their position with respect to each other can be more easily visualized. Since phase sequence is always assumed to be in a counterclockwise direction, the first subscript of the double-subscript notation determines the point of origin. Therefore, the phase sequence for this example is *A–B–C*.

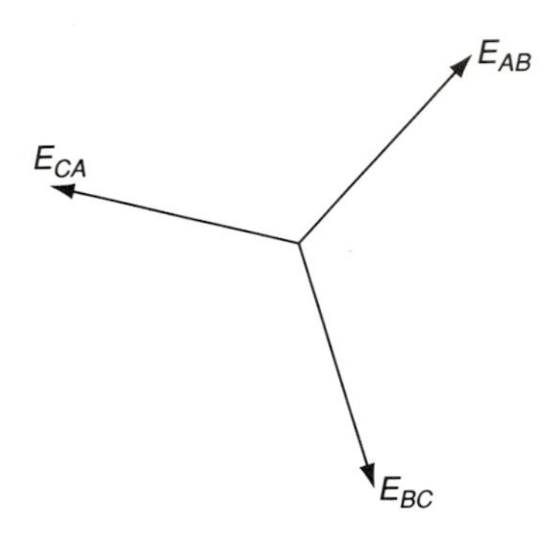

FIGURE 16-6 Phasor diagram for Example 16-1.

16-5 THREE-PHASE BALANCED WYE CONNECTION

The basic wye connection for a source is shown in Figure 16-7. In this method of connecting the windings, one of the leads from each phase is connected to form a common junction. This type of connection is alternatively known as the **star** or **tee connection**. If a wire is connected to the common junction it is considered to be the neutral conductor. This interconnection forms a **three-phase four-wire wye system**. In this system, all three of the currents return to the source through the neutral wire. When the currents in the three-line wires of a wye system have equal magnitudes and are displaced from one another by 120°, the system is said to be **balanced**.

If the loads and voltages of all three phases are balanced, the individual currents will have a phasor sum equal to zero, so there will in effect be no current flowing through the neutral conductor. If the system were to remain balanced, the neutral wire could be removed from the circuit without changing current and voltage values in the system. The network that is formed without using a neutral wire is called a **three-phase three-wire wye system.** This type of system is used extensively for transmitting power to balanced loads.

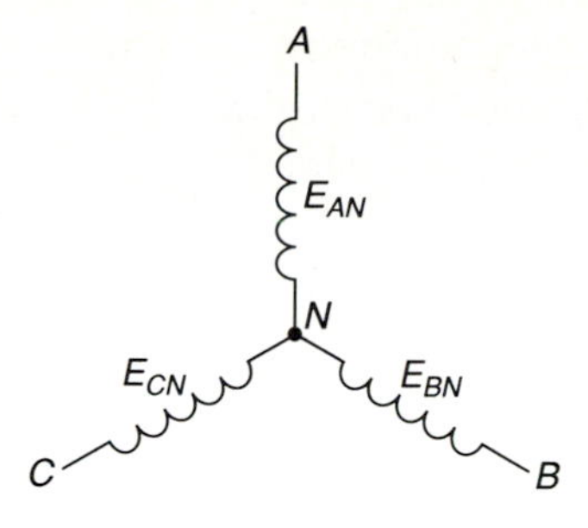

FIGURE 16-7 Three-phase three-wire wye-connected system.

In a balanced wye system, a definite relationship exists between the phase and line voltages. The voltages E_{AN}, E_{BN}, and E_{CN} found across the source windings in Figure 16-7 are phase voltages. To find the line voltage, the sum of two phase voltages combined subtractively must be determined. Figure 16-8 shows the relationship phase and line voltages using two different methods. Figure 16-8(a) is a **closed-face**, or **polygon**, **phasor diagram**. Figure 16-8(b) is called an **open-face**, or **polar**, **phasor diagram**. In either diagram the line voltages are calculated in exactly the same manner. For convenience, the magnitude of a phase voltage will be referred to as E_p. Assuming that phase voltage E_{AN} is the reference phasor, the relationship between the three-phase voltages of Figure 16-8 may be expressed in the following manner:

$$\begin{aligned} E_{AN} &= E_p \angle 0° \text{ V} \\ &= E_p(1 + j0) \end{aligned} \tag{16-1}$$

$$\begin{aligned} E_{BN} &= E_p \angle -120° \text{ V} \\ &= E_p(-0.5 - j0.866) \end{aligned} \tag{16-2}$$

$$\begin{aligned} E_{CN} &= E_p \angle -240° \text{ V} \\ &= E_p(-0.5 + j0.866) \end{aligned} \tag{16-3}$$

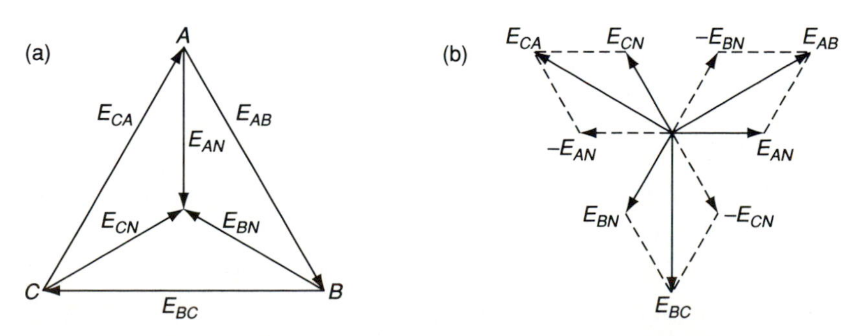

FIGURE 16-8 (a) Closed-face (polygon) phasor diagram; (b) open-face (polar) diagram.

The closed-face phasor diagram is sometimes referred to as a tip-to-tail diagram. To determine the line voltage using this type of diagram, think of the phase voltages as a series circuit. The line voltage is simply the phasor sum of two individual phase voltages.

$$E_{AB} = E_{AN} + E_{NB} \tag{16-4}$$

$$E_{BC} = E_{BN} + E_{NC} \tag{16-5}$$

$$E_{CA} = E_{CN} + E_{NA} \tag{16-6}$$

The polar-type diagram of Figure 16-8(b) also shows the relationship between phase and line voltages.

By substituting equations 16-1 and 16-2, the equation for E_{AB} is determined as

$$\begin{aligned} E_{AB} &= E_{AN} + E_{NB} \\ &= E_p(1 + j0) + E_p(0.5 + j0.866) \\ &= E_p(1.5 + j0.866) \\ &= E_p(1.73)\ \underline{/30^\circ}\ \text{V} \end{aligned}$$

Since $1.73 = \sqrt{3}$, the equation for line voltage E_{AB} is

$$E_{AB} = \sqrt{3}\, E_p\ \underline{/30^\circ}\ \text{V}$$

$$\begin{aligned} E_{BC} &= E_{BN} + E_{NC} \\ &= E_p(-0.5 - j0.866) + E_p(0.5 - j0.866) \\ &= \sqrt{3}\, E_p\ \underline{/-90^\circ}\ \text{V} \end{aligned}$$

$$\begin{aligned} E_{CA} &= E_{CN} + E_{NA} \\ &= E_p(-0.5 + j0.866) + E_p(-1 - j0) \\ &= \sqrt{3}\, E_p\ \underline{/150^\circ}\ \text{V} \end{aligned}$$

It can therefore be stated that the line voltage of a wye-connected three-phase system is $\sqrt{3}$ times the phase voltage.

$$E_{\text{line}} = \sqrt{3}\, E_p \tag{16-7}$$

Since each line-to-line connection of Figure 16-7 contains two series-connected windings, it is evident that the current in each line circuit must be the same as the current in the coil to which it is connected. That is:

> For a three-phase wye system, the line current is equal to the phase current.

$$I_{\text{line}} = I_p \tag{16-8}$$

From Kirchhoff's current law it can be stated that the phasor sum of the three coil currents in a Y system having no neutral conductor must be zero. This is true for both instantaneous and phasor values of current. Figure 16-9 shows a three-phase four-wire wye load connected to a three-phase four-wire wye source. Since the impedances of the wye-connected load are equal and the source voltages are balanced, the currents I_a, I_b, and I_c are also balanced and displaced from each other by 120°, the current in the neutral wire can be expressed in terms of the current in the three line wires by writing Kirchhoff's current for node N.

$$I_N = I_a + I_b + I_c$$

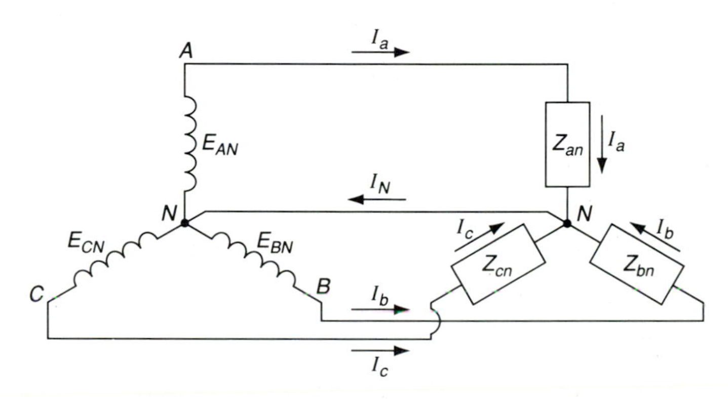

FIGURE 16-9 Three-phase wye-connected load and source.

EXAMPLE 16-2 The circuit shown in Figure 16-10 consists of a three-phase four-wire wye-connected source that supplies power to a balanced wye-connected load. Determine the line currents and draw the phasor diagram. Assume the phase sequence to be ABC.

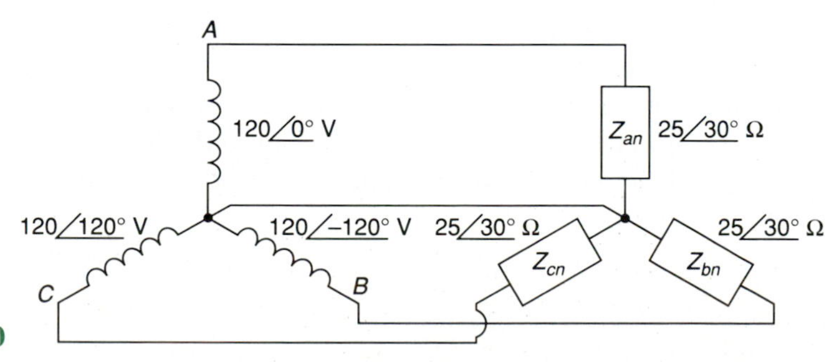

FIGURE 16-10

Solution $E_{AN} = V_{an} \qquad E_{BN} = V_{bn} \qquad E_{CN} = V_{bn}$

$$I_a = \frac{V_{an}}{Z_{an}} = \frac{120\,\angle 0^\circ \text{ V}}{25\,\angle 30^\circ\ \Omega} = 4.8\,\angle -30^\circ \text{ A}$$

$$I_b = \frac{V_{bn}}{Z_{bn}} = \frac{120\,\angle -120^\circ \text{ V}}{25\,\angle 30^\circ\ \Omega} = 4.8\,\angle -150^\circ \text{ A}$$

$$I_c = \frac{V_{cn}}{Z_{cn}} = \frac{120\,\angle 120^\circ \text{ V}}{25\,\angle 30^\circ\ \Omega} = 4.8\,\angle 90^\circ \text{ A}$$

To check:
$$\begin{aligned} I_N &= I_a + I_b + I_c \\ &= 4.8\,\angle -30^\circ \text{ A} + 4.8\,\angle -150^\circ \text{ A} + 4.8\,\angle 90^\circ \text{ A} \\ &= 0 \end{aligned}$$

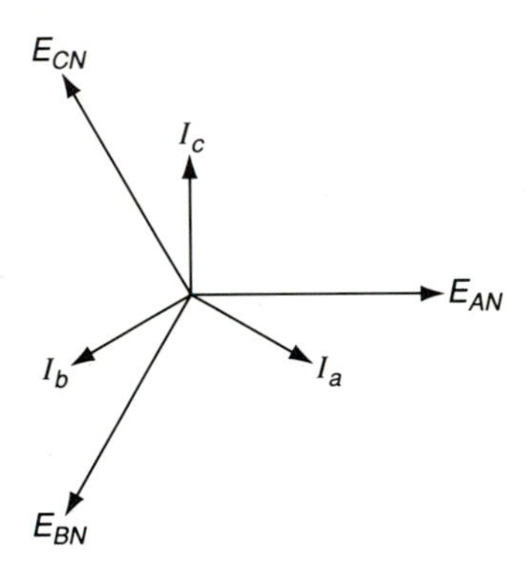

FIGURE 16-11 Phasor diagram for Example 16-2.

16-6 THREE-PHASE BALANCED DELTA CONNECTION

Another method of connecting either the phase voltages of the source or the phase voltages of the load is illustrated in Figure 16-12. This is called the *delta connection*, or alternatively, a *pi* or Π configuration. In a delta-connected ac generator, the start of one phase winding is connected to the end of the third winding. The start of the third phase winding is connected to the finish of the second phase winding. The delta is completed by connecting the start of the second phase winding to the finish of the fist winding. The three junction points are connected to the line wires leading to the load. Because the phases are connected directly across the line wires, phase voltage is equal to line voltage.

$$E_{\text{line}} = E_p \tag{16-9}$$

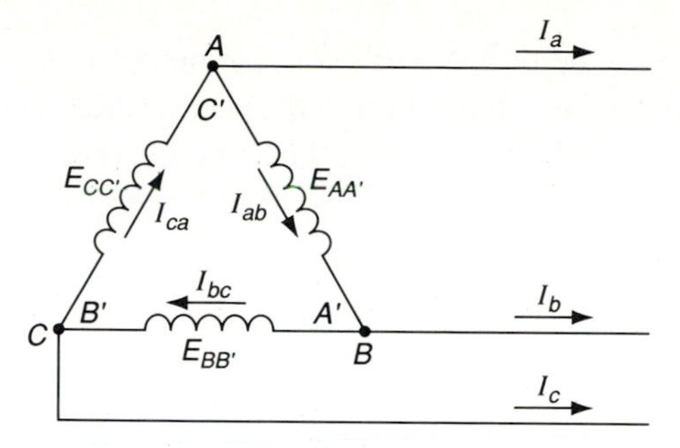

FIGURE 16-12 Delta-connected source.

Although the three delta coils form a closed circuit, when the magnitude of the phase voltages are equal and the angle between any two emfs is 120 electrical degrees, there is no circulating current around the delta. Therefore, even though the delta may appear to be a short circuit, the three voltages of the system will have a phasor sum of zero. If any one of the phases is reversed with respect to its correct connection, a short-circuit current will flow through the windings and damage the system.

Since the line voltages are the same as the phase voltages in a balanced delta system, the voltages for Figure 16-12 with a phase sequence of *ABC* are as follows:

$$E_{AB} = E_{AA'} = E \angle 0^\circ \text{ V}$$

$$E_{BC} = E_{BB'} = E \angle -120^\circ \text{ V}$$

$$E_{CA} = E_{CC'} = E \angle 120^\circ \text{ V}$$

Figure 16-13 shows the open- and closed-face phasor diagrams for voltages in a balanced delta system. Line voltage E_{AB} is chosen as the reference voltage and phase sequence is assumed to be *ABC*.

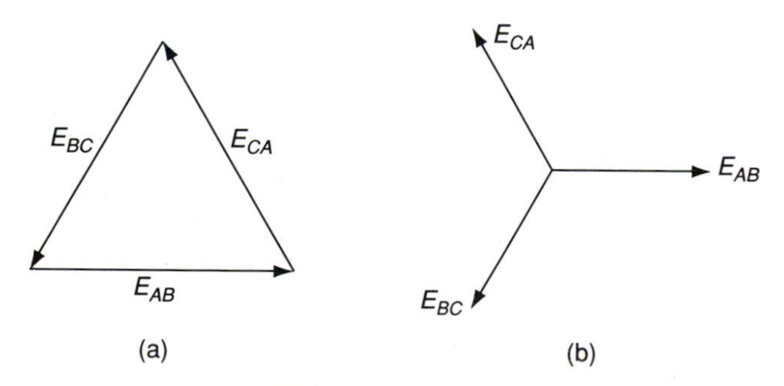

FIGURE 16-13 (a) Closed-face and (b) open-face phasor diagrams of voltages in a delta-connected system.

In Figure 16-14, a balanced delta load is shown connected to a delta source. Because each line connects at a junction of two phases, the line current is equal to the combination of two phase currents. The line current I_a may be expressed in terms of phase currents I_{ab} and I_{ca} by writing Kirchhoff's current law for node a at the delta load.

$$I_a = I_{ab} - I_{ca} \tag{16-10}$$

The equations for line currents I_b and I_c can be derived in a similar manner.

$$I_b = I_{bc} - I_{ab} \tag{16-11}$$

$$I_c = I_{ca} - I_{bc} \tag{16-12}$$

For convenience, the magnitude of the phase current will be called I_p, and phase current I_{ab} will be considered as the reference phasor.

$$\begin{aligned} I_{ab} &= I_p \angle 0^\circ \text{ A} \\ &= I_p(1 + j0) \end{aligned} \tag{16-13}$$

$$\begin{aligned} I_{bc} &= I_p \angle -120^\circ \text{ A} \\ &= I_p(-0.5 - j0.866) \end{aligned} \tag{16-14}$$

$$\begin{aligned} I_{ca} &= I_p \angle 120^\circ \text{ A} \\ &= I_p(-0.5 + j0.866) \end{aligned} \tag{16-15}$$

By combining equations 16-13, 16-14, and 16-15, the values of line currents I_a, I_b, and I_c are determined in the following manner:

$$\begin{aligned} I_a &= I_{ab} - I_{ca} \\ &= (1 + j0) - (-0.5 + j0.866) \\ &= \sqrt{3}\, I_p \angle -30^\circ \text{ A} \end{aligned}$$

$$\begin{aligned} I_b &= I_{bc} - I_{ab} \\ &= (-0.5 - j0.866) - (1 + j0) \\ &= \sqrt{3}\, I_p \angle -150^\circ \text{ A} \end{aligned}$$

$$\begin{aligned} I_c &= I_{ca} - I_{bc} \\ &= (-0.5 + j0.866) - (-0.5 - j0.866) \\ &= \sqrt{3}\, I_p \angle 90^\circ \text{ A} \end{aligned}$$

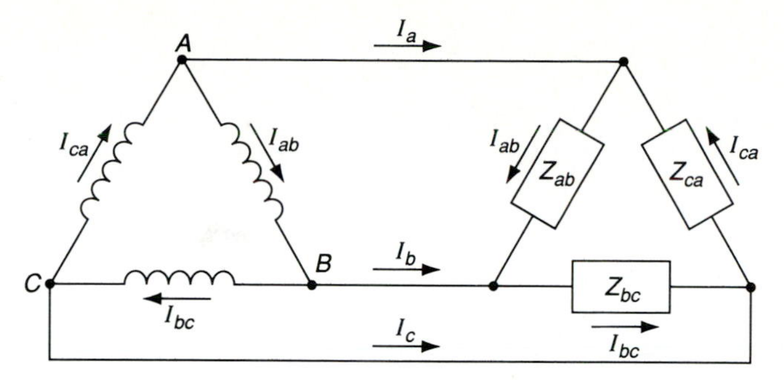

FIGURE 16-14 Current relationships in delta-connected load and delta-connected source.

The relationship that exists between the phase currents and the line current is illustrated in the phasor diagram of Figure 16-15. From this phasor diagram and the equations above it is obvious that in a balanced delta-connected, three-phase system, each line current is $\sqrt{3}$ times the phase current.

$$I_{\text{line}} = \sqrt{3}\, I_p \tag{16-16}$$

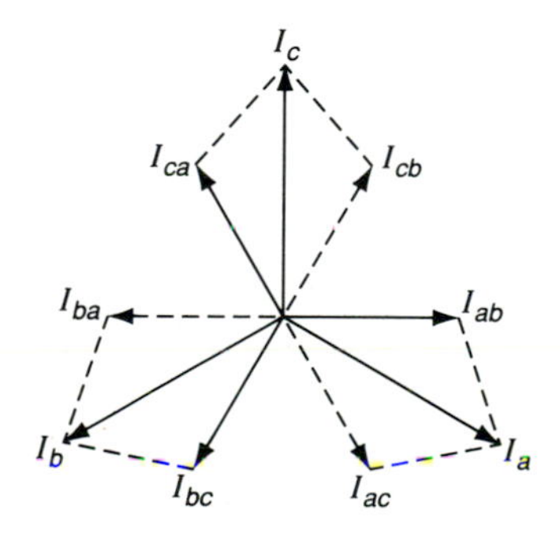

FIGURE 16-15 Phasor diagram of the currents in the circuit of Figure 16-14.

EXAMPLE 16-3 For the circuit shown in Figure 16-16, determine the values of line currents I_a, I_b, and I_c. Also, draw the phasor diagram for the circuit.

Solution $E_{AB} = V_{ab} \qquad E_{BC} = V_{bc} \qquad E_{CA} = V_{ca}$

$$I_{ab} = \frac{V_{ab}}{Z_{ab}} = \frac{120\,\underline{/0^\circ}\text{ V}}{8\,\underline{/30^\circ}\,\Omega} = 15\,\underline{/-30^\circ}\text{ A}$$

$$I_{bc} = \frac{V_{bc}}{Z_{bc}} = \frac{120\,\underline{/-120^\circ}\text{ V}}{8\,\underline{/30^\circ}\,\Omega} = 15\,\underline{/-150^\circ}\text{ A}$$

$$I_{ca} = \frac{V_{ca}}{Z_{ca}} = \frac{120\,\underline{/120^\circ}\text{ V}}{8\,\underline{/30^\circ}\,\Omega} = 15\,\underline{/90^\circ}\text{ A}$$

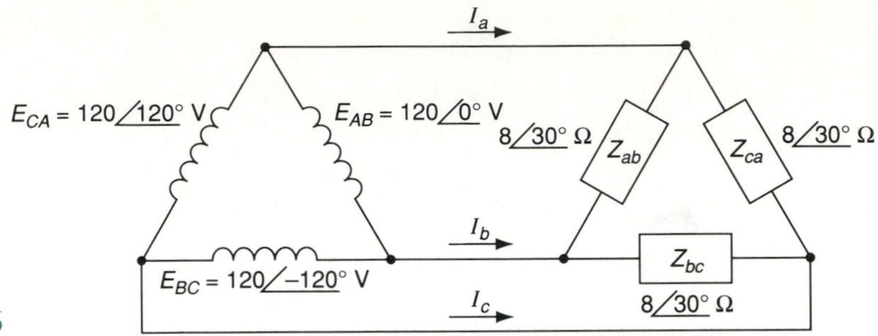

FIGURE 16-16

Next, apply Kirchhoff's current law at each junction of the load.

$$I_a = I_{ab} + I_{ac} = 15 \angle -30° \text{ A} + 15 \angle -90° \text{ A} = 26 \angle -60° \text{ A}$$

$$I_b = I_{bc} + I_{ba} = 15 \angle -150° \text{ A} + 15 \angle 150° \text{ A} = 26 \angle 180° \text{ A}$$

$$I_c = I_{ca} + I_{cb} = 15 \angle 90° \text{ A} + 15 \angle 30° \text{ A} = 26 \angle 60° \text{ A}$$

To check: $I_a + I_b + I_c = 0$

$$26 \angle -60° + 26 \angle 180° + 26 \angle 60° = 0$$

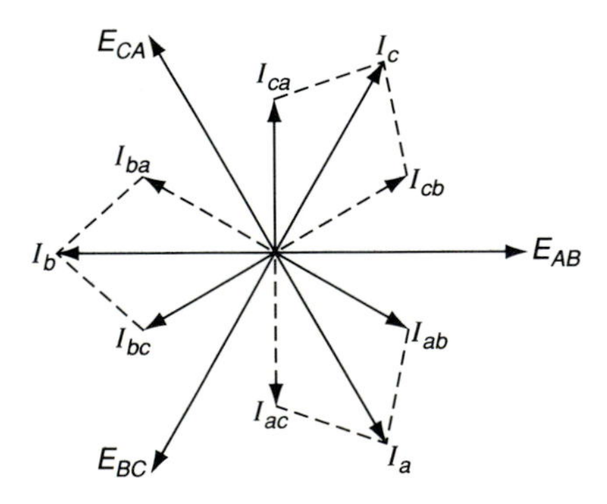

FIGURE 16-17 Phasor diagram for Example 16-3.

16-7 BALANCED WYE-DELTA AND DELTA-WYE SYSTEMS

Electric utility companies usually distribute power from a wye-connected source. When the load is delta connected, the voltage across each leg of the equals line voltage rather than phase voltage. A wye-connected source and a delta-connected load are shown in Figure 16-18. Since there is no neutral connection shown in this system, any change in impedance of the load will vary both the line and phase currents of the network. As long as the load remains balanced, the line currents will be equal in magnitude and the phase currents will also be equal in magnitude. Although the magnitude of the line and phase currents are proportionately equal, the line current is $\sqrt{3}$ times larger than the phase current of the load. When a wye-connected source is connected to a delta load, the phase voltages of the source are not used. The phase voltages of the source may be used to calculate the line voltage, but otherwise are not required.

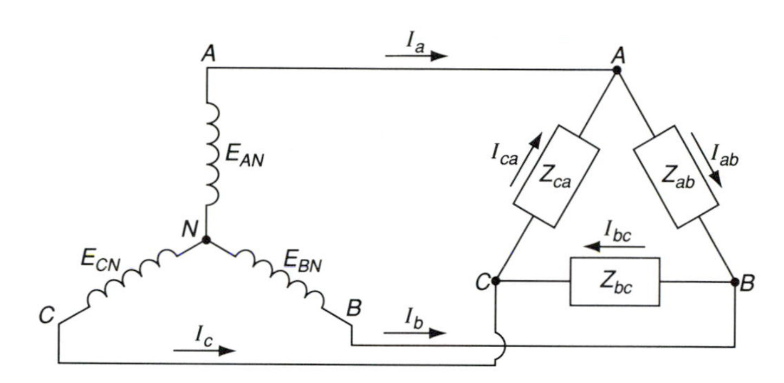

FIGURE 16-18 Wye-connected source and delta-connected load.

EXAMPLE 16-4 The balanced system shown in Figure 16-18 consists of the following voltages and impedances: $E_{AN} = 120\ \angle 0^\circ$ V, $E_{BN} = 120\ \angle -120^\circ$ V, $E_{CN} = 120\ \angle 120^\circ$ V, and $Z_{ab} = Z_{bc} = Z_{ca} = 20\ \angle 45^\circ\ \Omega$. Determine the line voltages, phase currents, and line currents.

Solution The line voltage is the phasor sum of the phase voltages.

$$E_{AB} = E_{AN} + E_{NB} = 120\ \angle 0^\circ\ \text{V} + 120\ \angle 60^\circ\ \text{V}$$
$$= 208\ \angle 30^\circ\ \text{V}$$

$$E_{BC} = E_{BN} + E_{NC} = 120\ \angle -120^\circ\ \text{V} + 120\ \angle -60^\circ\ \text{V}$$
$$= 208\ \angle -90^\circ\ \text{V}$$

$$E_{CA} = E_{CN} + E_{NA} = 120\ \angle 120^\circ\ \text{V} + 120\ \angle 180^\circ\ \text{V}$$
$$= 208\ \angle 150^\circ\ \text{V}$$

The phase current is found by dividing the line voltage by the phase impedance of the load.

$$E_{AB} = V_{ab} \qquad E_{BC} = V_{bc} \qquad E_{CA} = V_{ca}$$

$$I_{ab} = \frac{V_{ab}}{Z_{ab}} = \frac{208\ \angle 30^\circ}{20\ \angle 45^\circ} = 10.4\ \angle -15^\circ\ \text{A}$$

$$I_{bc} = \frac{V_{bc}}{Z_{bc}} = \frac{208\ \angle -90^\circ}{20\ \angle 45^\circ} = 10.4\ \angle -135^\circ\ \text{A}$$

$$I_{ca} = \frac{V_{ca}}{Z_{ca}} = \frac{208\,\underline{/150^\circ}}{20\,\underline{/45^\circ}} = 10.4\,\underline{/105^\circ}\text{ A}$$

To find the line currents, apply Kirchhoff's current law at each junction of the load.

$$I_a = I_{ab} + I_{ac} = 10.4\,\underline{/-15^\circ}\text{ A} + 10.4\,\underline{/-75^\circ}\text{ A}$$
$$= 18\,\underline{/-45^\circ}\text{ A}$$

$$I_b = I_{bc} + I_{ba} = 10.4\,\underline{/-135^\circ}\text{ A} + 10.4\,\underline{/165^\circ}\text{ A}$$
$$= 18\,\underline{/-165^\circ}\text{ A}$$

$$I_c = I_{ca} + I_{cb} = 10.4\,\underline{/105^\circ}\text{ A} + 10.4\,\underline{/45^\circ}\text{ A}$$
$$= 18\,\underline{/75^\circ}\text{ A}$$

To check: $I_a + I_b + I_c = 0$

$$18\,\underline{/-45^\circ}\text{ A} + 18\,\underline{/-165^\circ}\text{ A} + 18\,\underline{/75^\circ}\text{ A} = 0$$

Figure 16-19 shows a wye-connected load with a delta supply. To calculate the line currents for a load in this configuration, the simplest method is to convert the wye-connected circuit to its delta equivalent. Wye–delta conversion was illustrated in Chapter 15, and the same equations would apply to a problem of this nature.

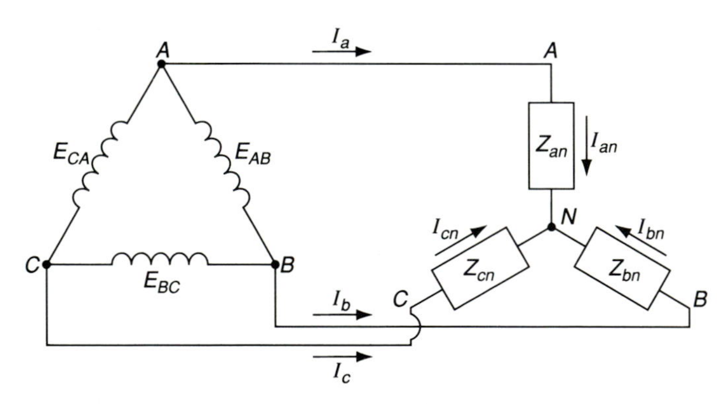

FIGURE 16-19 Delta-connected source with a wye-connected load.

EXAMPLE 16-5 The balanced system of Figure 16-19 has line voltages of $E_{AB} = 208\,\underline{/30^\circ}$ V, $E_{BC} = 208\,\underline{/-90^\circ}$ V, and $E_{CA} = 208\,\underline{/150^\circ}$ V. The load impedances are $Z_{an} = Z_{bn} = Z_{cn} = 15\,\underline{/30^\circ}\;\Omega$. Calculate the phase currents of the load.

Solution First, convert the wye connected load to its equivalent delta, as shown in Figure 16-20.

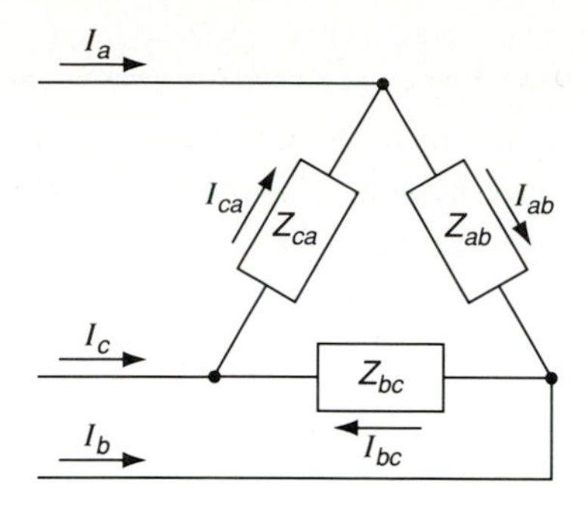

FIGURE 16-20 Delta equivalent circuit of wye-connected load.

$$Z_{ab} = \frac{Z_aZ_b + Z_aZ_c + Z_bZ_c}{Z_c}$$
$$= 45 \angle 30^\circ \ \Omega$$

Since the load is balanced, $Z_{ab} = Z_{bc} = Z_{ca} = 45 \angle 30^\circ \ \Omega$. The equivalent load phase currents can now be calculated:

$$E_{AB} = V_{ab} \qquad E_{BC} = V_{bc} \qquad E_{CA} = V_{ca}$$

$$I_{ab} = \frac{V_{ab}}{Z_{ab}} = \frac{208 \angle 30^\circ \text{ V}}{45 \angle 30^\circ \ \Omega} = 4.62 \angle 0^\circ \text{ A}$$

$$I_{bc} = \frac{V_{bc}}{Z_{bc}} = \frac{208 \angle -90^\circ \text{ V}}{45 \angle 30^\circ \ \Omega} = 4.62 \angle -120^\circ \text{ A}$$

$$I_{ca} = \frac{V_{ca}}{Z_{ca}} = \frac{208 \angle 150^\circ \text{ V}}{45 \angle 30^\circ \ \Omega} = 4.62 \angle 120^\circ \text{ A}$$

Next, the line currents are determined as the phasor sum of the phase currents.

$$I_a = I_{ab} + I_{ac} = 4.62 \angle 0^\circ \text{ A} + 4.62 \angle -60^\circ \text{ A}$$
$$= 8 \angle -30^\circ \text{ A}$$

$$I_b = I_{bc} + I_{ba} = 4.62 \angle -120^\circ \text{ A} + 4.62 \angle 180^\circ \text{ A}$$
$$= 8 \angle -150^\circ \text{ A}$$

$$I_c = I_{ca} + I_{cb} = 4.62 \angle 120^\circ \text{ A} + 4.62 \angle 60^\circ \text{ A}$$
$$= 8 \angle 90^\circ \text{ A}$$

In a wye-connected system, the phase currents and line currents are equal. Therefore, the phase currents for the load are

$$I_{an} = 8 \angle -30^\circ \text{ A} \qquad I_{bn} = 8 \angle -150^\circ \text{ A} \qquad I_{cn} = 8 \angle 90^\circ \text{ A}$$

16-8 POWER IN A BALANCED THREE-PHASE SYSTEM

The power in a three-phase system is calculated in the same manner as in any ac network. The total power in a three-phase system, regardless of whether it is balanced or unbalanced, is the sum of the power of the individual phases. The relationship for total power is expressed by the equation

$$P_T = E_pI_p \cos \theta_1 + E_pI_p \cos \theta_2 + E_pI_p \cos \theta_3 \qquad (16\text{-}17)$$

Total power in a balanced three-phase system is determined using the equation

$$P_T = 3E_pI_p \cos \theta \qquad (16\text{-}18)$$

For a balanced wye-connected load, the line voltage, E_L, is equal to the phase voltage, E_p, multiplied by $\sqrt{3}$, and the line current, I_L, equals I_p. Equation 16-17 can be written in terms of line voltage and current as

$$P_T = \sqrt{3}\, E_LI_L \cos \theta \qquad (16\text{-}19)$$

where
P_T = total average power, in watts
E_L = rms voltage
I_L = rms current
$\cos \theta$ = power factor

For a balanced delta-connected load, $E = E_p$ and $I_p = I_L/\sqrt{3}$. Therefore, regardless of whether the system is wye- or delta-connected, the total power in a balanced three-phase network is determined by equation 16-19. In equation 16-19, the term $\sqrt{3}\, E_LI_L$ represents the apparent power of a balanced three-phase circuit. Total apparent power is measured in volt-amperes and is represented by the symbol S_T.

$$S_T = \sqrt{3}\, E_LI_L \qquad (16\text{-}20)$$

The power factor of a balanced three-phase system is given by the equation

$$\text{PF} = \frac{P_T}{S_T} = \cos \theta \quad \text{(leading or lagging)} \qquad (16\text{-}21)$$

The total three-phase reactive power, Q_T, is the product of the apparent power and the sine of the phase angle, and is measured in VAR.

$$Q_T = S_T \sin\theta = \sqrt{3}\, E_L I_L \sin\theta \qquad (16\text{-}22)$$

A power triangle for a three-phase circuit can be constructed in the same manner as a single-phase power triangle. The real power is drawn as the reference phasor on the horizontal axis, the reactive power is drawn on the vertical axis, and the apparent power is the hypotenuse.

EXAMPLE 16-6 For the circuit shown in Figure 16-21, calculate the power dissipated by the load.

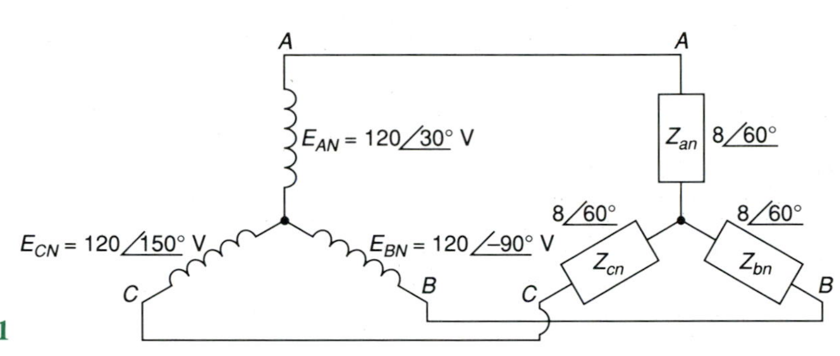

FIGURE 16-21

Solution Since each impedance has a power factor angle of $+60°$,

$$\cos\theta = \cos 60° = 0.5 \text{ lagging}$$

The phase current through any of the three loads is found by Ohm's law.

$$E_{AN} = V_{an}$$

$$I_{an} = \frac{V_{an}}{Z_{an}} = \frac{120\angle 30°\text{ V}}{8\angle 60°\ \Omega}$$

$$= 15\angle -30°\text{ A}$$

Therefore,

$$P_T = 3E_p I_p \cos\theta = (3)(120\text{ V})(15\text{ A})(0.5)$$
$$= 2700\text{ W}$$

EXAMPLE 16-7 A delta-connected load has an impedance per phase of 15 Ω resistance and 20 Ω capacitive reactance. When this load is connected to a 240-V three-phase delta source, calculate the total power dissipated by the load.

Solution $Z_p = \sqrt{R_p^2 + X_p^2} = \sqrt{15^2 + 20^2} = 25\ \Omega$

$$\cos\theta = \frac{R_p}{Z_p} = \frac{15\,\Omega}{25\,\Omega} = 0.6$$

Since the load is delta connected, $E_L = E_p$ and $I_L = \sqrt{3}\,I_p$.

$$I_p = \frac{E_p}{Z_p} = \frac{240\text{ V}}{25\,\Omega} = 9.6\text{ A}$$

$$I_L = \sqrt{3}\,I_p = (9.6\text{ A})/(\sqrt{3}) = 16.628\text{ A}$$

Therefore,

$$\begin{aligned} P_T &= \sqrt{3}\,E_L I_L \cos\theta = \sqrt{3}\,(240\text{ V})(16.628\text{ A})(0.6) \\ &= 4147.28\text{ W} \end{aligned}$$

16-9 MEASURING POWER IN A THREE-PHASE SYSTEM

In Section 16-8 it was shown how power in a balanced three-phase system can be calculated from line or phase values of voltage and current. However, if the load is unbalanced, the power calculations become more complicated. A more practical method of determining power dissipated in a system involves power measurement. The most common method of measuring ac power is by using the **wattmeter**. The wattmeter, which was discussed in Chapter 9, operates on the electrodynamic principle. The current coil of the wattmeter should be in series with each phase, and the voltage coil should be connected across each phase. Reversing the leads to one of the coils of a wattmeter will cause a negative torque, and the pointer will be deflected to the left and possibly damaged. For this reason, one terminal of each wattmeter coil is identified by a $\pm$ marking.

The proper connection of the wattmeter is shown in Figure 16-22, where the identified terminal of the current coil is connected to the source, and the identified terminal of the voltage coil is connected to the line containing the current coil. A wattmeter is rated in amperes and volts, rather than in watts, because the indicated watts of the scale do not indicate the amperes in the current coil or the voltage across the potential coil.

16-10 THREE-WATTMETER METHOD OF POWER MEASUREMENT

The power in a balanced or unbalanced three-phase system may be measured with three single-phase wattmeters or with one three-element polyphase wattmeter. Figure 16-22 shows three single-phase wattmeters connected for power measurement of a four-wire wye circuit. Each wattmeter measures the power dissipated by each phase. The current coil of each wattmeter is in series with the load, while the potential coil is in parallel with the load. The total average power is the sum of the three individual wattmeter readings.

$$P_T = P_1 + P_2 + P_3 \tag{16-23}$$

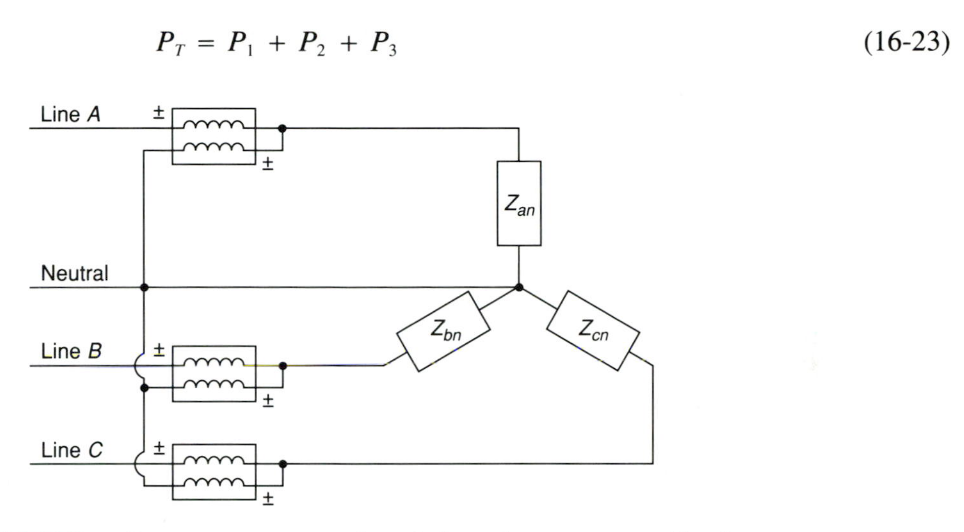

FIGURE 16-22 Power measurement for a balanced wye-connected load.

The total power delivered to a balanced or unbalanced delta-connected load can also be measured using three wattmeters, as shown in Figure 16-23. This method of measuring power in delta-connected loads would generally not be used unless the power in the individual phases was desired. Since the connection shown in Figure 16-23 requires that the current coil of each meter be connected inside the delta connection of the load, it is not a practical method of total power measurement.

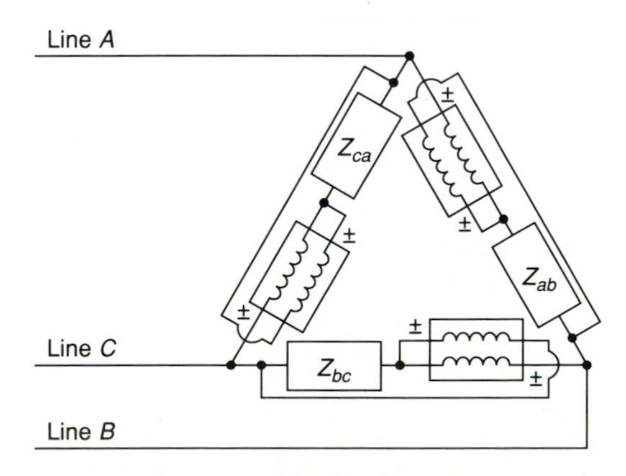

FIGURE 16-23 Power measurement in balanced or unbalanced delta-connected loads.

In many three-phase wye and delta loads, the junction points are not accessible for independent measurement of phase currents and voltages. When this type of situation occurs, the total power can be measured by the three-wattmeter method shown in Figure 16-24. The three-wattmeter method of power measurement is well suited for measuring power in a system where the power factor is constantly changing. If the three wattmeters have equal potential-circuit resistances, and the system is balanced, they will have the same readings regardless of power factor.

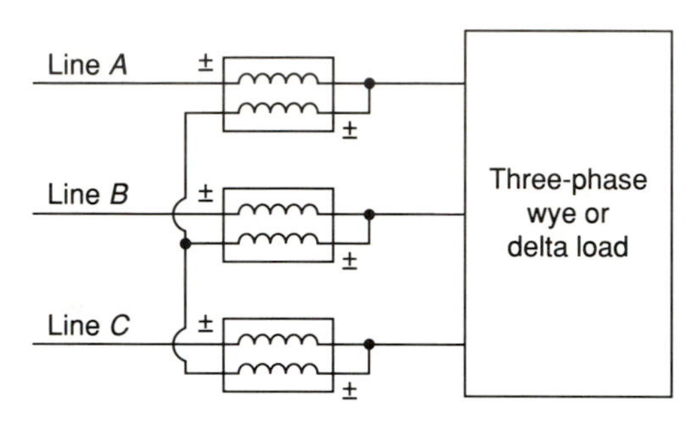

FIGURE 16-24 Power measurement for either wye or delta load.

16-11 TWO-WATTMETER METHOD OF POWER MEASUREMENT

The most commonly used method of measuring power in a three-phase load is the **two-wattmeter method**. In four-wire three-phase circuits, three wattmeters must be used, each connected as a single-phase meter between one line and neutral. In three-wire three-phase circuits, only two wattmeters are required. If the wattmeters are connected according to the circuit shown in Figure 16-25, they will measure the average power delivered to either a wye- or a delta-connected load. Also, provided that the load is balanced, the power factor of the load can be determined from the two wattmeter readings. The current coil of wattmeter W_1 is connected in line A and its potential coil is connected between lines A and B. Wattmeter W_2 is similarly connected with its current coil in conductor C and its potential coil connected between conductors B and C. It is assumed that each current coil lags its coil voltage by $\theta°$, so that the power factor of the system is $\cos \theta$.

Consider the circuit shown in Figure 16-26(a). The three voltages across the Y-connected balanced loads are V_{an}, V_{bn}, and V_{cn}. The load currents are shown as I_a, I_b, and I_c. The phase currents lag the phase voltages by $\theta°$. Since the line voltage in a three-phase system leads the phase voltage by 30°, the total angle between phase current I_{an} and line voltage E_{AB} is $(30° + \theta)$. Therefore, the power read by wattmeter W_1 is

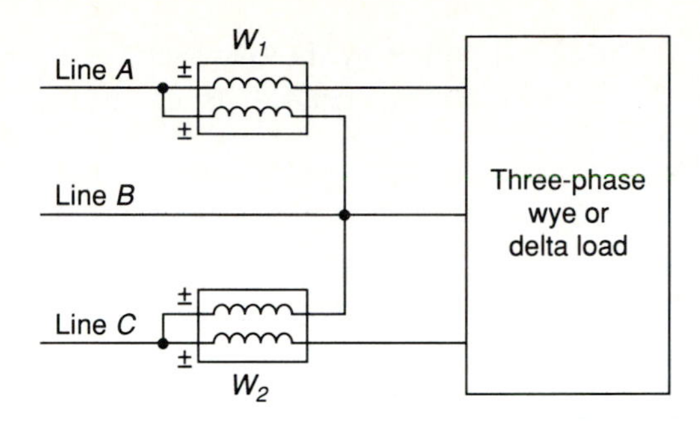

FIGURE 16-25 Two-wattmeter method of measuring power in a balanced wye or delta load.

$$P_1 = E_{AB}I_a \cos(30^\circ + \theta) \tag{16-24}$$

Similarly, P_2, the power measured by wattmeter W_2, is equal to the product of E_{CB}, I_c, and the cosine of their included angle.

$$P_2 = E_{CB}I_c \cos(30^\circ - \theta) \tag{16-25}$$

In general terms, equations 16-24 and 16-25 may be written in the following manner:

$$P_1 = E_LI_L \cos(30^\circ + \theta) \tag{16-26}$$

$$P_2 = E_LI_L \cos(30^\circ - \theta) \tag{16-27}$$

The phasor diagram of Figure 16-26(b) shows the relationship between the line and phase voltages and currents of the wye-connected load.

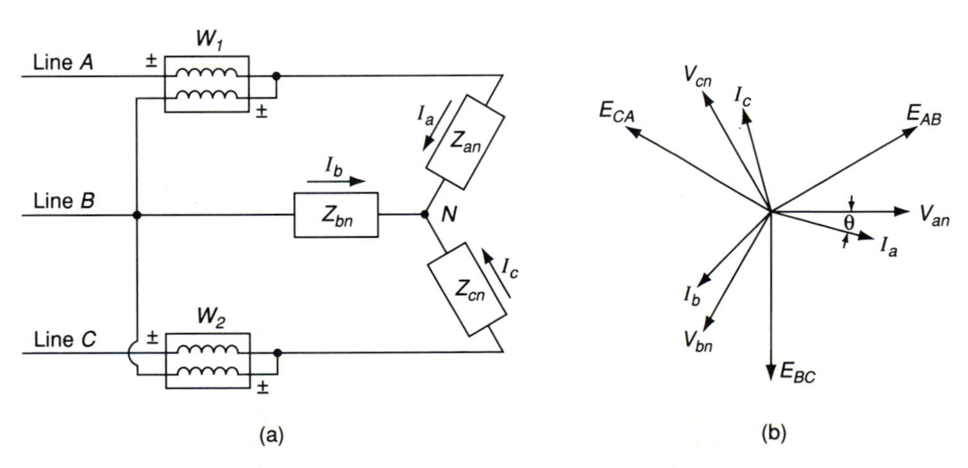

FIGURE 16-26 (a) Two-wattmeter power measurement in a balanced wye load; (b) resulting phasor diagram.

To prove the validity of finding total power by the two-wattmeter method, the trigonometric expansion for $\cos(A + B)$ and $\cos(A - B)$ can be substituted into equations 16-14 and 16-15.

$$\begin{aligned} P_T &= W_1 + W_2 \\ &= E_L I_L[\cos(30° + \theta) + \cos(30° - \theta)] \\ &= E_L I_L(\cos 30° \cos\theta + \sin 30° \sin\theta + \cos 30° \cos\theta \\ &\quad - \sin 30° \sin\theta) \\ &= E_L I_L\, 2 \cos 30° \cos\theta \end{aligned}$$

Since $2 \cos 30° = \sqrt{3}$,

$$P_T = \sqrt{3}\, E_L I_L \cos\theta$$

From the derivation above it can therefore be stated that:

The total power dissipated by any balanced or unbalanced three-phase load is the algebraic sum of the two wattmeter readings.

If the power factor is unity, both wattmeters will have the same reading,

$$P_1 = E_L I_L \cos(30° + 0°) = 0.866 E_L I_L$$

$$P_2 = E_L I_L \cos(30° - 0°) = 0.866 E_L I_L$$

In this situation, each wattmeter reads half of the total power. When the power factor is zero, that is, if $\theta = 90°$, the two cosine values are equal in magnitude, and both wattmeters will have the same reading.

$$P_1 = E_L I_L \cos(30° + 90°) = -0.5 E_L I_L$$

$$P_2 = E_L I_L \cos(30° - 90°) = 0.5 E_L I_L$$

Although the readings of both wattmeters will be equal when the power factor is zero, since the $\cos(30° + 90°)$ is negative, the total power indicated by the wattmeters is zero. Because this condition can occur only if the load is purely reactive, the total power indicated by the two meters is correct since no power is consumed by a purely reactive load.

If $\theta = 60°$, the power factor is 0.5, or 50%. Under these conditions wattmeter 2 will have a positive reading, since $\cos(-30°) = \cos(+30°)$ and is positive. W_1 must read zero, since $\cos 90°$ is zero. Under these conditions, W_2 indicates the entire power dissipated by the load. When the power factor is less than 50%, W_2 still has a positive reading, but W_1 reads negative, due to the cosine of the angle between 90° and 120° being negative. To read the power indicated on W_1 when the power factor is below 0.5, it is necessary to reverse the wattmeter potential or current coil leads so that the pointer may read upscale. The reading of W_1 is then subtracted from W_2 and the system power is found as

$$P_T = P_2 - P_1 \tag{16-28}$$

A specific ratio exists between P_1 and P_2 for any power factor of a balanced load. When the power factor is known, the ratio between the two wattmeter readings can be expressed by dividing equation 16-26 into equation 16-27, which is

$$\frac{P_1}{P_2} = \frac{\cos(30° + \theta)}{\cos(30° - \theta)}$$

A curve showing the power factor as a function of P_1/P_2 is illustrated in Figure 16-27. Note that this ratio is 1.0 at unity power factor since the wattmeter readings are equal; when P_1/P_2 is zero, the power factor is 0.5; and for negative values of P_1/P_2 the power factor is less than 0.5.

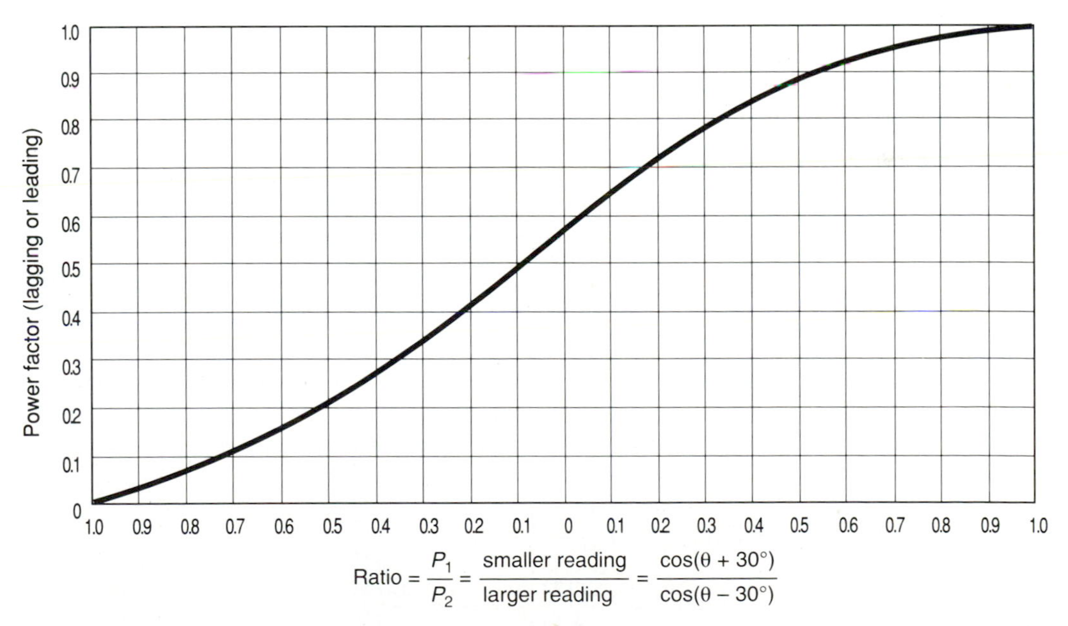

FIGURE 16-27 Power-factor diagram for a balanced three-phase system using the two-wattmeter method of measurement.

$$\text{Ratio} = \frac{P_1}{P_2} = \frac{\text{smaller reading}}{\text{larger reading}} = \frac{\cos(\theta + 30°)}{\cos(\theta - 30°)}$$

Another method of determining the power factor of a balanced three-phase load from two wattmeter readings is by mathematically solving the ratio of the difference between the wattmeter readings and the sum of the wattmeter readings. If the ratio between P_1 and P_2 is expanded and cross-multiplied, the result is found by the following equation:

$$P_1 \cos 30° \cos\theta + P_1 \sin 30° \sin\theta = P_2 \cos 30° \cos\theta - P_2 \sin 30° \sin\theta$$

$$0.866P_1 \cos\theta + 0.5P_1 \sin\theta = 0.866P_2 \cos\theta - 0.5P_2 \sin\theta$$

Dividing both sides of the equation above by $0.5 \cos\theta$ results in

$$\sqrt{3}\, P_1 + P_1 \tan\theta = \sqrt{3}\, P_2 - P_2 \tan\theta$$

Therefore,

$$\tan\theta = \sqrt{3}\,\frac{P_2 - P_1}{P_2 + P_1} \qquad (16\text{-}29)$$

or

$$\theta = \tan^{-1} \sqrt{3}\,\frac{P_2 - P_1}{P_2 + P_1}$$

In equation 16-29, P_2 represents the larger of the two wattmeter readings when the power factor is any value other than unity. As mentioned earlier, the two-wattmeter method cannot be used to measure power in a three-phase four-wire system unless the current in the neutral wire is zero.

EXAMPLE 16-8 The power input to a three-phase balanced load is measured by the two-wattmeter method. The wattmeter readings are 2400 W and −1100 W (meter potential coil reversed). Determine the power factor of the load.

Solution

$$\theta = \tan^{-1} \sqrt{3}\,\frac{P_2 - P_1}{P_2 + P_1} = \tan^{-1} \sqrt{3}\,\frac{2400\text{ W} - (-1100\text{ W})}{2400\text{ W} + (-1100\text{ W})} = 77.9°$$

$$\text{PF} = \cos\theta = \cos 77.9° = 0.21$$

EXAMPLE 16-9 For the circuit shown in Figure 16-28, find the readings of the two wattmeters used to determine the total system power. Phase voltages are $E_{AN} = 120\ \angle 0^\circ$ V, $E_{BN} = 120\ \angle -120^\circ$ V, and $E_{CN} = 120\ \angle 120^\circ$ V.

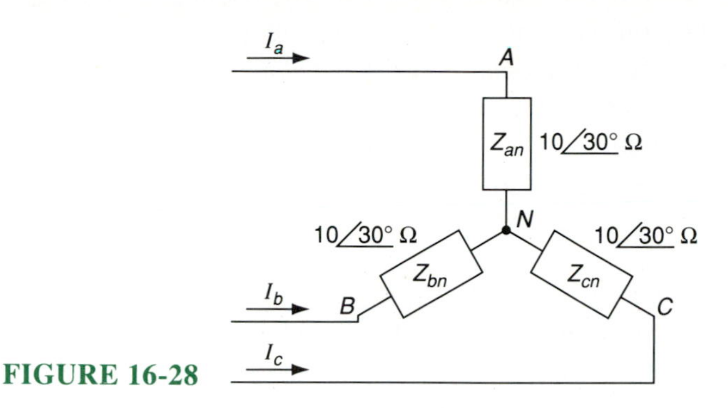

FIGURE 16-28

Solution First determine the line currents.

$$E_{AN} = V_{an} \qquad E_{BN} = V_{bn} \qquad E_{CN} = V_{cn}$$

$$I_a = \frac{V_{an}}{Z_{an}} = \frac{120\ \angle 0^\circ\ \text{V}}{10\ \angle 30^\circ\ \Omega} = 12\ \angle -30^\circ\ \text{A}$$

$$I_b = \frac{V_{bn}}{Z_{bn}} = \frac{120\ \angle -120^\circ\ \text{V}}{10\ \angle 30^\circ\ \Omega} = 12\ \angle -150^\circ\ \text{A}$$

$$I_c = \frac{V_{cn}}{Z_{cn}} = \frac{120\ \angle 120^\circ\ \text{V}}{10\ \angle 30^\circ\ \Omega} = 12\ \angle 90^\circ\ \text{A}$$

$$E_L = \sqrt{3}\,E_p = 208\ \text{V}$$

$$\theta = 30^\circ$$

With a balanced three-phase load, the wattmeter readings are

$$W_1 = E_L I_L \cos(30^\circ + \theta) = (208\ \text{V})(12\ \text{A})\cos(30^\circ + 30^\circ) = 1248\ \text{W}$$

$$W_2 = E_L I_L \cos(30^\circ - \theta) = (208\ \text{V})(12\ \text{A})\cos(30^\circ - 30^\circ) = 2496\ \text{W}$$

16-12 ONE-LINE EQUIVALENT CIRCUIT FOR BALANCED LOADS

According to the principle of delta–wye transformations, a delta-connected impedance network can be converted into a wye-connected impedance network, and vice versa. When the three-phase system is balanced, the line currents are of equal magnitude. Therefore, when the circuit is wye-connected, all the line currents can be obtained when only one line current is calculated and the phase sequence is known. This method of reducing a balanced three-phase wye circuit to its single-phase counterpart is known as a **one-line equivalent circuit.**

The one-line equivalent circuit shown in Figure 16-29(b) is one phase of the three-phase wye-connected circuit shown in Figure 16-29(a). The neutral is not required for a one-line equivalent circuit since the neutral current is zero for a balanced system. The line current for the one-line equivalent circuit is found by Ohm's law.

$$I_L = \frac{E_p}{Z_p}$$

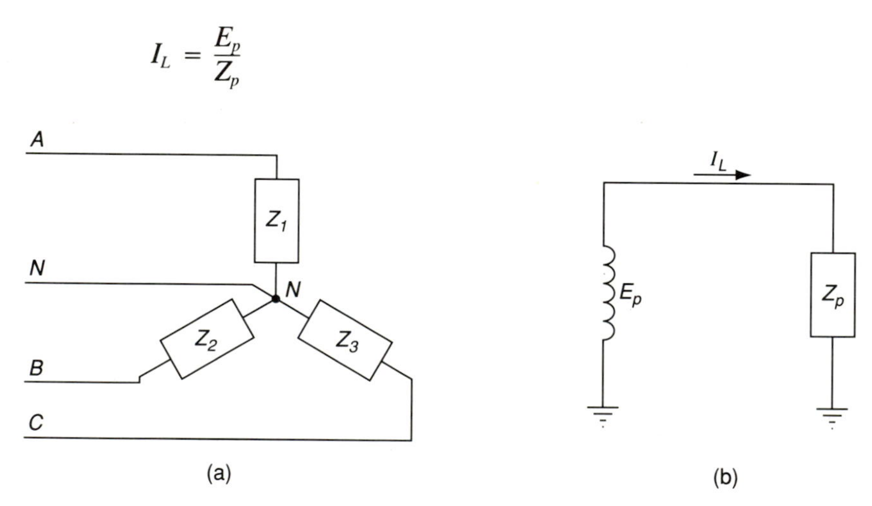

FIGURE 16-29 (a) Balanced wye-connected load; (b) one-line equivalent circuit.

EXAMPLE 16-10 Three identical impedances of $20 \angle 45° \ \Omega$ are connected in delta to a three-phase three-wire 240-V *ACB* system. Determine the line currents using the one-line equivalent method.

Solution First, convert the delta-connected impedances to an equivalent wye circuit.

$$Z_1 = Z_2 = Z_3 = \frac{Z\Delta}{3} = \frac{20 \angle 45°}{3} = 6.67 \angle 45°$$

$$E_p = \frac{E_L}{\sqrt{3}} = \frac{240}{\sqrt{3}} = 138.56\,\text{V}$$

Using voltage E_{AN} as a reference, the line current is found by Ohm's law:

$$I_L = \frac{138.56\,\angle 0^\circ\,\text{V}}{6.67\,\angle 45^\circ\,\Omega} = 20.77\,\angle -45^\circ\,\text{A}$$

The circuit is shown in Figure 16-30.

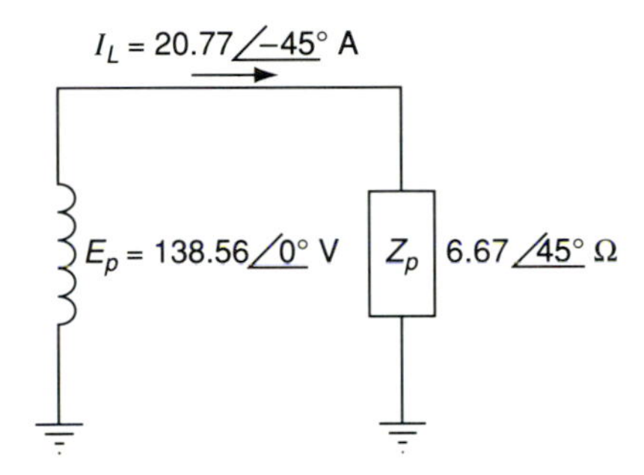

FIGURE 16-30 One-line circuit for Example 16-10.

16-13 UNBALANCED DELTA SYSTEM

Any polyphase load in which the impedance in one or more phases differs from those of other phases is said to be **unbalanced**. Also, if the voltages applied across the load are unequal and the phase angles are unequal, the system is said to be unbalanced.

If the three-phase line voltages across the terminals of an unbalanced delta are of a fixed value, the voltage drops across each phase impedance is known. The currents in each phase can be determined from the line (phase) voltage and load impedance. For a balanced delta system, $I_L = \sqrt{3}\, I_p$. However, no definite relationship exists between line and phase currents in an unbalanced delta system. The line currents in the unbalanced delta are found by phasor addition of the phase currents. The following example illustrates this method.

EXAMPLE 16-11 A three-phase three-wire 208-V *ACB* system supplies power to the delta-connected load shown in Figure 16-31. Find the line currents and the total power consumed by the system.

Solution With E_{AB} chosen as the reference phasor and the voltage sequence *ACB*, the line voltages have the following values:

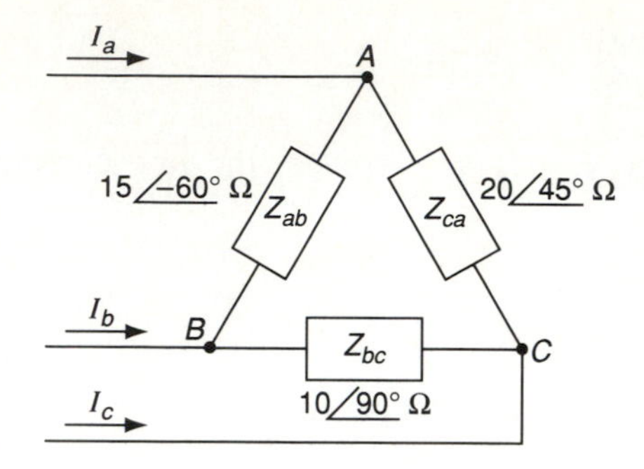

FIGURE 16-31

$E_{AB} = 208\ \angle 0°\ \text{V} \qquad E_{CA} = 208\ \angle -120°\ \text{V}$
$E_{BC} = 208\ \angle 120°\ \text{V}$

Each phase current is now solved for

$E_{AB} = V_{ab} \qquad E_{BC} = V_{bc} \qquad E_{CA} = V_{ca}$

$$I_{ab} = \frac{V_{ab}}{Z_{ab}} = \frac{208\ \angle 0°}{15\ \angle -60°\ \Omega} = 13.87\ \angle 60°\ \text{A}$$

$$I_{bc} = \frac{V_{bc}}{Z_{bc}} = \frac{208\ \angle 120°\ \text{V}}{10\ \angle 90°\ \Omega} = 20.8\ \angle 30°\ \text{A}$$

$$I_{ca} = \frac{V_{ca}}{Z_{ca}} = \frac{208\ \angle -120°\ \text{V}}{20\ \angle 45°\ \Omega} = 10.4\ \angle -165°\ \text{A}$$

The line currents are then found by applying Kirchhoff's current law to each node of the system.

$$I_a = I_{ab} - I_{ca} = 13.87\ \angle 60°\ \text{A} - 10.4\ \angle -165°\ \text{A}$$
$$= 22.46\ \angle 40.9°\ \text{A}$$

$$I_b = I_{bc} - I_{ab} = 20.8\ \angle 30°\ \text{A} - 13.87\ \angle 60°\ \text{A}$$
$$= 11.19\ \angle -8.3°\ \text{A}$$

$$I_c = I_{ca} - I_{bc} = 10.4\ \angle -165°\ \text{A} - 20.8\ \angle 30°\ \text{A}$$
$$= 30.96\ \angle -155°\ \text{A}$$

The total power is equal to the sum of the individual phase powers.

$$P_{AB} = E_{AB} \times I_{ab} \times \cos\theta_{ab}$$
$$= (208\ \text{V})(13.87\ \text{A})(\cos -60°)$$
$$= 1442.48\ \text{W}$$

$$
\begin{aligned}
P_{BC} &= E_{BC} \times I_{bc} \times \cos\theta_{bc} \\
&= (208\text{ V})(20.8\text{ A})(\cos 90^\circ) \\
&= 0\text{ W}
\end{aligned}
$$

$$
\begin{aligned}
P_{CA} &= E_{CA} \times I_{ca} \times \cos\theta_{ca} \\
&= (208\text{ V})(10.4\text{ A})(\cos 45^\circ) \\
&= 1529.61\text{ W}
\end{aligned}
$$

$$
\begin{aligned}
P_T &= P_{AB} + P_{BC} + P_{CA} \\
&= 1442.48\text{ W} + 0\text{ W} + 1529.61\text{ W} \\
&= 2972.09\text{ W}
\end{aligned}
$$

16-14 UNBALANCED FOUR-WIRE WYE-CONNECTED LOADS

When three impedances of different values are interconnected to form a four-wire wye system, phase currents of different magnitudes will flow in the system when a three-phase voltage is applied. When this situation exists, the system is said to be unbalanced, and current will flow in the neutral wire, as shown in Figure 16-32. The system will also be unbalanced if the ratios of resistance to reactance in the three impedances are not the same. When this condition occurs, the displacement angle between phase currents will not be 120°.

Since the neutral is a common point between the load and source, the phase current can be determined using the neutral as a reference regardless of the impedance of each phase. By applying Ohm's law, the phase current is simply

$$I_p = \frac{E_p}{Z_p}$$

In a balanced wye system, the line currents and phase currents are equal. However, in the unbalanced system, the algebraic sum of the individual line or phase currents is not zero. The phasor sum of either the line or phase currents is the current that will flow in the neutral wire. By applying Kirchhoff's current law, the equation for the current flowing in the neutral is found by the following equation:

$$I_n = I_a + I_b + I_c = I_{an} + I_{bn} + I_{cn} \tag{16-30}$$

EXAMPLE 16-12 A 208-V three-phase four-wire *ACB* system feeds the wye-connected system shown in Figure 16-32 with the following loads: $Z_{an} = 15\,\underline{/30^\circ}\ \Omega$, $Z_{bn} = 10\,\underline{/-60^\circ}\ \Omega$, and $Z_{cn} = 12\,\underline{/-30^\circ}\ \Omega$. Find the line currents and neutral current. Also, draw the phasor diagram.

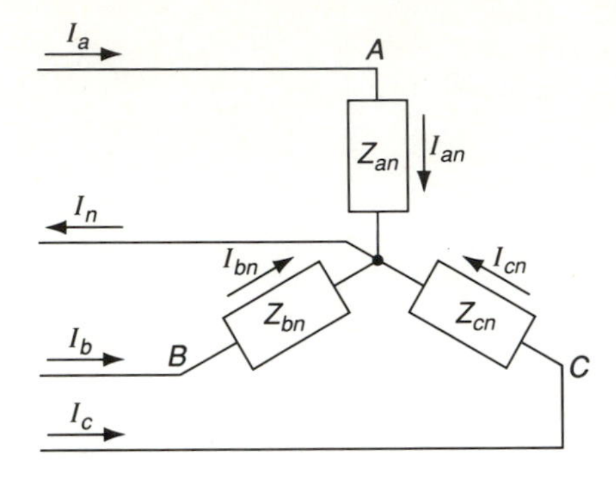

FIGURE 16-32 Direction of current flow in an unbalanced wye-connected load.

Solution First, determine the phase voltages of this *ACB* system with line voltage E_{ab} as the reference voltage.

$$V_{an} = \frac{208\,\underline{/0^\circ}}{\sqrt{3}} = 120\,\underline{/0^\circ}\text{ V}$$

$$V_{bn} = \frac{208\,\underline{/120^\circ}}{\sqrt{3}} = 120\,\underline{/120^\circ}\text{ V}$$

$$V_{cn} = \frac{208\,\underline{/-120^\circ}}{\sqrt{3}} = 120\,\underline{/-120^\circ}\text{ V}$$

Next, solve for the three line currents.

$$I_a = \frac{V_{an}}{Z_{an}} = \frac{120\,\underline{/0^\circ}\text{ V}}{15\,\underline{/30^\circ}\,\Omega} = 8\,\underline{/-30^\circ}\text{ A}$$

$$I_b = \frac{V_{bn}}{Z_{bn}} = \frac{120\,\underline{/120^\circ}\text{ V}}{10\,\underline{/-60^\circ}\,\Omega} = 12\,\underline{/180^\circ}\text{ A}$$

$$I_c = \frac{V_{cn}}{Z_{cn}} = \frac{120\,\underline{/-120^\circ}\text{ V}}{12\,\underline{/-30^\circ}\,\Omega} = 10\,\underline{/-90^\circ}\text{ A}$$

The neutral contains the phasor sum of line currents I_a, I_b, and I_c. Assuming the neutral current flow is toward the source,

$$\begin{aligned} I_n = I_a + I_b + I_c &= 8\,\underline{/-30^\circ}\text{ A} + 12\,\underline{/180^\circ}\text{ A} + 10\,\underline{/-90^\circ}\text{ A} \\ &= 14.9\,\underline{/-110^\circ}\text{ A} \end{aligned}$$

The complete phasor diagram is shown in Figure 16-33.

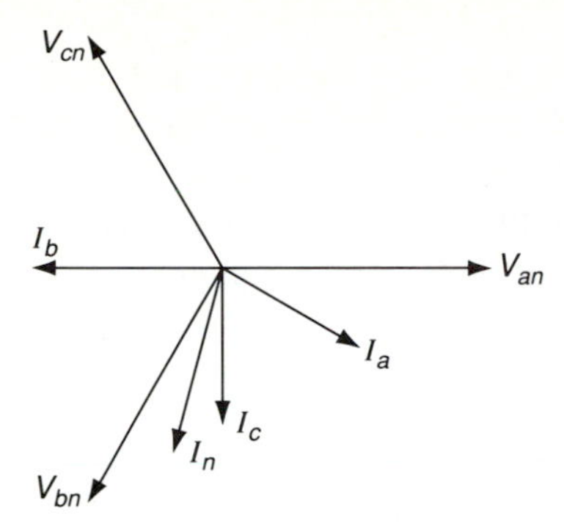

FIGURE 16-33 Phasor diagram for Example 16-12.

16-15 UNBALANCED THREE-WIRE WYE–CONNECTED LOADS

When the neutral wire of the unbalanced four-wire wye-connected load is removed, the phase voltages of the load will vary with the phase impedances of the load. This type of condition is referred to as a **floating neutral**. With the neutral wire open, the three-phase voltages are unbalanced in magnitude and in phase. In this situation, it is quite possible that one phase voltage can exceed the line voltage of the system. These unbalanced phase voltages make the floating neutral a very undesirable condition, and for this reason, three-wire wye connected loads are rarely used.

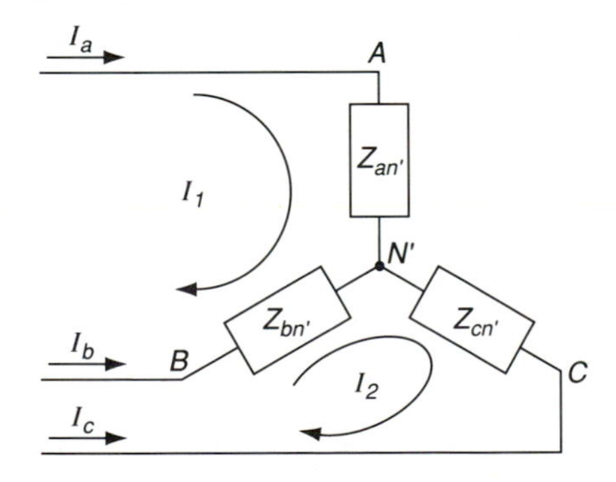

FIGURE 16-34 Three-wire wye-connected load.

The simplest method of solving for unknown values in a three-wire wye system consists of writing loop equations to determine the values of I_1 and I_2, as shown in Figure 16-34. The common junction between the three impedances is marked N' instead of N, since this value is not at the potential of the neutral itself.

$$E_{AB} = I_1(Z_{an'} + Z_{bn'}) - I_2(Z_{bn'})$$

$$E_{BC} = -I_1(Z_{bn'}) + I_2(Z_{bn'} + Z_{cn'})$$

The two loop equations are then changed into determinant form, and the values of I_1 and I_2 are solved for. The line currents I_a, I_b, and I_c can now be calculated.

$$I_a = I_1$$

$$I_b = I_2 - I_1$$

$$I_c = -I_2$$

The voltages across the three impedances are given by the products of the line currents and the corresponding impedances.

$$V_{an'} = I_a Z_{an'}$$

$$V_{bn'} = I_b Z_{bn}'$$

$$V_{cn'} = I_c Z_{cn'}$$

The floating neutral voltage $V_{nn'}$ is the displacement voltage between the floating neutral and the zero potential of the system. This voltage may be calculated using any of the three points A, B, or C and following the conventional double-subscript notation.

$$V_{nn'} = V_{an'} - V_{an} \quad (16\text{-}31)$$

$$V_{nn'} = V_{bn'} - V_{bn} \quad (16\text{-}32)$$

$$V_{nn'} = V_{cn'} - V_{cn} \quad (16\text{-}33)$$

From the equations above, a Kirchhoff voltage loop can be determined for the system which utilizes the floating neutral voltage.

$$V_{an'} + V_{n'n} + V_{bn'} + V_{n'n} + V_{cn'} + V_{n'n} = 0 \quad (16\text{-}34)$$

$$V_{an'} + V_{bn'} + V_{cn'} + 3V_{n'n} = 0$$

$$V_{nn'} = \frac{V_{an'} + V_{bn'} + V_{cn'}}{3}$$

EXAMPLE 16-13 A three-phase three-wire 208-V *ABC* system has a wye-connected load with $Z_{an'} = 12 \underline{/0^\circ}\ \Omega$, $Z_{bn'} = 8 \underline{/45^\circ}\ \Omega$, and $Z_{cn'} = 10 \underline{/30^\circ}\ \Omega$. Determine the line currents and phase voltages across each impedance. Also, construct an open-faced and closed-face phasor diagram and obtain the floating neutral voltage $V_{n'n}$. Use line voltage E_{BC} as a reference phasor.

Solution

$$E_{BC} = 208 \underline{/0^\circ}\text{ V}$$

$$E_{CA} = 208 \underline{/-120^\circ}\text{ V}$$

$$E_{AB} = 208 \underline{/120^\circ}\text{ V}$$

Figure 16-34 shows the three-wire wye system with loop currents I_1 and I_2.

$$E_{AB} = I_1(Z_{an'} + Z_{bn'}) - I_2(Z_{bn'})$$

$$208 \underline{/120^\circ}\text{ V} = I_1(12 \underline{/0^\circ}\,\Omega + 8 \underline{/45^\circ}\,\Omega) - I_2(8 \underline{/45^\circ}\,\Omega)$$

$$E_{BC} = -I_1(Z_{bn'}) + I_2(Z_{bn'} + Z_{cn'})$$

$$208 \underline{/0^\circ}\text{ V} = -I_1(8 \underline{/45^\circ}\,\Omega) + I_2(8 \underline{/45^\circ}\,\Omega + 10 \underline{/30^\circ}\,\Omega)$$

The loop currents can now be expressed in determinant form:

$$I_1 = \frac{(E_{AB})(Z_{bn'} + Z_{cn'}) - (E_{BC})(-Z_{bn})}{(Z_{an'} + Z_{bn'})(Z_{bn'} + Z_{cn'}) - (-Z_{bn})(-Z_{bn})}$$

$$= \frac{(208 \underline{/120^\circ}\text{ V})(8 \underline{/45^\circ}\,\Omega + 10 \underline{/30^\circ}\,\Omega) - (208 \underline{/0^\circ}\text{ V})(8 \underline{/-135^\circ}\,\Omega)}{281.34 \underline{/46.82^\circ}\,\Omega}$$

$$= 12.31 \underline{/83.3^\circ}\text{ A}$$

$$I_2 = \frac{(Z_{an'} + Z_{bn'})(E_{BC}) - (-Z_{bn})(E_{AB})}{(Z_{an'} + Z_{bn'})(Z_{bn'} + Z_{cn'}) - (-Z_{bn})(-Z_{bn})}$$

$$= \frac{(12 \underline{/0^\circ}\,\Omega + 8 \underline{/45^\circ}\,\Omega)(208 \underline{/0^\circ}\text{ V}) - (8 \underline{/-135^\circ}\,\Omega)(208 \underline{/120^\circ}\,\Omega)}{281.34 \underline{/46.82^\circ}\,\Omega}$$

$$= 9.3 \underline{/-8.9^\circ}\text{ A}$$

The line currents can now be found:

$$I_a = I_1 = 12.31 \underline{/83.3^\circ}\text{ A}$$

$$I_b = I_2 - I_1 = 15.71 \underline{/-60.4^\circ}\text{ A}$$

$$I_c = -I_2 = 9.3 \underline{/171.1^\circ}\text{ A}$$

The voltages across the three impedances are as follows:

$$V_{an'} = I_a Z_{an'} = (12.31 \underline{/83.3^\circ}\text{ A})(12 \underline{/0^\circ}\,\Omega)$$
$$= 147.72 \underline{/83.3^\circ}\text{ V}$$

$$V_{bn'} = I_b Z_{bn'} = (15.71 \angle -60.4^\circ \text{ A})(8 \angle 45^\circ \ \Omega)$$
$$= 125.68 \angle -15.4^\circ \text{ V}$$

$$V_{cn'} = I_c Z_{cn'} = (9.3 \angle 171.1^\circ \text{ A})(10 \angle 30^\circ \ \Omega)$$
$$= 93 \angle 201.1^\circ \text{ V}$$

Figure 16-35(a) and (b) show the open-face and closed-face phasor diagrams for the line and phase voltages.

$$V_{nn'} = \frac{V_{an'} + V_{bn'} + V_{cn'}}{3}$$
$$= \frac{147.72 \angle 83.3^\circ \text{ V} + 125.68 \angle -15.4^\circ \text{ V} + 93 \angle 201.1^\circ}{3}$$
$$= 31.7 \angle 57.1^\circ \text{ V}$$

To check: $V_{nn'} = V_{an'} - V_{an}$
$$= 147.72 \angle 83.3^\circ \text{ V} - 120 \angle 90^\circ \text{ V}$$
$$= 31.7 \angle 57.1^\circ \text{ V}$$

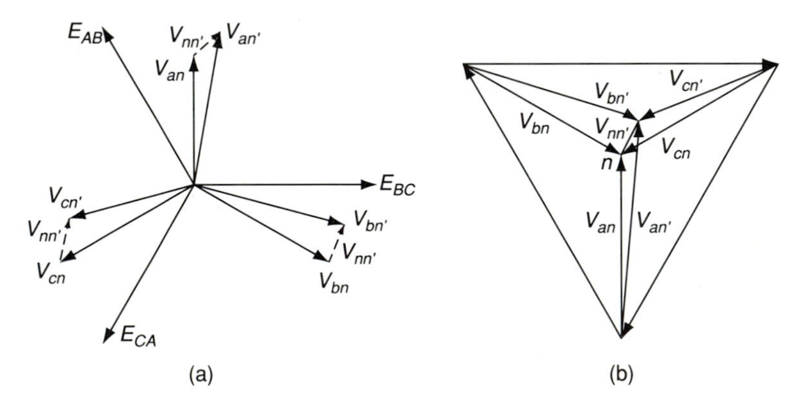

FIGURE 16-35 (a) Open-face and (b) closed-face phasor diagrams of line and phase voltages for Example 16-13.

16-16 POWER FACTOR IN UNBALANCED THREE-PHASE LOADS

Power factor in a single-phase system or in a balanced polyphase system is the ratio of real power to apparent power. In an unbalanced polyphase system, each phase has its own particular power factor. When phase loads are either purely inductive, or capacitive, the average of the individual phase power factors is a general indication of the ratio of total watts to total volt-amperes.

However, when an unbalanced system contains both inductive and capacitive loads, the compensating effect of capacitive reactance and inductive reactance is not taken into account when using this averaging method. Another limitation to this method is that it may be quite difficult to determine the individual phase power factors in many practical installations. For these reasons, average power factor is generally not considered as a feasible method of determining the total power factor of an unbalanced three-phase system.

A more accurate method of determining the power factor of this type of system involves constructing a power triangle for the unbalanced load. Using this method, the individual values of real, apparent, and reactive power are determined for each phase. The sums of each are then combined to form a total system ratio of the total power to apparent power. This method is illustrated in Example 16-14.

EXAMPLE 16-14 A three-phase four-wire *ABC* system supplies power to the wye-connected load shown in Figure 16-36. Determine the total power factor of the system. Phase voltage of the system is assumed to be 120 V.

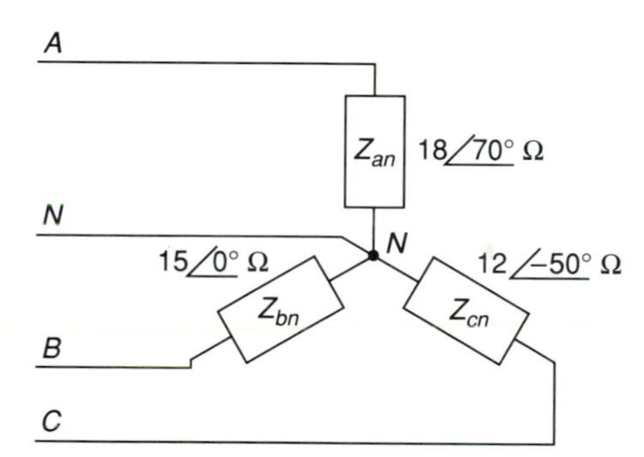

FIGURE 16-36

Solution With V_{an} chosen as the reference phasor and the voltages sequence *ABC*, the phase voltages are as follows:

$$V_{an} = 120\,\underline{/0^\circ}\text{ V}$$

$$V_{bn} = 120\,\underline{/-120^\circ}\text{ V}$$

$$V_{cn} = 120\,\underline{/120^\circ}\text{ V}$$

The phase currents are found by Ohm's law.

$$I_{an} = \frac{V_{an}}{Z_{an}} = \frac{120\,\underline{/0^\circ}\text{ V}}{18\,\underline{/70^\circ}\,\Omega} = 6.67\,\underline{/-70^\circ}\text{ A}$$

$$I_{bn} = \frac{V_{bn}}{Z_{bn}} = \frac{120\,\underline{/-120^\circ}\text{ V}}{15\,\underline{/0^\circ}\,\Omega} = 8\,\underline{/-120^\circ}\text{ A}$$

$$I_{cn} = \frac{V_{cn}}{Z_{cn}} = \frac{120\,\angle 120^\circ\ \text{V}}{12\,\angle -50^\circ\ \Omega} = 10\,\angle 170^\circ\ \text{A}$$

The power dissipated by each phase can now be determined.

For phase A:

$$P_a = E_{pa}I_{pa}\cos\theta_a = (120\ \text{V})(6.67\ \text{A})(\cos 70^\circ) = 273.75\ \text{W}$$

$$S_a = E_{pa}I_{pa} = (120\ \text{V})(6.67\ \text{A}) = 800.4\ \text{VA}$$

$$Q_a = E_{pa}I_{pa}\sin\theta_a = (120\ \text{V})(6.67\ \text{A})(\sin 70^\circ) = 752.13\ \text{VAR}$$

For phase B:

$$P_b = E_{pb}I_{pb}\cos\theta_b = (120\ \text{V})(8\ \text{A})(\cos 0^\circ) = 960\ \text{W}$$

$$S_b = E_{pb}I_{pb} = 960\ \text{VA}$$

$$Q_b = E_{pb}I_{pb}\sin\theta_b = (120\ \text{V})(8\ \text{A})(\sin 0^\circ) = 0\ \text{VAR}$$

For phase C:

$$P_c = E_{pc}I_{pc}\cos\theta_c = (120\ \text{V})(10\ \text{A})(\cos -50^\circ) = 771.35\ \text{W}$$

$$S_c = E_{pc}I_{pc} = (120\ \text{V})(10\ \text{A}) = 1200\ \text{VA}$$

$$Q_c = E_{pc}I_{pc}\sin\theta_c = (120\ \text{V})(10\ \text{A})(\sin -50^\circ) = -919.25\ \text{VAR}$$

The total power triangle shown in Figure 16-37 can now be found from the sum of the individual phase values.

$$\begin{aligned} P_T &= P_a + P_b + P_c \\ &= 273.75\ \text{W} + 960\ \text{W} + 771.35\ \text{W} \\ &= 2005.1\ \text{W} \end{aligned}$$

$$\begin{aligned} S_T &= S_a + S_b + S_c \\ &= 800.4\ \text{VA} + 960\ \text{VA} + 1200\ \text{VA} \\ &= 2960.44\ \text{VA} \end{aligned}$$

$$\begin{aligned} Q_T &= Q_a + Q_b + Q_c \\ &= 752.13\ \text{VAR} + 0\ \text{VAR} + (-919.25)\ \text{VAR} \\ &= -167.12\ \text{VAR} \end{aligned}$$

$$\text{PF} = \cos\theta = \frac{P_T}{S_T} = \frac{2005.1}{2960.44} = 0.68$$

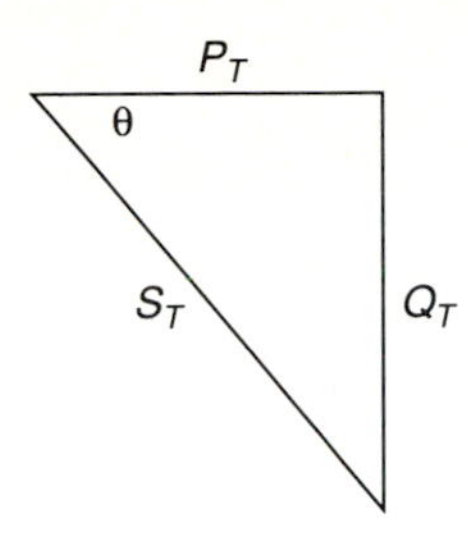

FIGURE 16-37 Power triangle for Example 16-14.

16-17 COMBINATION THREE-PHASE LOADS

Usually, a three-phase source supplies power to several wye and delta loads. Also, numerous single-phase loads may be connected between a pair of line wires, which would represent one leg of a delta system. The simplest method of solving these combination circuits is to convert all wye-connected impedances to equivalent delta impedances. The phase currents in each load can then be determined, and the line currents are obtained by using Kirchhoff's current law. The following example illustrates this method of solving combination three-phase systems.

EXAMPLE 16-15 For the circuit shown in Figure 16-38, calculate the line currents I_a, I_b, and I_c. The source is a three-phase three-wire ABC system with a line voltage of 460 V. The impedance values are as follows: $Z_a = 80\ \underline{/30°}\ \Omega$, $Z_b = 60\ \underline{/-20°}\ \Omega$, $Z_c = 90\ \underline{/45°}\ \Omega$, $Z_1 = 100\ \underline{/0°}\ \Omega$, $Z_2 = 50\ \underline{/60°}\ \Omega$, $Z_3 = 75\ \underline{/-30°}\ \Omega$.

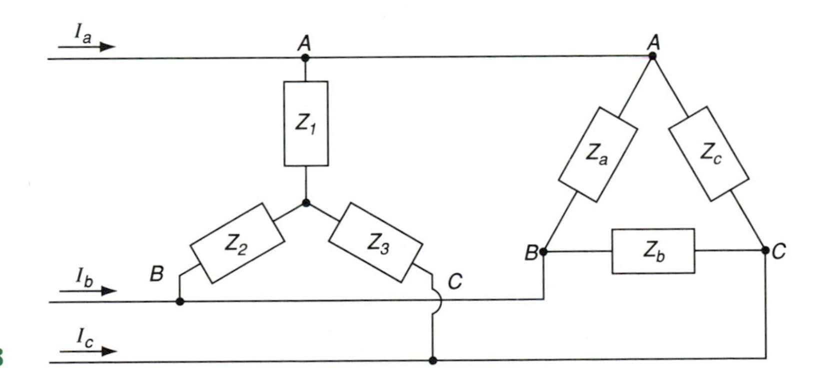

FIGURE 16-38

Solution The voltage phase sequence is ABC. Choosing E_{AB} as a reference phasor, the line voltages are

$$E_{AB} = 460 \angle 0^\circ \text{ V}$$

$$E_{BC} = 460 \angle -120^\circ \text{ V}$$

$$E_{CA} = 460 \angle 120^\circ \text{ V}$$

First, convert wye impedances to equivalent delta impedances.

$$Z_{a'} = \frac{Z_1Z_2 + Z_2Z_3 + Z_3Z_1}{Z_3}$$
$$= 166.49 \angle 41.34^\circ \ \Omega$$

$$Z_{c'} = \frac{Z_1Z_2 + Z_2Z_3 + Z_3Z_1}{Z_2}$$
$$= 249.73 \angle -48.7^\circ \ \Omega$$

$$Z_{b'} = \frac{Z_1Z_2 + Z_2Z_3 + Z_3Z_1}{Z_1}$$
$$= 124.87 \angle 11.34^\circ \ \Omega$$

The phase currents can now be calculated.

$$E_{AB} = V_{ab} \qquad E_{BC} = V_{bc} \qquad E_{CA} = V_{ca}$$

$$I_{ab} = \frac{V_{ab}}{Z_a} = \frac{460 \angle 0^\circ \text{ V}}{80 \angle 30^\circ \ \Omega} = 5.75 \angle -30^\circ \text{ A}$$

$$I_{bc} = \frac{V_{bc}}{Z_b} = \frac{460 \angle -120^\circ \text{ V}}{60 \angle -20^\circ \ \Omega} = 7.67 \angle -100^\circ \text{ A}$$

$$I_{ca} = \frac{V_{ca}}{Z_c} = \frac{460 \angle 120 \text{ V}}{90 \angle 45^\circ \ \Omega} = 5.11 \angle 75^\circ \text{ A}$$

$$I_{ab'} = \frac{V_{ab}}{Z_{a'}} = \frac{460 \angle 0^\circ \text{ V}}{166.49 \angle 41.34^\circ \ \Omega}$$
$$= 2.76 \angle -41.34^\circ \text{ A}$$

$$I_{bc'} = \frac{V_{bc}}{Z_{b'}} = \frac{460 \angle -120^\circ \text{ V}}{124.87 \angle 11.34^\circ \ \Omega}$$
$$= 3.68 \angle -131.34^\circ \text{ A}$$

$$I_{ca'} = \frac{V_{ca}}{Z_{c'}} = \frac{460 \angle 120^\circ \text{ V}}{249.73 \angle -48.7^\circ \ \Omega}$$
$$= 1.84 \angle 168.7^\circ \text{ A}$$

The line currents can now be determined by applying Kirchhoff's current law to the equivalent delta shown in Figure 16-39.

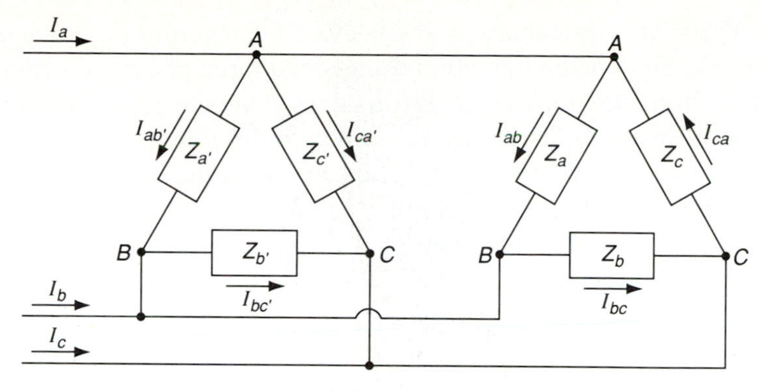

FIGURE 16-39 Equivalent delta circuit for Example 16-15.

$$\begin{aligned} I_a &= I_{ab'} - I_{ca'} + I_{ab} - I_{ca} \\ &= 2.76 \angle -41.34^\circ \text{ A} + 1.84 \angle 168.7^\circ \text{ A} \\ &\quad + 5.75 \angle -30^\circ \text{ A} + 5.11 \angle 75^\circ \text{ A} \\ &= 12.52 \angle -53^\circ \text{ A} \end{aligned}$$

$$\begin{aligned} I_b &= I_{bc'} - I_{ab'} + I_{bc} - I_{ab} \\ &= 3.68 \angle -131.34^\circ \text{ A} - 2.76 \angle -41.34^\circ \text{ A} \\ &\quad + 7.67 \angle -100^\circ \text{ A} - 5.75 \angle -30^\circ \text{ A} \\ &= 12.19 \angle -152.5^\circ \text{ A} \end{aligned}$$

$$\begin{aligned} I_c &= I_{ca'} - I_{bc'} + I_{ca} - I_{bc} \\ &= 1.84 \angle 168.7^\circ \text{ A} - 3.68 \angle -131.34^\circ \text{ A} \\ &\quad + 5.11 \angle 75^\circ \text{ A} - 7.67 \angle -100^\circ \text{ A} \\ &= 15.96 \angle 78.1^\circ \text{ A} \end{aligned}$$

16-18 PHASE SEQUENCE EFFECTS AND MEASUREMENT

When systems are connected in a wye or delta configuration, it is important to know the phase sequence in order to ensure a proper interconnection of coil windings. Phase sequence plays an important role in such factors as the direction of rotation of three-phase induction motors. Changing the phase sequence of a balanced system will have no effect on the magnitudes of the currents and voltages of that system. However, if the two-wattmeter method is being used to measure power, a change in phase sequence will cause the readings of both meters to reverse.

When the phase sequence is altered in an unbalanced three-phase system, the branch currents may change in magnitude as well as direction, although the changes caused will have no net effect on the total power consumed by the system. There are, essentially, two methods of determining voltage phase sequence. The first method involves using small three-phase induction motors which have been previously checked against a known phase sequence of a given system. In three-phase systems, only two different phase sequences are possible (*ABC* and *ACB*), so the direction in which the motor rotates can be used as an indicator of phase sequence.

The second method of checking phase sequence requires using an unbalanced circuit as shown in Figure 16-40. The three line wires whose voltage sequence is to be tested are labeled arbitrarily. Two incandescent lamps, *A* and *C*, form two phases of a wye, and an inductive reactance X_L forms the third phase. The resistances of the two lamps and the reactance of the coil have approximately the same magnitude. If lamp *A* is brighter than lamp *C*, the phase sequence of the line-to-line voltage is *ABC*. If lamp *C* is brighter than lamp *A*, the phase sequence is *ACB*.

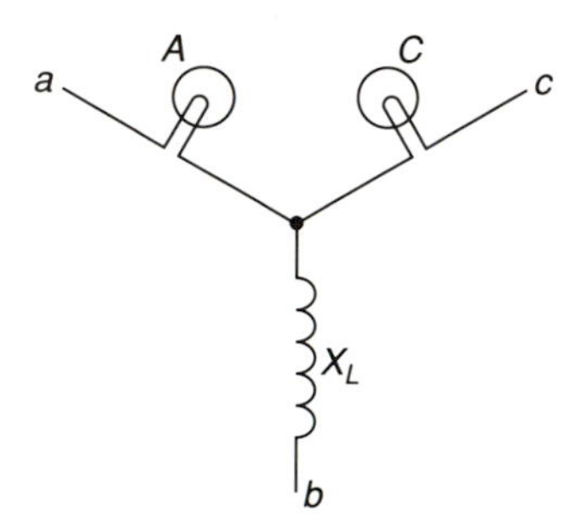

FIGURE 16-40 Phase sequence indicator.

16-19 HARMONICS IN THREE-PHASE SYSTEMS

The term **harmonics** in power systems refers to components of the system which have frequencies above the fundamental, or source, frequency. Voltage and current harmonics are usually found by means of Fourier series analysis. A harmonic of a given wave is another wave having a frequency equal to an integral multiple of the frequency of the given wave. For example, a harmonic with a frequency double that of the fundamental frequency is called the **second harmonic**. If the source frequency is 60 Hz, then

$$\begin{aligned} \text{Fundamental frequency} &= 60\ \text{Hz} \\ \text{Second harmonic frequency} &= 120\ \text{Hz} \\ \text{Third harmonic frequency} &= 180\ \text{Hz} \end{aligned}$$

Fourth harmonic frequency $= 240$ Hz

etc.

The relation between a fundamental harmonic and its second and third harmonics is shown in Figure 16-41.

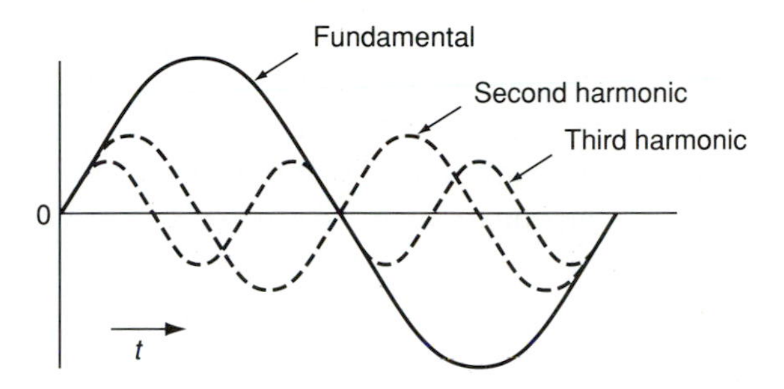

FIGURE 16-41 Relationship between fundamental, second, and third harmonics.

When a three-phase network has a balanced load, the neutral carries no current unless harmonics are present. Some loads, such as the ballast transformers of fluorescent lamps, generate currents that circulate between the phase and neutral conductors. Figure 16-42 shows the current waves in a three-phase network. Phase *A*, *B*, and *C* are drawn separately for a clearer representation of the harmonic wave. When third harmonics are produced in a balanced three-phase network, the third harmonics of all three phases are in phase with one another. For this reason, they are usually referred to as **zero-sequence currents**, since no phase difference exists between them.

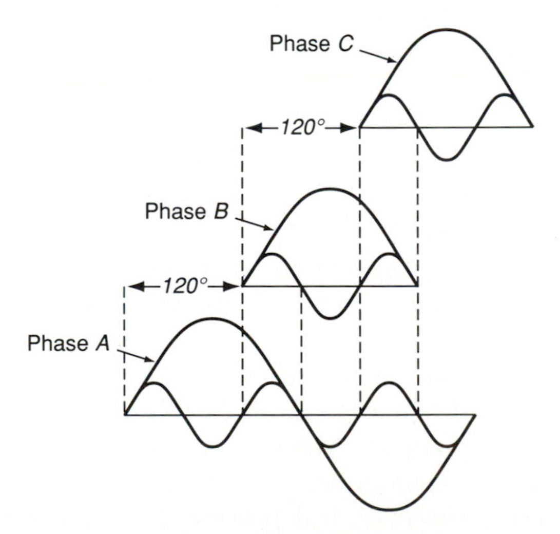

FIGURE 16-42 Presence of third harmonics in a three-phase circuit. The three phases are shown separately for clarity.

A voltage generated in a conductor will be sinusoidal only when the flux cutting the conductor varies according to a sine law. A purely sinusoidal waveform is considered to be perfectly symmetrical. In ac generators it is extremely difficult to obtain an exact sine wave of distribution of the field flux. The slots and teeth of the armature change the reluctance of the path for the flux and cause ripples in the flux wave. Also, when a load is connected to the generator, the distribution of the field flux is altered due to the effect of the armature reaction of the current in the armature. Since the generated emf wave is distorted, harmonic frequency components are induced in each phase of the circuit. Some of these harmonic components are canceled by the interconnection of windings in either a wye or delta configuration. If the angular velocity of a sine wave is given by $\omega = 2\pi f$, the equations for the fundamental voltages in each of the three phases is as follows:

$$E_a = E_{m1} \sin \omega t \text{ V} \tag{16-35}$$

$$E_b = E_{m1} \sin(\omega t - 120°) \text{ V} \tag{16-36}$$

$$E_c = E_{m1} \sin(\omega t - 240°) \text{ V} \tag{16-37}$$

The third-harmonic components of the voltages are defined by the equations

$$E_{a3} = E_{m3} \sin 3\, \omega t \text{ V}$$

$$E_{b3} = E_{m3} \sin(3\omega t - 360°) \text{ V}$$

$$E_{c3} = E_{m3} \sin(3\omega t - 720°) \text{ V}$$

If the phase voltages E_{bn} and E_{na} of Figure 16-43(a) are combined, the phasor diagrams shown in Figure 16-43(b) will result. The fundamental frequency has E_{ba} at 30° ahead of E_{na}. Since $E_{na1} = E_{m1} \sin \theta t$, $E_{ba1} = \sqrt{3}\, E_m \sin(\omega t + 30°)$. For the third harmonic, $E_{ba3} = 0$.

The phasor diagram of Figure 16-43(c) shows that the third harmonics in the two phases between any pair of terminals are in opposition and cancel each other out. The third harmonics cannot contribute anything to line voltage, although they do contribute to the phase voltage of the system. The ratio of line and phase voltage of a wye connection is $\sqrt{3}$ only when there is no third harmonic present.

In the wye-connected generator, there are no third-harmonic voltages, since the windings oppose each other and as a result are neutralized. The delta-connected generator has no third-harmonic voltages due to the third-harmonic being short-circuited by the delta connection, and appears in the system as a small internal impedance drop. However, it is possible for a third-harmonic

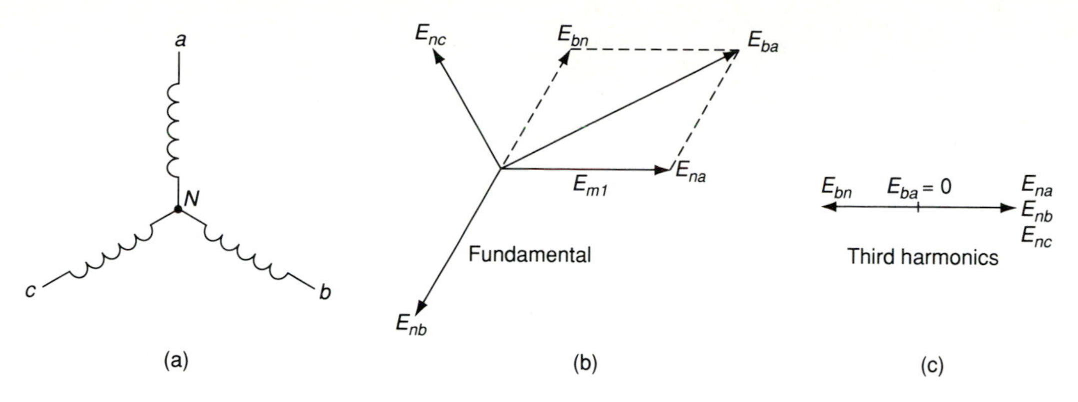

FIGURE 16-43 (a) Wye-connected generator; (b) phasor diagram; (c) phasor sum of third harmonics.

circulating current to develop in a delta-connected generator, which is one reason that wye-connected generators are preferred over delta-connected generators.

In a generator, a circulating current is regarded as an undesirable condition. However, in a transformer, this type of current is considered to be quite useful. A circulating current in a transformer acts as a component of the magnetizing current for the transformer's core. Some high-voltage transformers that have a wye-connected primary and secondary winding will also have a third winding which is delta-connected to allow a third-harmonic circulating current to flow. This type of delta-connected winding is referred to as a **tertiary winding.**

KEY TERMS

Polyphase system
Double-subscript method
Neutral
Line wires
Phase voltage
Line voltage
Phase sequence
Star (tee) connection
Three-phase four-wire wye system
Balanced system
Three-phase three-wire wye system
Closed-face (polygon) phasor diagram
Open-face (polar) phasor diagram
Wattmeter
Two-wattmeter method
One-line equivalent circuit
Unbalanced system
Floating neutral
Harmonic
Second harmonic
Zero-sequence currents
Tertiary winding

PROBLEMS

16-1 Determine the phase sequence of the following voltages: $E_{AB} = 120 \angle -30^\circ$ V, $E_{BC} = 120 \angle 90^\circ$ V, and $E_{CA} = 120 \angle -150^\circ$ V.

16-2 For the circuit shown in Figure 16-44, determine the three line currents I_a, I_b, and I_c.

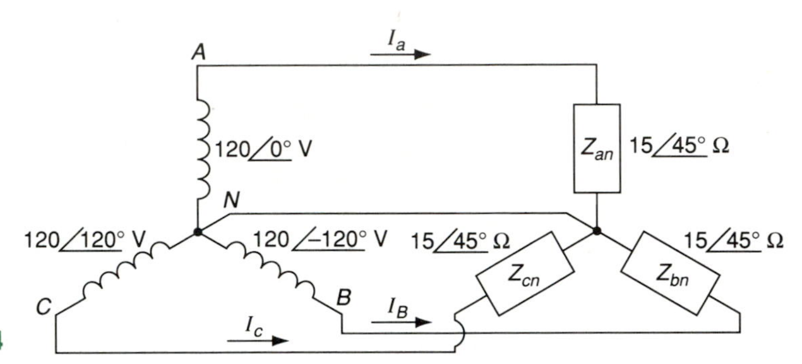

FIGURE 16-44

16-3 Determine the line voltages for the circuit of Figure 16-44.

16-4 For the circuit shown in Figure 16-45, calculate the values of line currents I_a, I_b, and I_c.

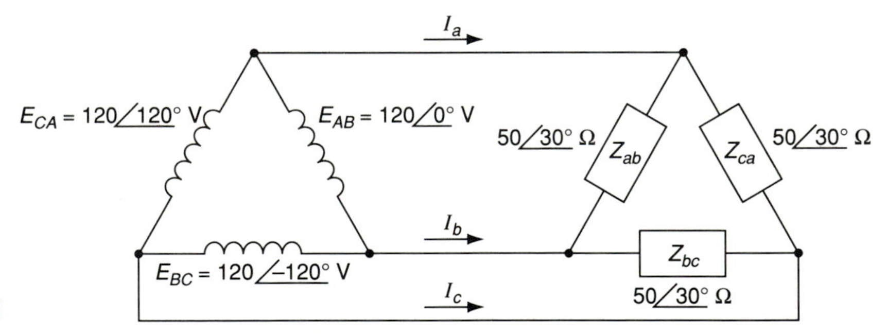

FIGURE 16-45

16-5 A balanced delta source with a reference phasor of $E_{AB} = 208 \angle 30^\circ$ V, and a phase sequence of *ACB* is connected to a balanced delta load. Each load has an impedance of $20 \angle 45^\circ$ Ω. Determine the three line currents for this circuit.

16-6 The balanced system shown in Figure 16-46 consists of the following voltages and impedances: $E_{AN} = 120 \angle 0^\circ$ V, $E_{BN} = 120 \angle -120^\circ$ V, $E_{CN} = 120 \angle 120^\circ$ V, and $Z_{ab} = Z_{bc} = Z_{ca} = 45 \angle 30^\circ$ Ω. Determine (a) the line voltages; (b) the phase currents; (c) the line currents.

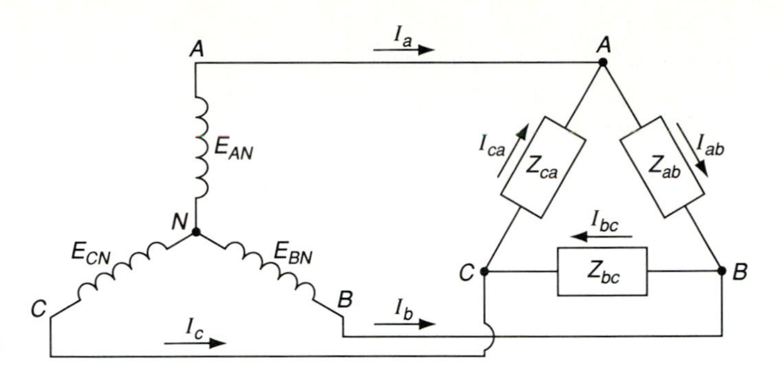

FIGURE 16-46

16-7 Repeat Problem 16-6 using the following voltages and impedances: $E_{AN} = 2400\ \underline{/0^\circ}$ V, $E_{BN} = 2400\ \underline{/120^\circ}$ V, $E_{CN} = 2400\ \underline{/-120^\circ}$ V, and $Z_{ab} = Z_{bc} = Z_{ca} = 600\ \underline{/45^\circ}\ \Omega$.

16-8 The balanced system of Figure 16-47 has line voltages of $E_{AB} = 208\ \underline{/30^\circ}$ V, $E_{BC} = 208\ \underline{/-90^\circ}$ V, and $E_{CA} = 208\ \underline{/150^\circ}$ V. The load impedances are $Z_{an} = Z_{bn} = Z_{cn} = 40\ \underline{/60^\circ}\ \Omega$. Determine the phase currents of the load.

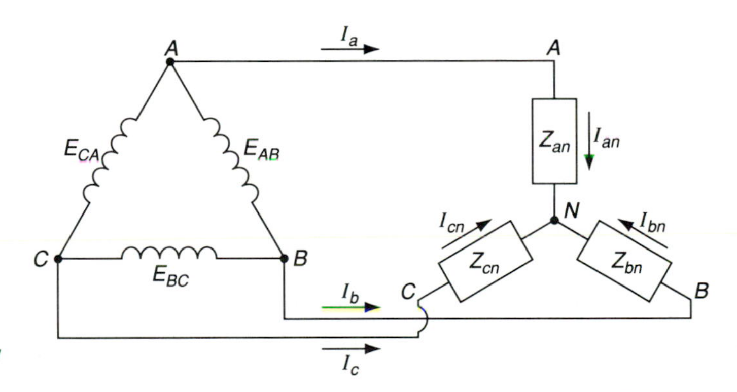

FIGURE 16-47

16-9 Repeat Problem 16-8 for a balanced load of $Z_{an} = Z_{bn} = Z_{cn} = 75\ \underline{/-30^\circ}\ \Omega$.

16-10 A balanced wye-connected load has an impedance per phase of $25\ \underline{/45^\circ}\ \Omega$. The load is connected to a balanced three-phase, wye-connected source with the following phase voltages: $E_{AN} = 120\ \underline{/0^\circ}$ V, $E_{BN} = 120\ \underline{/-120^\circ}$ V, and $E_{CN} = 120\ \underline{/120^\circ}$ V. Determine the total power dissipated by the load.

16-11 If the balanced wye-connected load of Problem 16-10 is connected to a balanced delta supply with a line voltage of 220 V, find the total power dissipated by the load.

16-12 A balanced delta-connected load has an impedance per phase of 65 Ω resistance and 45 Ω inductive reactance. When this load is connected to a 460-V three-phase delta source, determine the total power dissipated by the load.

16-13 The two-wattmeter method of power measurement is used to measure power to a balanced three-phase load. The wattmeter readings are 3500 W and 2500 W. Determine the power factor of the load.

16-14 A three-phase balanced load is metered by using two wattmeters. The wattmeter readings are 6.5 kW and −2.8 kW. Calculate the power factor of the load.

16-15 The balanced three-phase load shown in Figure 16-48 is metered by two wattmeters. The phase voltages are as follows: $E_{AN} = 120\ \underline{/0^\circ}$ V, $E_{BN} = 120\ \underline{/-120^\circ}$ V, and $E_{CN} = 120\underline{/120^\circ}$ V. Determine the readings of the two wattmeters.

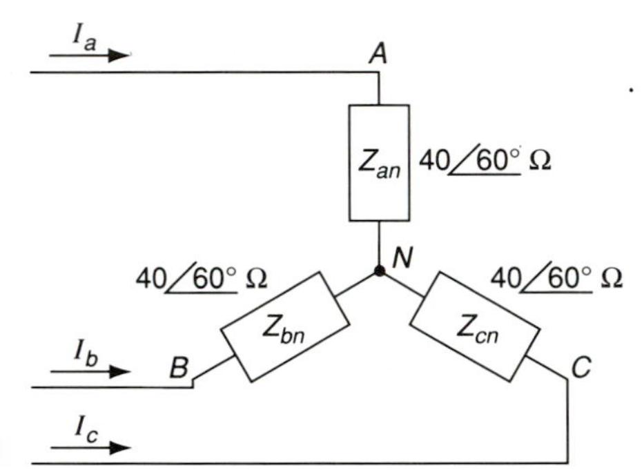

FIGURE 16-48

16-16 If two wattmeters in a balanced three-phase system read 70 kW and 30 kW for wattmeter 1 and wattmeter 2, respectively, determine (a) the system active power; (b) the system power factor; (c) the system apparent power; (d) the system reactive power.

16-17 Three identical impedances of $75\ \underline{/30^\circ}\ \Omega$ are connected in wye to a three-phase three-wire 208-V *ABC* system. Determine the line currents using the one-line equivalent method. E_{AB} is the reference phasor at $208\ \underline{/30^\circ}$ V.

16-18 Determine the line currents of the circuit in Problem 16-17 using the one-line equivalent method if the impedances are delta connected.

16-19 A three-phase three-wire 208-V *ABC* system supplies power to the delta-connected load shown in Figure 16-49. Determine the line currents.

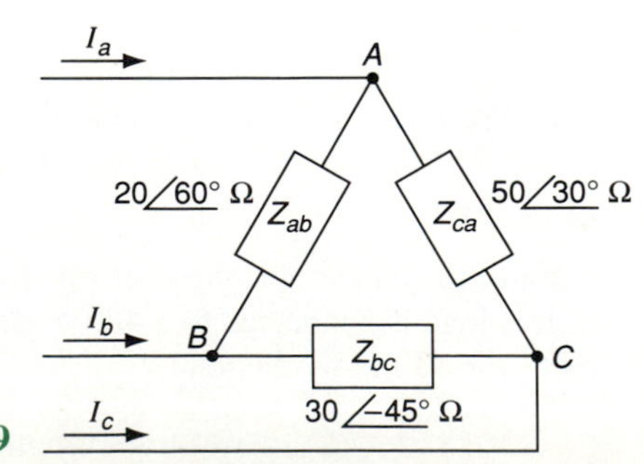

FIGURE 16-49

16-20 Determine the total power consumed by the circuit of Figure 16-49. Assume a three-phase three-wire 208-V *ABC* source.

16-21 A 208-V three-phase four-wire *ABC* system feeds the wye-connected load shown in Figure 16-50. Determine the line currents of the system if the load impedances are as follows: $Z_{an} = 65\,\underline{/20^\circ}\ \Omega$, $Z_{bn} = 40\,\underline{/-50^\circ}\ \Omega$, $Z_{cn} = 25\,\underline{/70^\circ}\ \Omega$.

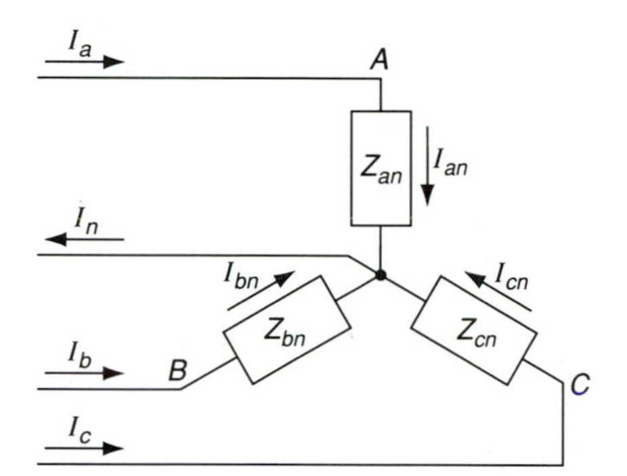

FIGURE 16-50

16-22 Determine the neutral current for the values given in Problem 16-21.

16-23 The wye-connected load shown in Figure 16-50 has the following impedances: $Z_{an} = 15\,\underline{/30^\circ}\ \Omega$, $Z_{bn} = 20\,\underline{/-10^\circ}\ \Omega$, and $Z_{cn} = 35\,\underline{/-40^\circ}\ \Omega$. The system supply is 208-V three-phase four-wire *ABC*. Determine the total power dissipated by the load.

16-24 The three-wire wye-connected load shown in Figure 16-51 consists of the following impedances: $Z_{an'} = 40\,\underline{/15^\circ}\ \Omega$, $Z_{bn'} = 70\,\underline{/30^\circ}\ \Omega$, and $Z_{cn'} = 65\,\underline{/-45^\circ}\ \Omega$. The circuit is supplied by a three-phase 208-V *ABC* system. Determine (a) the line currents; (b) the phase voltages; (c) the total power.

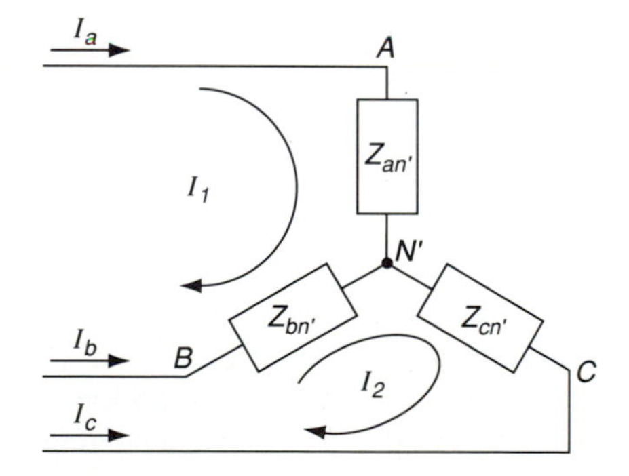

FIGURE 16-51

16-25 Repeat Problem 16-24 using the following load impedances: $Z_{an'} = 35\,\underline{/60^\circ}\ \Omega$, $Z_{bn'} = 20\,\underline{/-30^\circ}\ \Omega$, and $Z_{cn'} = 50\,\underline{/0^\circ}\ \Omega$.

16-26 A three-phase four-wire *ACB* system supplies power to the wye-connected load shown in Figure 16-52. Determine the total power factor of the system. Assume a phase voltage of 120 V.

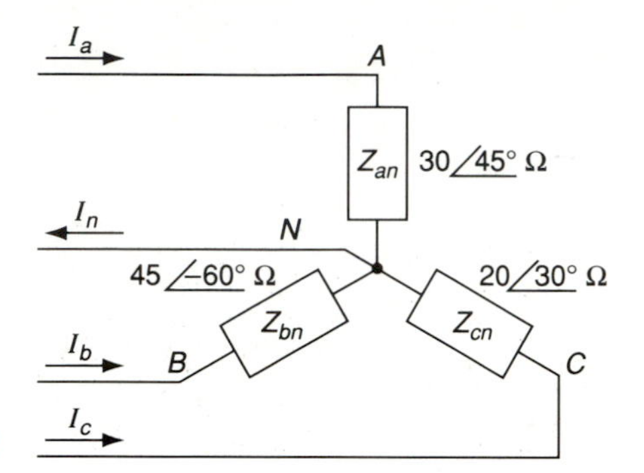

FIGURE 16-52

16-27 Repeat Problem 16-26 using the following load impedance values: $Z_{an} = 8\,\angle 30^\circ\ \Omega$, $Z_{bn} = 12\,\angle -45^\circ\ \Omega$, and $Z_{cn} = 15\,\angle 60^\circ\ \Omega$.

16-28 For the circuit shown in Figure 16-53, calculate the line currents I_a, I_b, and I_c. The system supply is three-phase three-wire *ABC*, with a line voltage of 208 V. The impedance values are as follows: $Z_a = 10\,\angle 30^\circ\ \Omega$, $Z_b = 6\,\angle 45^\circ\ \Omega$, $Z_c = 15\,\angle -20^\circ\ \Omega$, $Z_1 = 20\,\angle 60^\circ\ \Omega$, $Z_2 = 12\,\angle -40^\circ\ \Omega$, and $Z_3 = 16\,\angle 25^\circ\ \Omega$.

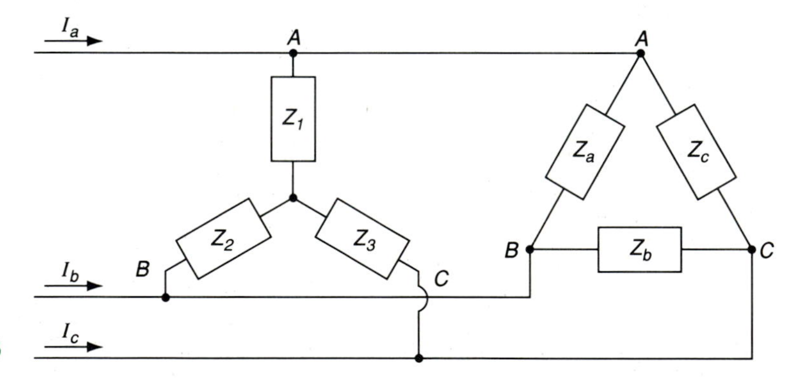

FIGURE 16-53

16-29 The circuit shown in Figure 16-54 consists of two single-phase loads and one three-phase load connected to a three-phase four-wire 120/208-V *ABC* supply. Load 1 draws 4.6 kW at unity power factor. Load 2 draws 6.5 kW at unity power factor. Load 3 is a three-phase 20-hp motor that operates at a lagging power factor of 0.766 and has an efficiency of 88%. Determine the line currents for the system. Assume that E_{AN} is the reference phasor at 0°.

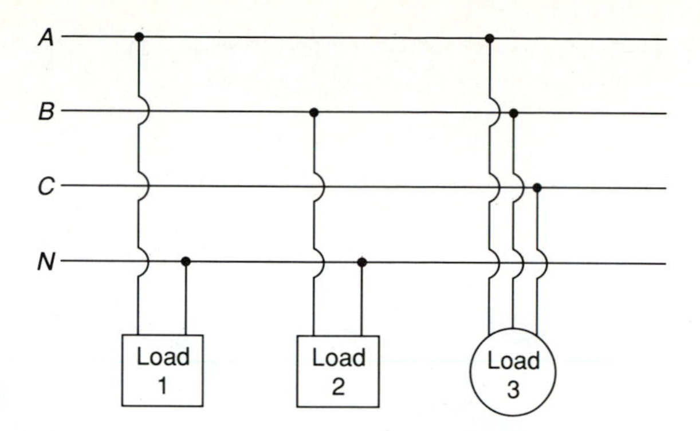

FIGURE 16-54

16-30 Repeat Problem 16-29 using the following load values: Load 1 draws 3.5 kW at 0.866 PF lagging, load 2 draws 5 kW at unity, load 3 is a three-phase 15-hp motor operating at a lagging power factor of 0.92 and has an efficiency of 90%.

CHAPTER

Transformers

LEARNING OBJECTIVES

Upon completion of this chapter you will be able to:

- Explain the basic operating principles of the transformer.
- List the standard types of transformers.
- Draw the schematic symbols for iron- and air-core transformers.
- Explain the standard markings used to identify transformer windings.
- Define leakage reactance and its effect on transformers.
- Explain the purpose of transformer tap changers.
- Understand the principles of reflected impedance.
- List the various losses associated with transformers.
- Perform open-circuit and short-circuit tests on a transformer.
- Determine the equivalent circuit for a transformer.
- Understand transformer phasor relations.
- Determine the voltage regulation and efficiency of a transformer.
- Express the significance of transformer polarity.
- Understand the basic operating characteristics of pulse transformers and instrument transformers.
- List the requirements for parallel operation of transformers.

17-1 INTRODUCTION

The transformer is a simple and reliable device that transfers energy from one circuit to another by electromagnetic induction. This transference is usually, but not always, accompanied by a change of voltage. The transformer is basically two coils wound on an iron core, which form a closed magnetic circuit. The only connection between the two coils is the common magnetic flux within the core of the transformer. When power is supplied to one coil at a specific frequency, power can be taken from the other coil at that same frequency. The two windings are referred to as the **primary** and **secondary**. The primary winding receives energy from the supply circuit, and the secondary receives energy by induction from the primary. In most transformers the two windings are magnetically linked by a closed core of laminated iron or steel.

A transformer that receives energy at one voltage and delivers it at the same voltage is called a **one-to-one transformer**. When the transformer receives energy at one voltage and delivers it at a higher voltage, it is called a **step-up transformer**. If the energy received by the transformer is delivered at a lower voltage, it is referred to as a **step-down transformer**.

17-2 TRANSFORMER CONSTRUCTION

Transformers are divided into two general types: core type and shell type. In the core type, the winding surrounds the iron core. The shell type of transformer has the core built around the coils. Figure 17-1 shows the two types of transformers. The windings may be cylindrical in form and placed one inside the other, with the necessary insulation between them, or they may be built up in thin flat sections called **pancake coils**.

The magnetic circuit, or core, of the transformer is usually made of high-grade silicon steel, containing about 3% silicon. By using silicon, the magnetic characteristics of the core are improved and the iron losses are reduced. The steel core is in the shape of a hollow rectangle and is made up of laminations that are either rectangular or L-shaped. The sheet steel is laminated to reduce eddy-current losses, which are generated by the alternating flux as it cuts through the iron core.

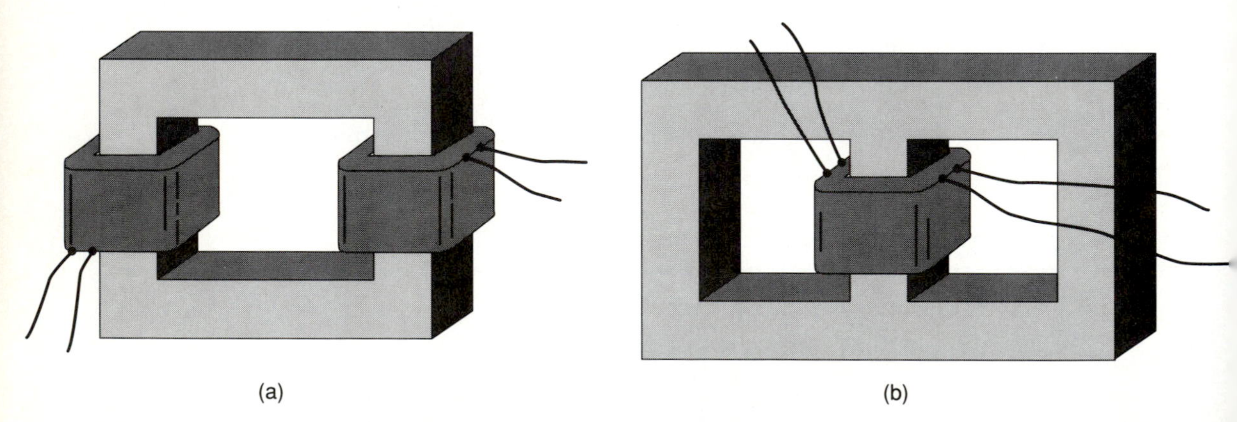

FIGURE 17-1 (a) Core-type and (b) shell-type transformers.

The primary and secondary windings may be placed on the same leg in either the core-type or shell-type transformer. The low-voltage winding is placed next to the core with the high-voltage winding placed around the low-voltage winding. In the shell-type transformer, the core has a middle leg upon which both primary and secondary windings are placed. This type of transformer provides a much shorter magnetic path, but the average length per turn of wire is much greater. The entire flux passes through the central leg of the core. Once the flux passes through the central core, it divides, with half of the flux going in each direction.

The complete core and coil assembly is placed in a steel tank. In some transformers, the complete assembly is immersed in a special mineral oil which provides a means of insulating and cooling the transformer. When a smooth tank surface does not provide a sufficient cooling area, the sides are either corrugated or constructed with cooling fins mounted on the tank. The oil also increases the dielectric strength of the insulation between the coils and therefore acts as an insulator in the tank. The rate at which the heat may be dissipated is dependent on the outside surface area of the tank. The volume, or cubical content, of the transformer and tank is approximately proportional to the kVA capacity of the transformer.

Transformers are sometimes cooled by either natural, or forced, air circulation. Generally, transformers that use natural air circulation are small, such as instrument transformers, and are installed in areas where the use of oil is considered a fire hazard. Transformers that are cooled by the forced circulation of air are usually power transformers which are used in substations located in densely populated areas. With this method of cooling, the air is forced up from the bottom of the transformer against the core and coils, thus carrying away the excess heat. One disadvantage of the forced-air method of cooling is the fact that an arc formed by the failure of insulation is quickly fanned into flame by the strong air current.

The schematic symbols used to represent transformers are shown in Figure 17-2. The symbol for an iron- or steel-core transformer is shown in Figure 17-2(a). A transformer with a nonmagnetic core material is represented by the air-core transformer symbol of Figure 17-2(b).

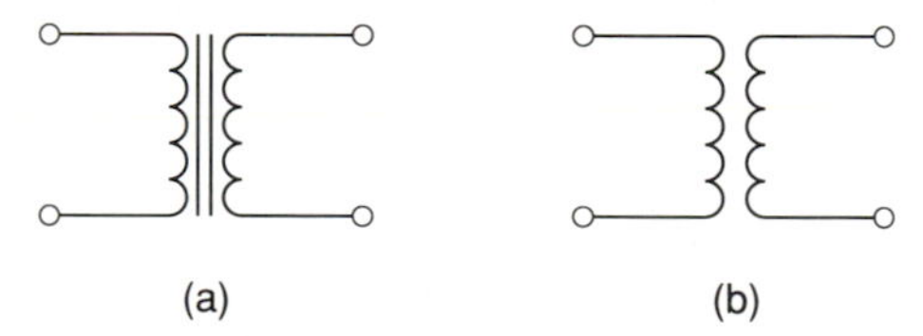

FIGURE 17-2 Schematic symbol for transformers: (a) iron or ferrite core; (b) air core.

17-3 TRANSFORMER PRINCIPLES

Figure 17-3 illustrates a basic transformer, consisting of primary and secondary windings wrapped around a laminated iron or steel core. The primary winding is connected to a source of ac electric power, and the secondary winding is connected to the load. When a transformer is used to step up voltage, the low-voltage winding is the primary. When a transformer is used to step down voltage, the high-voltage winding is the primary. The primary winding is always connected to the source of power, regardless of the application.

Although any transformer may be used as a step-up or step-down transformer, it is very important that the windings be properly connected. For this reason, standard markings have been adopted for transformer terminals. The winding that is to be connected to a high-voltage source or load is designated by H_1 and H_2. This winding is capable of handling high voltages and relatively low amounts of current, so its coil consists of a large number of turns of small-gage wire. The winding that is to be connected to the low-voltage source or load is labeled X_1 and X_2. This winding consists of a small number of turns of large-sized copper to carry the higher value of current. A step-up transformer would have its primary marked X_1 and X_2 and its secondary identified as H_1 and H_2. A step-down transformer would have its primary and secondary marked H_1, H_2 and X_1, X_2, respectively.

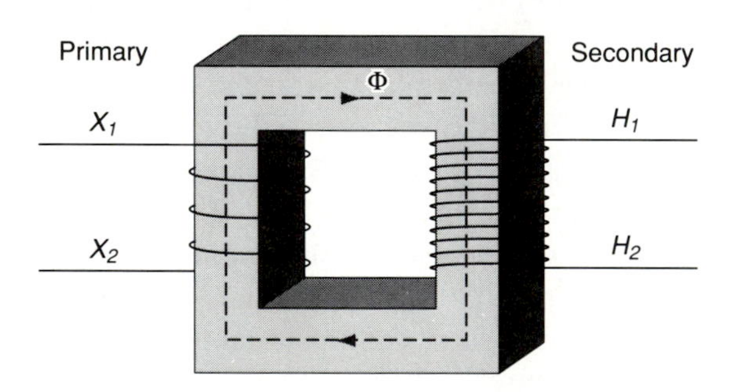

FIGURE 17-3 Transformer designed to step up voltage.

The transformer is based on the principle that energy may be efficiently transferred by magnetic induction from one winding to another by a varying magnetic flux. Therefore, when both windings are on the same magnetic circuit, **mutual inductance** exists between them. Since the primary is connected to an ac supply, the current and flux strength will rise to a maximum value and fall to zero twice in each cycle. During each cycle, the current and flux will flow in one direction for the first half-cycle and in the opposite direction for the second half-cycle. In Chapter 8 it was discussed that whenever changing flux links a coil of wire, an emf is developed in that coil. In a transformer, the magnetic lines of force from the primary coil link with the turns of the secondary coil and induce a voltage in the secondary winding. Voltage is induced in the secondary winding only if the current in the primary winding is alternating. A direct current in the primary winding has no effect on the secondary windings because the magnetic flux is stationary and does not cut, or link, the secondary turns.

The emf developed in the secondary coil is proportional to the rate of change of flux and is defined by the equation

$$E_{ave} = N \times \frac{\Phi_{pm}}{t} \tag{17-1}$$

where E_{ave} = average voltage induced into a coil, in volts

N = number of turns in the coil

Φ_{pm} = peak mutual flux, in webers

t = time required for the flux to rise from zero to its maximum value, in seconds

Equation 17-1 represents a steady value of magnetic flux, since the winding does not move in a transformer, the flux changes sinusoidally and rises from zero to Φ_{pm} in one-quarter of a cycle. Therefore,

$$E_{ave} = 4\Phi_{pm}fN$$

where 4 is the multiplier that one full cycle of flux variation requires and f is the frequency of the applied voltage, in hertz. The effective value of induced voltage becomes

$$\begin{aligned} E &= \frac{0.707}{0.637} 4\Phi_{pm}fN \\ &= 4.44\Phi_{pm}fN \end{aligned} \tag{17-2}$$

Equation 17-2 is called the **general transformer equation**. When the general transformer equation is applied to the primary winding of a transformer, the rms voltage of the primary winding is found by the equation

$$E_p = 4.44\Phi_{pm}fN_p \tag{17-3}$$

where E_p = effective voltage of primary winding

Φ_{pm} = peak mutual flux

f = frequency of applied voltage

N_p = number of turns in primary winding

The flux developed by the primary's magnetomotive force links the secondary winding. In the ideal transformer, where no losses occur, all the flux developed by the primary will link the secondary. In this situation, a voltage will be induced in the secondary winding based on the equation

$$E_s = 4.44\Phi_{pm}fN_s \tag{17-4}$$

EXAMPLE 17-1 The secondary of a 25-kVA 60-Hz transformer has 200 turns, and the flux in the core has a maximum value of 5.7×10^{-3} Wb. Determine the emf induced in the secondary.

Solution

$$E_s = 4.44\Phi_{pm}fN_s = 4.44(5.7 \times 10^{-3})(60)(200) = 303.7 \text{ V}$$

Since the flux shown in Figure 17-3 links each of the two windings, it follows that the same amount of emf per turn is induced in each winding, and the total induced emf in each winding must be proportional to the number of turns in that winding. To further illustrate this point, if we divide equation 17-4 into equation 17-3, the result is

$$\frac{E_p}{E_s} = \frac{N_p}{N_s} = a \tag{17-5}$$

where a is the transformation ratio. From equation 17-5 it can be stated that in an ideal transformer, the primary and secondary voltages are directly proportional to the number of turns in the windings. The ratio of secondary winding turns to primary winding turns is known as the **transformation ratio**, or **turns ratio**, and is given the letter symbol a. If the turns ratio is less than 1, the transformer would be called a step-up transformer, since E_p is less than E_s. Consequently, if the transformation ratio is greater than 1, it would be referred to as a step-down transformer.

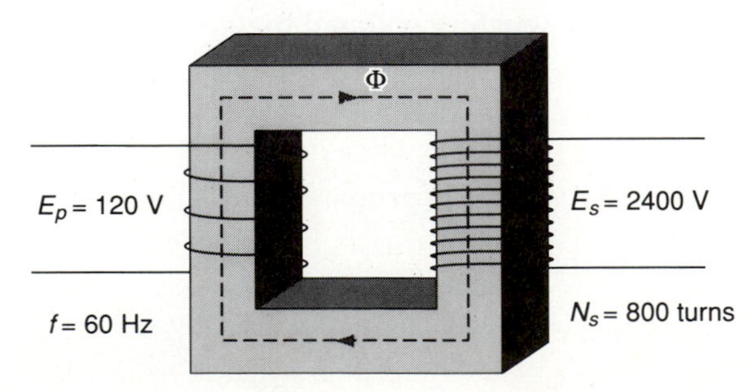

FIGURE 17-4

EXAMPLE 17-2 For the transformer shown in Figure 17-4, determine (a) the peak mutual flux; (b) the number of primary turns.

Solution (a) $E_s = 4.44\Phi_{pm}fN_s$. Therefore,

$$\Phi_{pm} = \frac{E_s}{4.44fN_s} = \frac{2400\ \text{V}}{4.44(60\ \text{Hz})(800)} = 11.26\ \text{mWb}$$

(b) $\frac{E_p}{E_s} = \frac{N_p}{N_s}$. Therefore,

$$N_p = \frac{E_pN_s}{E_s} = \frac{(120\ \text{V})(800)}{2400\ \text{V}} = 40\ \text{turns}$$

In previous discussions dealing with coils of negligible resistance, it was determined that the current lags the applied voltage by 90° and the current leads the induced voltage by 90°. This implies that the applied voltage is therefore 180° out of phase with the induced voltage. When an ac voltage is applied to the primary of a transformer with the secondary open, the primary current is very small, usually 2 to 5% of its rated value. This no-load current is called the **exciting current**, I_0, and it performs two functions. The exciting current supplies the no-load, or iron losses, in the core, and it also sets up the magnetic flux. The effect of the exciting current is illustrated in Figure 17-5.

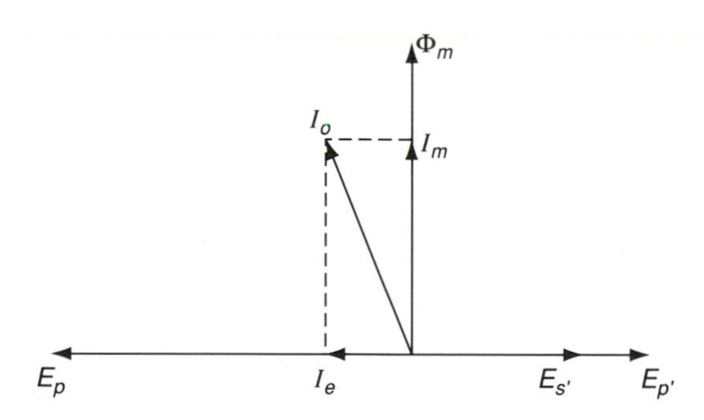

FIGURE 17-5 Effect of magnetizing and exciting current.

The exciting current is easiest understood when resolved into two components. The **magnetizing current**, I_m, lags the primary applied voltage, E_p, and produces the mutual flux, Φ_m, which must be in phase with it. In an actual transformer, because the flux is alternating, there must be hysteresis and eddy-current losses in the core. Even at no-load there is some copper loss in the primary winding. To supply these losses, an **energy current**, I_e, is required. Since such core losses are small, the primary on an open circuit is a very high inductive impedance, and the no-load power factor is 5 to 10%. The exciting current, which is the total no-load primary current, is the phasor sum of the magnetizing current and the energy current.

According to Lenz's law, as the current and flux decrease from a positive value at their maximum rate, which is when they are passing through their zero values in a negative direction, the induced voltage E_p' reaches a maximum positive value, which will oppose this change. Since the flux and the current, Φ_m and I_m, have reached their maximum positive values 90° earlier, the induced voltage E_p' lags 90° behind Φ_m and I_m.

When a load is connected to the secondary of a transformer, a current will flow through the load due to the induced voltage E_s'. The angle at which the secondary current I_s leads or lags E_s' depends on the resistance and reactance of the load. The secondary load current flowing through the secondary turns comprises a load component of magnetomotive force, which according to Lenz's law, is in such a direction as to oppose the flux that is producing it. This opposition tends to reduce the transformer flux by a slight amount. The reduction in flux is accompanied by a reduction in the counter emf induced in the primary winding of the transformer. Since the internal impedance of the primary winding is low and the primary current is mainly limited by the counter emf in the winding, the primary current in the transformer increases when the primary voltage is reduced.

In an ideal transformer, the current through the primary winding will continue to increase until the primary ampere-turns are equal to the secondary ampere-turns. The flux, by its rate of change, must balance the primary applied voltage, and since the primary counter emf varies only slightly from no-load to full-load current, the flux remains practically constant regardless of the load.

The exciting current is generally of a very small magnitude and can be ignored in most transformer calculations. If the no-load current is neglected in comparison with the total primary current, the primary and secondary ampere-turns are equal. Therefore,

$$N_p I_p = N_s I_s \quad \text{or} \quad \frac{N_p}{N_s} = \frac{I_s}{I_p} \tag{17-6}$$

By substituting equation 17-5 into equation 17-6, the following equation is derived:

$$\frac{E_p}{E_s} = \frac{N_p}{N_s} = \frac{I_s}{I_p} = a \tag{17-7}$$

EXAMPLE 17-3 A transformer delivers 3000 A at 240 V. The primary voltage is 2400 V. Determine the primary current.

Solution

$$\frac{E_p}{E_s} = \frac{I_s}{I_p}$$

$$I_p = \frac{E_s I_s}{E_p} = \frac{(240\text{ V})(3000\text{ A})}{2400\text{ V}} = 300\text{ A}$$

17-4 LEAKAGE REACTANCE

In the previous discussion of ideal transformers it was assumed that all the flux which links the primary winding will also link the secondary winding. This would imply that the magnetic coupling, or coefficient of coupling, is 100%. In practice, it is impossible to attain this condition. All the flux produced by the primary ampere-turns does not link the secondary, since a portion completes its magnetic circuit by passing through the air rather than around the core. This is shown in Figure 17-6 as the flux Φ_{LP}. The portion of the flux that goes through one of the transformer windings but not the other is called **leakage flux**. This situation can also occur in the secondary winding when some of the flux set up by the secondary current does not link the primary winding.

The total flux in the primary winding of the transformer can therefore be divided into two components:

1. The mutual flux, which is the flux that passes completely through the core and links both windings.
2. The leakage flux, which leaks out of the core and does not link both windings.

The mutual flux Φ_m is produced by the ampere-turns of both primary and secondary. The primary leakage flux Φ_{LP} is produced by the ampere-turns of the primary, which implies that it is proportional to the primary current. The primary leakage flux induces a voltage in the primary winding in the same manner that the secondary leakage flux induces a voltage in the secondary. The induced voltage in both windings is a counter emf which will oppose the flow of current in the respective winding. The leakage flux varies with frequency and will lag the current by 90°. The induced voltage caused by Φ_{LP} and Φ_{LS} is of a relatively small magnitude compared to the voltage induced by the mutual flux.

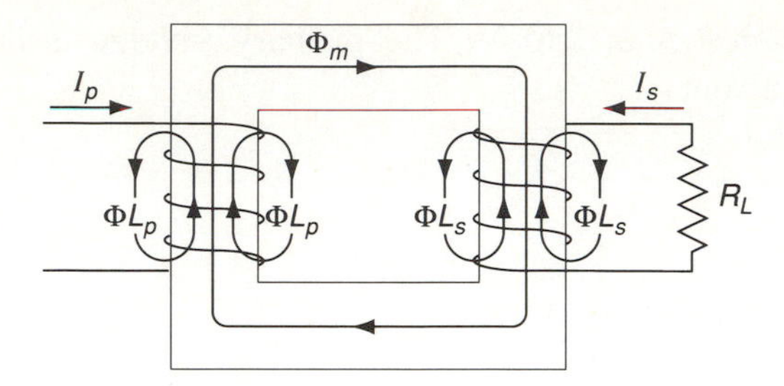

FIGURE 17-6 Leakage reactance of a transformer.

The leakage flux prevents the mutual flux from reaching its maximum value and is therefore a reactive voltage drop (Ix) which is caused by a **leakage reactance** (x) in series with the windings. The leakage reactance occurs in both the primary and secondary windings, and its overall effect is to reduce the output voltage at the secondary terminals of the transformer. Leakage reactance is, generally, an undesirable characteristic and manufacturers of small transformers attempt to reduce leakage flux by placing the primary and secondary windings as close together as possible. Transformers of very high kVA ratings usually require large values of leakage reactance to limit the short-circuit current.

17-5 TAP CHANGERS

If a transformer is supplied with power through a transmission circuit whose impedance is relatively high, the primary terminal voltage may vary drastically with changes in load, due to changes in the impedance drop in the transmission circuit. In such instances, it is highly desirable to adjust the turns ratio slightly in a transformer. This is most often accomplished by means of a **tap changer**, which selects different taps in the primary winding. The number of turns in use in the primary winding is determined by an arrangement of contactors so that the primary induced voltage per primary turn is maintained at a nearly constant value, which means that the flux is nearly constant. Large lap changing power transformers are also used frequently to control the flow of reactive power between two interconnected power systems, while at the same time allowing the voltages at specified points to be maintained at desired values.

The changes in transformer turns ratio are made on a deenergized transformer by means of a tap-changing switch. Figure 17-7 shows how a tap-changing switch is connected to a transformer winding. When the connector is in the position shown, the entire primary winding is connected in the circuit. When the connector is shifted to position 2, section A is removed from the circuit, which causes the secondary voltage to be raised. When the connector is consecutively shifted to positions 3, 4, and 5, sections *B*, *C*, and *D* are removed from the primary, and the secondary voltage increases in increments.

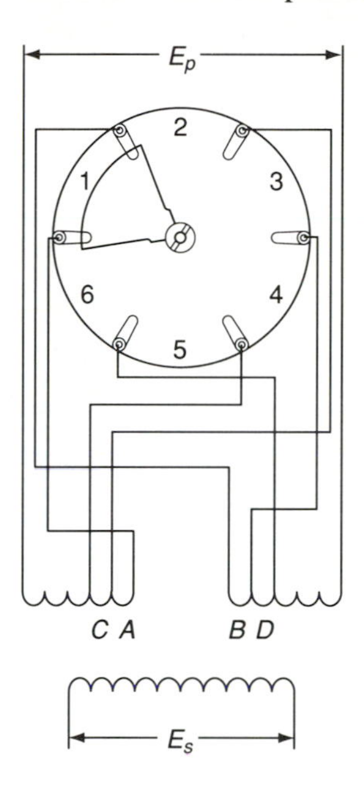

FIGURE 17-7 Transformer tap changer.

Quite often a tap-changing apparatus is required that is capable of adjusting the transformation ratio while the transformer is energized and under load. Figure 17-8 illustrates one method of accomplishing this by means of electric contactors. The coil *X* is a center-tapped iron-core reactor; contacts 1 and 2 are normally open, and contact 3 is normally closed. When the entire primary winding is in use, contactor 1 would be closed and 2 and 3 open. If an increase in load occurs and it is desired to reduce the turns on the primary winding, contactor 1 would open and 2 would close. However, if the primary circuit is to be uninterrupted, contactor 2 must close before contactor 1 is opened. The purpose of the reactor is to prevent the coils between taps 1 and 2 from being short-circuited while contactors 1 and 2 are both closed. In the normal operating position, contactor 3 is closed, short-circuiting the reactor. In this position, the primary current divides equally between both halves of the reactor. Since these currents are in opposite directions, their resultant flux is zero. When contactors 1 and 2 are closed during transition, 3 is open, and the current in the transformer coils between the taps is limited by the magnetizing reactance of the reactor. If the reactance is very high, the current in the transformer winding is not excessive.

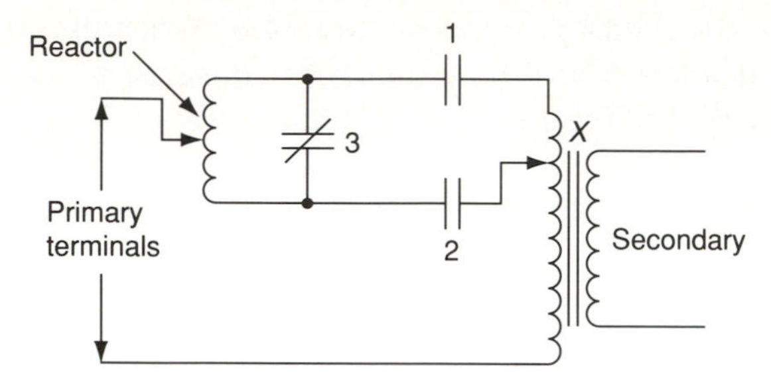

FIGURE 17-8 Tap-changing transformer.

17-6 REFLECTED IMPEDANCE

Any change in the load impedance connected to the transformer of Figure 17-4 will affect the current flowing through the secondary winding. Consequently, any variation in secondary current will cause the primary current to vary proportionately. This action is a result of Lenz's law. When a load is connected to the secondary, the current through the secondary develops a flux which is in opposition to the primary flux, and cancels, or nullifies, part of the primary flux. The primary draws more current from the source to compensate for the flux canceled by the secondary. If we take the transformation ratio into account, the amount of current flow caused by an impedance load would be the same, regardless of whether it was connected to the primary or secondary of the transformer. This modified load impedance which appears as an input impedance at the primary is called the **reflected impedance**, and is defined as:

> The value of a load impedance reflected from the secondary into the primary of a transformer is equal to the load impedance multiplied by the square of the transformation ratio.

In equation form,

$$Z_R = a^2 Z_L \tag{17-8}$$

where Z_R = reflected impedance of primary, in ohms

a = transformation ratio

Z_L = load impedance of secondary, in ohms

Reflected impedance is generally used for simplifying transformer circuits when performing impedance-matching calculations to ensure maximum power transfer. When used as a step-down transformer, the primary impedance must be a^2 times greater than the secondary impedance. Therefore, when used as a step-up transformer, the secondary impedance must be a^2 times greater than the primary, or $Z_L = a^2 Z_R$.

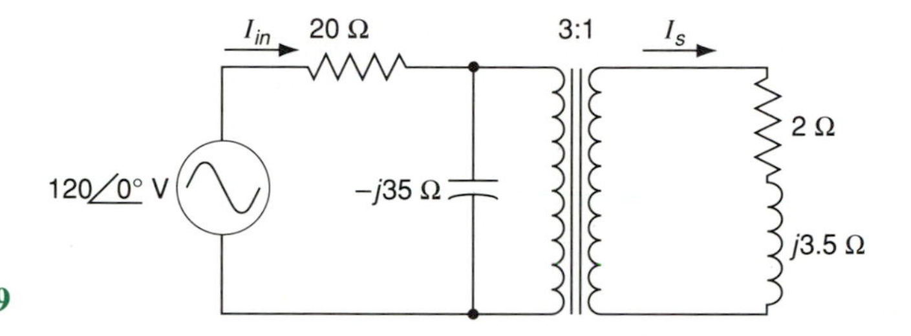

FIGURE 17-9

EXAMPLE 17-4 For the circuit shown in Figure 17-9, determine the input current I_{in} and the current through the load connected to the secondary I_s by the reflected impedance method.

Solution

$$a = \frac{N_p}{N_s} = 3$$

$$a^2 = 9$$

$$Z_L = (2 + j3.5)\ \Omega$$

$$Z_R = 9(2 + j3.5) = 36.28\ \underline{/60.3^\circ}\ \Omega$$

The circuit can now be redrawn with the transformer removed, as shown in Figure 17-10. The total impedance of the circuit is calculated as a series–parallel combination.

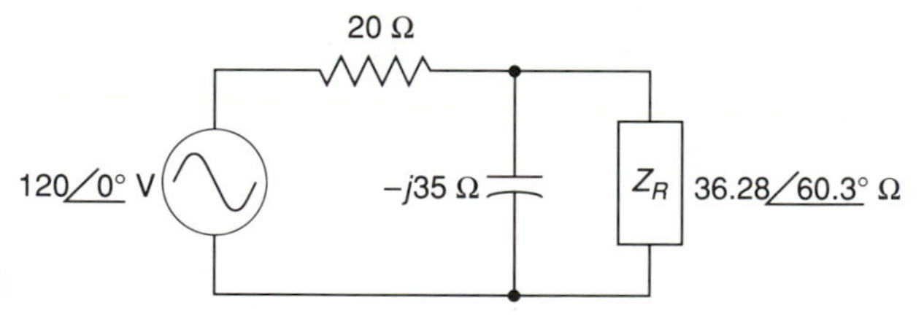

FIGURE 17-10

$$\begin{aligned} Z_T &= 20\ \Omega + \frac{(35\ \underline{/-90^\circ}\ \Omega)(36.28\ \underline{/60.3^\circ}\ \Omega)}{35\ \underline{/-90^\circ}\ \Omega + 36.28\ \underline{/60.3^\circ}\ \Omega} \\ &= 88.525\ \underline{/-14.6^\circ}\ \Omega \end{aligned}$$

The input current is determined by Ohm's law.

$$\begin{aligned} I_{in} = \frac{E_{in}}{Z_T} &= \frac{120\ \underline{/0^\circ}\ \text{V}}{88.525\ \underline{/-14.6^\circ}\ \Omega} \\ &= 1.36\ \underline{/14.6^\circ}\ \text{A} \end{aligned}$$

Since the transformation ratio is given, the secondary voltage is

$$E_s = \frac{120\angle 0^\circ \text{ V}}{3} = 40\angle 0^\circ \text{ V}$$

and the current through the load is

$$I_s = \frac{E_s}{Z_L} = \frac{40\angle 0^\circ \text{ V}}{(2 + j3.5)\,\Omega}$$

$$= 9.92\angle -60.3^\circ \text{ A}$$

17-7 TRANSFORMER LOSSES

Losses in transformers are due to eddy currents and hysteresis in the core and are known as iron, or core, losses. Losses that occur due to the resistance of the windings are referred to as copper losses. Since the magnetic flux in a transformer is changing direction many times a second, heat is developed because of the **hysteresis** of the magnetic material. Low hysteresis losses are obtained by using **grain-oriented steel**. The sheets of steel are cut so that the magnetic flux flows in the direction of the structural grain of the material. By using grain-oriented steel, the molecular friction is reduced and the hysteresis of the magnetic material is lowered.

The main purpose of laminating the core is to reduce eddy currents. Eddy-current loss is an I^2R loss that is set up by emfs induced in the core by the changing magnetic flux. These emfs are in a direction perpendicular to the flux path, so the layers of the laminations are laid parallel to the direction of the flux. After the sheets of steel are annealed, a coat of oxide is formed on the surface of the sheets, which increases the resistance in the path of the eddy currents. To insulate one sheet from another more effectively, a coat of insulating varnish is put on the steel sheets.

Even when no load is connected to the transformer, there is some loss of energy due to the magnetizing current flowing in the primary winding. This I^2R loss also appears as heat generated in the winding and must be dissipated to prevent too great a temperature rise. When a load is connected to a transformer and current is flowing in both the secondary and primary windings, further losses of electrical energy occur. These losses, called **copper losses**, are also considered to be I^2R losses, since the loss in each winding is proportional to the resistance of the winding and the square of the current flowing in it. The copper losses may be calculated if the primary and secondary resistances are known. If R_H and R_X are the resistances of the high- and low-voltage windings, respectively, the copper loss is calculated using the equation

$$\text{copper loss} = I_H^2 R_H + I_X^2 R_X \tag{17-9}$$

From equation 17-9 it is obvious that copper losses vary with the load, being practically negligible at no load and at a maximum at full load. Transformers with a high kVA rating require conductors of large cross section to minimize copper loss. Unfortunately, a transformer requires a large number of turns to produce a high percentage of flux linkage, so a compromise between the size of conductors and the number of conductors must be made. The amount of core and copper losses in a transformer may be determined by the open-circuit and short-circuit tests.

17-8 OPEN-CIRCUIT AND SHORT-CIRCUIT TESTS

The **open-circuit test** of a transformer is, as the name implies, a test in which one of the windings is open-circuited. When one side of a transformer is connected to a source of alternating current at rated frequency and voltage, and the other side is left open-circuited, the core loss in a transformer can be determined. When rated voltage is applied across the winding, rated flux exists in the core, and therefore the core loss is at its normal value. Since the flux remains virtually constant at all loads, the core loss also remains virtually constant at all loads. Figure 17-11 shows the proper connections for making an open-circuit test.

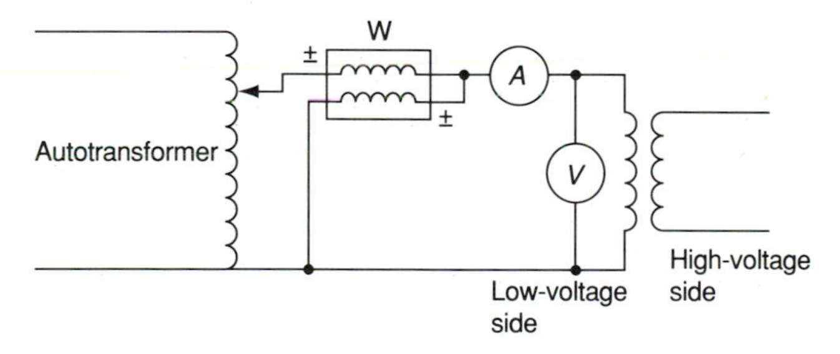

FIGURE 17-11 Open-circuit transformer test.

During the open-circuit test, there are, essentially, no copper losses in the primary winding and none whatsoever in the secondary. As a result, a wattmeter connected as shown in Figure 17-11 will only measure the **core losses**. To adjust the voltage to its rated value and measure the I^2R loss, an autotransformer, or variac, is used. When the circuit is energized, a great deal of precaution should be taken to prevent anyone coming in contact with the high-voltage side. Although the open-circuit test can be performed on either winding, it is much safer to work with as low a value of voltage as possible.

When the adjustable voltage source is raised up to the rated value of the transformer winding, the open-circuit power is recorded from the wattmeter. The magnetic core loss of the transformer is the wattmeter reading. From the voltmeter and ammeter readings, the total impedance of the magnetizing circuit can be determined. This information may be used for both the efficiency calculation of the transformer as well as determining the equivalent circuit.

The **short-circuit test** is used to determine the copper losses in a transformer. Since the copper loss increases with the square of the current, the primary current increases in the same ratio as the secondary current. As mentioned previously, the copper loss can be determined by measuring the ohmic resistance of each winding and calculating the loss in each winding for a given value of current. However, the resistance of the windings will vary with temperature, which makes the short-circuit method of testing a more accurate means.

In Figure 17-12, the transformer of Figure 17-11 is reversed and the low-voltage winding is short-circuited. The windings are reversed for this test since the applied voltage is less than 5% of the rated voltage of the winding. This makes it quite difficult to obtain an appreciable voltmeter deflection when the low-voltage winding is used.

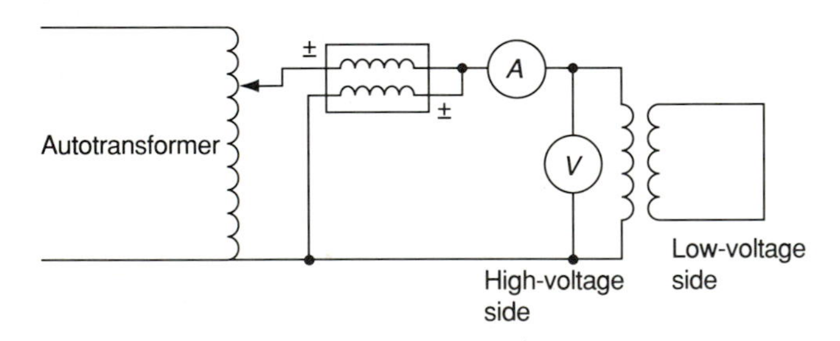

FIGURE 17-12 Short-circuit transformer test.

An autotransformer, or rheostat, is once again used on the input side of the transformer, except that this time it is adjusted until rated current flows in the primary winding. When rated current flows in the primary, rated current will also flow in the secondary by means of transformer action. This current can be measured by an ammeter of proper range placed either in the primary circuit as shown, or in the secondary. The wattmeter will indicate the total copper loss in both windings, including a negligible core loss. From this wattmeter reading, the equivalent resistance of the transformer can be determined. The following equation is used to find the equivalent transformer resistance:

$$R_e = \frac{W}{(I_p)^2} \tag{17-10}$$

where R_e = equivalent transformer resistance, in ohms

W = wattmeter reading, in watts

I_p = current in primary winding, in amperes

R_e is referred to as the equivalent transformer resistance since it is not the arithmetic sum of the resistance of the two windings. The total internal impedance of the transformer can be found by applying Ohm's law to the voltmeter and ammeter readings of the short-circuit test.

$$Z_e = \frac{V}{I} \tag{17-11}$$

where Z_e = internal impedance of transformer, in ohms

V = actual applied voltage obtained in test (approximately 50% of rated value)

I = rated current, in amperes

The equivalent reactance is now found by applying the Pythagorean theorem.

$$Z_e = \sqrt{R_e^2 + X_e^2}$$
$$\therefore X^2 = \sqrt{Z_e^2 - R_e^2} \tag{17-12}$$

17-9 TRANSFORMER EQUIVALENT CIRCUIT

To determine the performance characteristics of a practical transformer properly, equivalent circuits containing the losses that occur in a transformer are used. These losses include leakage flux, copper (I^2R) losses, eddy current, and hysteresis losses. An equivalent circuit of a practical iron-core transformer is shown in Figure 17-13(a). In this circuit the secondary winding resistance, a^2R_s, and the inductive reactance, a^2X_{LS}, have been reflected back, along with the reflected impedance of the load. The primary, secondary, and magnetization circuits of the resulting network are shown in a series–parallel combination. The primary current, I_p, is the sum of a magnetizing current component, I_m, and the load current component, I_p'. The magnetizing reactance of the core and the core losses due to hysteresis and eddy currents are represented by the inductive reactance, X_{Lm}, and the resistance, R_m, respectively. The values of X_{Lm} and R_m could be found by the open-circuit test.

In Figure 17-13(a), the current I'_p flows through the inductive reactance and resistance of the secondary winding as well as through the reflected impedance of the load. If the load were disconnected, current I'_p would be zero and negligible losses would occur in that portion of the circuit.

When the transformer is under load conditions, the magnetizing current, I_m, is considered negligible. The circuit of Figure 17-13(a) can be simplified by discarding the I_m branch and grouping the primary and secondary resistances and reactances together. The simplified equivalent circuit of Figure 17-13(b) illustrates how the primary and reflected values can be grouped together in the following combinations:

$$R_{ep} = R_p + a^2R_s \tag{17-13}$$

$$X_{ep} = X_{Lp} + a^2X_{Ls} \tag{17-14}$$

The equivalent impedance is found by the Pythagorean theorem.

$$Z_{ep} = \sqrt{R_{ep}^2 + X_{ep}^2} \tag{17-15}$$

Assuming that value of I_m is minimal, the primary current and the load current component are approximately equal.

$$I_p \approx I'_p$$

The resistive and reactive components of the load can be combined and the value of input current determined.

$$I_p = \frac{V_p}{\sqrt{(R_{ep} + a^2R_L)^2 + (X_{ep} \pm a^2X_L)^2}} \tag{17-16}$$

where $+X_L$ is the reactance of an inductive load and $-X_L$ is the reactance of a capacitive load.

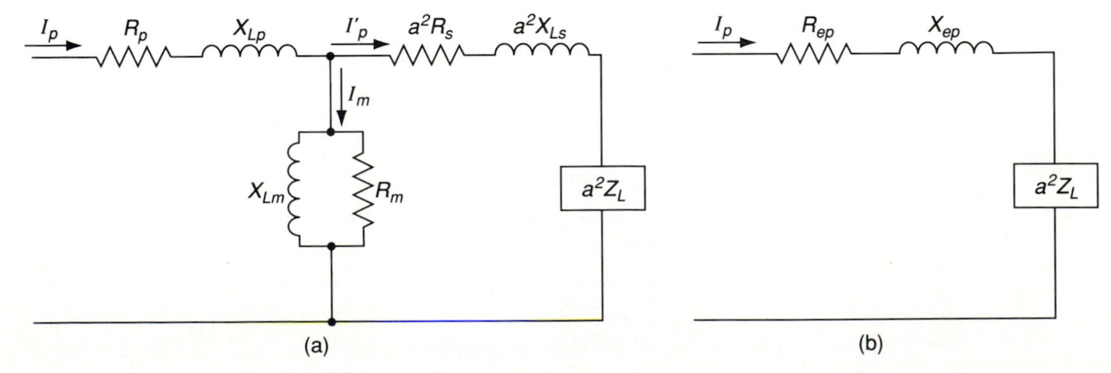

FIGURE 17-13 (a) Equivalent circuit for power transformer; (b) simplified equivalent circuit assuming that $I_m = 0$.

EXAMPLE 17-5 For the circuit shown in Figure 17-14, determine (a) the internal resistance R_e, referred to the primary; (b) the internal reactance X_e, referred to the primary; (c) the internal impedance Z_e, referred to the primary; (d) the secondary load impedance Z_L, referred to the primary; (e) the primary load current at rated primary voltage.

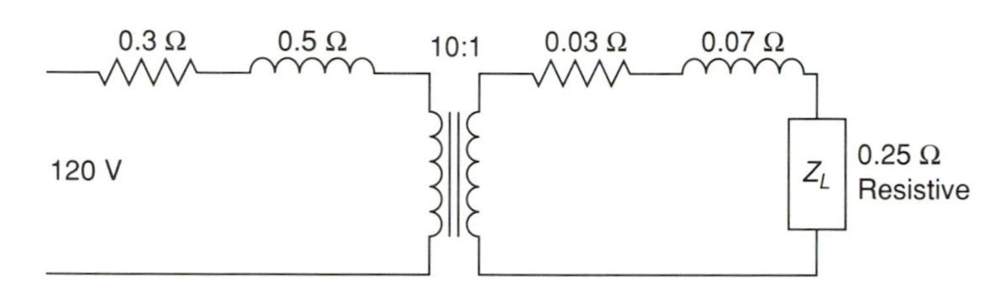

FIGURE 17-14

Solution

(a) $R_{ep} = R_p + a^2R_s = 0.3\ \Omega + (10)^2(0.03\ \Omega) = 3.3\ \Omega$

(b) $X_{ep} = X_{Lp} + a^2X_{Ls} = 0.5\ \Omega + (10)^2(0.07\ \Omega) = 7.5\ \Omega$

(c) $Z_{ep} = \sqrt{R_{ep}^2 + X_{ep}^2} = \sqrt{3.3^2\ \Omega + 7.5^2\ \Omega} = 8.2\ \Omega$

(d) $a^2Z_L = (10)^2(0.25\ \Omega) = 25\ \Omega$

(e)
$$I_p = \frac{E_p}{(R_{ep} + a^2R_L) + (X_{ep} \pm a^2X_L)}$$
$$= \frac{120\ \text{V}}{\sqrt{(3.3\ \Omega + 25\ \Omega)^2 + 7.5^2\ \Omega}}$$
$$= 4.1\ \text{A}$$

17-10 TRANSFORMER PHASOR RELATIONS

In Section 17-9 the magnetizing current, I_m, of the primary current, I_p, was neglected. By doing this it is possible to reflect impedances to the secondary and determine the secondary power factor. Impedances are reflected to one side of a transformer for simplification purposes. Just as it is possible to simplify the actual transistor circuit by means of an equivalent network, it is also possible to simplify the phasor diagram of the transformer.

The secondary current of the transformer creates a leakage flux that is varying sinusoidally. This leakage flux will induce a voltage that is lagging by 90°. The voltage drop caused by leakage reactance, I_sX_s, will be out of phase with the voltage induced in the secondary winding, E_s. The other voltage drop that occurs in the secondary is the result of the winding resistance, I_sR_s. For this reason, the voltage that is induced into the secondary winding must not only supply the rated output voltage of the secondary, V_s, it must also supply the two voltage drops I_sR_s and I_sX_s. If the power factor of the load is lagging, the phasor diagram for the secondary would be similar to that shown in Figure 17-13(a). The same voltage drops I_pR_p and I_pX_p exist in the primary winding of the transformer, except they are subtracted from the terminal voltage, V_p, which results in the induced voltage E_p. The phasor diagram for the primary of a transformer with a lagging power factor is shown in Figure 17-15(b).

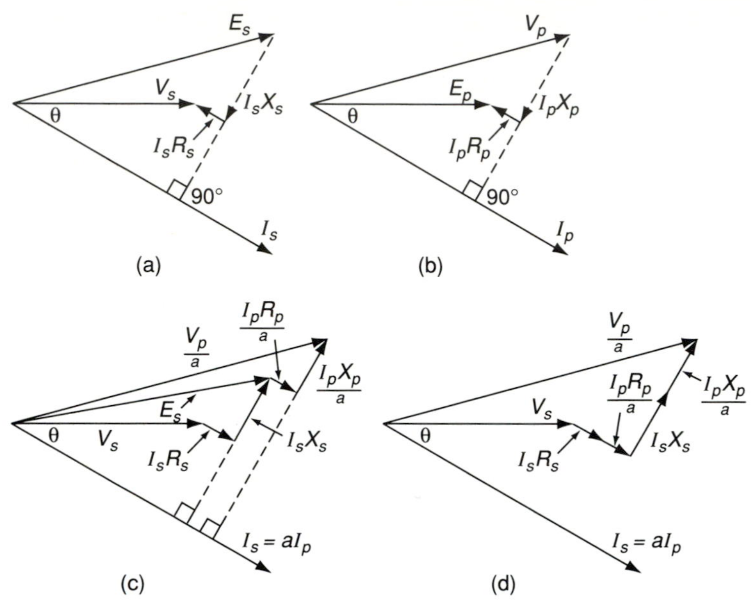

FIGURE 17-15 (a) Phasor diagram of secondary winding with lagging PF load; (b) phasor diagram of primary winding with lagging PF load; (c) resulting phasor diagram of combined primary and secondary windings; (d) phasor diagram with a transformation ratio of 1:1.

If the phasor diagrams of Figure 17-15(a) and b) are combined, the resulting phasor diagram is as shown in Figure 17-15(c). Since the transformer is assumed to be a step-down type, it is necessary to divide the primary voltages by the transformation ratio. This simplifies the phasor diagram, as the transformer now has an equivalent transformation ratio of 1:1. The transformer phasor diagram now appears as shown in Figure 17-15(d).

The primary and secondary values of resistance can now be reflected to the secondary, and the equivalent resistance and reactance can be found.

$$R_{es} = R_s + \frac{R_p}{a^2} \tag{17-17}$$

$$X_{es} = X_s + \frac{X_p}{a^2} \tag{17-18}$$

Once again, the equivalent impedance is found by the Pythagorean theorem.

$$Z_{es} = \sqrt{R_e^2 + X_e^2}$$

The phasor diagram shown in Figure 17-16 is the result of reducing the transformation ratio to 1:1, and reflecting the primary impedance to the secondary.

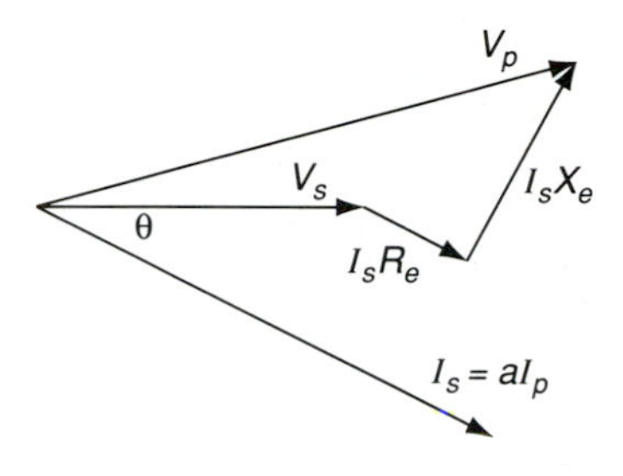

FIGURE 17-16 Phasor diagram with primary impedance reflected to the secondary.

Since impedance may be reflected to either the primary or the secondary, a proportional relationship exists between the equivalent ohmic values on the primary and secondary sides. By using the transformation ratio, the equivalent values are easily converted from one side of the transformer to the other.

$$Z_{es} = \frac{Z_{ep}}{a^2} \tag{17-19}$$

$$R_{es} = \frac{R_{ep}}{a^2} \tag{17-20}$$

$$X_{es} = \frac{X_{ep}}{a^2} \tag{17-21}$$

The power factor of the secondary is easily determined from a phasor diagram of a transformer. Since the phase angle between V_s and I_s is considered to be the power-factor angle, the equation for finding the power factor of the secondary is

$$\theta_s = \tan^{-1} \frac{V_s \sin \theta_L + I_s X_s}{V_s \cos \theta_L + I_s X_s} \tag{17-22}$$

where θ_s = power factor of the secondary of the transformer

θ_L = power factor angle of the load connected to the secondary

V_s = output voltage at the terminals of the secondary

17-11 VOLTAGE REGULATION

The **voltage regulation** of a transformer may be determined either by phasor relations or by data obtained in a short-circuit test. From the preceding two sections it should be apparent that the secondary current, I_s, causes voltage drops I_sR_s and I_sX_s, and the primary current produces the voltage drops I_pR_p and I_pX_p in the primary circuit. For this reason, the induced voltage is less than the terminal voltage on the primary, and the output voltage is less than the induced voltage on the secondary.

When there is no load connected to the secondary, the output voltage is the same as the induced voltage in the winding. As load is applied, the resistive and reactive voltage drops occur. Generally, the output voltage at no load will be higher than at full load. However, if the power factor is relatively low and the load is capacitive, it is possible to achieve a higher output voltage at full-load than at no-load. The voltage regulation of a transformer is defined in the following manner: Voltage regulation is the difference between the no-load and rated full-load values of the secondary terminal voltage, expressed in percent of the rated full-load secondary voltage. This is expressed in equation form as

$$\text{voltage regulation} = \frac{E_{\text{nl}} - E_{\text{fl}}}{E_{\text{fl}}} \times 100 \qquad (17\text{-}23)$$

where E_{nl} is the transformer no-load output voltage and E_{fl} is the transformer full-load output voltage.

Provided that a short-circuit test has been performed on a transformer, if the load current is known, the induced voltage, E_{nl}, can be determined. If the phasor diagram of Figure 17-16 is redrawn in the shape of a right-angle triangle with the secondary current as a reference phasor, the induced voltage, E_{nl}, can be solved for as the hypotenuse of the larger of the two triangles shown in Figure 17-17.

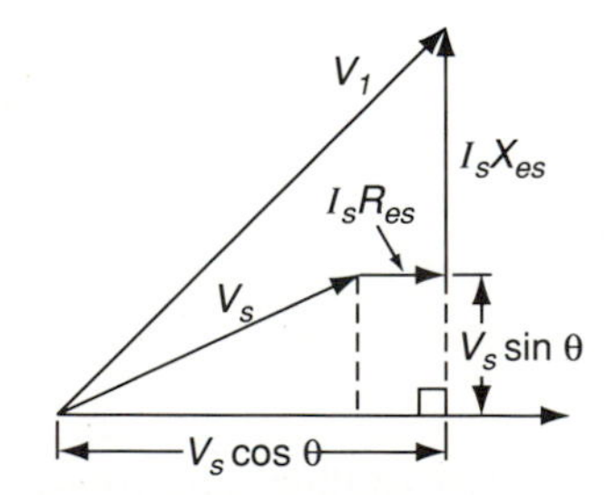

FIGURE 17-17

For the lagging power factor circuit shown in Figure 17-17, the no-load voltage is calculated by the following equation:

$$E_{nl} = V_1 = \sqrt{(V_s \cos\theta + I_s R_{es})^2 + (V_s \sin\theta + I_s X_{es})^2} \qquad (17\text{-}24)$$

If the PF of the load is leading, the equation is expressed as follows:

$$E_{nl} = \sqrt{(V_s \cos\theta + I_s R_{es})^2 + (V_s \sin\theta - I_s X_{es})^2} \qquad (17\text{-}25)$$

Voltage regulation can also be expressed in terms of the primary voltage by the following equations:

Lagging PF:

$$E_{nl} = V_2 = \sqrt{(V_p \cos\theta + I_p R_{ep})^2 + (V_p \sin\theta + I_p X_{ep})^2} \qquad (17\text{-}26)$$

Leading PF:

$$E_{nl} = V_2 = \sqrt{(V_p \cos\theta + I_p R_{ep})^2 + (V_p \sin\theta - I_p X_{ep})^2} \qquad (17\text{-}27)$$

EXAMPLE 17-6 A 30-kVA 2400/240-V 60-Hz transformer is short- circuited on the low-voltage side, and the following data are obtained from a short-circuit test: $V = 81$ V, $W = 340$ W, and $I =$ rated current. Determine the voltage regulation with a 0.92 lagging PF load connected (a) in terms of the primary winding; (b) in terms of the secondary winding.

Solution (a) $I_p = \dfrac{\text{rated volt-amperes}}{\text{rated primary volts}} = \dfrac{30{,}000\text{ VA}}{2400\text{ V}} = 12.5\text{ A}$

The equivalent ohmic values can now be obtained using equations 17-10, 17-11, and 17-12.

$$Z_{ep} = \frac{V}{I_p} = \frac{81\text{ V}}{12.5\text{ A}} = 6.48\,\Omega$$

$$R_{ep} = \frac{W}{I_p^2} = \frac{340\text{ W}}{(12.5\text{ A})^2} = 2.176\,\Omega$$

$$X_{ep} = \sqrt{Z_{ep}^2 - R_{ep}^2} = \sqrt{6.48^2\,\Omega - 2.176^2\,\Omega} = 6.104\,\Omega$$

$$\theta = \cos^{-1} 0.92 = 23.1°$$

transformation ratio, $a = 10$

$$E_{nl} = V_2 = \sqrt{(V_p \cos\theta + I_p R_{ep})^2 + (V_p \sin\theta + I_p X_{ep})^2}$$

$$= \sqrt{[2400 \cos 23.1° + (12.5)(2.176)]^2 + [2400 \sin 23.1° + (12.5)(6.104)]^2}$$

$$= 2455.68 \text{ V}$$

$$\% \text{ regulation} = \frac{E_{nl} - E_{fl}}{E_{fl}} \times 100 = \frac{2455.68 \text{ V} - 2400 \text{ V}}{2400 \text{ V}}$$
$$= 2.32\%$$

(b) $$I_s = \frac{\text{rated volt-amperes}}{\text{rated secondary volts}} = \frac{30{,}000 \text{ VA}}{240 \text{ V}} = 125 \text{ A}$$

Equations 17-19, 17-20, and 17-21 can now be used to find the equivalent ohmic values of the secondary.

$$Z_{es} = \frac{Z_{ep}}{a^2} = \frac{6.48\,\Omega}{100} = 0.065\,\Omega$$

$$R_{es} = \frac{R_{ep}}{a^2} = \frac{2.176\,\Omega}{100} = 0.022\,\Omega$$

$$X_{es} = \frac{X_{ep}}{a} = \frac{6.104\,\Omega}{100} = 0.061\,\Omega$$

$$E_{nl} = V_1 = \sqrt{(V_s \cos\theta + I_s R_{es})^2 + (V_s \sin\theta + I_s X_{es})^2}$$

$$= \sqrt{[240 \cos 23.1° + (125)(0.022)]^2 + [240 \sin 23.1° + (125)(0.061)]^2}$$

$$= 245.59 \text{ V}$$

$$\% \text{ regulation} = \frac{E_{nl} - E_{fl}}{E_{fl}} \times 100$$
$$= \frac{245.59 \text{ V} - 240 \text{ V}}{240} \times 100$$
$$= 2.32\%$$

17-12 EFFICIENCY

The efficiency of a transformer is a ratio of the output power to the input power. In equation form,

$$\text{efficiency} = \frac{\text{output}}{\text{input}} \times 100 = \eta$$

Since the output power is less than the input power due to the losses that occur in the transformer, it is necessary to use these losses in the efficiency calculation. The copper and core losses may be determined experimentally by means of the open-circuit and short-circuit tests. The sum of these two losses is always very small, so the efficiency of a transformer is usually above 95%. As long as the supply frequency remains constant, the core losses are virtually constant for all conditions of loading as well as at no-load. The copper losses vary as the square of the kVA output of the transformer, and are considered variable for that reason. When expressed in terms of losses, the equation for the efficiency of a transformer becomes

$$\eta = \frac{P_{\text{out}}}{P_{\text{out}} + P_{\text{losses}}}$$
$$= \frac{V_s I_s \cos\theta}{V_s I_s \cos\theta + (\text{core loss} + I_s R_{es})} \times 100 \qquad (17\text{-}28)$$

where

V_s = rated output voltage

I_s = rated output current

$\cos\theta$ = decimal equivalent of the power factor

core loss = core loss power in watts from the open-circuit test

R_{es} = equivalent total winding resistance reflected to the secondary, as determined by the short-circuit test

EXAMPLE 17-7 The core loss in a 20-kVA 2200/220-V 60-Hz transformer is 180 W, as indicated by the wattmeter when an open-circuit test is performed with the high-side winding open. On short-circuit test with full-load current and the low side shorted, the wattmeter reads 540 W. The power factor is 0.8 lagging. Calculate (a) the efficiency at rated kVA output; (b) the efficiency at half-load output; (c) the efficiency at one-fourth load output.

Solution (a)

$$\begin{aligned}\text{Total losses} &= \text{core loss at all loads} + \text{full-load copper loss}\\ &= 180\text{ W} + 540\text{ W}\\ &= 720\text{ W}\end{aligned}$$

$$\begin{aligned}\text{Efficiency} &= \frac{V_s I_s \cos\theta}{V_s I_s \cos\theta + \text{total losses}} \times 100\\ &= \frac{(20{,}000\text{ VA})(0.8)}{(20{,}000\text{ VA})(0.8) + 720\text{ W}} \times 100\\ &= 95.69\%\end{aligned}$$

(b) At half-load, the currents in both windings are one-half of their full-load values. Therefore, the copper losses will be one-fourth of their full-load value since they vary as the *square* of the current.

$$\text{Total losses} = 180 + \frac{1}{4}(540) = 315\text{ W}$$

$$\begin{aligned}\text{Efficiency} &= \frac{V_s I_s \cos\theta}{V_s I_s \cos\theta + \text{total losses}} \times 100 \\ &= \frac{(10{,}000\text{ VA})(0.8)}{(10{,}000\text{ VA})(0.8) + 315\text{ W}} \times 100 \\ &= 96.21\%\end{aligned}$$

(c) At 1/4 load the copper losses are now 1/16 of their full-load values.

$$\text{Total losses} = 180 + \frac{1}{16}(540) = 213.75\text{ W}$$

$$\begin{aligned}\text{Efficiency} &= \frac{V_s I_s \cos\theta}{V_s I_s \cos\theta + \text{total losses}} \times 100 \\ &= \frac{(5000\text{ VA})(0.8)}{(5000\text{ VA})(0.8) + 213.75\text{ W}} \times 100 \\ &= 94.93\%\end{aligned}$$

From Example 17-7 it is seen that as the transformer goes from full-load toward no-load, the efficiency reaches a maximum value and proceeds to drop off. This maximum value, which is known as **maximum efficiency**, will occur when the copper losses are equal to the core losses. In equation form,

$$I_s^2 R_{es} = P_c \tag{17-29}$$

where I_s = secondary, or load current

R_{es} = equivalent resistance, in terms of the secondary

P_c = core loss, in watts

Equation 17-29 can be rewritten in terms of I_s as

$$I_s = \sqrt{\frac{P_c}{R_{es}}}$$

The values of current and equivalent resistance in the equations above are taken from the short-circuit test and would be almost exclusively high-voltage-side values. Transformers are usually designed so that their maximum efficiency occurs at somewhat less than full load.

EXAMPLE 17-8

A 2400/240-V transformer has the following characteristics: $R_p = 4.2\ \Omega$, $X_p = 7\ \Omega$, $R_s = 0.42\ \Omega$, and $X_s = 1.5\ \Omega$. If the open-circuit test indicates core losses of 70 W, determine the level of output current at which maximum efficiency occurs.

Solution

$$a = \frac{2400\text{ V}}{240\text{ V}} = 10$$

$$R_{es} = R_s + \frac{R_p}{a^2} = 0.42\ \Omega + \frac{4.2\ \Omega}{(10)^2}$$
$$= 0.462\ \Omega$$

$$I_s = \sqrt{\frac{P_c}{R_{es}}} = \sqrt{\frac{70\text{ W}}{0.462\ \Omega}}$$
$$= 12.31\text{ A}$$

The **all-day efficiency** of a transformer is the ratio of the watthour output to the watthour input, which is generally calculated over a 24-hour period. Distribution transformers are always energized, although the secondaries may supply power only on an intermittent basis. Since the core losses are being supplied continuously, transformers are usually designed so that the maximum efficiency takes into account the constant losses over 24 hours.

$$\text{All-day efficiency} = \frac{\text{W} \cdot \text{h}_o}{\text{W} \cdot \text{h}_o + \text{W} \cdot \text{h}_{\text{core loss}} + \text{W} \cdot \text{h}_{\text{copper loss}}} \times 100 \quad (17\text{-}30)$$

where $\text{W} \times \text{h}_o$ is the watthour output.

EXAMPLE 17-9

Find the all-day efficiency of a 20-kVA 2200/220-V 60-Hz transformer with normal core loss of 175 W and a copper loss of 300 W, determined by the open- and short-circuit tests, respectively. Calculate the all-day efficiency based on the transformer operating for 10 hours a day at full load and 14 hours a day at zero load.

Solution During the 24-hour period:

$$\text{Energy output} = 20{,}000 \times 10 = 200\text{ kWh}$$

$$\text{Energy expended in core losses} = 175 \times 24 = 4.2\text{ kWh}$$

$$\text{Energy expended in copper losses} = 300 \times 10 = 3\text{ kWh}$$

$$\text{All-day efficiency} = \frac{200 \times 10^3}{200{,}000 + 4200 + 3000} \times 100$$
$$= 96.5\%$$

17-13 AUTOTRANSFORMER

The **autotransformer** is a device that accomplishes the desired transformer action within one coil, as compared to two or more coils of a standard transformer. In a standard transformer, the primary and secondary coils are magnetically coupled with no physical connection between the primary and secondary. In an autotransformer, the coils are magnetically coupled as well as physically connected, since one of the coils is common to both the primary and secondary circuits. The autotransformer is used extensively for reduced voltage starting of ac motors, balance coils for three-wire dc generators, variacs, and for stepping voltage up or down for transmission lines.

Figure 17-18 illustrates the circuit diagram of an autotransformer. Winding bc, which current I_c flows through, is called the **common winding** because its voltage appears on both sides of the transformer. The other winding, *ab*, is referred to as the **series winding**, since it is in series with the common winding. Voltage V_x on the low-voltage side is the terminal voltage V_2 of the common winding. From this diagram it is seen that the voltage V_H on the high-voltage side is the phasor sum of the terminal voltages of the series and common windings. In equation form,

$$V_X = V_2$$
$$V_H = V_1 + V_2 \tag{17-31}$$

The voltages induced in the windings of the autotransformer are very close to the applied voltage values since the autotransformer is more efficient than a standard transformer. The induced voltage in the low-voltage circuit is identified as E_2, while the induced voltage in the high-voltage circuit is E_1. The induced voltages E_1 and E_2 are proportional to the number of turns in the series and common windings, respectively. Therefore,

$$\frac{E_2}{E_1} = \frac{N_2}{N_1}$$

and

$$\frac{V_H}{V_X} = \frac{N_1 + N_2}{N_2} \tag{17-32}$$

If Kirchhoff's current law is applied to the junction of the common winding and the series winding, the following equation is developed:

$$I_2 = I_1 + I_c \tag{17-33}$$

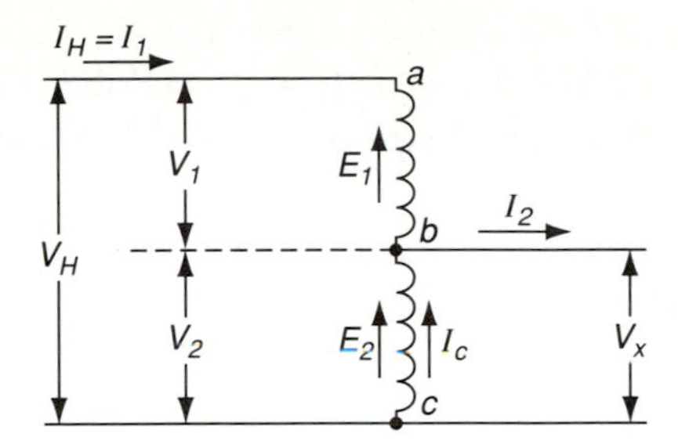

FIGURE 17-18 Circuit diagram of autotransformer.

where I_c is the common current in the transformer.

The autotransformer can be used either as a step-up or a step-down transformer. When used as a step-up transformer, the equation for the current relationship is as follows:

$$I_1 = I_2 + I_C \qquad (17\text{-}34)$$

The common current, I_C, is the arithmetic difference between I_1 and I_2 in both applications. If the magnetizing current is neglected, the currents through each winding cause equal and opposing mmfs. The relationship between the opposing winding currents I_1 and I_C is

$$N_{ab}I_1 \approx N_{bc}I_C \qquad (17\text{-}35)$$

When used as a step-down transformer, the difference in current between the series coil and common coil is $I_2 = I_1 + I_C$. It can therefore be said that all the current through winding *ab* is conducted to I_2. So, the power which is transferred conductively from the input to the output is found by the equation

$$\text{power}_{\text{conductive}} = V_2I_1 \qquad (17\text{-}36)$$

Since $V_H = V_1 + V_2$, the difference between the input terminal voltage, V_H, and the output terminal voltage, V_X, is a ratio of the power that is induced, or transformed. Consequently, the power that is transferred from primary to secondary by means of transformer action is

$$\text{power}_{\text{transformed}} = V_1I_1 \qquad (17\text{-}37)$$

where V_1 is the voltage applied across the series winding of a step-down transformer.

When used as a step-up transformer, the difference between the current in the series coil and the common coil is found by $I_2 = I_1 - I_c$, and the relationship between conducted power and transformed power can be determined in a similar manner to that shown above.

The power transformed in a step-down autotransformer can also be found by using the transformation ratio, as applied to the equation

$$\begin{aligned} P_{\text{trans}} &= \left(V_H - \frac{V_H}{a}\right) I_1 \\ &= V_H I_1 \left(1 - \frac{1}{a}\right) \\ &= \text{power input} \times \left(1 - \frac{1}{a}\right) \end{aligned} \tag{17-38}$$

For either a step-up or a step-down transformer, the total amount of power transferred from the primary to the secondary is the sum of the conductive power and the transformed power.

$$P_{\text{total}} = P_{\text{trans}} + P_{\text{cond}} \tag{17-39}$$

EXAMPLE 17-10 When used as a step-down autotransformer, the transformation ratio is 2, and the input voltage is 240 V. Determine the current in the common winding when there is a load of 80 Ω connected to the output terminal.

Solution

$$V_X = \frac{V_H}{a} = \frac{240\text{ V}}{2} = 120\text{ V}$$

$$I_2 = \frac{V_X}{R_L} = \frac{120\text{ V}}{80\,\Omega} = 1.5\text{ A}$$

$$I_1 = \frac{I_2}{a} = \frac{1.5\text{ A}}{2} = 0.75\text{ A}$$

$$\begin{aligned} I_C = I_2 - I_1 &= 1.5\text{ A} - 0.75\text{ A} \\ &= 0.75\text{ A} \end{aligned}$$

EXAMPLE 17-11 An autotransformer having a primary voltage of 2850 V and a secondary of 2400 V delivers a load of 25 kW at unity power factor. Determine the power transformed and the power conducted directly from primary to secondary.

Solution

$$a = \frac{2850\text{ V}}{2400\text{ V}} = 1.188$$

$$P_{\text{trans}} = \text{power input} \times \left(1 - \frac{1}{a}\right)$$

$$= (25 \times 10^3 \text{ W})\left(1 - \frac{1}{1.188}\right)$$

$$= 3956.23 \text{ W}$$

$$P_{\text{cond}} = P_{\text{total}} - P_{\text{trans}}$$
$$= 25{,}000 \text{ W} - 3956.23 \text{ W}$$
$$= 21.04 \text{ kW}$$

17-14 TRANSFORMER POLARITY

The primaries and secondaries of transformers usually consist of two or more windings instead of a single winding. This is done so that the windings on each side of the transformer may be connected in either series or parallel, and several voltage and current ratings may be obtained from one transformer. Single-phase transformers are also interconnected to supply three-phase power, discussed later in this chapter. The winding connections must be made so that the correct phase relations exist among the winding voltages. **Transformer polarity** is defined as the relative direction of the induced voltages in the primary and secondary windings with respect to the winding terminals.

There are two methods used for identifying the winding terminals of a transformer. One method, which was mentioned earlier in this chapter, involves using a set of letters and numbers established by the American Standards Association. The high-voltage terminals are labeled H_1 and H_2, and the low-voltage terminals are marked X_1 and X_2. The subscripts 1 and 2 denote the instantaneous polarity of the terminals with respect to each other. The terminals H_1 and X_1 have the same instantaneous polarity, and the terminals H_2 and X_2 have the same instantaneous polarity. When transformers have two coils on each side, they would be labeled with H_3 and H_4 also on the high-voltage side, and X_3 and X_4 also on the low-voltage side. Terminals H_3 and X_3 would have the same instantaneous polarity, as would terminals H_4 and X_4.

The other method of identifying the terminals of a transformer is called the **dot convention**. When this type of marking is used, terminals with the

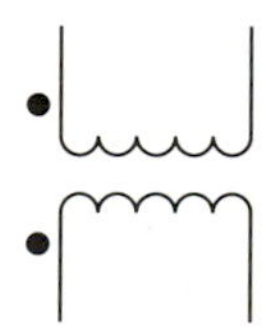

FIGURE 17-19 Transformer terminals identified using the dot convention.

same instantaneous polarity are identified by dots and the other terminals are left blank. An example of the dot convention is shown in Figure 17-19. The primary windings in Figure 17-20 are interconnected so that the direction of the magnetizing current in both coils at any instant tends to establish a flux in the same direction in the core. Since the induced voltages in both windings are caused by the same sinusoidal flux, the direction of the counter emf in both windings will be the same. Therefore, the windings shown in Figure 17-20 are connected in **additive polarity**.

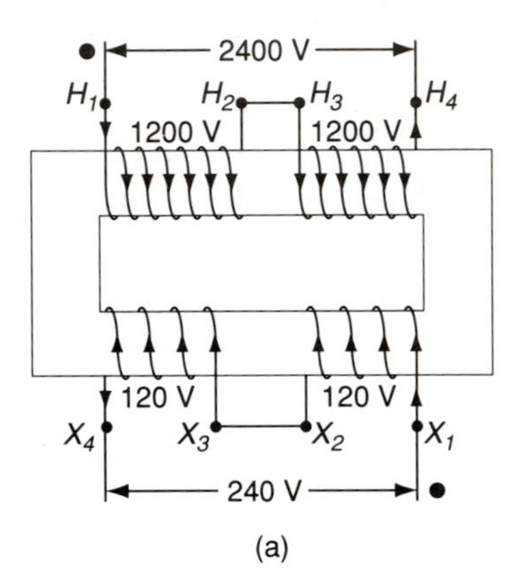

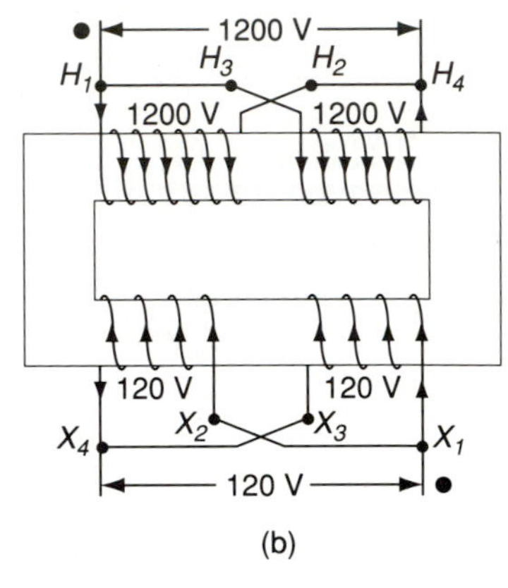

FIGURE 17-20 Additive polarity transformer with (a) primary and secondary in series and (b) primary and secondary in parallel.

When the primary coils are connected so that the mmfs are set up in opposite directions in the core, the resulting flux is zero. The coils are then said to be connected in **subtractive polarity**, as shown in Figure 17-21. Additive polarities are generally required in large transformers, up to 200 kVA, while small transformers use subtractive polarities.

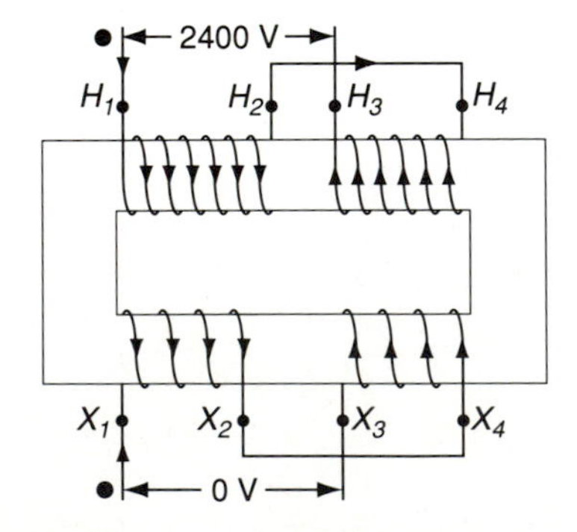

FIGURE 17-21 Transformer connected with primary and secondary in opposite, or subtractive, polarity.

Whether the external polarity is additive or subtractive is determined entirely by the manner in which the leads from the windings are connected to the external terminals. Since the windings in a transformer are sealed and not visible, it is necessary to rely on the manufacturers external polarity markings for proper **phasing** of the windings. However, if a transformer has markings that are difficult to decipher, or nonexistent, a simple voltage test can be performed to determine the external polarity. Although any safe voltage is suitable for this test, Figure 17-22 shows 120 V being used, and the voltmeter should have a range of about twice the testing voltage. Assuming that there are no markings on the transformer, the right- and left-hand terminals, as viewed from the high-voltage side of the transformer, may be marked H_1 and H_2, respectively. A connection is made between H_1 and the low-voltage terminal directly opposite H_1. A voltmeter is then connected between H_2 and the low-voltage terminal directly opposite H_2.

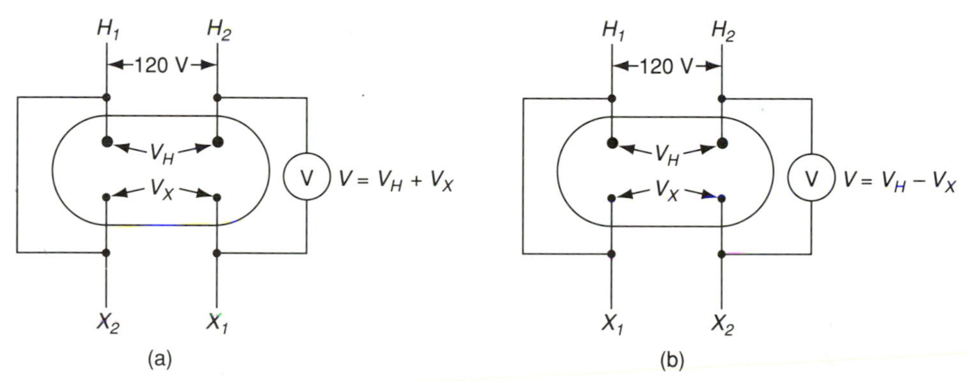

FIGURE 17-22 Method of testing polarity of transformer using a voltmeter: (a) voltage reading over 120 V = additive polarity; (b) voltage reading under 120 V = subtractive polarity.

If the voltmeter reading is greater than the input voltage, as in Figure 17-22(a), the transformer is said to have additive polarity, and the low-voltage terminal directly opposite H_1 is marked X_2. The other low-voltage terminal is marked X_1. If the voltmeter reads below the input value, the transformer is considered to be subtractive, and its low-voltage terminals are marked as shown in Figure 17-22(b).

The distribution transformers used to supply domestic loads usually have two 120-V secondary windings which may be connected in either series or parallel. Figure 17-23 shows three common connections for this type of application with the primary connected to 4160 V and the secondary windings interconnected for various voltage outputs. The most common connection for domestic supply is the circuit shown in Figure 17-23(a). The three wires provide both 120 V and 240 V, with the center wire shown as the neutral conductor.

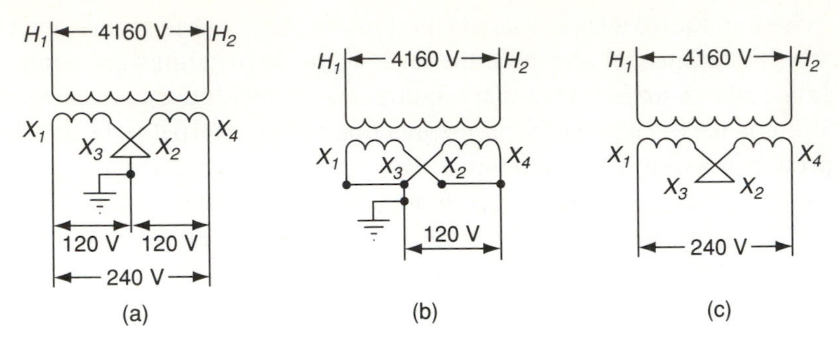

FIGURE 17-23 Three main types of distribution transformer connection: (a) 240 V and 120 V secondary; (b) 120 V secondary; (c) 240 V secondary.

17-15 PULSE TRANSFORMERS

Pulse transformers are widely used in electronic circuits. The pulse transformer technique was originally developed for use in radar systems. However, the increased use of square-wave voltages has led to extending pulse transformer theory to a variety of applications, including video signals and inverters. Figure 17-24 shows the output of a pulse transformer for a rectangular input pulse. The pulse width is shown as a value at approximately one-half of the output amplitude. The **pulse width** of the input signal is defined as the time between the start of the rise and the start of the fall of the input pulse. The pulse width of the input pulse and the output pulse are not necessarily equal. The rise and fall times of the pulse are often determined by the time elapsed between 10 and 90% of the pulse amplitude.

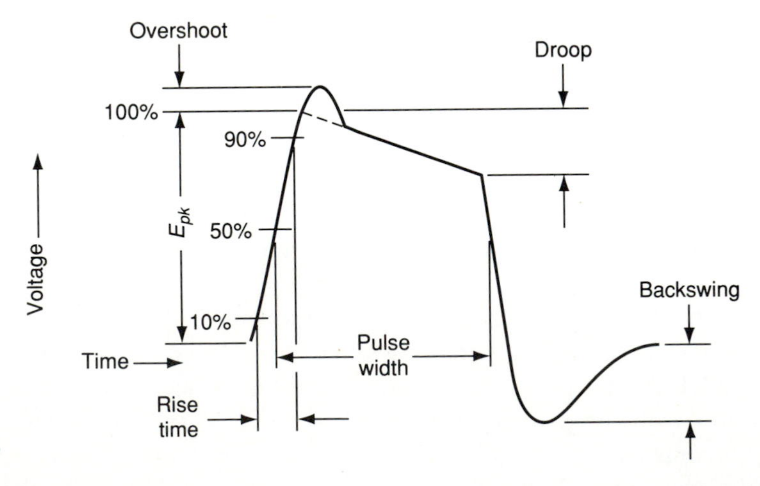

FIGURE 17-24

Pulse transformers are often used to couple a trigger pulse to a thyristor, such as an SCR, to obtain electrical isolation between two circuits. Pulse transformers used for thyristor control are generally either a 1:1 ratio (two windings) or a 1:1:1 ratio (three windings). Figure 17-25 shows an SCR connected to a 1:1 ratio pulse transformer. Diode D_1 is connected in series with the gate of the SCR to prevent reversal of the pulse transformer output voltage. Figure 17-26 shows how a 1:1:1 pulse transformer is used to drive an inverse-parallel pair of SCRs. In this circuit, the two SCRs and the trigger pulse generator are fully isolated.

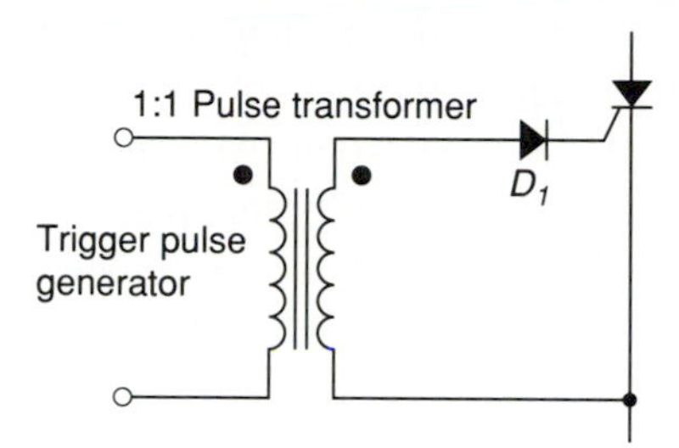

FIGURE 17-25 Basic pulse transformer coupling.

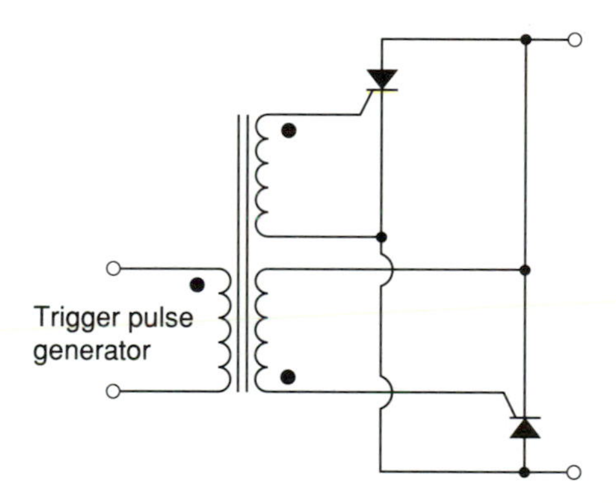

FIGURE 17-26 1:1:1 pulse transformer controlling two SCRs.

17-16 INSTRUMENT TRANSFORMERS

Measuring instruments and relays for the operation of protective and control devices are usually connected to ac circuits through **instrument transformers** if the circuit voltage or current is of a large value. Instrument transformers are classified as **potential transformers** or **current transformers** according to whether they are used in measurement of potential difference or current. The primary of the transformer is connected to the circuit in which the voltage or current is to be measured, and the secondary is connected to the instruments or relays. These transformers differ from power or distribution transformers in that they are of comparatively small capacity and are designed to maintain a much higher degree of accuracy in their ratios of transformation. To differentiate from other transformers, the load carried by instrument transformers is referred to as their **burden**. Both potential and current transformers may be of the dry-type or oil-filled. They may be equipped with either solid porcelain or oil-filled high-voltage bushings, through which the high-voltage leads are brought out of their case.

Potential transformers (PTs) have a small output and a high accuracy. PTs of several **accuracy classes** may be purchased depending on the degree of accuracy required for a given application. The windings of a PT are compensated to produce an exact ratio at a specified rated burden. Although the output of a potential transformer is typically only a few volt-amperes, its physical size can be quite large, due to the high-voltage insulation requirements. Figure 17-27 shows a potential transformer connected between a single-phase source and load. PTs are used with wattmeters, voltmeters, watthour meters, relays, and other control devices. Several instruments may be connected to the same PT provided that their sum does not exceed the burden for which the transformer is designed.

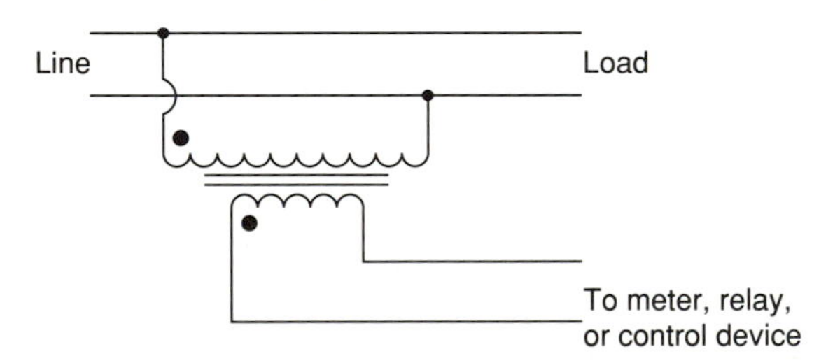

FIGURE 17-27 Potential transformer connection.

The primary of a current transformer (CT) is connected in series with the circuit whose current is to be measured, and the secondary is connected to the current-measuring instrument or relay. The primary winding may consist of one or few turns and must be insulated to sustain the line voltage. An example of a typical CT is shown in Figure 17-28. In this circuit the insulated line conductor acts as the primary winding, passing through the core around which the secondary windings are wound. The secondary of a current transformer should never be left open, since extremely high voltages can appear across its open secondary terminals. In Figure 17-28, a switch, or **shorting interlock**, is connected in parallel with the meter. This interlock must be closed in order for the meter to be removed for inspection or adjustment.

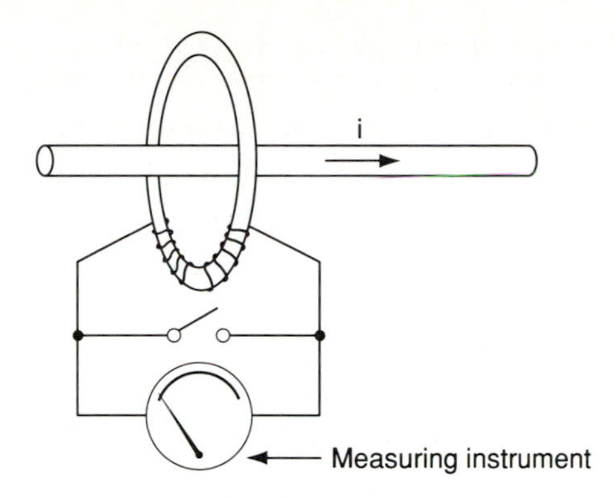

FIGURE 17-28 Current transformer.

Current transformers are also used for comparison of two currents in different circuits or parts of the same circuit. In Figure 17-29, the current through one winding of a polyphase motor is measured by two identical current transformers, CT′ and CT″. A relay coil is shown connected across the secondaries of the two current transformers. I_2' and I_2'' indicate the direction of current flow in the secondary windings. The secondaries of the two CTs are connected to each other with a relative polarity such that the current through the relay coil is the difference between the secondary currents in the two transformers.

If the motor winding is in good condition, the currents in the two transformers are equal and there is no current in the relay coil. If a fault occurs in the winding, such as insulation breakdown, the currents at either end of the winding will be different. Consequently, a proportional current flows in the relay coil, and a control device, such as a circuit breaker, would actuate. This configuration is the basis for **differential relaying**, which is commonly used to protect machines, short feeders, parallel circuits, and other installations.

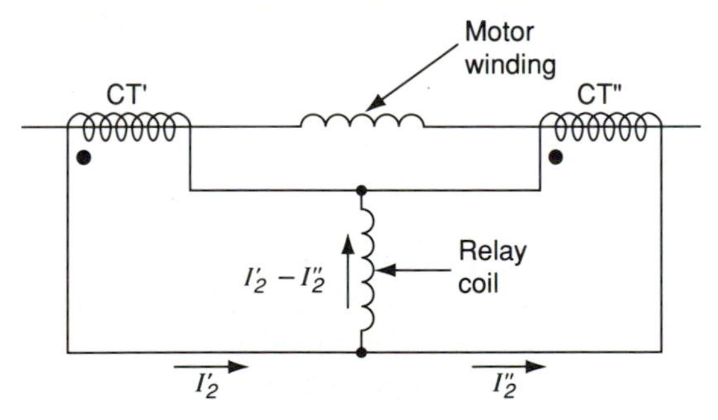

FIGURE 17-29 Two CTs used for differential-current relaying.

17-17 PARALLEL OPERATION OF TRANSFORMERS

Transformers are said to be connected in parallel when both their primary and secondary windings are connected in parallel. Transformers are frequently operated in parallel in single-phase or three-phase combinations when an existing system is required to be expanded to increase its kVA capacity. In some situations, such as hospitals, it may be desirable to supply an important load from several groups of transformers to maintain continuity of service despite failure of one of the transformers or its associated circuits. For transformers to be paralleled successfully, several important conditions must be fulfilled:

1. The voltage ratings of the primaries and secondaries must be the same. That is, the transformation ratio of both transformers must be identical.
2. Before connecting the secondary windings in parallel, their terminals must be checked for proper polarity.
3. To avoid circulating currents and operation at different power factors, the ratio of equivalent reactance to equivalent resistance should be the same.
4. When paralleling transformers with different kVA ratings, the equivalent impedances should be *inversely proportional* to the individual kVA ratings. In other words, if two transformers of the same kVA rating are to divide the load equally, they should have the same equivalent impedance. If one has twice the kVA rating of the other, it should supply twice the current. A proportional division of the load therefore requires that its equivalent impedance shall be one-half that of the other, since the impedance voltage drops must be the same.

When the primaries of two or more transformers are connected in parallel across a supply system, and the secondaries are connected to the same load, as in Figure 17-30, the foregoing criteria must be met to avoid circulating currents. These circulating currents are caused by unequal secondary voltages and can be quite large since they are limited only by the small impedance of the transformer windings.

Although transformers connected in parallel should ideally have the same transformation ratio, it is sometimes necessary to parallel transformers, which have different ratios. In this type of situation, such as in an emergency or when transformers have tap-changing equipment operating on different taps, the magnitude of the circulating current should be calculated to determine the feasibility of this type of connection. This method of load sharing should not be used on transformers for extended periods if the value of circulating current exceeds 10% of the rated current.

To perform an exact calculation of the circulating current between two

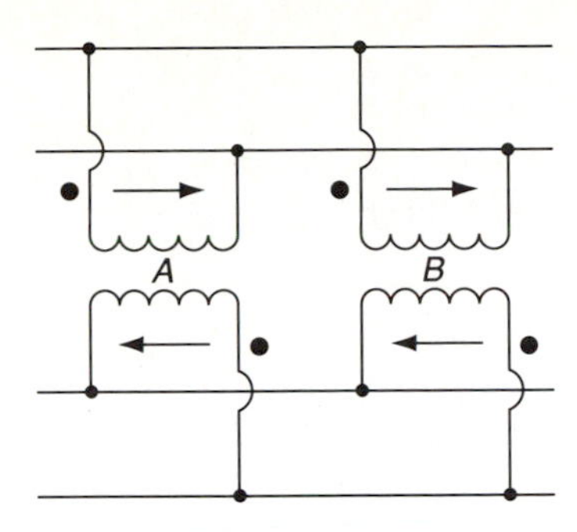

FIGURE 17-30 Correct phasing of parallel-connected transformers. Arrows indicate direction of current.

transformers, it would be necessary to use the phasor quantities of the equivalent impedances. In other words, the phase displacement between the transformers must be taken into consideration. However, equation 17-40 shows a simple method of arriving at the value of circulating current by using Z_e as a scalar quantity. Although not as precise as the phasor analysis method, the value determined by using this equation is extremely close, and for most applications, well suited.

$$I_{cc} = \frac{V_s(a_1 - a_2)}{a_1 Z_{e1} + a_2 Z_{e2}} \tag{17-40}$$

where I_{cc} = circulating current, in amperes

V_s = rated secondary voltage

a_1, a_2 = turns ratio of two transformers

Z_{e1}, Z_{e2} = equivalent of impedance of two transformers in terms of the secondary

EXAMPLE 17-12 A 20-kVA 2500/240-V transformer with a secondary impedance of $Z_e = 0.095\ \Omega$ is to be paralleled with a 30-kVA 2400/240-V transformer with a secondary impedance of $Z_e = 0.08\ \Omega$. Calculate the circulating current when no load is connected.

Solution

$$a_1 = \frac{2500\text{ V}}{240\text{ V}} = 10.417$$

$$a_2 = \frac{2400\text{ V}}{240\text{ V}} = 10$$

$$I_{cc} = \frac{V_s(a_1 - a_2)}{a_1 Ze_1 + a_2 Ze_2}$$

$$= \frac{(240\ \text{V})(10.417 - 10)}{(10.417)(0.095\ \Omega) + (10)(0.08\ \Omega)}$$

$$= 55.92\ \text{A}$$

When two transformers of unequal transformation ratios are connected to a load, the total load does not divide inversely as compared to the transformer impedances. The division of current between the transformers can be determined from the equations

$$I_1 = \frac{V_s(a_2 - a_1) + I_L a_2 Z_{e2}}{a_1 Z_{e1} + a_2 Z_{e2}} \tag{17-41}$$

$$I_2 = \frac{V_s(a_1 - a_2) + I_L a_1 Z_{e1}}{a_1 Z_{e1} + a_2 Z_{e2}} \tag{17-42}$$

where I_L is the value of load current to be supplied, in amperes.

EXAMPLE 17-13 Determine the individual load currents for the transformers of Example 17-12 if they are connected to a 60-kVA load.

Solution $a_1 = 10.417 \quad a_2 = 10 \quad Z_{e1} = 0.095\ \Omega \quad Z_{e2} = 0.08\ \Omega$

$$I_L = \frac{60{,}000\ \text{VA}}{240\ \text{V}} = 250\ \text{A}$$

$$I_1 = \frac{V_s(a_2 - a_1) + I_L a_2 Z_{e2}}{a_1 Z_{e1} + a_2 Z_{e2}}$$

$$= \frac{(240\ \text{V})(10 - 10.417) + (250\ \text{A})(10)(0.08\ \Omega)}{(10.417)(0.095\ \Omega) + (10)(0.08\ \Omega)}$$

$$= 55.83\ \text{A}$$

$$I_2 = \frac{V_s(a_1 - a_2) + I_L a_1 Z_{e1}}{a_1 Z_{e1} + a_2 Z_{e2}}$$

$$= \frac{(240\ \text{V})(10.417 - 10) + (250\ \text{A})(10.417)(0.095\ \Omega)}{(10.417)(0.095\ \Omega) + (10)(0.08\ \Omega)}$$

$$= 194.17\ \text{A}$$

17-18 THREE-PHASE CONNECTIONS

Virtually all electric energy is generated in three-phase generators and transmitted over three-phase transmission lines. Since a transformer is used every time a voltage is changed, a vast number of transformers are presently in use throughout the world. Three-phase voltage transformation is accomplished either by a bank of three single-phase transformers, or by one three-phase transformer in which the magnetic circuits of the three phases are interconnected. During this discussion of three-phase connections, we shall use banks of single-phase transformers to illustrate the proper connections.

To avoid unnecessary complications, it shall be assumed that all three transformers in a bank have the same parameters. That is, the transformation ratios, kVA ratings, and internal impedances will be the same for each individual single-phase transformer. The methods of connecting the transformers for three-phase power shall be in accordance with the American Standards Association using H_1–H_2 and X_1–X_2, with the odd-numbered subscripts instantaneously positive. When phasor diagrams are used to illustrate a transformer connection, the subscript E is assumed to be a primary voltage and V a secondary voltage.

There are four standard methods of connecting a three-phase transformer bank: Y–Y, Δ–Δ, Δ–Y, and Y–Δ. Two other methods of connecting transformers, the open-delta and Scott-tee, are discussed later in this section.

The Y–Y connection

Transformer windings are connected in wye when three terminals of like polarity are joined to form the neutral point of the wye, as shown in Figure 17-31(a). The phasor diagram of Figure 17-31(b) shows the relationship between the primary line and phase voltages. The secondary line and phase voltages are in-phase with their respective primaries. As in any Y-connected circuit, the phasor relations among the line-to-neutral voltages E_{AN}, E_{BN}, and E_{CN}, and the line-to-line voltages E_{AB}, E_{BC}, and E_{CA} are as follows:

$$E_{AB} = E_{AN} + E_{NB} = E_{AN} - E_{BN}$$

$$E_{BC} = E_{BN} - E_{CN}$$

$$E_{CA} = E_{CN} - E_{AN}$$

When the phase voltages are balanced, the effective value of the line voltage equals $\sqrt{3}$ times the effective value of the phase voltage in either the primary or secondary. On both the primary and secondary sides, the phase currents and line currents are equal. Under balanced conditions, the phasor sum of the secondary line currents equals zero, and no current flows in the neutral. However, if the load is unbalanced, the neutral will float and cause the three-phase voltages to become unequal. The Y–Y connection is rarely used on large power systems, but is used occasionally for local power distribution, such as within industrial plants.

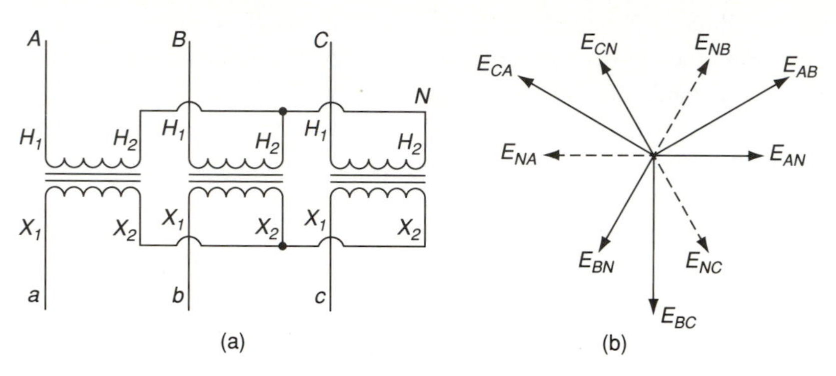

FIGURE 17-31 (a) Y–Y connection; (b) phasor diagram.

The Δ–Δ connection

In Figure 17-32(a), three single-phase transformers are shown in their primaries and secondaries connected in delta. On both the primary and secondary sides, terminals of opposite polarity are connected together to form the Δ. In this type of connection, it is extremely important that the transformers have the same transformation ratio; otherwise, a circulating current will be present in the transformers even when the bank is under no-load conditions. The Δ–Δ system does not experience the problems of unbalanced loads, as in the Y–Y connection, which makes this combination better suited for that type of application.

When the transformers are identical and the circuit is balanced, each phase is of equal magnitude and displaced by 120°. Figure 17-32(b) shows the relationship between line and phase circuits. Phasors I_{ab}, I_{bc}, and I_{ca} represent the currents in the primary windings. If the power is unity, these primary currents are in phase with the corresponding primary voltages. When the currents are balanced, the effective value of the line currents equals $\sqrt{3}$ times the effective values of the currents in the Δ-connected windings.

When transformers are connected in the configuration shown in Figure 17-32(a), there are parallel paths between every pair of line terminals on both the primary and secondary sides. Therefore, the distribution of current among transformers of a Δ–Δ bank depends on their equivalent impedance. If one Z_e is different from the other, the current in the parallel circuit will take the path of least resistance. For this reason, even if the line currents are balanced, the individual transformers of a Δ–Δ connection will be loaded unequally unless their equivalent impedances are the same.

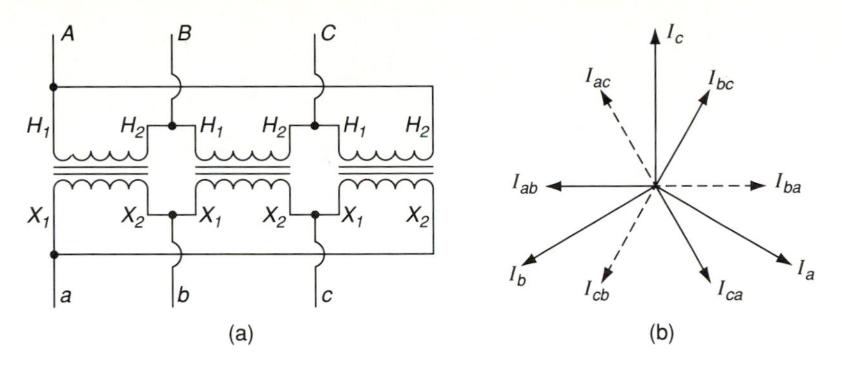

FIGURE 17-32 (a) Δ–Δ connection; (b) phasor diagram.

The Δ–Y connection

The Δ–Y connection is generally used to step-up voltages, although it is occasionally applied in a step-down configuration. The neutral of the Y-connected high-voltage side is often grounded when the Δ–Y bank is used for high-voltage transmission of power. By grounding the neutral, the voltage distribution between the lines and ground is balanced, and the voltages between the transformer coils and core are reduced. One advantage of the Δ–Y connection over the Δ–Δ is that insulation requirements for the secondaries are reduced, which is particularly advantageous for very high secondary voltages. For example, in a 100,000-V system, the Y-connected secondaries only require insulation for 58,000(100,000/$\sqrt{3}$) volts.

Until this point, there has been no mention of the phase shift that can occur between the primary and secondary voltages of a three-phase transformer connection In the Y–Y and Δ–Δ connections, no phase shift occurs between the secondary. However, in the Δ–Y and Y–Δ connections, a phase displacement of 30° occurs. When H_1 and X_1 of the transformer are considered to have the same instantaneous polarity, the voltage across the windings X_1–X_2 and H_1–H_2 are considered to be in-phase with each other. Consider the Δ–Y step-up transformer shown in Figure 17-33(a). Since the voltages across the windings X_1–X_2 and H_1–H_2 are always in phase, in this configuration the primary line voltage E_{AB} is in phase with the secondary phase voltage V_{AN}. Assuming that this is a step-up transformer, V_{AN} and E_{AB} are shown as the reference phasor in Figure 17-33(b). This 30° phase displacement occurs between the primary and secondary line voltages; the line voltage for V_{AB} in the wye-connected bank is the sum of V_{AN} + V_{NB}. In the circuit shown in Figure 17-33, the secondary line voltage was seen to lead the primary line voltage by 30°. By reversing the winding polarities, this type of connection could also produce a 30° lagging line voltage.

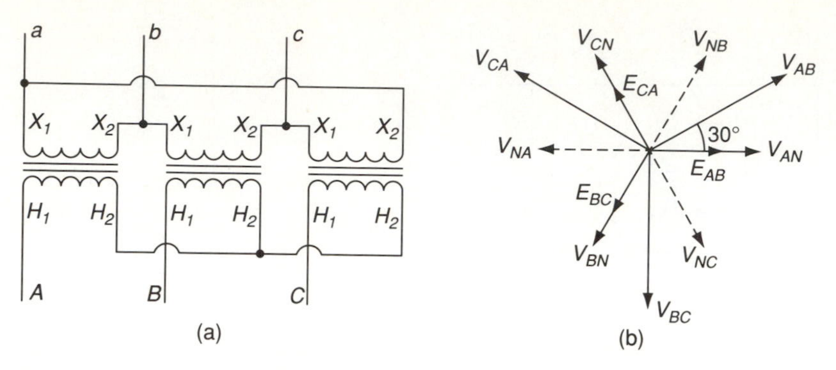

FIGURE 17-33 (a) Δ–Y step-up transformer; (b) phasor diagram.

The Y–Δ connection

This connection is the reverse of the Δ–Y system and is shown in Figure 17-34(a). The Y–Δ connection is generally used to step down the high voltage of a transmission line to a moderate voltage for distribution. As mentioned previously, there is also a 30° phase displacement on the Y–Δ system. The phasor diagram of Figure 17-34(b) shows the relationship between the primary and secondary line voltages when used as a step-down transformer. As was the case with the Δ–Y system, there is a 30° phase shift, except this time the primary line voltage is leading the secondary by 30°. This is an ideal situation for paralleling a Δ–Y and Y–Δ bank of transformers since the phase shift between the two banks would cancel each other out using this combination.

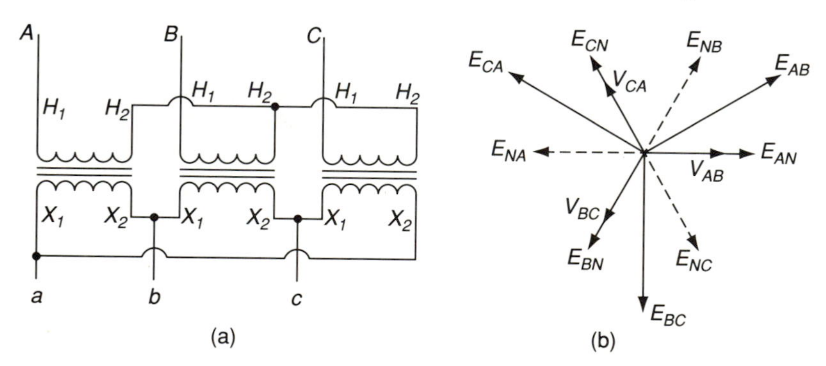

FIGURE 17-34 (a) Y–Δ step-down transformer; (b) phasor diagram.

The Δ–Y, Y–Δ connection

This particular transformer configuration is a very popular means of high-voltage transmission. However, there are two important considerations to take into account before attempting to parallel these two connections: first, both banks of transformers require the same voltage ratios, and second, both combinations require the same phase shift.

As mentioned earlier, a wye-delta connection is generally used in a step-down voltage transformation. The wye connection takes advantage of the fact that the one leg of a Y, or the line-to-neutral voltage, is less than the line-to-line voltage by the $\sqrt{3}$ factor. This is especially important when the primary voltage is a few hundred thousand volts. The Y–Δ transformer bank has a 30° phase shift between the primary and secondary voltages, as was shown in the phasor diagram of Figure 17-35(b). This displacement poses no problems for the Δ–Y transformer bank which is connected in parallel, but would be disastrous if a Y–Y or Δ–Δ bank were connected to it. The transformer bank connection shown in Figure 17-35 illustrates one method of connecting a Δ–Y and Y–Δ bank to produce a 0° phase shift between the input and output voltages.

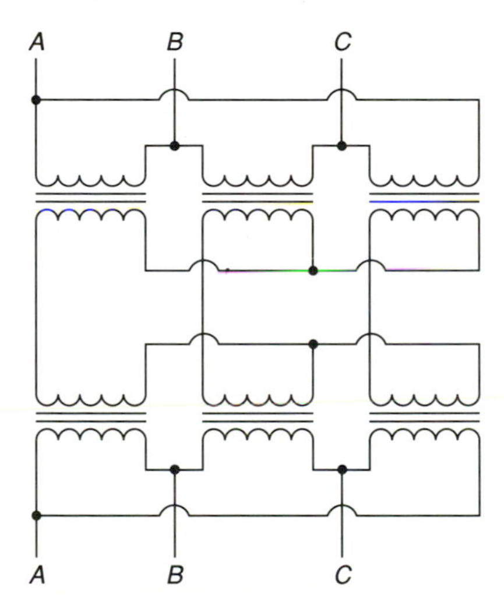

FIGURE 17-35 Δ–Y and Y–Δ connection with 0° phase shift.

The open-delta, or V–V connection

When one transformer of a Δ–Δ bank is removed, the other two transformers will continue to supply three-phase power to the secondary circuit. This particular connection is quite useful in situations where if one of a group of Δ–Δ-connected transformers becomes defective, it can be removed from the circuit without the interruption of three-phase power being distributed.

The power-handling capability of an open-delta connection is not two-thirds of a full Δ–Δ as one may initially assume. Since both transformers operate at a reduced power factor when connected in open-delta, the kVA rating of the V–V connection is less than two-thirds of the delta connection having three individual transformers of equal rating. The ratio of the open-delta rating to the delta rating is $1/\sqrt{3}$, or 58% of the full-load value. An open-delta connection is shown in Figure 17-36. The three-phase source supplies the primaries of the two transformers and the secondaries supply a three-phase voltage to the load. In the open-delta connection, the line current is equal to the transformer phase current. In a normal delta connection, the line current is equal to $\sqrt{3}$ times the phase current. When a unity power factor load is connected to a Δ–Δ system, the power is determined by

$$P\Delta = \sqrt{3}\, E_{\text{line}} I_{\text{line}}$$

In an open-delta connection, the line current is equal to the phase current. Since $I_{\text{phase}} = I_{\text{line}}/\sqrt{3}$ in an open-delta, the equation for power when connected to a unity power factor load becomes

$$P_v = \sqrt{3}\,\frac{E_{\text{line}} \times I_{\text{line}}}{\sqrt{3}} = E_L I_L \qquad (17\text{-}43)$$

Therefore, the ratio of rated power in an open delta to the rated power in a closed delta is determined as

$$\frac{P_v}{P_\Delta} = \frac{E_L I_L}{\sqrt{3}\, E_L I_L} = \frac{1}{\sqrt{3}} = 0.577 \quad \text{or} \quad 57.7\% \qquad (17\text{-}44)$$

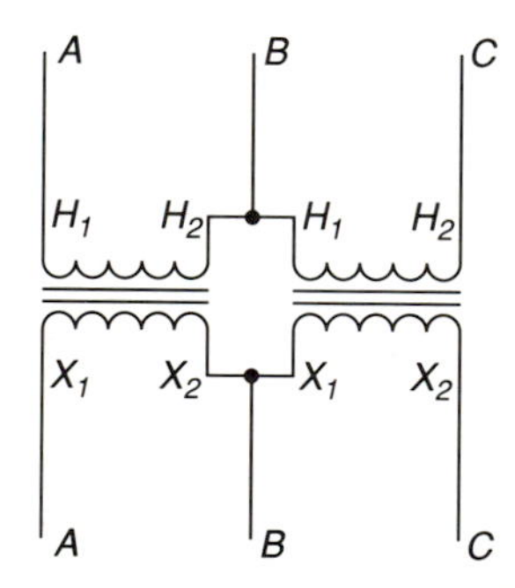

FIGURE 17-36 Open-delta transformer connection.

If an open-delta bank consisting of two transformers is connected to a balanced three-phase load having a power factor of cos θ, one of the transformers will have a power factor of $\cos(30° - \theta)$, and the other transformer will have a power factor of $\cos(30° + \theta)$. When the power factor of a balanced load is less than 1, or cos θ, the two transformers will carry the same values of current but do not supply the same amount of power. The equations for determining the power supplied by each transformer are as follows:

$$P_1 = E_L I_L \cos(30^\circ - \theta) \tag{17-45}$$

$$P_2 = E_L I_L \cos(30^\circ + \theta) \tag{17-46}$$

where P_1 = power supplied by transformer 1

P_2 = power supplied by transformer 2

θ = power factor angle of the load

EXAMPLE 17-14 Two transformers connected in open-delta supply a 200-kW 600-V balanced load operating at a power factor of 0.8 lagging. Determine the power delivered by each transformer.

Solution

$$I_L = \frac{P_T}{E_L\sqrt{3}\cos\theta} = \frac{200{,}000 \text{ W}}{600 \text{ V} \times \sqrt{3} \times 0.8}$$

$$= 240.563 \text{ A}$$

$$\theta = \cos^{-1}(0.8) = 36.9^\circ$$

$$\begin{aligned} P_1 &= E_L I_L \cos(30^\circ - \theta) \\ &= (600 \text{ V})(240.563 \text{ A}) \cos(30^\circ - 36.9^\circ) \\ &= 143.29 \text{ kW} \end{aligned}$$

$$\begin{aligned} P_2 &= E_L I_L \cos(30^\circ + \theta) \\ &= (600 \text{ V})(240.563 \text{ A}) \cos(30^\circ + 36.9^\circ) \\ &= 56.63 \text{ kW} \end{aligned}$$

The open-Y open-Δ connection

The open-wye open-delta connection shown in Figure 17-37(a) is very similar to the open-delta, or V–V connection. The main difference is that the primary voltages are derived from two phase voltages and a neutral. This type of connection is very useful in situations where three-phase power is required and only two phases are available.

A major disadvantage of this type of connection is that a very large current must flow in the neutral of the wye-connected primary. As a given instant, the neutral current on the primary side, I_n, is equal to the current flowing from points N to B, and N to A. Assuming a per unit basis of 1 and following Kirchhoff's current law, the following statement can be made:

> At the neutral point on the primary side of the two transformers connected in open-wye open-delta, the currents entering the transformers and the currents leaving must equal zero.

This statement can be expressed in equation form as

$$-I_n = I_{an} + I_{bn} \tag{17-47}$$

From equation 17-47 and taking into account the additive polarity marking between *B* and *N* on the primary, and *B* and *C* on the secondary, I_{bn} must equal the secondary line current, $-I_c$. The phasor diagram of Figure 17-37(b) shows the relationship between the primary currents with subscript *I*, and the secondary currents, subscripted *I′*. From the phasor diagram and equation 17-47, using a per unit basis of 1, the equation for the negative neutral current becomes

$$-I_n = I_{an} + (-I_c) = 1\,\underline{/0^\circ} + (-1\,\underline{/-60^\circ})$$
$$= 1.732\,\underline{/-30^\circ} \qquad (17\text{-}48)$$

By reversing this phasor by 180°, the neutral current is determined as

$$I_n = 1.732\,\underline{/150^\circ}$$

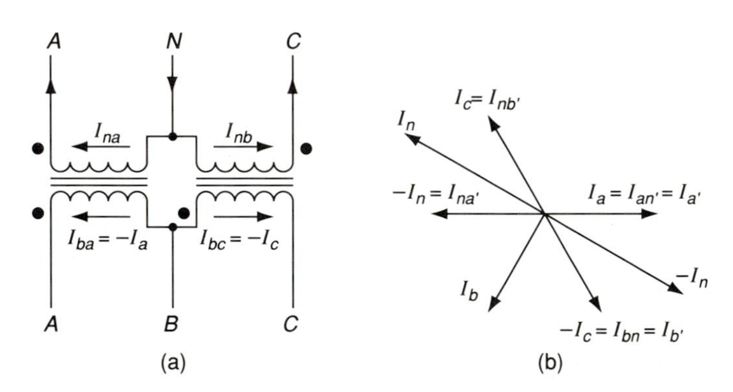

FIGURE 17-37 (a) Open-Y open-Δ transformer connection; (b) phasor diagram.

The Scott, or T, connection

Although this type of connection may be used for three-phase to three-phase voltage transformation, or for transforming two-phase voltages to three-phase values, its most common application is in transforming three-phase voltages to two phases. The method of connection for three-phase to two-phase shown in Figure 17-38(a) is named after its inventor, C.F. Scott. The primary of one transformer on the three-phase side is connected across two line voltages and its center tap is connect to the second transformer, known as the *teaser*. The primary of the teaser is tapped at 86.6% of its turns. The windings of the secondary transformers are connected in the same manner as the secondary of the open-Y open-Δ, except in this case the center is a neutral wire.

The phasor diagram of Figure 17-38(b) shows the voltage relationship for a 1:1 Scott connection. Using voltage E_{AB} as the reference phasor, voltage E_{AC} is the sum of E_{AD} and E_{EC}. On a per unit basis of 1,

$$E_{AC} = E_{AD} + E_{EC}$$

Therefore,

$$E_{EC} = E_{AC} - E_{AD} = 1 \angle -60^\circ - 0.5 \angle 0^\circ$$
$$= 0.866 \angle -90^\circ$$

Since the transformer has a transformation ratio of 1:1, voltages E_{AB} and $E_{A'B'}$ are equal, as are voltages E_{EC} and $E_{B'C'}$. The kVA ratings of the main and teaser transformers will be identical, even though the voltage across the teaser is only 86.6% of the main transformer voltage. This is due to the phase displacement of the two halves of the main transformer. Phasors E_{AD} and E_{DB} are 60° out of phase with each other, so when they are added vectorially, the kVA total of the main is equal to that of the secondary.

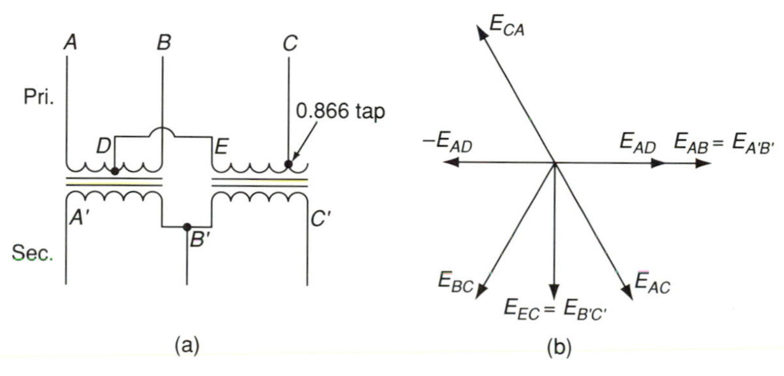

FIGURE 17-38 (a) Scott, or T, transformer connection; (b) phasor diagram.

17-19 THREE-PHASE TRANSFORMERS

One three-phase transformer is considerably smaller than three single-phase transformers having the same kVA rating. So, in situations where weight and physical size are important considerations, three-phase transformers can be quite advantageous. Three-phase transformers are constructed in a similar manner to their single-phase counterparts. There are two basic types: the shell type, in which magnetic circuit is a shell encircling the windings, and the core type, in which the magnetic circuit is a core surrounded by the windings.

Figure 17-39 shows the winding connections of a core-type transformer in a Y–Δ configuration. If the induced voltages are balanced and have a sinusoidal waveform, their instantaneous sum is zero. If the windings are arranged so that the positive directions of their core fluxes are the same in the adjoining legs of the cores, the instantaneous sum of the fluxes in these legs is zero. The primary and secondary of each phase are wound over one of the three legs of the transformer. Each of the three legs may act in turn as a return path for the fluxes of the other two.

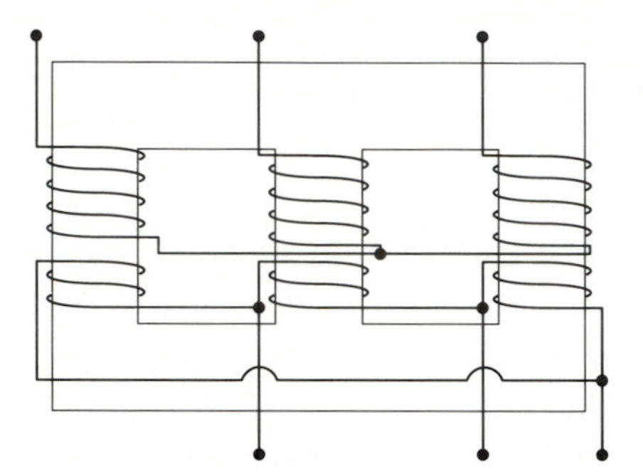

FIGURE 17-39 Three-phase core-type transformer with Y–Δ connection.

Figure 17-40 show the primary winding of a three-phase shell-type transformer with three separate magnetic circuits. When the primary windings are symmetrically connected to a three-phase supply, each of the combined yoke sections, marked x, carries half of the fluxes Φ_1 and Φ_2 at 120°. Therefore, the flux in these sections equals one-half of the core flux. The reluctance of the magnetic path for all windings is the same in the shell-type transformer, so there is no unbalance in the magnetizing currents.

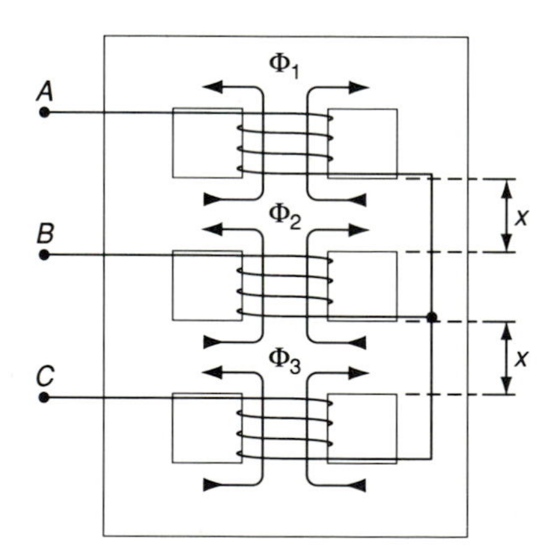

FIGURE 17-40 Three-phase shell-type transformer.

KEY TERMS

Primary
Secondary
One-to-one transformer
Step-up transformer
Step-down transformer
Pancake coils
Mutual inductance
General transformer equation
Transformation ratio (turns ratio)
Exciting current
Magnetizing current
Energy current
Leakage flux
Leakage reactance
Tap changer
Reflected impedance
Hysteresis
Grain-oriented steel
Copper losses
Open-circuit test
Core loss
Short-circuit test
Voltage regulation
Maximum efficiency
All-day efficiency
Autotransformer
Common winding
Series winding
Transformer polarity
Dot convention
Additive polarity
Subtractive polarity
Phasing
Pulse transformers
Pulse width
Instrument transformer
Potential transformer
Current transformer
Burden
Accuracy classes
Shorting interlock
Differential relaying

PROBLEMS

17-1 An ideal transformer with a turns ratio of 20 has an ac voltage of 120 V applied to its primary winding. Determine the secondary voltage.

17-2 The primary and secondary currents of a transformer were measured as 6.5 A and 202.6 A, respectively. Determine the primary voltage if the secondary is 120 V.

17-3 The 2400-V primary of a 60-Hz transformer has 960 turns. Calculate the peak mutual flux developed.

17-4 A 4160/230-V 60-Hz transformer produces a maximum flux of 0.012 Wb. Calculate the number of turns on the secondary winding.

17-5 A 120/2400-V 60-Hz transformer has 600 turns on the secondary winding. Determine the peak mutual flux of the transformer.

17-6 The secondary of a 100-kVA 60-Hz transformer has 295 turns. The flux in the transformer's core has a maximum value of 0.0048 Wb. Calculate the voltage induced in the secondary winding.

17-7 A 250-kVA 17,300/4160-V 60-Hz transformer has 40 turns on the secondary winding. Determine (a) the current in the primary and secondary windings; (b) the maximum flux in the core.

17-8 A 2400/460-V transformer supplies 2500 A to a load connected to the secondary. Determine the primary current.

17-9 A 120/24-V 200-Hz transformer is required to be used in a circuit operating at 60 Hz. Determine the primary and secondary voltages at 60 Hz.

17-10 An ideal transformer having a turns ratio of 20 and a load impedance of 8 Ω connected to the secondary. Use reflected impedance to calculate the impedance looking into the primary winding.

17-11 An ideal transformer is supplying a secondary current of 8 A. The transformation ratio is 6, and the impedance at the primary is 100 Ω. Determine (a) the primary voltage; (b) the secondary voltage; (c) the impedance connected to the secondary winding.

17-12 An impedance-matching transformer is to be used in a circuit to allow for maximum power transfer. The load impedance of the secondary is 600 Ω and the line impedance is 75 Ω. Determine the transformation ratio that will allow for maximum power transfer.

17-13 The output impedance of an audio power amplifier is measured as 2500 Ω. An impedance-matching transformer with 1000 turns on its primary is to be used to match the amplifier output to an 8-Ω load. Determine the number of turns on the secondary winding of the impedance matching transformer.

17-14 Repeat Problem 17-13 for a 4-Ω load.

17-15 For the circuit shown in Figure 17-41, calculate the input current, I_{in}, and the output current, I_S, by the reflected impedance method.

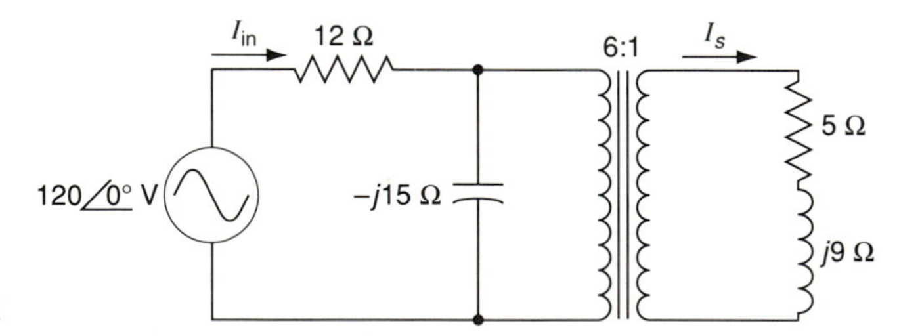

FIGURE 17-41

17-16 A 4160/230-V 25-kVA step-down transformer is connected as shown in Figure 17-42. The test data obtained for the high-voltage side are as follows: W = 325 W, I_p = 6 A, and V = 72 V. Determine (a) the equivalent impedance, resistance, and reactance, referred to the high-voltage side; (b) the equivalent impedance, resistance, and reactance, referred to the low-voltage side.

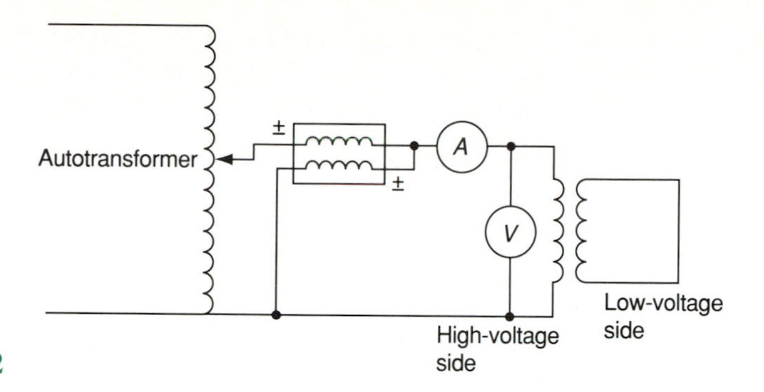

FIGURE 17-42

17-17 A 2300/230-V 10-kVA transformer has the following resistance and reactance values: $R_p = 0.65\ \Omega$, $X_p = 2.5\ \Omega$, $R_s = 0.0036\ \Omega$, and $X_s = 0.017\ \Omega$. Determine the equivalent transformer values (a) in terms of primary; (b) in terms of the secondary.

17-18 A short-circuit test was performed on a 4160/240-V 20-kVA transformer, and the following data obtained: $W = 980$ W, $I_p = 4.8$ A, and $V = 83$ V. Calculate the copper losses when the load is (a) 10 kVA; (b) 30 kVA.

17-19 For the circuit shown in Figure 17-43, determine the following values: (a) internal resistance, R_e, referred to the primary; (b) internal reactance, X_e, referred to the primary; (c) internal impedance, Z_e, referred to the primary; (d) secondary load impedance, Z_L, referred to the primary; (e) primary load current at rated primary voltage.

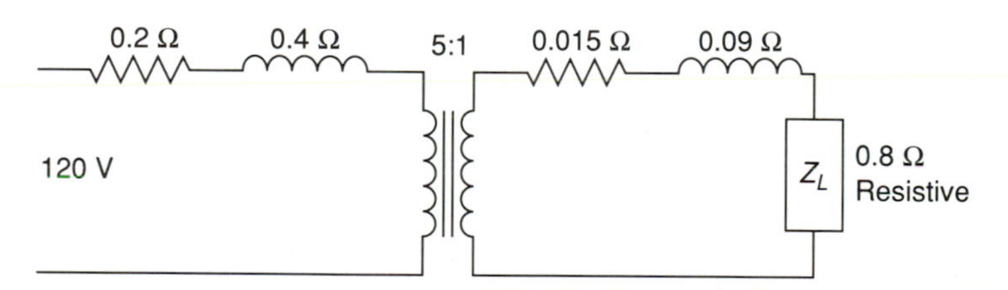

FIGURE 17-43

17-20 Repeat Problem 17-19 for a load of $Z_L = 3 + j5\Omega$ and a transformation ratio of 10.

17-21 Determine the percent regulation of a 2400/240-V transformer with a no-load voltage measured at 245 V.

17-22 A 2300/115-V transformer has a percent regulation of 3.47%. Determine the no-load voltage of the secondary winding.

17-23 A short-circuit test is performed on the low-voltage side of a 40-kVA 4800/240-V 60-Hz transformer and the following data are obtained: $W = 240$ W, $V = 66$ V, and I = rated current. Determine the percent voltage regulation with a 0.88 lagging PF load connected: (a) in terms of the primary winding; (b) in terms of the secondary winding.

17-24 A short-circuit test was performed on a 25-kVA 2400/240-V transformer and the following data obtained: $W = 232$ W, $I_p = 10.42$ A, and $V = 63$ V. Determine the voltage regulation of the transformer for 0.85 lagging PF load.

17-25 A 4800/240-V transformer has a primary current of 6.2 A and a secondary current of 96.3 A. Determine the efficiency of the transformer.

17-26 An open-circuit test is performed on a 2300/230-V 20-kVA transformer with the high-side winding open, and the core loss was found to be 220 W. When a short-circuit test was performed with the low-side shorted, the wattmeter indicated 472 W. The power factor of the load is 0.87 lagging. Determine the efficiency at rate kVA output.

17-27 Repeat Problem 17-26, solving for the efficiency at one-fourth kVA output.

17-28 A 2300/230-V 10-kVA transformer is subjected to a short-circuit test and an open-circuit test and the following data are obtained: open-circuit test, high side open, $W = 175$ W. Short-circuit test, low side shorted, $W = 255$ W. Determine the value of load current at which maximum efficiency occurs.

17-29 A distribution transformer is rated at 60 kVA, 2300/230 V. The resistance of the primary winding is 0.35 Ω, and the secondary winding resistance is 0.002 Ω. The core losses are measured as 275 W. Determine the all-day efficiency based on the transformer operating at full load for 8 hours, one-half for 4 hours, and no-load for 12 hours.

17-30 When used as a step-down autotransformer, the transformation ratio is 5 and the input voltage is 120 V. Determine the current in the common winding when there is a load of 50 Ω connected at the output.

17-31 An autotransformer with a primary voltage of 330 V and a secondary of 230 V delivers a load of 2 kW at unity power factor. Determine (a) the power transformed; (b) the power conducted directly from primary to secondary.

17-32 A 20-kVA 4800/240-V transformer with a secondary impedance of 0.063 Ω is to be paralleled with a 25-kVA 4600/240-V transformer with a secondary impedance of 0.05 Ω. Determine the circulating current when no load is connected.

17-33 Determine the individual load currents for the transformers in Problem 17-32 if they are connected to a 50-kVA load.

17-34 Two transformers connected in an open-delta configuration supply a 150-kW 480-V balanced load operating at a power factor of 0.85 lagging. Determine the power delivered by each transformer.

CHAPTER

Polyphase Induction Motors

LEARNING OBJECTIVES

Upon completion of this chapter you will be able to:

- List the main parts of a polyphase induction motor.
- Explain the rotating magnetic field produced by the windings of a three-phase induction motor.
- Determine the slip and rotor speed of a three-phase motor.
- Define rotor frequency and explain its effect on the performance of a motor.
- Explain how torque is developed in a three-phase induction motor.
- List the three factors affecting the torque, regulation, and efficiency of a squirrel-cage motor.
- List the four conditions that cause breakdown torque at heavy loads.
- Explain the operating characteristics of the wound-rotor motor.
- List the five different types of induction motors classified by NEMA.
- Determine the losses and efficiency of the induction motor.
- List the four methods of reduced voltage starting.
- Name the three factors that determine the speed of an induction motor.
- Explain the basic operating principles of the inverter and cycloconverter.

18-1 INTRODUCTION

The induction motor is the most common type of alternating-current motor and is also the simplest in construction. It consists of two main parts: the primary winding, or **stator**, and the secondary winding, or **rotor**. The rotor and stator are not electrically connected; instead, the currents are induced in the rotor, which is why this type of motor is called the induction motor. The induction motor has no commutator or any moving contacts between the rotor and stator. The transfer of power from the stator to the rotor is similar in principle to the transfer from the primary to the secondary of a transformer. However, the rotor of the induction motor is free to rotate and instead of supplying electric power to a stationary circuit, it supplies power to drive a rotating machine. Therefore, the induction motor combines the principles of a motor and a transformer, rotation being produced by the revolving unit following a rotating magnetic field, which is the resultant of two or more alternating magnetic fields set up by the polyphase current flowing through the stator windings.

18-2 CONSTRUCTION

The induction motor is similar in construction to the direct-current shunt motor, the essential difference being that the armature current of the dc motor utilizes brushes, whereas the armature current of the induction motor is an induced, or transformer, current in the rotor windings produced by the alternating field set up by the currents flowing in the stator windings. The primary, or stator, of the induction motor is made up of a stationary frame and a laminated steel core. In one manufacturing process, the armature winding is first wound in slots and properly connected. It is then dipped in varnish, baked, and pressed into its frame. The windings are connected in groups corresponding to the number of phases and poles, one group per phase per pair of poles. The stator is supplied with polyphase alternating currents and a revolving magnetomotive force is produced. This mmf develops a magnetic field revolving at a constant speed, which is referred to as the **synchronous speed** of the motor.

Due to the presence of the air gap in the magnetic circuit, the magnetizing current in the induction motor is much larger than in the transformer, the leakage reactances are larger, and the power factor tends to be quite low. Usually in small and medium-sized motors, the stator slots are partially closed, as shown in Figure 18-1. This helps to reduce the reluctance of the magnetic circuit, the flux density, and the iron losses in the teeth. The main slot insulation is in the form of a trough, which is folded over the slot wires, and the entire contents of individual slots are firmly held in place by means of a slot wedge. In large motors, the slots are generally of the open type, as the comparative reluctance of the flux paths is considerably lower. Large low-voltage motors have stator currents of such a high value that the conductors are in the form of rectangular bars. The main insulation is applied to the conductors, which are then pushed through semiclosed slots.

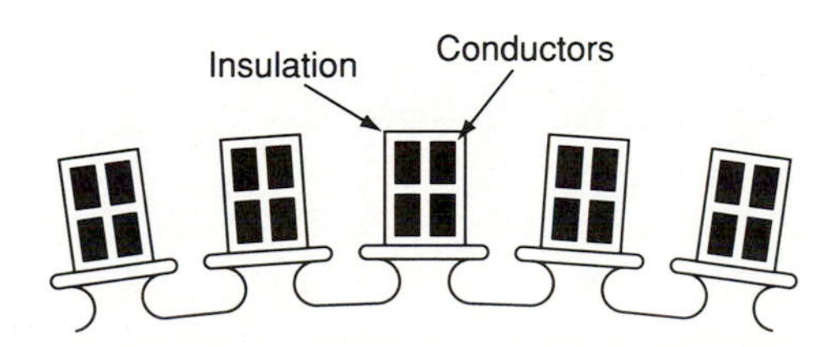

FIGURE 18-1 Stator core of an induction motor with partially closed slots.

The secondary, or rotor, is made in two basic types: the wound rotor and the squirrel-cage rotor. Both designs use a slotted laminated core pressed tightly on a shaft. The slots are not parallel to the shaft axis, but are skewed. This slanting of the rotor slots results in a smoother, quieter operation. The wound rotor consists of a laminated iron core with slots carrying the winding, which must have the same number of poles as the stator winding, but may have a different number of phases. It is usually wound for three phases and the ends of the windings are brought out to slip rings, so that resistances may be inserted in series with the winding for starting and the terminals short-circuited under running conditions.

The squirrel-cage rotor winding consists of a number of heavy copper bars short-circuited at the two ends by two heavy brass rings. In practically all squirrel-cage rotors, the slots are either partially closed, as in Figure 18-2, or totally enclosed. With very small machines of output up to 1 or 2 hp, the bars, end rings, and fan blades can be of aluminum cast in one piece. This type of rotor is virtually indestructible under service conditions.

When the rotor with its closed windings is placed in the revolving magnetic field produced by the stator currents, the flux cuts across the conductors on the rotor and generates emfs in them. Currents flow in the rotor equal to the induced emfs divided by the rotor impedances. These currents reacting on the magnetic field produce torque and the rotor revolves in the direction of the field. At no load, the rotor runs almost as fast as the field, and very small emfs and currents are induced in its conductors. When the motor is loaded, the rotor lags behind the field in speed and large currents are induced and the required torque is developed.

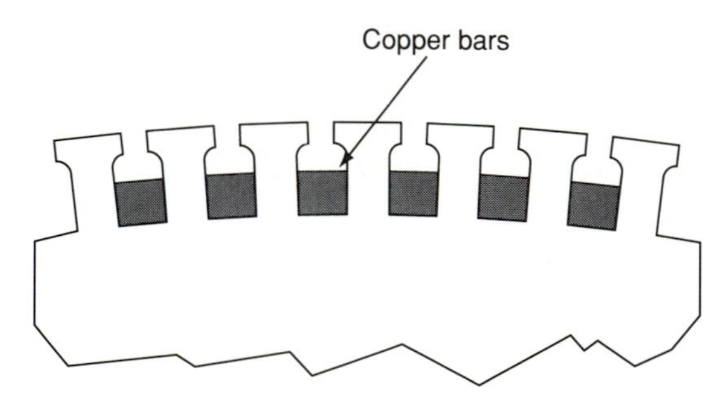

FIGURE 18-2 Rotor of squirrel-cage motor with semiclosed slots.

18-3 ROTATING MAGNETIC FIELD

In the induction motor, the current in the stator's distributed winding produces a magnetic field similar to the field coils on the pole piece of a direct-current machine. The magnetic field will rotate in a polyphase machine due to the action between the magnetic fields set up by two or more phases. When an electric current flows through a conductor, it sets up a magnetic field that rotates around the conductor, as shown in Figure 18-3. When the current is flowing away from the reader, as indicated by the cross is Figure 18-3(a), the magnetic field rotates in a clockwise direction. Consequently, when current flows toward the reader, as in Figure 18-3(b), the direction of the field is counterclockwise. From Figure 18-3 it can be seen that the direction of the magnetic field is downward. If the positions of the two conductors were interchanged, the direction of the magnetic field would be upward.

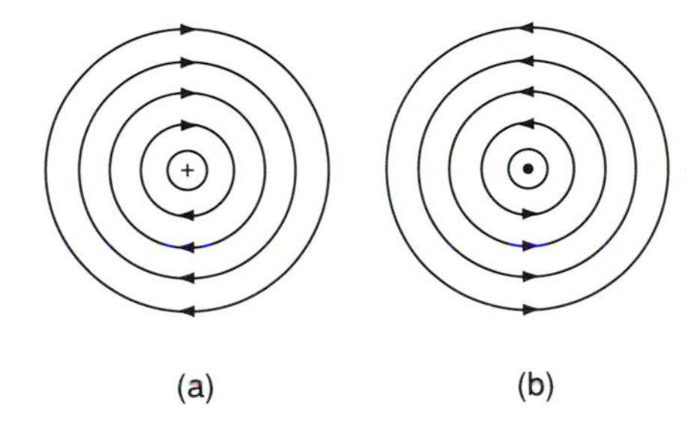

FIGURE 18-3 Magnetic field set up by current flowing in electrical conductors: (a) current flows away from reader; (b) current flows toward reader.

In Figure 18-4(a), the direction of the flux between the two conductors is in the opposite direction. If these conductors are moved close together, the two fields combine and surround the two conductors, as in Figure 18-4(b). It is on this simple principle that the induction motor is based. From the direction of the magnetic fields indicated in Figure 18-3(a) and (b), it can be seen that the current in the two groups of conductors in Figure 18-4(c) will produce a field having a direction as indicated.

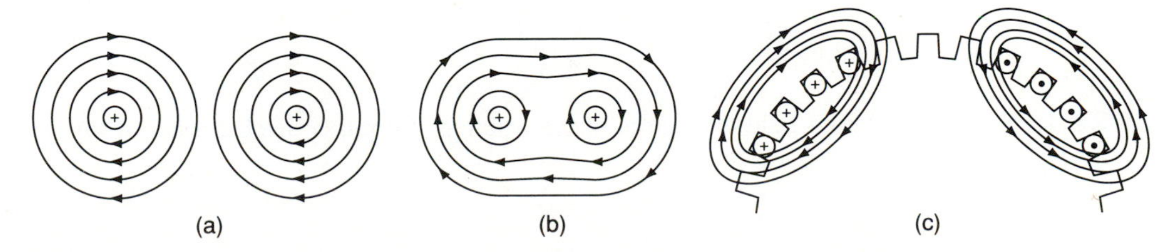

FIGURE 18-4 (a) Two conductors with current flow in the same direction; (b) magnetic fields combine when conductors are placed close together; (c) direction of magnetic field in stator of induction motor.

Figure 18-5 (a) shows a three-phase four-pole single-layer winding induction motor. The three-phase input current waveforms I_A, I_B, and I_C are shown in Figure 18-5(b). These currents, when flowing into the three-phase machine, produce a four-pole rotating field. In the windings marked (+), such as $+B$, the current direction is away from the reader when the waveform of Figure 18-5(b) is in the positive direction. In the windings marked (−), the current flow is toward the reader when the current waveform is negative.

At 0° on the horizontal axis of Figure 18-5(b), the current in phase *A* is zero. Since phase *B* lags by 120° it is negative and equal to 0.866 of the maximum value. Phase *C*, which leads phase *A* by 120°, is positive and also 0.866 of the maximum value. Since no current is flowing in phase *A* at this instant, there is no current in the A conductors of Figure 18-5(a), and the current flow is outward, or toward the reader, in the $+B$ and $-C$ conductors, and inward, or away from the reader, in the $-B$ and $+C$ conductors.

By applying Fleming's left-hand rule to the 0° point of the waveforms, the paths of the fluxes are determined as indicated. North poles are formed at the top and bottom of Figure 18-5(a) and south poles at either side. The arrow points in the direction of the upper north pole.

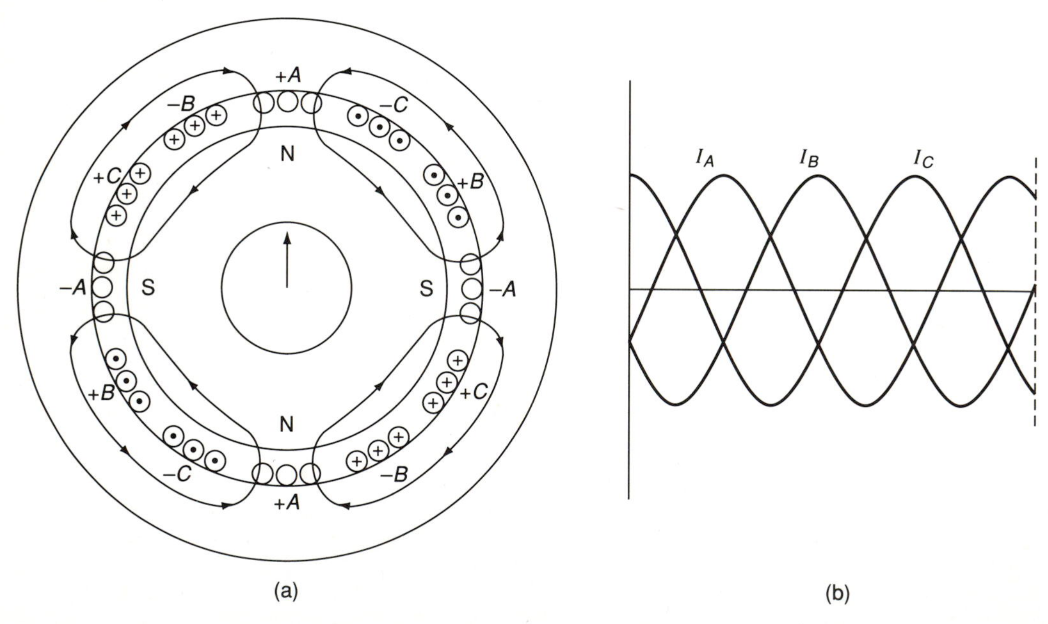

FIGURE 18-5 (a) Rotating field with three-phase winding; (b) phase displacement of three-phase voltages.

After 30° the current in I_A is now positive and equal to 0.5 of its maximum value, while the current I_B is negative and equal to its maximum value, and I_C is positive and equal to 0.5 of its maximum value. At 30° in Figure 18-5(a), the currents in the *B* and *C* conductors are still flowing in the same direction, but the current in the *B* conductors has increased while the current in the *C* windings has decreased. Also, since phase *A* is now carrying current, the currents in the $+A$ conductors are inward and the $-A$ current flow is outward. By applying Fleming's left-hand rule, the direction and path of the fluxes are determined as indicated.

After 30°, all four poles have advanced 15° in a clockwise direction, which is one-half the angle of advance of the current waves. Figure 18-6 shows this 15° advancement, as well as the new path of the flux around the *A* conductors which are energized after 30 electrical degrees. It can therefore be stated that in a four-pole induction machine, the field revolves 90°, or one pole span, as the current I_A changes from a positive maximum to a negative maximum, or 180°. Consequently, the speed of the revolving field is inversely proportional to the number of pairs of poles.

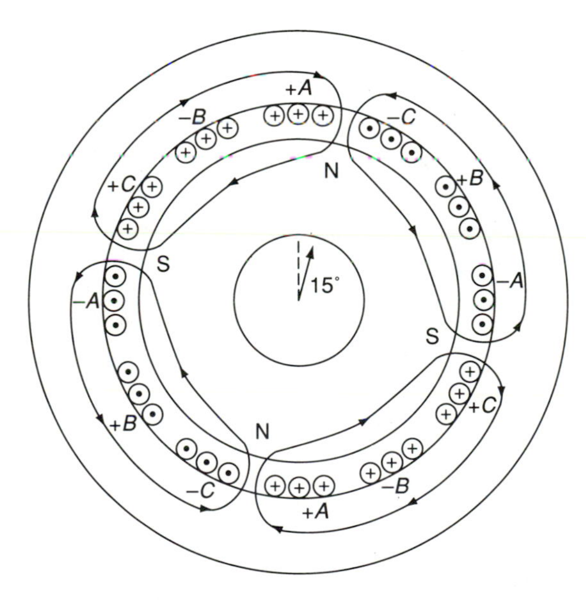

FIGURE 18-6 Effect of advancing poles 15 mechanical degrees.

18-4 SLIP AND ROTOR SPEED

The speed at which the rotating field moves around the stator is called the synchronous speed. The angular speed of an ac rotating field depends on two factors: the frequency of the current and the number of poles for which the motor is wound. The relation among speed, frequency, and poles is given by

$$N_s = \frac{120 \times f}{P} \tag{18-1}$$

where N_s = synchronous speed, in rpm

f = frequency, in hertz

P = number of poles for which the stator is wound

EXAMPLE 18-1 Calculate the synchronous speed of a six-pole induction motor operating at a frequency of 60 Hz.

Solution $$N_s = \frac{120 \times f}{\text{P}} = \frac{120 \times 60}{6} = 1200 \text{ rpm}$$

Equation 18-1 is for the synchronous speed of the revolving field and not the speed of the rotor. The rotor always operates at a slightly lower speed, usually 2 to 5%, depending on the mechanical load on the motor. The difference between the synchronous speed, or the speed of the stator field, and the speed of the rotor is called the **slip** and is expressed as a percent of synchronous speed:

$$\text{percent slip} = \frac{N_s - N_r}{N_s} \times 100 \tag{18-2}$$

where N_s is the synchronous speed, in rpm, and N_r is the rotor speed, in rpm.

EXAMPLE 18-2 A four-pole 60-Hz squirrel-cage induction motor has a full-load speed of 1720 rpm. Determine the percent slip of this motor.

Solution $$N_s = \frac{120 \times f}{P} = \frac{120 \times 60}{4} = 1800 \text{ rpm}$$

$$N_r = 1720 \text{ rpm}$$

$$\% \text{ slip} = \frac{N_s - N_r}{N_s} \times 100 = \frac{1800 - 1720}{1800} \times 100 = 4.4\%$$

Since the percent slip is dependent on the ratio of synchronous speed to rotor speed, it stands to reason that the rotor speed must differ from the synchronous speed by some proportion of the slip. The equation for rotor speed in terms of synchronous speed and slip is as follows:

$$N_r = \frac{120 \times f}{P} \times (1 - s) \qquad (18\text{-}3)$$

where N_r = rotor speed, in rpm

f = frequency of stator, in hertz

P = number of poles

s = slip, in decimal form

18-5 ROTOR FREQUENCY

A very important characteristic of the motor is the rotor frequency and its effect on the performance of the motor. When the rotor of a two-pole 60-Hz induction motor is at a standstill, and voltage is applied to the stator, the flux is cutting each turn in the stator at exactly the same rate as in the rotor. This means that each conductor in both the stator and rotor will be cut by a north pole 60 times a second and by a south pole 60 times a second. Therefore, if the rotor is locked so that it cannot move, the rotor and the stator will have exactly the same frequency. In this situation the motor is acting like a transformer in which the stator winding is the primary and the rotor is the secondary.

If the rotor of a 60-Hz induction motor turns at one-half synchronous speed, or with 50% slip, the relative speeds of the stator flux and the rotor are reduced by one-half. The stator poles will cut across the rotor conductors at half synchronous speed, and the frequency of the rotor currents will be reduced to 30 Hz. If the rotor turns at ¾ synchronous speed, or 25% slip, the stator flux will cut the rotor conductor at ¼ synchronous speed, and the rotor frequency will be ¼ that of the stator frequency. Consequently, if the rotor turns at synchronous speed, the frequency on the rotor will be zero. This implies that during normal operation, the rotor frequency is very low. Furthermore, the rotor frequency varies every time the speed and slip varies. This relationship is expressed by the equation

$$f_r = sf \qquad (18\text{-}4)$$

where f_r = rotor frequency, in hertz

s = slip

f = stator frequency, in hertz

EXAMPLE 18-3 Determine the frequency of the currents in the rotor of a six-pole 60-Hz induction motor if the rotor speed is 1172 rpm.

Solution

$$N_s = \frac{120 \times 60}{6} = 1200 \text{ rpm}$$

$$s = \frac{1200 - 1172}{1200} = 0.023$$

$$f_r = sf = (0.023)(60) = 1.4 \text{ Hz}$$

18-6 TORQUE OF THE INDUCTION MOTOR

When starting an induction motor, the stator winding is connected to the line and a current flows through the winding and produces a magnetic field that as long as the rotor is at a standstill, revolves by the conductors at a speed equal to the synchronous speed of the machine. This sets up a large value of current in the rotor's conductors having a frequency equal to the line frequency. The reactance of the rotor on starting is much higher than the resistance, due to the relatively high frequency. This causes the power factor of the rotor circuit to be low, and the rotor current lags the voltage by a large phase angle. At the instant the motor is started, the percent slip is 100%, or unity, due to the fact that the rotor is at standstill. At this point, a definite torque is developed by the motor. This torque is called the **starting torque**. The starting torque of an induction motor is generally between 1½ and 2½ times the full-load torque, but the starting current may be between 5 and 10 times the full load value.

As the rotor comes up to speed, the current and frequency decrease in the rotor's conductors until a speed is reached where the current in the rotor is just sufficient to produce the necessary torque to carry the load. If the increase in load is too great, the rotor will come to a standstill, or **stall.** The torque developed when the rotor is stalled is referred to as the **breakdown torque**, and is usually between 2½ and 3½ times the rated load torque.

The torque of an induction motor is dependent on the strength of the rotor and stator magnetic fields and the phase relations between them. The equation for calculating the total starting torque of an induction motor is as follows:

$$T_s = K\phi_s I_r \cos\theta_r \tag{18-5}$$

where T_s = total starting torque

K = proportionality constant, depending on the system of units used for the other terms and the dimensions of the rotor

ϕ_s = effective value of stator flux

I_r = effective value of rotor current

$\cos \theta_r$ = rotor power factor

At the instant of starting, the current through the rotor is equal to the induced voltage of the rotor at standstill divided by the impedance of the rotor at standstill. In equation form, the value of rotor current at starting is

$$I_r = \frac{E_{br}}{\sqrt{R_r^2 + X_{br}^2}} \tag{18-6}$$

where E_{br} = effective value of rotor induced emf at standstill

R_r = effective combined rotor resistance at standstill

X_{br} = combined rotor reactance at standstill

If I_r from equation 18-6 is substituted into equation 18-5, the combined equation for starting torque is as follows:

$$T_s = \frac{K\phi_s E_{br} R_r}{R_r^2 + X_{br}^2} \tag{18-7}$$

Torque and current are both linear functions of slip between synchronous speed and full-load speed. For most normal operating conditions, the induction motor operates at synchronous speed with a small value of slip. When torque increases, the speed of the motor decreases slightly, and the slip increases. Consequently, the torque of an induction motor is directly proportional to the slip. The following equation expresses the torque–slip characteristic for any value between no-load and full-load:

$$\frac{T_{\text{nl}}}{T_{\text{fl}}} = \frac{s_{\text{nl}}}{s_{\text{fl}}} \tag{18-8}$$

The flux in a transformer is proportional to the applied voltage. In an induction motor, the flux of the stator is proportional to the voltage applied across the rotor at standstill. If the stator voltage is doubled, the flux doubles, and twice as much rotor voltage is induced, which would double the rotor current. Then, according to equation 18-5, the torque becomes four times as great. Therefore, *the torque of an induction motor varies as the square of the voltages*. From this reasoning, it can be stated that the line voltage, E_L, is directly proportional to the blocked rotor voltage, E_{br}. Also, the effective combined rotor resistance and reactance of a squirrel-cage induction motor is constant and can be combined into a new starting torque constant, K'. Using these symbols, E_L and K', equation 18-7 reduces to

$$T_s = K'E_L^2 \qquad (18\text{-}9)$$

and

$$T_s' = K'(E_L')^2 \qquad (18\text{-}10)$$

where

K' = constant including combined rotor resistance and reactance

E_L = rated line voltage applied to stator winding

T_s' = torque at starting for any value of line voltage except rated

E_L' = any other value of line voltage applied to stator winding

By combining equations 18-9 and 18-10, the following equation is derived:

$$T_s' = T_s\left(\frac{E_L'}{E_L}\right)^2$$

EXAMPLE 18-4 A four-pole 110-hp 550-V three-phase induction motor has a starting torque of 300 ft·lb. When a reduced three-phase voltage of 480 V is applied to the stator, determine the starting torque.

Solution

$$T_s' = T_s\left(\frac{E_L'}{E_L}\right)^2$$

$$= (300\text{ lb-ft})\left(\frac{480\text{ V}}{550\text{ V}}\right)^2$$

$$= 228.5\text{ lb-ft}$$

The basic equation involving torque, speed, and power in English units is

$$\text{hp} = \frac{T \times S}{5252}$$

In the SI system, power is a product of torque, in newton-meters, and speed, in radians per second:

$$W = T \times \omega$$

Since 1 hp = 746 W, to convert torque, in ft·lb, and speed, in rpm, to watts, the equation is

$$W = \frac{T \times S(746)}{5252}$$

Torque, in ft·lb, expressed in terms of watts and rpm, is

$$T = 7.04 \times \frac{W}{S}$$

If RPD is the rotor power developed in watts, and RPO is the mechanical power output of the rotor, the difference between RPO and RPD is the rotational losses of the rotor. Therefore, the equation for determining the net output torque, in newton-meters, is

$$T = \frac{\text{RPO}}{\omega_r}$$

where T = output torque, in newton-meters

RPO = rotor mechanical power output, including rotor losses

ω_r = rotor speed, in radians per second

18-7 SQUIRREL-CAGE MOTOR OPERATING CHARACTERISTICS

The starting torque, starting current, maximum torque, percent regulation, and efficiency of a **squirrel-cage motor** depend on the following three factors:

1. Rotor resistance

2. Air gap between the stator and rotor

3. Shape of the teeth and slots in the stator and rotor

The resistance of the rotor is dependent on the type of material used in its construction and the area of cross section of the end rings. The core, or armature, of the rotor is constructed in a similar manner to that of the dc armature and is usually built up of slotted steel punchings. As mentioned earlier, large motors have windings that are actually copper bars placed in the slots, with their ends connected by conducting rings. These conducting rings are referred to as **end rings**. The copper bars and end rings are either brazed together or the rotor may be placed in a mold and the ends of the copper bars cast in a ring of solid copper.

The squirrel-cage motor is considered to operate at a constant speed. At zero load, the slip is a very small value, and the current that flows in the rotor is just enough to develop the required torque to overcome the no-load losses of the motor. As the load on the rotor increases, a larger value of rotor current is needed to develop the necessary torque in order to carry the increased load. Consequently, the rotating magnetic field must cut the rotor conductors at a greater rate in order to produce the required increase in emf. Therefore, the slip of the rotor must increase so that the rotor speed drops. The squirrel-cage motor has a very low rotor resistance, so a small increase in slip produces a large increase in current. When load is placed on the squirrel-cage motor, the following reactions occur:

1. The torque that is developed by the rotor will continue to increase until it reaches a value that is sufficient to carry the load placed on the motor.

2. The rotor current increases.

3. The speed of the rotor decreases.

4. As the rotor speed decreases, the slip increases, since the stator field is operating at synchronous speed.

5. The induced emf across the rotor increases, due to the rotor bars being cut more frequently.

The variation of torque with slip for a typical squirrel-cage motor is shown in Figure 18-7. From this curve it can be seen that when the percent slip is nearly zero, the rotor is running at its maximum speed. As the load is increased, the torque initially increases almost in proportion to the slip. Therefore, as the percent slip increases, the induced rotor currents increase, and the torque also increases. At larger values of percent slip, the torque increases less rapidly than the percent slip, and eventually the breakdown torque is

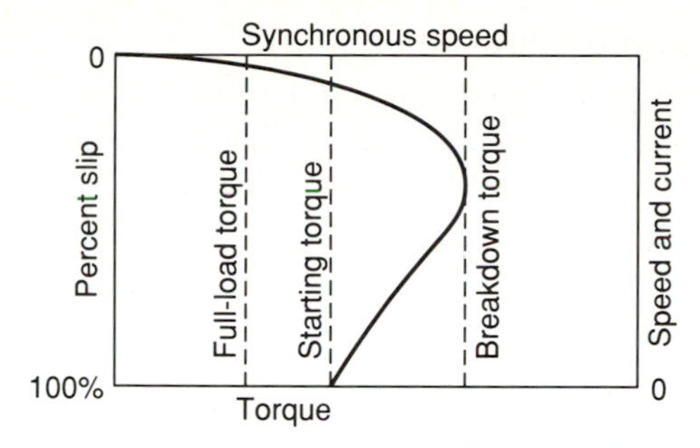

FIGURE 18-7 Induction motor torque–slip characteristic curve.

reached. Because slip is directly proportional to rotor reactance and rotor frequency, an increase in slip beyond pull-out, or breakdown, torque results in an increase in rotor frequency and a decrease in torque. If the reactance of the rotor increases, the power factor of the rotor decreases, causing a net decrease in the torque of the motor. The slip that occurs at breakdown torque is a ratio of the rotor resistance to the blocked rotor reactance. In equation form,

$$s_{br} = \frac{R_r}{X_{br}}$$

There are, essentially, four conditions that in sequence cause breakdown torque to occur at heavy loads:

1. When a heavy load is applied to the rotor, the leakage flux increases, which reduces the mutual flux across the air gap. Consequently, a greater percent slip is required to induce the necessary rotor emf and current to develop the required torque.

2. An increase in slip results in an increase in the rotor frequency and reactance, which causes an increase in the rotor impedance. Because of the increased rotor impedance, additional slip is required to induce a greater value of rotor emf. This increase in rotor emf in necessary to produce the rotor current and torque needed to overcome the increase in rotor impedance.

3. Since rotor resistance remains constant, an increase in the rotor reactance will cause an increase in the rotor power factor. This in turn reduces the torque so that a greater percent slip is required to produce the necessary torque.

4. By applying load to the motor, a point is reached where an increase in slip and rotor current fails to develop additional torque. If the loading of the

motor continues beyond this point, the torque will decrease and the motor stalls. This breakdown torque occurs at a slip well beyond the normal range of operation of a squirrel-cage induction motor. The breakdown torque occurs when the rotor reactance is equal to the rotor resistance.

One disadvantage of the squirrel-cage motor is that, on starting, it requires a large value of current to develop a low power factor. Also, even with this large inrush current the squirrel-cage motor will develop only a moderate starting torque. This low value of starting torque is further diminished when it is necessary to use reduced voltage for starting the motor. For example, if 50% of the rated line voltage is applied across the stator, only 25% of the normal starting torque is obtained. Therefore, the squirrel-cage motor is simply not practical in situations where large starting torque is required. There are, however, many applications for which the squirrel-cage motor is well suited. Since the rotor resistance is low, its operating characteristics for constant-speed applications are excellent. Also, with a small percent slip, the motor has very good speed regulation. It should be noted that the speed of the ordinary squirrel-cage motor is not adjustable.

18-8 DOUBLE-SQUIRREL-CAGE ROTORS

As mentioned in the preceding section, the main disadvantage of the squirrel-cage motor is that it requires a large starting current to develop a relatively low starting torque. If the resistance of the rotor were increased, the starting torque would improve, but the net result would be a reduction in efficiency as well as a decrease in speed regulation. If it is necessary to have a small slip and also a high starting torque, the best solution is a **double-squirrel-cage rotor**. In this motor there are two distinct rotor windings: one having rotor bars close to the surface, and the other with bars located away from the surface.

The two sections of the rotor slots are shown in Figure 18-8. The outer winding, placed near the surface, has a small cross section and a relatively high resistance. The outer winding is usually made of brass. The inner winding has large, low-resistance bars of copper or aluminum alloy. The leakage flux linking the inner winding is much greater than the outer winding, and consequently, this inner winding has a larger amount of self-inductance. The contracted section of the slot, which connects the inner and outer sections, has a relatively low reluctance, allowing it to be traversed by considerable slot-leakage flux. Since the inductance of the inner winding is greater than that of the outer winding, on starting the reactance of inner winding is high. Therefore, most of the starting current flows in the outer, high-resistance low-reactance windings, which gives good starting torque and low starting current. As

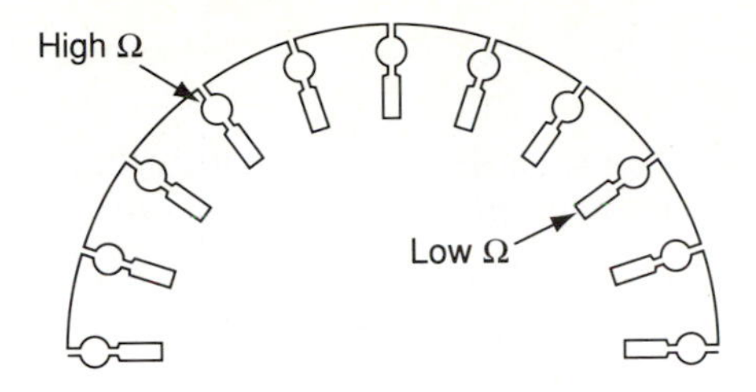

FIGURE 18-8 Double-squirrel-cage rotor.

the rotor gains speed, the rotor frequency decreases and the reactance of both windings decrease. This causes a larger proportion of the total rotor current to flow in the low-resistance winding.

When the motor approaches synchronous speed, the frequency of the rotor becomes very low, the division of current in the two windings is almost the inverse of the winding resistances, and nearly all the rotor current now flows in the low-resistance winding. This results in the motor having the running characteristics of the ordinary squirrel-cage rotor.

18-9 WOUND-ROTOR MOTOR OPERATING CHARACTERISTICS

In the squirrel-cage motor, copper conducting bars are placed around the rotor core. The wound-rotor motor has a rotor that is made up of copper windings, and the rotor core is placed over these windings. The currents are induced in the windings in the same manner as they would be if copper bars were used. The main advantage of using windings is that the conductors can be brought out through slip rings and the resistance of the rotor can be varied. The three-phase rotor winding is wye-connected with the open ends of each winding connected to one of three slip rings.

In an induction motor, the slip for any given value of torque is proportional to the rotor resistance. If resistance is added in the rotor circuit, the slip for any given value of torque will increase. This fact is used to control both the speed and starting torque of the induction motor.

The air-gap flux of the induction motor is practically constant, since terminal voltage and counter emf are also constant. If the resistance of the rotor is increased, the rotor impedance would also increase. As the motor usually operates at low values of slip, the armature reactance is small compared with the armature resistance, and the power factor is almost unity due to the rotor impedance being almost a purely resistive component. If the slip remains at a constant value, the induced emf of the rotor does not change. By applying Ohm's law, the armature current, which is equal to the emf divided by the rotor impedance, decreases due to the increased rotor resistance. Consequently, the torque will decrease.

For the torque to return to its original value, the armature current must increase. This is accomplished by increasing the emf induced across the armature. However, since the air-gap flux is constant, the induced armature emf may only be raised by increasing the rate at which the flux cuts the rotor conductors. Therefore, the slip must increase when resistance is inserted into the rotor circuit.

For maximum torque to occur when starting an induction motor, the rotor resistance must equal the blocked rotor reactance. In equation form,

$$R_r + R_{\text{ex}} = X_{br} \tag{18-11}$$

where R_{ex} is the external resistance to be added to result in maximum starting torque. When the external resistance is adjusted to make the rotor resistance and reactance equal, the full-voltage starting current of the wound-rotor motor can be reduced to a relatively moderate value of about 250%. Also, by varying the external resistance it is possible to obtain a certain degree of variable speed control.

Typical speed–torque curves for different values of external resistance are shown in Figure 18-9. The curve labeled R_r is the speed versus torque curve for the rotor with its slip ring terminals short-circuited. As the rotor resistance is increased, the slip is also increased to develop the same torque. Curve R_2 shows the characteristic curve when the rotor resistance doubled, and consequently, the slip doubled. Curve R_3 shows the performance of the wound-rotor motor with four times the original external resistance applied, resulting in a quadrupling of the slip of the motor. Curve R_4 is the characteristic curve when eight times the external resistance is applied, and eight times the slip occurs. From this family of curves it can be seen that as the rotor resistance is increased, full-load torque is obtained at a greater slip and a lower value of speed. The breakdown torque is not affected by the change in rotor resistance except that it takes place at an increased value of rotor slip. When sufficient resistance is added to the rotor, the maximum torque is obtained at 100% slip.

From the speed–torque curves of Figure 18-9, it is evident that the speed of the motor can be controlled by introducing resistance in the rotor circuit. From the curves it can also be seen that the starting torque increases as the rotor resistance increases, up to the point of maximum breakdown torque, and then decreases. If more resistance is added to the rotor circuit, the resistance of the rotor circuit will be quite high compared to the reactance of the rotor. Therefore, the power factor of the wound-rotor motor is considered to be high when the motor is starting. This is an advantage of the wound-rotor motor compared with the squirrelcage. That is, the wound-rotor motor is capable of

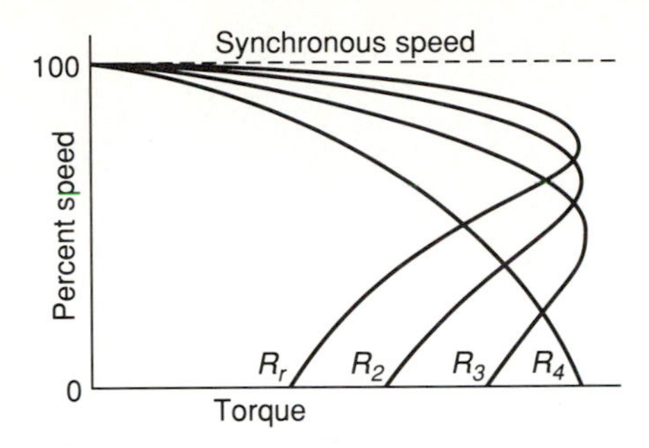

FIGURE 18-9 Speed–torque characteristic curves for different values of external resistance in the rotor circuit.

developing a large starting torque at a high power factor, with less starting current than the squirrel-cage motor. The main disadvantage of the wound-rotor motor is that the speed regulation of this type of motor is not as good as that of the squirrel-cage. Consequently, the wound-rotor motor is considered to have better starting characteristics and poorer running characteristics than these of the squirrel-cage motor.

18-10 INDUCTION MOTOR CLASSIFICATION

By varying the rotor characteristics of induction motors, it is possible to produce a large cross section of speed–torque curves. The National Electrical Manufacturers Association, NEMA, has classified polyphase induction motors in accordance with these rotor characteristics, and identifies the various types by using code letters. There are five types of induction motors classified by NEMA: class A, B, C, D, and F. The speed–torque curves for these five motors are shown in Figure 18-10. The construction of the rotor differs in each classification, which affects the resistance and reactance of each rotor design.

NEMA Class A. These squirrel-cage motors are designed for general-purpose duty, requiring low starting torque. At starting, the rotor has a high resistance value and low reactance, which results in a starting torque of approximately 150% of that of the rated full-load torque. The starting current of a class A motor is 5 to 7 times the rated full-load value. Due to its high starting current, the type of motor is generally used for smaller applications, such as fans, flowers, and centrifugal pumps.

NEMA Class B. Class B motors are also designed for general-purpose applications, and are usually started directly from line voltage. The speed–

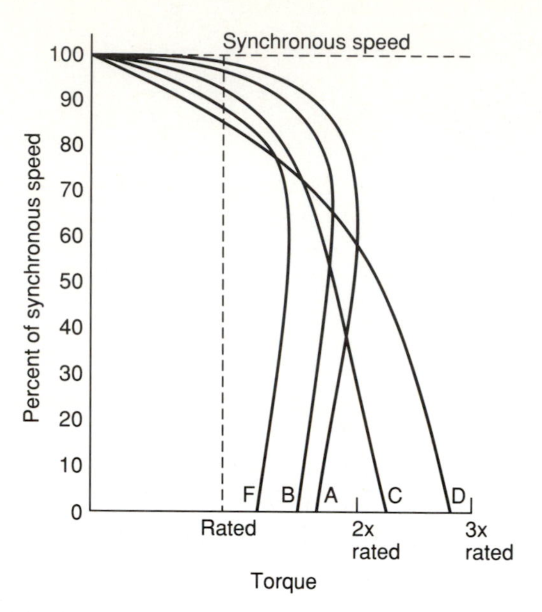

FIGURE 18-10 Speed–torque characteristics for NEMA class A, B, C, D, and F induction motors.

torque curves of class A and B motors are very similar, but have a larger value of reactance on starting. The starting current for the class B motor is approximately 4½ to 5 times the full-load current. The starting torques of class A and B motors are virtually identical. Due to its lower starting current, the class B motor is found in many applications that the class A motor was once used for. Its principal areas of use are in those mentioned for the class A, as well as machine tools, motor-generator sets, and others.

NEMA Class C. These motors are the double-squirrel-cage design and are usually started by applying full voltage. Compared to classes A and B, this type of motor develops a higher starting torque, between 2 and 2½ times rated torque, as well as a lower starting current, from 3½ to 5 times rated value. Frequent applications are reciprocating pumps, crushers, mixers, air compressors, refrigerating machines, and conveyor equipment.

NEMA Class D. This type of motor has a very high starting torque, over 275% of rated torque, as well as a high value of rotor resistance. The efficiency of the class D motor is lower than that of the preceding three classes, but it does have a relatively low starting current at 4 to 5 times the rated value. The high starting torque and low starting current make this motor ideal for heavy starting duty, although it is not recommended for frequent starting due to its poor thermal dissipating ability. Usually, the class D motor is found in equipment such as bulldozers, punch presses, small hoists, flywheel machinery, and foundry equipment.

NEMA Class F. The class F motor is also a double-squirrel-cage motor, except that it has a relatively low-torque capability. The rotor resistance is above the value that provides maximum torque at starting, so the starting torque is approximately 125% of the full-load torque. This type of motor has the lowest starting current, at 2 to 4 times the rated value. It is generally used where starting current limitations are severe and both starting and maximum torque requirements are low, and in equipment with high-inertia loads. Some applications of the class F motors are large flywheels, punch presses, shears, fans, centrifugal pumps, and compressors.

18-11 LOSSES AND EFFICIENCY OF THE INDUCTION MOTOR

The calculation of the efficiency of an induction motor is done in basically the same manner as it is for the transformer. That is,

$$\text{percent efficiency} = \frac{P_{\text{input}} - \text{losses}}{P_{\text{input}}} \times 100 \qquad (18\text{-}12)$$

The percent efficiency could also be found in terms of the output power by using equation 17-28. However, equation 18-12 is better suited for the percent efficiency of an induction motor since the input quantities are usually easier to obtain. The input power to an induction motor is in the form of three-phase currents and voltages. The same losses occur in the induction motor as in the transformer. These losses include copper losses (I^2R losses in the stator) and core losses (I^2R losses due to hysteresis and eddy currents in the stator). Also, there are copper, friction, and windage losses in the rotor.

Another loss that should be taken into account is the air-gap power, P_g. This is the power transferred by the air-gap magnetic field from the stator windings to the rotor. Neglecting the stator core losses, the air-gap power is determined by subtracting the stator copper loss (SCL) from the power input to the stator. In equation form

$$P_g = P_{\text{in}} - \text{SCL}$$

The efficiency of an induction motor may be determined by performing the following four tests:

1. *No-load test.* The no-load, or open-circuit, test determines the rotational losses of the motor. The induction motor is connected to its rated line voltage and is run with no load connected to its shaft. Since there is no power output, the power supplied to the stator supplies the copper, core,

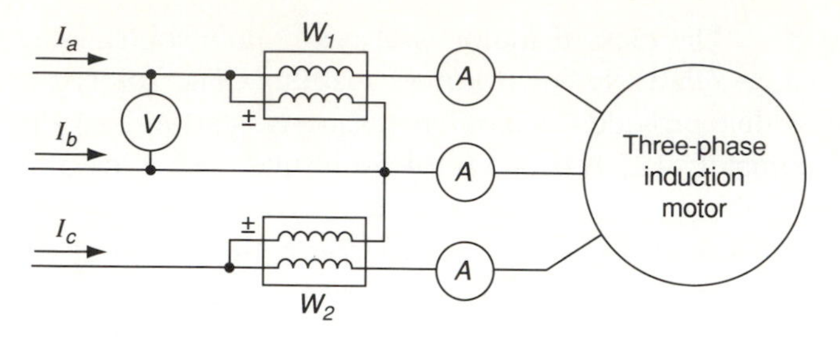

FIGURE 18-11 Circuit for no-load test of an induction motor.

friction, and windage losses in the rotor. This test is the equivalent of the open-circuit test in transformer testing. The circuit for no-load testing is shown in Figure 18-11. The motor is generally assumed to have a wye-connected stator winding. When the two-wattmeter method is used to measure the power input, the total power is the difference between the wattmeter readings, since the no-load power factor is less than 0.5. The difference between the total power and the I^2R losses of the stator is the rotational losses. In equation form,

$$P_{\text{ROT}} = \sqrt{3}\, E_L I_s \cos\theta - 3I_s^2 R_s \tag{18-13}$$

where I_s is the ac current through a stator winding and R_s is the ac resistance of a stator winding.

2. *Load test.* The load test uses the same test circuit as that shown in Figure 18-11. The motor is connected to its normal load, and power, current, and power factor are measured. In the load test, the input power is the sum of the two wattmeters.

3. *Dc stator resistance test.* The dc stator resistance test, shown in Figure 18-12, is a simple Ohm's law test performed on the stator windings to reveal the dc stator resistance. The dc voltage and current are obtained from the meter readings. Since two stator windings are being measured in this test, it is necessary to divide the result of the Ohm's law calculation by 2.

$$R_{\text{dc}} = \frac{V_{\text{dc}}}{2I_{\text{dc}}} \tag{18-14}$$

This test is to be performed after the stator windings have warmed up to operating temperature. To convert this dc resistance to an effective ac resistance, a multiplier of 1.25 is used to compensate for temperature and skin effect.

$$R_s = 1.25 \times R_{dc} \tag{18-15}$$

where R_s is the stator resistance.

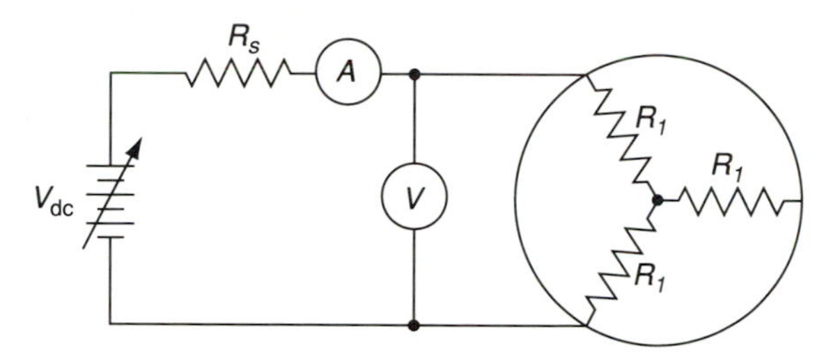

FIGURE 18-12 Circuit for dc resistance test.

4. *Blocked-rotor test.* The fourth test that can be performed on an induction motor to determine its circuit parameters is the blocked rotor test. This test is performed with the rotor mechanically locked against rotation, and a low voltage of between 10 and 20% of rated value is applied to the stator. This test corresponds to the short-circuit test on a transformer, and the wiring diagram of Figure 18-11 is used. Extreme caution should be taken when performing the blocked rotor test since the motor temperature will rise very quickly unless the meter readings are obtained quickly and the value of current is low. Also, if the rotor is not properly blocked, there is the danger that whatever is used to prevent the rotor from turning may become loose, and possibly cause damage or injury. Because the blocked rotor test and the transformer test are the same in principle, the same equations for equivalent impedance, reactance, and resistance apply. Assuming a wye connection, we have

$$Z_{es} = \frac{V_{br}/\sqrt{3}}{I_{br}} \tag{18-16}$$

$$R_{es} = \frac{P_{br}}{3 \times I_{br}^2} \tag{18-17}$$

$$X_{es} = \sqrt{Z_{es}^2 - R_{es}^2} \tag{18-18}$$

where Z_{es} = equivalent impedance reflected to the stator per phase, in ohms

V_{br} = line-to-line voltage across stator terminals with rotor blocked

I_{br} = stator current with rotor blocked

R_{es} = equivalent resistance reflected to the stator per phase, in ohms

P_{br} = total power input with rotor blocked

X_{es} = equivalent reactance reflected to the stator per phase, in ohms

The rotor resistance per phase in terms of the stator can now be obtained.

$$R_{er} = R_{es} - R_s \tag{18-19}$$

and

$$X_{br} = X_s = \frac{X_{es}}{2} \tag{18-20}$$

where R_{er} = resistance of the rotor per phase reflected to the stator

X_{br} = inductive reactance of the rotor under blocked rotor conditions

X_s = inductive reactance of the stator

EXAMPLE 18-5 A three-phase induction motor was subjected to a no-load, dc stator resistance, and load test. The following data were obtained from these three tests:

No-load test: $E_{nl} = 240$ V, $P_1 = 850$ W, $P_2 = -375$ W, $I_{nl} = 15$ A

dc stator resistance per phase = 0.08 Ω

Load test: $E_L = 240$ V, $P_1 = 3805$ W, $P_2 = 2210$ W, $I_L = 41.5$ A

Determine the percent efficiency of this motor.

Solution $R_s = 1.25 \times R_{dc} = 1.25 \times 0.08 = 0.1\ \Omega$

No-load copper loss $= I_{nl}^2 R_s = (15)^2(0.1) = 22.5$ W

$$\begin{aligned}\text{Rotational losses} &= P_{total} - P_{Cu}\\ &= (850\text{ W} - 375\text{ W}) - 22.5\text{ W}\\ &= 452.5\text{ W}\end{aligned}$$

$$\%\ \text{Efficiency} = \frac{P_{input} - \text{losses}}{P_{input}} \times 100$$

$$= \frac{(3805\text{ W} + 2210\text{ W}) - 452.5\text{ W}}{3805\text{ W} + 2210\text{ W}}$$

$$= 92.48\%$$

EXAMPLE 18-6 A 480-V 40-hp three-phase induction motor has the following data obtained from various testing:

Blocked-rotor test: wattmeter 1 = 1400 W
wattmeter 2 = 800 W
$I_{br} = 37\text{ A}, E_L = 90\text{ V}$

Load test: wattmeter 1 = 9800 W
wattmeter 2 = 8450 W
$I_L = 37\text{ A}, E_L = 480\text{ V}$

No-load test: wattmeter 1 = 1850 W
wattmeter 2 = −1200 W
$I_{\text{nl}} = 14.2\text{ A}, E_{\text{nl}} = 480\text{ V}$

Determine the percent efficiency.

Solution Full-load copper losses are obtained by the blocked rotor test.

$$P_{\text{Cu}} = W_1 + W_2 = 1400\text{ W} + 800\text{ W} = 2200\text{ W}$$

The equivalent resistance can now be determined.

$$R_{es} = \frac{P_{br}}{3 \times I_{br}^2} = \frac{2200\text{ W}}{3(37\text{ A})^2} = 0.536\ \Omega$$

Next, the no-load copper loss is calculated.

$$P_{\text{Cu0}} = I_{\text{nl}}^2 R_s = (14.2\text{ A})^2(0.536\ \Omega)(1.25) = 135.1\text{ W}$$

The no-load total power input is the difference between the two wattmeter readings.

$$P_{\text{in0}} = W_1 - W_2 = 1850\text{ W} - 1200\text{ W} = 650\text{ W}$$

The total no-load power input minus the no-load copper losses results in the core, friction, and windage losses.

$$P_{\text{core}} = 650\text{ W} - 135.1\text{ W} = 514.9\text{ W}$$

The total full-load losses are the sum of the core losses plus the full-load copper losses.

$$P_{\text{losses}} = P_{\text{Cu}} + P_{\text{core}} = 2200\text{ W} + 514.9\text{ W} = 2714.9\text{ W}$$

The percent efficiency can now be calculated.

$$\% \text{ Efficiency} = \frac{P_{in} - \text{losses}}{P_{in}} \times 100$$
$$= \frac{(9800\text{ W} + 8450\text{ W}) - 2714.9\text{ W}}{9800\text{ W} + 8450\text{ W}} \times 100$$
$$= 85.12\%$$

18-12 CIRCLE DIAGRAM OF THE INDUCTION MOTOR

In Section 18-11 it was shown that the parameters of small induction motors are readily determined by performing load tests. However, these tests are considerably more complicated and expensive when they are performed on large polyphase motors. Consequently, it is more advantageous to obtain the performance characteristics of such a motor without having to perform a load test on the machine. By using the no-load, stator resistance, and blocked rotor test, sufficient data can be obtained to construct a **circle diagram**. By constructing a circle diagram it is possible to determine the input, output, slip, torque, power factor, and efficiency for any value of assumed stator current.

The three-phase motor is always considered to be wye-connected when constructing a circle diagram, and the values of voltage, current, and power per phase should be used. Figure 18-13 shows the circle diagram of a three-phase induction motor. The following procedure is used to plot this type of diagram:

1. The first line to be plotted is the rated voltage, E_p, per phase, which is obtained by dividing the line voltage by $\sqrt{3}$.
2. The next line to plot is the no-load current per phase, OI_e, which is obtained from the no-load test. A convenient scale should be chosen; otherwise, the circle diagram will be extremely large. The power factor angle between I_e and E_p is determined by dividing the total power obtained by the two wattmeters in the no-load test by $\sqrt{3}\, E_L I_e$. Therefore,

$$\theta_0 = \frac{P_{in}}{\sqrt{3}\, E_L I_e}$$

3. Next, the blocked rotor current, OI_B, is drawn to the same scale as the no-load current. OI_B lags E_p by the blocked rotor power factor angle θ_B.
4. From the tip of I_e, draw a line out to point K. This line will be parallel to the X-axis.
5. Also, from the tip of I_e to the tip of I_B, draw another line, I_2B. Measure I_2B to find its magnitude in amperes.
6. I_e and I_B are two points on a circle. Since I_2B has been measured, one-half

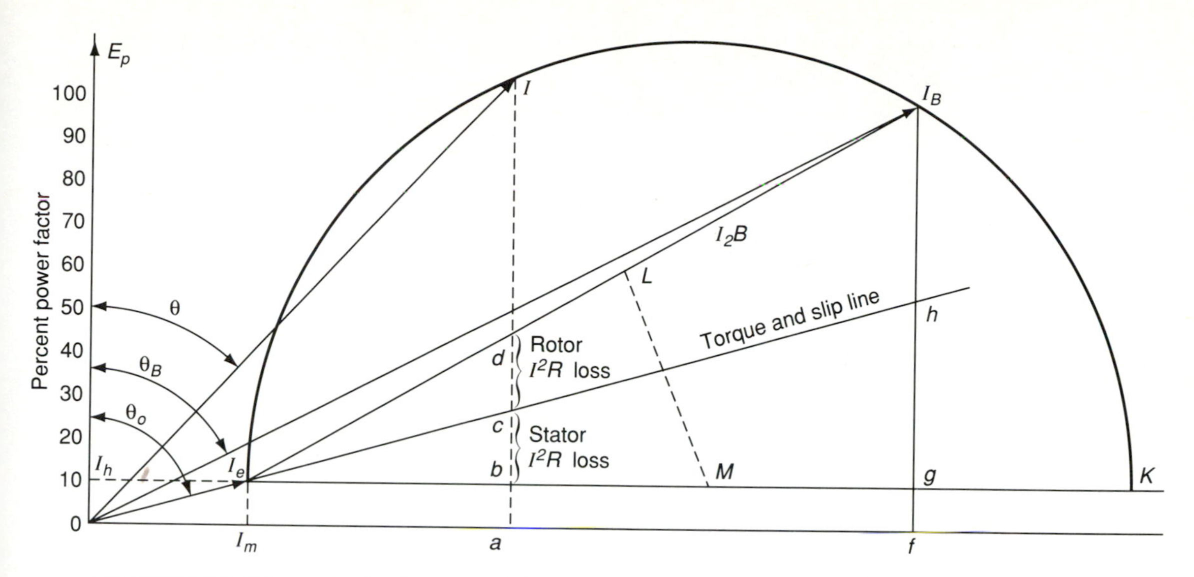

FIGURE 18-13 Circle diagram of three-phase induction motor.

its length is point L. The center of the circle, M, is found by drawing a perpendicular line, LM, from the center of I_2B. With I_eM as a radius and M as a center, the semicircle I_eI_BK is drawn.

7. Draw a line from point f to I_B. This line represents the component of phase current which is in phase with voltage E_p when rated voltage is applied during the blocked-rotor test. Its length represents, and is proportional to, the power input per phase. Since the rotor is blocked, the power output is zero, and fI_B represents the losses per phase of the motor. fg is equal to the no-load losses. The position of point h is found by dividing the stator I^2R losses per phase (with the rotor blocked) by the voltage per phase; or

$$gh = \frac{I_{br}^2 R_s}{E_p}$$

8. With point h determined, a line is now drawn from the tip of I_e through point h. This is the torque and slip line.

9. If the rated line current of the motor is known, the value of this current is changed to the scale of the drawing, and is plotted by scribing an arc, centered at point O. The point where the arc intersects the semicircle is labeled I. A line is now drawn from O to I.

10. Finally, a dashed line is drawn straight down from I to a, and the points of intersection with the other lines are labeled d, c, and b, respectively.

From the circle diagram, the following data are obtained by measuring the scaled lengths of lines:

$$\text{Total power input, watts} = aI \times 3E_p$$

$$\text{Rotational losses, watts} = ab \times 3E_p$$

$$\text{Stator copper loss, watts} = bc \times 3E_p$$

$$\text{Rotor copper loss, watts} = cd \times 3E_p$$

$$\text{Output power, watts} = dI \times 3E_p$$

$$\text{Output horsepower} = \frac{dI \times 3E_p}{746}$$

$$\text{Output torque, newton-meters} = \frac{cI \times 3E_p}{\omega_s}$$

$$\text{Starting torque, newton-meters} = \frac{hI \times 3E_p}{\omega_s}$$

$$\text{Slip} = \frac{cd}{cI}$$

$$\text{Speed} = N \times \frac{dI}{cI}$$

$$\text{Efficiency} = \frac{di}{aI}$$

$$\text{Power factor} = \frac{aI}{OI}$$

18-13 STARTING INDUCTION MOTORS

Very small induction motors, up to 10 hp, may be started by connecting them directly across the source of supply. However, motors of large ratings should not be started in this manner, since the large starting current at low power factor may disturb the voltage regulation of the system. When a motor is connected to the system directly, it uses what is known as an **across-the-line starter**. This method of starting a motor can be accomplished simply by using a hand-operated three-pole switch. The more common across-the-line starter uses a magnetic switch, or contactor, which can be operated from one or more remote stations by pushbuttons. The connections of an across-the-line starter, adapted to either a squirrel-cage or a wound-rotor motor, are shown in Figure 18-14. When the start button is pushed, operating coil M is energized, contacts M_1, M_2, and M_3 close. Contact M_4, which is the maintaining contact, also closes, so that when the normally open start button is depressed and released, the circuit remains energized through contact M_4. When the stop button is pressed, the operating coil, M, is deenergized and the normally open contacts return to their open positions, shutting off power to the induction motor.

The fuses shown, F_1, F_2, and F_3, are generally selected with a rating of 300% of the full-load motor current, although the inrush current may be much higher than this value. The three overload relays, marked OL on Figure 18-14, are connected in series with the operating coil M and are usually thermally actuated. The thermal overload relays are set to protect the motor against continuous overload but are not affected by the large starting current or by any brief overloading of the motor. A thermal overload element, or heater, is connected in series with each of the motor leads. If the motor overheats, the thermal element opens, which in turn causes one of the overload contacts to open, which results in power being interrupted in the control circuit. The selection of the rating of disconnect switches, fuses, and overload protection is given by the National Electrical Code. There are, essentially, four methods of reduced voltage starting: (1) the compensator method, (2) the primary resistor method, (3) the wye–delta method, and (4) the part-winding method.

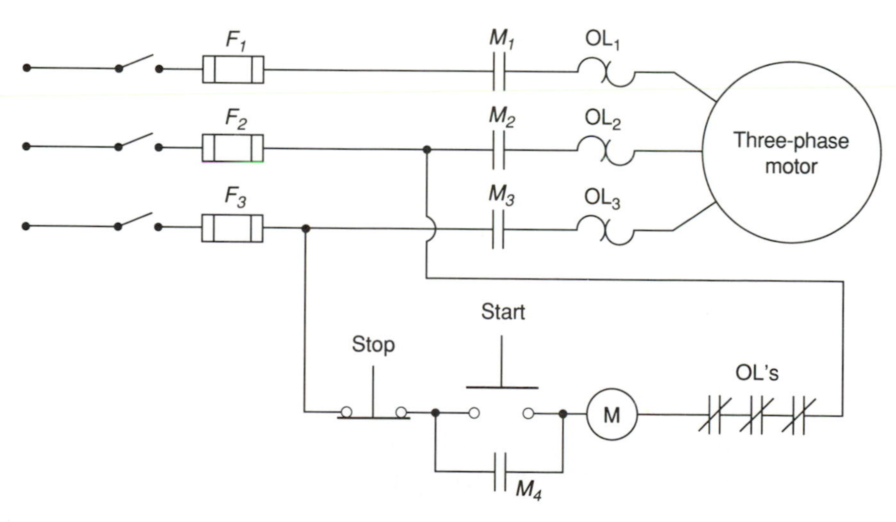

FIGURE 18-14 Across-the-line starter for a three-phase induction motor.

Compensator, or autotransformer, starting. Figure 18-15 shows two delta-connected autotransformers connected to a three-phase motor. The taps on the transformers vary from 50 to 80% of the rated voltage. In the starting position, the three S_2 contacts are closed, and the motor starting current is reduced due to the transformation ratio of the autotransformer. When the motor is being brought up to speed, different values of taps are selected. Once in the run position, the start contacts open and the run contacts close, thereby eliminating the autotransformers from the circuit. The circuit shown in Figure 18-15 illustrates a closed transition method of starting, since the voltage is uninterrupted between the start and run modes.

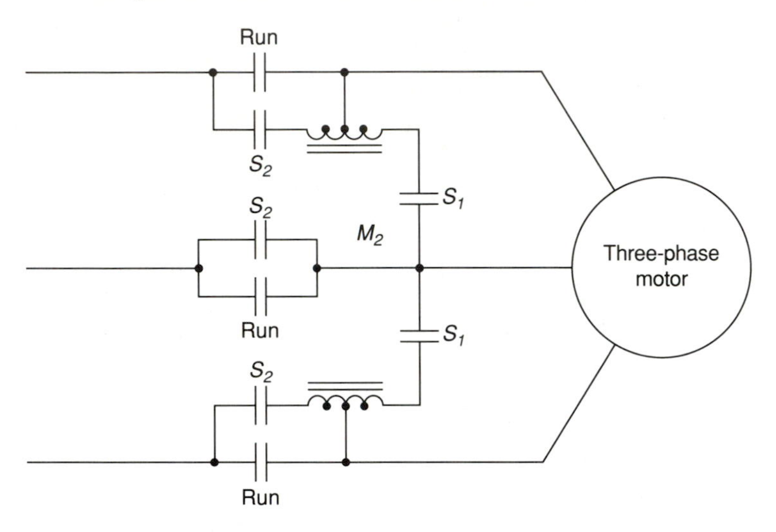

FIGURE 18-15 Autotransformer starting of three-phase motor.

Primary resistor starting. When resistance is inserted in series with the line conductors, it is equivalent to the procedure used to start dc motors. By inserting resistance in the primary, or stator, winding, the motor starting speed and current are reduced. Figure 18-16 shows a pushbutton stop/start as well as a time-delay run wiring diagram. The motor is started in the same manner as an across-the-line starter. The start button energizes the operating coil, M, which causes the four M contacts to close. Resistors R_1, R_2, and R_3 are now in series with the stator windings of the motor. After a certain amount of time elapses, time-delay relay T_R times out and seals in the three T_R contacts, effectively removing the line resistance. When the stop button is pressed, both coil M and coil T_R deenergize, which causes the M contacts to open and the T_R contacts to return to their normally open positions.

Wye–delta method of starting. This method of starting is used when the motor is designed for delta operation at its rated voltage. When the stator windings are wye-connected, the voltage across each winding is $1/\sqrt{3}$, or 57.7%, of the rated voltage. Since the voltage across the windings is only 57.7% of its rated value, the current in each phase is only 57.7% of what it

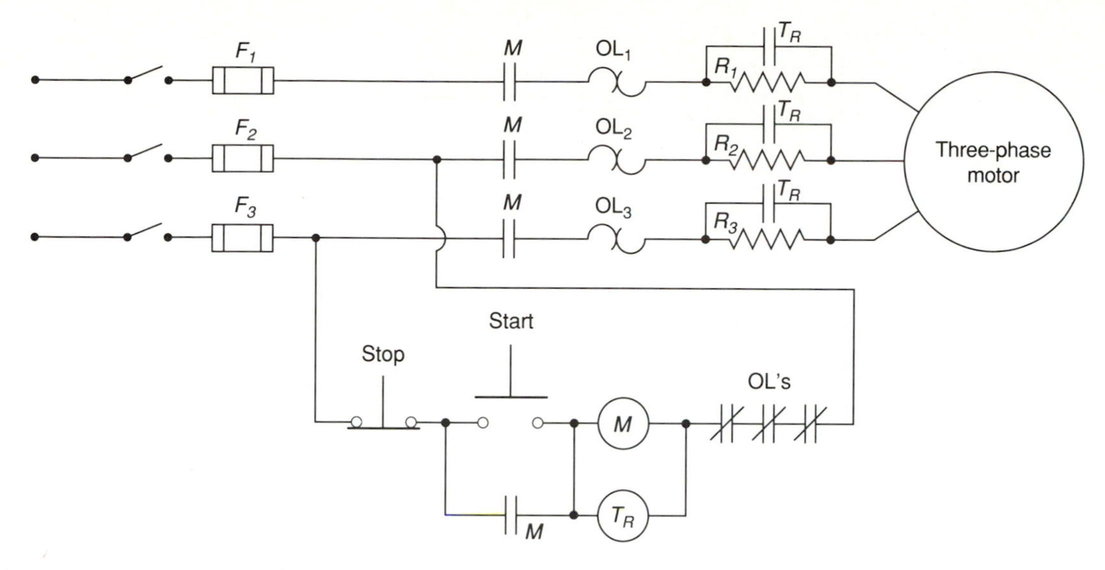

FIGURE 18-16 Primary resistor starter for reduced-voltage starting.

would be if full line voltage were applied. Also, the starting torque varies as the square of the applied voltage. Therefore, the starting torque is reduced to 0.577^2, or 33% of that at normal voltage.

The wiring diagram shown in Figure 18-17 is that of a wye–delta automatic starter. When the start button is pressed, operating coil CR is energized, which closes the four CR contacts. The time-delay relay T_R is now energized and begins its timing cycle. Coils *M* and *S* are simultaneously energized and seal in the three *M* contacts and the two *S* contacts, respectively. The windings marked *a′*, *b′*, and *c′* are now connected together to form a *Y*. After a specified period of time, T_R activates the normally open and normally closed T_R contacts. The n/c contact in series with coil *S* opens, which causes the two *S* contacts to open, and the n/o contact that is in series with coil *R* closes. This action closes the three *R* contacts, and the windings are now connected in delta and full-load current flows through the windings.

Part-winding starting. Squirrel-cage induction motors are often designed with two identical windings, which can be series connected for high-voltage systems, or connected in parallel for lower-voltage systems. These motors, sometimes referred to as **dual-voltage motors**, have nine leads brought out of the motor and are labeled as shown in Figure 18-18. This type of motor may have a dual-voltage rating, such as 220/440 V. If the dual-voltage motor is connected for low-voltage operation, it is possible to use part-winding starting. Part-winding starters are not suitable for use with delta-wound dual-voltage motors.

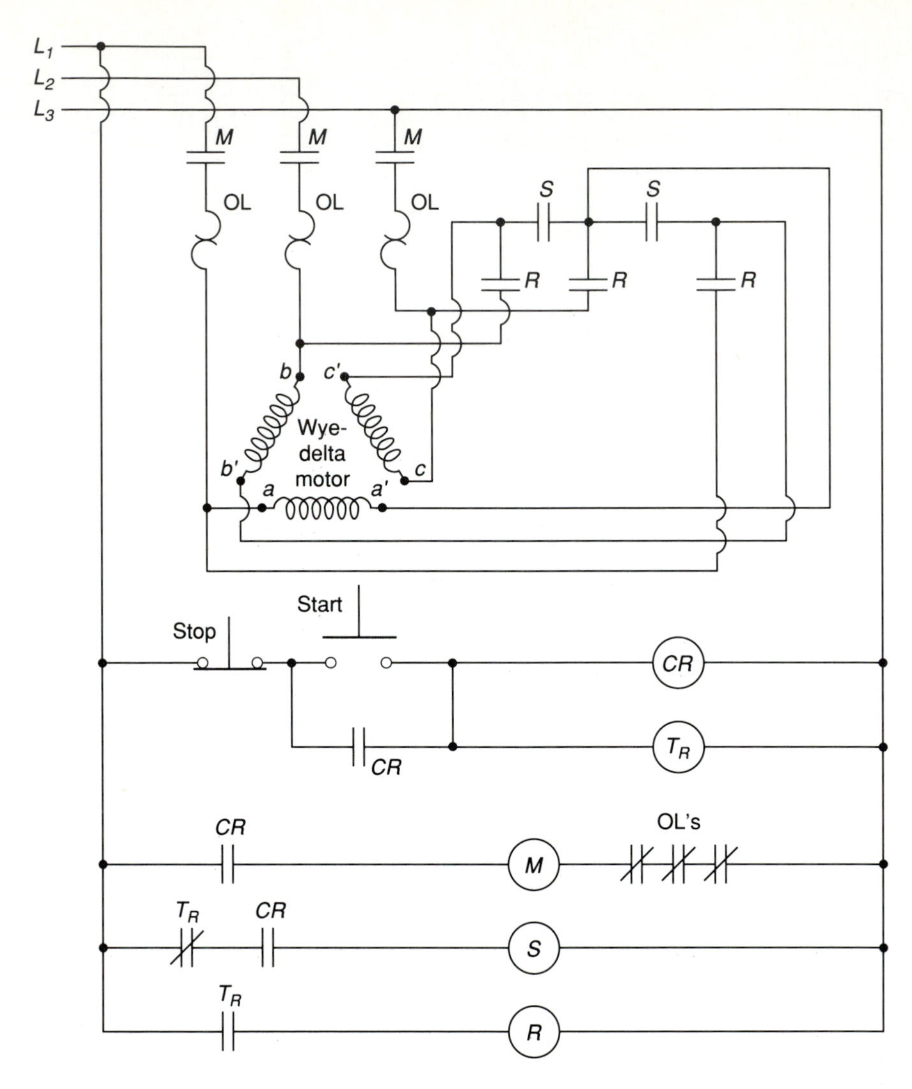

FIGURE 18-17 Wiring diagram for wye–delta motor starting.

Figure 18-19 shows a two-step part-winding starting method. As shown in the diagram, terminals T_4, T_5, and T_6 are permanently joined together to form a second wye connection out of the windings. When the motor is started, the S contacts close, and the wye-connected windings, $T_1 - T_4$, $T_2 - T_5$, and $T_3 - T_6$, are connected to the three-phase source. After a time delay of approximately 5 s, the three R contacts also close, placing the two wye-connected windings in parallel. When this type of starting is used, the inrush current is limited to about 65% of the normal full-voltage starting value. The main disadvantage of the part-winding starter is that the starting torque is poor.

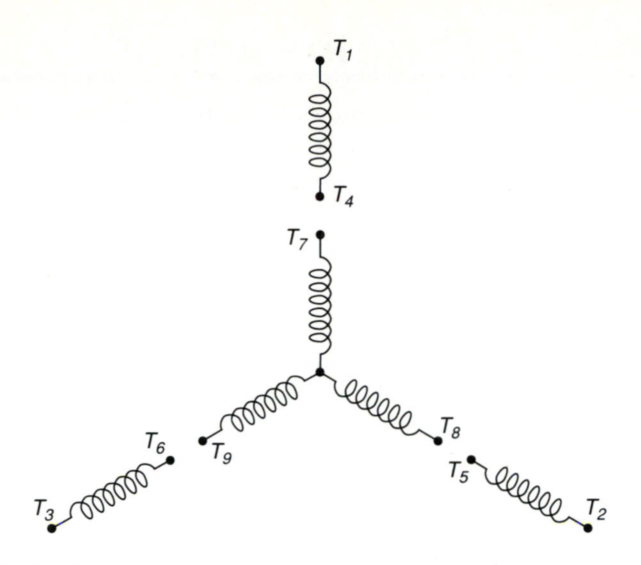

FIGURE 18-18 Terminal markings and connections for dual-voltage motor.

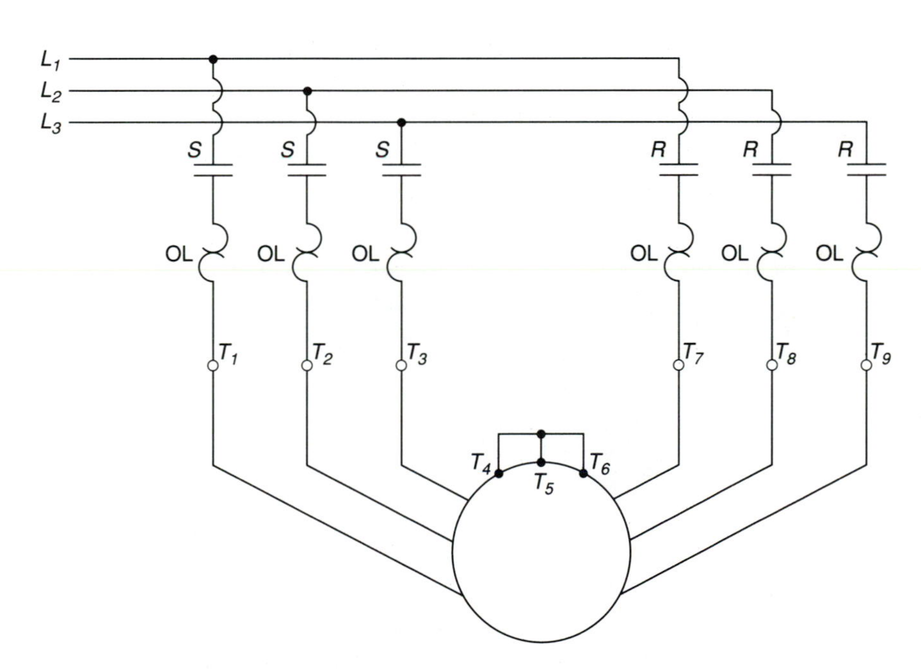

FIGURE 18-19 Two-step part-winding starting method.

18-14 SPEED CONTROL OF INDUCTION MOTORS

The speed of the rotor of an induction motor is found by equation 18-3:

$$N = \frac{f \times 120}{P} \times (1 - s)$$

The three factors in the equation above, frequency, slip, and the number of poles, determine the speed of an induction motor. Any one or any combination of these factors may be changed to affect the speed of the induction motor. The following methods are a few of the more popular means employed to control the speed of induction motors.

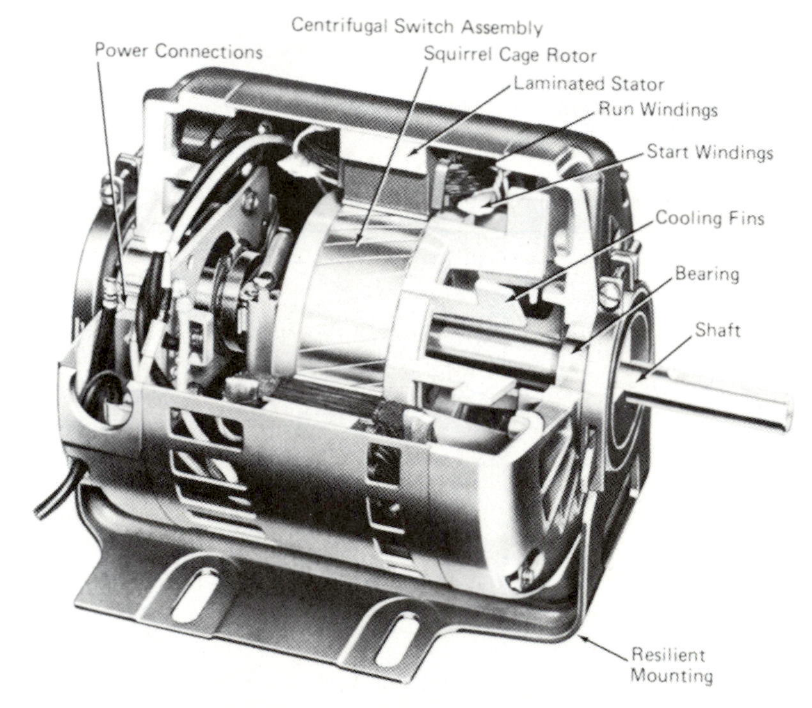

Cutaway view of an alternating-current induction motor

Source: Prentice Hall Inc. / Marathon Electric

Wound-rotor method. This method involves changing the slip and can only be used on wound-rotor motors. This type of control corresponds to the use of resistance in series with the armature of a dc shunt motor. The disadvantages of this method are the decrease in efficiency, due to the additional I^2R losses in the rotor circuits, and the poor voltage regulation obtained. These disadvantages can be overcome by using counter emf instead of resistance in the rotor circuit. This can be done by using a commutator on the rotor, which due to the high inherent cost is feasible only on very large machines.

Changing the number of poles. With a specially designed and suitably connected stator winding, it is possible to change the number of stator poles and, consequently, the synchronous speed of the induction motor. If the number of poles in a 60-Hz machine changes from eight to four, the synchronous speed is changed from 900 rpm to 1800 rpm. When this is done, the best possible design is not usually obtainable at both speeds. For example, it is impossible for a motor to have both high power factor and high efficiency at two different speeds simply by changing the number of poles. Since this method of speed control is quite costly, due to the complexity of the switching arrangement required to change poles in a motor, this type of speed control is not popular when more than two speeds are required.

Concatenation method. It is possible to obtain several operating speeds by connecting two induction motors in tandem. This connection requires one of the motors to be a wound-rotor motor, while the other can be either a squirrel-cage or a wound rotor. The stators of both motors should be wound for the same voltage. The stator winding of the wound-rotor motor is connected to the source, and its rotor winding is connected to the stator wind of the second motor. The rotor shafts are directly coupled or rigidly connected using gears. The rotor winding of the wound-rotor motor may be connected to the stator winding of the second motor in such a manner that the motors both turn in the same direction. The resulting speed will be determined by the sum of the poles in both machines. When both motors turn in the same direction, this is known as **direct concatenation**. When the motors are connected so that they turn in opposite directions, the resulting speed is determined by the difference in the number of poles of both machines. This method is known as **differential concatenation**.

EXAMPLE 18-7 Two 60-Hz motors are connected in direct concatenation. The wound-rotor motor has four poles, and the second motor also has four poles. Calculate the speed of the rotors if the slip is 0.105.

Solution Number of poles = 4 + 4 = 8

$$N_r = \frac{f \times 120}{8}(1 - s) = \frac{60 \times 120}{8}(1 - 0.105)$$
$$= 805.5 \text{ rpm}$$

Change of frequency. This method is very popular due to recent advances in solid-state circuitry. By changing the frequency of the stator, the speed of the magnetic field will change. Consequently, the no-load point on the motor's speed–torque curve will move. To prevent the magnetizing current from becoming too large in value, the line voltage must be reduced in direct proportion to the line frequency.

18-15 SOLID-STATE VARIABLE-SPEED DRIVES

The most modern solid-state variable-frequency drives are self-contained portable units which are relatively inexpensive and highly reliable. The original variable-frequency drives required one or more ac generators to supply power to a synchronous or induction motor. The variations in motor speed would be accomplished by varying the generator frequency. This method of control was, and still is, particularly adaptable to the electric propulsion of ships, in which a high-voltage, high-speed alternator is used to drive a slow-speed propeller though a slow-speed synchronous motor. The ac generator provides the proportional decrease in both the stator frequency and voltage of the induction motor. The main disadvantage of this type of frequency variation is that it is very expensive.

There are two basic methods of producing a high-power, variable-frequency source for controlling the speed of an ac induction motor. These two methods are:

1. Convert a dc source into ac by switching a group of SCRs or transistors in a specific sequence. This type of circuit is called an **inverter**.
2. Convert an ac source into a lower-frequency ac source by controlling the firing angles of SCRs. A circuit that does this is called a **cycloconverter**.

An ac inverter accepts a fixed-frequency ac input signal and converts it to a variable output frequency. Many inverters use pulse width modulation (PWM) to produce the varying frequency. By precisely varying the width of the pulse and the time between pulses, the motor ''sees'' a frequency. The power in the dc pulses produced by PWM is equivalent to the power in an ac sine-wave frequency.

A typical single-phase inverter is shown in Figure 18-20(a). In this circuit, only one SCR in each leg can be on at any one time. If SCR_1 is on, SCR_2 must be off, in order to prevent a short-circuit between positive and negative input terminals. The diodes in parallel with the SCRs act as freewheeling diodes. The gate-firing circuits and commutation circuitry have been omitted to simplify the circuit explanation.

Figure 18-20(b) shows the phase and line output voltages for the load. When SCR_1 conducts, point A will be positive with respect to the negative input terminal. When SCR_1 switches off and SCR_2 is triggered, the potential at point A is equal to the negative terminal. When SCR_3 conducts, point B is at the potential of the positive terminal. When SCR_3 is off and SCR_4 is triggered, point B is equal to the negative terminal voltage. Therefore, by alternately switching SCR_1 and SCR_2 on and off, a series of positive voltage pulses appear at V_{an}. Cycling SCR_3 and SCR_4 results in a series of positive voltage pulses at V_{bn} which are 180° out of phase with V_{an}.

When the inverter voltage is controlled by varying the width of the output square wave, it is referred to as **pulse width control**. In Figure 18-20(b), SCR_1 and SCR_4 are on for one half-cycle, and SCR_2 and SCR_3 are on for the other half-cycle. Voltage control is achieved by varying the phase of SCR_3 and SCR_4 with respect to SCR_1 and SCR_2. Figure 18-20(c) shows the resulting output voltage waveform when the conduction intervals are retarded by an angle of 90°. The angle of retard is defined as the delay time, or the time after voltage starts to go positive at which the SCR is fired. The inverter output voltage can be smoothly adjusted from a maximum to zero by retarding the control signals for one pair of SCRs with respect to the other. In actual fact, because of the commutation interval, the resulting waveform is not quite square.

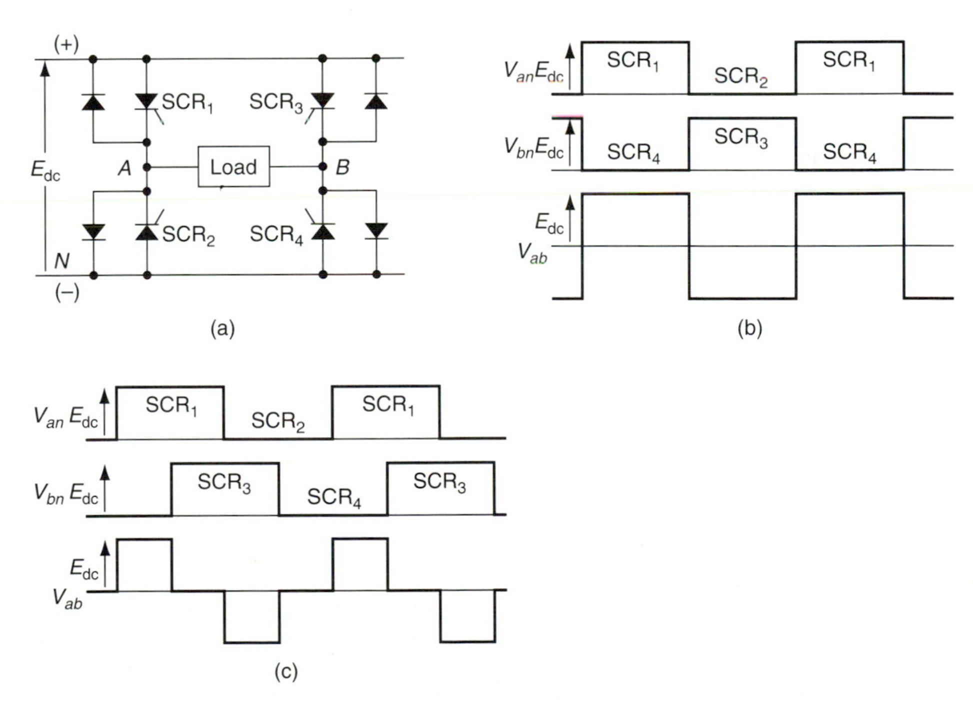

FIGURE 18-20 Voltage control using pulse width control: (a) full bridge inverter (commutating circuits are not shown); (b) voltage waveforms with 0° retard; (c) voltage waveforms with 90° retard.

Figure 18-21 shows a three-phase six-step inverter driving a wye-connected motor. Once again the SCR gate-triggering circuits are not shown. By switching the SCRs on and off in the proper sequence, the dc supply is switched across the stator windings in such a way that a rotating magnetic field is created which duplicates the action of a three-phase ac source.

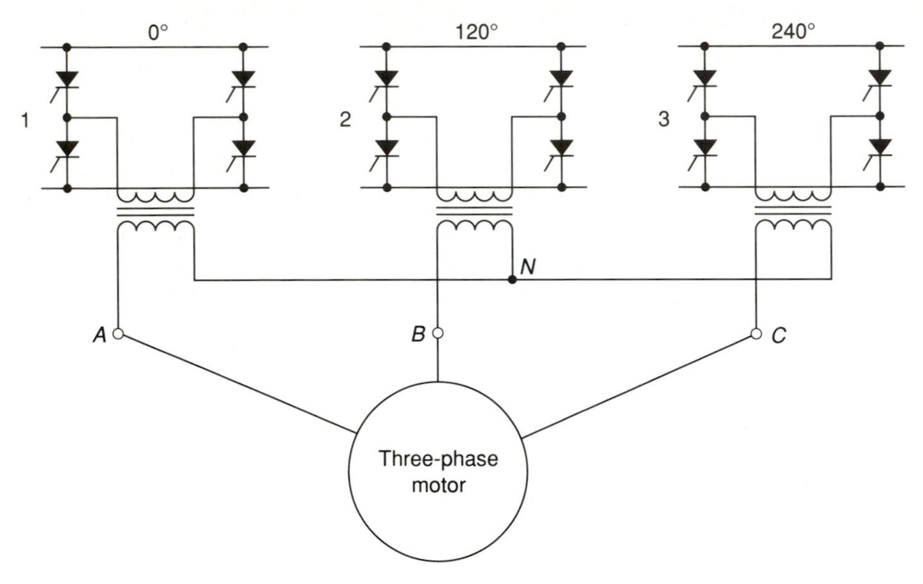

FIGURE 18-21 Six-step inverter consisting of three single-phase inverters.

If the gating signals for inverter 2 are delayed 120° from those of inverter 1, and if the gating signals for inverter 3 are delayed 180° from those of inverter 2, the resulting line-to-neutral voltages would appear as shown in Figure 18-22(a). The line-to-line secondary voltage waveforms shown in Figure 18-22(b) are obtained by summation of the square waves of Figure 18-22(a).

$$V_{ab} = V_{an} + (-V_{bn})$$

$$V_{bc} = V_{bn} + (-V_{cn})$$

$$V_{ca} = V_{cn} + (-V_{an})$$

Figure 18-23 shows an inverter circuit using power transistors to create a pulse-width-modulated approximation of a three-phase ac. There are certain advantages to using transistors instead of SCRs in inverter circuits. The main advantage is that motor **cogging** (pulsating torque) is greatly reduced. The waveshape produced by the inverter circuit of Figure 18-23 more closely approximates a sine wave than the waveshape shown in Figure 18-22.

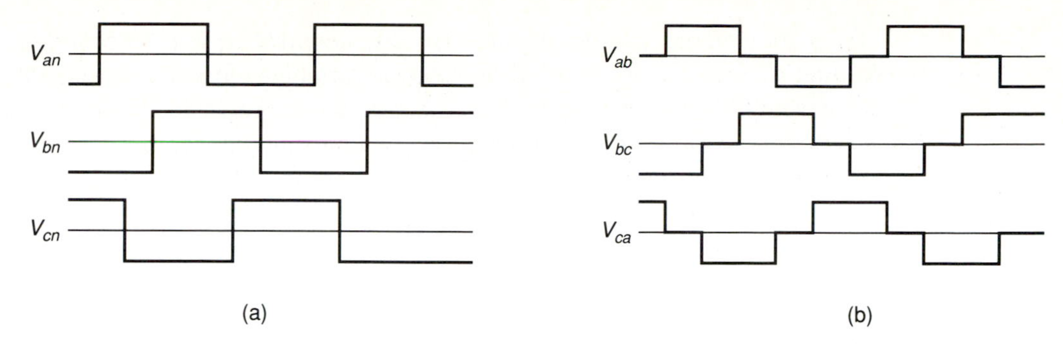

FIGURE 18-22 Waveforms of six-step inverter: (a) line-to-neutral output voltages (for 180° pulses); (b) line-to-line output voltages.

In Figure 18-23, each leg of the inverter bridge has two transistors connected in series across the dc bus. A diode is connected in parallel with each transistor. The junction between the two transistors (i.e., Q_1 and Q_4) connects to a motor terminal. Since the inverter drives a motor terminal, which is an inductive load, the current lags the voltage. When transistor Q_1 shuts off, there is a brief delay before Q_4 turns on. This delay prevents a short circuit across the dc bus. During this delay interval, the diode in parallel with Q_4 conducts the lagging current. Without these diodes in parallel with the transistors, a high voltage would develop across the transistor that is turning off.

The most modern types of PWM inverters use insulated gate bipolar transistors (IGBTs). These transistors are extremely fast switching devices that can accommodate more complex switching and modulation techniques. IGBTs are voltage controlled, which means that the drive currents are very low and the drive circuits are very compact and highly efficient.

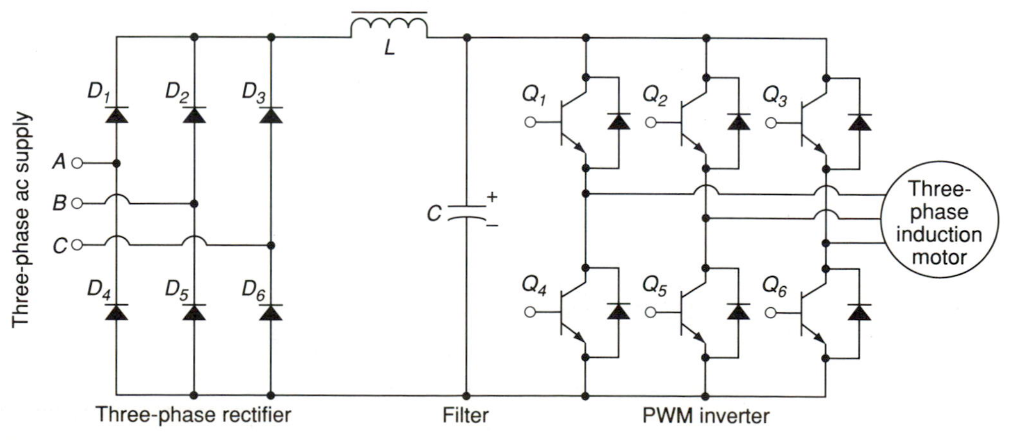

FIGURE 18-23 Inverter circuit using power transistors.

A cycloconverter uses SCRs and associated control circuits to change the frequency of the incoming line directly to a variable-frequency output. The basic principle of the single-phase cycloconverter is shown in Figure 18-24. In this circuit, two two-pulse midpoint phase-controlled converters are used. One converter uses two SCRs to form a positive group, and the other converter forms a negative group. The output current from each of these dual converters flows in only one direction. Consequently, to produce an alternating current in the motor, the positive and negative groups must be connected in inverse-parallel.

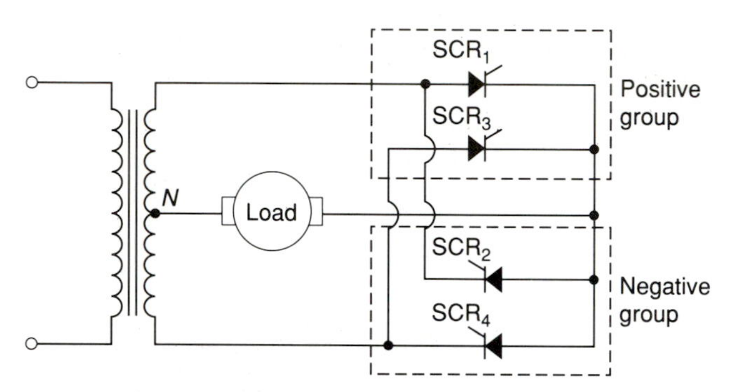

FIGURE 18-24 Single-phase to single-phase two-pulse cycloconverter.

In Figure 18-24, if SCR_1 and SCR_3 are triggered, the dc output voltage would be as shown on the left side of Figure 18-25(b). If SCR_2 and SCR_4 are triggered, the output voltage would be below the zero axis shown in Figure 18-25(b). The triggering circuit for the SCRs in the cycloconverter can change the firing points by applying a control voltage. Varying the firing delay angles in a sinusoidal manner at the desired output frequency will result in an output load voltage that is sinusoidal at the desired frequency. When the firing delay angle is 90°, the output voltage is zero. When the firing delay is 0°, the output voltage is at its maximum. To eliminate the ripple voltage produced by the

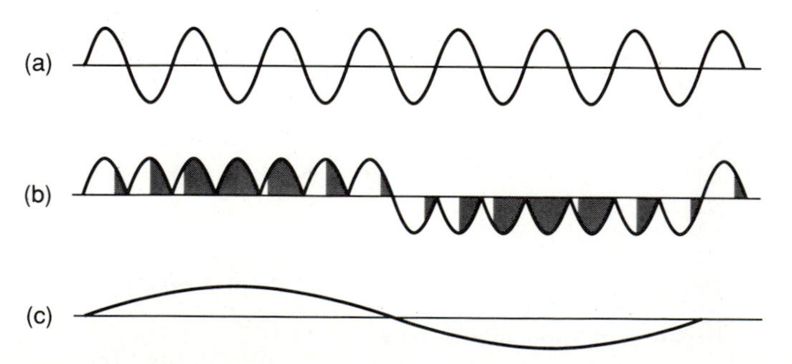

FIGURE 18-25 Single-phase cycloconverter waveforms: (a) input; (b) output; (c) output after filtering.

output of the SCRs, a filtering device such as a capacitor would be needed to eliminate the ripple.

Figure 18-26 outlines the basic cycloconverter components required for the speed control of a three-phase induction motor. Three dual converters are used for the power electronics, one for each motor phase. A waveform generator is also required to supply the ac reference to each dual converter. The waveform generator must control the phase sequence to set the motor rotation, as well as the amplitude, to control the motor torque, and the frequency to control the motor speed. The output frequency is determined by the number of gate pulses per half-cycle of the output waveform.

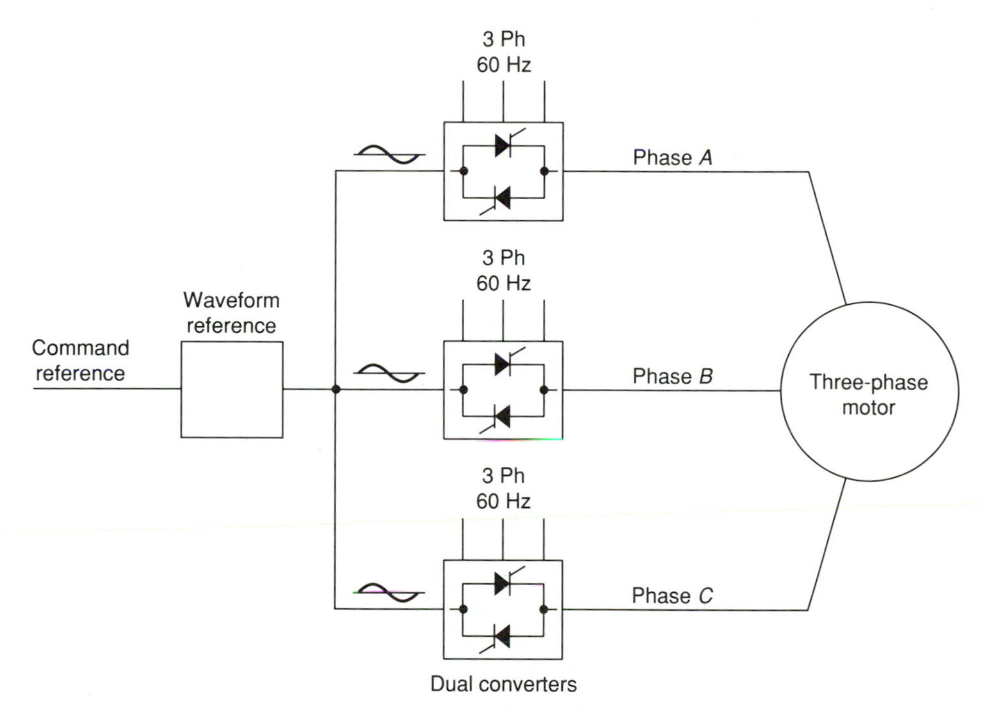

FIGURE 18-26 Typical cycloconverter equipment.

Figure 18-27 shows a typical three-phase commercial cycloconverter using 36 SCRs to provide phase-angle control of an induction motor. This type of speed control system is highly efficient since the large number of SCRs increase the pulse number, which causes a smaller-amplitude ripple content in the load voltage waveform.

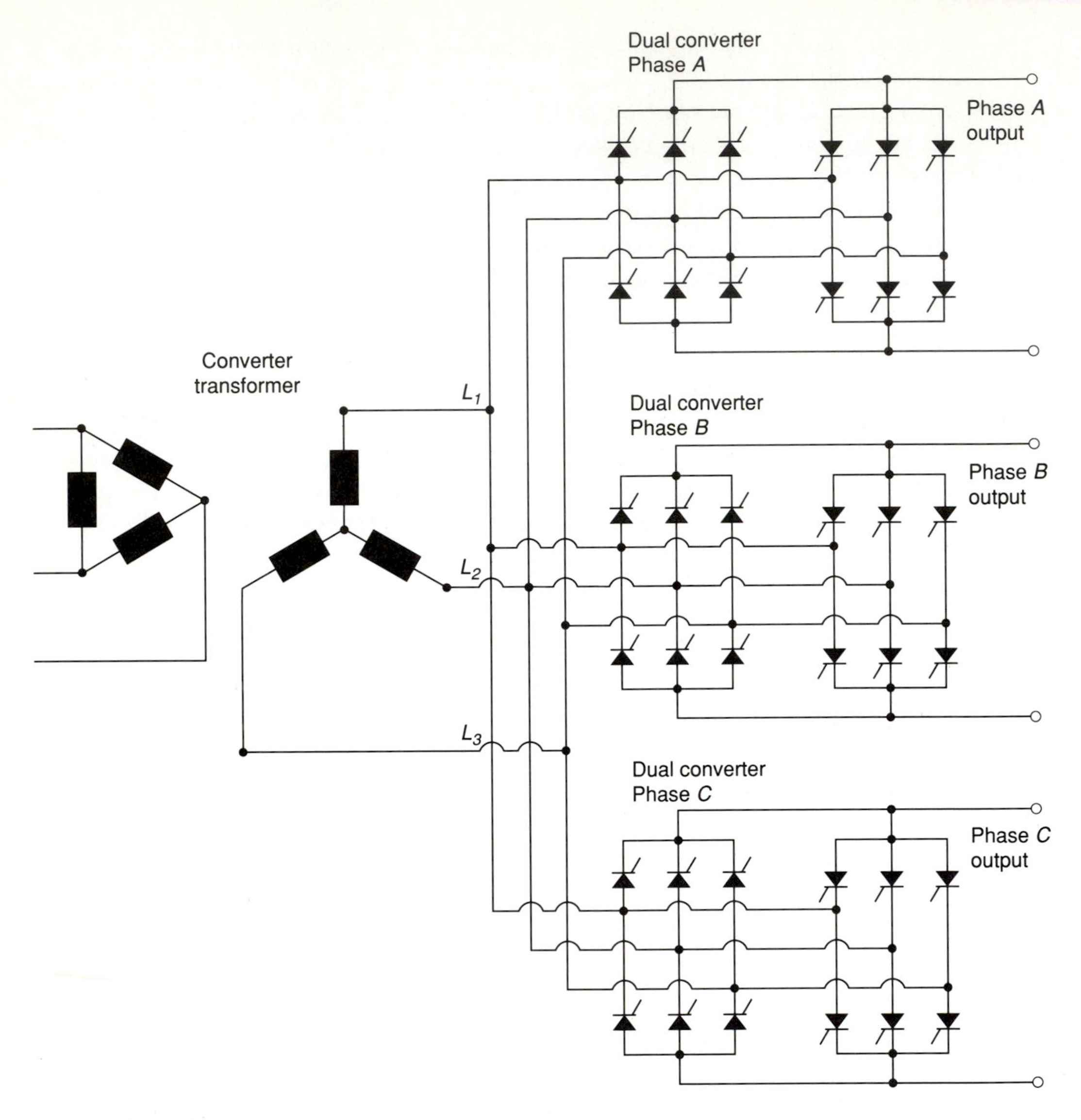

FIGURE 18-27 Cycloconverter, three-phase, 36 legs.

18-16 INDUCTION GENERATOR

If the rotor of an induction motor is driven above synchronous speed, the rotor's conductors will be cutting the flux from the stator in an opposite direction, and the slip will become negative. Consequently, the direction of the motor's inducted torque will reverse and the motor will act like a generator. That is, it will deliver power to the line instead of receiving it. The stator flux is not affected by the increase in speed of the rotor, and it will revolve in the same direction as when the machine operated as a motor. However, since the slip has reversed, the emfs and currents induced in the rotor are reversed.

The power transferred from the rotor to the stator depends on the slip, just as in the induction motor. Therefore, to increase the power delivered by the induction generator, its speed must be increased. Because of this, if an induction generator is connected to a prime mover of variable speed, it will supply power almost in proportion to the increase of its speed above synchronous speed.

The induction generator has two very serious disadvantages: It is not self-exciting, and it cannot supply reactive currents to an inductive load. One application of the induction generator is in elevators; if the load in the car tends to overspeed the motor, it becomes an induction generator and supplies power back into the system and acts as a brake to keep the car and its contents under control. Since its speed is not in synchronism with line frequency, the induction generator is also known as an **asynchronous generator**.

KEY TERMS

Stator
Rotor
Synchronous speed
Slip
Starting torque
Stall
Breakdown torque
Squirrel-cage motor
End rings
Double-squirrel-cage rotor
NEMA motor classification
Circle diagram
Across-the-line starter
Dual-voltage motor
Direct concatenation
Differential concatenation
Inverter
Cycloconverter
Pulse width control
Cogging
Asynchronous generator

PROBLEMS

18-1 Determine the synchronous speed of a four-pole 60-Hz squirrel-cage motor.

18-2 A machine is to be driven at 900 rpm by a three-phase induction motor. How many poles would the motor have if it operated at 60 Hz?

18-3 Determine the frequency required to operate a four-pole induction motor at 1500 Hz.

18-4 Determine the full-load speed of a six-pole induction motor if its slip is 4% when operating on a 60-Hz supply.

18-5 A four-pole 60-Hz induction motor has a full-load speed of 1760 rpm. Determine the percent slip at full load.

18-6 An eight-pole three-phase 240-V 60-Hz induction motor has a percent slip of 3.1% when operating under full-load conditions. Determine the full-load speed of the motor.

18-7 Determine the frequency of the currents in the rotor of a four-pole 60-Hz induction motor if the rotor speed is 1690 rpm.

18-8 A six-pole three-phase 230-V 60-Hz induction motor runs at a speed of 1125 rpm when fully loaded. Determine (a) the per unit slip; (b) the rotor frequency at rated speed.

18-9 A four-pole induction motor operating from a 60-Hz supply develops 15 hp at 1730 rpm. Determine the operating speed if the load torque is reduced to one-half.

18-10 Determine the horsepower of the motor in Problem 18-9 if the torque is reduced by 50%.

18-11 A 20-hp 60-Hz four-pole induction motor operates at a slip of 4% when fully loaded. The slip is reduced to 2.5% when the load on the motor is decreased. Determine the torque developed by the motor, in newton-meters, when the slip is 2.5%.

18-12 A six-pole 460-V motor has a starting torque of 250 lb-ft. Calculate the starting torque if the line voltage is reduced by 15%.

18-13 An induction motor develops 55.3 hp when operating at 1640 rpm. Determine the output torque in lb-ft.

18-14 A four-pole 550-V three-phase 60-Hz induction motor draws 35 A at a power factor of 0.88 lagging. The rotor copper loss was measured at 400 W and the rotational losses are 600 W. Determine the output torque in newton-meters if the slip is 0.014.

18-15 A four-pole 10-hp 60-Hz induction motor has a full-load slip of 0.06. Determine (a) the synchronous speed of the motor; (b) the rotor speed at full load; (c) the rotor frequency at full load; (d) the output torque at rated load, in lb-ft.

18-16 An eight-pole 60-Hz induction motor is loaded down until it stalls at 550 rpm. If the rotor has a resistance per phase of 0.35 Ω, calculate the blocked rotor reactance.

18-17 A 480-V three-phase induction motor is drawing 30 A at a power factor of 0.82 lagging. The stator copper losses are 900 W and the rotor copper losses are 650 W. The rotational losses are 300 W. Determine (a) the total input power; (b) the air-gap power; (c) the output power; (d) the efficiency of motor.

18-18 A three-phase induction motor was subjected to a no-load, dc stator resistance, and load test. The following data were obtained from these tests:

No-load test: $E_{nl} = 230$ V, $P_1 = 725$ W, $P_2 = -450$ W, $I_{nl} = 12$ A

dc stator resistance per phase $= 0.065\ \Omega$

Load test: $E_L = 230$ V, $P_1 = 3650$ W, $P_2 = 1825$ W, $I_L = 37$ A

Determine the percent efficiency based on the data above.

18-19 Determine the percent efficiency of a 480-V 50-hp three-phase induction motor which has been tested and the following data obtained:

Blocked rotor test: wattmeter 1 = 1800 W

wattmeter 2 = 1100 W

$I_{br} = 45$ A, $V_L = 110$ V

Load test: wattmeter 1 = 11,350 W

wattmeter 2 = 10,500 W

$I_L = 45$ A, $E_L = 480$ V

No-load test: wattmeter 1 = 2700 W

wattmeter 2 = −1650 W

$I_{nl} = 16.7$ A, $E_{nl} = 480$ V

18-20 Two 60-Hz motors are connected in direct concatenation. One motor has six poles and the other motor has four poles. Determine the speed of the rotors if the slip is 0.091.

18-21 Determine the individual speeds of the two motors in Problem 18-20 if operated separately.

CHAPTER

Alternating-Current Generators

LEARNING OBJECTIVES

Upon completion of this chapter you will be able to:

- Name the two types of rotors used in alternating-current generators.
- Explain the difference between a concentrated winding and a distributed winding.
- Define pitch factor and distribution factor.
- Determine the generated voltage in an alternator.
- Understand the effects of load power factors on alternator regulation.
- Define synchronous impedance.
- Perform open-circuit and short-circuit tests on an ac generator.
- List the five losses associated with an ac generator.
- Explain the basic principles of electronic voltage regulators.
- List four advantages of operating ac generators in parallel.
- Explain the procedure for synchronizing alternators.
- Define the term *hunting* and list three methods used to reduce hunting.
- Explain the rating and load capacity of an alternator.

19-1 INTRODUCTION

The alternating-current generator and the direct-current generator are of similar construction. Both machines have a field winding excited by direct current, and both machines have an armature in which the emf is generated. The dc and ac generator both develop alternating emfs. However, in the dc generator, the alternating voltage is rectified by a commutator and brushes. The ac generator does not require a commutator, and delivers ac power by means of slip rings. Since the terms **alternating-current generator, synchronous generator, synchronous alternator**, and **alternator** are virtually identical in meaning, they are used interchangeably throughout this chapter.

19-2 CONSTRUCTION

The most common type of ac generator is the rotating field type. Since the armature of an alternator has no commutator, there is no need for the armature to rotate. All that is required for voltage generation is that there is relative motion between a conductor and a field. There are two distinct advantages in constructing a machine with a stationary armature and a moving field. First, a rotating armature requires at least three slip rings for carrying the current from the armature to an external circuit. These rings, which are exposed, are difficult to insulate, particularly for the high voltages that alternators are required to supply. Also, when slip rings are required to carry high voltages, they are subject to arc-overs and short circuits. A stationary armature requires no slip rings, and the high-voltage insulation required for the armature conductors is easier to install on a stationary armature. When the armature is rotating, the conductors are subject to centrifugal force and vibration resulting from rotation.

In a revolving field alternator, two slip rings are required for the rotating field. These slip rings provide the dc voltage for exciting the field, and the voltage is typically 125 V in small machines and 250 V in large machines. In ac generators, the stationary armature is called the stator, and the rotating field is referred to as the rotor. The revolving field components are made in two forms: the **salient pole rotor** and the **nonsalient**, or **cylindrical, rotor**. The term *salient* means *protruding*, and a salient pole is a magnetic pole that protrudes from the surface of the rotor. The rotor with salient poles, shown in Figure 19-1, is used for all machines except those driven by steam turbines. Cylindrical rotors are utilized with high-output turbo-alternators which must run at extremely high speeds to obtain the maximum rated output. The rotors are made of steel alloy or have slots milled out to carry the field windings. A cylindrical rotor can be precisely balanced, and the noise and windage factors are very low with this type of rotor.

Salient pole rotor of alternator

Source: Prentice Hall Inc. / Siemens

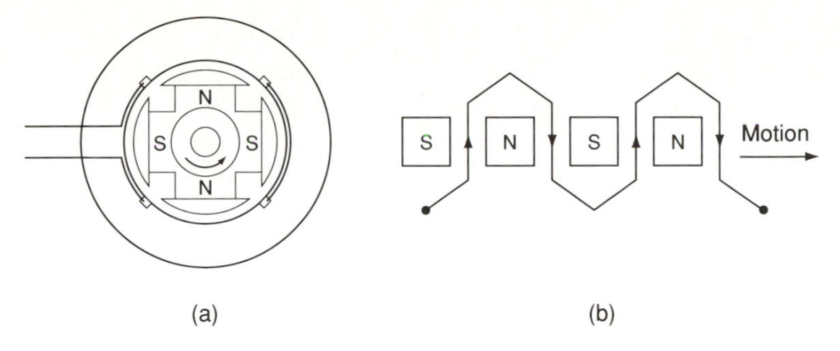

FIGURE 19-1 (a) Salient pole rotor; (b) armature winding.

The stator of an alternator consists of a slotted steel core, as shown in Figure 19-2. When the alternator is in operation, the armature core is continuously cut by the flux of the rotating field and must be laminated to reduce eddy-current losses. As alternators are usually large machines, the laminations are in sections, since it is not economical to cut complete circles of greater than about 40 in. in diameter. In order that the heat generated by the iron losses in the stator core, primarily the hysteresis and eddy-current losses, may be dissipated, ventilating ducts must be provided. A general rule of thumb is to allow one duct about 1.5 cm wide for each 7 cm of core length. In the turbo-alternator, a prominent characteristic is the large axial length compared with the diameter, so that the completed stator is almost in the form of a tunnel.

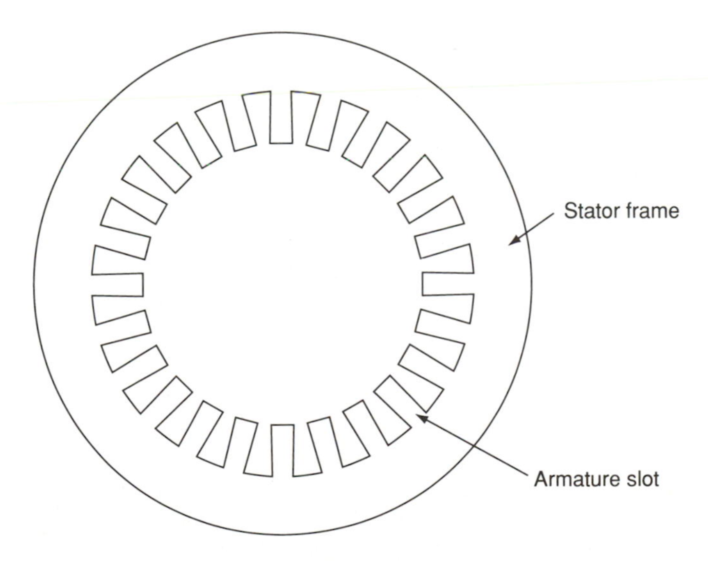

FIGURE 19-2 Stator, or armature, of an alternator.

Armature slots are divided into two basic classes, the open slot and the semiclosed, or overhung, slot. The open slot is more common because the coils can be form-wound and insulated prior to being placed in the slots. The semiclosed slot is rarely used in ac generators, but it is very common in induction motors. In both types of slots, the conductors are usually held in position by a fiber wedge. The effect of the semiclosed slot may be obtained by using open slots in conjunction with magnetic wedges.

Armature windings for alternators may be either the delta-connected type or the wye-connected type. When the windings are wye-connected, one end of each of the three separate sets of coils is brought out through the frame to stationary insulated terminals. The general principles that apply to the windings of dc generators are also used for the windings of ac generators. The span of each coil must be approximately one pole pitch, and consequently, the two sides of any coil must lie under adjacent poles. Also, the windings must be connected so that their emfs are additive.

Even though many different type of windings may be used on the stator of an alternator, only two types are found on most commercial applications: the lap winding and the chain (or basket) winding. These windings are much simpler in design than those used on dc machines because no commutator is required. The lap winding has a double layer, made up of form-wound coils with a twist between the coil sides so that the coils can be lapped.

A winding that has only one slot per pole per phase is called a **concentrated winding**. Large ac machines rarely use concentrated windings since only a small portion of the stator surface is utilized in this type of winding. A more practical method is to use two or more windings per pole per phase. This is known as a **distributed winding**. Figure 19-3 shows a distributed winding for a two-pole machine. Each pole is 180° apart, and there are six slots between each pole. Therefore, the phase difference between the adjacent slots is 180/6, or 30 electrical degrees.

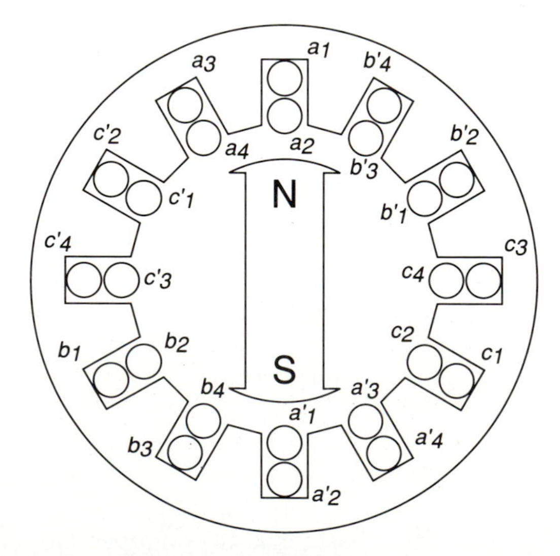

FIGURE 19-3 Distributed winding for a two-pole ac machine.

19-3 FREQUENCY

In North America, alternators are designed to supply power at a frequency of 60 or 25 cycles per second. Sixty-hertz alternators generally supply power to loads situated a short distance from the generating station. Twenty-five-hertz alternators are occasionally used to supply power to loads at a considerable distance from the source of supply. This lower frequency is used because it gives a correspondingly lower reactance drop, Ix, in the transmission line.

The frequency of the alternator voltage depends on the speed of rotation of the rotor and the number of poles. The more poles there are on the rotor, the higher the frequency is for a certain speed. In a given coil, one complete cycle of emf is generated when a north and south rotor pole is moved past one winding. Then in an alternator with P number of poles, $P/2$ cycles of voltage will be generated for each revolution of the field poles. The number of cycles per second, or frequency, equals the number of cycles per revolution, $P/2$, times the number of revolutions per second, or times the number of revolutions per minute, N, divided by 60.

$$f = \frac{P}{2} \times \frac{N}{60} = \frac{PN}{120} \tag{19-1}$$

where f = frequency, in hertz

P = number of poles

N = rotor speed, in rpm

EXAMPLE 19-1 What is the frequency of the emf generated in an eight-pole alternator operating at 900 rpm?

Solution $f = \frac{PN}{120} = \frac{(8)(900)}{120} = 60 \text{ Hz}$

19-4 PITCH FACTOR

The pole pitch of an alternator is the distance between the centers of adjacent poles, equal to 180°. In a **full-pitch winding**, the span of the coil equals the pole pitch. The emfs in the two sides of a full-pitch coil combine in phase with each other. Armatures in most commercial alternators are wound with a **fractional-pitch winding**. In a fractional-pitch winding, the span of the coil is less than the pole pitch and is generally measured in slots. From example, a $^7/_8$-pitch coil spans $^7/_8$ of the distance between two adjacent poles. The fractional pitch can also be thought of as a ratio of the slots the windings span to the number of slots per pole. If an alternator has 12 slots per pole, and the coil span is nine slots, a 9/12- or 3/4-pitch coil results. The pitch of a coil can also be given in electrical degrees.

The **pitch factor** of an alternator is a ratio of the voltage generated by a fractional-pitch coil to the voltage generated by a full-pitch coil. The pitch factor of a coil is determined by

$$k_p = \sin\frac{p^\circ}{2} \qquad (19\text{-}2)$$

where k_p is the pitch factor, a value always less than 1, and p° is the coil span in electrical degrees, where full pitch is considered to be 180 electrical degrees.

EXAMPLE 19-2 Determine the pitch factor for a polyphase alternator whose coils are wound with a 5/6 coil pitch.

Solution

$$p^\circ = \frac{5}{6} \times 180^\circ = 150^\circ$$

$$k_p = \sin\frac{p^\circ}{2} = \sin\frac{150^\circ}{2}$$

$$= 0.97$$

EXAMPLE 19-3 Find the pitch factor of a four-pole 36-slot three-phase alternator. The windings span eight slots.

Solution First, find the slots per pole.

$$\frac{36}{4} = 9 \text{ slots per pole}$$

The fractional pitch is a ratio of the slots the windings span to the slots per pole.

$$\text{Fractional pitch} = \frac{8}{9}$$

$$p^\circ = \frac{8}{9} \times 180^\circ = 160^\circ$$

$$k_p = \sin\frac{p^\circ}{2} = \sin\frac{160^\circ}{2}$$

$$= 0.98$$

19-5 DISTRIBUTION FACTOR

If a number of coils in a pole group are series connected, the voltage generated by that pole group is the sum of the individual voltages, provided that two or more coils lie in the same slot. The induced emfs of each of these coils will be displaced by the same degree that the slots have been distributed.

The total voltage in any phase of an alternator will be the phasor sum of the individual coil voltages. The phasor sum of the coil voltages is less than their arithmetic sum due to the phase differences. The factor by which these phase differences reduce the total voltage is called the **distribution factor**. The distribution factor is a convenient way to summarize the decrease in voltage caused by the spatial distribution of the coils in the armature winding. The distribution factor is defined by

$$k_d = \frac{E_\Phi}{E_c} \tag{19-3}$$

where k_d = distribution factor

E_Φ = phasor sum of coil voltages per phase

E_c = arithmetic sum of coil voltages per phase

The calculation of the distribution factor in terms of voltages is very impractical. Therefore, k_d is usually determined by the number of electrical degrees between the adjacent slots and by the number of slots per pole per phase. Using this method, the distribution factor is given by

$$k_d = \frac{\sin(nd/2)}{n \times \sin(d/2)} \tag{19-4}$$

where n is the number of slots per pole per phase, and d is the number of electrical degrees between adjacent slots.

EXAMPLE 19-4 Calculate the distribution factor for a 48-slot eight-pole three-phase alternator.

Solution

$$\text{Slots per pole} = \frac{48}{8} = 6$$

$$d = \frac{180}{6} = 30°$$

$$n = \frac{48}{8 \times 3} = 2$$

$$k_d = \frac{\sin(nd/2)}{n \times \sin(d/2)} = \frac{\sin(2 \times 30/2)}{2 \times \sin(30/2)}$$
$$= 0.966$$

19-6 GENERATED VOLTAGE IN AN ALTERNATOR

When a coil is made up of two conductors, the equation for the effective generated voltage per conductor is

$$E = 4.44\Phi f N_p \qquad \text{volts} \tag{19-5}$$

where E = effective, or rms, voltage per phase

f = frequency, in hertz

N_p = total number of turns per phase

The total number of series turns per phase, N_p, is the total number of turns in the armature divided by the number of phases for which the armature is wound.

Equation 19-5 is not representative of the effective voltage developed in a per pole per phase group of coils in an alternator. When calculating the value of voltage generated in each phase of an ac generator, it is necessary to take the pitch factor, k_p, and distribution factor, k_d, into account. The equation becomes

$$E_{gp} = 4.44\Phi f N_p\, k_p k_d \qquad \text{volts} \tag{19-6}$$

where E_{gp} = voltage generated per pole per phase

k_p = pitch factor

k_d = distribution factor

The pitch factor, k_p, is required in equation 19-6 because the wave form of the emf per conductor is identical in shape with the curve of flux density against distance around the air gap. This curve is made as nearly sinusoidal as possible in the case of salient-pole machines by design considerations such as a gradual increase in the length of the air gap toward the pole tips, or by the skewing of the pole face.

The distribution factor takes into account the distribution of the winding over several slots per pole per phase. Since the coils in a group are spread out, the emfs induced in the individual coils are not in phase with one another, so the resultant emf is not quite equal to the emf per coil multiplied by the number of coils. Consequently, the distribution factor will be a value less than 1, where its actual value will depend on the number of slots per pole per phase.

EXAMPLE 19-5 In a 7500-kVA 13,800-V 60-Hz three-phase four-pole synchronous generator, there are 60 slots and the armature is lap wound, 5/6 pitch, with two coil sides per slot and four turns per coil. The flux per pole is 0.51 Wb distributed sinusoidally along the air gap. Determine (a) the number of slots per pole per phase; (b) the distribution factor; (c) the pitch factor; (d) the total number of turns per phase; (e) the total generated voltage per phase.

Solution

(a) $$\frac{60}{4} = 15 \text{ slots per pole}$$

$$n = \frac{15}{3} = 5 \text{ slots per pole per phase}$$

(b) $$d = \frac{180}{15} = 12^\circ$$

$$k_d = \frac{\sin(nd/2)}{n \times \sin(d/2)} = \frac{\sin(5 \times 12^\circ/2)}{5 \times \sin(12^\circ/2)} = 0.957$$

(c) $$\text{Fractional pitch} = \frac{5}{6}$$

$$p^\circ = \frac{5}{6} \times 180 = 150^\circ$$

$$k_p = \sin\frac{p^\circ}{2} = \sin\frac{150^\circ}{2} = 0.966$$

(d) $$N_p = \frac{60 \times 4}{3} = 80 \text{ turns/phase}$$

(e) $$\begin{aligned} E_{gp} &= 4.44\Phi f N_p k_p k_d \\ &= 4.44(0.51 \text{ Wb})(60 \text{ Hz})(80)(0.966)(0.957) \\ &= 10{,}048.1 \text{ V/phase} \end{aligned}$$

19-7 ALTERNATOR REGULATION

The field windings of the ac and dc generators are excited by direct current to produce a steady magnetic field flux. However, the alternator and dc generator differ in that the alternator must be driven at its synchronous speed by a prime mover so that the proper ac frequency may be delivered. The definition of regulation for the ac and dc generator is the same; that is, it is the change in the terminal voltage when rated load is reduced to zero, divided by the voltage at rated load.

If there is no load on the ac or dc generator, the terminal voltage and the generated voltage are the same. When load is applied to the dc generator, the difference between terminal voltage and generated voltage is caused by resistance drop in the armature circuit and armature reaction. In the ac generator, the difference between terminal and generated voltage is also caused by the I_aR_a drop and armature reaction, but it is also due to the reactance of the armature. The armature reactance, I_aX_a, and the armature reaction will usually cause the terminal voltage to increase when the load is increased. The amount of terminal voltage rise or drop will depend on the magnitude of the load and the actual over all power factor of the load. Therefore, the terminal voltage of an alternator may drop by 8 to 20% from no-load to full-load, but the voltage may drop from 30 to 40% when the power factor is lagging. For this reason, the regulation of the ac generator is worse than that of the dc generator.

Figure 19-4 shows the effects of different load power factors on the change in the terminal voltage with changes of load on an alternator. From these characteristic curves of different power factors, it can be seen that the voltage regulation varies to such an extent that it even becomes negative for a leading power factor. The percent regulation formula for ac generators is the same equation as that used to determine the percent regulation for dc generators. The only difference is a change in subscripts.

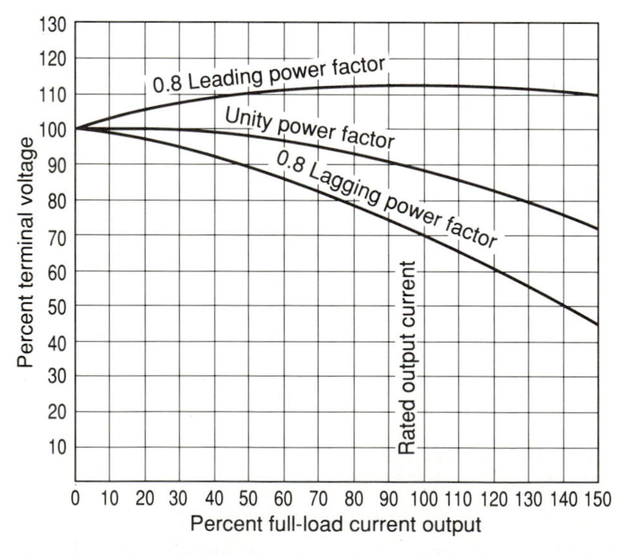

FIGURE 19-4 Output voltage and current characteristics of an alternator at three different power factors.

$$\text{Percent regulation} = \frac{E_{gp} - V_p}{V_p} \times 100$$

where E_{gp} is the internal generated voltage per phase at no load and V_p is the terminal voltage at rated load.

EXAMPLE 19-6 Calculate the percent regulation of the leading PF load shown in Figure 19-4.

Solution

$$\% \text{ Regulation} = \frac{E_{gp} - V_p}{V_p} \times 100 = \frac{100\text{ V} - 112.5\text{ V}}{112.5\text{ V}} \times 100$$
$$= -11.1\%$$

The negative regulation in Example 19-6 indicates that the full-load voltage is greater than the no-load voltage. A condition caused when the armature reaction aids the main-field flux. The effects of the three possible power factors—unity, leading, and lagging—all have a substantial effect on the operating characteristics of an alternator. The following discussion deals with these three types of power factors.

Unity power factor load. A simplified phasor diagram of an alternator operating at unity power factor is shown in Figure 19-5. The generated voltage in the phase winding is equal to the phasor sum of the terminal voltage for the phase V_p and the internal voltage drop in the armature resistance I_aR_a. Because the armature reactance drop, I_aX_a, is 90° out of phase with the current, the terminal, or voltage generated per phase, E_{gp}, is approximately equal to the generated voltage, minus the I_a drop in the armature winding. At unity power factor, the armature reaction drop, I_aX_{ar}, leads the armature current and is always in phase with the armature reactance drop, I_aX_a, regardless of the power factor. Because of the relationship between I_aX_a and I_aX_{ar}, it is often more convenient to consider them as one reactance drop, I_aX_s. The equation for determining the induced voltage caused by dc excitation alone can be stated as follows:

$$E_{gp} = \sqrt{(V_p + I_aR_a)^2 + (I_aX_s)^2} \tag{19-7}$$

or, in complex form,

$$E_{gp} = V_p + I_aR_a + jI_aX_s$$

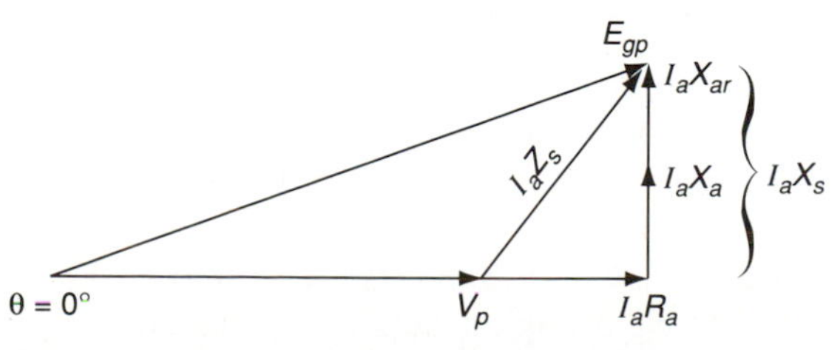

FIGURE 19-5 Phasor diagram of alternator operating at unity power factor.

Lagging power factor load. The voltage phasors for a lagging power factor load are shown in Figure 19-6. The armature phase current and the I_aR_a drop lag the terminal voltage by angle θ, and both are still in phase with each other. The armature reactance and reaction drops once again lead the armature current by 90°. The relationship between E_{gp} and V_p is more easily visualized when $V_p \cos\theta$ is made parallel to the original I_aR_a, and $V_p \sin\theta$ is drawn parallel to I_aX_s. A second I_aR_a is made parallel to the original I_aR_a so that it is an extension of $V_p \cos\theta$. By drawing a second $V_p \sin\theta$ in parallel with the first $V_p \sin\theta$, a right-angle triangle has now been formed with E_{gp} as the hypotenuse. The Pythagorean theorem yields the following equation for a lagging power factor:

$$E_{gp} = \sqrt{(V_p \cos\theta + I_aR_a)^2 + (V_p \sin\theta + I_aX_s)^2} \quad (19\text{-}8)$$

or, in complex form,

$$E_{gp} = V_p \cos\theta + I_aR_a + j(V_p \sin\theta + I_aX_s)$$

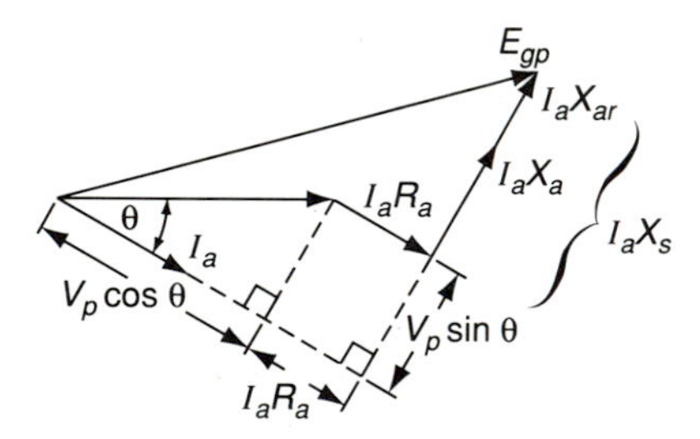

FIGURE 19-6 Phasor diagram of an alternator operating at a lagging power factor.

Leading power factor load. The voltage phasors for a leading power factor load are shown in Figure 19-7. The I_aR_a voltage drop now leads the phasor V_p by the load power factor angle. This condition may result in an increase in terminal voltage above the value of E_{gp}, depending on the power factor angle. When this negative regulation occurs, it is due to the combination of the rotationally induced voltage, E_{gp}, and the self-induced voltage. As in any ac circuit, the self-induced voltage is caused by the varying field linking the armature conductors. The self-induced voltage always lags the current by 90°. Consequently, when I_a leads V_p, the self-induced voltage aids E_{gp}, and V_p increases in value.

The magnitude of voltage E_{gp} is once again determined by the Phythagorean theorem. $V_p \sin\theta$ and $V_p \cos\theta$ are drawn in relation to the phasor V_p in a manner similar to that of Figure 19-6. Once again, I_aR_a is drawn parallel to the original as an extension of $V_p \cos\theta$. Since I_aX_s is a part of $V_p \sin\theta$, it is subtracted from this side of the right-angle triangle. Therefore,

$$E_{gp} = \sqrt{(V_p \cos\theta + I_a R_a)^2 + (V_p \sin\theta - I_a X_s)^2} \tag{19-9}$$

or, in complex form,

$$E_{gp} = V_p \cos\theta + I_a R_a + j(V_p \sin\theta - I_a X_s)$$

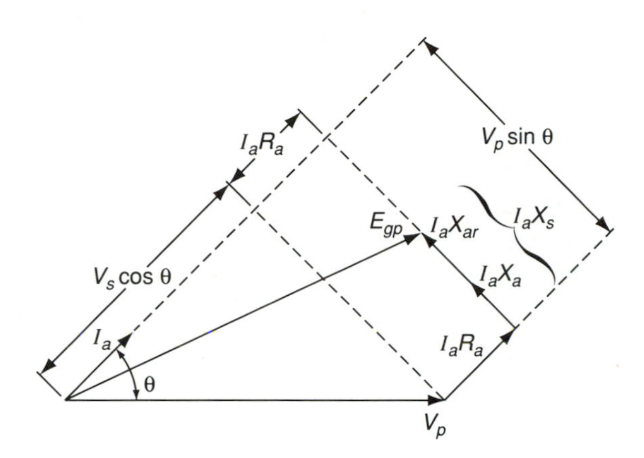

FIGURE 19-7 Phasor diagram of an alternator operating at a leading power factor.

EXAMPLE 19-7 A 4000-kVA 13,800-V 60-Hz three-phase wye-connected alternator has an effective armature resistance of 5.2 Ω per phase and a combined armature reactance, X_s, of 54 Ω per phase. Determine the full-load generated voltage per phase at (a) unity power factor; (b) 0.8 lag power factor; (c) 0.72 leading power factor.

Solution

$$V_p = \frac{V_L}{\sqrt{3}} = \frac{13{,}800\ \text{V}}{\sqrt{3}} = 7967.43\ \text{V}$$

$$I_a = I_p = \frac{V_A}{3V_p} = \frac{4000 \times 10^3\ \text{VA}}{3 \times 7967.43\ \text{V}} = 167.35\ \text{A}$$

$$I_a R_a = (167.35\ \text{A})(5.2\ \Omega) = 870.22\ \text{V}$$

$$I_a X_s = (167.35\ \text{A})(54\ \Omega) = 9036.9\ \text{V}$$

(a) $$E_{gp} = \sqrt{(V_p + I_a R_a)^2 + (I_a X_s)^2} = \sqrt{(7967.43\ \text{V} + 870.22\ \text{V})^2 + (9036.9\ \text{V})^2} = 12{,}640\ \text{V/phase}$$

(b) $$\theta = \cos^{-1} 0.8 = 36.87°$$

$$E_{gp} = \sqrt{(V_p \cos\theta + I_a R_a)^2 + (V_p \sin\theta + I_a X_s)^2}$$

$$= [(7967.43\ \text{V} \cos 36.87° + 870.22\ \text{V})^2 + (7967.43\ \text{V} \sin 36.87°)^2 + 9036.9\ \text{V}]^{1/2}$$

$$= 15{,}601.2\ \text{V/phase}$$

(c) $\theta = \cos^{-1} 0.72 = 43.95°$

$$E_{gp} = \sqrt{(V_p \cos\theta + I_a R_a)^2 + (V_p \sin\theta + I_a X_s)^2}$$
$$= \sqrt{[(7967.43\text{ V} \cos 43.95°) + 870.22\text{ V}]^2 + (7967.43\text{ V} \sin 43.95°) + 9036.9\text{ V}^2]}$$
$$= 7479.6\text{ V/phase}$$

19-8 SYNCHRONOUS IMPEDANCE

The simplest method of determining the voltage regulation of an alternator is to connect an external load of known power factor to the terminals. However, in large alternators, it would be extremely difficult to obtain the voltage regulation by connection of a load. For these types of alternators, several methods have been devised to determine the regulation. The **synchronous impedance method** is one of the methods used for predetermining the operating characteristics of an alternator without actually placing the alternator under load conditions. The synchronous impedance method consists of an open-circuit test and a short-circuit test.

As mentioned previously, the armature resistance, R_a, is the effective ac resistance of the winding per phase. The combined armature reactance and armature reaction, X_s, is called the **synchronous reactance** of the alternator, which is also a per phase value. The synchronous impedance, Z_s, of the armature winding is the phasor sum of R_a and X_s, which is also on a per phase basis. The phasor diagram of Figure 19-8(a) shows the voltage drops of the armature winding with the armature current, I_a, as the reference phasor. Dividing each of the voltage drops by the armature current yields the impedance triangle of Figure 19-8(b).

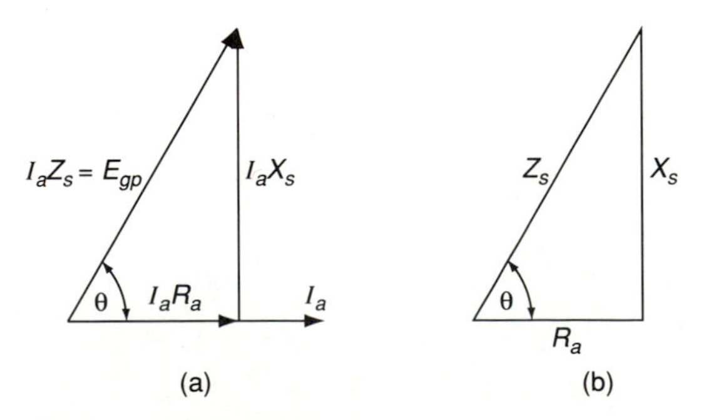

FIGURE 19-8 (a) Phasor diagram and (b) impedance triangle for synchronous impedance of an alternator.

In addition to the open-circuit and short-circuit tests, it is also necessary to perform a resistance test on an ac generator to determine the effective armature resistance. The circuit for a resistance test is shown in Figure 19-9. When performing the resistance test, it is assumed that the alternator is wye-connected if it is a three-phase machine. The dc field winding is opened, and the dc resistance is measured between each pair of terminals. The average of the three sets of resistance values is called R_t. R_t is then divided by 2 to obtain the resistance on a per phase basis, R_{dc}. With the alternator stationary, and ideally with its armature windings at the normal operating temperature, connect the circuit as shown, using a high-current rheostat to adjust the current to approximately the rated value. The value of R_t is then obtained by applying Ohm's law to the ammeter and voltmeter readings. Once the average of R_t is found, R_{dc} is determined. To convert R_{dc} to a value approximately equal to the resistance when ac current flows through the winding, a multiplier of 1.25 is used.

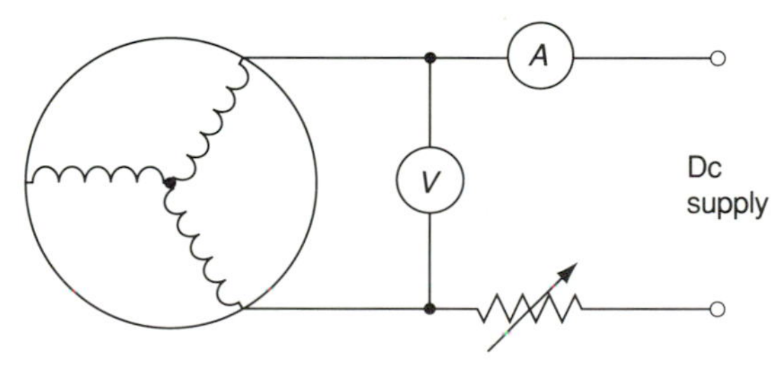

FIGURE 19-9 Resistance test of an ac generator.

The open-circuit test is shown in Figure 19-10. The armature winding is open and the alternator is operated at synchronous speed. A dc ammeter is connected in the field circuit, and an ac voltmeter is connected across any two of the armature windings to record the line voltage. The purpose of this test is to plot an **open-circuit saturation curve**. A dc source is connected to the field with a rheostat to adjust the magnitude of field current. Readings of current and voltage are taken at increments beginning at zero amperes through the field and increased until the voltage between any pair of stator windings is slightly above rated value. These readings are then plotted on the saturation curve of Figure 19-12. Before plotting the curve, it is necessary to divide the voltage reading by $\sqrt{3}$ to obtain the phase voltage.

The connection for the short-circuit test on an alternator is shown in Figure 19-11. The alternator is driven at rated speed with some dc excitation. By adjusting the rheostat, the field current is set at zero, and the alternator is brought up to synchronous speed. The field current is gradually increased, and the value of I_f and the average of the three ac ammeters are recorded at different increments.

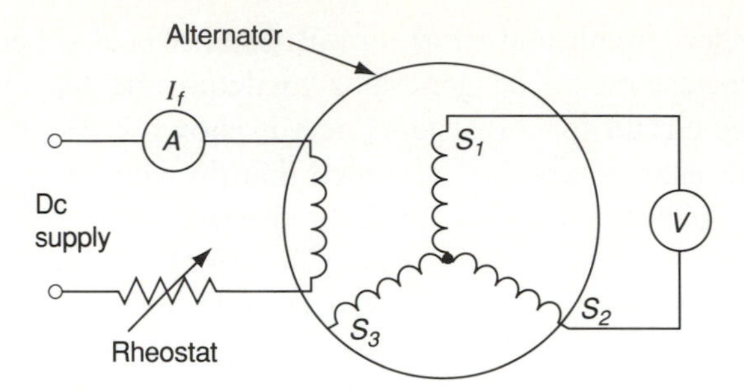

FIGURE 19-10 Open-circuit test of an ac generator.

The curve obtained by the short-circuit test is known as the **short-circuit characteristic** (SCC). Note that the short-circuit characteristic is a straight line. This is due to the fact that under short-circuit conditions, the stator magnetic opposes the rotor field. Consequently, the magnetic field does not saturate. If the synchronous machine stays unsaturated, the excitation voltage will increase linearly with the excitation current and the armature current will increase linearly with the field current.

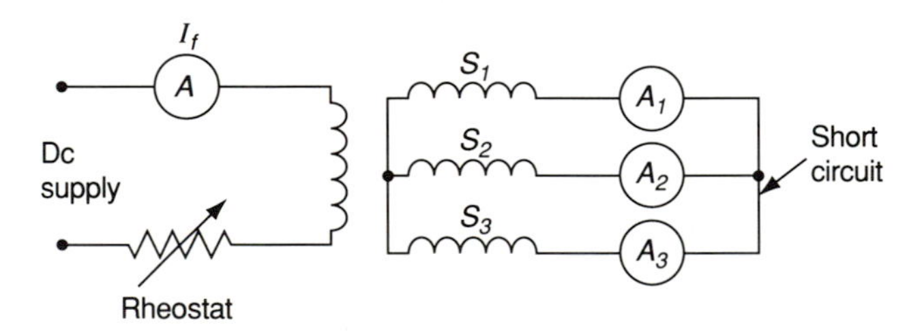

FIGURE 19-11 Short-circuit test of an ac generator.

The synchronous impedance for the alternator can now be found from the curves plotted in Figure 19-12. The value of I_{sc} at the field current that gives rated alternator voltage per phase is located on the graph. Z_s is then found by applying Ohm's law to the open-circuit voltage and the short-circuit current.

$$Z_s = \frac{\text{open-circuit voltage per phase}}{\text{short-circuit current}}$$

$$= \frac{E_{gp}}{I_{sc}} \tag{19-10}$$

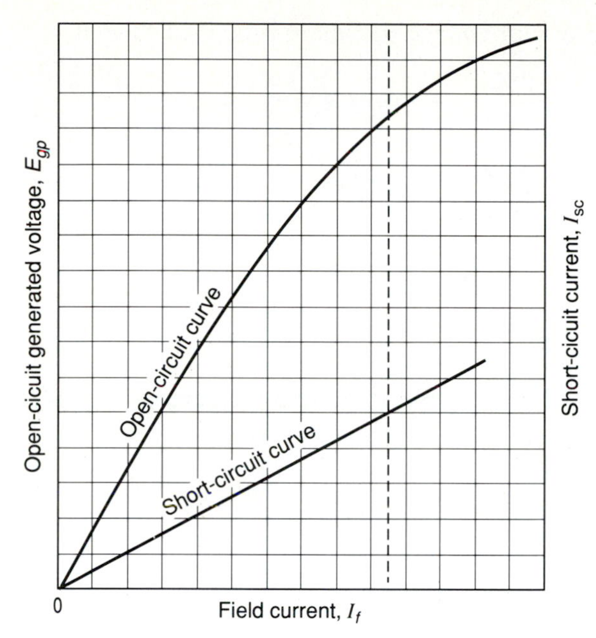

FIGURE 19-12 Open-circuit and short-circuit characteristic curves for an alternator, on a per phase basis.

EXAMPLE 19-8 A 750-kVA 2300-V wye-connected three-phase alternator has had open-circuit and short-circuit tests performed and the following data obtained:

Short-circuit test	*Open-circuit test*
Field current = 31.5 A	Field current = 31.5 A
Line current = rated	Line voltage = 1050 V

The dc resistance of the windings is averaged at 0.38 Ω. Calculate the values of R_a, Z_s, and X_s.

Solution

$$I_a = \frac{\text{kVA} \times 1000}{\sqrt{3} \times E_L}$$
$$= \frac{750{,}000\ \text{VA}}{\sqrt{3} \times 2300\ \text{V}}$$
$$= 188.27\ \text{A}$$

$$R_t = 0.38\ \Omega$$

$$R_{dc} = \frac{0.38\ \Omega}{2} = 0.19\ \Omega$$

$$R_a = 1.25 \times 0.19 = 0.238\ \Omega$$

The open-circuit voltage per phase is now calculated.

$$E_{gp} = \frac{E_L}{\sqrt{3}} = \frac{1050\text{ V}}{\sqrt{3}} = 606.22\text{ V}$$

$$Z_s = \frac{E_{gp}}{I_{sc}} = \frac{606.22\text{ V}}{188.27\text{ A}} = 3.22\ \Omega$$

$$\begin{aligned} X_s &= \sqrt{Z_s^2 - R_a^2} \\ &= \sqrt{3.22^2\ \Omega - 0.238^2\ \Omega} \\ &= 3.21\ \Omega \end{aligned}$$

EXAMPLE 19-9 Using the values obtained in Example 19-8, find the percent regulation when the alternator is connected to a 0.72 lagging PF load.

Solution

$$V_p = \frac{E_L}{\sqrt{3}} = \frac{2300\text{ V}}{\sqrt{3}} = 1327.9\text{ V}$$

$$I_aR_a = (188.27\text{ A})(0.238\ \Omega) = 44.81\text{ V}$$

$$I_aX_s = (188.27\text{ A})(3.21\ \Omega) = 604.35\text{ V}$$

$$\theta = \cos^{-1} 0.72 = 43.95°$$

$$\begin{aligned} E_{gp} &= \sqrt{(V_p \cos\theta + I_aR_a)^2 + (V_p \sin\theta + I_aX_s)^2} \\ &= \sqrt{(1327.9\text{ V} \cos 43.95° + 44.81\text{ V})^2 + (1327.9\text{ V} \sin 43.95° + 604.35\text{ V})^2} \\ &= 1824.88\text{ V} \end{aligned}$$

$$\begin{aligned} \%\text{ regulation} &= \frac{E_{gp} - V_p}{V_p} \times 100 \\ &= \frac{1824.88\text{ V} - 1327.9\text{ V}}{1327.9\text{ V}} \times 100 \\ &= 37.43\% \end{aligned}$$

The synchronous impedance method gives a value of regulation which is worse than the actual regulation of the alternator. For this reason, the synchronous impedance method is often referred to as the pessimistic method of determining voltage regulation. The value given by using this method is basically a worst-case scenario of the percent regulation of the machine. The greatest error in the synchronous impedance method is that there is no saturation effect. The value of the synchronous reactance, particularly that part which replaces armature reaction, is too large, for it is determined at short circuit, when the iron is not saturated. In the percent regulation calculation, a constant value of

Z_s is assumed despite the fact that the impedance decreases with increasing values of field current. Also, the effect of the armature mmf is much greater at short circuit, where the value of X_s is determined than under the usual conditions of load. When the armature is short circuited, the current lags the induced voltage by nearly 90°, and the armature reaction is almost totally demagnetizing. This demagnetizing effect further reduces the degree of saturation.

19-9 EFFICIENCY

The efficiency of large alternating-current generators is very high, sometimes reaching 98%. High efficiency is of very great importance, as it means lower losses and lower temperature rise and, therefore, longer life for the machine. The efficiencies of an ac generator and a dc generator are calculated in exactly the same manner. As mentioned previously, the only real difference between an ac generator and a dc generator is the fact that in the alternator, the armature is stationary and the field is rotating at a constant speed. The losses in an ac generator are the same as in its dc counterpart:

1. Rotor copper (I^2R) losses
2. Stator copper (I^2R) losses
3. Core losses
4. Windage and friction losses
5. Stray loss or load loss

Rotational losses of an alternator may be found by performing the same no-load test that was used on the induction motor. Since an alternator is strictly a one-speed machine, most of the rotational loss is a constant. The windage and friction loss are affected by the size and shape of the rotating components, fan design, bearing design, and type of enclosure.

A very simple method of determining the rotational losses in an alternator is to use a calibrated dc motor as a prime mover for the alternator. The alternator is driven at synchronous speed but with no field excitation. By using a calibrated motor, the losses are known, and the output is easily determined. The dc motor output, which is known, is therefore the input to the alternator and is considered to consist of the friction and windage losses of the alternator. The next step involves bringing the alternator up to synchronous speed again, but this time with field excitation. If the normal operating value of field excitation is unknown, it is necessary to perform the synchronous impedance test to determine the internal voltage drops, and then adjust the open-circuit voltage using the dc motor, so that the open-circuit voltage equals the rated voltage plus the internal voltage drop. The difference between the output of the dc motor with field excitation and without field excitation is the core loss of the alternator. The last step involves short-circuiting the armature and adjusting the field current to obtain rated line current. The difference between the dc motor output when the armature is short-circuited compared to when it is open-circuited is the armature copper loss and stray loss.

The dc field copper losses are easily found by simple dc measurement using a voltmeter and an ammeter.

$$P_f = I_f^2 R_f \tag{19-11}$$

where $P_f = I^2R$ losses in field winding

I_f = adjusted field amperes

R_f = rotating terminal to terminal field resistance, including slip rings

EXAMPLE 19-10

A 2000-kVA 2300-V 60-Hz three-phase alternator is driven by a calibrated dc motor. The following data are obtained: dc motor output with no field excitation, P_1 = 64 kW, dc motor output with normal field excitation, P_2 = 92.5 kW, output of dc motor with short-circuited armature and rated current flowing, P_3 = 108 kW. The field excitation of the alternator is 22 A at 120 V dc. Determine (a) the core losses; (b) the copper and stray losses; (c) the field losses; (d) the full-load efficiency at 0.8 lagging power factor.

Solution

(a) Core losses $= P_2 - P_1 = 92.5\text{ kW} - 64\text{ kW} = 28.5\text{ kW}$

(b) Copper losses $= P_3 - P_1 = 108\text{ kW} - 64\text{ kW} = 44\text{ kW}$

(c) Field losses $= 22\text{ A} \times 120\text{ V} = 2.64\text{ kW}$

Total losses = friction + windage losses + core loss + copper loss + field loss
$= 64\text{ kW} + 28.5\text{ kW} + 44\text{ kW} + 2.64\text{ kW}$
$= 139.14\text{ kW}$

(d) $$\%\ \text{Efficiency} = \frac{\text{output}}{\text{output} + \text{losses}} \times 100 = \frac{1600\text{ kVA}}{1600\text{ kVA} + 139.14\text{ kW}} \times 100 = 92\%$$

19-10 ELECTRONIC VOLTAGE REGULATORS

The regulation of an alternator is poor, since it is dependent on the power factor of the load. The terminal voltage of the alternator also varies with the magnitude of the load. If the load increases, the terminal voltage of the ac generator is reduced and resistance in the field rheostat must be decreased. Consequently, a decrease in load will cause an increase in the terminal voltage which requires the resistance of the field rheostat to be increased. A common method of controlling the terminal voltage is by using a **voltage regulator** to control the amount of dc field excitation. The voltage regulator is an auxiliary device which monitors the alternator output voltage. When the output changes, the regulator causes a corresponding change in the field current of the exciter that supplies field current to the generator. The exciter is usually on the same shaft as the prime mover and the alternator. When the alternator terminal voltage drops, the voltage regulator will automatically increase the field excitation.

Figure 19-13 shows the block diagram of a solid-state voltage regulator. The regulator monitors the alternator voltage, compares a rectified portion of that voltage with the reference voltage, and supplies the field current necessary to maintain the ratio between the alternator voltage and the reference voltage. The sensing circuit consists of one or two sensing transformers as well as a dc rectifier. This circuit monitors the output of the alternator, steps-down and rectifies the voltage, and feeds the resultant dc signal back to the error detector and error amplifier. The error detector consists of a zener diode which holds the reference voltage at a steady value. The zener voltage is proportional to the signal from the sensing circuit. The difference between the two signals is sent to the error amplifier. The error amplifier supplies phase-angle control of the SCRs in the power controller. The amount of output current from the power controller depends on the conduction time of the SCRs and the exciter field resistance. Essentially, the stabilization network is made up of *RC* timing constants, which provide a stabilizing signal from the power stage to the error amplifier, which prevents oscillations, or hunting.

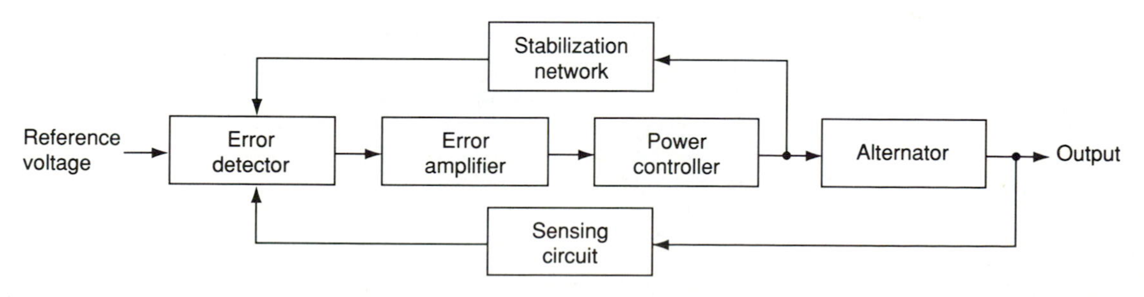

FIGURE 19-13 Block diagram of a regulator circuit for an ac generator.

The output voltage produced by a three-phase alternator may be regulated to control the charging of a battery, as shown in Figure 19-14. In this circuit an automobile voltage regulator is used to charge a 12-V battery. Diode *Z* is a **zener diode**, which is a two-terminal semiconductor often found in voltage regulator circuits. The zener diode operates as a regulating device when it is reverse-biased. It will hold the anode-to-cathode voltage at a fixed value and allow relatively large values of current to flow through the device.

When the ignition switch is closed, switch *S* closes and the automobile engine rotates the field winding, which is connected through the slip rings in the SCR circuit. The SCR will provide pulses of current that energize the field winding, and the three stationary windings of the ac generator produce up to 20 V ac.

Diodes D_1, D_2, D_3, D_4, D_5, and D_6 form a three-phase full-wave rectifier. The purpose of this rectifier is to convert the alternating current from the generator into a direct current for the car battery. The automobile-battery voltage is applied across the top of the circuit, point 1, and the bottom of the circuit, point 3. Variable resistor R_1 acts as voltage divider network across the battery. The voltage at point 2 is the **reference voltage**. If the voltage at point 2 is 10 V or less, Q_1 will conduct.

If we assume that the voltage at point 2 is 10 V, the emitter–base junction of Q_1 is forward-biased. The zener diode, *Z*, holds the emitter voltage at a steady value of 11 V. Since the base of the PNP transistor is more negative than the emitter, Q_1 conducts and the SCR is fired. When the SCR is on, current through the field winding is supplied by the car battery. As the current in the rotating field increases, the voltage generated by the ac generator increases. If the voltage peaks generated by the alternator are greater than the voltage of the battery, the diodes in the rectifier are forward-biased and the battery begins to charge.

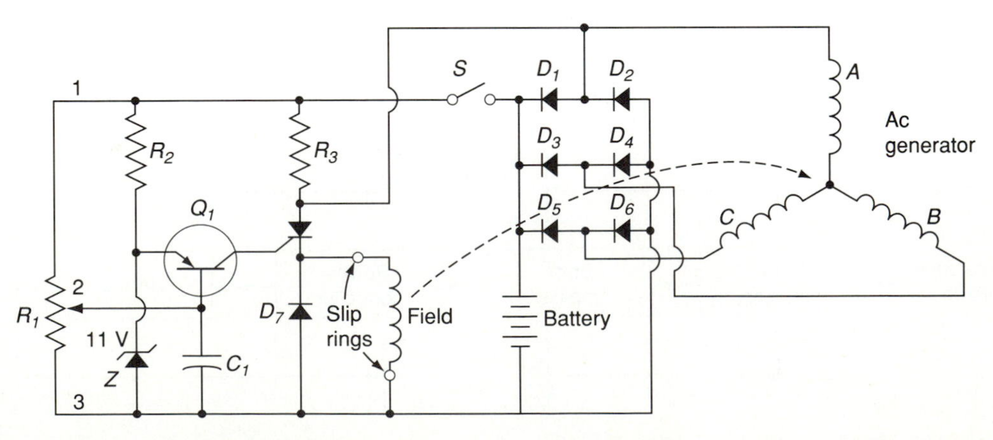

FIGURE 19-14 Voltage regulator circuit for automobile alternator.

When the anode of the SCR becomes negative during each cycle of the ac generated voltage, the SCR becomes reverse-biased and shuts off. The SCR is fired again when the anode returns positive (provided that the battery voltage is low). When the battery is fully charged, the voltage at the base of Q_1 equals the zener voltage and the transistor becomes reverse-biased and the SCR stops firing.

19-11 PARALLEL OPERATION OF AC GENERATORS

Alternating-current power systems usually consist of several generating units connected in parallel to a common bus line. The same basic reasons for paralleling dc generators are also applicable for the parallel operation of ac generators. The main advantages of parallel operation are as follows:

1. Having many generators in parallel increases the reliability of the power system. If only one large alternator supplies an entire power grid, in the event of that machine failing, the generating station is no longer functional.
2. If only one large generator is used and it is not operating at full load, the system becomes uneconomical. Several smaller generators may be removed or added in accordance with the demands placed on the system. The smaller generators that are in operation are then operating near full load, and the system is far more efficient.
3. Increases in consumer demand, such as new factories and housing, can be handled by adding generators without disturbing the original installation.
4. The power demands placed on a single alternator may exceed the capabilities of that machine. There are circumstances where electric power loads exceed the sizes to which a single alternator can be built. For example, Niagara Falls routinely exports 1 million kilowatts of power to New York City—it would not be feasible to build a single generator capable of handling this amount of power.

When attempting to operate ac generators in parallel, or for the paralleling of an alternator to a common bus line, it is necessary to observe the same precautions as in the case of the direct-current generator. These precautions are as follows:

1. When two alternators are to be paralleled, both machines must have the same voltage rating.
2. The frequency of both alternators must be the same, even though the speed of each machine may be different. For example, a four-pole alternator

driven at 1800 rpm may be operated in parallel with a six-pole alternator driven at 1200 rpm, since both machines generate 60 Hz.

3. The phase sequence of polyphase alternators must be the same.
4. For alternators to operate satisfactorily in parallel, they both must have a drooping speed characteristic. That is, the speed must fall as the load is increased.
5. The individual phase voltages of each alternator must be connected to each other so that the phase angles are equal. In other words, phase *A* of alternator 1 is connected to phase *A* of alternator 2.

19-12 SYNCHRONIZING ALTERNATORS

The first condition for paralleling, or synchronizing, alternators was stated in the preceding section: The terminal voltages must be equal. This can be determined by connecting voltmeters across the machine terminals. The frequency, phase relations, and phase sequence of the two alternators that are to be paralleled are checked by using either synchronizing lamps or a **synchroscope**. There are two simple connections using synchronizing lamps which will determine whether the phase sequence is correct. These circuits are the **dark lamp** and the **bright lamp**. Both connections will also provide a means of testing for the phase relations of the alternators.

Figure 19-15 shows the bright lamp method of testing phase sequence. Two lamps are connected in series on each phase, due to the high voltages which may be encountered with this type of connection. The prime mover of the incoming machine brings the alternator up to approximately its rated speed. The field current of the incoming alternator is adjusted so that its terminal voltage is now the same as that of the running alternator. If the phases are properly connected, the three lamps will go bright and then dark at the same time. If the phase sequence is not correct, the lamps will go bright and dark alternately. If the three lamps flicker, it is an indication that there is a difference in frequency between the two alternators. Unfortunately, the flicker of lamps does not indicate which machine has the higher frequency. However, if the incoming generator has its speed increased slightly and the rate of lamp flicker increases, the incoming alternator is above rated frequency and should have its speed adjusted accordingly.

The two bright/one dark lamp phase sequence is shown in Figure 19-16. Using this method, by noting the sequence of brightness of the lamps, it can be determined whether the incoming generator is fast or slow. Assume that the incoming generator is out of synchronism, and lagging, the running generator. All three lamps will appear to glow steadily. As the lagging alternator is

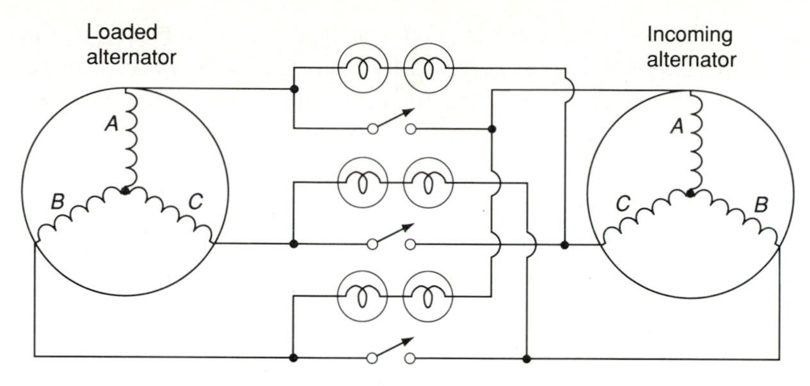

FIGURE 19-15 Connections for synchronizing lamps using the bright lamp method.

accelerated, the lamp, or differential, frequency decreases until the lights begin to flicker noticeably. The flickering of the lamps will have a rotating sequence if the connections are correct, and will indicate which generator is faster. At a point approaching synchronism, the lamps connected between phase *C* of the two alternators will darken. This is because the potential difference between phase *C* of the incoming alternator and phase *C* of the loaded alternator is almost zero. Under perfectly synchronized conditions, the phase voltages across the other sets of lamps are 120° apart, due to their cross-connection, and will glow brightly. If the phase sequence is incorrect, the lamps connected to either phase *A* or *B* will become brighter as the lamps connected to phase *C* become darker.

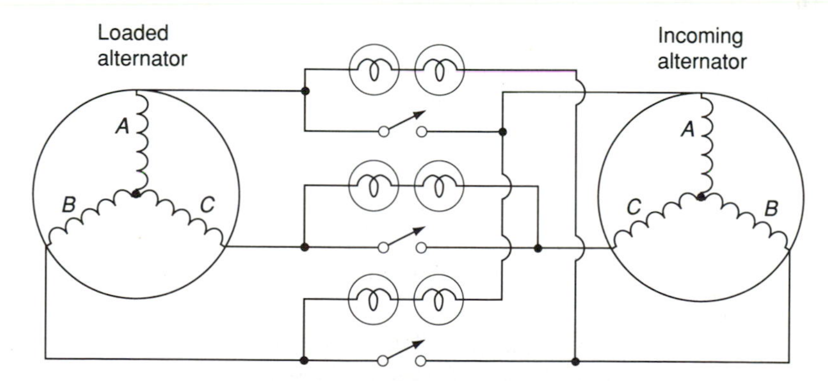

FIGURE 19-16 Connections for synchronizing lamps using the rotating lamp (two bright, one dark) method.

The dark lamp phase sequence method is shown in Figure 19-17. If the alternator is connected properly, the three lamps should all become bright and dark simultaneously. If they brighten and darken in sequence, it means that the phase rotation of the incoming alternator is opposite to that of the existing alternator. This problem is corrected by reversing one phase. Since the lamps flicker at a frequency equal to the difference in the frequency between the two machines, as the alternators approach synchronism, the flicker becomes less and less rapid. When all the lamps are dark, the switches may be closed. The fact that the lamps are dark indicates that the potential difference across the lamps is virtually zero. One disadvantage of this type of method is that the lamps do not indicate whether the incoming alternator is running fast or slow.

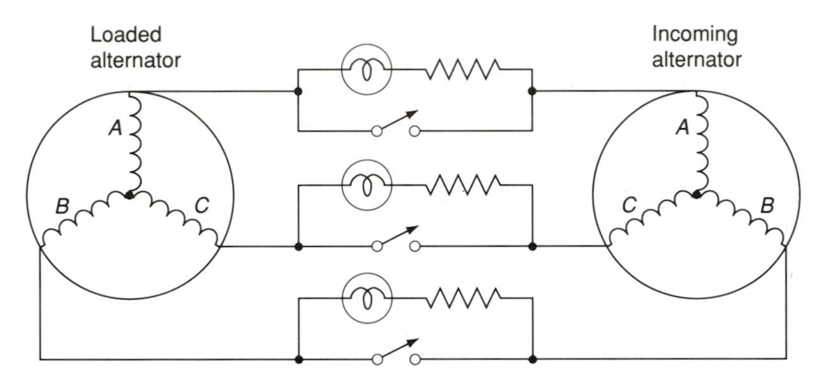

FIGURE 19-17 Connections for synchronizing lamps using the dark lamp method.

19-13 SYNCHROSCOPE

The synchronizing methods mentioned in the preceding section are not foolproof, and the possibility of error when using these methods does exist. For this reason, most utility companies will use a synchroscope when paralleling alternators. As shown in Figure 19-18, a synchroscope is an instrument with a rotating pointer which indicates whether the incoming alternator is running fast or slow. When the synchroscope pointer is in the vertical position, the voltages are in phase and the two alternators can be paralleled. The one disadvantage of the synchroscope is that it cannot detect phase sequence, something that the synchronizing lamp method is capable of.

The synchroscope pointer is polarized to the frequency of the running machine by means of a rotor coil. The stator winding consists of two coils, one connected across two lines of the incoming machine. Therefore, the synchroscope is actually a single-phase induction motor, where the rotating field of the stator rotates at the frequency of the incoming machine, and the pointer is magnetized at the running alternator's frequency. If the incoming machine is fast, the pointer will rotate clockwise over a circular scale. Consequently, if the incoming alternator is slow, the pointer rotates counterclockwise. The

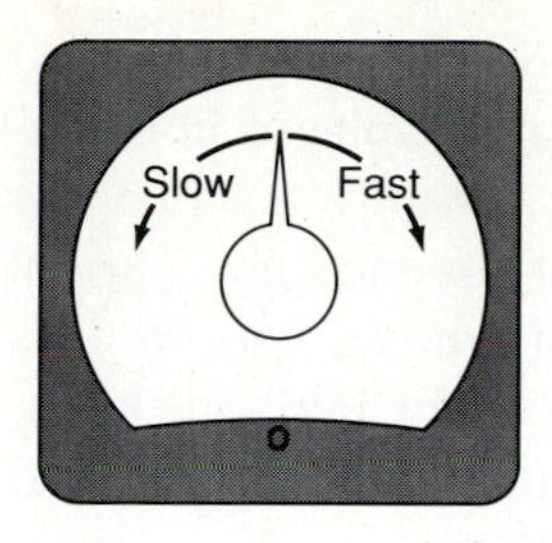

FIGURE 19-18 Synchroscope.

number of revolutions per second of the pointer is equal to the difference in frequency between the two machines.

19-14 HUNTING

If two alternators are operating in parallel and delivering a common load, they are said to be in a stable equilibrium. That is, both are in synchronism and operating at the same frequency, provided that the speed of the prime mover is constant over a full cycle of revolution. Unfortunately, the driving torque of a steam, diesel, or gasoline engine is of a reciprocating type and is not uniform during a revolution of the flywheel. Even with a heavy flywheel, this variation of torque results in the alternator being driven ahead of synchronism on the power stroke, and it falls behind synchronism on the return stroke of the engine.

Should one alternator slow down, due to a sudden decrease in the speed of its prime mover, the second machine supplies power to pull it into synchronism again. The impulse received by the rotor causes it to oscillate with respect to its normal operating speed. The oscillations are called **hunting**. Even a slight variation of the rotor's angular position can adversely affect the phase angle of the voltage. For example, if the rotor of a 60-pole alternator is displaced by 1°, it will make a difference of 30° in the phase angle of the emf.

If the reciprocating action that produces the speed pulsation is repeated periodically, the oscillations, instead of dying out, will increase in amplitude. Each successive oscillation will increase over the previous one, and if no protective devices are present, serious damage may result. It is therefore necessary to reduce this compounding effect of oscillations. The following techniques are some of the methods used to reduce hunting:

1. Changing the natural period of vibration of the machine by changing the flywheel. A large, heavy flywheel will increase the inertia of the prime mover, which will result in a more constant speed.

2. **Amortisseur**, or **damper**, **windings** will also reduce hunting. Essentially, the amortisseur winding is a squirrel-cage winding placed in the pole faces of the alternator. At synchronous speed, the armature reaction flux is stationary, relative to the fields, and does not produce any current in the windings. If the machine falls below, or runs above, synchronous speed, the flux sweeps across the windings and produces emfs in them. Consequently, large values of current will flow, which will react on the field and tend to hold the machine in exact synchronism.
3. Dashpots are used on the throttles, or governors, of fuel engines. The dashpot serves to dampen the governor to prevent its immediate response to sudden changes in power demand from the alternator.

19-15 RATING AND LOAD CAPACITY OF THE ALTERNATOR

Alternators are usually rated in kVA, at a definite voltage and frequency. Occasionally, the alternator is rated in kilowatts and a unity power factor is assumed. The maximum output of an alternator is limited by its operating temperature. The temperature rise is due to the losses in the machine. This loss depends on the magnitude of the current and is independent of the power factor. For example, a 200-A 480-V alternator will produce the same I^2R loss regardless of the power factor. However, the power factor will affect the output power of the machine. The output power of an alternator is determined by the following equation:

$$P = nEI \cos \theta \qquad \text{watts} \tag{19-12}$$

where

E = voltage per phase

I = full-load current per phase

$\cos \theta$ = power factor of the load

n = number of phases

From equation 19-12 it is apparent that the power output of an alternator depends on the voltage, which is a fixed quantity, the current, which is variable, and the copper losses and other losses in the machine, as well as the power factor of the load. Because of this, alternators should not be rated in watts, but rather, in volt-amperes, or kVA.

EXAMPLE 19-11 A 13,800-V three-phase 60-Hz alternator is rated at 12,550 kW at 0.8 power factor and has a full-load efficiency of 0.96 excluding the field loss. Determine (a) the kVA rating of the alternator; (b) the current rating; (c) the power delivered by the prime mover at full load.

Solution (a) $\text{kVA} = \dfrac{12{,}550}{0.8} = 15{,}687.5\text{ kVA}$

(b) $I = \dfrac{15{,}687.5\text{ kVA} \times 10^3}{\sqrt{3} \times 13{,}800\text{ V}} = 656.32\text{ A}$

(c) $\dfrac{12{,}550\text{ kVA}}{0.96} = 13{,}072.92\text{ kW}$

KEY TERMS

Alternating-current generator
Synchronous generator
Synchronous alternator
Alternator
Salient pole rotor
Nonsalient (cylindrical) rotor
Concentrated winding
Distributed winding
Full-pitch winding
Fractional-pitch winding
Pitch factor
Distribution factor
Synchronous impedance method
Synchronous reactance
Open-circuit saturation curve
Short-circuit characteristic (SCC)
Voltage regulator
Zener diode
Reference voltage
Synchroscope
Dark lamp
Bright lamp
Hunting
Amortisseur (damper) windings

PROBLEMS

19-1 Determine the frequency of a four-pole alternator operating at 1800 rpm.

19-2 Calculate the speed required for a six-pole ac generator to develop 50 Hz.

19-3 Determine the pitch factor for a three-phase alternator which has coils wound with a $^7/_8$ coil pitch.

19-4 Repeat Problem 19-3 for a $^3/_4$ coil pitch.

19-5 Determine the pitch factor of a four-pole 32-slot three-phase alternator. The windings span five slots.

19-6 Calculate the distribution factor for a 36-slot four-pole three-phase alternator.

19-7 Find the distribution factor for a 48-slot eight-pole three-phase alternator.

19-8 Determine the effective voltage generated in one phase of an alternator given the following data: $F = 60$ Hz, turns per phase = 320, and maximum flux per pole = 0.025 Wb.

19-9 A three-phase alternator has eight poles, 120 slots, and 120 eight-turn coils. The coil span is 12 slots. The rotor is driven at a speed of 900 rpm, and the rotor flux is 0.045 Wb/pole. Determine (a) the number of series turns per phase; (b) the frequency; (c) the distribution factor; (d) the pitch factor; (e) the total generated voltage per phase.

19-10 A 10,000-kVA 13,800-V 60-Hz three-phase eight-pole synchronous generator has a flux per pole of 0.368 Wb. There are 96 slots, the armature is lap wound, $^7/_8$ pitch with two coil sides per slot and four turns per coil. Determine (a) the slots per pole per phase; (b) the distribution factor; (c) the pitch factor; (d) the total number of turns per phase; (e) the total generated voltage per phase.

19-11 The terminal voltage of a three-phase alternator rises from 550 V to 690 V when the load is removed. Calculate the voltage regulation.

19-12 A 20,000-kVA 13,800-V 60-Hz three-phase wye-connected alternator has an effective armature resistance of 0.3 Ω/phase and a combined armature reactance of 5.5 Ω/phase. Determine the voltage regulation at a power factor of 0.85 lagging.

19-13 Repeat Problem 19-12 for a leading power factor of 0.91.

19-14 A 500-kVA 1100-V wye-connected three-phase alternator has had open-circuit and short-circuit tests performed and the short-circuit current was found to be 175 A. The dc resistance of the windings is averaged at 0.29 Ω. Calculate the values of R_a, Z_s, and X_s.

19-15 Determine the percent regulation of the alternator of Problem 19-14 if it is connected to a load with a unity power factor.

19-16 A 200-kVA 2300-V 60-Hz wye-connected synchronous generator has a measured armature resistance of 0.15 Ω/phase and a synchronous reactance of 1.75 Ω/phase. Determine the percent voltage regulation if the generator is connected to a 0.88 lagging PF load.

19-17 A 20-kVA 480-V 60-Hz three-phase wye-connected synchronous generator has a per phase armature resistance of 0.4 Ω and a field-winding resistance of 5.2 Ω. The excitation current is 12.5 A for a 0.82 lagging PF load. Determine the efficiency of the generator if the mechanical losses are 280 W and the core losses are 400 W.

19-18 Repeat Problem 19-17 for a 0.87 leading PF and an excitation current of 11.3 A.

19-19 A 2500-kVA 2300-V 60-Hz three-phase synchronous generator is driven by a calibrated dc motor. The following data are obtained: dc motor output with no field excitation, $P_1 = 72$ kW. Dc motor output with normal field excitation, $P_2 = 100$ kW; output of dc motor with short-circuited armature and rated current flowing, $P_3 = 122$ kW. The field excitation of the generator is 24 A at 120 V dc. Determine (a) the core losses; (b) the copper and stray losses; (c) the field losses; (d) the full-load efficiency at 0.85 lagging power factor.

19-20 A 2300-V three-phase 60-Hz alternator is rated at 800 kW at 0.82 PF. The full-load efficiency is 0.93, excluding field loss. Determine (a) the kVA rating of alternator; (b) the current rating; (c) the power delivered by prime mover at full load.

CHAPTER

20 Synchronous Motors

LEARNING OBJECTIVES

Upon completion of this chapter you will be able to:

- Explain the basic operating principles of the synchronous motor.
- List the three different torques associated with the synchronous motor.
- Define the synchronous-induction motor and the supersynchronous motor.
- Plot the V-curves for a synchronous motor.
- Describe how the power factor of a synchronous motor can be controlled.

20-1 INTRODUCTION

When two alternators are operating in parallel and the prime mover is disconnected from one, the alternator with no prime mover will continue to rotate by drawing power from the other alternator. An alternator operating in this manner is considered to be a **synchronous motor**. Any alternator will operate as a synchronous motor, in the same manner as any dc generator may operate as a dc motor. A synchronous motor moves in synchronism with the rotating magnetic field that is created by its stator windings. Consequently, the motor will operate at an average constant speed, regardless of the load.

The stator of a synchronous motor is similar to that in an induction motor. One main difference is that the synchronous motor has field poles that are excited from a direct-current source, either from a small generator driven by the motor or from an outside supply. Also, the power factor of the induction motor is always lagging, while the synchronous motor can be varied over a wide range by adjusting the dc field excitation. This permits using the motor for power factor correction as well as for driving mechanical loads.

20-2 PRINCIPLES OF OPERATION

The operation of the synchronous motor is due to the reaction between a magnetic field of a fixed polarity produced by a direct current and a field of constantly changing polarity set up by an alternating current. Therefore, two magnetic fields are present in the synchronous motor, and the stator field will tend to line up with the rotor field.

When the rotor winding and the stator winding are energized as the poles of the rotating magnetic field move toward rotor poles of opposite polarity, as shown in Figure 20-1(a), the force of attraction will tend to turn the rotor in a direction opposite to that of the rotating field. In the next half-cycle, the frequency reverses the direction of current in the coil and the rotor attempts to turn in a clockwise direction, as shown in Figure 20-1(b). Therefore, the rotating field tends to pull the rotor poles first in one direction and then in the other, and as a result, the starting torque is zero. However, if the armature and coil can, by some means, be rotated or advanced one pole pitch during the time the current in the coil is reversed, the conductors are brought under poles of opposite polarity and the torque will still be in a clockwise direction, which will produce rotation.

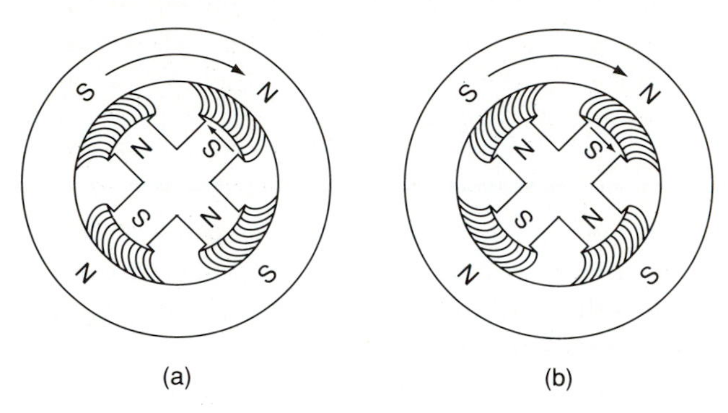

FIGURE 20-1 Operating principles of a synchronous motor: (a) tendency of rotor to turn counter-clockwise; (b) tendency of rotor to turn clockwise.

In a synchronous motor, either the armature or field structure must rotate or advance a distance equal to one pole pitch each half-cycle, if the motor is to operate. The synchronous motor must then operate at constant speed if the frequency is constant. If this average speed varies by even a slight amount, the average torque will ultimately become zero and the motor will come to a standstill. Therefore, a synchronous motor will either run at synchronous speed or not at all. The speed of any synchronous motor can be determined by the same equation as for the synchronous alternator, which is

$$N = \frac{120f}{P}$$

where N = motor speed, in rpm

f = frequency, in hertz

P = number of poles

EXAMPLE 20-1 Determine the rated-load speed and no-load speed of a 13,800-V 500-kVA 60-Hz 12-pole synchronous motor.

Solution With constant frequency, the motor speed must be the same at all loads.

$$N = \frac{120 \times 60}{12} = 600 \text{ rpm}$$

20-3 STARTING SYNCHRONOUS MOTORS

The single-phase synchronous motor has no starting torque, although the polyphase synchronous motor when operated without load is self-exciting, as there is always some turning effort exerted on the rotor due to the fact that as the current in the stator coils of one phase reaches zero the current is increasing in the coils of the other phase or phases, resulting in a revolving field around the surface of the armature. With the field circuit open, this rotary flux sets up eddy currents in the pole faces and reacts with them to develop torque. To assist the starting of polyphase synchronous motors, the polar faces of the rotating field have copper bars embedded in parallel slots in the rotor core, the bars being connected to end rings forming an auxiliary cage winding. In this way, a complete squirrel-cage winding is formed. This type of winding is referred to as an amortisseur, or damper, winding. Amortisseur windings will also damp out any tendency of the motor to hunt or oscillate. To start the motor, the dc field winding is short-circuited, so the rotor is left deenergized and a polyphase voltage is applied to the stator. Therefore, the motor is simply started as an induction motor and brought up to synchronous speed. The rotor is then excited from the dc supply and the field rheostat adjusted for minimum line current. The opposite polarity poles on the stator and rotor will attract each other, causing the rotor to lock in step with rotating field and be pulled around at synchronous speed.

When the motor is running at synchronous speed, if the armature has the correct polarity, the stator current will decrease when the excitation voltage is applied. As the field circuit is energized, it may be found that the armature polarity is incorrect, and the rotor will then drop back in phase by 180°. This change in phase will also be accompanied by a sudden increase in current. To avoid this scenario, it is better to excite the fields through a large resistance just before synchronism is achieved, then increase the field to normal and raise the applied voltage to full value. When a synchronous motor is started using this method, the motor will draw a large value of lagging current, since the applied voltage must be consumed by the impedance of the armature, and the power factor is very low. The applied starting voltage should be reduced to about one-third of its full value in order to reduce the starting current.

A simple method of reducing the applied starting voltage is to connect a resistor of low ohmic value across the rotor dc field winding during the starting period. The dc field winding is disconnected from the source during starting, and the resistor is connected across the field terminals. This allows alternating current to flow in the dc field winding. Since the impedance of this winding is high compared with the inserted external resistance, the internal voltage drop limits the terminal voltage to a safe value. When extremely large synchronous motors are started as induction motors, the same methods for reducing voltage and current are used as the methods employed for starting induction motors. These methods include autotransformer, wye–delta, line-resistance starting, and others.

20-4 STARTING SYNCHRONOUS MOTORS UNDER LOAD

If a load is applied to a dc shunt motor, the speed will decrease slightly, which reduces the counter emf and allows more current to enter the armature, and the motor is able to carry the increased load. When load is applied to a synchronous motor, the speed cannot decrease, since the motor must operate at constant speed. Because the field does not change, the counter emf remains relatively constant during this loading process. Therefore, the synchronous motor cannot draw additional current from the line, as the counter emf does not decrease. Also, the motor is separately excited and its counter emf is controlled exclusively by the value of the direct current in its field windings. Under normal conditions, the field current is usually adjusted so that the counter emf is practically equal to the terminal, or line, voltage. This is referred to as the **normal excitation** of the machine. Even under no-load conditions, the counter emf cannot reach its maximum value at exactly the same instant as the line voltage, since they are in direct opposition to each other, as shown in the phasor diagram of Figure 20-2. In this diagram, E_R is the resultant of V_t and E_{gp}. The armature current in a synchronous motor can be defined by the following equation:

$$I_a = \frac{V_t - E_{gp}}{R_a + jX_s} = \frac{E_R}{Z_s} \tag{20-1}$$

where I_a = armature current

V_t = applied terminal voltage

E_{gp} = voltage generated per phase in armature

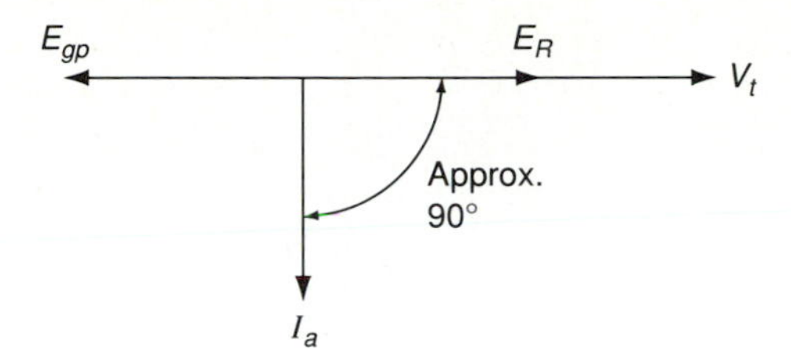

FIGURE 20-2 Resultant phase current in a synchronous motor.

R_a = effective armature resistance per phase conductors

X_s = synchronous armature reactance per phase

E_R = phasor difference between the applied armature voltage and the generated voltage

Z_s = synchronous armature impedance per phase

By rearranging equation 20-1, the generated voltage per phase, E_{gp}, is calculated as follows:

$$E_{gp} = V_t - I_a(R_a + jX_s)$$

In a normal stator winding, the effective armature resistance is maintained at as low a value as possible to reduce I^2R losses. The synchronous armature impedance is predominately a reactive component. The angle between E_R and I_a is virtually 90°. Since the input to the motor has a 90° phase angle, the input power must be zero and there is no torque developed by the motor to maintain rotation. Under conditions of normal excitation, the dc excitation is adjusted so that the armature current is rotated approximately 90° and is in phase with the terminal voltage, V_t.

Figure 20-3 shows the phasor diagram of a synchronous motor under normal conditions of excitation. When load is applied to the motor, it causes the poles to be pulled α degrees behind their no-load position, and the counter emf occurs α degrees later. The resultant voltage, E_r, causes the armature current, I_a, to lag behind V_t by angle θ. Because the speed is constant, if the field excitation is decreased, the counter emf, E_{gp}, decreases, and the resultant voltage, E_R, becomes greater. The armature current will increase in value and lag the terminal voltage, V_t, by a greater amount. On the other hand, if the field excitation is increased until the armature current and terminal voltage are in phase, as in Figure 20-3, the power factor of the motor becomes unity for a given load.

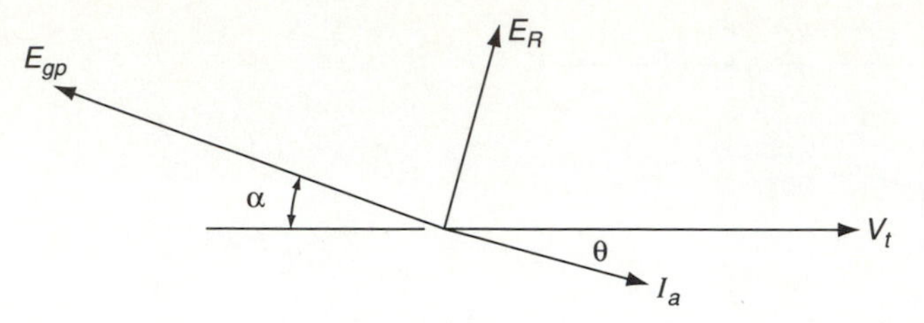

FIGURE 20-3 Phasor diagram of a synchronous motor under normal conditions of excitation.

The resultant voltage, E_R, may be calculated for any value of α by using the equation

$$E_R = (V_t - E_{gp} \cos \alpha) + j(E_{gp} \sin \alpha) \tag{20-2}$$

Angle α is given in electrical degrees. In some situations it is necessary to convert from mechanical degrees to electrical degrees. This is accomplished by using

$$\alpha = P\frac{\beta}{2} \tag{20-3}$$

where β is the mechanical degrees and P is the number of poles.

EXAMPLE 20-2 A 12-pole 480-V 60-Hz wye-connected three-phase synchronous motor is operating at no-load with its generated voltage per phase and terminal voltage at equal values of magnitude. The rotor is shifted 0.75 mechanical degree from its synchronous position, and the synchronous reactance and effective resistance are 12 Ω and 1.6 Ω per phase, respectively. Determine (a) the rotor shift in electrical degrees from its synchronous position; (b) the resultant emf across the armature; (c) the armature current.

Solution

(a) $\alpha = P\frac{\beta}{2} = 12\frac{0.75}{2} = 4.5°$

(b) $V_t = \frac{E_L}{\sqrt{3}} = \frac{480\text{ V}}{\sqrt{3}} = 277.13\text{ V}$

$$E_{gp} = V_t = 277.13\text{ V}$$

$$\begin{aligned}E_R &= (V_t - E_{gp} \cos \alpha) + j(E_{gp} \sin \alpha)\\ &= (277.13\text{ V} - 277.13\text{ V} \cos 4.5°) + j(277.13\text{ V} \sin 4.5°)\\ &= 0.854 + j21.743\\ &= 21.76 \underline{/87.8°}\text{ V}\end{aligned}$$

$$\text{(c) } Z_s = R_a + jX_s$$
$$= 12.11 \angle 82.4^\circ\ \Omega$$

$$I_a = \frac{E_R}{Z_s} = \frac{21.76 \angle 87.8^\circ \text{ V}}{12.11 \angle 82.4^\circ\ \Omega} = 1.8 \angle 5.4^\circ \text{ A}$$

As load is applied to a synchronous motor, the rotor shifts backward in phase position and draws a power component of current from the line, which supplies the additional power necessary to carry the increase in load. The angle by which the rotor shifts backward at any load is called it **synchronous position**, and the angle by which it shifts backward at full load is called the **torque angle**.

20-5 SYNCHRONOUS MOTOR TORQUE

There are three different torques that must be taken into consideration when specifying the characteristics of a synchronous motor for a given application. These three torques are as follows:

1. *Starting torque.* This torque is an indication of the motor's ability to start the load from rest.
2. *Pull-in torque.* This represents the ability of the motor to maintain operation during the change from induction to synchronous operation.
3. *Pull-out torque.* This torque specifies the motor's ability to maintain operation under peak-load conditions.

The motor's starting and pull-in torques are characteristics of the synchronous motor when it is operating initially as an induction motor and are determined by the design of the starting winding. The pull-out torque is a characteristic of the motor when it is operating in its synchronous mode, and is dependent directly on the field strength of the machine. The starting torque of a synchronous motor can range from about 50% to in excess of 200% of full-load torque. Pull-in torque can vary from 35 to 125%, while pull-out torques can be as high as 400% of the full-load value.

When a motor is started, sufficient torque must be developed to start and accelerate the load under the most difficult conditions. The starting torque varies as the square of the applied voltage, and if the supply voltage is low, the starting torque may be too low to start the motor or it may accelerate so slowly that the windings will become overheated.

If the resistance of the squirrel-cage starting winding is increased, the starting characteristic of the motor will improve, although this alone may not be sufficient to accelerate the motor to 95% of synchronous speed. The method of connecting the field circuit will also have an effect on the starting characteristic of the synchronous motor. The field circuit may be either open-circuited or short-circuited, depending on the application of the motor.

Figure 20-4 shows the full-voltage starting characteristic of a large 25-Hz synchronous motor with the field of curve (a) open-circuited, and curve (b) represents the field short-circuited through a resistance. With the field open-circuited, the starting torque is about 350% of full load, and the starting kVA or current is about 525% of normal. When the field is close through a suitable resistance, the starting torque is decreased to 300% and the starting kVA is increased to 575%.

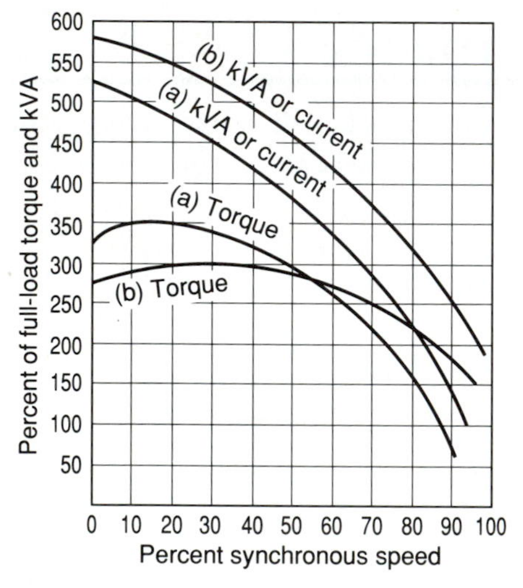

FIGURE 20-4 Full-voltage starting characteristics of synchronous motor: (a) with field open-circuited and (b) with field closed through a resistance.

The maximum torque of a synchronous motor occurs when the torque angle, α, is 90°. When the torque on the synchronous motor's shaft exceeds the maximum, or pull-out, torque, the rotor will no longer be locked in synchronism with the stator and magnetic fields. Consequently, the rotor starts to slip. As the speed of the rotor decreases, the magnetic field of the stator continuously passes the rotor, which results in the induced torque of the rotor reversing each time it is passed. Severe vibration occurs in the synchronous motor when the torque surges first in one direction and then in the other. This loss of synchronization after maximum torque is exceeded is known as **slipping poles**. Maximum torque is determined as follows:

$$T_{max} = \frac{3V_tE_{gp}}{\omega_sX_s}$$

The torque developed by a synchronous motor is found by the following equations:

$$\text{English units: } T_D = \frac{7.04}{N_s}(\sqrt{3}\,E_LI_L\cos\theta - 3I_a^2R_a) \quad (20\text{-}4)$$

$$\text{SI units: } T_D = \frac{\sqrt{3}\,E_LI_L\cos\theta - 3I_a^2R_a}{\omega_s} \quad (20\text{-}5)$$

where N_s is the synchronous speed, in rpm, and ω_s is the synchronous speed, in radians/second. The maximum power developed, P_D, is equal to the input power minus the stator copper loss. In equation form

$$P_D = \sqrt{3}\,E_LI_L\cos\theta - 3I_a^2R_a \quad (20\text{-}6)$$

Since the torque angle, α, of a synchronous machine is the angular displacement between the voltage generated per phase, E_{gp}, and the terminal voltage, V_t, it may be calculated using equation 20-4. The armature current and synchronous reactance are *added* in the numerator when the power factor is leading, and *subtracted* when the power factor is lagging.

$$\alpha = \theta - \tan^{-1}\frac{V_t\sin\theta \pm I_aX_s}{V_t\cos\theta \pm I_aR_a} \quad (20\text{-}7)$$

The angle between the resultant voltage, E_R, and the terminal voltage, V_t, is represented by the symbol δ, and is calculated as follows:

$$\delta = \tan^{-1}\frac{E_{gp}\sin\alpha}{V_t - E_{gp}\cos\alpha} \quad (20\text{-}8)$$

In situations where the generated voltage per phase and terminal voltage are equal, the resultant voltage, E_R, would bisect the angle between E_{gp} and V_t. The angle between E_R and V_t would therefore be calculated as

$$\delta = \frac{180 - \alpha}{2} \quad (20\text{-}9)$$

When $E_{gp} = V_t$, the resultant voltage is determined as follows:

$$E_R = 2V_t\cos\delta \quad (20\text{-}10)$$

The synchronous impedance angle, β, is the angle between the resultant voltage, E_R, and the armature current, I_a. When the motor is operating at a lagging power factor, the angle between E_R and V_t is equal to β minus the phase angle θ. In equation form,

$$\text{lagging PF: } \delta = \beta - \theta$$

When the motor is operating at a leading power factor, the angle between E_R and V_t is the sum of β and θ, or

$$\text{leading PF: } \delta = \beta + \theta$$

When the synchronous motor is operating at unity power factor, θ is zero since there is no angular displacement between V_t and I_a. Therefore,

$$\text{unity PF: } \delta = \beta$$

EXAMPLE 20-3 A 20-pole 850-kW 13,800-V three-phase wye-connected synchronous motor has a synchronous reactance of 6.5 Ω/phase and an effective armature resistance of 0.75 Ω. The efficiency of the motor at the rated load, 0.85 leading, is 94% Determine the torque angle.

Solution

$$\text{Input power} = \frac{850 \times 10^3}{0.94} = 904.3 \text{ kW}$$

$$I_a = \frac{\text{input power}}{\sqrt{3}\,E_L \cos\theta} = \frac{904.3 \times 10^3}{\sqrt{3} \times 13{,}800 \times 0.85} = 44.5 \text{ A}$$

$$V_t = V_p = \frac{E_L}{\sqrt{3}} = 7967.4 \text{ V}$$

$$\theta = \cos 0.85 = 31.8°$$

$$\begin{aligned}\alpha &= \theta - \tan^{-1}\frac{V_t \sin\theta + I_a X_s}{V_t \cos\theta - I_a R_a} \\ &= 31.8° - \tan^{-1}\frac{7967.4 \sin 31.8° + (44.5 \times 6.5)}{7967.4 \cos 31.8° - (44.5 \times 0.75)} \\ &= -1.86°\end{aligned}$$

The torque angle of a synchronous machine is also referred to as the **power angle**. When speed is constant, torque and power are linearly proportional. Since the power developed by a synchronous motor is dependent on the torque angle, the following equation may be used to determine power:

$$P = \frac{3V_t E_{gp} \sin \alpha}{X_s} \tag{20-11}$$

20-6 V-CURVES OF THE SYNCHRONOUS MOTOR

When the load on a synchronous motor is held constant and the terminal voltage does not change, the power factor is determined by the field excitation. A weak field causes a lagging current, while a strong field results in a leading current. The power factor at which the synchronous motor operates, and consequently the armature current, can be controlled by adjusting the field excitation. By reducing the field current, a lagging armature current results which will exceed the minimum current at normal excitation. The curve that indicates the relation between field current and armature current at a constant load and terminal voltage is referred to as a **V-curve**, due to its characteristic shape.

The performance of a synchronous motor under different levels of field excitation can be determined by performing a V-curve test, as shown in Figure 20-5. A direct-current ammeter and a field rheostat are connected in series with the field wiring and the alternating-current instruments are inserted in the line. Since the motor is a balanced three-phase load, the two-wattmeter method of power measurement is shown, although any polyphase wattmeter method discussed in Chapter 16 would be applicable. The power input is held constant by adjusting the mechanical load on the motor. A constant terminal voltage is also required for this test. The motor is tested by applying a load, and by using the rheostat, the field current is varied in increments from the lowest to the highest possible value. For each setting of the rheostat, the values of the line currents, field current, and power input are obtained. This allows the power factor of the motor to be calculated for each individual field current setting. The results of these readings are then plotted as V-curves.

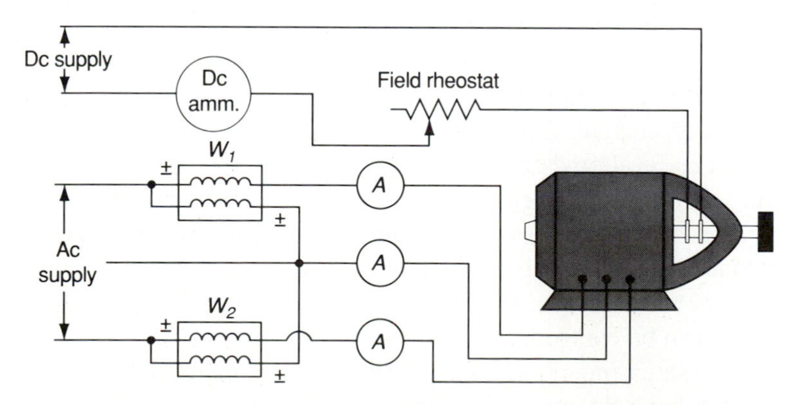

FIGURE 20-5 Circuit connections for V-curves of synchronous motor.

A family of V-curves is shown in Figure 20-6 at no-load, half-load, and full-load. Note that these curves are not symmetrical in shape, and the minimum values of armature current for the three load conditions occur at different values of field current. The armature current is minimum at unity power factor and increases as the power factor becomes poor, either leading or lagging. If the motor is heavily overloaded, the field poles are pulled out of step with the rotating field and a heavy current will flow through the armature. The power factor of the synchronous motor can be determined from the wattmeter readings, as shown in Figure 20-6(b), and the results are plotted against the field current for the three given loads. In both diagrams it can be seen that as the load is increased, each of the curves in the family will shift to the right.

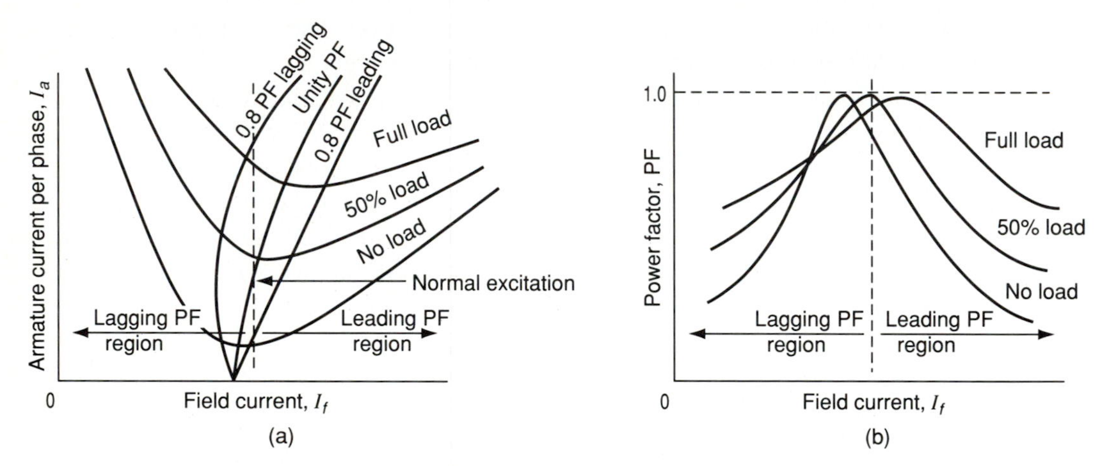

FIGURE 20-6 Synchronous motor V-curves: (a) armature current versus field current; (b) power factor versus field current.

20-7 SYNCHRONOUS MOTOR POWER FACTOR CONTROL

The power factor of an induction motor when connected to a given load cannot be altered without changing the motor design. Also, the induction motor must take a lagging current. The power factor of the synchronous motor at any given load can be altered at will, and the current can be changed from lagging to leading simply by changing the field excitation. As mentioned previously, the rotor of a synchronous motor shifts backward in phase position with respect to its stator as load is applied, but continues to run at a constant average speed in synchronism with the line voltage. The power factor of the synchronous motor can be controlled by increasing or decreasing the field current.

In the dc shunt motor, when the field current is increased, the motor slows down. When the field current is increased in the synchronous motor, it cannot slow down, but instead must run at a constant speed. Since the speed is constant, the counter emf is controlled exclusively by the field excitation, and is increased when the field current is increased. Therefore, the counter emf of the synchronous motor is independent of the load, and the motor will operate even when its counter emf is greater than the line voltage. Under this condition, the motor takes a leading current from the line and is said to be **overexcited**, as shown in Figure 20-7. Overexcitation produces a demagnetizing effect because of increased armature reaction with increased load. The result of this demagnetizing action is a reduction in the air-gap flux to a value just sufficient to set up the necessary induced, or counter emf, in the armature. Consequently, a leading current in a synchronous motor has the same effect on the air-gap flux as a lagging current if the same machine is used as an alternator. The net result of this demagnetizing, or weakening, of the field, is to produce a leading power factor.

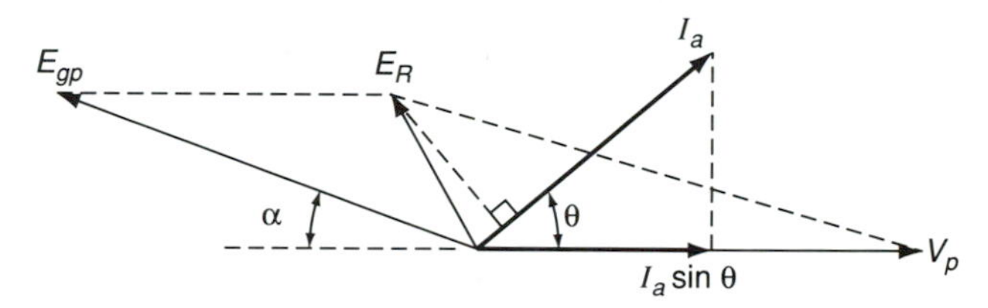

FIGURE 20-7 Phasor diagram of a synchronous motor with an overexcited field, resulting in a leading power factor.

When the field current in the synchronous motor decreases, the counter emf is reduced, but again the motor cannot increase in speed as is the case with the dc shunt motor, since the synchronous motor must run at a constant speed. As the armature current is increased, the motor takes a lagging current from the line and is said to be **underexcited**, as shown in Figure 20-8. In a

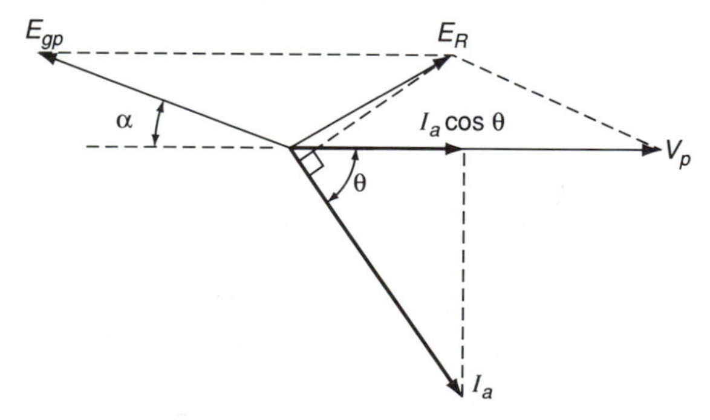

FIGURE 20-8 Phasor diagram of a synchronous motor with an underexcited field, resulting in a lagging power factor.

synchronous motor, a lagging current strengthens the field by armature reaction and opposes the effect of decreased field current. This is because a lagging current in the synchronous motor produces a magnetizing action that increases the air-gap flux and raises the counter emf to oppose the change in field. The net result of the field being strengthened by armature reaction is a lagging power factor being produced.

EXAMPLE 20-4 A small industrial plant operating at a voltage of 600 V has the following three loads connected to a common bus bar: load 1 is a 150-kW 0.8-PF lagging induction motor; load 2 is a 250-kW 0.82-PF lagging induction motor; load 3 is a 200-kW synchronous motor. Determine the overall power factor of these three loads if the synchronous motor has a power factor of (a) 0.86 lag; (b) 0.92 lag.

Solution

$$\text{Load 1} = 150\text{ kW},\ \theta = \cos^{-1} 0.8 = 36.87°$$

$$Q_1 = P_1 \tan\theta = (150\text{ kW}) \tan 36.87° = 112.38\text{ kVAR}$$

$$\text{Load 2} = 250\text{ kW},\ \theta = \cos^{-1} 0.82 = 34.92°$$

$$Q_2 = P_2 \tan\theta = (250\text{ kW}) \tan 34.92° = 174.53\text{ kVAR}$$

(a) $$\text{Load 3} = 200\text{ kW},\ \theta = \cos^{-1} 0.86 = 30.68°$$

$$Q_3 = P_3 \tan\theta = (200\text{ kW}) \tan 30.68° = 118.66\text{ kVAR}$$

$$\begin{aligned}\text{Total real power} &= P_1 + P_2 + P_3 \\ &= 150\text{ kW} + 250\text{ kW} + 200\text{ kW} \\ &= 600\text{ kW}\end{aligned}$$

$$\begin{aligned}\text{Total reactive power} &= Q_1 + Q_2 + Q_3 \\ &= 112.38\text{ kVAR} + 174.53\text{ kVAR} + 118.66\text{ kVAR} \\ &= 405.57\text{ kVAR}\end{aligned}$$

The overall power factor can now be determined:

$$\begin{aligned}\text{PF} = \cos\theta &= \cos\left(\tan^{-1}\frac{Q_T}{P_T}\right) \\ &= \cos[\tan^{-1}(405.57\text{ kVAR}/600\text{ kW})] \\ &= 0.83\text{ lag}\end{aligned}$$

(b) $$\text{Load 3} = 200\text{ kW},\ \theta = \cos^{-1} 0.92 = 23.07°$$

$$Q_3 = P_3 \tan\theta$$
$$= (200 \text{ kW}) \tan 23.07°$$
$$= 85.18 \text{ kVAR}$$

$$P_T = 600 \text{ kW}$$

$$Q_T = Q_1 + Q_2 + Q_3$$
$$= 112.38 \text{ kVAR} + 174.53 \text{ kVAR} + 85.18 \text{ kVAR}$$
$$= 372.09 \text{ kVAR}$$

$$\text{PF} = \cos\theta = \cos\left(\tan^{-1}\frac{Q_T}{P_T}\right)$$
$$= \cos\left(\tan^{-1}\frac{372.09 \text{ kVAR}}{600 \text{ kW}}\right)$$
$$= 0.85$$

The fact that the armature current in the synchronous motor may be made to lead or lag the applied voltage over a wide range, simply by overexciting or underexciting the field, makes it very useful for improving the power factor of a system. Synchronous motors operating without mechanical load may be connected to a power system for the sole purpose of power factor correction. When used in this manner, a synchronous motor is referred to as a **synchronous capacitor**, or **synchronous condenser**. In this day and age, conventional static capacitors are more economical to buy and use than are synchronous capacitors, although a large number of synchronous capacitors are still in service in older plants.

The construction of a synchronous capacitor is slightly different from that of a synchronous motor. Since the synchronous capacitor has no mechanical output, it has no shaft extension. Also, the dc excitation is considerably higher than for ordinary synchronous motors, so the construction of the rotating poles and windings is somewhat heavier.

When a synchronous capacitor is placed in parallel with an inductive load, the total power is found by the phasor sum of the individual power of each machine. For example, Figure 20-9 shows the phasor diagram of a parallel-

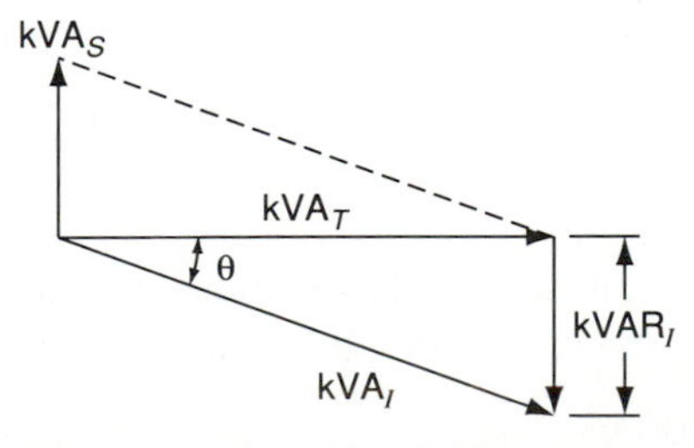

FIGURE 20-9 Power factor corrected to unity. The reactive power of the network is now equal to the apparent power.

connected synchronous capacitor and induction motor. For unity power factor to be achieved, the reactive powers of both machines must be equal. Therefore, the apparent power of the synchronous capacitor is simply the product of the apparent power of the inductive motor multiplied by $\sin\theta$, or

$$\text{kVA}_S = \text{kVA}_I \sin\theta$$

Neglecting all losses in the synchronous capacitor, the input power is considered to be strictly a vertical component. This is why the apparent power of the synchronous motor must equal the reactive power of the inductive load to produce a unity power factor.

EXAMPLE 20-5 An induction motor load on a factory takes 1000 kW from a 2300-V three-phase supply line at 0.6 lagging power factor. A synchronous capacitor is to be installed in parallel with the load to improve the power factor to unity. Calculate the required kVA rating of the synchronous capacitor.

Solution

$$\theta = \cos^{-1} 0.6 = 53.13°$$

$$\text{kVA}_I = \frac{1000}{0.6} = 1666.67 \text{ kVA}$$

$$\text{kVA}_S = \text{kVA}_I \sin\theta = 1666.67 \text{ kVA} \sin 53.13° = 1333.17 \text{ kVA}$$

Occasionally, a synchronous motor is used for the dual purpose of supplying a mechanical load as well as correcting a lagging power factor. In this situation the synchronous motor has an in-phase, or power, component as well as a reactive power component. The total power of the system is the arithmetic sum of the inductive and synchronous motor loads, and the total reactive power is the difference of the two loads.

EXAMPLE 20-6 An industrial plant has an average load of 600 kW at 0.65 lagging power factor. The supply line is a 2300-V three-phase wye-connected system. A synchronous motor is to be installed to take an additional load of 110 kW and to raise the overall power factor to 0.92. Determine (a) the kVA rating of the synchronous motor; (b) the power factor at which the synchronous motor will operate. Also, draw a phasor diagram of the system.

Solution Since the system is wye connected, phase values are used. Phase voltage, E, is chosen as the reference phasor in Figure 20-10.

$$\text{Phase voltage, } E = \frac{2300 \text{ V}}{\sqrt{3}} = 1327.9 \text{ V}$$

$$\text{Original kW load, } Ob = 600 \text{ kW}$$

$$\text{Original kVA load, } Od = \frac{600\,\text{kW}}{0.65} = 923.08 \text{ kVA}$$

$$\text{Original kVAR load, } bd = \sqrt{923.08^2 - 600^2} = 701.48 \text{ kVAR}$$

$$\text{Additional load, } bc = 110 \text{ kW}$$

$$\begin{aligned}\text{Total power of both loads, } Oc &= 110 \text{ kW} + 600 \text{ kW} \\ &= 710 \text{ kW}\end{aligned}$$

$$\text{Desired PF} = 0.92, \theta = \cos^{-1} 0.92 = 23.07°$$

$$\begin{aligned}\text{Desired kVAR, } cf = Oc \tan 23.07° &= 710 \text{ kW} \times 0.426 \\ &= 302.46 \text{ kVAR}\end{aligned}$$

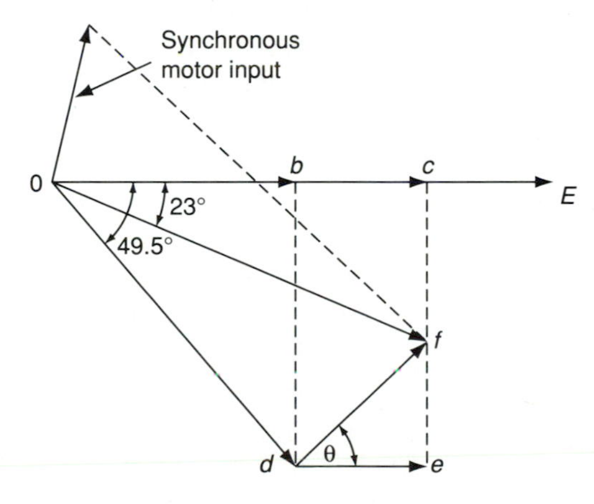

FIGURE 20-10 Phasor diagram for Example 20-6.

The kVAR required to be supplied by the synchronous motor is the difference between the original kVAR and the desired kVAR.

$$\begin{aligned}\text{kVAR of synchronous motor, } ef = ce - cf &= 701.48 \text{ kVAR} - 302.46 \text{ kVAR} \\ &= 399.02 \text{ kVAR}\end{aligned}$$

(a) The kVA input of the synchronous motor can now be found by applying the Pythagorean theorem:

$$\begin{aligned}\text{kVA}_S = \sqrt{\text{kW}_S^2 + \text{kVAR}^2} &= \sqrt{110^2\,\text{kW} + 399.02^2\,\text{kVAR}} \\ &= 413.9\,\text{kVA}\end{aligned}$$

(b) $$\begin{aligned}\text{Power factor of synchronous motor} = \frac{de}{df} &= \frac{110\,\text{kW}}{413.9\,\text{kVA}} \\ &= 0.27 \text{ leading}\end{aligned}$$

20-8 SYNCHRONOUS-INDUCTION MOTOR

The **synchronous-induction motor** is a small polyphase motor that requires no dc field excitation, although it has the constant-speed characteristics of the synchronous motor. This type of motor is, essentially, a wound-rotor slip-ring induction motor, which has a squirrel-cage winding on the periphery of the rotor. The motor is started with resistance in the squirrel-cage winding, which is gradually cut out as the motor accelerates, in the same manner as the wound-rotor machine.

The salient pole rotor of the synchronous-induction motor allows the motor to be pulled into synchronism quite easily. In effect, the lower reluctance of the protruding poles of the rotor causes it to be excited by stator flux. Therefore, the synchronous-induction motor develops reluctance torque and is occasionally incorrectly referred to as a polyphase reluctance motor. The main difference between a synchronous-inductance motor and a reluctance motor is the rotor design and construction. The rotor of the synchronous induction motor is designed in such a way that the flux entering from the stator is guided, or steered, through low-reluctance paths with effective flux barriers. This results in the formation of distinct magnetic circuits within the rotor. The flux barriers, which are composed of high-reluctance air gaps, oppose any tendency of the rotor to slow down and therefore contribute to the synchronous operation of the motor. The paths of minimum reluctance allow the rotor poles to lock in synchronism with the poles of the rotating stator field. The synchronous-induction motor has a relatively high starting torque of up to 400% of full-load torque. However, the high-resistance rotor winding reduces efficiency, and the pull-out torque is about 30% of the normal induction motor.

20-9 SUPERSYNCHRONOUS MOTOR

In some situations extremely high starting torque is required of the synchronous motor. The **supersynchronous motor** is a special type of motor that is designed to produce a starting torque that is equal to its pull-out torque. This type of motor is relatively expensive, due to the machine's armature structure. The armature, which is stationary in the conventional synchronous motor, is mounted in a cradle supported by bearings and is free to rotate. The armature may be restrained or blocked by tightening a band brake around the outside of the yoke frame. The field structure, or rotor, is connected directly to the load. The rotor is a standard squirrel-cage type with its field winding brought out to slip rings on the rotor shaft. Since the stator also rotates, it is excited in the same manner, through slip rings, and is usually started at a reduced voltage.

Upon starting the supersynchronous motor, the band brake is released and a reduced voltage is applied to the stator. Since the rotor is coupled to the load, it remains stationary and the stator accelerates. The induction motor torque produced by the rotor poles reacts in opposition to the stator conductors, which causes a torque in the stator that is opposite in direction to the direction of rotation required by the load. As the stator approaches synchronous speed, the field switch is closed and full ac stator voltage is applied and the motor pulls into synchronism.

The band brake is now gradually tightened and the stator is brought to a standstill. The stator slows down as the rotor speeds up. Since the stator and rotor are now operating in synchronism, their relative speeds must be equal to synchronous speed. For example, if the synchronous speed is 2400 rpm, a stator speed of 1600 rpm in a clockwise direction would imply that the rotor is turning at 800 rpm in a counterclockwise rotation. Therefore, when the stator is completely stopped, the rotor will be turning at synchronous speed. This relationship between revolving stator and rotor allows the motor to start under no-load conditions of torque and inrush current, and then on starting, the load develops a torque that will be equal to the pullout value. The average starting torque of a supersynchronous motor is 325% of the full-load torque.

20-10 SYNCHRONOUS CONVERTER

In some situations it is necessary to convert alternating current to direct current. This can be accomplished mechanically by the use of a **synchronous converter**. This type of machine is utilized in situations where dc motors are required to drive equipment such as elevators, printing presses, and machine tools. Many applications of synchronous converters have gradually been replaced by solid-state rectifiers, although synchronous converters are still used in industry.

The synchronous converter, or rotary converter, is a combination of a synchronous motor and a direct-current generator. It receives alternating current and converts it to direct current. The converter has fixed poles, a rotating armature, a commutator, a shunt field, and usually a series field.

The fields are excited by a shunt winding connected between the dc brushes or from a separate exciter. In some cases a series winding is added for compounding. If the dc brushes are open-circuited or removed, the machine becomes a synchronous motor of the rotating armature type. Consequently, if direct current is supplied to the brushes and commutator and the slip-ring brushes are disconnected, the machine becomes a shunt, or compound, motor. Synchronous converters are almost always of the three-phase type. Single-phase synchronous converters are very impractical because they are inefficient and have a tendency to hunt.

KEY TERMS

Synchronous motor
Normal excitation
Synchronous position
Torque angle
Slipping poles
Power angle
V-curve
Overexcited
Underexcited
Synchronous capacitor (synchronous condenser)
Synchronous-induction motor
Supersynchronous motor
Synchronous converter

PROBLEMS

20-1 Determine the rated load speed and no-load speed of a 2300-V 25-kVA 60-Hz eight-pole synchronous motor.

20-2 A synchronous motor with a rated speed of 600 rpm is to be started with a smaller 60-Hz induction motor. Determine the number of poles that the induction motor must have.

20-3 A 20-hp 240-V 60-Hz three-phase wye-connected synchronous motor delivers full load at a power factor of 0.82 leading. The rotational losses are 350 W and the field losses are 100 W. Calculate the armature current.

20-4 A 50-hp 440-V 60-Hz 12-pole wye-connected three-phase synchronous motor is operating at a relatively light load condition. The rotor is shifted 0.6 mechanical degree from it synchronous position. The excitation has been adjusted for a generated armature voltage per phase of 220 V. Determine the resultant armature voltage per phase.

20-5 If the synchronous impedance is $7.4\ \underline{/88^\circ}\ \Omega$ for the motor in Problem 20-4, determine the armature current.

20-6 A 440-V three-phase wye-connected synchronous motor has a full-load armature current of 15 A. The synchronous armature reactance per phase is 10 Ω, and the effective armature resistance is 1.5 Ω. Determine (a) the generated voltage per phase; (b) the torque angle when the motor is fully loaded at unity power factor.

20-7 Repeat Problem 20-6 for a power factor of 0.85 leading.

20-8 A 460-V three-phase delta-connected synchronous motor has a synchronous reactance per phase of 8.3 Ω and an effective armature resistance of 1.2 Ω. When operating at full load the armature current is 22 A at a leading power factor of 0.82. Determine the maximum power developed by the motor if the field excitation does not change.

20-9 A 150-hp 2300-V three-phase wye-connected synchronous motor has an effective armature resistance of 1.35 Ω and a synchronous reactance per phase of 8.75 Ω. The motor efficiency, excluding field loss, is 88%. If the motor is operating at rated load with a power factor of 0.85, determine the power developed.

20-10 If the motor in Problem 20-9 operated under load at 202 rad/s, determine the torque developed.

20-11 A 2300-V 500-hp eight-pole 60-Hz wye-connected synchronous motor is operating under light load with a torque angle of 4°. The excitation has been adjusted so that E_{gp} = 1150 V. Determine the resultant voltage E_R.

20-12 If the motor in Problem 20-11 has its field excitation adjusted so that E_{gp} = 1328 V, determine the resultant voltage.

20-13 The synchronous motor in Problem 20-11 has an effective armature resistance per phase of 1.4 Ω and a synchronous reactance per phase of 9.2 Ω. Determine (a) the armature current; (b) the power factor angle if E_{gp} = 1150 V and α = 4°.

20-14 Determine the total input power for the motor of Problem 20-13.

20-15 A 4000-V 60-Hz 12-pole three-phase wye-connected synchronous motor has a synchronous reactance per phase of 4.7 Ω. The input power is 1100 kW and the field excitation has been adjusted so that the counter emf is 4600 V. Determine the torque angle of this motor.

20-16 What would be the maximum torque developed by the motor of Problem 20-15?

20-17 A manufacturing plant operating at a voltage of 550 V has the following three loads connected to a common bus bar:

Load 1 = 200-kW 0.85-PF lagging induction motor.
Load 2 = 300-kW 0.8-PF lagging induction motor.
Load 3 = 250-kW 0.9-PF lagging synchronous motor.

Determine the overall power factor of the three loads.

20-18 If load 3 in Problem 20-17 is changed to 250 kW at 0.95 lagging, determine the overall PF.

20-19 An industrial plant draws 6000 kW at 0.82 PF lagging. A 250-hp induction motor with an efficiency of 97% and a power factor of 0.85 lagging is to be replaced. If the induction motor is replaced with a synchronous motor that is capable of operating at 0.83 PF leading, determine the new power factor of the plant. Assume that the efficiency of the synchronous motor is the same as that of the induction machine.

20-20 For the plant in Problem 20-19, what percent change in line current will result from replacing the induction motor with a synchronous motor?

20-21 An induction motor used in a manufacturing plant draws 1200 kW from a 2300-V three-phase supply at a power factor of 0.71 lagging. A synchronous capacitor is to be installed in parallel with the induction motor to improve the power factor to unity. Determine the kVA rating of the synchronous capacitor.

20-22 An industrial plant draws 25 kVA at a power factor of 0.65 lagging. A synchronous motor is to be connected in the plant to improve the overall power factor. If the motor draws 2 kW at 0.08 PF leading, what is the new power factor of the plant?

CHAPTER

Single-Phase Motors

LEARNING OBJECTIVES

Upon completion of this chapter you will be able to:

- List the three basic classes of single-phase motors.
- Explain the double revolving field theory and cross-field theory.
- State the basic operating principles of the split-phase motor.
- Understand the principle of electronic switching of capacitor-start induction motors.
- Discuss the operating characteristics of the shaded-pole induction motor and the ac servomotor.
- List four design modifications to reduce limitations of the ac series motor.
- Explain the principles of electronic speed control of universal motors.
- List the three basic types of single-phase synchronous motors.

21-1 INTRODUCTION

Single-phase motors, as their name implies, operate on a single-phase power supply. This type of motor is generally built in fractional-horsepower sizes, classified as **small motors.** The National Electrical Manufacturers Association (NEMA) defines a small motor as a motor built in a frame smaller than that having a continuous rating of 1 hp, open-type, at 1700 to 1800 rpm. According to this definition, a 3/4-hp 1200-rpm motor would not be classified as fractional, because a similar frame, designed for 1800 rpm, would have a larger horsepower rating than 1 hp. Single-phase motors are also manufactured in standard integral horsepower sizes: 1.5, 2, 3, 5, 7.5, and 10 hp for both 120- and 220-V applications. There are three basic classes of single-phase motors:

1. Induction motors
2. Commutator motors
3. Synchronous motors

21-2 SINGLE-PHASE INDUCTION MOTOR

Torque is developed in the polyphase induction motor by currents in the stator winding which set up a rotating magnetic field. After the polyphase induction motor has been started, if one phase is disconnected, the motor will continue to run but at a greatly reduced torque. Under this condition, the induction motor would develop no starting torque. A single-phase induction motor with only one stator winding and a squirrel-cage rotor is similar to a three-phase induction motor except that the single-phase motor has no magnetic revolving field on starting and, therefore, no starting torque. For this reason, the single-phase induction motor is not self-starting.

However, once the rotor begins turning, an induced torque will be produced in the motor. There are two basic theories to explain why a torque is produced in the rotor once it is turning: the **cross-field theory** and the **double revolving field theory**.

Cross-field theory

According to the cross-field theory, when the rotor is turning, a voltage is generated in the rotor as a result of its rotation in the stationary field. A single-phase induction motor with a single stator winding, AA', is shown in Figure 21-1(a). The rotor consists of a squirrel-cage winding, which is shown as being divided into two equal sections, the coils aa', which magnetize along the axis of the stator winding, and the coils bb', in quadrature. For simplicity, we shall assume that each section of the rotor winding has the same number of effective turns as that of the stator.

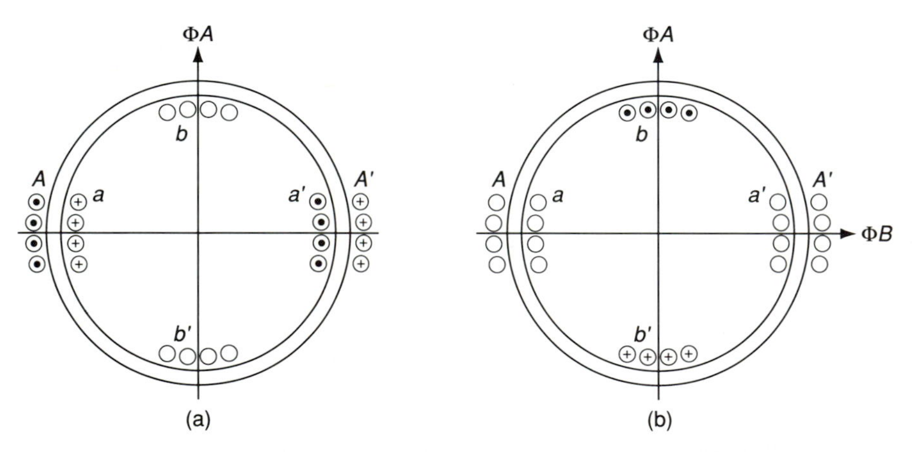

FIGURE 21-1 (a) Single-phase induction motor when voltage is initially applied across stator; (b) rotor turning at synchronous speed.

If a single-phase voltage is applied across the stator, a magnetizing current will flow and a flux, Φ_A, is produced, which links the stator winding, AA', and the rotor winding, aa'. The alternating flux will consequently induce an emf in winding AA'. This induced emf will be equal in magnitude to the voltage that was originally applied across the stator, although it will have a phase displacement that is exactly opposite to the stator voltage. An emf will also be induced in winding aa', which, neglecting voltage drops due to resistance and leakage reactance, will also be equal in magnitude to the applied emf.

The two windings act as the primary and secondary of a transformer. Since the secondary winding is initially short-circuited, a large value of current will flow, although no torque is developed to turn the rotor. To develop torque, flux is required in the quadrature, or horizontal, direction.

Figure 21-1(b) shows the direction of currents and flux if the rotor were turning at synchronous speed. The rotor coils, bb', cut the stator flux, Φ_A, and an emf is generated in these coils which is proportional to the flux Φ_A and to the speed. The emf generated from the speed of the rotor is in phase with the flux Φ_A, and since the rotor winding is closed, a magnetizing current flows that produces a flux Φ_B in quadrature behind Φ_A. Since the rotor sets up a flux, Φ_B, which is at right angles to the stator flux, Φ_A, it is referred to as the cross field. At synchronous speed, the flux Φ_B is equal to the flux Φ_A. Therefore, when the single-phase induction motor is operating at synchronous speed, a revolving field of constant value rotating at synchronous speed is produced.

Double revolving field theory

The double revolving field theory is based on the principle that a stationary pulsating magnetic field can be resolved into two magnetic fields that rotate in opposite directions but are equal in magnitude. Since the stator winding has only one field, the magnetic field does not rotate. Instead, the magnetic field pulsates up and down with time. The magnitude of this field is determined as follows:

$$B = B_m \cos \omega t \tag{21-1}$$

where B_m is the maximum flux density of motor.

As the stator current varies sinusoidally, the B_m varies sinusoidally with time. The flux density, B, is resolved into two components, B_1 and B_2. These two revolving fields are equal in magnitude but rotate in opposite directions. Each magnetic field tends to produce torque in its direction of rotation. At standstill, the two torques are of equal strength and opposite direction. Therefore, the net result is the torque developed by the motor is zero. If the clockwise and counterclockwise components of flux density remain equal when the rotor is revolving, each of the component fields would produce a torque–speed characteristic similar to that of a polyphase induction motor. In three-phase motors, the slip was determined as the ratio between the angular velocity of the rotor, ω_r, and the rotating magnetic field, ω_s. If the motor is rotating in a clockwise direction, the slip is found as follows:

$$s = \frac{\omega_s - \omega_r}{\omega_s} \tag{21-2}$$

Consequently, the slip of a motor producing torque in a counterclockwise direction would be found by

$$s_r = \frac{-\omega_s - \omega_r}{-\omega_s} = 1 + \frac{\omega_r}{\omega_s} = 2 - s$$

Since the torque developed by an induction motor is proportional to its effective rotor resistance per phase, if the slip is less than unity, the rotor resistance is higher when the rotor turns clockwise than when it turns counterclockwise. In other words, the torque developed by B_1 will be higher than the torque developed by B_2. The resultant torque is clockwise, which will tend to maintain rotation of the rotor in a clockwise direction.

Due to the fact that the single-phase induction motor is not self-starting, several different methods are used to provide the motor with starting torque. These methods identify the motor as split-phase, capacitor, shaded pole, repulsion, and so on.

21-3 SPLIT-PHASE MOTOR

The **split-phase**, or **resistance-start**, **motor** is a very popular single-phase induction motor. To make it self-starting, the motor is constructed with two stator windings. These windings are referred to as an auxiliary, or starting winding, and a main, or running, winding. These two windings are displaced by 90 electrical degrees along the stator of the motor. The auxiliary winding is constructed so that at a predetermined speed, it will be switched out of the circuit by a centrifugal switch. The starting winding has a comparatively high resistance and low inductance, while the main winding has a comparatively low resistance and high inductance. This ratio allows the current in the auxiliary winding to lead the current in the main winding.

Figure 21-2 shows a conventional diagram of the split-phase induction motor. The windings are drawn at right angles to each other to represent the 90° displacement. The main winding occupies the lower half of the slots, and the starting winding occupies the upper half. The starting winding is connected is series with a switch operated by a centrifugal mechanism in the rotor. On starting, the centrifugal switch is closed, and the two windings are connected in parallel across the line. When the rotor accelerates to about 75% of synchronous speed, the centrifugal switch opens and the start winding is removed from the circuit.

The phase relations of the locked-rotor currents at the instant of starting are shown in Figure 21-3. The phase angles shown are considered typical but

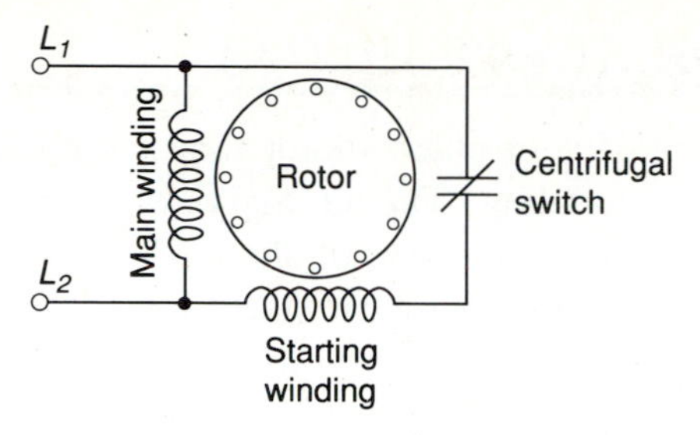

FIGURE 21-2 Circuit diagram of a split-phase induction motor.

may vary slightly among different sizes and depending on the manufacturer. The phase difference of the currents of the run and start windings is approximately 30°. In the phasor diagram, I_S is the current in the start winding, I_M the current in the main winding, and I_L is the total motor current. The current in the start winding lags the line voltage, V_L, by approximately 15°.

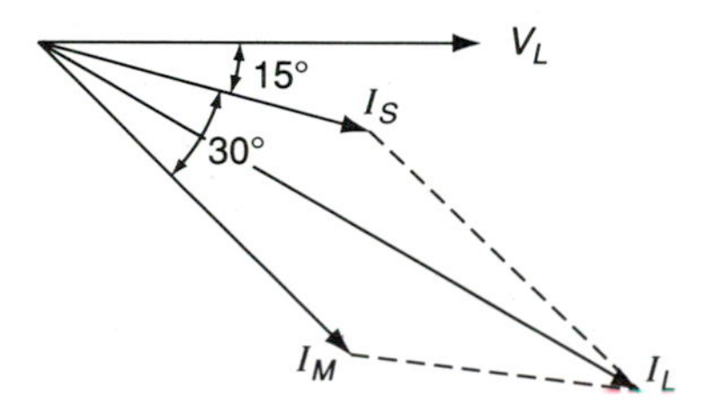

FIGURE 21-3 Phasor diagram of starting currents in a split-phase induction motor.

The direction of rotation of the split-phase motor is changed by reversing the connections of the starting winding to the line. Many of these motors are designed to operate on either 120 or 240 V, and for this reason they are referred to as **dual**, or **two-voltage, motors**. When operating on the lower voltage, the stator coils are divided into two equal groups, and these groups are connected in parallel. The higher voltage rating has the two groups of stator coils connected in series. The split-phase induction motor has a starting torque between 150 and 200% of the full-load torque, and the starting current is between 600 and 800% of the full-load current. Since this type of motor has a relatively high starting current, the split-phase induction motor is used only where the starting requirements are moderate.

21-4 SPLIT-PHASE CAPACITOR-START MOTOR

A **capacitor-start motor** is a form of split-phase motor that has a capacitor connected in series with the starting winding. The auxiliary circuit is opened when the motor has attained a predetermined speed of about 75% of synchronous speed. A conventional diagram of the capacitor-start motor is shown in Figure 21-4.

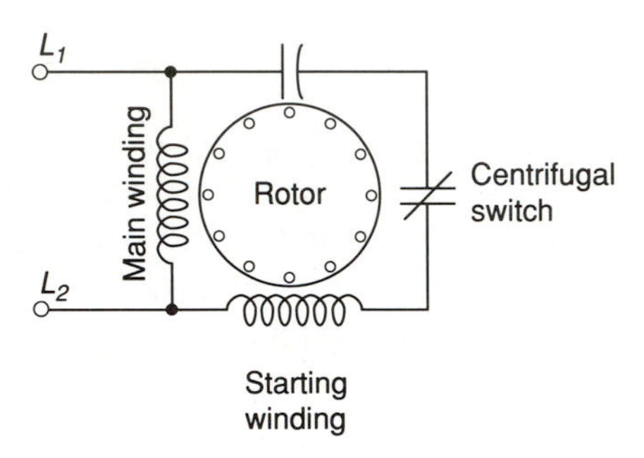

FIGURE 21-4 Circuit diagram of a capacitor-start motor.

A capacitor-start motor has windings that are constructed differently from the standard split-phase motor. The starting winding is made of small-gage wire but has considerably more turns than the main winding. The larger number of turns increases the inductive reactance and is offset by the capacitive reactance of the series-connected capacitor. This results in a net increase in the current and flux in the auxiliary winding. By connecting a capacitor in the starting winding, the current through the auxiliary winding will lead the voltage across it, as shown in Figure 21-5. Since the current is shown to lead the

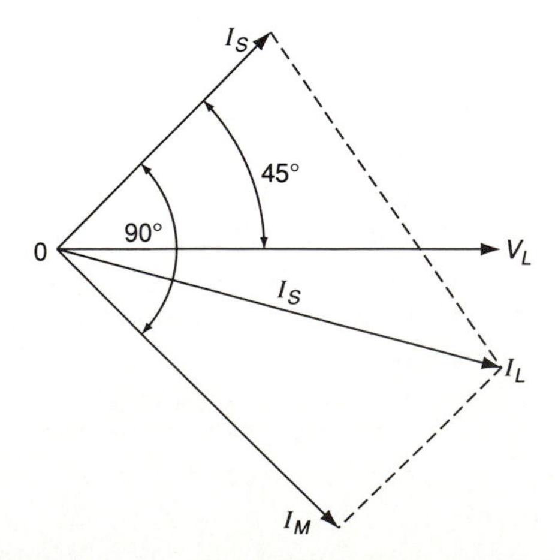

FIGURE 21-5 Phasor diagram of starting currents in a capacitor-start motor.

voltage in the starting winding by about 45°, a phase displacement of almost 90° is now obtained between the currents in the main and auxiliary windings. The net result of an increased phase displacement between the two windings is an improvement in the starting torque of the motor, due to the fact that starting torque is dependent on the sine of the angle between the currents in the start and run windings.

The split-phase capacitor-start motor has a starting torque of 3.5 to 4.5 times the rated torque, which makes it well suited for a wide variety of applications. This direction of this type of motor may be reversed by changing the connections of the auxiliary winding. The capacitor-start motor has an electrolytic capacitor which varies in size from about 80 μF for a ⅛-hp motor to 400 μF for a 1-hp motor.

21-5 PERMANENT-SPLIT-CAPACITOR MOTOR

Some split-phase motors are constructed so that the auxiliary winding and its series capacitor are connected permanently to the line. This type of motor is referred to as a **permanent-split-capacitor motor** and has windings that are virtually identical to the capacitor-start motor discussed in the preceding section. Since the motor runs continuously as a permanent-split-phase motor, no centrifugal switch is required. In place of the electrolytic capacitor, which will deteriorate under continuous duty, an oil-filled capacitor is used. The size of the capacitor is based on its maximum running, rather than its starting, characteristics. Consequently, the value of capacitance required is considerably lower than that of the capacitor-start motor.

Since there is less capacitance in the starting winding, this type of motor has a lower starting torque than that of the capacitor-start motor. Due to the fact that the capacitor is sized to balance the currents in the start and run windings at normal load conditions, a capacitor that balances the windings under running conditions will leave them very unbalanced when the machine is starting. However, this disadvantage is relatively minor compared to the higher full-load efficiency, higher power factor at full load, and lower full-load line current which the permanent-split-capacitor motor has compared to the conventional split-phase motor.

21-6 CAPACITOR-START, CAPACITOR-RUN MOTOR

When both the maximum starting torque and the best running conditions are required, two capacitors can be used with the starting winding, as shown in Figure 21-6. These types of motors are called **capacitor-start, capacitor-run**

motors, and have windings which are very similar to the capacitor motors mentioned in the preceding two sections. This particular motor is also referred to as a **two-value capacitor motor**, since it starts with one value of capacitance in series with the auxiliary winding and runs with a capacitor of smaller value.

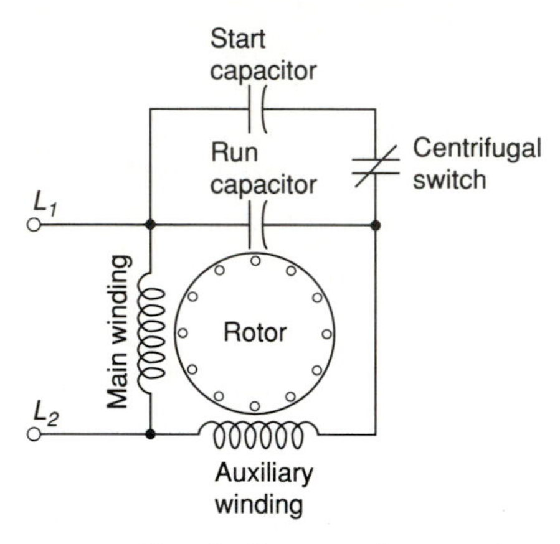

FIGURE 21-6 Circuit diagram of a capacitor-start, capacitor-run motor.

The centrifugal switch shown in Figure 21-6 will automatically cut out the larger-value capacitor when the motor reaches about 75% of synchronous speed. The motor then continues to accelerate as a capacitor motor, having achieved optimum starting torque by using both capacitors on starting.

Another method of utilizing a capacitor-start, capacitor-run system involves using only one capacitor and an autotransformer, as shown in Figure 21-7. This method is based on the principle of reflected impedance, which was discussed in Chapter 17. The transformer is connected to the circuit by a

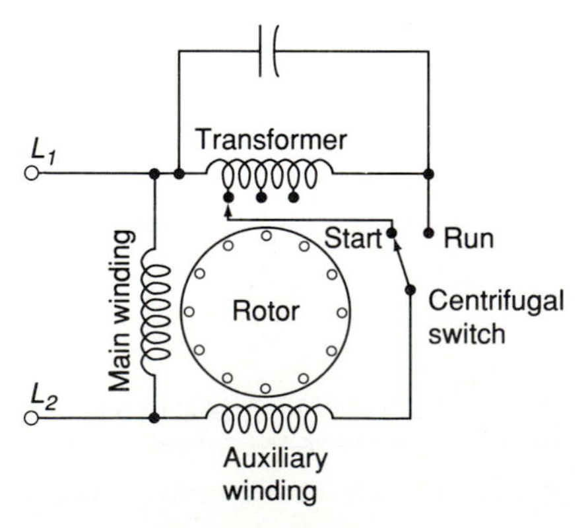

FIGURE 21-7 Capacitor-start, capacitor-run motor using a single capacitor and an autotransformer.

centrifugal switch. The value of capacitive reactance in the primary winding will depend on which tap is used in the autotransformer. For example, if the circuit had an autotransformer with 200 turns which was tapped at the 40-turn point, and the running capacitor was rated at 12 μF, the amount of capacitance in the start winding would be $(200/40)^2 \times 12\ \mu$F, or 300 μF.

21-7 ELECTRONIC SWITCHING OF CAPACITOR-START INDUCTION MOTOR

Solid-state devices have gradually begun to replace mechanical devices such as centrifugal switches in motor starting circuits. By using electronic switching, more precise timing and greater reliability are obtained. Figure 21-8 shows a capacitor-start motor with a **triac** in series with the starting capacitor. A triac is a semiconductor device that is capable of conducting current in both directions once it has been triggered, or it can block current in both directions before triggering. The gate, *G*, of the triac is used to switch the device into a conducting state.

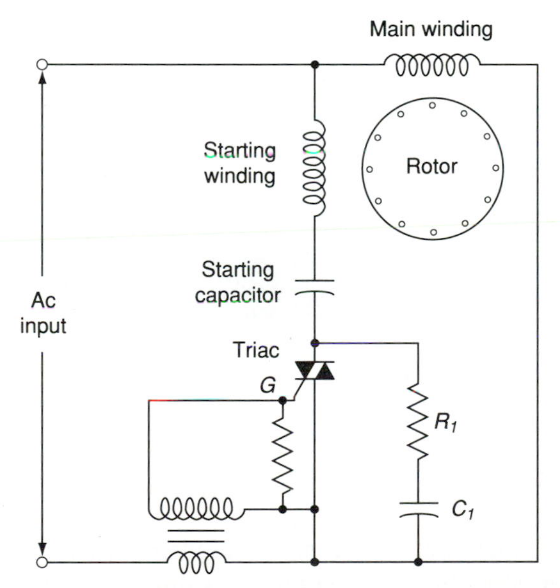

FIGURE 21-8 Electronic switch for capacitor-start induction motor.

In Figure 21-8 the gate circuit of the triac is coupled to the ac line by means of a current transformer. The primary of the transformer consists of a few turns of a relatively large conductor, and the secondary has a large number of turns, so the triac is triggered by the inrush of current through the main winding of the motor. Capacitor C_1 and resistor R_1 form an ***RC* snubber**, which prevents the triac being triggered by voltage spikes caused by the starting winding.

When the triac is switched into a conducting state, the starting winding is energized. As the speed of the motor increases, the current in the primary winding of the transformer decreases until the triac is no longer being triggered by the secondary winding. Consequently, the triac switches off and the start winding is removed from the circuit.

21-8 SHADED-POLE INDUCTION MOTOR

The **shaded-pole motor** is a type of single-phase induction motor which has a squirrel-cage rotor and a stator containing salient poles. Each salient pole is constructed with its own exciting coil, similar to a dc motor. Each pole contains a shallow slot that houses a short-circuited copper strap called a **shading coil**. The portion of the main pole that is encircled by the shading coil is referred to as a shaded pole. A four-pole and a two-pole shaded coil motor are shown in Figure 21-9.

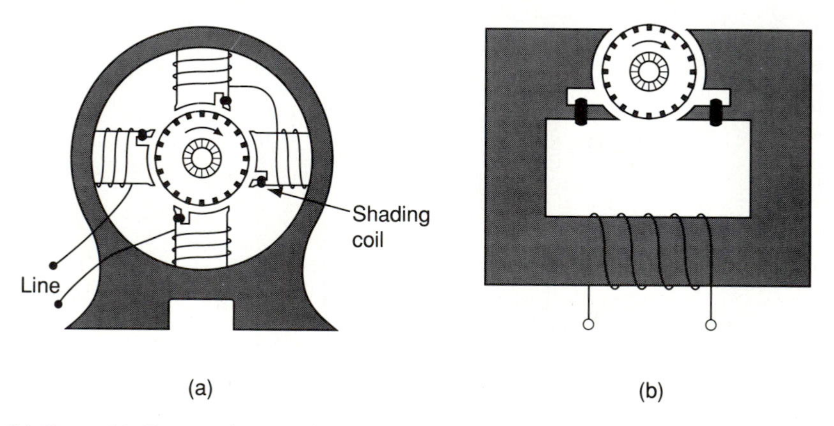

FIGURE 21-9 (a) Four-pole and (b) two-pole shaded coil motor.

In Figure 21-9(a), the single-phase current flowing in the exciting winding produces a four-pole alternating flux. Some of the flux through each pole links with the short-circuited shading coil and induces current in the shading coil. According to Lenz's law, this induced current opposes the force and the flux producing it. Therefore, initially, the greater part of the flux passes through the unshaded section of the poles, and the flux buildup in the shaded section is delayed.

As the main flux reaches its maximum value, the rate of change of flux is zero, and the voltage and current in the shading coil are also zero. This means that the flux is now more evenly distributed over the entire pole face, and the effective center of the total flux is basically in the center of the pole. As the main flux begins to decrease, the voltage and current induced in the shading coil reverse polarity and set up a flux that tends to oppose this decrease. The effective flux now shifts from the center of the pole to the shaded pole. The net result is that the flux in the shaded pole is displaced in time from that in the main pole. As the flux drops to zero and begins increasing in the opposite polarity, this displacement of flux again occurs between the main pole and the shaded pole. This action is equivalent to a sweeping movement of the field across the pole face from the unshaded pole to the shaded pole. The overall effect is that the rotor is turned in a clockwise direction from the main section of the pole toward the shaded section.

The shaded-pole motor does not produce a circular rotating field, that is, a field in which the flux is constant with respect to both angular position and time. The amplitude of the field of the shaded-pole motor varies considerably during rotation, and as a result the starting torque is poor, and the motor is only suitable for very small power applications. These motors are built in sizes ranging from $^1/_{250}$ to $^1/_{20}$ hp, where very low starting torque is required.

Since the torque of shaded pole motors is very low, the torque is often measured in terms of pound-inches (lb-in.) or ounce-inches (oz-in.). Torque is found in terms of ounce-inches by the equation

$$T = \frac{\text{hp}}{\text{rpm}} \times 10^6 \qquad \text{oz-in.}$$

21-9 AC SERVOMOTOR

Most **ac servomotors** are two-phase induction motors with high-resistance, low-inertia rotors. The main difference between a standard split-phase motor and an ac servomotor is that the servomotor has narrower conducting bars in the squirrel-cage rotor, and consequently the rotor resistance is greater. The two windings of an ac servomotor are referred to as the **control winding** and the **main winding**. A typical connection for an ac servomotor is shown in Figure 21-10. The voltage V_c, which is applied across the control winding, is taken from the output of an electronic amplifier. The magnitude of V_c is a function of the degree of action required of the motor.

The main winding voltage, V_m, and control winding voltage, V_c, are derived from the same voltage source and operate in synchronism with each other. The capacitor in series with the main winding is a phase shifting capacitor that introduces a 90° phase difference between V_m and V_c. When V_m leads

V_c, rotation in one direction is obtained. When V_m lags V_c, the motor rotates in the opposite direction. In some servo systems, the phase shifting is accomplished by the electronic amplifier instead of using a capacitor. In these situations, the V_c output is 90° out of phase with the input voltage, V_{in}, to the amplifier.

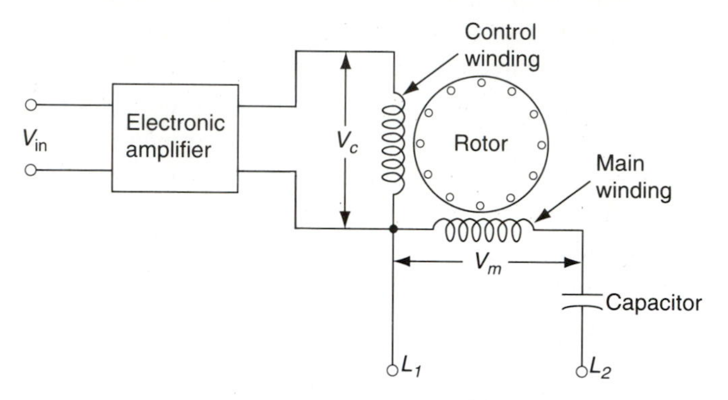

FIGURE 21-10 Circuit of a conventional ac servomotor.

The input voltage, V_{in}, is also called the **error voltage** since it is the signal that determines how far the device connected to the rotor shaft is from its desired position. For example, if the difference between the desired position and the actual position is small, V_{in} will be small and the shaft will turn at a slow speed. When the device controlled by the motor is in its desired position, the error voltage will be zero, and the motor will stop.

Servomotors are designed to produce large values of torque at low speeds. Since these types of motors are widely used in positional control devices, such as robots, it is important that the torque be reduced at high speeds to prevent the motor from **overshooting** its desired position. This is one reason why the ac servomotor has high-resistance rotor bars. As the speed of the motor

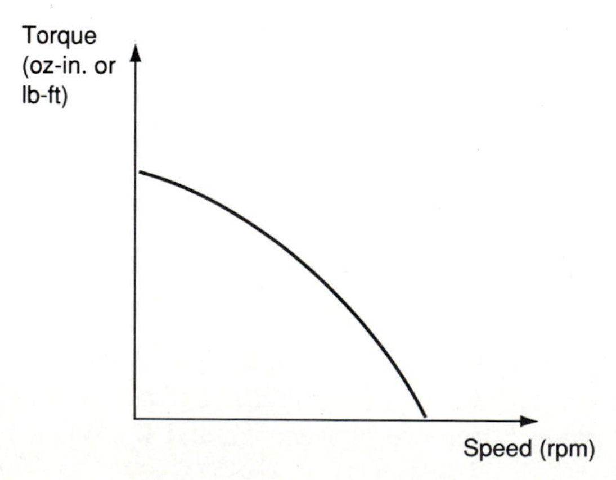

FIGURE 21-11 Torque–speed curves for an ac servomotor.

increases, the resistance of the rotor bars prevents the inductive reactance of the motor affecting the torque. As a result, the torque of the motor decreases as the speed increases. The torque–speed characteristic of an ac servomotor is shown in Figure 21-11.

21-10 REPULSION MOTOR

The stator of a **repulsion motor** is a simple single-phase stator without any auxiliary winding. The rotor is very similar to a standard dc armature. That is, the rotor is lap-wound and has one or more pairs of brushes brought to a commutator. The brushes are permanently short-circuited. By allowing the brush rigging to rotate, the position of the short-circuited brushes with respect to the polar axis may be changed. Repulsion motors have an inherently high starting torque with low starting current. At starting, the torque developed and the current drawn by the motor both depend on the brush position.

The principle of the repulsion motor's operation can perhaps be better understood by reference to Figure 21-12. Assuming that the stator current is increasing and is in the direction shown in Figure 21-12(a), the flux produced

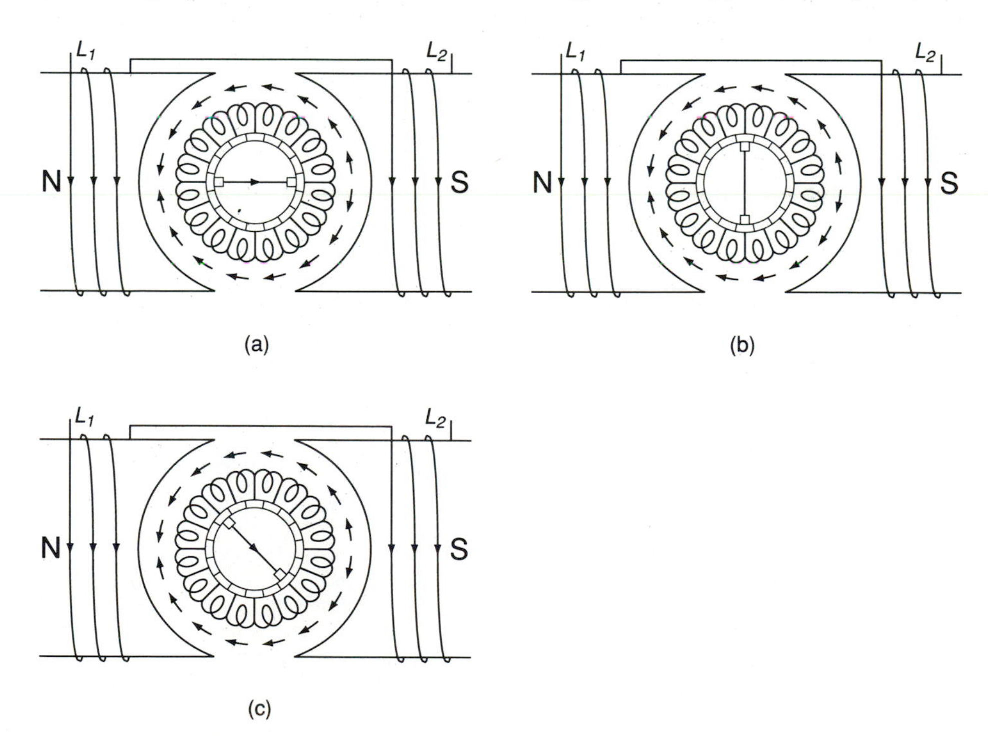

FIGURE 21-12 Three different brush positions of the repulsion motor.

will induce voltages in the armature conductors as indicated. Since these voltages are additive on each side of the brushes, a high current is forced through the armature and short-circuited brushes. However, because half of the conductors under each pole carry current in one direction and the other half carry current in the opposite direction, the net result is that no torque is developed. In Figure 21-12(b), the brushes have been shifted by 90°. This results in the voltage in each path being neutralized, and consequently no torque is produced since no current flows in the armature.

If the brushes are shifted 45° from their original position, as shown in Figure 21-12(c), a resulting voltage will exist in each path and current will flow through the armature as indicated. In this situation, all conductors under one pole carry current in one direction, while the conductors under the opposite pole will carry current in the opposite direction. Therefore, when the brushes are shifted from the position shown in Figure 21-12(a) to the position shown in Figure 21-12(c), the torque gradually increases. At some point in the shifting of brushes, a maximum torque will be developed, and if the brushes are shifted beyond this point the torque will decrease.

Since repulsion motors usually have lap-wound armatures, a four-pole motor would have two pairs of short-circuited brushes, and a six-pole motor would have three pairs. Instead of a single stator winding, the motor generally has two windings, as shown in Figure 21-13. The two series-connected windings are displaced by 90 electrical degrees on the stator. The brushes are positioned along the axis of one of these windings, called the transformer field, Φ_{Tx}, which produces current in the armature by transformer action. The other winding, called the torque field, Φ_{Tr}, is placed 90° away from the first winding and develops torque, as in the dc motor.

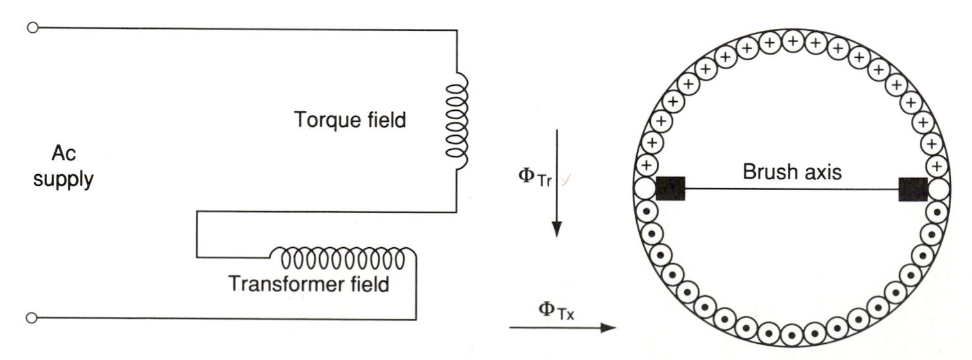

FIGURE 21-13 Stator and armature fluxes for repulsion motor.

The two fields at 90° create a resultant field, as shown in Figure 21-14. This resultant field is at 45° from the main field, which is in effect the same as shifting the brushes 45°. By using two separate windings, a larger value of

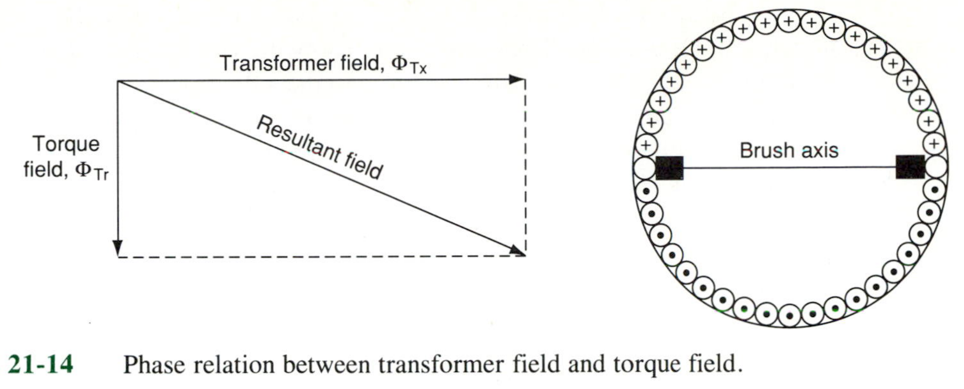

FIGURE 21-14 Phase relation between transformer field and torque field.

torque flux results, which reduces the armature current required to produce a given torque. The power factor of the motor is also improved and becomes unity at synchronous speed.

The advantages of the repulsion motor, such as excellent starting torque and low starting current, are outweighed by the disadvantages, such as poor speed regulation, noisy operation, and a tendency to run above synchronous speed at light loads. For these reasons, very few repulsion motors are in use today. However, the repulsion principle is often used to start certain types of motors that have little or no starting torque.

21-11 REPULSION-START INDUCTION MOTOR

The simple repulsion motor described in the preceding section can develop starting torques of four or five times full-load torque, but it has several disadvantages which were discussed, including the tendency to run at excessive speed at light load. To combine the high-starting-torque, low-starting-current features of the repulsion motor with the constant operating speed of the induction motor, the **repulsion-start induction motor** was developed.

As the name implies, the repulsion-start motor operates as a repulsion motor on starting but switches to an induction motor when brought up to speed. This type of motor has a form-wound rotor with commutator and brushes. The windings of the rotor are connected to a commutator, and the commutator segments are shorted out to convert the rotor to an induction-motor rotor. The commutator segments are short-circuited by a centrifugal device which also removes the brushes from the commutator when the motor reaches approximately 75% of synchronous speed. In most repulsion-start induction motors, the brushes are removed and the commutator segments are shorted at exactly the same time, which reduces brush wear and noise.

A typical speed–torque curve for a repulsion-start induction motor is shown in Figure 21-15. The dashed lines show the characteristic curves for an induction motor and a repulsion motor, and the solid lines represent the actual characteristic curve of a typical repulsion-start induction motor. The motor starts as a repulsion motor with its brushes set to the maximum torque position. As a repulsion motor, it accelerates until reaching the crossover point of the two curves, which is about ¾ synchronous speed. When it reaches this point, the commutator segments are short-circuited, and the motor begins to operate as an induction motor.

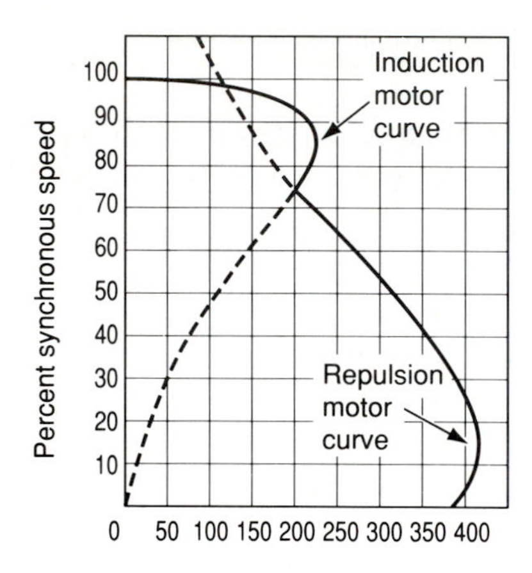

FIGURE 21-15 Speed–torque characteristic curves for a repulsion-start motor.

In fractional-horsepower applications, the capacitor-start motor has replaced the repulsion-start induction motor. The repulsion-start motor is limited since its direction of rotation cannot be changed unless the angle of brush shift is reversed. Also, this type of motor is relatively noisy on starting, and the commutator arcing produces radio-frequency interference (RFI). However, in larger-horsepower applications the repulsion-start induction motor is still in use due to its high starting torque, low starting current, and high pull-out torque.

21-12 REPULSION-INDUCTION MOTOR

The **repulsion-induction motor** is virtually identical to the standard repulsion motor except that it has a squirrel-cage rotor winding in addition to the repulsion motor armature winding. The motor does not have a centrifugal mechanism or brush lifting system, and the two rotor windings are completely isolated from each other. Both windings are continuously operating, resulting in a characteristic curve in which the motor torque is the sum of both the repulsion and induction torques. The repulsion-induction motor has a double-cage rotor which has the upper winding connected to the commutator, and the lower winding is a high-reactance induction-type squirrel-cage winding. The squirrel-cage winding has a high value of leakage reactance on starting as well as at low speeds. Consequently, the starting torque produced in the squirrel-cage winding is extremely small when the motor is starting. This means that the motor starts as a repulsion motor, and as the rotor accelerates the rotor frequency and reactance decrease, causing the current in the squirrel-cage winding to increase, and the torque required by the motor is eventually developed by *both* windings. This is the main advantage of the repulsion-induction motor over the repulsion-start induction motor. For any given load, the motor operates as a repulsion motor as well as an induction motor. Figure 21-16 shows the combined characteristics of the repulsion-induction motor.

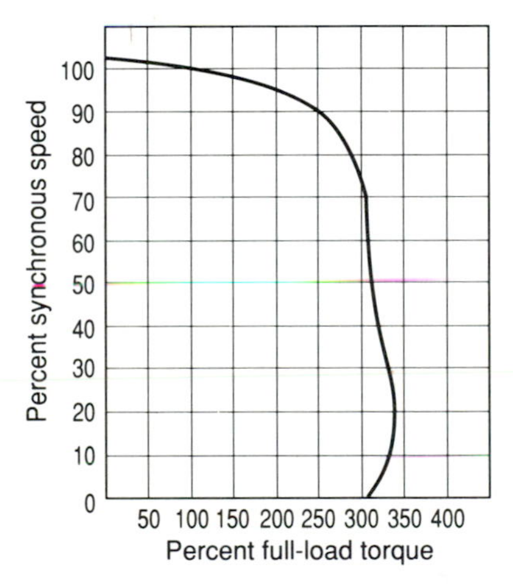

FIGURE 21-16 Speed–torque characteristic curve for a repulsion-induction motor.

On very light loads, the repulsion-induction motor runs slightly above synchronous speed. In this situation the induction winding acts as a generator, since it is now cutting flux in a direction opposite to when it was operating below synchronous speed. Therefore, the torque developed is opposite to the direction of rotation. This implies that the torque is negative and will exert a braking action on the commutator winding. Consequently, a repulsion-induction motor cannot rise more than 2 or 3% above synchronous speed.

The power factor of this type of motor is quite high, typically about 95% at full load. This is due to both windings being in operation simultaneously, so they are actually operating in parallel, resulting in a low effective rotor resistance with correspondingly low rotor copper losses. The repulsion-induction motor operates at an almost constant speed and can also develop torque under sudden heavy applied loads. Due to its constant-speed characteristic, this type of motor is well suited for lathes and milling machinery.

21-13 AC SERIES MOTOR

When the polarity of a dc series motor is reversed, the direction of rotation remains the same. Consequently, if an ac voltage is applied to a dc series motor, it has no effect on the motor's direction. However, an ac voltage will cause a substantial amount of sparking at the brushes of a dc series motor, and the efficiency and power factor of the motor will be quite low. To overcome these limitations, the ac series motor has the following design modifications:

1. The entire field structure and armature core are laminated. This reduces the large eddy-current and hysteresis losses caused by alternating flux.
2. The reactance of the field windings is kept low by using shallow pole pieces, fewer series field turns, low frequency, and low reluctance in the form of a short air gap.
3. To reduce armature reactance drop and improve commutation, the motor is equipped with a compensating winding, similar to that used in some dc machines. If the compensating winding is connected in series with the armature, the armature is said to be *conductively compensated*. If the compensating winding is short-circuited on itself, the armature is *inductively compensated*. Figure 21-17 shows the two types of compensating winding connections.

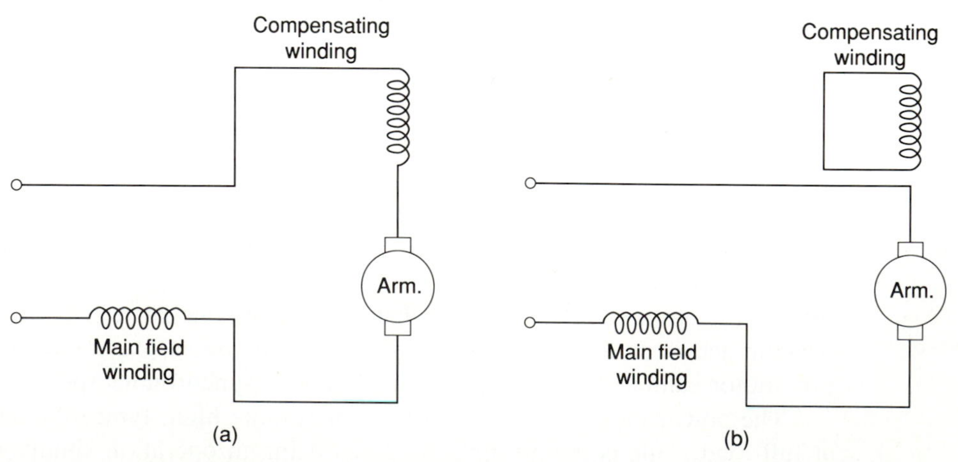

FIGURE 21-17 (a) Series motor with conductively compensated armature; (b) series motor with inductively compensated armature.

4. When the armature coils are short-circuited by brushes during commutation, they are linked by the alternating flux from the field poles. This flux induces a transformer emf in these coils, which are instantaneously short-circuited by the brushes. Unless this is prevented, large short-circuit currents and severe sparking result. This difficulty is overcome in part by using resistance, or *preventive*, leads, as shown in Figure 21-18. These leads are connected between the coils and the commutator segments, which increases the impedance of these coils on short-circuit. Therefore, resistors such as R_1 and R_2 act in series to limit the current in any short-circuited coil, but act in parallel to the main armature current.

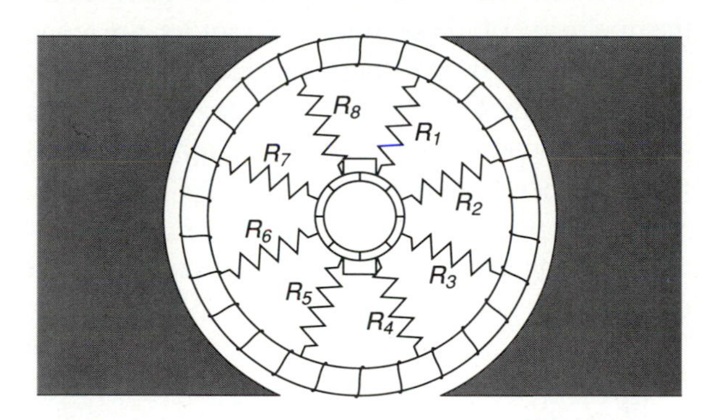

FIGURE 21-18 Series motor with preventive leads.

Because of the modifications that must be made to reduce eddy currents and reactance, ac series motors are more complex in structure, heavier per horsepower, and consequently more expensive than dc motors of the same rating. The operating characteristics of the ac and dc series motors are very similar. At light loads, the speed is high. As the load is applied, the torque increases and the speed decreases without a large increase in current. However, if the load is removed completely from the motor, the speed will increase to a dangerously high level. At full load, the power factor of the ac series motor is about 90%. Fractional-horsepower ac series motors are called universal motors.

21-14 UNIVERSAL MOTORS

The **universal motor** is a series-wound motor of fractional horsepower rating, which is designed to operate with either ac or dc voltage sources. This type of motor is very popular in appliances which are sold internationally and may be used where the frequencies vary. Different supply voltages in various countries are compensated for by using a tapped transformer or resistor in conjunction with a motor of this type. The universal motor is, as the name implies, used in a wide variety of applications, including sewing machines, vacuum cleaners, food processors, portable drills, and hair dryers.

In addition to being adaptable to a wide variety of frequency and voltage sources, the universal motor also has a high starting torque. Its speed is controlled by using a variable series resistance, or solid-state devices consisting of triacs or SCRs in wave-chopping circuits. To have a high power factor and output, the universal motor must run at high speeds, usually from 4000 to 10,000 rpm.

Small, inexpensive universal motors, up to ¼ hp, are occasionally uncompensated, although the majority of these types of motors are conductively compensated, which improves the overall performance characteristics of the motor. To prevent the universal motor from running away at light loads, they are often equipped with an automatic centrifugally operated switch, or governor. When a predetermined speed is attained, a centrifugal device actuated by the armature inserts resistance in series with the motor circuit, as shown in Figure 21-19. To prevent arcing and burning of the governor contacts, they are shunted by a capacitor.

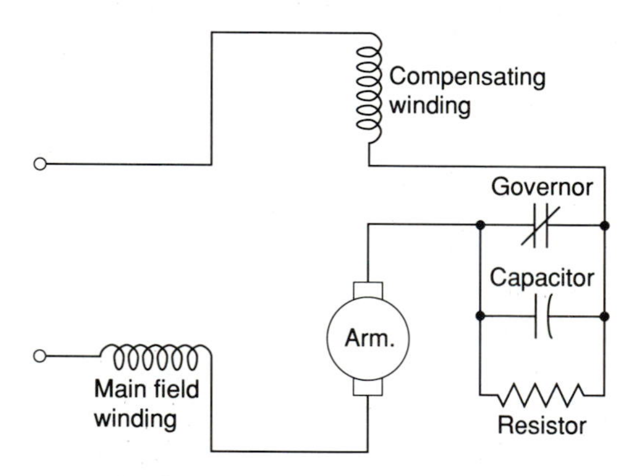

FIGURE 21-19 Conductively compensated universal motor.

21-15 ELECTRONIC SPEED CONTROL OF UNIVERSAL MOTORS

The speed control of universal motors is easily accomplished by use of thyristors such as triacs or SCRs. Figure 21-20 shows a circuit that uses a triac to control the speed of a universal motor. A **diac** is shown connected in series with the gate of the triac. A diac is a two-terminal semiconductor device that must have a specific voltage applied to it in order for the device to switch into a conducting state. This voltage is referred to as the **breakover voltage**. Once breakover occurs, current flow is in a direction depending on the voltage across the terminals. The diac switches off when the current drops below the holding value. For example, if the breakover voltage is 30 V and the holding current is 2 mA, the diac will switch on when the applied voltage reaches 30 V, and switch off when the current through the device falls below 2 mA.

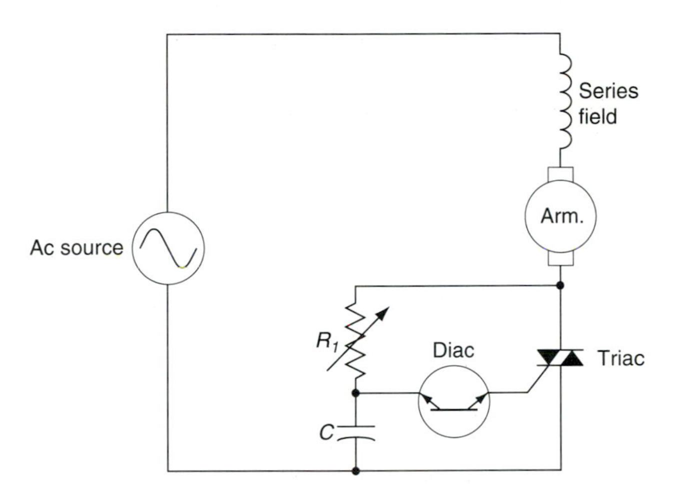

FIGURE 21-20 Speed-control circuit for a universal motor.

Since the diac acts as an ac voltage-sensitive switch, when the capacitor in the circuit of Figure 21-20 rises to the breakover voltage of the diac, the diac turns on and discharges the capacitor through the gate of the triac. This discharge pulse fires the triac. Once the diac conducts and discharges the capacitor, the capacitor will not begin to charge again until the triac has been shut off near the zero-crossing point of the ac cycle.

When the capacitor begins to charge during the second 180° of the cycle, it will charge at a polarity opposite that of the first 180°. When the breakover voltage of the diac is reached, the triac conducts again. The length of time required to charge the capacitor can be varied by changing the amount of resistance at R_1.

Figure 21-21 shows the speed of a universal motor controlled by one SCR. This circuit provides **half-wave control** due to the fact that only 180° of the ac cycle is used. During the other 180°, the SCR is reverse-biased, so no current flows. The advantage of using this circuit is that it provides good **speed regulation**. If the motor load changes, the control circuit can detect this change and compensate for an increase or decrease in load.

The circuit of Figure 21-21 operates by comparing the residual counter emf, V_2, with a circuit generated reference voltage, V_1. The reference voltage

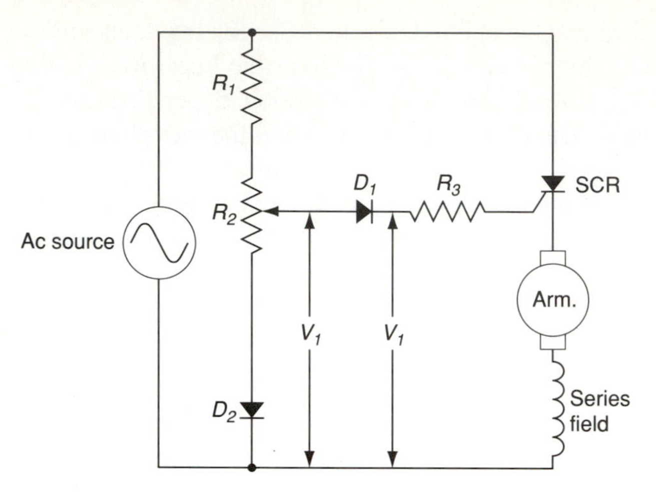

FIGURE 21-21 Half-wave speed control for universal motor.

is set by a voltage divider network consisting of resistor R_1 and potentiometer R_2. Current will flow in this branch only during the positive half-cycle due to diode D_2. When the motor is turning at the desired rpm, a counter emf is developed by the motor. If the voltages V_1 and V_2 are equal, diode D_1 is reverse-biased and does not conduct. If the motor is loaded down so that its speed decreases and counter emf decreases, V_1 becomes greater than V_2 and diode D_1 conducts. A gate voltage is now applied to the SCR and the SCR triggers earlier in the cycle, supplying more power to the motor.

The motor control circuit of Figure 21-22 provides the full-wave speed control of a universal motor. In addition to providing manual speed control, this circuit also incorporates a sensing circuit that allows the torque of the motor to increase when the load increases. In an electric drill, the motor naturally slows down when the load increases. As the motor tries to develop more torque with an increasing load, the current drawn by the motor increases. In the circuit of Figure 21-22, the voltage applied to the motor automatically increases as the current demand increases.

Diodes D_1, D_2, D_3, and D_4 form a single-phase bridge rectifier. The rectifier converts the incoming ac into a pulsating dc voltage which is used by the control circuit for the electric drill. Phase control of the triac results from the charging of capacitor C_1 through resistors R_2 and R_3 from the voltage level set by the zener diode, Z_1. Resistor R_3 is a variable resistor that would be the speed control trigger on the electric drill.

Transistor Q_1 is a unijunction transistor (UJT). This transistor, in addition to R_2, R_3, and C_1, forms a relaxation oscillator. When C_1 develops a sufficient voltage at the emitter, E, of the UJT, Q_1 conducts and the triac is fired by the

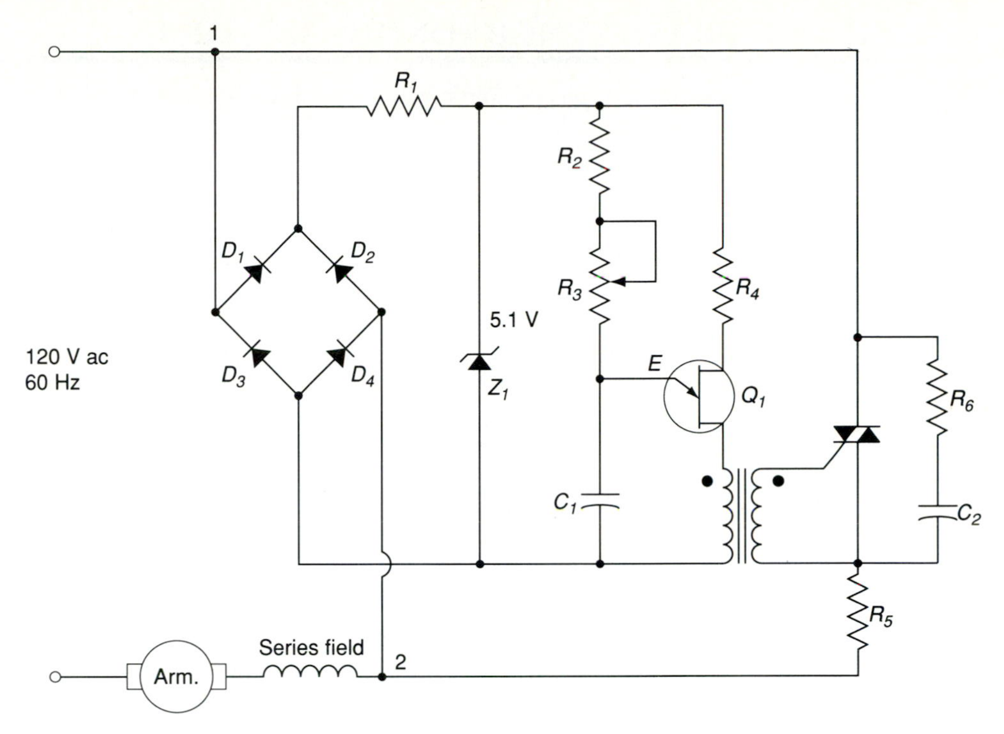

FIGURE 21-22 Full-wave speed control for electric drills.

pulse from the transformer. The pulse transformer provides electrical isolation between the control circuit and the power circuit. Once the UJT is fired, capacitor C_1 quickly discharges and the UJT shuts off. The cycle begins again as C_1 recharges until it reaches the voltage level required to fire the UJT.

During the time that the triac is conducting, the voltage between circuit points 1 and 2 is reduced to a value that is below the level required to fire Q_1. Resistor R_5 is called a **sensing resistor** due to the fact that the actual voltage applied across C_1 depends on the motor current and the voltage dropped across R_5. Consequently, as the motor current increases, the voltage drop across R_5 increases, and the reference voltage for C_1 will be higher when it begins its charge cycle. As a result, the triggering time of the triac will be advanced more as the load on the motor increases and the current demand increases.

Capacitor C_2 and resistor R_6 form an RC snubber circuit. The purpose of the snubber is to limit the rate of rise of voltage (dV/dt) across the triac during switching. Snubber networks are a necessary precaution when inductive loads, such as motors, are used. Otherwise, the inductive kickback pulse developed when the triac switches to its off state can have a rate of rise that is sufficient to retrigger the triac regardless of gating conditions.

21-16 SINGLE-PHASE SYNCHRONOUS MOTORS

There are three basic types of single-phase synchronous motors:

1. Reluctance motor
2. Hysteresis motor
3. Subsynchronous motor

These types of motors are used to drive phonograph motors, electric clocks, timing devices, and any other mechanism that requires an absolutely constant speed.

Reluctance motor

Single-phase salient-pole synchronous motors are usually referred to as **reluctance motors**. A reluctance motor depends on reluctance torque for its operation. To produce reluctance torque, an object must be elongated along axes at angles corresponding to the angles between adjacent poles of an external magnetic field. If the field coils are excited from an ac supply, the flux will build up in the core, which results in two flux components that are out of phase with each other. This principle is what a compass is based on. If a needle is placed in a magnetic field, the needle will line up in the magnetic field. The reluctance torque of a compass is the torque that appears between the external magnetic field and the magnetic field inducted in the needle. Anything that makes the reluctance of the air gap a function of the angular position of the rotor with respect to the stator coil axis will produce a reluctance torque when the rotor is revolving at synchronous speed.

The reluctance motor is started as an induction motor, and when it is brought up to synchronous speed the rotor poles are pulled into synchronism with the revolving poles of opposite polarities and the motor operates at a constant speed. This pulling-in process is accomplished by the principle of reluctance torque. Each time a salient pole passes through the rotating synchronous speed field, there is a reduction of reluctance in that area. This reduction in reluctance causes an increase in the circulating flux, which in turn develops a reluctance torque and pulls the rotor into synchronism.

A typical reluctance motor's speed–torque characteristic curve is shown in Figure 21-23. The starting torque varies with the rotor pole position and is usually between 300 and 400% of full-load torque. At about 75% of synchronous speed, a centrifugal switch opens the starting winding, and the motor continues developing torque by its running winding only.

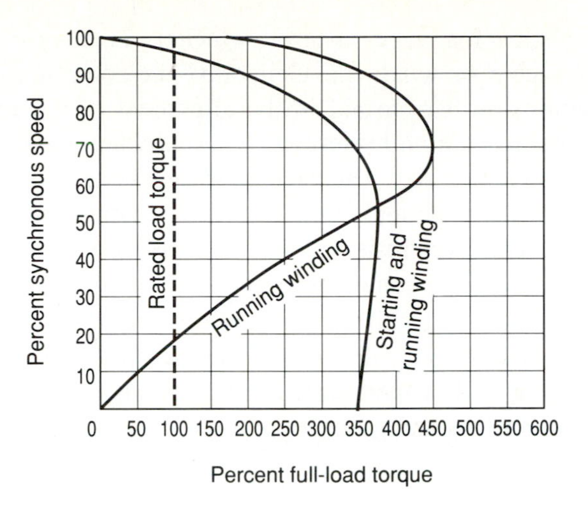

FIGURE 21-23 Characteristic curves for reluctance motor.

Hysteresis motor

The **hysteresis**, or **shading-pole synchronous, motor** is very popular for use in clocks and industrial timing devices. Where the reluctance motor operates on reluctance torque, the hysteresis motor operates on the principle of hysteresis torque. Hysteresis torque is, essentially, the residual magnetism that remains in the field poles after the magnetizing force is removed. When the motor approaches synchronous speed, the rotating stator flux, as it sweeps across the field poles, will have a decreasing tendency to leave the poles. If this hysteresis effect is strong enough, the rotor will pull into synchronism without the aid of the field current.

Figure 21-24 shows a hysteresis motor with shading coils. The rotor is

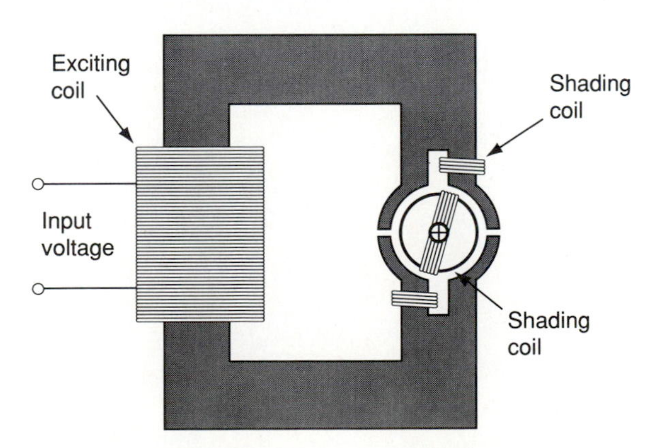

FIGURE 21-24 Hysteresis motor with shading coils.

made of a permanent-magnet magnetic material which is in the form of a smooth cylinder with no teeth or windings. Generally, the rotor is made of hardened steel rings with cross bars to aid in the alignment of the retained magnetism. Hardened steel has a high hysteresis loss, and when the rings are acted on by the rotating magnetic field, a substantial torque is developed. The rotor begins to turn due to the rotating action of the flux established by the exciting and shading coils. This hysteresis action causes the rotor to accelerate to almost synchronous speed and is then pulled into synchronism.

Hysteresis torque is relatively small and requires the rotor to be geared down through a gear train. However, the torque developed by this type of motor is extremely constant and is well suited for a variety of equipment, including tape deck capstan drives and gyroscopic rotors.

Subsynchronous motor

This type of hysteresis motor has a salient pole rotor, as shown in Figure 21-25. The synchronous speed of the motor is determined by the frequency and number of salient poles on the rotor. The subsynchronous motor is started and accelerated by hysteresis torque. Since it operates at a speed lower than that determined by the field poles, it is called a **subsynchronous motor.**

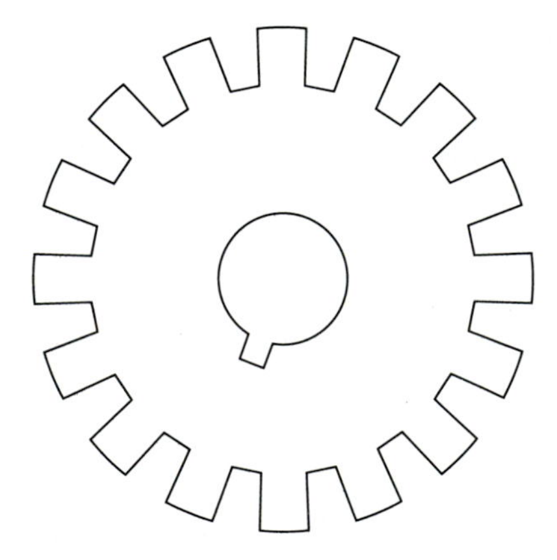

FIGURE 21-25 Subsynchronous motor.

KEY TERMS

Small motors
Double revolving field theory
Cross-field theory
Split-phase (resistance-start) motor
Dual (two-voltage) motor
Capacitor-start motor
Permanent-split-capacitor motor
Capacitor-start, capacitor-run motor
Two-value capacitor motor
Triac
RC snubber
Shaded-pole motor
Shading coil
Ac servomotor
Control winding
Main winding
Error voltage
Overshooting
Repulsion motor
Repulsion-start induction motor
Repulsion-induction motor
Universal motor
Diac
Breakover voltage
Half-wave control
Speed regulation
Sensing resistor
Reluctance motor
Hysteresis (shading-pole synchronous) motor
Subsynchronous motor

PROBLEMS

21-1 Determine whether the following motors are fractional horsepower or integral: (a) 1½ hp, 3600 rpm; (b) ¾ hp, 1200 rpm; (c) 1½ hp, 2000 rpm; (d) ⅔ hp, 900 rpm.

21-2 A 120-V 60-Hz four-pole single-phase induction motor is operating at a speed of 1720 rpm. Determine the slip of the motor.

21-3 If the motor of Problem 21-2 were rotated in the opposite direction, what would its slip be?

21-4 A ½-hp 120-V split-phase motor is connected to a load and draws a starting current of 6.5 A in its starting winding, and 8.2 A in its main winding. The current in the starting winding lags the supply voltage by 15° and the main winding current lags V_L by 45°. Determine the line current and power factor of the motor.

21-5 A ¼-hp 120-V 60-Hz capacitor-start motor has a main winding impedance of $6.2 + j5.1\ \Omega$ and an auxiliary winding impedance of $5.6 + j2.7\ \Omega$. What value of starting capacitor will place the main and auxiliary winding currents in quadrature at starting?

21-6 Calculate the full-load torque, in pound-feet and ounce-inches, of a 125-hp 1600-rpm shaded-pole motor.

21-7 A 150-hp 1700-rpm shaded-pole motor draws 60 W when operating under full-load conditions. Determine the efficiency of the motor.

21-8 A 120-V two-pole universal motor operates at a full-load speed of 7500 rpm, and draws a current of 15 A at a lagging PF of 0.95. The impedance of the series field winding is $0.6 + j1.15\ \Omega$, and the impedance of the armature winding is $1.25 + j1.5\ \Omega$. Determine the generated voltage.

21-9 If the motor of Problem 21-8 has a rotational loss of 90 W, determine the efficiency.

APPENDIX A

CONVERSION FACTORS

To convert from	To	Multiply by
Btus	Calorie-grams	251.996
	Ergs	1.054×10^{10}
	Foot-pounds	777.649
	Hp-hours	0.000393
	Joules	1054.35
	Kilowatthours	0.000293
Centimeters	Angstrom units	1×10^{8}
	Feet	0.0328
	Inches	0.3937
	Meters	0.01
	Miles	6.214×10^{-6}
Circular mils	Square centimeters	5.067×10^{-6}
	Square inches	7.854×10^{-7}
Cubic inches	Cubic centimeters	16.387
	Gallons (U.S.)	0.00433
Cubic meters	Cubic feet	35.315
Days	Hours	24
	Minutes	1440
	Seconds	86 400
Dynes	Newtons	0.00001
	Pounds	2.248×10^{-6}
Electronvolts	Ergs	1.60209×10^{-12}
Ergs	Dyne-centimeters	1.0
	Electronvolts	6.242×10^{11}
	Foot-pounds	7.376×10^{-8}
	Joules	1×10^{-7}
	Kilowatthours	2.777×10^{-14}
Feet	Centimeters	30.48
	Meters	0.3048
Foot-candles	Lumens/square foot	1.0
	Lumens/square meter	10.764

Foot-pounds	Dyne-centimeters	1.3558×10^{7}
	Ergs	1.3558×10^{7}
	Horsepower-hours	5.050×10^{-7}
	Joules	1.3558
	Newton-meters	1.3558
Gallons (U.S.)	Cubic inches	231
	Liters	3.785
	Ounces	128
	Pints	8
Gauss	Maxwells/square centimeter	1.0
	Lines/square centimeter	1.0
	Lines/square inch	6.4516
Gilberts	Ampere-turns	0.7958
Grams	Dynes	980.665
	Ounces	0.0353
	Pounds	0.0022
Horsepower	Btus/hour	2547.16
	Ergs/second	7.46×10^{9}
	Foot-pounds/second	550.221
	Joules/second	746
	Watts	746
Inches	Angstrom units	2.54×10^{8}
	Centimeters	2.54
	Feet	12
	Meters	0.0254
Joules	Btus	0.000948
	Ergs	1×10^{7}
	Foot-pounds	0.7376
	Horsepower-hours	3.725×10^{-7}
	Kilowatthours	2.777×10^{-7}
	Wattseconds	1.0
Kilograms	Dynes	980 665
	Ounces	35.2
	Pounds	2.2
Lines	Maxwells	1.0
Lines/square centimeter	Gauss	1.0
Lines/square inch	Gauss	0.1550
	Webers/square inch	1×10^{-8}
Liters	Cubic centimeters	1000.028

	Cubic inches	61.025
	Gallons (U.S.)	0.2642
	Quarts (U.S.)	1.0567
Lumens	Candle power	0.0796
Lumens/square centimeter	Lamberts	1.0
Lumens/square foot	Foot-candles	1.0
Maxwells	Lines	1.0
	Webers	1×10^{-8}
Meters	Angstrom units	1×10^{10}
	Centimeters	100
	Feet	3.2808
	Inches	39.370
	Miles	0.000621
Miles/hour	Kilometers/hour	1.609344
Newton-meters	Dyne-centimeters	1×10^{7}
	Kilogram-meters	0.10197
Oersteds	Ampere-turns/inch	2.0212
	Ampere-turns/meter	79.577
	Gilberts/centimeter	1.0
Quarts (U.S.)	Cubic centimeters	946.353
	Cubic inches	57.75
	Gallons (U.S.)	0.25
	Liters	0.9463
	Ounces (U.S.)	32
	Pints (U.S.)	2
Radians	Degrees	57.2958
Slugs	Kilograms	14.5939
	Pounds	32.1740
Watts	Btus/hour	3.4144
	Ergs/second	1×10^{7}
	Horsepower	0.00134
	Joules/second	1.0
Webers	Lines	1×10^{8}
	Maxwells	1×10^{8}
Years	Days	365
	Hours	8760
	Minutes	525 600
	Seconds	3.1536×10^{7}

APPENDIX B

TABLE OF STANDARD RESISTOR VALUES

Resistance Tolerance (± %)

0.1% 0.25% 0.5%	1%	2% 5%	10%	0.1% 0.25% 0.5%	1%	2% 5%	10%	0.1% 0.25% 0.5%	1%	2% 5%	10%	0.1% 0.25% 0.5%	1%	2% 5%	10%	0.1% 0.25% 0.5%	1%	2% 5%	10%	0.1% 0.25% 0.5%	1%	2% 5%	10%
10.0	10.0	10	10	14.7	14.7	—	—	21.5	21.5	—	—	31.6	31.6	—	—	46.4	46.4	—	—	68.1	68.1	68	68
10.1	—	—	—	14.9	—	—	—	21.8	—	—	—	32.0	—	—	—	47.0	—	47	47	69.0	—	—	—
10.2	10.2	—	—	15.0	15.0	15	15	22.1	22.1	22	22	32.4	32.4	—	—	47.5	47.5	—	—	69.8	69.8	—	—
10.4	—	—	—	15.2	—	—	—	22.3	—	—	—	32.8	—	—	—	48.1	—	—	—	70.6	—	—	—
10.5	10.5	—	—	15.4	15.4	—	—	22.6	22.6	—	—	33.2	33.2	33	33	48.7	48.7	—	—	71.5	71.5	—	—
10.6	—	—	—	15.6	—	—	—	22.9	—	—	—	33.6	—	—	—	49.3	—	—	—	72.3	—	—	—
10.7	10.7	—	—	15.8	15.8	—	—	23.2	23.2	—	—	34.0	34.0	—	—	49.9	49.9	—	—	73.2	73.2	—	—
10.9	—	—	—	16.0	—	16	—	23.4	—	—	—	34.4	—	—	—	50.5	—	—	—	74.1	—	—	—
11.0	11.0	11	—	16.2	16.2	—	—	23.7	23.7	—	—	34.8	34.8	—	—	51.1	51.1	51	—	75.0	75.0	75	—
11.1	—	—	—	16.4	—	—	—	24.0	—	24	—	35.2	—	—	—	51.7	—	—	—	75.9	—	—	—
11.3	11.3	—	—	16.5	16.5	—	—	24.3	24.3	—	—	35.7	35.7	—	—	52.3	52.3	—	—	76.8	76.8	—	—
11.4	—	—	—	16.7	—	—	—	24.6	—	—	—	36.1	—	36	—	53.0	—	—	—	77.7	—	—	—
11.5	11.5	—	—	16.9	16.9	—	—	24.9	24.9	—	—	36.5	36.5	—	—	53.6	53.6	—	—	78.7	78.7	—	—
11.7	—	—	—	17.2	—	—	—	25.2	—	—	—	37.0	—	—	—	54.2	—	—	—	79.6	—	—	—
11.8	11.8	—	—	17.4	17.4	—	—	25.5	25.5	—	—	37.4	37.4	—	—	54.9	54.9	—	—	80.6	80.6	—	—
12.0	—	12	12	17.6	—	—	—	25.8	—	—	—	37.9	—	—	—	56.2	—	—	—	81.6	—	—	—
12.1	12.1	—	—	17.8	17.8	—	—	26.1	26.1	—	—	38.3	38.3	—	—	56.6	56.6	56	56	82.5	82.5	82	82
12.3	—	—	—	18.0	—	18	18	26.4	—	—	—	38.8	—	—	—	56.9	—	—	—	83.5	—	—	—
12.4	12.4	—	—	18.2	18.2	—	—	26.7	26.7	—	—	39.2	39.2	39	39	57.6	57.6	—	—	84.5	84.5	—	—
12.6	—	—	—	18.4	—	—	—	27.1	—	27	27	39.7	—	—	—	58.3	—	—	—	85.6	—	—	—
12.7	12.7	—	—	18.7	18.7	—	—	27.4	27.4	—	—	40.2	40.2	—	—	59.0	59.0	—	—	86.6	86.6	—	—
12.9	—	—	—	18.9	—	—	—	27.7	—	—	—	40.7	—	—	—	59.7	—	—	—	87.6	—	—	—
13.0	13.0	13	—	19.1	19.1	—	—	28.0	28.0	—	—	41.2	41.2	—	—	60.4	60.4	—	—	88.7	88.7	—	—
13.2	—	—	—	19.3	—	—	—	28.4	—	—	—	41.7	—	—	—	61.2	—	—	—	89.8	—	—	—
13.3	13.3	—	—	19.6	19.6	—	—	28.7	28.7	—	—	42.2	42.2	—	—	61.9	61.9	62	—	90.9	90.9	91	—
13.5	—	—	—	19.8	—	—	—	29.1	—	—	—	42.7	—	—	—	62.6	—	—	—	92.0	—	—	—
13.7	13.7	—	—	20.0	20.0	20	—	29.4	29.4	—	—	43.2	43.2	43	—	63.4	63.4	—	—	93.1	93.1	—	—
13.8	—	—	—	20.3	—	—	—	29.8	—	—	—	43.7	—	—	—	64.2	—	—	—	94.2	—	—	—
14.0	14.0	—	—	20.5	20.5	—	—	30.1	30.1	30	—	44.2	44.2	—	—	64.9	64.9	—	—	95.3	95.3	—	—
14.2	—	—	—	20.8	—	—	—	30.5	—	—	—	44.8	—	—	—	65.7	—	—	—	96.5	—	—	—
14.3	14.3	—	—	21.0	21.0	—	—	30.9	30.9	—	—	45.3	45.3	—	—	66.5	66.5	—	—	97.6	97.6	—	—
14.5	—	—	—	21.3	—	—	—	31.2	—	—	—	45.9	—	—	—	67.3	—	—	—	98.8	—	—	—

APPENDIX C

MAGNETIC PARAMETER CONVERSIONS

SI (MKS)	CGS	English
Φ webers (Wb) 1 Wb	maxwells $= 10^8$ maxwells	lines $= 10^8$ lines
B Wb/m^2 1 Wb/m^2	gauss (maxwells/cm^2) $= 10^4$ gauss	lines/in.^{2} $= 6.452 \times 10^4$ lines/in^2
A 1 m^2	$= 10^4$ cm^2	$= 1550$ in.^{2}
μ_o $4\pi \times 10^{-7}$ Wb/Am	= 1 gauss/oersted	= 3.20 lines/Am
$\mathcal{F}$ NI(At) 1 At	$= 0.4\pi$NI (gilberts) = 1.257 gilberts	NI (At) 1 gilbert = 0.7958 At
H NI/l (at/m) 1 At/m	0.4πNI/l (oersteds) $= 1.26 \times 10^{-2}$ oersted	NI/l (At/in.) $= 2.54 \times 10^{-2}$ At/in.
H_g 7.97×10^5 B_g (At/m)	B_g (oersteds)	$0.313B_g$ (At/in.)

APPENDIX D

THE GREEK ALPHABET AND COMMON MEANINGS

Name	Capital	Lower case	Commonly used to designate
Alpha	A	α	Angles, area, coefficients
Beta	B	β	Angles, flux density, coefficients
Gamma	Γ	γ	Conductivity, specific gravity
Delta	Δ	δ	Variation, density
Epsilon	E	ϵ	Base of natural logarithms
Zeta	Z	ζ	Impedance, coefficients, coordinates
Eta	H	η	Hysteresis, coefficients, efficiency
Theta	Θ	θ	Phase angle, temperature
Iota	I	ι	
Kappa	K	κ	Dielectric constant, susceptibility
Lambda	Λ	λ	Wavelength
Mu	M	μ	Micro, amplification factor, permeability reluctivity
Nu	N	ν	
Xi	Ξ	ξ	
Omicron	O	o	
Pi	Π	π	Ratio of circumference to diameter = 3.1416
Rho	P	ρ	Resistivity
Sigma	Σ	σ	Sign of summation
Tau	T	τ	Time constant, time phase displacement
Upsilon	Υ	ν	
Phi	Φ	ϕ	Magnetic flux, angles
Chi	X	χ	
Psi	Ψ	ψ	Dielectric flux, phase difference
Omega	Ω	ω	Capital, ohms; lowercase, angular velocity

APPENDIX E

ANSWERS TO ODD NUMBERED PROBLEMS

CHAPTER 1

1-1 18.288 cm
1-3 95.24 Wh
1-5 (a) 10^{-4}
(b) 10^{5}
(c) 10^{-8}
(d) 10^{8}
1-7 3.64×10^{-4} m
1-9 (a) 6 770 000
(b) 0.00325
(c) 846 000 000
(d) 0.000173
1-11 (a) 10^{3}
(b) 10^{5}
(c) 10^{-9}
(d) 10^{12}
(e) 10^{3}
1-13 (a) 3.40×10^{10}
(b) 1.08×10^{-2}
(c) 2.17×10^{-8}
(d) 2.76×10^{16}
1-15 20.9 mN
1-17 604.8 N

CHAPTER 2

2-1 24.6 $\mu\Omega$
2-3 2.02 Ω
2-5 0.58 Ω
2-7 (a) 0.0164 in.
(b) 0.0742 in.
(c) 0.0346 in.
(d) 0.0286 in.
2-9 0.217 Ω
2-11 5.15 Ω
2-13 2.08 Ω
2-15 154.97 $\mu\Omega$

2-17	(a) 28.35 Ω
	(b) 20.17 Ω
	(c) 26.43 Ω
2-19	2.31 S
2-21	61%
2-23	1.61 Ω
2-25	No.10 AWG
2-27	689.6 ft
2-29	52.2 ft
2-31	47 k Ω ±10%
2-33	Brown, green, gold, silver

CHAPTER 3

3-1	55.8 Ω
3-3	6 V
3-5	9 V
3-7	The current also doubles.
3-9	1226.25 J
3-11	(a) 5.27 lb·ft
	(b) 7.11 J
3-13	40 W
3-15	100 V, 50 mA
3-17	24.57 W
3-19	0.83 hp
3-21	1554.2 W
3-23	(a) 0.23 kWh
	(b) 8.4×10^5 J
3-25	\$1728
3-27	\$244.13

CHAPTER 4

4-1	0.12 A
4-3	25 Ω
4-5	5.6 W
4-7	900 mW
4-9	P_1 = 117.76 W, P_2 = 80.64 W, P_3 = 57.6 W, P_T = 256 W
4-11	500 Ω, 200 mW
4-13	4:1
4-15	15.45 Ω
4-17	413.6 Ω
4-19	I_1 = 0.055 A, I_2 = 0.026 A, I_T = 0.081 A
4-21	I_1 = 22.37 mA, I_2 = 10.17 mA, I_3 = 7.46 mA

4-23	$V_T = 9$ V, $I_1 = 13.5$ mA, $I_2 = 45$ mA, $I_3 = 22.5$ mA
4-25	(a) 3333.3 Ω
	(b) 300 μS
4-27	80 Ω
4-29	(a) 571.43 Ω
	(b) 42 mA
	(c) 6 mA
	(d) 1.008 W
	(e) $P_1 = 0.576$ W, $P_2 = 0.288$ W, $P_3 = 0.144$ W
4-31	18.07 mA

CHAPTER 5

5-1	973.91 Ω
5-3	11.667 Ω
5-5	0.69 A
5-7	0.75 A
5-9	10.8 W
5-11	5.61 kΩ
5-13	7.41 V
5-15	$I_1 = 4.29$ A, $I_2 = 5.73$ A
5-17	$I_1 = 2.24$ A, $I_2 = 1.09$ A, $I_3 = 1.94$ A
5-19	−25.2 mA
5-21	30.82 V
5-23	$I_1 = 20.84$ A, $I_2 = 10.39$ A, $I_N = 10.45$ A
5-25	$I_a = 29.61$ A, $I_b = 38.06$ A, $I_c = 41.233$ A
5-27	$P_1 = 97.44$ W, $P_2 = 45.03$ W, $P_3 = 5.99$ W
5-29	1.39 A
5-31	1.22 mA
5-33	6 V, 3 Ω
5-35	(a) 0.41 mA, 19.27 V
	(b) 0.33 mA, 22.44 V
5-37	7.72 mA
5-39	1.01 A
5-41	(a) 2.8 Ω
	(b) 12.86 W
5-43	(a) 0.8 Ω
	(b) 4 Ω
	(c) 0.8 Ω
	(d) 83.33%

CHAPTER 6

6-1	816 μC
6-3	16.25 V
6-5	4.68×10^{-9} C
6-7	0.35×10^{-9} F
6-9	400 V/m
6-11	1.215 pF
6-13	0.55 mm
6-15	0.68 μF $\pm$10% 100 V
6-17	22×10^{-6} s
6-19	220 μF
6-21	0.89 s
6-23	13.33 μF
6-25	(a) 2.4 mA
	(b) 0.88 mA
6-27	63.2 V
6-29	37.44 V
6-31	28.2 ms
6-33	$Q_1 = 111.12$ μC, $Q_2 = 44.4$ μC, $Q_3 = 66.6$ μC

CHAPTER 7

7-1	0.03 T
7-3	1.74×10^{-9} Wb
7-5	2.75 T
7-7	300 At

CHAPTER 8

8-1	60 At, 750 At/m
8-3	36×10^{5} At/Wb
8-5	2700 At
8-7	2.97 A
8-9	1.87 T
8-11	409.4 At
8-13	(a) 75 At
	(b) 272.73 At/m
8-15	(a) 1428.57 At/m
	(b) 0.48 T
	(c) 3 μWb
8-17	(a) 25 At
	(b) 500 At/m
	(c) 15 T
8-19	0.133 A

8-21	8.25 A
8-23	278.75 mA
8-25	34.6 kg
8-27	4091.3 N. The magnet can lift 417 kg.

CHAPTER 9

9-1	0.3535 Ω
9-3	99.8 mA
9-5	184.8 mA
9-7	130.34 A
9-9	29 980 Ω
9-11	85 Ω/V
9-13	3333.33 Ω/V
9-15	6 MΩ
9-17	45.45 V
9-19	99.66%
9-21	256.57 kΩ
9-23	857.14 Ω
9-25	1608 m

CHAPTER 10

10-1	2.6 V
10-3	21.18 V
10-5	22.6 Wb/s
10-7	0.59 H
10-9	355 μH
10-11	3.5 H
10-13	(a) 169.4 mWb
	(b) 26.14 H
10-15	0.74
10-17	307.2 H
10-19	8.33 H
10-21	(a) 0.75 H
	(b) 0.125
10-23	299.72 mH
10-25	3.86 ms
10-27	4.4 mA
10-29	2.75 ms
10-31	4.17 ms
10-33	2.3 J

CHAPTER 11

11-1	16
11-3	2
11-5	1080 V
11-7	120 Hz
11-9	1800 rpm
11-11	2.5 A
11-13	4.78%
11-15	230.2 V
11-17	2437.5 V
11-19	(a) 2285.71 At
	(b) 3200 At
11-21	1298.69 At/series coil
11-23	(a) 233.8 V
	(b) 1180.49 W
	(c) 6345.79 W
	(d) 107.53 kW
	(e) 4.09%
11-25	0.77

CHAPTER 12

12-1	0.63 N
12-3	1.525×10^{-3} N
12-5	83.34 N·m
12-7	11.19 kW
12-9	18.95 hp
12-11	67.83 ft·lb
12-13	9.33 A
12-15	1903 rpm
12-17	1600 rpm
12-19	(a) 0.84 Ω
	(b) 1705 rpm
12-21	36 V

CHAPTER 13

13-1	75.40 rad
13-3	99.7°
13-5	30.53 μs
13-7	7539.82 rad/s
13-9	2.61 A
13-11	8.485 V
13-13	108.12 V

13-15	66.15 V
13-17	215 sin(2200 + 10°)
13-19	$208 \angle 60°$ V
13-21	−15 +j12
13-23	(a) 9 − j14
	(b) −9 − j14
13-25	(a) $86.84 \angle -54.8°$
	(b) $15.3 \angle 101.3°$
	(c) $47.17 \angle -122°$
13-27	$3.67 \angle 21.1°$ A
13-29	$365.74 \angle 15°$
13-31	(a) ·30 V
	(b) 15 V
	(c) 10.6 V
13-33	30°

CHAPTER 14

14-1	(a) 54.63 mA
	(b) 137.54 V
14-3	863.94 Ω
14-5	1.59 H
14-7	1.88 A
14-9	(a) 175 Ω
	(b) 0.137 A
	(c) 10.3 V
	(d) 13.7 V
14-11	195 Ω
14-13	10 kΩ
14-15	(a) 5.1 Ω
	(b) I_{C1} = 9.96 A, I_{C2} = 13.57 A
	(c) I_{CT} = 23.53 A
14-17	$752.9 \angle 15°$ Ω
14-19	(a) $2234.5 \angle 57.5°$ Ω
	(b) V_R = 111.76 V, V_L = 175.43 V
14-21	R = 17.21 Ω, X_L = 21.64 Ω
14-23	2.2 μF
14-25	$2235 \angle 10.2°$ Ω
14-27	V_R = 92.2 V, V_L = 76.8 V
14-29	$I_1 = 5.29 \angle 56.44°$ A, $I_2 = 3.46 \angle -54.73°$ A, $Z_T = 15.48 \angle -31.86°$ Ω

14-31	$R_s = 8.3\ \Omega$, $X_s = 5.54\ \Omega$
14-33	62.35 W
14-35	(a) 6.25 W
	(b) 12.15 VA
14-37	(a) 650 W
	(b) 100 VAR (lagging)
	(c) 657.6 VA
	(d) 0.988 (leading)
14-39	272.95 Hz
14-41	6.03
14-43	$f_1 = 78.08$ kHz, $f_2 = 128.08$ kHz
14-45	18.06 dB
14-47	−7 dB
14-49	2.52 W
14-51	1496 Hz
14-53	$L = 1.06$ mH, $C = 66.3\ \mu F$
14-55	4.42 nF

CHAPTER 15

15-1	$5.8 \angle -37.2^\circ$ A, $4.14 \angle 37^\circ\ \Omega$
15-3	$I_1 = 1.24 \angle -64.8^\circ$ A, $I_2 = 0.59 \angle 135.4^\circ$ A
15-5	$I_1 = 0.76 \angle -67.5^\circ$ A, $I_2 = 0.64 \angle 46.8^\circ$ A,
	$I_3 = 0.77 \angle -17.6^\circ$ A
15-7	$V_{R2} = 9.34 \angle 27^\circ$ V
15-9	$I_{Z3} = 5.88 \angle -24.3^\circ$ A
15-11	$I_L = 1.42 \angle -35.6^\circ$ A
15-13	$V_{TH} = 5.62 \angle -20.6^\circ$ V
15-15	$V_{TH} = 71.4 \angle -164.3^\circ$ V, $Z_{TH} = 11.5 \angle 5.4^\circ\ \Omega$
15-17	$I_L = 0.32 \angle 65.3^\circ$ A
15-19	$I_L = 0.83 \angle -8^\circ$ A
15-21	$R_x = 68.2\ k\Omega$, $C_x = 1.47\ \mu F$
15-23	$R_x = 3.3\ k\Omega$, $L_x = 669$ mH
15-25	(a) $48.26 \angle -56^\circ\ \Omega$
	(b) $2.22 \angle 0^\circ$ A
	(c) Apparent power = 235.7 VA Active power = 131.8 W Reactive power = 195.4 VAR (leading)
15-27	612.5 mW
15-29	$5.19 \angle 22^\circ\ \Omega$
15-31	$Z_{ab} = 42.8 \angle -60^\circ\ \Omega$, $Z_{bc} = 30.8 \angle -26.3^\circ\ \Omega$,
	$Z_{ca} = 38.05 \angle 42^\circ\ \Omega$

CHAPTER 16

16-1 Phase sequence = A-C-B

16-3 $E_{ab} = 208 \angle 30^\circ$ V, $E_{bc} = 208 \angle -90^\circ$ V,
$E_{ca} = 208 \angle 150^\circ$ V

16-5 $I_a = 18 \angle 15^\circ$ A, $I_b = 18 \angle 135^\circ$ A, $I_c = 18 \angle -105^\circ$ A

16-7 (a) $E_{ab} = 4160 \angle -30^\circ$ V, $E_{bc} = 4160 \angle 90^\circ$ V, $E_{ca} = 4160 \angle -150^\circ$ V
(b) $I_{ab} = 6.93 \angle -75^\circ$ A, $I_{bc} = 6.93 \angle 45^\circ$ A, $I_{ca} = 6.93 \angle -195^\circ$ A
(c) $I_a = 12 \angle -45^\circ$ A, $I_b = 12 \angle 75^\circ$ A, $I_c = 12 \angle -165^\circ$ A

16-9 $I_a = 1.6 \angle 30^\circ$ A, $I_b = 1.6 \angle -90^\circ$ A, $I_c = 1.6 \angle 150^\circ$ A

16-11 1368.84 W

16-13 0.96

16-15 $W_1 = 0$ W, $W_2 = 540.4$ W

16-17 $I_a = 1.6 \angle -30^\circ$ A, $I_b = 1.6 \angle -150^\circ$ A, $1.6 \angle 90^\circ$ A

16-19 $I_a = 14.15 \angle -68.4^\circ$ A, $I_b = 4.11 \angle 145.9^\circ$ A, $I_c = 11 \angle 99.4^\circ$ A

16-21 $I_a = 1.85 \angle -20^\circ$ A, $I_b = 3 \angle -70^\circ$ A, $I_c = 4.8 \angle 50^\circ$ A

16-23 1856.5 W

16-25 (a) $I_a = 6.48 \angle -30.2^\circ$ A, $I_b = 8.13 \angle 166.4^\circ$ A, $I_c = 2.67 \angle 30.3^\circ$ A
(b) $E_{an} = 226.8 \angle 29.8^\circ$ V, $E_{bn} = 162.6 \angle 136.4^\circ$ V, $E_{cn} = 133.5 \angle 30.3^\circ$ V
(c) 2236.1 W

16-27 0.73

16-29 $I_a = 94.09 \angle -24.8^\circ$ A, $I_b = 108.66 \angle -141.3^\circ$ A, $I_c = 61.44 \angle 80^\circ$ A

CHAPTER 17

17-1 6 V

17-3 9.38 mWb

17-5 15.02 mWb

17-7 (a) $I_p = 14.45$ A, $I_s = 60.1$ A
(b) 0.39 Wb

17-9 $E_p = 36$ V, $E_s = 7.2$ V

17-11 (a) 133 V
(b) 22.2 V
(c) 2.78 Ω

17-13 56.6

17-15 $I_{in} = 6.05 \angle 51.61^\circ$ A, $I_s = 1.52 \angle -98.3^\circ$ A

17-17 (a) $R_{ep} = 1.01\ \Omega$, $X_{ep} = 4.2\ \Omega$
(b) $R_{es} = 0.01\ \Omega$, $X_{es} = 0.042\ \Omega$

17-19 (a) 0.57 Ω
(b) 2.65 Ω
(c) 2.7 Ω
(d) 20 Ω
(e) 5.79 A

17-21	2.08%
17-23	(a) 1.27%
	(b) 1.27%
17-25	77.66%
17-27	94.58%
17-29	98.36%
17-31	(a) 606.27 W
	(b) 1393.7 W
17-33	$I_1 = -0.12$ A, $I_2 = 208.45$ A

CHAPTER 18

18-1	1800 rpm
18-3	50 Hz
18-5	2.22%
18-7	3.67 Hz
18-9	1765 rpm
18-11	51.53 N·m
18-13	177.09 lb·lft
18-15	(a) 1800 rpm
	(b) 1692 rpm
	(c) 3.6 Hz
	(d) 31.04 lb·ft
18-17	(a) 20 452 W
	(b) 19.552 W
	(c) 18 602 W
	(d) 90.95%
18-19	82.53%
18-21	$rpm_6 = 1090.8$ rpm, $rpm_4 = 1636.2$ rpm

CHAPTER 19

19-1	60 Hz
19-3	0.98
19-5	0.832
19-7	0.966
19-9	(a) 320 turns/pole
	(b) 60 Hz
	(c) 0.957
	(d) 0.951
	(e) 3491.3 V/phase
19-11	25.45%
19-13	−4.45%
19-15	84.41%

19-17	88.23%
19-19	(a) 28 kW
	(b) 50 kW
	(c) 2.88 kW
	(d) 93.3%

CHAPTER 20

20-1	900 rpm
20-3	45.09 A
20-5	5.02 −66.1° A
20-7	(a) 343.54 V/phase
	(b) 24°
20-9	121.45 kW
20-11	197.71 $\angle 24°$ V
20-13	(a) 21.25 $\angle -57.35°$ A
	(b) 57.35°
20-15	16.32°
20-17	0.85 lagging
20-19	0.84
20-21	1190.3 kVA

CHAPTER 21

21-1	(a) Fractional
	(b) Integral
	(c) Fractional
	(d) Integral
21-3	1.96
21-5	278.86 μF
21-7	24.9%
21-9	70.4%

GLOSSARY

A

Absolute permittivity The flux produced with a vacuum as dielectric; also known as absolute capacitivity.

Ac bridge A method of measuring the separate resistive and reactive components of an impedance; also used for phase shifting, filtering undesired signals, and frequency measurement.

Accuracy class A method of classifying potential transformers depending on the degree of accuracy required for a given application.

Ac generator *See* Alternator.

Ac resistance The effective resistance of a conductor; includes many factors such as skin effect, radiation loss, etc.

Across-the-line starter A switching device generally using a magnetic switch, or contactor, that can be operated from one or more remote stations by pushbuttons.

Ac servomotor Typically, a two-phase induction motor with a high-resistance, low-inertia rotor.

Active device A device, such as a transistor, capable of controlling voltage or current.

Active filter A filter network that uses an active device to obtain the desired filtering effect.

Active network A circuit containing at least one active element.

Active power The average power input to the resistance of an ac circuit.

Additive polarity A winding connection that causes the direction of counter emf in both windings to be the same.

Admittance The ease with which an ac current flows in a circuit; the reciprocal of impedance, measured in siemens.

Admittance triangle A right-angled triangle relating conductance, susceptance, and admittance.

Air-core inductor An inductor that contains no magnetic iron, and is generally wound on a tubular insulating material.

Air gap Part of the magnetic circuit of rotating electrical machines and electromagnets, requiring the largest proportion of the magnetomotive force.

All-day efficiency Ratio of watt-hours output to watt-hours input.

Alternating current A current in which the magnitude and direction varies with time.

Alternating emf A voltage in which the magnitude and direction varies with time.

Alternator An ac generator that delivers ac power by means of slip rings; also known as synchronous generator and synchronous alternator.

American Wire Gage (AWG) A standard by which wires are manufactured and numbered according to.

Ammeter A measuring instrument used to indicate electrical current in amperes.

Ampere The base SI unit of electric current. It is the rate of electric charge flow when 1C of charge passes a given point in 1s.

Ampere's circuit law A law that states the algebraic sum of the rises and drops of the mmf around a closed loop of a magnetic circuit is equal to zero.

Amplification The process of increasing the voltage, current, or power of an electrical signal.

Amplitude The maximum positive or negative value of an alternating current, voltage, or power; also known as peak value.

Analog-to-digital converter A device that compares an unknown voltage with a reference voltage and indicates which of the two voltages is higher.

Angular velocity The rate of change of a quantity, such as voltage, in an ac circuit.

Anode Positive element of an electrical device or battery.

Antiresonance A term that describes the condition in which the impedance of an LC circuit is maximum.

Apparent power The product of the total rms voltage and current in a circuit, expressed in voltamperes.

Arc A column of ionized gas, appearing as a luminous discharge.

Arc chute A device that creates a deionizing effect to reduce the arcing of contacts.

Arcing A phenomenon caused by interrupting current, such as opening a switch, that produces a very high induced voltage due to the rapidly collapsing field.

Argument The angular displacement between two quantities, such as current and voltage.

Armature Part of the magnetic path through a rotating machine.

Armature reaction The reduction in flux in an electrical machine caused by current flow in the armature conductors.

Armature windings The windings in which a voltage is induced.

Armortisseur winding A squirrel-cage winding placed in the pole faces of a synchronous machine to reduce hunting.

Asynchronous generator *See* Induction generator.

Atom The smallest particle of an element that can exist alone or in combination.

Attenuation A reduction in the magnitude of a quantity, such as voltage or current.

Autotransformer A transformer having one winding that is common to both primary and secondary sides.

Average power The average of instantaneous power over a complete cycle, expressed in watts.

Average value The mean value over an integral number of repetitions. When calculating average values of sine waves only one-half of the cycle is used.

B

Balanced delta A delta-connected circuit having impedances of equal values.

Balanced system When the currents in each line of a polyphase network have equal magnitudes and are displaced by 120°.

Balanced wye A wye-connected circuit having impedances of equal value.

Bandpass The width of a band of frequencies.

Bandpass filter A circuit used to allow a specific range of frequencies to pass and to attenuate all other frequencies.

Bandstop filter A circuit designed to attenuate a specific range of frequencies; also known as band-reject filter and notch filter.

Bandwidth The total number of cycles below and above resonant frequency for which the current is equal to or greater than 70.7% of its resonant value.

Battery A series connection of voltaic primary cells or secondary cells.

Beamfinder An oscilloscope function that is used to locate a waveform that has been shifted off the screen.

Bel The unit of measure for the logarithm of the ratio of two voltages, currents or power levels.

B/H curve A curve that illustrates the various stages of magnetization of ferrous material; also illustrates the demagnetization of material.

Bipolar junction transistor (BJT) A three-terminal device whose emitter-to-collector conduction is controlled by its base terminal current.

Blocked-rotor test A test performed on rotating machines to determine circuit parameters.

Bode plot A straight-line approximation of the magnitude and phase angle response to frequency.

Bound electrons The electrons very close to the nucleus and tightly held in their orbit.

Breakdown The point at which current begins to flow in an insulator.

Breakdown torque The torque developed when the rotor is stalled.

Breakover voltage The amount of voltage required to switch a device into a conducting state.

Breakpoints The points at which the gain drops to 3 dB below the mid-frequency gain.

Bridge rectifier An electronic circuit that converts alternating current to direct current.

Bright-lamp method A technique used for synchronizing two alternators.

Brush A device that makes a rubbing contact on the commutator of a rotating machine to conduct current to and from an outside circuit.

Brushless motor A motor that uses an electronic commutator.

Brush rigging A device for holding brushes firmly against a commutator.

Brush shift The movement of the brushes on a commutator away from the neutral position.

Burden The load carried by instrument transformers.

C

Candela The base SI unit of luminous intensity.

Capacitance The property of an electric circuit to oppose any change in voltage across the circuit.

Capacitance comparison bridge An ac bridge used to measure the impedance of a capacitive circuit.

Capacitive reactance The opposition of capacitance to alternating current.

Capacitor A device capable of storing electrical energy. It is constructed of two conductor materials separated by an insulator.

Capacitor motor A single-phase motor with a split stator winding and an external capacitor in the auxiliary circuit to increase starting torque and decrease starting current.

Capacitor-start, capacitor-run motor A motor with two capacitors that starts with one value of capacitance in series with the auxiliary winding and runs with a capacitor of smaller value.

Cathode The negative terminal of an electrical device or battery. The part of an electronic device from which electrons are emitted.

Cell A chemical source of electrical energy consisting of two electrodes in an electrolyte.

Centrifugal switch A centrifugally operated automatic switching mechanism used to change circuits in a single-phase motor.

CGS system A system of physical units based on the centimeter, gram, and second.

Charge An accumulation or deficiency of electrons.

Charge curve An exponential curve showing the increase in charge across an inductor or capacitor from zero to maximum.

Chopper Essentially an on-off switch that connects and disconnects the load to a dc voltage source.

Chorded winding An armature winding with a fractional pitch.

Circle diagram A term used to describe a diagram that reveals the input, output, slip, torque, power factor and efficiency for any value of assumed stator current in a rotating machine.

Circular mil The area of a circular cross-section having a diameter of 1 mil.

Closed-face phasor diagram A polygon phasor diagram showing the relationship between voltages and currents in a system.

Coefficient of coupling The degree of closeness with which the primary and secondary windings are coupled.

Coercive force The demagnetizing force necessary to remove the residual flux from the magnetic material.

Cogging A pulsating torque caused by motor control circuits such as electronic inverters.

Color code band A method of determining resistance, capacitance, and tolerance of electrical and electronic components such as resistors and capacitors.

Common winding The winding of an autotransformer that has a physical connection between the primary and secondary.

Commutator A mechanical means of converting alternating current to direct current.

Complex circuit A circuit that contains two or more sources in different branches; or a circuit which combines series and parallel elements in various interconnections.

Complex number A method of expressing the real and imaginary components of a quantity.

Complex waveform A sustained periodic waveform that cannot be completely described in terms of a single sine function.

Compound field A series-parallel combination of field windings.

Compound generator A machine with each pole having two coils. One coil is part of the series winding and the other is part of the shunt winding.

Compounding A method of adjusting the characteristics of an electrical machine in accordance with its loading.

Concatenation A term used to describe the connection of two induction machines where the rotor winding of one motor is connected to the stator winding of the other motor for the purpose of speed control.

Concentrated winding A winding that has only one slot per pole per phase.

Conductance The ability of resistance of pass alternating current; the reciprocal of resistance, measured in siemens.

Conductivity The reciprocal of resistivity; the conductance per unit length and cross-section of a material.

Conductor A material that offers a low resistance to the passage of electric current.

Conjugate When the imaginary part of a complex number is multiplied by -1.

Controller A motor starter capable of regulating the electric power delivered to a machine.

Control winding The winding of a servomotor that controls the rotor speed.

Conventional current A current flow adopted by the IEEE in which current is defined as the direction in which positive charge carriers flow through a circuit.

Conversion factor A constant that relates a measurement in one measurement system to an equivalent in another system.

Converter A circuit that converts alternating current to direct current.

Copper loss A term loosely applied to the I^2R loss that occurs in a conductor due to the resistance of the conductor.

Coulomb The unit of charge in SI units. A coulomb represents the quantity of electric charge possessed by 6.24×10^{18} electrons.

Coulomb constant A proportionality constant whose value depends on what material fills the volume of space in which the bodies are located.

Coulomb's law The force of attraction or repulsion between two charged bodies is directly proportional to the square of the distance between them.

Counter emf (cemf) The emf that is generated in an electric circuit by a change of current within the circuit itself.

Counter emf starter A method of starting a motor that is dependent on the counter emf developed across the armature for operation.

Coupling capacitor A capacitor that will pass ac signals and block dc signals.

Critical field resistance A value of resistance in a generator that will not allow the machine to build beyond a certain voltage due to residual magnetism.

Critical frequency *See* cutoff frequency.

Cross field theory A theory that states when a rotor is turning, a voltage is generated in the rotor as a result of its rotation in the stationary field.

Crystal An assembly of molecules having a definite internal structure and the external form of a solid enclosed by a number of symmetrical plane faces.

Cumulative compound generator A machine in which the current in the series coil sets up a flux in the same direction as that set up by the shunt coil.

Current A movement of electric charge. The direction of a current is taken arbitrarily as that of the movement of positive charges and opposite to that of negative charges.

Current divider Parallel electrical circuit providing current to more than one circuit.

Current divider rule A rule that states the amount of current in one of two parallel resistances is calculated by multiplying their total current by the other resistance and dividing by their sum.

Current-limit starter A motor starting circuit using contacts that remain open as long as a minimum value of current flows through them.

Current magnification factor *See* Q factor.

Current per volt The admittance of an ac circuit; the ratio of I/V.

Current source A device capable of supplying current.

Current transformer An instrument transformer used for the transformation of current.

Cutoff frequency A specific frequency where the signal is attenuated.

Cycle One complete set of positive and negative values of alternating current or voltage; the smallest nonrepetitive portion of a periodic wave.

Cycloconverter Converts an ac source into a lower frequency ac source by controlling the firing angles of SCRs.

D

Damping (1) Any means employed to keep moving parts from oscillating. (2) The reduction in Q-factor by reducing the resistance of the parallel resonant circuit.

Damping effect A method of automatically slowing down a motor to prevent overshoot.

Damping resistor A variable resistor used in parallel resonant circuits to reduce the Q of the circuit.

Dark-lamp method A technique used for synchronizing two alternators.

D'Arsonval galvanometer A very sensitive electrical instrument used in detecting extremely small currents.

D'Arsonval principle A principle that states when current is fed through a moving coil, the resulting magnetic field reacts with the magnetic field of a permanent magnet and causes the coil to rotate.

Dc chopper Electronic control circuit that provides variable dc output voltage.

Decade A tenfold change in frequency.

Decibel A logarithmic unit used to express gain or loss in signal level; it is equal to ten times the common logarithm of a power ratio.

Degree of compounding A factor determined by the number of series-field ampere-turns with respect to the number of shunt-field ampere-turns.

Delta network A network of components arranged in the configuration of the Greek capital letter delta; also called a pi network.

Delta-wye conversion An equivalent circuit method of network analysis where delta-connected components are converted to an equivalent wye-connected network.

Dependent node A node connected directly to a current or voltage source.

Dependent source An electrical source that produces a voltage or current as a function of a voltage or current at some other point in a circuit.

Derived units The units that are not selected as fundamental but are derived from the fundamental.

Detent torque A residual torque that is present when a motor is not turning.

Determinant A matrix, or square array of elements arranged between two vertical lines.

Developed torque The electromagnetic torque developed by the armature of a motor.

Diac A two terminal bidirectional thyristor capable of conducting current in both directions.

Diamagnetic material A material that exhibits a very slight opposition to magnetic lines of force.

Dielectric Insulating material used to store electrical charges.

Dielectric constant The property which determines the electrostatic energy stored per unit volume for unit potential gradient.

Dielectric flux The total strength of an electric field.

Dielectric hysteresis An effect in a dielectric material caused by changes in orientation of electron orbits in the dielectric.

Dielectric strength The voltage per unit thickness at which breakdown occurs.

Differential compound generator A machine in which the current in the series coil sets up a flux in the opposite direction to that of the shunt field winding.

Differential relaying A method commonly used to protect machines, circuit feeders, parallel circuits, and other installations.

Digital counter A sophisticated frequency meter that is capable of measuring frequencies over an extremely wide range.

Digital meter A solid state electronic instrument that measures electrical quantities and displays the measured value in decimal numeric form.

Diode An electronic device that permits current flow in one direction only.

Direct concatenation A method of obtaining several operating speeds of a motor by connecting two motors in tandem, both turning in the same direction.

Direct current An unchanging, unidirectional current.

Discharge curve An exponential curve showing the decrease in charge across an inductor or capacitor from maximum to zero.

Dissipation factor The ratio of energy dissipated to the energy stored in an element for one cycle.

Distributed winding A winding that has two or more windings per pole per phase.

Distribution factor Used in the calculation of the emf generated in the winding of an ac machine.

Diverter A low-resistance shunt connected in parallel with the series field of a machine to control the degree of compounding.

Domain A microscopic needle-shaped crystal that contains a large number of spinning electrons.

Dot convention (1) A notation used to represent the direction of flux produced by a coil. (2) A method of identifying the terminals of transformer windings where terminals with the same instantaneous polarity are identified as dots and the other terminals are left blank.

Double revolving field theory A theory that states a stationary pulsating magnetic field can be resolved into two magnetic fields that rotate in opposite directions but are equal in magnitude.

Double squirrel-cage rotor A motor with two distinct rotor windings having the characteristics of small slip and high starting torque.

Double-subscript notation A notation in which the first subscript designates the point at which a value is measured with respect to a reference point, designated by a second subscript.

Drum controller A device that utilizes a drum switch as the main switching element.

Dual A parallelism between the equations and theorems of electric circuits.

Dual-slope integration A common analog-to-digital technique used in digital meters.

Dual-voltage motor A motor with two voltage ratings, i.e. 220/440 V.

Duplex winding Consists of two complete sets of windings which are independent of each other.

Duty cycle A characteristic of a pulse waveform that is a ratio of the conducting time to the period of one cycle.

DVM Abbreviation for digital voltmeter.

Dynamic braking A method of bringing a motor to a stop by inducing emf in the rotating windings.

Dynamo A machine that may be used to either convert mechanical energy into electrical energy, or vice-versa.

Dyne The CGS unit of force.

E

Eddy current A circulating current caused by the induced emf within a magnetic material.

Effective resistance The total of all resistive effects for an ac circuit; obtained by dividing the total losses caused by current (i.e. hysteresis loss) by the square of the rms value of current.

Effective value *See* rms value.

Efficiency A measure of how completely the power put into a device is used as output.

Elastance The opposition to the setting up of electric lines of force in an electric insulator or dielectric.

Electric circuit A network that contains at least one closed path.

Electric current *See* current.

Electric flux density *See* flux density.

Electric potential The amount of work required to move a unit charge from one point to another.

Electrodynamometer A meter that is capable of measuring both ac and dc voltages and currents.

Electromagnet A magnet excited by a current in a coil surrounding a ferromagnetic core.

Electromagnetic induction The process by which an electromotive force is induced, or generated, in an electric circuit when there is a change in the magnetic flux linking the circuit.

Electromechanical device A device that converts electrical energy into mechanical energy and vice-versa.

Electromotive force (emf) The force that tends to cause a movement of charge carriers in a conductor. Electromotive force is commonly known as voltage.

Electron An elementary particle containing the smallest negative electric charge.

Electron flow The movement of electrons past a given point from the negative terminal of the power source, through the load, to the positive terminal of the source.

Electron shell A group of several closely spaced permissible energy levels that may be occupied by orbiting electrons.

End rings Conducting rings that connect the windings of a motor.

Energy The ability to do work.

Energy current Supplies energy for hysteresis and eddy-current losses in a transformer.

Engineering notation A form of scientific notation in which a number is shown with a power of 10 that is divisible by 3 so the number can be directly related to an SI prefix.

Equivalent circuit A circuit that, under certain conditions of use, may replace another circuit without substantial effect on electrical performance.

Error voltage The difference between the measured voltage and the desired voltage.

Exchange interaction An electrostatic force that maintains a magnetic alignment up to a certain temperature.

Exciting current Supplies the iron losses in the core and sets up the magnetic flux in the transformer winding.

F

Farad The unit of capacitance.

Faraday's law A law that states the generation of voltage increases with additional flux or a faster rate of change of the flux field.

Ferromagnetic materials Materials possessing pronounced magnetic properties.

Ferromagnetism A phenomenon exhibited by materials having a permeability that is considerably greater than unity, and that varies with flux density.

Field intensity A measurement of the magnetomotive force needed to establish a certain flux density in a unit length of the magnetic circuit.

Field poles Laminated steel poles that are connected to the yoke of a dynamo to support field windings.

Field windings Series connected windings in a dynamo that produce alternate north and south poles to obtain the correct direction of emf in the armature conductors.

Filter A circuit designed to pass certain frequencies and block other frequencies.

Flat-compound generator A machine that operates with its terminal voltage equal to the no-load voltage.

Fleming's left hand rule A rule expressing the relationship between current direction, field direction, and the developed force on the conductors.

Flemings right-hand rule A rule that expresses the relation between the direction of motion of a conductor, the direction of the magnetic field, and the direction of the induced emf.

Floating neutral A condition caused by the removal of the neutral wire in a polyphase load, resulting in the phase voltage varying with the phase impedance.

Flux density The charge-inducing capability of an electric field; the number of lines of force per unit area.

Flux linkage The product of the number of lines of magnetic flux and the number of turns of a coil through which they pass. The unit of linkage is one unit of magnetic flux passing through one turn of a coil.

Focus control An oscilloscope adjustment that allows the focus of the waveform to be modified.

Form factor A ratio between the average and effective values of a signal.

Forward-biased A condition where the internal resistance of a diode is very low.

Fourier series A method of mathematical analysis of a recurring nonsinusoidal waveform.

Fractional pitch The span of the coil is less than the pole pitch.

Fractional-pitch winding A winding that is accommodated in a number of slots not divisible by the product of the number of phases and poles.

Free electron The valence electrons of a material that are free to move about among the positive material ions forming a crystal lattice structure.

Free-wheeling diode A diode used in motor control circuits to suppress high values of induced voltage when equipment is deenergized.

Frequency The number of cycles per second.

Friction loss A loss that includes bearing and windage losses caused by a rotating armature.

Fringing The spreading of magnetic lines of force as they cross an air gap in a ferromagnetic material.

Frog-leg winding A combination of the lap and wave winding in electric machines.

Full-pitch coil A coil that covers a span of 180 electrical degrees.

Fundamental frequency The lowest frequency component in the harmonic or Fourier expansion of a nonsinusoidal quantity.

Fundamental units The units selected to serve as a basis, in terms of which the other units of the system may be conveniently derived.

G

Gain A ratio of output to input; also called amplification.

Galvanometer A very sensitive electrical instrument used in detecting and measuring extremely small currents.

Gauss CGS Unit of flux density in magnetic circuit; equal to one magnetic line of force per square centimeter.

Gauss's law A law that states the net number of electric lines of force crossing any closed surface in an outward direction is numerically equal to the net total charge within that surface.

General bridge equation An equation that states the phasor product of one pair of opposite impedances must equal the phasor product of the other pair of opposite impedances.

General transformer equation The equation describing the voltage induced into a transformer winding in terms of the peak mutual flux, frequency, and number of turns.

Generated voltage An induced voltage that is developed by mechanical motion between a conductor and a magnetic field.

Generator A machine that converts mechanical energy into electrical energy.

Generator action The induction of voltages in a conductor moving in a magnetic field.

Generator magnetization curve The relation between the field ampere-turns and the flux per pole in a generator driven at a constant speed; also known as generator saturation curve.

Germanium A tetravalent semiconductor element having low conductivity at room temperature and increasing conductivity with rising temperature.

German silver A resistance alloy containing copper, nickel, and zinc.

Gilbert The CGS unit of magnetomotive force (mmf).

Grain-oriented steel Sheets of steel that are cut so that the magnetic flux flows in the direction of the structural grain of the material.

Graticule A transparent ruled screen mounted in front of the fluorescent screen in an oscilloscope.

Ground An electrical connection between a circuit and the earth, or between a circuit and a metal object that takes the place of the earth.

H

Half-power point The point(s) on a response curve that represents half the power intensity of that of the maximum point; also known as the 3 dB point.

Half-wave control Phase angle control for up to 180 electrical degrees.

Hall-effect sensor A device that produces an output voltage in the presence of a magnetic field.

Harmonic An integral multiple of the repetition (or fundamental) frequency of the waveform.

Harmonic distortion A change in waveform due to the inclusion of additional frequency components.

Hay bridge An ac bridge used to measure the impedance of an inductive circuit where the inductive reactance is quite large compared to the resistance.

Heat losses Losses that occur in electric circuits and machines due primarily to eddy currents and hysteresis losses.

Henry The SI unit of inductance.

Hertz The SI unit of frequency.

High-pass filter A circuit designed to pass high frequencies and attenuate low frequencies.

Holding current The minimum value of current required to maintain conduction.

Hole Positive electrical charge existing in a semiconductor material.

Hooke's law A law that states that a spiral spring's angular deflection is proportional to the amount of torque applied; the force exerted by the spring is a restoring force that always points toward the origin.

Horsepower A practical unit of power, equal to 746 watts.

Hunting An oscillation phenomenon associated with synchronous machines.

Hypotenuse The longest side of a right-angled triangle.

Hysteresis The lagging of the magnetization of a ferromagnetic material behind the magnetomotive force that produces it.

Hysteresis loop The loop formed when a complete magnetization cycle is plotted as a B/H curve.

Hysteresis loss A heat loss that occurs in any magnetic material in which the magnetic field continually reverses direction.

Hysteresis motor A synchronous motor with no salient poles and no direct current excitation; also known as shading-pole synchronous motor.

I

Ideal independent voltage source A voltage source that provides a constant voltage across its terminals.

Imaginary number A number out of phase by 90°; in electric circuits the letter *j* is used to represent this value.

Impedance The total opposition to the flow of alternating current.

Impedance matching The process of matching the output impedance of one circuit to the input impedance of another circuit to achieve maximum power transfer.

Impedance triangle A vector triangle formed by the resistance, reactance, and impedance of a circuit.

Independent node A node that is not directly connected to a voltage or current source.

Independent source An electrical source whose electrical operation depends only on its own electrical characteristics.

Inductance The property of a circuit that opposes any change in the amount of current in that circuit.

Induced emf The emf associated with a changing magnetic field.

Induction generator A generator similar in construction to an induction motor; also known as an asynchronous generator.

Induction motor An ac motor in which the current in the stator winding produces a rotating flux that induces a current in the winding of the rotor.

Inductive reactance The opposition of an inductor, such as a coil, to the flow of alternating current.

Inductive time constant *See* Time constant.

Inductor A device that introduces inductance into an electric circuit.

Inferred zero-resistance temperature The temperature at which the extrapolation of a resistance versus temperature curve intersects the temperature axis.

In phase A condition where there is no angular displacement between quantities; i.e. voltage and current.

Insertion loss The difference between power received at the load before a device is inserted and the power after insertion.

Instantaneous value The magnitude of a voltage, current, or power at an exact instant in time.

Instrument transformer A transformer that is used to reproduce very accurate proportional values of voltage or current for measuring instruments.

Insulators Materials that are poor conductors and have electrons that are tightly bound to individual atoms.

Intensity control An oscilloscope adjustment that varies the brightness of the displayed waveform.

Internal resistance The resistance to the flow of an electric current within a source.

Internal sync An oscilloscope function that applies the output of the vertical amplifier to the sweep generator.

Internal voltage drop The voltage drop occurring inside a voltage or current source.

Interpole A method of improving commutation in electrical generators.

Inverter A circuit that converts direct current into alternating current by triggering a group of SCRs or transistors in a specific sequence.

Ion An atom having an electrical charge.

Iron core inductor An inductor with a core made of various alloys of iron and other materials to give the required characteristics for a particular application.

J

Jones chopper circuit An electronic control circuit using two SCRs to provide variable dc output voltage.

***j* operator** Imaginary number used in the analysis of ac circuits that rotates a phasor by 90° counterclockwise.

Joule The SI unit of work or energy. It is given by the work done by a force of one newton acting through the distance of one meter; or the work done by the transfer of an electric charge of one coulomb through a potential difference of one volt.

K

Kelvin The base SI unit of temperature.

Kilogram The base SI unit of mass.

Kilowatthour The amount of work done in a specific period of time.

Kinetic energy The energy that a mechanical system possesses by virtue of its motion.

Kirchhoff's current law (KCL) A law that states the algebraic sum of the currents entering and leaving a node is zero.

Kirchhoff's voltage law (KVL) A law that states that, in any closed loop, the algebraic sum of the voltage drops and rises equals zero.

L

Lagging A term used to describe a waveform that is delayed in reference to another waveform.

Lagging power factor The power factor of an inductive load.

Lap winding A winding in which the coil ends are connected to adjacent commutator segments.

Laws of magnetic attraction and repulsion Like magnetic poles repel each other. Unlike magnetic poles attract each other.

Law of conservation of electric charge The algebraic sum of all electric charges in any isolated system is a constant.

LCD Liquid crystal display.

Leading A term used to describe a waveform that is advanced in reference to another waveform.

Leading power factor The power factor of a capacitive load.

Leak-off The process by which a capacitor discharges through its leakage resistance.

Leakage current The dc current that flows through a capacitor due to imperfections in the dielectric or to surface paths from one plate to the other.

Leakage flux The portion of the flux that goes through one of the transformer windings but not the other.

Leakage reactance Occurs in both the primary and secondary windings of a transformer; reduces the output voltage at the secondary terminals.

Leakage resistance The relatively high resistance representing the dielectric in a capacitor.

LED *See* Light emitting diode.

Length The SI unit of length is the meter.

Lenz's law A law that states the current that flows as a result of an induced emf is in such a direction that the magnetic field established by the current reacts to stop the motion that generates the emf.

Light-emitting diode A diode that radiates energy in the form of light when conducting a current.

Linear circuit A circuit made up of resistors and driven by sources of constant voltage and current.

Linear resistance A resistance that has a linear voltage versus current relationship.

Line of force (1) The path along which an electric charge moves in an electric field. (2) The path along which a theoretically isolated magnetic pole moves from one pole of a magnet to another.

Line voltage The voltage between two lines of a single-phase system, or the voltage between two lines of a symmetrical three-phase system.

Liquid crystal display A low-power indicating device used to produce alphanumeric displays.

Loading effect (1) A condition producing an inaccurate reading by an ammeter caused by the meter resistance of an ammeter being too high. (2) Produces inaccurate voltmeter readings due to the voltmeter acting as a shunt resistor.

Logarithm The exponent to which a base number must be raised to produce a given number.

Logarithmic scale A graph scale in which the linear displacement along the axis is proportional to the logarithm of the numbers represented.

Loop Any closed path in an electric circuit.

Loop current The current flowing in a closed path.

Loop analysis A circuit analysis method in which Kirchhoff voltage law equations are written to solve for loop currents in a circuit.

Loose coupling The coefficient of coupling between two magnetic circuits that produces a small value of coupling.

Low-pass filter A circuit designed to pass low frequencies and attenuate high frequencies.

M

Magnetic blowout A method of arc control that sets up a magnetic field across the arc.

Magnetic circuit The closed path around which magnetic flux passes.

Magnetic field The space near a magnet in which magnetic forces are present.

Magnetic field intensity *See* Field intensity.

Magnetic flux The total number of lines of force in a given region.

Magnetic hysteresis Phenomenon observable in the magnetization of some materials in which their magnetic state at an instant is related to their previous state. *See* Hysteresis, B/H curve, Hysteresis loss.

Magnetic leakage The part of the magnetic flux that follows a path in which it is ineffective for the desired application.

Magnetic pole The regions where a magnet's strength is concentrated.

Magnetic saturation A condition where a magnetic material is saturated to the point that to produce an increase in the flux density, an extremely large increase in magnetizing force is necessary.

Magnetism A property associated with materials that attract iron and iron alloys.

Magnetization curve *See* B/H curve.

Magnetizing current Lags the primary applied voltage in a transformer winding and produces mutual flux.

Magnetomotive force (mmf) The force that causes a magnetic flux to exist in any magnetic circuit.

Mass The property that determines the acceleration the body will have when acted upon by a given force.

Matched circuit A circuit where the source and load impedances are equal.

Matter A substance of which all physical objects are composed.

Maximum efficiency Occurs in transformers when the copper losses are equal to the core losses.

Maximum power transfer theorem A procedure for determining the load resistance or impedance that when connected to a two-terminal network will receive maximum power from the network.

Maximum value The greatest value a signal may obtain; also known as peak value.

Maxwell CGS unit of magnetic flux.

Maxwell bridge *See* Maxwell-Wein bridge.

Maxwell-Wein bridge An ac bridge used to measure the impedance of an inductive circuit.

Mean An average value obtained by adding successive, equally spaced values and then dividing the sum by the number of values.

Megger An instrument for measuring very high resistance values.

Meter The base SI unit of length or distance.

Millman's theorem A procedure that combines the source transformation theorem with both the Thévenin and Norton theorems.

MKS system A system of physical units based on the meter, kilogram, and second.

Modulus The magnitude of a phasor.

Mole The base SI unit of molecular substance.

Motor A machine that converts electrical energy into mechanical energy.

Motor action The force exerted on a confining structure due to the motion of a conductor moving across a magnetic field.

Multimeter *See* VOM.

Multiplex winding A winding that consists of multiple independent sets of wave windings on the armature.

Multiplier The resistance connected in series with the moving coil of a voltmeter to extend the basic range of measurement.

Murray loop *See* Slide wire bridge.

Mutual flux A flux produced by the ampere-turns of both the primary and secondary of a transformer.

Mutual inductance The inductance between two separate coils where the change of current in one coil will induce a voltage in the other coil.

N

NEMA National Electrical Manufacturers' Association.

Neutral The common return conductor in a single-phase of polyphase system.

Neutral plane A position of brushes in an electrical machine.

Neutron An elementary particle with no electrical charge.

Nodal analysis A circuit analysis method in which Kirchhoff current law equations are written to solve for unknown voltages in a circuit.

Node Any point in a circuit where two or more circuit paths intersect.

Node voltage The voltage at a node with respect to a common reference point.

No-load test An open-circuit test performed on rotating machines to determine the rotational losses.

Nonideal source A source that contains internal resistance.

Non-salient rotor A rotor that is cylindrical in shape with no protrusions.

North pole The region on a magnet that magnetic lines of force exit from.

Norton's equivalent circuit A circuit determined by applying Norton's theorem.

Norton's theorem A procedure for reducing a complex electrical circuit to one having a single current source and a parallel resistance.

Nucleus The core of an atom composed mainly of protons and neutrons.

Null detector Instrument for comparing an unknown quantity with a known quantity.

O

Octave A doubling of frequency.

Oersted The CGS unit of magnetizing force, equal to $4\pi/10$ times the ampere-turns per centimeter.

Ohm The unit of measurement applied to resistance and impedance. A conducting path has a resistance of 1Ω when the passage of a current of 1A requires a potential difference across the path of 1V.

Ohmic resistance The resistance offered by a conductor to the free flow of direct current.

Ohmmeter A measuring instrument used to indicate resistance in ohms.

Ohm's law A law that states the current produced in a given conductor is directly proportional to the difference of potential between its end points.

One-line equivalent circuit A method of reducing a balanced three-phase wye circuit to its single-phase equivalent.

One-to-one transformer A transformer that receives energy at one voltage and delivers it at the same voltage.

Open-circuit characteristic curve The saturation curve of a machine with an open-circuited armature winding.

Open-circuit test A test performed with one winding of a transformer open-circuited; used to determine copper losses in a transformer.

Open-face phasor diagram A polar phasor diagram that shows the relationship between voltages and currents in a system.

Operational amplifier (Op-amp) Integrated circuit amplifier.

Oscilloscope A measuring instrument consisting of a cathode ray tube (CRT) and various associated electronic circuitry; commonly used to automatically plot a particular voltage variation vs. time.

Out-of-phase A condition where there is an angular displacement between quantities; i.e. voltage and current.

Overcompounded When the rated voltage is higher than the no-load voltage.

Overexcited A condition where the motor takes a leading current from the line.

Overload protection A device that protects a system against excessive current flow.

Overshoot Exceeding, or overtravelling, a predetermined stopping point.

P

Pancake coils Windings constructed in thin, flat sections.

Parallax error An error in reading measuring instruments caused by looking at a meter from an angle that will cause the pointer to appear left or right of the true position.

Parallel circuit A circuit with two or more common points so that the same voltage is across all elements.

Parallel resonant circuit A circuit in which the inductance and capacitance have equal values of reactance.

Paramagnetic material A material that becomes only slightly magnetized when under the influence of a strong magnetic field.

Passband The band of frequencies that are passed through a circuit such as a filter.

Passive device A device, such as a resistor, whose value does not change as a result of the application of voltage or current.

Passive filter A filter network consisting of passive devices.

Passive network A network containing no sources.

Peak-to-peak value The amplitude of a waveform from its positive peak to its negative peak value.

Peak value The maximum instantaneous positive or negative value of a voltage, current, or power during a cycle.

Percentage overcompounding The voltage rise between no-load and full-load currents, calculated as a percentage of the no-load voltage.

Percent voltage regulation A ratio expressed in percentage form, between the no-load voltage of a machine or device and the full load voltage.

Period The duration of one cycle of a sustained oscillation or alternation.

Periodic wave A wave that repeats itself after given time intervals.

Permanent magnet stepper motor A motor that operates on the reaction between a permanent magnet rotor and an electromagnetic field.

Permeability Permeance per unit length and cross-sectional area of magnetic materials.

Permeance The property of a magnetic circuit that permits the passage of magnetic flux.

Permittivity The ratio in an insulating material, of the electric flux density to the electric field intensity, compared with the same ratio for free space.

Phase angle The angle by which a sinusoidally varying quantity is displaced in time from another quantity of the same frequency.

Phase sequence The order in which polyphase voltages pass through their respective maximum values.

Phase voltage The voltage between any line of a single-phase or polyphase system and the neutral point of the system.

Phasing A method of establishing the proper polarity connections for a transformer.

Phasor Term describing the representation of a steady-state sine-varying quantity as a complex number.

Phasor diagram A two-dimensional drawing that shows the magnitude and phase relationships of two or more sinusoidally varying quantities.

Pi network A network of components that are arranged in the configuration of the Greek letter pi; also called a delta network.

Pitch factor A ratio between the voltage generated by a fractional-pitch coil and the voltage generated by a full-pitch coil.

Plex A method of classifying armature windings.

Plugging A method of bringing a motor to a stop by reversing the developed torque in the armature.

Polar coordinates The representation of a complex number, such as a phasor quantity, in terms of the magnitude and phase angle of the number. Also known as polar form.

Pole (1) The boundary where magnetic lines of force enter or leave a ferromagnetic material. (2) A term used to describe how an electrical contact arrangement operates. (3) An RC network that produces a −20 dB/decade rate of attenuation.

Pole piece The part of the magnetic circuit of a machine, the yoke, and the air gap.

Pole pitch The distance between identical points on adjacent poles, equal to 180°.

Pole shoe The separable part of a pole piece that faces the armature of a machine.

Polyphase system Two or more equal voltages with fixed-phase differences supplying power to various ac loads.

Positive ion An atom that has lost one or more valence electrons and has a net positive charge.

Potential energy Energy possessed by a system by virtue of position or condition.

Potential transformer An instrument transformer used for the transformation of voltage.

Potentiometer A variable resistor.

Power The rate at which work is done or energy is converted from one form to another.

Power angle *See* torque angle.

Power factor Ratio of the active power to the apparent power in an electrical system.

Power factor angle The angular displacement between the active power and apparent power in an electrical system.

Power factor correction The addition of capacitors or inductors to a circuit to reduce the total current drawn from an ac source by reducing the power factor angle of the circuit.

Power supply A device that converts one type of electric potential or current to another.

Power triangle A right-angled triangle relating apparent power, active power, and reactive power.

Practical source *See* Nonideal source.

Primary cell An electrolytic cell in which two electrodes of different conducting materials associated with an electrolyte generate an emf.

Primary winding The winding of a transformer that receives energy from the supply circuit.

Principle node A node connecting three or more branches.

Product-over-sum rule When only two resistances are connected in parallel, the resistance of the parallel combination of two resistances is equal to the product of the individual resistances divided by their sum.

Prony brake A method of making direct measurements of efficiency in motors.

Proton Component of the nucleus of an atom. It has a positive charge equal in magnitude to that of an electron.

Pull-in torque The maximum constant torque under which the synchronous motor will pull its connected load into synchronism.

Pull-out torque The maximum sustained torque that the synchronous machine will develop at synchronous speed with rated voltage applied.

Pulse A brief but sharp rise or fall of electrical current or voltage.

Pulse transformers A transformer that develops a signal of short duration for control circuit applications such as coupling a trigger pulse to an SCR.

Pulse width The time between the start of the rise and the start of the fall of the input pulse.

Pulse width control Voltage control by using semiconductor devices such as SCRs or transistors.

Pulse-width modulation (PWM) A method of voltage control in which the amplitude and repetition frequency are held constant but the width of the pulses is varied in accordance with the amplitude of the modulating signal.

Pushbutton A momentary contact device.

Pythagorean's theorem A method of solving for an unknown in a right-angle triangle when two quantities are known.

Q

Q factor Used to describe a figure of quality or merit of a resonant circuit.

R

Radian The SI unit of angular measure, approximately equal of 57.3°.

Radiation loss A power loss associated with high-frequency ac circuits, such as antennas.

Ramp voltage A linearly increasing voltage.

Ratio-meter Measures the ratio of two quantities such as voltage to current, or kVAR to kW.

RC snubber A surge suppression technique using a resistor and capacitor to prevent voltage spikes.

Reactive factor The ratio of reactive volt-amperes to total volt-amperes.

Reactive power The rate at which energy is stored and alternately returned to the source by a reactive component such as an inductor or capacitor.

Reactive volt-ampere (VAR) The unit of reactive power.

Real power *See* average power.

Reciprocity theorem A theorem that states if any voltage source located at one point in a circuit produces a current at a second point in the circuit, the same source of voltage acting at the second point will produce the same current at the first point.

Rectangular coordinates The representation of a complex number, such as a phasor quantity, in terms of the real component and imaginary component of the number.

Rectifier An electronic circuit that converts ac into pulsating dc.

Reference node Common point for all sources in a network.

Reference phasor A phasor lying on the reference axis.

Reference voltage A voltage used as a standard of reference.

Reflected impedance The impedance appearing at the primary terminals of the transformer as a direct result of the impedance connected to the secondary terminals.

Refresh A method of maintaining a charge across a capacitor.

Regenerative braking A method of bringing a motor to a stop by using the load to apply a negative torque on the motor and drive it as a generator so that power is returned to the supply.

Relative motion The relation between the motion of a conductor cutting lines of force in a field that remains stationary.

Relative permittivity The property that determines the electrostatic energy stored per unit volume for unit potential gradient. Relative permittivity is dimensionless since it is a ratio.

Relaxation oscillator An electronic circuit that is used to create a repetitive signal voltage.

Relay An electromagnetic device that will open or close one or more sets of electrical contacts.

Reliability The probability that a device, such as a resistor, will function without failure under stated conditions for a stated period of time.

Reliability factor An indication of the failure rate of devices such as resistors.

Reluctance The opposition that a magnetic path offers to magnetic flux when a magnetomotive force is applied.

Reluctance motor A synchronous machine with a wound stator and a cage rotor constructed to provide two axes of substantially different magnetic reluctance.

Repulsion-induction motor A single-phase motor possessing the starting characteristics of a repulsion motor and the running characteristics of an induction motor.

Repulsion motor A single-phase motor in which the field and armature fluxes repel each other to produce a torque in the rotor.

Repulsion-start induction motor A motor that has high starting torque, low starting current, and a constant operating speed.

Residual flux density *See* Residual magnetism.

Residual magnetism The amount of flux density remaining in the material after the magnetizing force has been removed.

Resistance The property of a device or a circuit that opposes the movement of current through it. The unit of resistance is the ohm.

Resistivity (specific resistance) The resistance of a conductor having unit length and unit cross-sectional area.

Resistor A device used to insert electrical resistance in a circuit.

Resonance A condition in an ac circuit when the inductive reactance equals the capacitive reactance.

Resonant frequency The frequency at which the inductive reactance is numerically equal to the capacitive reactance of an ac circuit.

Response curve A plotted curve indicating the reaction of a circuit to a given input.

Restoring torque A very small percentage of the maximum torque of a motor; occurs when rotor is moved from its position of minimum reluctance.

Reversal When the imaginary part of a complex number is multiplied by 1.

Reverse-biased A condition where the internal resistance of a diode is very high.

Rheostat Controls circuit current by varying the amount of resistance in the resistance element.

Right-hand rule A rule in which the right hand determines the direction of the magnetic field surrounding a conductor and the current flowing through the conductor.

RMS value The value of a sine wave that indicates its heating effect; represents 0.707 of the peak value.

Root-mean-square *See* RMS value.

Rotor The rotating component of a machine.

S

Salient pole A pole piece that projects beyond the magnetic yoke towards the armature.

Sawtooth waveform A waveform in which the magnitude increases uniformly with time for a period and then falls rapidly to zero.

Scalar A quantity that has only magnitude.

Schmitt trigger An electronic switching device.

Scientific notation A format of representing very large or very small numerical values by using powers of 10.

Scott connection A three-phase/two-phase connection of transformers.

Second The base SI unit of time.

Secondary cell A cell that is designed to be charged after discharge.

Secondary winding The winding of a transformer that receives energy by induction from the primary winding.

Second harmonic A harmonic with a frequency that is double that of the fundamental frequency.

Selectivity A measure of the bandwidth of a resonant circuit in terms of the Q factor and resonant frequency.

Self-inductance The ability of an inductor to induce a voltage into itself.

Semiconductor A material having electrical properties intermediate between those of good electrical conductors and insulators.

Semilogarithmic scale A graph scale in which linear graduations are indicated on the vertical axis and logarithmic graduations are indicated on the horizontal axis.

Sensing resistor A resistor used in feedback circuits.

Series circuit A circuit in which there is only one current path and all components are connected end-to-end along this path.

Series generator A machine with the field winding in series with the armature circuit.

Series motor A motor with the field coils connected in series with the armature.

Series-parallel circuit A circuit containing both series and parallel connected devices.

Series winding The winding of an autotransformer that is in series with the common winding.

Servomechanism A closed-loop amplifying system employing negative feedback to control positioning automatically and accurately.

Servo motor A small dc or ac motor used in a servomechanism.

Seven-segment display An indicating device made up of segments that can be controlled to produce alpha-numeric displays.

Shaded-pole motor A single-phase induction motor with a main winding and a permanently short-circuited auxiliary winding displaced in magnetic position from the main winding.

Shading coil A short-circuit copper strap used in shaded-pole motors.

Shoe The outer end of a laminated pole.

Short-circuit characteristic (SCC) The curve obtained from the short-circuit test.

Short-circuit test A test performed on an electrical machine with its output terminals short-circuited and full-load current flowing; used to determine the equivalent resistance and reactance of transformer windings.

Shorting interlock A safety device used with current transformers to allow the meter to be removed for inspection or adjustment.

Shunt The branch connected in parallel with another branch of a circuit; used in current meters to provide a path around meter movement.

Shunt generator A machine with the field winding in parallel with the armature circuit.

Shunt motor A motor with its field winding connected in parallel with the armature winding.

Shunt resistor A low value resistor usually made of manganin strips that are brazed to heavy copper blocks to ensure a low temperature coefficient of resistance.

SI Abbreviation for International System of Units (Le Système International d'Unités).

Siemens The SI unit of conductance; the reciprocal of the ohm.

Silicon A tetravalent semiconductor element used in the manufacture of semiconductor devices.

Silicon controlled rectifier (SCR) A three terminal pnpn device that acts as a gated diode. The gate terminal switches the device on, allowing current to flow from anode to cathode.

Simplex winding A single, complete, closed winding wound on an armature.

Sine wave An electrical waveform created when magnitude varies in proportion to the sine function of the angles of rotation.

Single-phase circuit An electric circuit energized by a single alternating voltage.

Single-phase motor A motor that operates on a single-phase power supply.

Single-pole filter A filter with one RC network.

Skewed rotor A rotor with slanted slots to produce a smoother, quieter operation.

Skin effect A phenomenon associated with high-frequency ac circuits where the current flow is on the outside, or skin, of the conductor.

Slide wire bridge A method of locating ground faults in cables and wires.

Slip The fractional speed-difference between the stator and rotor of an induction motor.

Slipping poles The loss of synchronization after maximum torque is exceeded.

Slip-ring A conducting ring connected with a winding and rotating with it, allowing electrical conduction to be maintained.

Small motor A motor built in a frame smaller than that having a continuous rating of 1 hp, open-type at 1700 to 1800 rpm.

Solenoid An electromagnet constructed by winding a coil of wire around a cylindrical form.

Source conversion A procedure for replacing a constant-current source in a circuit diagram with an equivalent constant-voltage source, and vice-versa.

South pole The region on a magnet that magnetic lines of force enter.

Space permeability The permeability of air or a vacuum.

Speed regulation The ability of a motor to maintain its speed when a load is applied.

Split-phase motor A single-phase induction motor equipped with an auxiliary stator winding connected in parallel with the main stator winding.

Square wave The waveform of alternating current or voltage that is approximately square or rectangular.

Squirrel-cage motor An induction motor with a squirrel-cage rotor.

Squirrel-cage rotor The rotor of an induction motor with a squirrel-cage winding.

Squirrel-cage winding A type of rotor winding consisting of a series of bars in rotor slots, connected at each end to a common conducting ring.

Stabilization winding A small number of series turns added to a motor to increase the flux with an increase of load.

Stall An increase in load on a motor that causes the rotor to come to a standstill.

Star connection *See* Wye network.

Starting torque The torque exerted by a synchronous motor during the starting period.

Static charge Charges that are transferred by causing friction between materials.

Static electricity The movement of static charges between materials.

Stator The fixed or stationary component of a machine.

Steady state A condition where the signal is constant and does not change with time.

Step The angular rotation produced by the output shaft of a motor each time the motor receives a step pulse.

Step angle The rotation of the output shaft of a motor caused by each step pulse, measured in degrees.

Step-down transformer A transformer that receives energy at one voltage and delivers it at a lower voltage.

Stepper motor Converts electrical pulses into a proportional mechanical movement.

Step-pulse A voltage signal of a fixed duration.

Step-up transformer A transformer that receives energy at one voltage and delivers it at a higher voltage.

Stopband The band of frequencies that are attenuated.

Storage element A device that receives electrical energy, stores it in the form of a magnetic field, and returns the energy to the circuit at a later point in time.

Stray capacitance Capacitance arising from proximity of component parts, wire, and ground.

Subsynchronous motor A type of hysteresis motor that has a salient pole rotor and operates at a speed lower than that determined by the field poles.

Subtractive polarity A winding connection that causes the direction of counter emf in each winding to be opposite.

Superposition theorem A procedure that is used to analyze networks containing multiple power sources by isolating sources, solving for individual currents, and adding the results.

Supersynchronous motor A special type of motor designed to produce a starting torque equal to its pull-out torque.

Susceptance The ability of a reactive component, such as a coil or capacitor, to permit current flow; the reciprocal of reactance, measured in siemens.

Sweep generator Used in oscilloscopes to develop a voltage at the horizontal deflection plate that increases linearly with time.

Symmetrical components A method of analyzing unbalanced conditions in an electrical network or machine.

Synchronous alternator *See* Alternator.

Synchronous capacitor A synchronous motor used for power factor correction; also known as a synchronous condensor.

Synchronous converter A combination of a synchronous motor and dc generator; used in some situations to convert ac to dc.

Synchronous generator *See* Alternator.

Synchronous impedance The ratio of the short-circuit field current to the open-circuit current of a synchronous machine.

Synchronous impedance angle The angle between the resultant voltage and the armature current in a rotating machine.

Synchronous impedance method A method of predetermining the operating characteristics of an alternator without actually placing the machine under load conditions.

Synchronous induction motor A small polyphase motor that requires no dc field excitation, but having the constant speed characteristics of a synchronous motor.

Synchronous motor A constant-speed machine that transforms electrical power into mechanical power.

Synchronous position The angle by which the rotor shifts backward at any load.

Synchronous reactance The combined armature reactance and armature reaction in a synchronous machine.

Synchronous speed The speed at which the rotating magnetic field moves around the stationary field.

Synchroscope An instrument with a rotating pointer that is used for paralleling alternators.

T

Tap changer A method of controlling the voltage ratio of a power transformer by tapping the windings to change the turns ratio.

Tank circuit *See* Parallel resonant circuit.

Tantalum Component used in the construction of certain types of electrolytic capacitors.

Teaser transformer Used in a Scott connection for three-phase/two-phase transformation.

Tee network *See* Wye network.

Temperature coefficient of resistance The change in resistance per ohm of a material for each degree of change in temperature from the reference temperature of 20°C.

Tertiary winding An auxiliary winding used to allow a circulating current to flow.

Tesla The SI unit of flux density.

Thévenin equivalent circuit The reduction of a circuit to its Thévenin values.

Thévenin's theorem A procedure for reducing a complex electrical circuit to one having a single voltage source and a series resistance.

Three-phase system Three voltages with fixed phase differences supplying power to various ac loads.

Three-wattmeter method A three-phase power measurement using three wattmeters.

Three-wire distribution A system of electric supply consisting of three conductors, one of which is maintained at a potential midway between the potential of the other two conductors.

Tight coupling When the coefficient of coupling between two magnetic circuits is 1, or unity.

Time The SI unit of time is the second.

Time constant The time required for the voltage across a capacitor or inductor to rise to 63.2 percent of maximum, or fall to 36.8 percent of its maximum value.

Time delay relay An electromagnetic device that, when the coil is energized, has an adjustable time delay before its contacts change state.

Time element starter A motor-starting circuit with a variable resistance in series with the armature that is gradually reduced during a specified time interval.

Time period The amount of time required for one cycle to change.

Tolerance Allowable deviation from the marked value of a component.

Toroid A coil wound on a circular core.

Torque The effectiveness of any force to produce rotation.

Torque angle The electrical angle between the stator and rotor mmfs in an electrical machine; also known as power angle.

Tractive force The force of attraction that exists across the air gap of a magnetic circuit.

Transfer impedance The ratio of the voltage in one branch to the current in another branch.

Transformer A device that transfers energy from one circuit to another by electromagnetic induction.

Transformer polarity The relative direction of the induced voltages in the primary and secondary windings with respect to the winding terminals.

Transient The part of the change in a variable that disappears when going from one steady-state condition to another. Also known as transient state.

Transistor A three-terminal semiconductor device used for amplification or switching.

Triac A gated diac capable of conducting a current in both directions when the gate terminal is triggered.

Trigger control Oscilloscope adjustment to provide a stable waveform display.

True power *See* Active power.

Tuned circuit A resonant circuit.

Turns ratio The ratio of the number of secondary turns to the number of primary turns in a transformer winding.

Two-pole filter A filter using two RC networks to produce a roll-off rate of −40 dB/decade.

Two-wattmeter method A three-phase power measurement using two wattmeters.

U

UJT relaxation oscillator *See* Relaxation oscillator.

Unbalanced load Any polyphase load in which the impedance in one or more phases differs from those of the other phases.

Undercompounded When the rated voltage is less than the no-load voltage.

Underexcited A machine condition where if the armature current is increased, the motor takes a lagging current from the line.

Unijunction transistor (UJT) A three terminal device that acts as a diode with its own internal voltage divider biasing circuit.

Unit hypotenuse A right-angle triangle used for deriving sine values.

Universal motor A series-wound motor of fractional horsepower rating that is designed to operate with either ac or dc voltage sources.

V

Valence orbit The outermost shell occupied by orbiting electrons.

VAR The unit of reactive power.

Variable-reluctance stepper motor A stepper motor with no permanent magnet, and consequently, no reluctance torque.

V-curve The load characteristic of a synchronous machine giving the relationship between the field current and armature current for constant values of load, armature voltage, and power.

Vector A quantity that has both magnitude and direction.

Vertical controls The selector switches on an oscilloscope that allows for an appropriate volts/div range to be selected; also moves the waveform up or down by using the position knob.

Volt The SI unit of voltage or potential difference.

Volt-ampere-reactive *See* VAR.

Voltage divider A device to permit a fixed or variable fraction of a given supply voltage to be obtained.

Voltage divider rule A rule that states the ratio between any two voltage drops in a series circuit is the same as the ratio of the two resistances across which these voltage drops occur.

Voltage drop The difference of voltages at the two terminals of a passive element.

Voltage regulation The voltage variation of a source as it changes from no load to full load as a ratio of full-load voltage.

Voltage regulator A device that, supplied a given input voltage, will provide an adjustable output voltage.

Voltampere (VA). Unit of apparent electrical power.

Voltmeter A measuring instrument used to indicate the magnitude of voltage in a circuit.

Voltmeter loading The effect of connecting a voltmeter in a circuit, resulting in a lower-than-normal reading by the meter.

Voltmeter sensitivity An indication of the shunting action of a voltmeter, measured in ohms/volt.

VOM Instrument used to measure voltage, resistance, and current.

W

Ward-Leonard system of speed control A motor control circuit that controls the motor via a dc generator.

Watt The power expended when one ampere of direct current flows through a resistance of one ohm; unit of active or average power.

Watt-hour meter An instrument for measuring the energy supplied to the residential or commercial user of electricity.

Wattless power Power that is received from the supply during the first 180° and returned to the supply during the second 180° of each cycle.

Wattmeter An instrument used for measuring power in watts.

Waveform The pattern of variations of a current or voltage.

Wave winding A winding used on four-pole machines to increase the number of conductors in series between brushes without increasing the number of coils on the armature.

Weber The SI unit magnetic flux.

Weber's theory A theory that states the molecules of a magnetic material are tiny magnets each with a north pole and a south pole, and a surrounding magnetic field.

Wheatstone bridge An instrument that measures resistance with a high degree of accuracy.

Wiper The movable contact in a potentiometer.

Wire table A table of sizes and properties of round copper wire.

Work The amount of energy converted from one form to another, as a result of motion or conversion of energy.

Wound-rotor motor An induction motor having a wound rotor with connections made to slip-rings; also known as slip-ring motor.

Wye-delta conversion An equivalent circuit method of network analysis where wye-connected components are converted to an equivalent delta-connected network.

Wye network A three-phase source or load connection in which one of the leads from each element is connected to form a common junction.

Y

Yoke The frame of a motor or generator.

Z

Zener diode A diode that is designed to conduct in the reverse direction when its value of breakdown voltage is reached.

Zero-sequence currents Currents with no phase difference that flow as a result of harmonics in a system.

I N D E X

N

O

P

Q

R

T

U